# Progress in Drug Research

Volume 77

Progress in Drug Research is a prestigious book series which provides extensive expert-written reviews on highly topical areas in current pharmaceutical and pharmacological research.

Founded in 1959 by Ernst Jucker, the series moved from its initial focus on medicinal chemistry to a much wider scope. Today it encompasses all fields concerned with the development of new therapeutic drugs and the elucidation of their mechanisms of action, reflecting the increasingly complex nature of modern drug research. Invited authors present their biological, chemical, biochemical, physiological, immunological, pharmaceutical, toxicological, pharmacological and clinical expertise in carefully written reviews and provide the newcomer and the specialist alike with an up-to-date list of prime references.

Starting with volume 61, Progress in Drug Research is continued as a series of monographs and contributed volumes.

Virat Khanna • Shweta Dang • Alka Ahuja
Editors

# Nanotechnologies for Drug Delivery and Biopharmaceutical Development

Springer

*Editors*
Virat Khanna
University Centre for Research
and Development
Chandigarh University
Mohali, Punjab, India

Shweta Dang
Department of Biotechnology
Jaypee Institute of Information Technology
Noida, Uttar Pradesh, India

Alka Ahuja
College of Pharmacy
National University of Science
and Technology
Muscat, Oman

ISSN 0071-786X ISSN 2297-4555 (electronic)
Progress in Drug Research
ISBN 978-3-032-00795-7 ISBN 978-3-032-00796-4 (eBook)
https://doi.org/10.1007/978-3-032-00796-4



This Springer imprint is published by the registered company Springer Nature Switzerland AG
The registered company address is: Gewerbestrasse 11, 6330 Cham, Switzerland

# Contents

## Part III Nanotechnology in Biopharmaceutical Formulation

# About the Editors

**Virat Khanna** is serving as Associate dean, Research and Development, Chandigarh University, Punjab, India. He is an accomplished academic and researcher with over 15 years of experience in higher education and scientific research. He has authored more than 80 publications in high-impact, peer-reviewed journals indexed in SCI, Scopus, and Web of Science. His research contributions span nanotechnology, materials science, and sustainable development. He has served as an editor for multiple books and conference proceedings with leading international publishers, including Springer, CRC Press, IGI Global, and Bentham Science. He is actively involved in scholarly publishing and currently serves on the Editorial Board of *Scientific Reports* (Nature Group) and *Discover Nano* (Springer). Additionally, he holds the position of Academic Editor for *Journal of Nanotechnology* and *Advanced Materials Science and Engineering*, both published by Wiley. As a recognized leader in his field, Dr. Khanna is the Book Series Editor for *Recent Advances in Materials for Nanotechnology*, published by CRC Press (Taylor & Francis). He has also guest-edited several special issues for renowned international journals, including those published by Springer and De Gruyter. He has delivered numerous invited seminars, keynote addresses, and workshops at prestigious international conferences. His contributions to research and academia have been recognized with multiple best research paper awards.

**Shweta Dang** is a Professor in Biotechnology, Jaypee Institute of Information and Technology, with 19 plus years of experience in Teaching and Research. She has worked as a research scientist in Jubilant Life Sciences for the US regulatory market. She has worked on eight major research projects financed and supported by Government Agencies like ICMR, DBT, and CSIR as PI and Co-PI/Mentor. She has been an invited speaker at national and international forums, Seminars, and Conferences globally. Her research article "Nose-to-brain delivery of lamotrigine-loaded PLGA nanoparticles" was featured in the India Science Wire, Vigyan Prasar, 2019. Her research paper, published in the *Indian Journal of Pharmaceutical*

*Sciences*, was awarded the Prof. M.L. Khorana IJPS Best Paper award for the year 2018. She has more than 80 research publications in journals of repute with close to 5500 citations and an h-index of 29. She has more than 30 research theses supervised at graduate, postgraduate, and doctoral levels. She has three Indian patents granted, and she has contributed 18 book chapters.

**Alka Ahuja** is a Professor and Assistant Dean for Academic Affairs at the College of Pharmacy, National University of Science and Technology, Muscat, Sultanate of Oman (in academic partnership with West Virginia University, USA). She chairs the college's Research and Innovation Committee. She is an editorial board member of several peer-reviewed journals and an invited speaker at national and international forums. Her research has been cited in leading scientific texts, including *Bioadhesive Drug Delivery Systems: Fundamentals, Novel Approaches, and Development*, edited by Edith Mathiowitz (Marcel Dekker, New York), and numerous high-impact journals. She has led major research initiatives funded by ICMR, UGC, DST, AICTE (India), and MOHERI (Oman). Her contributions have earned her several prestigious honors, including the IDMA-G.P. Nair Award, AICTE Career Award for Young Teachers, Asian Education Leadership Award, Global Education Leadership Award, and the Academic Excellence Award, to name a few. With over 250 research publications, she has guided undergraduate, postgraduate, and doctoral research. She has also worked on consultancy projects funded by a Scottish company and is the author or co-author of 11 books in various areas of pharmaceutical sciences.

# Part I
# Fundamentals of Nanotechnology in Biopharmaceuticals

# Chapter 1
# Nanotechnology in Biopharmaceuticals: Enhancing Drug Delivery and Development

**Swapnadipa Patra, Jyoti Bhattacharjee, Arko De, Pritam Roy Chowdhury, and Subhasis Roy**

**Abstract** Nanotechnology has evolved as a pivotal tool in advancing biopharmaceutical research and development. Researchers have unlocked new drug delivery and therapeutic efficacy frontiers by leveraging nanoscale materials and devices. This chapter explores the intersection of nanotechnology and biopharmaceuticals, focusing on its role in enhancing targeted delivery, improving bioavailability, and enabling novel therapeutic strategies. This chapter also delves into the transformative potential of nanotechnology within biopharmaceuticals, focusing on its role in enhancing drug stability, bioavailability, and targeting specificity. Key topics include designing and engineering nanocarriers for high-precision drug delivery, applying nanoscale imaging techniques for diagnostics, and developing innovative nanosensors for real-time monitoring of therapeutic responses. Furthermore, the regulatory considerations and prospects of nanotechnology in biopharmaceuticals are explored, underscoring its profound impact on personalized medicine and the future of healthcare. Key advancements such as nanoparticle-based drug carriers, nanoscale imaging techniques, and nanosensors for diagnostics are discussed, highlighting their potential to overcome current drug formulation and delivery challenges. Furthermore, nanotechnology's regulatory landscape and prospects in biopharmaceutical development are explored, emphasizing its transformative impact on personalized medicine and healthcare delivery.

**Keywords** Nanoscale materials · Diagnostic instruments · Nanomedicine · Nanotechnology · Drug delivery

---

S. Patra · J. Bhattacharjee · A. De · P. R. Chowdhury · S. Roy (✉)
Department of Chemical Engineering, University of Calcutta, Kolkata, West Bengal, India

 
V. Khanna et al. (eds.), *Nanotechnologies for Drug Delivery and Biopharmaceutical Development*, Progress in Drug Research 77,
https://doi.org/10.1007/978-3-032-00796-4_1

## 1.1 Introduction to Nanotechnology in Biopharmaceuticals

### *1.1.1 Fundamentals of Nanotechnology in Biopharmaceuticals*

Nanotechnology is the research of working with matter on the particle and subatomic level. Nanotechnology alters or produces materials within the 1–100 nanometers range in at least one dimension. An experimental and theoretical field of applied science and technology is called nanotechnology. It makes the material stronger, more durable, faster, very small, and light in weight. In nanotechnology, the term "Nano" refers to a billionth ($1 \times 10^{-9}$) (Nikolova et al. 2023; Bhattacharjee and Roy 2024a). One of the most popular areas of nanotechnology research is nanomedicine (Mansoori and Fauzi Soelaiman 2005). It uses nanotechnology to develop highly targeted pharmaceutical interventions to diagnose, prevent, and treat diseases. The field of nanomedicine has witnessed significant development over the last two decades, and this is gaining pace in its commercialization into various medicines; i.e., drug delivery medicines sales have already exceeded 75% of the entire nanomedicine business (Mazayen et al. 2022). A vital contribution from nanotechnology is being made towards fighting cancer and other life-threatening diseases. It helps detect many diseases, such as diabetes mellitus, Parkinson's, and Alzheimer's disease, as well as germs and viruses (Kaur et al. 2024). Pharmaceutical sciences, such as tissue engineering, diagnostics, creation of nanomedicines, biomarkers, development of biosensors, and targeted drug delivery, can all be enhanced with nanotechnology (Bhattacharjee and Roy 2024a). Pharmaceutical sciences utilize several nano-based technologies, such as carbon nanotubes, quantum dots, dendrimers, liposomes, polymeric micelles, metallic nanoparticles, nanocomposites, and many more.

In the conventional drug delivery system, drugs are less efficient in reaching the dead cells, and they exhibit uncontrolled bursting mechanisms in drug release, which can attack nearby healthy cells and organs, generating a toxic effect; further, their pharmacokinetic & physical characteristics are inferior leading to inactivation or degradation during its delivery process (Huang and Ding 2022). While at the same time, nanoparticle-based DDSs vary from 10 to 1000 nm, and due to their small sizes and active surface area, pharmacokinetic & physical properties enable nanoparticles to penetrate the targeted cells effectively (Yusuf et al. 2023; Khan et al. 2019; Banerjee et al. 2016). Nanoparticles, despite being minute, exhibit specific adverse effects because some of them can be engineered with less clearance to have a prolonged presence in the body, but an unwelcome prolonged residence of them, e.g., cytotoxic type, leads to toxicity during the "First-pass" time (Yusuf et al. 2023). It is essential in nanoparticle-based DDS for design-controlled administration.

Intensive research has been conducted on DDSs to design different drugs to fight against diseases such as cancer and neurodiseases (Wang et al. 2020; Liu et al. 2022). The subsequent section concerns some of the common nanoparticle DSSs being researched.

#### 1.1.1.1 Polymeric DSSs

Polymeric nanoparticle DSSs have made notable progress in medicine (Bhattacharjee and Roy 2024b; Begines et al. 2020). Several copolymers of PLGA [Poly-lactic-co-glycolic acid], PLA [poly lactic acid], chitosan, and PEG [polyethylene glycol] were prepared using DSSs (Fraczek et al. 2023; Nelson Sathy et al. 2010; Xiao et al. 2010). However, the research on PEG, PLGA, and PLA has been studied more extensively than that of chitosan. It is receiving increased interest because of its low toxicity, immunogenicity, and biocompatibility. PEGylated drugs are the most developed polymeric DSSs, with some FDA-approved products for clinical use. However, PLGA and PLA both have been found to show an initial burst release behavior (within 24 h), which may lead to uncontrolled drug delivery at undesired locations (Su et al. 2021; Yoo and Won 2020; Vlachopoulos et al. 2022). Some polymer-based DSSs are pH-sensitive, temperature-sensitive, ROS-sensitive, and enzyme-sensitive, and can be engineered to change their response (Bhattacharjee and Roy 2024b). By increased metabolism, cancer cells produce more ROS than normal cells; such parameters may represent the state of tissue physiology (Perillo et al. 2020). Polyamines that are pH-sensitive and self-assembling have been demonstrated to exhibit varied delivery properties. Polyamines that bear doxorubicin and are folate- and transcriptional transactivator ligand (TAT) ligand-coated on irremediable cancer cells have been effective (Yusuf et al. 2023). These pH-sensitive polyamines release the entrapped doxorubicin when exposed to an acidic pH condition. This approach can effectively target the tumor microenvironment because it is predominantly acidic and must be targeted with appropriate targeting agents (Xu et al. 2022).

#### 1.1.1.2 Lipid-Based DSSs

Lipid-based DSSs come in different sizes and compositions in two large forms, liposomes and micelles (Bhattacharjee et al. 2025). A lipid monolayer spontaneously arranges to become a nanovesicle with sizes ranging from 5 to 50 nm in the aqueous phase to create a micelle (Kus-Liśkiewicz et al. 2021). They effectively deliver molecules by binding with the hydrophobic tails and creating an interior that interacts with the drug, and the hydrophilic phospholipid moiety is outside, facing the aqueous component (Kus-Liśkiewicz et al. 2021; Bhattacharjee and Roy 2025a).

Liposomes are bilayered nano-vesicles similar to cell membranes and range from 10 nm to some microns in size, unlike micelles (Bhattacharjee and Roy 2025a). Liposomes are four: unilamellar vesicles (ULVs) made of a single layer of lipid membrane, with sizes ranging from 10 to 100 nm; multilamellar vesicles (MLVs); oligolamellar vesicles (OLVs); and multivesicular liposomes (MLVs). OLVS and MLVs consist of more than one bilayer of phospholipid spherical moiety, layered onion-like and delineated by an aqueous layer. In contrast, MLVs contain one lipid bilayer containing several water chambers (Roschenko et al. 2023). As the outer phospholipid component comes into contact with the aqueous phase, hydrophobic drugs are trapped by the hydrophobic tails of the bilayer, and hydrophilic drugs are generally trapped by the aqueous core (Bhattacharjee and Roy 2025a). Since liposomes can transport water-soluble and insoluble drugs, they are some of the most effective delivery systems for drugs against cancer. Conventional drugs provide toxic side effects to healthy cells because they are not localized in the tumor microenvironment when they diffuse back into the bloodstream. Conventional drugs are usually small molecules with low specificity for cancer cells (Banerjee et al. 2016). However, through an enhanced permeability and retention (EPR) effect, liposomes can be used to improve the delivery of such drugs to the cancer microenvironment, where the junctions are tight with gaps of 100–800 nm, in contrast to the normal epithelial junctions of 5–10 nm. Liposomes accumulate between cancer cell tight junctions to deliver entrapped drugs, cross the blood vessel walls, and penetrate the microenvironment of tumor tissues by the EPR effect. Liposomes are metabolically labile; nonetheless, drug delivery sciences have reached milestones like using liposomes with an increased half-life upon PEGylation. Upon preparation of DSPE-PEG, for example, the amide function of DSPE is attached to the carboxyl end of PEG through a covalent bond to change PEG into phosphatidyl ethanolamine of DSPE. Similar to targeted drug delivery, various moieties can be attached to the liposome through the PEG linker at the surface of the liposome.

Liposomes can entrap hydrophobic drugs within the lipid bilayer using hydrophobic interaction with the phospholipid fatty acid tail. On the other hand, the water-like interior of the liposome can encapsulate water-soluble substances like DNA or crystalline drugs. Like PEG, the surface modifications on the surface of the liposomes offer increased bioavailability. Different linkers such as avidin-biotin complexes, PEG, or peptide linkers chemically conjugated to the protein or drug of interest and the head of phospholipid are usually used in surface coating of drugs using electrostatic or ionic binding or conjugation of antibodies, chemotherapeutic agents, peptides, and other proteins can prove beneficial for targeted delivery (Wang et al. 2020).

### *1.1.2 Nanomaterials for Drug Delivery*

Novel drug delivery systems encapsulated in liposomes provide new strategies to improve therapeutic effectiveness due to site-specific targeting and controlled release of medication. Such spherical vesicles consisting of biocompatible phospholipid bilayers are well-suited to protect the encapsulated drug from premature degradation and improve its bioavailability. Surface modifications may ensure site-specific targeting by ligands or antibodies and minimize off-target effects or toxicity. At the tissue of interest, liposomes are incorporated into cells from which drug release occurs intracellularly. Its metabolism mainly accounts for hepatic and reticuloendothelial system clearance. This intelligent delivery system is changing treatment paradigms in diseases such as cancer, infectious diseases, and more chronic disorders, allowing one to create a specific platform for personalized and precision medicine in therapeutics, as shown in Fig. 1.1 (Liu et al. 2022).

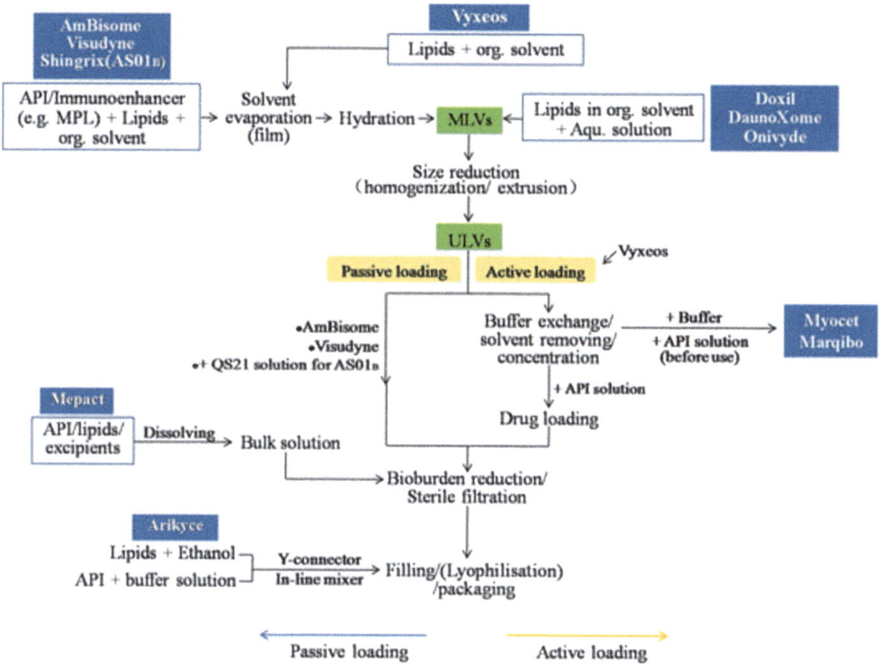


**Fig. 1.1** Schematic representation of potential manufacturing processes for marketed liposomal products compiled and summarized from relevant patents and scientific publications. (Reproduced with permission from Liu et al. 2022)

#### 1.1.2.1 Peptide Nanoparticle DSSs

Another essential component of the currently existing nanoparticle DSSs is linear and cyclic peptides, which can be synthesized or obtained from pre-existing fragments of naturally occurring proteins. The cell surface receptors often use peptides as their ligands because most proteins bind to such receptors by recognizing a specific peptide sequence fragment (Mansoori and Fauzi Soelaiman 2005; Liu et al. 2022). These include ease of preparation and low immunogenicity, making peptides invaluable as potential DSSs. Peptides have been used alone or with other nanoparticles for surface modification and enhanced delivery. EGFR signaling was inhibited in a lung cancer cell line by Kim et al. using an encapsulated peptidomimetic of the EGFR ligand in a cationic liposome. To facilitate the targeted killing of specific cancer cells, doxorubicin, methotrexate, and paclitaxel have been coupled with somatostatins, a class of peptide hormones that ligands abundantly expressed in several cancer cells. Short and designed to bind the serine protease hepsin, which is extensively described on the surface of prostate cancer and hepatoma cells, is the RIPL peptide. In contrast to traditional liposomes lacking the targeting peptide, Kang et al.'s RIPL-conjugated liposome produced higher selectivity and cellular uptake by targeting a panel of cancer cells that overexpress hepsin (Bhattacharjee and Roy 2024b).

## 1.2 Biocompatibility and Safety Assessment of Nanomaterials

Their cytotoxicity and overall biocompatibility should be assessed in nanomaterials used for biomedical purposes. Their influence on cell viability and operation is termed cytotoxicity. On the other hand, biocompatibility involves a proper study of NMs both in vitro and in vivo. They should be checked through standard procedures to ensure their biosafety. For example, NMs should be examined for (a) cytotoxicity, (b) genotoxicity, carcinogenicity, and reproductive toxicity, (c) immunotoxicity, irritation, and sensitization, (d) hemocompatibility, (e) pyrogenicity, (f) systemic toxicity, and (g) implantation. These examinations are necessary to safeguard the workers in the industry and the patients undergoing NM-based treatment. Toxicology analyses are crucial in preventing harmful effects, but to extend NM's applications, it is imperative to comprehend NM's physicochemical properties and their specific bio-effects (Begines et al. 2020).

Here, we target and evaluate the most relevant tests after the specific application requirements. As these consequences are most common in the literature and associated with exposures in work and factory environments, we focused on how NMs influence cytotoxicity, immunomodulation, fibrosis, and genotoxicity (Fraczek et al. 2023).

The most common features of NM that influence these processes are illustrated in Fig. 1.2.

### *1.2.1 Nanomaterial Immunomodulation*

Immunomodulation by nanomaterials is an emerging area of research that focuses on the capability of nanoscale materials to modulate the immune system. Nanomaterials can interact with biological systems in ways impossible for larger particles due to their tiny size and special physicochemical characteristics. Nanomaterials can be engineered to directly target drugs, antigens, or genes to immune cells, thus improving immunotherapies' specificity and efficacy. Take nanoparticles, for instance; these can bind to certain classes of immune cells, e.g., dendritic cells or macrophages, and transport controlling factors. This could mean more effective vaccines and therapies for cancer and autoimmune conditions, since nanomaterials induce or dampen targeted immunity as the case may be (Nelson Sathy et al. 2010).

Tailoring surface features of nanomaterials also allows them to communicate with the immune system. Modifications such as the deposition of polyethylene glycol (PEG) onto the surface of nanomaterials can make them invisible to the immune system and allow for temporary circulation within the body, facilitating drug delivery to target organs. Alternatively, nanomaterials that will be engulfed by immune cells and then stimulated could require modification methods different from those

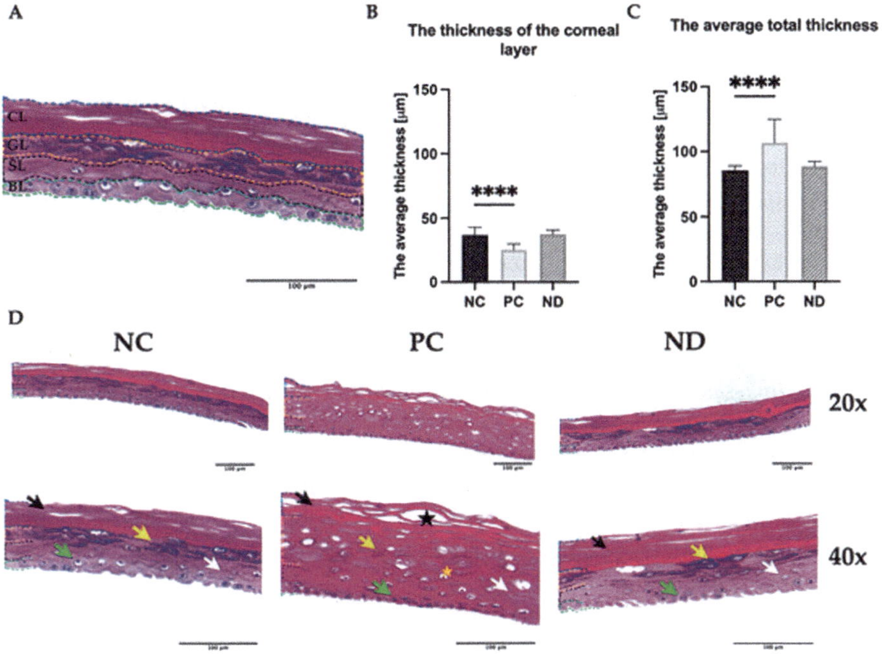


**Fig. 1.2** Histological study of the EpiDerm™ tissue model. (**a**) Schematic tissue structure with the different strata marked: corneal layer (CL, blue), granular layer (GL, yellow), spinous layer (SL, black), and basal layer (BL, green); (**b**) mean stratum corneum thickness; (**c**) total tissue thickness; (**d**) hematoxylin-eosin stained tissues at 20× and 40×. Black arrows—stratum corneum cells; yellow arrows—lamellar bodies with keratohyalin granules; white arrows—flattened spinous layer cells; green arrows—basal cells. A black star indicates damage; necrosis by an orange star. Data are presented as mean ± standard deviation (SD). $p < 0.05$, $p < 0.0001$: statistics. NC Negative Control (Dulbecco's Phosphate Buffered Saline, DPBS), PC Positive Control (1% Sodium Dodecyl Sulfate, SDS), ND Nanodiamond (25 µg/mL) treated. (Reproduced with permission from Fraczek et al. 2023)

above. The prospect of harnessing these interactions with control leaves room for tailored medications and the enhancement of several current immunotherapies. That being said, substantial studies have to be performed because of the complexity of immunity and the unknown implications in manufacturing the nanostructures able to initiate therapeutic effects (Xiao et al. 2010).

### *1.2.2 Nanomaterial Cytopathogenicity and Genopathogenicity*

Due to their unique properties and widespread uses in different fields, nanomaterials have raised serious issues of cytotoxicity and genotoxicity. Cytotoxicity is the ability of such materials to harm living cells, which can ultimately lead to cell death.

Due to their nanoscale size, they can easily penetrate cellular membranes, interact with intracellular components, and disrupt normal cellular functions. Mechanisms of cytotoxicity include oxidative stress, inflammation, and disruption of cell signaling pathways. For example, metal-based nanomaterials like silver nanoparticles can cause the formation of reactive oxygen species (ROS), which will cause oxidative damage to cell components such as lipids, proteins, and DNA. This consequently leads to apoptosis or necrosis, affecting tissue health and function (Su et al. 2021).

Genotoxicity, however, is a quantification of whether or not nanomaterials possess the capability of causing genetic mutation, breaking chromosomes, or otherwise modifying the DNA of genes in cells. The genotoxic activity may be through the direct action of nanomaterials on DNA or indirectly through ROS formation.

DNA damage causes mutations and may result in carcinogenesis and other genetic illnesses when repair does not occur. Some studies have indicated that carbon nanomaterials, such as carbon nanotubes, can cause DNA strand breaks and chromosomal aberrations. Shape, size, surface charge, and chemical composition are also essential factors in the genotoxicity of nanomaterials. Therefore, cytotoxicity and genotoxicity need to be explored in detail during the production and application of nanomaterials for safety purposes with low risk to human health and the environment (Yoo and Won 2020).

### *1.2.3 Therapeutic Applications of Nanomaterials*

Due to NMs' ability to penetrate more than any other material through the biological membrane, drug delivery and photodynamic treatment have been possible. More exciting is the capability of treating diseases related to abnormal cell development using the cytotoxicity of NPs. For instance, NPs were utilized in experiments to treat hypertrophic scarring (Vlachopoulos et al. 2022). Solid cancer may also be cured by employing some NPs. For example, a gold nanoparticle biosynthesized was cytotoxic to a breast cancer cell. NMs are also ROS quenchers and have been used to decrease genome damage. Some of them include Au, quantum dots, and fullerenes. Topical and systemic effects of these drugs must be tested because accurate targeting is difficult (Perillo et al. 2020).

## 1.3 Nanotechnology-Based Drug Delivery Systems

### *1.3.1 Lipid-Based Nanoparticles for Drug Delivery*

Lipid-based nanocarriers that leverage their characteristic properties to transport therapeutic agents effectively are a recent drug delivery system. Nanocarriers, such as liposomes, lipid-polymer hybrid nanoparticles, solid lipid nanoparticles (SLNs), and nanostructured lipid carriers (NLCs), are endowed with multiple benefits because of their biocompatibility, potential to encapsulate hydrophilic as well as hydrophobic drugs, and the ability to be targeted. Liposomes are vesicles with a spherical shape that have one or more phospholipid bilayers. They are widely utilized in drug encapsulation due to their superior biocompatibility. They have been studied widely in gene therapy, cancer treatment, and vaccine delivery (Xu et al. 2022). Due to their solid lipid core matrix surfactant-stabilized, SLNs have improved stability of the drug and controlled drug release. Thus they can be used in parenteral, topical, and oral drug delivery. Increased drug loading capacity and decreased drug ejection during storage characterize NLCs, a blend of SLNs and liquid lipids suitable for the delivery of poorly water-soluble drugs and cosmetic purposes. Lipid-polymer hybrid nanocarriers are investigated for cancer therapy, gene delivery, and vaccination. They have a core-shell structure with a polymer core and lipid shell, capitalizing on the best of both polymers and lipids, such as improved stability and extended release, as shown in Fig. 1.3 (Bhattacharjee et al. 2025).

These lipid nanocarriers, which are lipid-based, exhibit three principal mechanisms of drug delivery: controlled release, active targeting, and passive targeting. Because malignant tumors are often invaded by leaky vasculature, which will allow

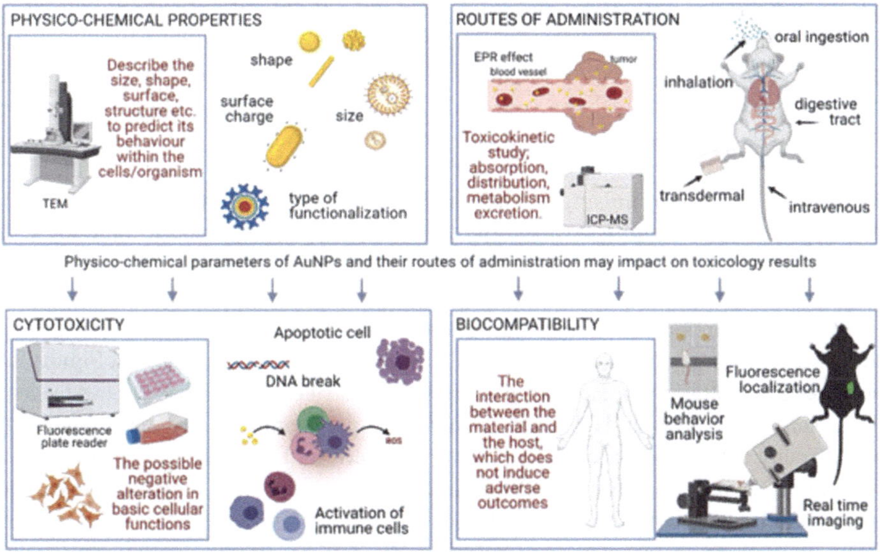


**Fig. 1.3** Generally, a schematic layout of the multiple factors that have a role in influencing the cytotoxicity and biocompatibility of gold nanoparticles (AuNPs). Their physicochemical factors may include size, shape, surface charge, route, and method of administration, which can be oral and transdermal. These parameters have significant biological interactions that lead to adverse outcomes such as apoptosis, generation of reactive oxygen species, and behavioral changes. Evaluation can be carried out at either the cellular or organism level with the help of techniques like transmission electron microscopy and inductively coupled plasma mass spectrometry. Enhanced Permeability and Retention effect. Created with BioRender (https://biorender.com, accessed on 6 October 2021). (Reproduced with permission from Kus-Liśkiewicz et al. 2021)

nanoparticles to accumulate in the tumor tissue, passive targeting frequently capitalizes on this effect, referred to as the Enhanced Permeability and Retention (EPR) effect. To enhance precision in drug delivery, active targeting involves surface-modifying nanocarriers with ligands that bind selectively to receptors on target cells. The therapeutic agent can be released more selectively and on schedule using controlled release mechanisms to respond to stimuli such as pH, temperature, or enzymes. Lipid nanocarriers have numerous advantages, including, first and foremost, that they are both biocompatible and biodegradable, which reduces the risk of toxic effects and allergic reactions often associated with other forms of drug delivery. Such biocompatibility ensures safety for patients as well as effectiveness in treatment protocols. Such nanocarriers can even enhance a medication significantly in both solubility and stability, particularly when the molecule is low in water solubility. This improves the bioavailability and therapeutic activity of the drug (Kus-Liśkiewicz et al. 2021). Their appeal is also enhanced by the potential for targeted delivery, allowing site-specific drug release, minimizing off-target effects, and maximizing the therapeutic effect at the target site of action. Notwithstanding these promising advantages, several issues must be addressed before lipid-based nanocarriers realize their full potential. To ensure the mass production of these innovative drug delivery systems at a low cost, scalable and economical manufacturing methods must be developed. The stability of the nanocarriers in storage and upon drug administration remains a significant concern because any instability could compromise the efficacy and safety of the medication. Regulatory barriers are also serious challenges because clinical approval under stringent regulations requires extensive testing and verification to ensure efficacy and safety. In addition, personalized medicine, where therapies are tailored for individuals based on their genetic and molecular analysis, is very promising for lipid-based nanocarriers in the future. Although this approach can minimize side effects and maximize therapeutic outcomes, it requires a well-understood concept of the underlying biology and sophisticated diagnostic tools (Bhattacharjee and Roy 2025a).

Research into lipid-based nanocarriers remains a highly active area with ongoing studies attempting to overcome such issues and uncover new applications. Figure 1.4 provides a schematic representation of a monolayer of curcumin-based phot osensitizer. Nanotechnology and advances in material science make next-generation lipid-based nanocarriers that are more potent, targeted, and capable of transporting a broader array of therapeutic agents possible. For instance, these nanocarriers can be optimized further for controlled release by adding stimuli-responsive elements, allowing even more targeted drug delivery. In addition, more targeted approaches can be developed with the aid of advanced imaging techniques and molecular diagnostics, ensuring that the nanocarriers very specifically target their intended sites of action. In short, lipid-based nanocarriers hold immense potential to revolutionize drug delivery systems by offering a versatile and effective platform for delivering therapeutic molecules. Their unique attributes, such as their biocompatibility, ability to encapsulate a wide range of drugs, and potential for customized delivery, make them extremely attractive to a wide range of medicinal applications. Although there are continuing challenges of scale-up, stability, regulatory acceptance, and

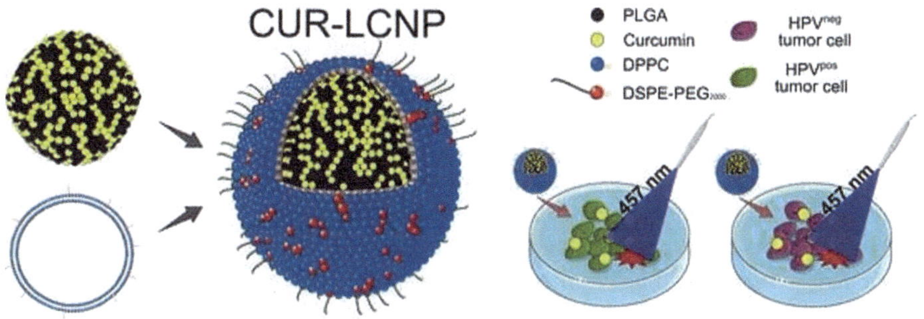


**Fig. 1.4** Schematic representation of a monolayer of curcumin-based photosensitizer (CUR-LCNPs) LPN [lipid polymer hybrid nanoparticle] preparation using a single-step approach. (Reproduced with permission from Roschenko et al. 2023)

individualized medicine, these issues ought to be addressed by ongoing research and technological advances, paving the way for lipid-based nanocarriers to be employed broadly in clinical medicine. Lipid-based nanocarriers promise to revolutionize medicine in the future by making drug therapy more effective and safe and enhancing the outcomes of patients as the science advances (Roschenko et al. 2023).

### *1.3.2 Polymeric Nanoparticles in Bio-Pharmaceuticals Development*

As pioneers in biopharmaceutical formulation, polymeric nanoparticles are a robust and versatile tool for therapeutic agent delivery, including proteins, nucleic acids, and low-molecular-weight compounds. Poly [lactic-co-glycolic acid] [PLGA], polylactic acid [PLA], and polycaprolactone [PCL], which are several advantages in medication delivery uses, are typically used to formulate the nanoparticles. One of the primary benefits of polymeric nanoparticles is their ability to provide controlled and sustained release of drugs, which can enhance therapeutic effectiveness and reduce the frequency of dosing. This is particularly useful for long-term therapies of chronic diseases. In addition, it is easy to attach various functional groups, ligands, or antibodies to the surface of polymeric nanoparticles, which enables targeted delivery to specific cells or tissues. This targeted approach lowers off-target activities and increases the drug's therapeutic index (Liu et al. 2022).

A higher dose of the drug arrives at the desired site of action due to encapsulation of drugs in polymeric nanoparticles, in addition to protecting the chemicals from degradation and premature release. This is especially important for sensitive molecules that the body's enzymes could easily degrade, like proteins and nucleic acids. Polymeric nanoparticles are extremely nontoxic and possess superior biocompatibility, thus making them suitable for a wide range of therapeutic interventions. Due to their size, these particles are more likely to penetrate into tissues and traverse the body than larger ones, which increases drug distribution and absorption (Bhattacharjee and Roy 2024c). The use of polymeric nanoparticles in cancer therapy is among the most significant advancements in the technology. Scientists can target cancer cells specifically with the help of nanoparticles by functionalizing their surface with targeting ligands such as folic acid, antibodies, or peptides. This permits sparing of healthy tissues and minimizing side effects. Targeted delivery of the particles is achieved via ligand-receptor interactions, which facilitate the cancer cells to bind to and internalize the nanoparticles more efficiently. In addition, polymeric nanoparticles can be engineered to respond to specific stimuli, including pH, temperature, or enzymes, in the tumor microenvironment, leading to the release of the medicine precisely where it is needed. Gene therapy applications for polymeric nanoparticles are also in development. Such nanoparticles are capable of encapsulating nucleic acids, like DNA, RNA, or siRNA, that may protect genetic materials from degradation and facilitate its transportation into target cells. Such therapy is likely feasible for treating infectious diseases, tumors, and hereditary issues. To allow for the escape of nucleic acids to the cytoplasm, where they can exercise their therapeutic functions, polymeric nanoparticles, for instance, can be designed to release their contents in response to the acidic nature of endosomes.

In terms of vaccine development, polymeric nanoparticles are a new approach to antigen delivery. Antigens and adjuvants can be formulated in nanoparticles for enhanced immunity and controlled antigen release that imitates natural infection patterns (Bhattacharjee and Roy 2024d). This can lead to sustained and enhanced immunity. In addition, overall vaccine effectiveness can be enhanced by co-delivering numerous antigens or adjuvants with antigens in a single nanoparticle delivery. Despite their multiple advantages, there are a few challenges involved in producing and applying polymeric nanoparticles. Reproducibility and scalability in the production of nanoparticles are two key issues. A uniform size, shape, and drug loading from one batch to another are necessary for clinical use. Improvements in production methods such as scalable emulsification processes and microfluidics partially overcome these issues. The potential for immunogenicity, where the immune system of the body recognizes the nanoparticles and attacks them, is another challenge. This could be minimized by selecting the polymer components with care and modifying the surface to reduce immune system detection. Regulatory approval is another significant challenge to polymeric nanoparticles in drug development. Due to their nature, these systems need to be well-characterized and tested to ensure their quality, safety, and effectiveness. Regulatory agencies such as the FDA and EMA establish standards to deal with these matters, yet the process is tedious and time-consuming (Silva et al. 2002).

In conclusion, polymeric nanoparticles have immense potential to advance biopharmaceutical research by making medicinal compounds targeted, controlled, and sustained release distribution possible. Their flexibility, biocompatibility, and ability to protect and deliver a wide range of drugs make them a very attractive option for treating an extensive range of diseases, such as genetic issues, infectious diseases, and cancer. Continued research and technological advancement overcome the challenges associated with their preparation and application, paving the way for their extended use in clinical applications. Polymeric nanoparticles are also poised to feature prominently in the future of medicine as the science continues to advance further, opening new avenues for effective and safe therapies (Silva 2017).

### *1.3.3 Dendrimers and Nanogels Future Trends in Drug Delivery*

There are many new drug delivery systems, such as dendrimers and nanogels, that offer unique advantages due to their novel structural features and versatility. Dendrimers are tree-like macromolecules with a clearly defined, monodisperse, three-dimensional structure. They possess numerous branches. Their capacity to control their size, shape, and surface functionality with great accuracy makes them ideal for drug administration applications (Bhattacharjee and Roy 2023). Multivalent surface of dendrimers can be functionalized with targeting ligands, imaging probes, or other bioactive compounds to enhance the therapeutic potency and selectivity of the encapsulated agents, such as small molecules, proteins, and nucleic acids. Designing dendrimer-based nanomedicines for cancer treatment, which allow for targeted delivery of chemotherapeutics to cancer cells while minimizing systemic toxicity and maximizing therapeutic outcomes, is one of the prominent themes in dendrimer research. Since dendrimers can bind to nucleic acids in stable complexes that protect them from degradation, they are also being explored for gene delivery shown in Fig. 1.5 (Shaikh et al. 2024).

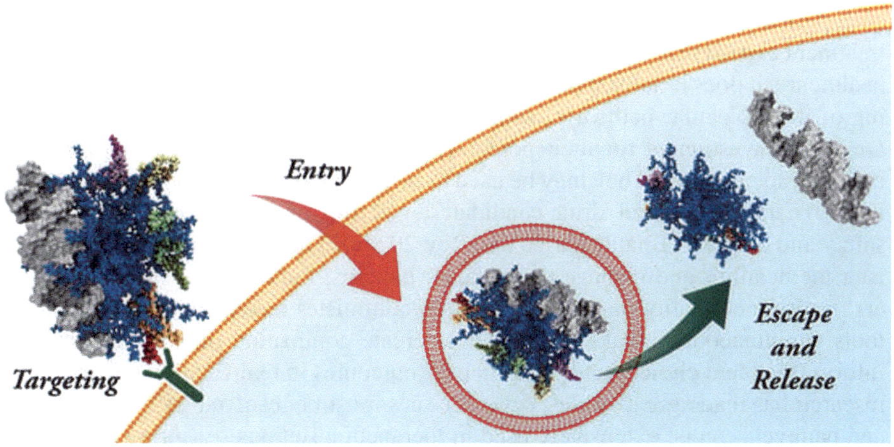


**Fig. 1.5** Schematic to show the chemical structure of a Generation 5 PAMAM dendrimer and the proposed self-assembly mechanism of dendrimers with 20 nm gold nanoparticles to generate one-dimensional ‘nano chains’, driven by electrostatic interactions. (Reproduced with permission from Laurini et al. 2019)

On the other hand, therapeutic chemicals can be encapsulated inside the matrix of three-dimensional, hydrophilic polymer networks called nanogels, which can expand in water settings. Their soft, flexible form and high water content make them biocompatible and appropriate for various drug delivery applications. Controlled and on-demand drug release is made possible by nanogels' ability to react to various physiological stimuli, including pH, temperature, and enzyme activity. Delivering medications to certain locations inside the body, such as tumors or inflammatory tissues, where the local environment is different from that of healthy tissues, is made easier by this stimuli-responsive behavior. The construction of multifunctional nanogels, or theranostics, which combines therapeutic and diagnostic properties, is one of the latest developments in nanogel research. These nanogels can co-deliver numerous medications. To monitor medication distribution and therapy response in real-time, for example, nanogels can be designed to contain both imaging probes and chemotherapeutic drugs (Laurini et al. 2019). Dendrimers and nano gels are two materials that are being combined with other nanotechnology platforms more and more to develop hybrid systems that use the advantages of different materials. Dendrimer-based nanocomposites, for instance, might include inorganic nanoparticles, like gold or iron oxide, to improve imaging properties or offer further therapeutic features, such as photothermal treatment. Likewise, liposomes or polymeric nanoparticles can be paired with nanogels to enhance drug loading ability, stability, and targeting effectiveness. The development of hybrid systems in drug delivery research is a promising avenue that holds the promise of overcoming the drawbacks of single-component systems and achieving more individualized and effective medicines.

Moreover, new avenues for personalized treatment are made possible by the molecular engineering of dendrimers and nanogels. Maximizing therapeutic success and reducing side effects is conceivable by customizing these nanocarriers to each patient's unique needs, including their genetic profile, illness status, and past treatment experiences. Although this method has immense potential for the future of healthcare, it does require sophisticated diagnostic tools and a thorough understanding of the molecular pathways behind various diseases. Dendrimers and nanogels are being investigated for their potential in drug discovery and development and their therapeutic uses. They may be used to research the pharmacokinetics and pharmacodynamics of novel drug candidates, offering important insights into their safety and efficacy (Bhattacharjee and Roy 2024e). They can also precisely administer medications and manage their release patterns. Furthermore, these nanocarriers' multifunctionality makes it possible to administer medications and diagnostic tools simultaneously, making it easier to create companion diagnostics that may inform treatment choices and track therapy outcomes in real-time. Even though the research has made great strides, several issues must be resolved before dendrimers and nanogels can be extensively used in therapeutic settings. Ensuring these nanocarriers are safe and biocompatible is one of the main problems, especially in light of their potential for accumulation in the body and long-term impacts. In-depth research is required to assess the toxicological profiles of various dendrimers and nanogels and to comprehend how they interact with biological systems. Regulatory approval procedures must also be shortened to make it easier for these nanocarriers to go from the lab to the clinic. The repeatability and scalability of dendrimer and nanogel fabrication provide another difficulty. For its therapeutic use, various batches must exhibit consistent quality and function. While developments in synthetic chemistry and nanofabrication are assisting in resolving these problems, further investigation is required to create reliable and affordable production procedures (Bhattacharjee and Roy 2024f). Even while there are still obstacles to overcome, continued research and technical developments are making it possible for these novel nanocarriers to be successfully incorporated into clinical practice, possibly completely changing how we identify, treat, and track illnesses. Dendrimers and nanogels are set to play a major role in drug delivery in the future, providing new opportunities for efficient, secure, and customized treatments as the area develops (Percec et al. 2010).

## 1.4 Nanotechnology in Biopharmaceutical Formulation

Nanotechnology plays an important role in biopharmaceutical manufacturing as it advances drug delivery, stability, and efficacy. It allows the formation of nanoscale carriers such as liposomes, nanoparticles, and micelles that improve drug solubility, protect therapeutic agents from degradation, and allow targeted delivery to specific tissues or cells based on This specificity reduces side effects and increases bioavailability. In addition, nanotechnology allows for controlled drug release, resulting in extended therapeutic action and reduced doses. As a result, nanotechnology has become a useful tool in the biopharmaceutical industry to enable generic medicine and improved patient care (Bhattacharjee et al. 2024).

### *1.4.1 Nanoscale Formulation Technique for Biotherapeutics*

Nanoscale formulation methods for biotherapeutics are emerging as feasible means to improve stability, bioavailability, and targeted delivery of therapeutic drugs. By particle size diminution in the nanometer, these formulations are able to confer water solubility, augment drug insolubility, and stabilize proteins, peptides, and other fragile biotherapeutic drugs against enzymatic degradation. Emulsification, nanoprecipitation, and lipid or polymer-based nanoparticles are typically utilized for formulating such formulations for controlled release. These nanocarriers are typically designed to possess certain surface properties enabling them to deliver selectively to cells or tissues and reduce and provide off-target effects to maximize clinical effects (Herdiana et al. 2024). The main advantage of nanoscale formulations in biomedicine is the capacity to circumvent biological barriers, e.g., the blood-brain barrier, that limit the effectiveness of many conventional drug delivery systems and control drug kinetics and biodelivery with accuracy through nanoparticle properties such as size tunability, charge, and surface functionalization Industrial Progress also facilitated the creation of multifunctional nanocarriers that can co-deliver multiple drugs or incorporate medical diagnostic functions [theranostics] For all these advantages, however, challenges await large-scale operation, regulatory approval, and long-term safety for more research and clinical validation the need is emphasized (Mat Yeh et al. 2024).

### 1.4.2 Nanoencapsulation of Bioactive Compounds

The majority of bioactive compounds, considered to be the key to human health, are hydrophobic and poor solubility. Carotenoids, essential fatty acids, essential oils, phenolic compounds, and insoluble vitamins are just a few among many. The greatest problems facing the use of such chemicals in food and pharmaceutical sectors are low stability and bioavailability.

The principle of encapsulation is the encasement of a bioactive molecule in any state—liquid, solid, or gas—into a matrix or inert substance, usually a polymer.

Encapsulated food or flavor molecules/ingredients within a wall can be stabilized, enhancing the stability of the substance in the proper environment. By being released more effectively at the active physiological site, encapsulation can enhance the stability of bioactive in food. It finds many uses in business and research applications, including foods, pharmaceuticals, cosmetics, and medicine. One of the most frequent uses is covering up bad tastes or smells in food recipes or simply changing liquid phases to solid phases (Dikici 2024). Less than 100 nm-sized particulates are formed using nanoencapsulation, and for Jafari and Nikmaram et al., the encapsulated particles have three forms: nanocapsules (less than 0.2 μm), microcapsules (from 0.2 to 5000 μm), and microcapsules (more than 5000 μm). One of the greatest advantages is homogeneity which nanoencapsulation offers and through which it causes improved encapsulation efficiency and proper physical and chemical properties.

This strategy offers a seductive substitute for reformulation of functional food ingredients, which can significantly influence the processability, bioavailability, and stability of the products (Patel and Haemmerich 2024). Today, increased focus is on the controlled release of bioactive raw materials under specific conditions, resulting in a longer shelf life and optimal nutrient consumption. There are several key factors that should be considered, such as the final particle size, ingredient activity, discharge mode, and strength of capsules.

Moreover, cost factors and wall material also become crucial factors while selling the products manufactured through this new technology.

Moreover, antioxidant activity and bioaccessibility have been shown to be desirable in recent times for delivery systems developed through nanotechnology. Establishing a food's bioaccessibility—or potential to be digested into free nutritional molecules—should come before establishing its physiological absorption. Tests for bioaccessibility are thus required to secure functional meals and determine the potential health impacts of certain BACs (Teron et al. 2024).

#### 1.4.2.1 Nanoencapsulation System Classification

The two general approaches to the fabrication of nanomaterials in nanoencapsulation technologies have classically been referred to as "top-down" and "bottom-up." The "top-down" strategy claims that the application of precise instruments during encapsulation reduces particle size.

The "bottom-up" technique constructs larger particle sizes.

Top-down is emulsification, whereas techniques such as complexation, coacervation, self-assembly and self-organization of molecules, supercritical fluids, and nanoprecipitation are bottom-up techniques. The two methods can be blended to accomplish the best outcomes. When not in use, nano-encapsulated systems are generally liquid-liquid, solid-liquid, and solid-solid systems. An O/W system (Ganguly and Margel 2023) is most commonly a liquid-liquid method used for solubilization, as shown in Fig. 1.6.

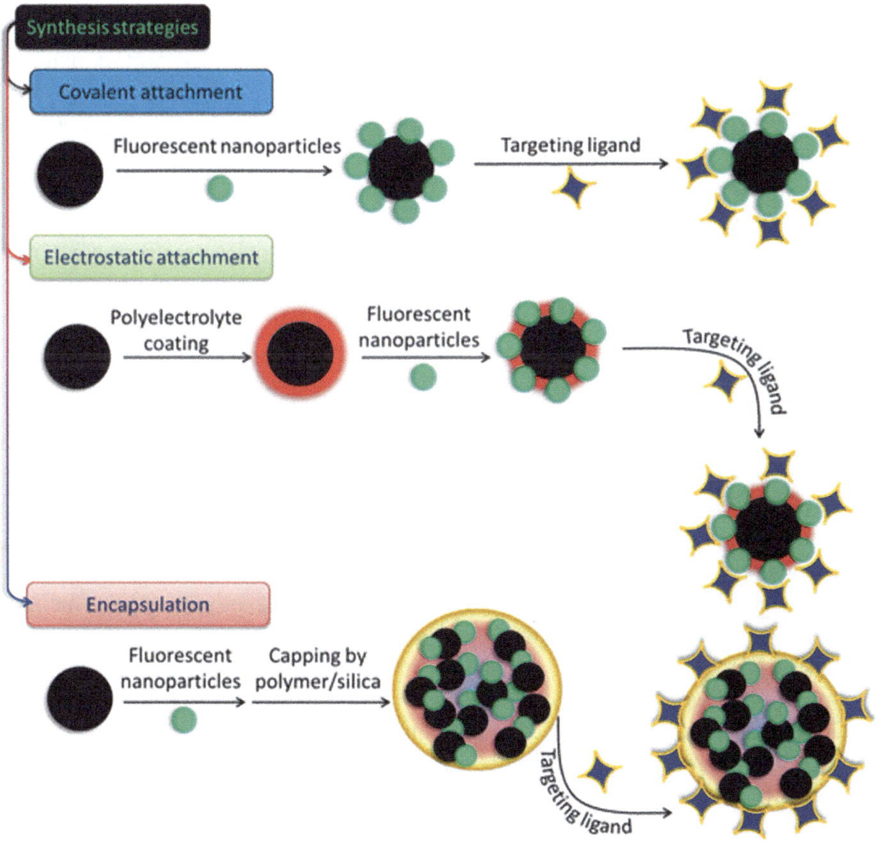


**Fig. 1.6** Presentation of bioimaging structures. (Reproduced with permission from Ganguly and Margel 2023)

### *1.4.3 Nanoparticle-Based Vaccines and Immunotherapies*

Nanoparticle-based vaccines and immunotherapy represent a promising approach to improving the efficacy and treatment of various diseases, including infectious diseases and cancer. These innovations use nanoparticles—particles on the nanometer scale —to deliver antigens drugs to specific cells or tissues in the body or administer immunosuppressants.

Nano carrier-based delivery methods have several benefits over traditional vaccination strategies, including better adjuvant properties, superior stability, and increased protection against premature degradation. Researchers can investigate alternative administration routes by using nanocarriers to encapsulate or coat the surface of an antigen, which can shield the immunogen from premature proteolytic destruction.

Besides their protective properties, nanocarriers can also lengthen the time antigen is shown to APCs and other critical immune cells, which is necessary to develop long-term immunity. They can also enhance the specificity of antigen delivery to APCs (Scattolin et al. 2024).

Many different types of nanoparticles, such as self-assembled protein nanoparticles, virus-like particles (VLPs), inorganic and polymeric nanoparticles, and liposomes, have been investigated as possible antigen carriers for vaccination applications. Viral antigens have been effectively delivered using biocompatible inor]ganic nanoparticles like carbon, silica, and gold.

By merging with the target cell membrane, liposomes can deliver vaccinations and stimulate the immune system. Liposomes exhibit great versatility since their aqueous core facilitates the easy incorporation of hydrophilic molecules into the nanoparticles, while hydrophobic compounds can potentially be enclosed within the phospholipid bilayer.

Due to their capacity to elicit strong host immunological responses, gold nanoparticles have demonstrated exceptional efficacy in transporting bacterial and viral antigens. In vivo, vaccinations against influenza, HIV, foot and mouth disease, and tuberculosis have all been employed with gold nanoparticles. Inorganic nanoparticles such as silica and gold have several benefits for vaccine development procedures, including being inexpensive, highly repeatable, and linked to favorable safety profiles (Osman et al. 2024).

Studies using nanoparticles as a platform for vaccine development have included a variety of liposome forms, such as unilamellar and multilamellar vesicles made of biodegradable phospholipids including cholesterol, phosphatidylcholine, and phosphatidylserine.

Numerous liposome forms, including unilamellar and multilamellar vesicles made of biodegradable phospholipids such as phosphatidylserine, phosphatidylcholine, and cholesterol, have been used in nanoparticle-based vaccination experiments.

## 1.5 Nanotechnology in Diagnostics and Imaging

### *1.5.1 Nanosensors for Biomedical Applications*

Nanosensors are transforming biomedical applications because they offer extraordinary sensitivity and specificity for illness detection, monitoring, and treatment. Utilizing the special qualities of nanomaterials, such as high surface area-to-volume ratios, quantum effects, and tunable physical and chemical properties, these minuscule sensors function at the molecular or cellular level as shown in Fig. 1.7 (Zahra et al. 2024).

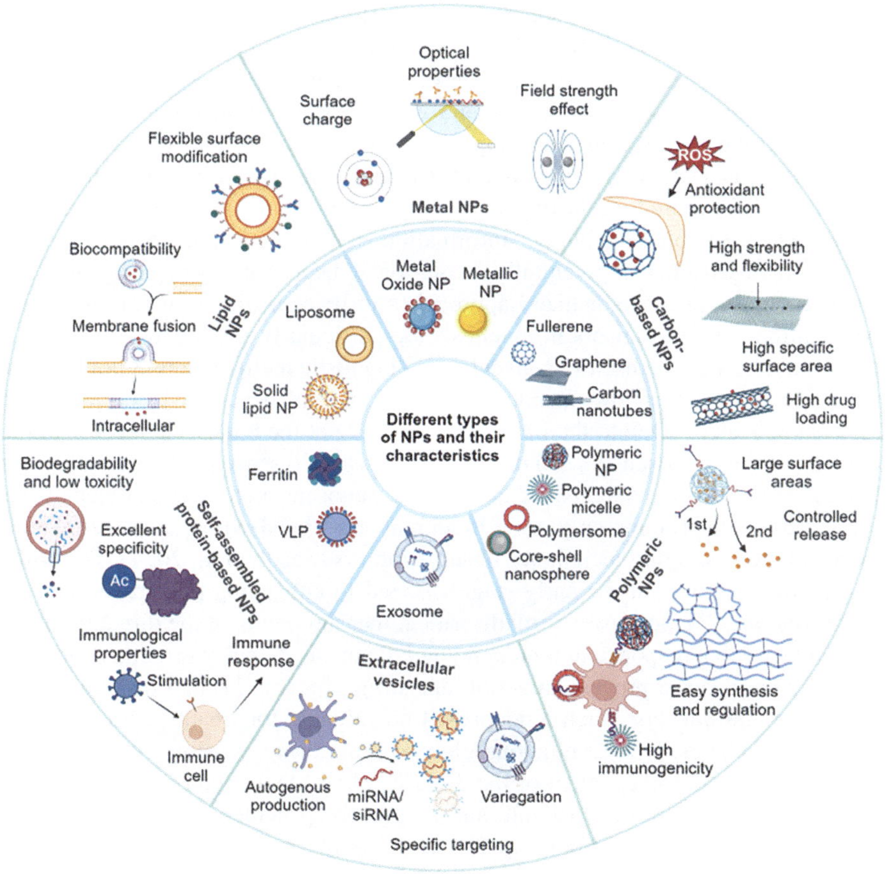


**Fig. 1.7** Schematic of six common nanomaterials and their essential properties are as follows: lipid nanoparticles (biocompatibility, surface modifiability), metallic nanoparticles (optical/electronic properties), carbon-based nanomaterials (high tensile strength, drug loading), polymer nanoparticles (structural diversity), self-assembled nanoparticles (biodegradability, immune response), and exosomes (natural vesicles for intercellular communication). (Reproduced with permission from Huang et al. 2024)

Nanosensors are beneficial for diagnostics since they can detect biomolecules at very low concentrations, facilitating early detection of illness and tracking the progression of a disease or the effectiveness of treatment in real time. To deliver rapid and reliable diagnosis, they can detect viruses, genetic aberrations, and cancer biomarkers. Nanosensors are crucial to drug delivery systems used in therapy because they ensure therapeutic chemicals reach the disease location precisely and minimize the effects of side reactions while optimizing treatment efficacy. Due to the ability to track physiological information continually and adjust treatment depending on every specific patient, sensors are equally vital to personal medicine. Usefulness is enhanced for nanosensors by wearable and implanted devices with continuous health monitoring and rapid medical intervention. Smart delivery of drugs, where sensors could dispense medicines in response to particular physiological changes to enhance therapy precision, is achieved because of breakthroughs in the nanosensor field (Huang et al. 2024). In addition, nanosensors in regenerative medicine can monitor the microenvironment within artificial organs or tissues and supply feedback to enhance growth conditions and ensure tissue engineering success. This field's ongoing research and development endeavor to resolve several problems, including sensor stability, biocompatibility, and data integration with healthcare systems to offer more long-lasting and reliable biomedical applications. Finally, the convergence of biotechnology, information technology, and nanotechnology through nanosensors can potentially revolutionize healthcare by offering more efficient, accurate, and personalized approaches to disease diagnosis and treatment. Applications of nanosensors in diagnosis are particularly groundbreaking. During the initial phases of a disease, conventional diagnostic methods might not be sensitive enough to detect low-abundance biomarkers and occasionally require quite large sample sizes. Due to their ability to interface at the nanoscale level with biological entities, nanosensors can detect trace levels of disease markers to allow earlier and more accurate diagnosis. For example, nanosensors for cancer detection can sense particular proteins or genetic mutations associated with cancer cells, many months before imaging techniques observe tumors (Saadh et al. 2024). The ability to diagnose diseases at the early stages is vital to enhancing patient survival and prognosis rates. Nanosensors provide critical measurements of the direction of disease progression, response to treatment, and diagnosis. Sensors provide continuous feedback on disease process direction and drug efficacy through disease activity biomarkers monitoring. Such a talent will be valuable for treating chronic conditions like diabetes, with continuous glucose monitoring possibly revolutionizing patient status and management of the disease. Wearable technology equipped with nanosensors may provide patients and healthcare providers access to real-time health information, enabling faster and more informed decision-making. Nanosensors enhance the accuracy and effectiveness of therapy in therapeutic devices. Systemic drug delivery via pharmaceutical administration is prevalent in conventional drug delivery devices, which might lead to adverse side effects and off-target effects (Saadh et al. 2024). By restricting the release of drugs to the exact sites where they are needed, nanosensors enable customized drug delivery. In cancer therapy, chemotherapeutic agents can be delivered specifically to tumor cells by nanosensors, reducing damage to normal organs and decreasing side effects. This targeted strategy improves the patient's quality of life during therapy and enhances the efficacy of therapies. Another area where nanosensors are very powerful is personalized medicine, which tailors treatment for individual patients based on their biological characteristics. Nanosensors provide accurate patient health data by monitoring physiological parameters, such as blood pressure, glucose levels, and other biomarkers. Physicians can adapt therapies to the needs of each patient using this real-time information, leading to more effective and tailored care. For example, heart rate and blood pressure fluctuations are monitored by nanosensors to aid in the control of cardiovascular disorders. This ensures that the medicine dosage is optimized and undesirable effects are avoided. The list of applications and availability of nanosensors continues to grow as they are combined with wearable and implantable devices. Wearable health monitors are equipped with nanosensors, which can analyze several health metrics such as body temperature, heart rate, and respiration rate. This information is beneficial for early diagnosis and preventive medicine. On the other hand, implantable nanosensors track continuous physiological status, particularly for patients with chronic medical conditions or those undergoing treatment (Ahmad et al. 2024). These implanted sensors enable timely and proactive medical interventions by alerting healthcare providers to any issues before their deterioration. Nanosensors are critical for monitoring and guiding tissue engineering efforts in regenerative medicine. The microenvironment has to be finely tuned for engineered tissues and organs to grow and function correctly. Researchers can fine-tune tissue development conditions by integrating nanosensors within these tissues to get real-time feedback on pH levels, oxygenation, and nutrient supply. The establishment of functional artificial organs that can replace patients' diseased or damaged tissues is contingent upon this feedback mechanism, and this is central to the success of tissue engineering research. Despite positive trends, many challenges remain to be met before nanosensors in biological use can achieve their full potential (Bhattacharjee and Roy 2025b). Maintaining the stability and reliability of nanosensors in biological situations is a major challenge. Long-term operation and tolerance of the intricate and dynamic conditions present in the human body are specifications of nanosensors. Researchers are studying different methods to improve nanosensors' stability and longevity, including using biocompatible materials and protective coatings. Biocompatibility is another major factor, in that when incorporated into the body, nanosensors should prevent toxicity or unhelpful immune responses. Biocompatible functional materials for nanosensors form a key research area of this study. Technical and administrative challenges are faced while merging nanosensor information with healthcare infrastructures. Advanced data structuring and analysis tools must be utilized for efficient collection, management, and understanding of the enormous amounts of data produced by nanosensors. It is also important to safeguard data confidentiality and security as sensitive health information needs to be protected from unauthorized access. Addressing these concerns requires interdisciplinary collaboration between scientists, engineers, and clinicians. By integrating biotechnology, information technology, and nanotechnology using nanosensors, healthcare can be revolutionized entirely. Nanosensors can significantly improve patient outcomes and healthcare quality by providing more accurate, quick, and personalized ways of managing and treating disease. The extensive utilization of nanosensors in healthcare practice is envisioned as research and development in the field proceeds, ushering a new era of precision medicine and innovative healthcare interventions (Munawar et al. 2019).

### *1.5.2 Nanoparticles in Molecular Imaging*

With their ground-breaking capacities for the visualization, diagnosis, and treatment of illnesses at the molecular and cellular levels, nanoparticles have completely transformed the area of molecular imaging. High surface area-to-volume ratios, customizable sizes, and adaptable surface functions are just a few of their distinctive physical and chemical characteristics. These make them perfect options for improving imaging methods' sensitivity, specificity, and resolution. Due to their limits in resolution and contrast, traditional imaging methods sometimes have trouble identifying minor clinical alterations or early-stage illnesses (Ding et al. 2024). On the other hand, contrast agents such as magnetic resonance imaging (MRI), computed tomography (CT), positron emission tomography (PET), and optical imaging can be significantly enhanced by engineering nanoparticles. For example, because of their superparamagnetic characteristics, iron oxide nanoparticles are frequently utilized as contrast agents in magnetic resonance imaging (MRI) to improve signal contrast and enable fine-grained visualization of tissues and organs.

Similarly, because of their high atomic number, which raises X-ray attenuation, gold nanoparticles can improve contrast in CT imaging. Furthermore, one form of semiconductor nanoparticle known as quantum dots has special optical characteristics, including strong photostability and size-tunable fluorescence, which makes them very useful for optical imaging applications like in vivo imaging and fluorescence microscopy. Nanoparticles in molecular imaging offer a notable benefit: they may be functionalized with diverse biomolecules, including peptides, small molecules, and antibodies (Ruiz-Sanchez et al. 2017). The particular targeting of sick cells or tissues by nanoparticles is made possible by this functionalization, which improves imaging precision and specificity. To precisely localize and characterize malignancies, for instance, nanoparticles coated with antibodies that recognize cancer-specific antigens can attach to tumor cells preferentially. By minimizing background signals from non-targeted locations, this tailored technique enhances illness identification and produces more precise and more informative pictures, as given in Fig. 1.8.

One of the most important benefits of employing nanoparticles for molecular imaging is that nanoparticles can be functionalized with various biomolecules, i.e., antibodies, small molecules, and peptides. Functionalization with biomolecules improves imaging sensitivity and specificity, and provides for nanoparticle targeting of sick cells or tissue. For instance, antibody-antigen specificity in nanoparticles makes it possible to selectively target malignant cells by binding tumor-specific

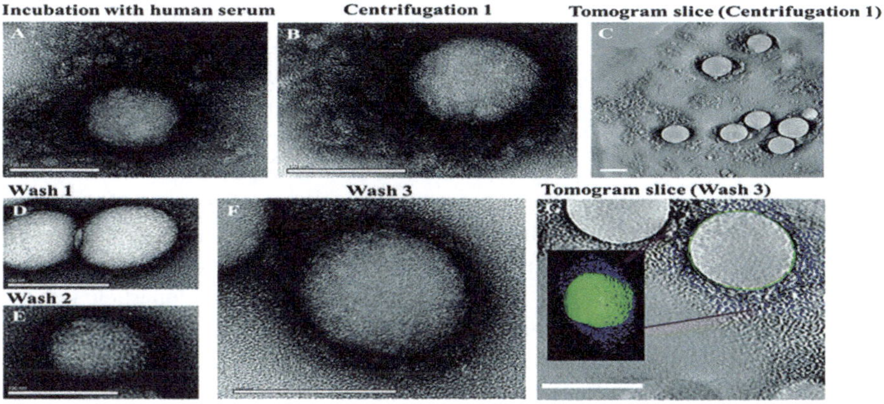


**Fig. 1.8** TEM micrographs of the protein corona surrounding un-functionalized polystyrene nanoparticles. (**a**) Hard and soft corona after incubation of un-functionalized PS-NPs in human serum, and (**b**) after centrifugation and removal of the supernatant, but before washing steps. (**c**) Tomogram slice of an area such as [**b**], scale bar: 100 nm. (**d**) Hard corona after first wash, (**e**) after second wash, and (**f**) after third wash. (**g**) Electron tomogram slice of the area presented in [**f**], superimposed by a 3D reconstruction. Scale bar: 100 nm. (Reproduced with permission from Kokkinopoulou et al. 2017)

antigens on these cells (Kokkinopoulou et al. 2017). This tailored approach lowers background signals in non-target locations, enhancing the capacity to recognize diseases while generating clearer and more insightful images. Nanoparticles are also key to enhancing the performance of advanced imaging techniques such as multimodal imaging, where two or more imaging modalities are combined to provide complementary and comprehensive information regarding biological systems. Scientists can overcome the limitations of single-modality imaging and obtain a more accurate and detailed view of disease mechanisms by designing nanoparticles accessible to multiple imaging modalities. For dual optical and PET imaging, fluorescent dye- and radioactive isotope-labeled nanoparticles can offer high-resolution anatomical details and functional and metabolic information. This multimodal approach is highly effective for cancer diagnosis and treatment as it allows precise tumor localization, tumor metabolism evaluation, and assessment of treatment response in a single imaging session. In addition, target sites inaccessible to conventional imaging probes are accessible by nanoparticles because of their small size and large surface area, which permit them to traverse biological barriers (Bhattacharjee and Roy 2024g). Given the challenges in delivering diagnostic and therapeutic agents to the brain because of the blood-brain barrier (BBB), this ability is significant for applications involving imaging in the brain. Neurological diseases such as Parkinson's disease, Alzheimer's disease, and tumors in the brain are easier to diagnose owing to the ability of nanoparticles to cross the blood-brain barrier and target imaging probes to specific brain regions. Neurological diseases can be correctly diagnosed and detected at an early stage with the help of nanoparticles that are functionalized with ligands targeting BBB transport mechanisms. Nanoparticles can cross the barrier efficiently and enhance imaging contrast in the brain. Extensive examination of a nanoparticle's pharmacokinetics, safety, and biocompatibility is also important while developing it for molecular imaging. Effective application of nanoparticles to clinical practice is ensured if they do not exhibit toxicity or negative immunological effects. Researchers are now examining several ways to enhance nanoparticle biocompatibility, such as incorporating biodegradable materials, coating them with biocompatible polymers, and optimizing their size and surface charge. Tailoring the nanoparticle design and ensuring their efficient and safe use in vivo also needs to consider their pharmacokinetics, meaning their dispersion, metabolism, and excretion. Dynamic contrast-enhanced imaging and nanoparticle tracking analysis are two sophisticated imaging technologies applied to investigate nanoparticles' activity in biological systems and tailor their properties. Nanoparticles are driving the development of new imaging techniques and technology and improving existing imaging modalities (Li et al. 2018). For example, applying nanoparticles as a contrast agent has enhanced photoacoustic imaging, combining ultrasonic and optical imaging. High-resolution imaging of tissues with deep penetration is facilitated by amplification of the photoacoustic signal by gold nanoparticles and other nanomaterials with high optical absorption properties. Likewise, investigations on the use of nanoparticles in magnetic particle imaging (MPI) have just begun. With the magnetic properties of nanoparticles, MPI is a new imaging modality that generates highly contrasted, sensitive images free from ionizing radiation. Nanoparticles are ideal for MPI applications due to their unique magnetic behavior and response to alternating magnetic fields. Such behavior can benefit imaging performance and safety. Personalized and precision medicine is being pushed forward by integrating nanoparticles with molecular imaging technologies. Nanoparticles enable the identification of specific disease markers and monitoring each patient's response to treatment by providing detailed molecular and cellular information (Bhattacharjee and Roy 2024h). Due to this personalized modality, it is also possible that better and customized treatments are attainable, enabling personalized treatment strategies based on the unique aspects of the patient's disease. Additionally, simultaneous administration of therapeutics and imaging probes on the nanoparticle will make personalized theranostic platforms more achievable, in which real-time treatment adjustment can be made per imaging input. Nanoparticles in molecular imaging hold promise in the future due to ongoing investigations and technological advances. The number of available imaging agents is being expanded, and the development of new nanomaterials like polymeric nanoparticles, metal-organic frameworks, and carbon-based nanoparticles is enhancing their performance. Advances in nanoparticle production and functionalization techniques enable the production of highly selective and multifunctional imaging probes. In addition, advances in data analysis methods and imaging instruments increase the quantitative sensitivity and resolution of nanoparticle imaging (Bhattacharjee and Roy 2024i).

### *1.5.3 Quantum Dots and Nanoprobes for Bioimaging*

The physicochemical and optical peculiarities of quantum dots [QDs) Nanoprobes make them valuable tools for bioimaging on the molecular and cellular levels. Size-tunable fluorescence is one of the characteristics of semiconductor nanocrystals referred to as quantum dots, making it possible to have precise control over the emission wavelength via size adjustment. Due to tunability, different QDs with different sizes and, therefore, different colors can be used in combination to target different cellular constituents or chemical analytes, facilitating multiplexed imaging. Their higher photostability than traditional organic dyes ensures prolonged imaging time without significant signal deterioration, making them ideal for long-term studies. In addition, the narrow and symmetric emission band of QDs and the large excitation band reduce overlap among different fluorescence signals, making them more valuable for demanding imaging applications (Bhattacharjee et al. 2024). QDs are often combined with biomolecules such as peptides, antibodies, or nucleic acids in bioimaging to generate nanoprobes that can specifically bind to target molecules and image specific proteins, nucleic acids, or other biological structures. The nanoprobe-QD conjugates can be utilized to track the distribution and dynamics of biomolecules in living organisms through different imaging techniques like flow cytometry, fluorescence microscopy, and in vivo imaging. In addition to basic research, QDs and nanoprobes are increasingly used in the clinical environment for disease diagnosis and other similar work. Applications include the early and specific detection of biomarkers associated with cancers. For example, QDs can specifically label cancer cells, diagnose early tumors, and potentially improve patient outcomes, as shown in Fig. 1.9 (Nikolaev et al. 2024).

In addition, the surface chemistry of QDs can be altered to reduce toxicity and enhance biocompatibility, two areas of their clinical translation that have been very challenging. More recently, advances in QD synthesis have yielded heavy metal-free, non-toxic QDs better adapted for biological use. The synthesis of near-infrared (NIR) QDs, capable of emitting light in the NIR spectral range where tissue absorption and autofluorescence are minimized, and allowing for more distinct, deeper imaging of tissue, is the focus of another area of research. Combining QDs with other nanomaterials, including gold or magnetic nanoparticles, has also been explored to create multifunctional nanoprobes capable of integrating therapeutic functions, such as drug delivery or photothermal therapy, into imaging (Bhattacharjee and Roy 2024j; Aslam et al. 2023; De et al. 2023). These multiuse probes, or theranostic agents, bring personalized medicine within reach, uniting therapy and diagnostics in a single targeted approach.

Finally, quantum dots and nanoprobes open new, previously unimaginable prospects for the visualization and understanding of complex biological systems, representing a quick progression toward bioimaging. They are highly valuable tools for clinical and research applications because of their versatility, tunability, and ability to be functionalized for specific purposes. In the coming years, they may significantly impact therapy, diagnostics, and personalized medicine fields (De et al. 2023; Liga et al. 2024; Halder et al. 2025).

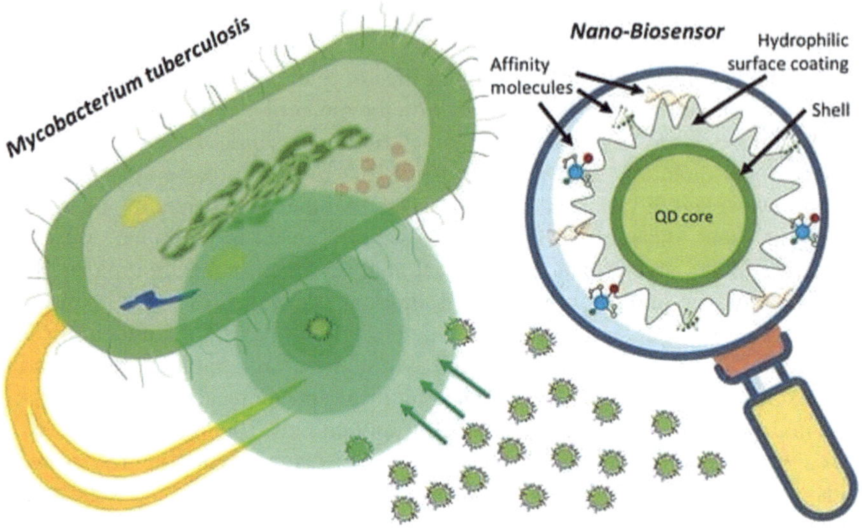


**Fig. 1.9** Quantum dots application in vitro imaging. (Reproduced with permission from Nikolaev et al. 2024)

## 1.6 Conclusions

Several developed and even emerging nations have shown concern with the increasing cost of healthcare in recent times. Governments need to better understand nanopharmaceuticals' cost-effectiveness to maximize cost efficiency. The initial step towards establishing this market, which should be emphasized, is a standardized cost-effectiveness analysis that indicates whether the additional health benefits of nanopharmaceuticals compared to traditional formulations would be worth the extra cost. Regulatory authorities now intend to set guidelines for this development and estimate the monetary advantages of cultivating this business. A few hindrances interfere with the profitable commercialization of nanomedicines, well beyond the prevailing scarcity of such rudimentary data. Patients' and citizens' trust in nanomedicines' efficacy and safety, as well as that of doctors, is essential for the market's development. To hurdle this challenge, the future market must increase public awareness of the benefits, risks, and safety issues related to nanopharmaceuticals. Despite the aforementioned challenges, the nanopharmaceuticals and nanomedicines market will expand during the next 5 years. These drivers will be mainly due to innovations in nanoengineering and bionanotechnology, strict regulations for nanotechnology-based new medicines, more government funding, greater consensus on environmental matters, and the partnership between the dominant pharmaceutical and medical companies and nanomedicine start-ups or R&D units. In September 2016, of the 17 FDA-approved drugs, five were related to cancer.

This, together with the sheer number of other cancer drugs that have been approved over the past few years, indicates both the scale of the cancer therapy market and the patients' urgent need for better cancer therapies. Also promising is the combination of photothermal or photodynamic treatment with chemotherapy. These are drug-device combinations; thus, obtaining full approval can be more difficult. Immune system diseases are also of extreme importance to the pharmaceutical industry. To fight infections and cancer, such applications may involve up- and down-regulating the immune system to combat allergies and autoimmune diseases. The increasing number of deaths from cancer is what drives the projected growth in the size of the global nanomedicine market in the coming years.

**Acknowledgement** Author, S. Roy would like to acknowledge 'Scheme for Transformational and Advanced Research in Sciences [STARS]' [MoE-STARS/STARS-2/2023-0175] by the Ministry of Education, Govt. of India for promoting translational India-centric research in sciences implemented and managed by the Indian Institute of Science [IISc], Bangalore, for their support.

**Conflict of Interests** The authors declare no conflict of interest.

## References

Ahmad S, Ahmad S, Ali S, Esa M, Khan A, Yan H (2024) Recent advancements and unexplored biomedical applications of green synthesized Ag and Au nanoparticles: a review. Int J Nanomedicine 19:3187–3215. https://doi.org/10.2147/IJN.S453775

Aslam M, Ahmad T, Manzoor MH, Laiba, Verpoort F (2023) MXenes as theranostics: diagnosis and therapy including in vitro and in vivo applications. Appl Mater Today 35:102002. https://doi.org/10.1016/j.apmt.2023.102002

Banerjee A, Qi J, Gogoi R, Wong J, Mitragotri S (2016) Role of nanoparticle size, shape and surface chemistry in oral drug delivery. J Control Release 238:176–185. https://doi.org/10.1016/j.jconrel.2016.07.051

Begines B et al (2020) Polymeric nanoparticles for drug delivery: recent developments and future prospects. Nano 10(7):1403. https://doi.org/10.3390/nano10071403

Bhattacharjee J, Roy S (2023) Green remediation of microplastics using bionanomaterials. In: Remediation of plastic and microplastic waste. CRC Press, pp 240–260

Bhattacharjee J, Roy S (2024a) Synthesis of thermoelectric nanocomposites by incorporating reduced graphene oxide. In: Archives of advanced engineering science. Bon View Publishing PTE. https://doi.org/10.47852/bonviewaaes42023514

Bhattacharjee J, Roy S (2024b) Synergistic insights: electro-organic photocatalysis and nanostructures. Chem Pap 78:8077–8105. https://doi.org/10.1007/s11696-024-03657-2

Bhattacharjee J, Roy S (2024c) Density functional waltz: molecular theory of spin dynamics. J Phys Opt Sci 6:1–4. https://doi.org/10.47363/jpsos/2024(6)242

Bhattacharjee J, Roy S (2024d) A review on photocatalysis and nanocatalysts for advanced organic synthesis. Hybrid Adv 6:100268. https://doi.org/10.1016/j.hybadv.2024.100268

Bhattacharjee J, Roy S (2024e) Review on green resources and AI for biogenic solar power. Energ Stor Conv 2(1):457

Bhattacharjee J, Roy S (2024f) Exploration of molybdenum oxide compounds—a review. Catal Res 4(3):1–19. https://doi.org/10.21926/cr.2403011

Bhattacharjee J, Roy S (2024g) Smart materials for sustainable energy. In: Natural resources conservation and research, vol 7. EnPress Publisher, p 5536

Bhattacharjee J, Roy S (2024h) Significance of renewable energy in water management and irrigation. In: Water resources development and management. Springer Nature Singapore, pp 235–252

Bhattacharjee J, Roy S (2024i) Utilizing a variable material approach to combat climate change. Mater Sci Res India 20(3):141–145

Bhattacharjee J, Roy S (2024j) Nature-inspired catalysts: a new era for water-splitting technology. Oriental Scientific Publishing Company. https://doi.org/10.13005/msri/210201

Bhattacharjee J, Roy S (2025a) A comprehensive review on integrated photo rechargeable batteries- supercapacitors, and their techno-economic feasibility. J Photochem Photobiol 25:100257. https://doi.org/10.1016/j.jpap.2024.100257

Bhattacharjee J, Roy S (2025b) Soft computing techniques in wind conversion systems: a review. In: Soft computing in renewable energy technologies. CRC Press, pp 121–142

Bhattacharjee J, Bhakta D, Roy S (2024) Characteristics, applications, and limitations of nanocomposites in biosensing. Adv Chem Mater Eng:161–191. https://doi.org/10.4018/979-8-3693-1094-6.ch006

Bhattacharjee J, Roy S, Shaikh AA, Datta P (2025) Efficient fabrication and characterization of doped nanocomposites for thermoelectric materials. In: Nano trends, vol 10. Elsevier BV, p 100109. https://doi.org/10.1016/j.nwnano.2025.100109

De A, Bhattacharjee J, Chowdhury SR, Roy S (2023) A comprehensive review on third-generation photovoltaic technologies. J Chem Eng Res Updates 10:1–17. https://doi.org/10.15377/2409-983X.2023.10.1

Dikici S (2024) Enhancing wound regeneration potential of fibroblasts using ascorbic acid-loaded decellularized baby spinach leaves. Polym Bull 81(11):9995–10016. https://doi.org/10.1007/s00289-024-05185-1

Ding L, Agrawal P, Singh SK, Chhonker YS, Sun J, Murry DJ (2024) Polymer-based drug delivery systems for cancer therapeutics. Polymers 16(6):843. https://doi.org/10.3390/polym16060843

Fraczek W, Kregielewski K, Wierzbicki M, Krzeminski P, Zawadzka K, Szczepaniak J, Grodzik M (2023) A comprehensive assessment of the biocompatibility and safety of diamond nanoparticles on reconstructed human epidermis. Materials 16(16):5600. https://doi.org/10.3390/ma16165600

Ganguly S, Margel S (2023) Bioimaging probes based on magneto-fluorescent nanoparticles. Pharmaceutics 15(2):686. https://doi.org/10.3390/pharmaceutics15020686

Halder S, Bhattacharjee J, Roy S, Das BC, Biswas AB (2025) Surface modification of lignin by Cetyl Trimethyl ammonium bromide. In: Jawaid M, Ahmad A, Meraj A (eds) Handbook of lignin. Springer, Singapore. https://doi.org/10.1007/978-981-97-2664-6_38-1

Herdiana Y, Febrina E, Nurhasanah S, Gozali D, Elamin KM, Wathoni N (2024) Drug loading in chitosan-based nanoparticles. Pharmaceutics 16(8):1043. https://doi.org/10.3390/pharmaceutics16081043

Huang S, Ding X (2022) Precise design strategies of nanotechnologies for controlled drug delivery. J Funct Biomater 13(4):188. https://doi.org/10.3390/jfb13040188

Huang Y, Guo X, Wu Y et al (2024) Nanotechnology's frontier in combatting infectious and inflammatory diseases: prevention and treatment. Signal Transduct Target Ther 9:34. https://doi.org/10.1038/s41392-024-01745-z

Kaur H et al (2024) Unravelling role of crocin for the treatment of Alzheimer's and Parkinson's disease: sojourn from food to nanomedicine. Phytochem Rev 23:1737–1760. https://doi.org/10.1007/s11101-024-09936-w

Khan I, Saeed K, Khan I (2019) Nanoparticles: properties, applications and toxicities. Arab J Chem 12(7):908–931. https://doi.org/10.1016/j.arabjc.2017.05.011

Kokkinopoulou M, Simon J, Landfester K, Mailänder V, Lieberwirth I (2017) Visualization of the protein corona: towards a biomolecular understanding of nanoparticle-cell-interactions. Nanoscale 9(25):8858–8870. Royal Society of Chemistry (RSC). https://doi.org/10.1039/c7nr02977b

Kus-Liśkiewicz M, Fickers P, Ben TI (2021) Biocompatibility and cytotoxicity of gold nanoparticles: recent advances in methodologies and regulations. Int J Mol Sci 22(20):10952. https://doi.org/10.3390/ijms222010952

Laurini E, Marson D, Aulic S, Fermeglia M, Pricl S (2019) Evolution from covalent to self-assembled PAMAM-based dendrimers as nanovectors for siRNA delivery in cancer by coupled in silico-experimental studies. Part II: Self-Assembled siRNA nanocarriers. Pharmaceutics 11(7):324. https://doi.org/10.3390/pharmaceutics11070324

Li BS, Wang TG, Ding J, Cai Y, Shi J, Zhang X (2018) Influence of $N_2$/Ar flow ratio on microstructure and properties of the AlCrSiN coatings deposited by high-power impulse magnetron sputtering. Coatings 8:3. https://doi.org/10.3390/coatings8030003

Liga S, Paul C, Moacă E-A, Péter F (2024) Niosomes: composition, formulation techniques, and recent progress as delivery systems in cancer therapy. Pharmaceutics 16(2):223. https://doi.org/10.3390/pharmaceutics16020223

Liu P, Chen G, Zhang J (2022) A review of liposomes as a drug delivery system: current status of approved products, regulatory environments, and future perspectives. Molecules 27(4):1372. https://doi.org/10.3390/molecules27041372

Mansoori G, Fauzi Soelaiman T (2005) Nanotechnology — an introduction for the standards community. J ASTM Int 2(6):1–22. https://doi.org/10.1520/JAI13110

Mat Yeh RM, Taha BA, Bachok NN, Sapiee NM, Othman AR, Karim NHA, Arsad N (2024) Advancements in detecting porcine-derived proteins and DNA for enhancing food integrity: taxonomy, challenges, and future directions. Food Control 161:110399. https://doi.org/10.1016/j.foodcont.2024.110399

Mazayen ZM, Ghoneim AM, Elbatanony RS, Basalious EB, Bendas ER (2022) Pharmaceutical nanotechnology: from the bench to the market. Futur J Pharm Sci 8(1):12. https://doi.org/10.1186/s43094-022-00400-0

Munawar A, Ong Y, Schirhagl R, Tahir MA, Khan WS, Bajwa SZ (2019) Nanosensors for diagnosis with optical, electric and mechanical transducers. RSC Adv 9(12):6793–6803. Royal Society of Chemistry (RSC). https://doi.org/10.1039/c8ra10144b

Nelson Sathy B, Anitha A, Nair SV, Furuike T, Tamura H, Jayakumar R (2010) Preparation of poly[lactic acid]/chitosan nanoparticles for anti-HIV drug delivery applications. Carbohydr Polym 80:833–838. https://doi.org/10.1016/j.carbpol.2009.12.040

Nikolaev VV, Lepekhina TB, Alliluev AS, Bidram E, Sokolov PM, Nabiev IR, Kistenev YV (2024) Quantum dot-based nanosensors for in vitro detection of mycobacterium tuberculosis. Nano 14(19):1553. https://doi.org/10.3390/nano14191553

Nikolova MP, Joshi PB, Chavali MS (2023) Updates on biogenic metallic and metal oxide nanoparticles: therapy, drug delivery, and cytotoxicity. Pharmaceutics 15(6):1650. https://doi.org/10.3390/pharmaceutics15061650

Osman AI, Zhang Y, Farghali M, Rashwan AK, Eltaweil AS, Abd El-Monaem EM, Mohamed IMA, Badr MM, Ihara I, Rooney DW, Yap P-S (2024) Synthesis of green nanoparticles for energy, biomedical, environmental, agricultural, and food applications: a re. Environ Chem Lett 22(2):841–887. https://doi.org/10.1007/s10311-023-01682-3

Patel PR, Haemmerich D (2024) Re on electrospray nanoparticles for drug delivery: exploring applications. Polym Adv Technol 35(7). https://doi.org/10.1002/pat.6507

Percec V, Wilson DA, Leowanawat P, Wilson CJ, Huges AD, Kaucher MS, Hammer DA, Levine DH, Kim AJ, Bates FS et al (2010) Self-assembly of Janus dendrimers into uniform dendrimersomes and other complex architectures. Science 328:1009–1014

Perillo B et al (2020) ROS in cancer therapy: the bright side of the moon. Exp Mol Med 52(2):192–203. https://doi.org/10.1038/s12276-020-0384-2

Roschenko V, Ayoub AM, Engelhardt K, Schäfer J, Amin MU, Preis E, Mandic R, Bakowsky U (2023) Lipid-coated polymeric nanoparticles for the photodynamic therapy of head and neck squamous cell carcinomas. Pharmaceutics 15(10):2412. https://doi.org/10.3390/pharmaceutics15102412

Ruiz-Sanchez AJ, Parolo C, Miller BS, Gray ER, Schlegel K, McKendry RA (2017) Tuneable plasmonic gold dendrimer nanochains for sensitive disease detection. J Mater Chem B 5(35):7262–7266. Royal Society of Chemistry (RSC). https://doi.org/10.1039/c7tb01394a
Saadh MJ, Rasheed YS, Sdiq ZM, Chandra S, Lateef SR, Ghnim ZS, Zain Al-Abdeen SH, Hawas MN, Elmasry Y (2024) BN-biphenyl nanosheet as a potential drug carrier for 5-fluorouracil: a DFT investigation. Inorg Chem Commun 160:111857. https://doi.org/10.1016/j.inoche.2023.111857
Scattolin T, Tonon G, Botter E, Canale VC, Hasanzadeh M, Cuscela DM, Buschini A, Zarepour A, Khosravi A, Cordani M, Rizzolio F, Zarrabi A (2024) Synergistic applications of cyclodextrin-based systems and metal–organic frameworks in transdermal drug delivery for skin cancer therapy. J Mater Chem B 12(16):3807–3839. https://doi.org/10.1039/D4TB00312H
Shaikh AA, Bhattacharjee J, Datta P, Roy S (2024) A comprehensive review of the oxidation states of molybdenum oxides and their diverse applications. Sustain Chem Environ 7:100125
Silva FJG (2017) Nanoindentation on tribological coatings. In: Tiwari A, Natarajan S (eds) Applied nanoindentation in advanced materials. Wiley, pp 111–133
Silva FJG, Baptista APM, Pereira E, Teixeira V, Fan QH, Fernandes AJS, Costa FM (2002) Microwave plasma chemical vapour deposition diamond nucleation on ferrous substrates with Ti and Cr interlayers. Diam Relat Mater 11(10):1617–1622. https://doi.org/10.1016/S0925-9635(02)00065-7
Su Y et al (2021) PLGA-based biodegradable microspheres in drug delivery: recent advances in research and application. Drug Deliv 28(1):1397–1418. https://doi.org/10.1080/10717544.2021.1938756
Teron C, Bhuyan A, Saikia P, Dutta SR, Gogoi H, Rongpi S (2024) A comprehensive review on niosomes in drug delivery and recent advancements. J Drug Deliv Ther 14(6):262–273. https://doi.org/10.22270/jddt.v14i6.6651
Vlachopoulos A et al (2022) Poly[lactic acid]-based microparticles for drug delivery applications: an overview of recent advances. Pharmaceutics 14(2):359. https://doi.org/10.3390/pharmaceutics14020359
Wang K et al (2020) Therapeutic nanomaterials for neurological diseases and cancer therapy. J Nanomater 2020:1–18. https://doi.org/10.1155/2020/2047379
Xiao RZ, Zeng ZW, Zhou GL, Wang JJ, Li FZ, Wang AM (2010) Recent advances in PEG–PLA block copolymer nanoparticles. Int J Nanomedicine 5:1057–1065. https://doi.org/10.2147/IJN.S14912
Xu Z et al (2022) pH-Sensitive nanoparticles based on amphiphilic imidazole/cholesterol modified hydroxyethyl starch for tumor chemotherapy. Carbohydr Polym 277:118827. https://doi.org/10.1016/j.carbpol.2021.118827
Yoo J, Won Y-Y (2020) Phenomenology of the initial burst release of drugs from PLGA microparticles. ACS Biomater Sci Eng 6(11):6053–6062. https://doi.org/10.1021/acsbiomaterials.0c01228
Yusuf A, Almotairy ARZ, Henidi H, Alshehri OY, Aldughaim MS (2023) Nanoparticles as drug delivery systems: a review of the implication of nanoparticles' physicochemical properties on responses in biological systems. Polymers 15(7):1596. https://doi.org/10.3390/polym15071596
Zahra M, Abrahamse H, George BP (2024) Green nanotech paradigm for enhancing Sesquiterpene lactone therapeutics in cancer. Biomed Pharmacother 173:116426. https://doi.org/10.1016/j.biopha.2024.116426

# Chapter 2
# Nanomaterials for Drug Delivery

**Rajkamal, Sunil Dutt, Rajesh Kumar, and Ankit Awasthi**

**Abstract** Nanotechnology has revolutionized medicine by providing precise control over therapeutic delivery at the molecular level. It has significantly advanced drug delivery systems, gene therapy, and medical imaging. Nanomaterials like liposomes, dendrimers, polymeric nanoparticles, metal nanoparticles, and quantum dots each have unique properties that make them ideal for specific medical applications. They employ various drug delivery mechanisms, including passive diffusion, endocytosis, and receptor-mediated uptake, to effectively target cancer cells and enhance treatment efficacy. Despite these benefits, several challenges impede the widespread use of nanotechnology in medicine. These include biocompatibility issues, complex manufacturing processes, and regulatory hurdles that must be overcome to ensure safety and efficacy. Future advancements are expected to focus on improving targeting precision, achieving controlled release of therapeutic agents, and integrating artificial intelligence to enable personalized medicine. These innovations could enhance treatment effectiveness, reduce side effects, and lead to better patient outcomes. As research progresses, nanotechnology's potential to transform healthcare and provide tailored therapeutic solutions is becoming increasingly clear.

**Keywords** Nanomaterials · Nanoparticles · Biomedical application · Drug delivery

Rajkamal
School of Pharmacy, Desh Bhagat University, Mandi Gobindgarh, Punjab, India

S. Dutt · R. Kumar
Maharaja Agrasen School of Pharmacy, Maharaja Agrasen University, Atal Shiksha Kunj, Solan, Himachal Pradesh, India

A. Awasthi (✉)
Chitkara College Pharmacy, Chitkara University, Rajpura, Punjab, India

 
V. Khanna et al. (eds.), *Nanotechnologies for Drug Delivery and Biopharmaceutical Development*, Progress in Drug Research 77,
https://doi.org/10.1007/978-3-032-00796-4_2

## 2.1 Introduction

Nanotechnology is increasingly recognized as a transformative field, especially in medicine. It enables scientists to design systems at the molecular level, offering precise control at the atomic scale. This precision brings unique physical, optical, and electronic properties invaluable in fields like drug delivery, gene therapy, and medical imaging (Milan Singh et al. 2025). Researchers are synthesizing innovative materials that advance disease prevention, diagnosis, and treatment by combining nanoparticles with biological molecules. These developments hold great promise for improving drug effectiveness and patient outcomes. Additionally, nanoscale drug delivery systems improve drug release, stability, and solubility, addressing some of the biggest challenges in pharmaceutical science (Sajid 2022). Since the 1970s, when liposomes first demonstrated targeted drug delivery (Liu et al. 2022), the field has made significant strides, including developing polymeric nanoparticles and dendrimers in the 1980s and 1990s and exploring gold nanoparticles and carbon nanotubes in the 2000s (Ajith et al. 2023). Recently, multifunctional nanoparticles have emerged, combining therapeutic and diagnostic capabilities in a single system, ushering in a new era of theranostic applications in medicine. Nanomaterials offer distinct advantages for drug delivery, including their nanoscale size, which enables prolonged circulation in the bloodstream, immune system evasion, and precise tissue targeting (Shekho et al. 2024). Their high surface-area-to-volume ratio supports efficient drug loading and facilitates the attachment of targeting molecules, allowing drugs to be delivered specifically to diseased cells while minimizing impact on healthy ones. By enhancing solubility and controlling drug release, nanomaterials increase therapeutic efficacy and reduce side effects, offering versatility across various delivery routes, such as oral, injectable, and transdermal. Additionally, some nanomaterials are designed to respond to specific stimuli—such as pH or temperature—that trigger drug release exactly when and where needed (Pham et al. 2020). In bioimaging, they play a critical role by targeting and highlighting malignant cells, aiding in disease diagnosis and monitoring. Various types, including liposomes, dendrimers, metallic nanoparticles, and polymers, are already in clinical use or advanced trials, demonstrating the transformative potential of nanomaterials in personalized and targeted therapies.

## 2.2 Types of Nanomaterials

Nanomaterials can be categorized into nanostructured and nanocrystalline types. Nanostructured materials, such as carbon-based materials, liposomes, dendrimers, and quantum dots, are notable for their unique shapes and specialized functions. On the other hand, nanocrystalline materials are nonpolymers characterized by their nanoscale size, high surface area, and biocompatibility (Mekuye and Abera 2023). Liposomes, micelles, and polymer-based nanocarriers have progressed significantly, with several formulations already approved for medical use. On the other

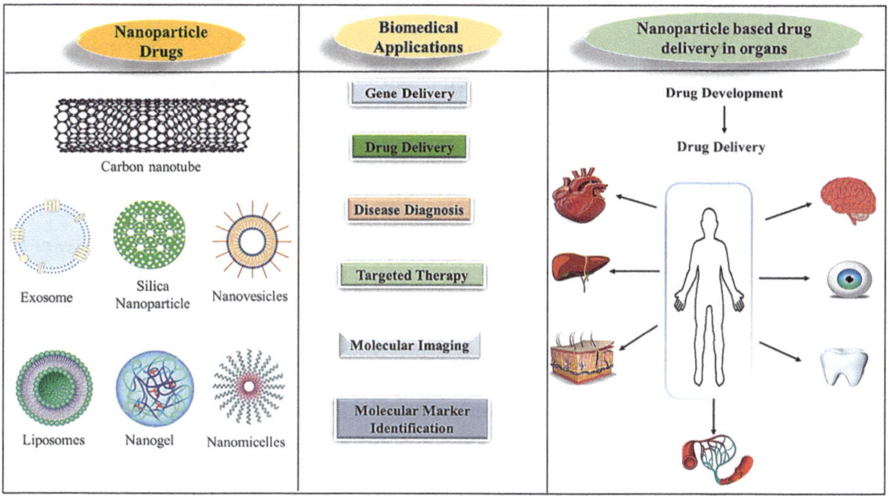


**Fig. 2.1** Illustrating the different types of nanoparticles, and their biomedical application and various organs for drug delivery

hand, many inorganic-based drug delivery systems are still in preclinical stages. With a high surface-to-volume ratio and capacity for surface modifications, nanomaterials integrate aspects of chemistry, materials science, and biology, supporting applications in multiplexed detection, biosensing, bioimaging, gene and drug delivery, and cancer therapies (Fig. 2.1). These materials also address challenges in conventional chemotherapy, such as toxicity and multidrug resistance (V. Singh et al. 2020). Various types of nanoparticles along with their advantages and disadvantages are summarised in Table 2.1.

## 2.3 Liposomes

Liposomes are small spherical vesicles composed of a lipid bilayer that mimics cell membranes, making them compatible with the human body. Their structure features a double layer of molecules, where hydrophilic (water-attracting) heads face outward and hydrophobic (water-repelling) tails point inward, creating a protective sphere (Yin et al. 2022). This unique arrangement enables liposomes to encapsulate both hydrophilic drugs within their water-filled core and hydrophobic drugs within the lipid bilayer, allowing for the transport of a diverse range of therapeutic agents. The encapsulation protects against degradation by enzymes and other biological factors, ensuring that a greater amount of the drug reaches its intended target. By modifying the surface of liposomes, such as through polyethylene glycol (PEG) attachment, their circulation time in the bloodstream can be prolonged, helping them avoid detection and clearance by the immune system (Nsairat et al. 2022).

**Table 2.1** Different types of nanoparticles along with their advantages and disadvantages

| Nanoparticles | Properties | Advantages | Disadvantages | References |
|---|---|---|---|---|
| Liposomes | Encapsulation of hydrophilic and hydrophobic drugs<br>Flexibility in surface modification | Targeted drug delivery<br>Reduced systemic toxicity<br>Enhanced drug solubility | Limited stability<br>Batch-to-batch variability<br>Potential immunogenicity | Hawthorne et al. (2022) |
| Dendrimers | Well-defined structure<br>Precise drug loading and surface functionalization | Enhanced drug stability<br>Tailored surface modifications<br>Potential for targeted delivery | Cytotoxicity concerns<br>Immunogenicity<br>Complex synthesis methods | Kaurav et al. (2023) |
| Polymeric nanoparticles | Biodegradability<br>Versatile drug loading and surface modification | Sustained drug release<br>Protection of drugs from degradation<br>Reduced toxicity | Potential toxicity of polymers<br>Challenges in large-scale production<br>Limited drug loading capacity | Bhardwaj and Jangde (2023) |
| Micelles | Self-assembled from amphiphilic block copolymers<br>Small size for improved tissue penetration | Targeted drug delivery<br>Enhanced stability and bioavailability<br>Utilization of EPR effect | Limited drug loading capacity<br>Challenges in achieving uniform size distribution<br>Potential instability in biological environments | Ghezzi et al. (2021) |
| Gold nanoparticles | Unique optical and physicochemical properties<br>Biocompatibility<br>Surface functionalization | Targeted drug delivery<br>Photothermal therapy<br>Imaging applications | Potential toxicity at high doses<br>Challenges in large-scale production<br>Limited drug loading capacity | Huang et al. (2023) |
| Carbon nanotubes | High surface area<br>Unique electronic properties; mechanical strength | Drug delivery<br>Photothermal therapy<br>Imaging applications | Potential toxicity<br>Challenges in functionalization<br>Concerns regarding biocompatibility | Murjani et al. (2022) |
| Quantum dots | Size-tunable emission properties; high photostability | Imaging applications<br>Multiplexed detection<br>Long-term tracking *in vivo* | Potential toxicity due to heavy metal content<br>Limited tissue penetration<br>Clearance concerns | Reshma and Mohanan (2019) |
| Graphene oxide | High surface area<br>Functionalization capabilities<br>Biocompatibility | Drug delivery<br>Targeted therapy<br>Imaging applications | Potential toxicity<br>Challenges in large-scale production<br>Limited stability in physiological conditions | Shafiee et al. (2022) |

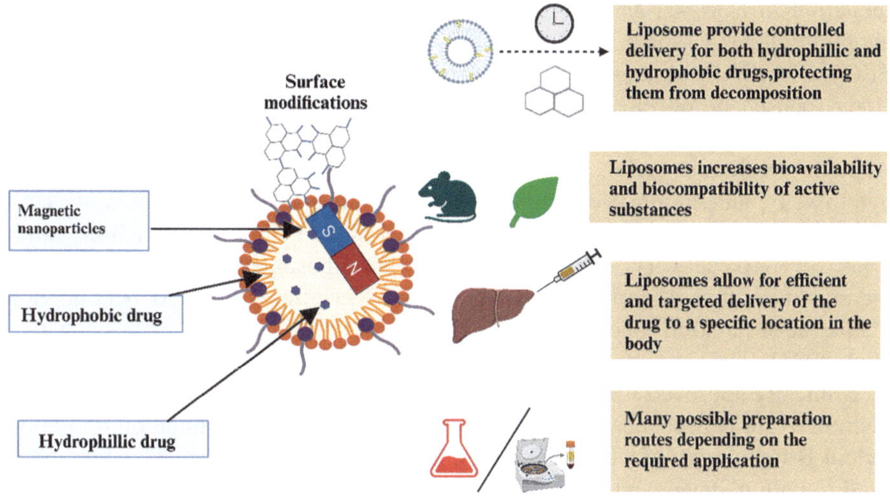


**Fig. 2.2** Magnetically guided liposomal drug delivery system encapsulates hydrophilic and hydrophobic drugs, enabling targeted, controlled release, enhanced bioavailability, biocompatibility, and versatile preparation for medical applications

This characteristic enhances their effectiveness in delivering drugs to specific tissues. Liposomes are highly versatile and have become a key component in clinical applications, particularly in cancer therapy, where they facilitate targeted delivery of chemotherapy drugs directly to tumour cells. This targeted approach not only minimizes harm to healthy cells but also reduces side effects, making liposomes one of the most widely approved nanomaterials for medical use due to their proven safety and efficacy (Fig. 2.2). Maritim et al. examined the formulation and characterization of BIIB093 (glibenclamide) liposomes. PEGylated liposomes containing BIIB093 were formulated using ethanol injection method. Variables such as lipid chain length, unsaturation, phase transition temperature (Tc), and cholesterol content were tested. The liposomes achieved an average size of 130 ± 20 nm, PDI of 0.15 ± 0.1, and ZP of −2 ± 1 mV. Lipids with longer chains showed higher drug loading and retention, while those with greater unsaturation and lower Tc increased release rates. Additional cholesterol enhanced PDI and release rate but lowered drug incorporation (Maritim et al. 2021). In another study, Badivi et al. developed an optimized liposomal formulation loaded with bee venom (BV) and coated with PEG (BV-Lipo-PEG) for potential cancer treatment. Among the formulations, F3 displayed the narrowest size distribution with a low PDI (193.72 ± 7.35), indicating uniform particle size and stability against agglomeration. BV-Lipo-PEG showed remarkable stability over 3 months at 4 °C and exhibited pH-dependent drug release. The formulation achieved high entrapment efficiency (96.74 ± 1.49%) and demonstrated enhanced endocytosis, leading to improved anticancer activity. In A549 lung cancer cell lines, BV-Lipo-PEG achieved a lower IC50 and higher apoptotic rate than both the free drug and uncoated BV-Lipo. Gene expression analysis revealed a reduction in MMP-2, MMP-9, and Cyclin E, alongside increased expression of apoptotic genes Caspase3 and Caspase9, highlighting the therapeutic promise of PEGylated BV liposomes for lung cancer treatment (Badivi et al. 2024). Another study focused on developing liposomal formulations for the nose-to-brain (NtB) delivery of donepezil, memantine, BACE-1 siRNA, and their combination to treat Alzheimer's disease (AD). Liposomes were evaluated for cellular internalization, cytotoxicity, and suppression of beta-amyloid plaque and pro-inflammatory cytokine expressions. The Calu-3 Transwell model simulated nasal epithelium conditions for NtB delivery. Results indicated that liposomes effectively delivered therapeutics intracellularly, crossed BBB tight junctions, reduced beta-amyloid plaque accumulation, and decreased pro-inflammatory cytokine expression (Lee et al. 2024).

Clinically approved liposomal formulations include Doxil, a liposomal doxorubicin used for cancer therapy, and AmBisome, a liposomal formulation of amphotericin B for fungal infections (Gu et al. 2023). Liposomes have also been widely used in pain management, such as DepoDur, a liposomal morphine for postoperative pain relief (Angst and Drover 2006). Additionally, LNPs have played a critical role in delivering mRNA-based vaccines, including the Pfizer-BioNTech and Moderna COVID-19 vaccines, demonstrating their potential in gene therapy and immunization (Khurana et al. 2021).

## 2.4 Polymeric Nanoparticles

Polymeric nanoparticles are tiny carriers crafted from polymers—materials made of long chains of repeating molecules—which can be either naturally derived or synthetically made. Their structure allows them to transport a wide variety of therapeutic agents, including small drug molecules, proteins, and even genetic material like DNA or RNA (Zielińska et al. 2020). One key advantage of polymeric nanoparticles is their ability to control and sustain drug release. Scientists can design these particles to break down gradually over a chosen timeframe, releasing the drug steadily, which helps in achieving long-term therapeutic effects without needing repeated doses. Additionally, polymeric nanoparticles can be "functionalized," meaning specific molecules, known as targeting ligands, are attached to their surface (Elmowafy et al. 2023). These ligands can seek out and bind to specific cells or tissues, such as tumour cells, to deliver the drug directly to the target site, thereby reducing side effects in healthy tissues. Due to these features, polymeric nanoparticles have become important tools in cancer treatment, where precise targeting is crucial. They are also being applied in vaccine development and gene therapy, providing new ways to enhance immune responses and deliver genetic material to specific cells (Fig. 2.3) (Gagliardi et al. 2021). Almoustafa et al. examined the effectiveness of hyaluronic acid (HA)-coated PEG-PLGA NPs for delivering doxorubicin (DOX) in the treatment of breast cancer metastasis using an athymic mouse model. The HA facilitated targeting by binding to CD44 receptors, which are often overexpressed in

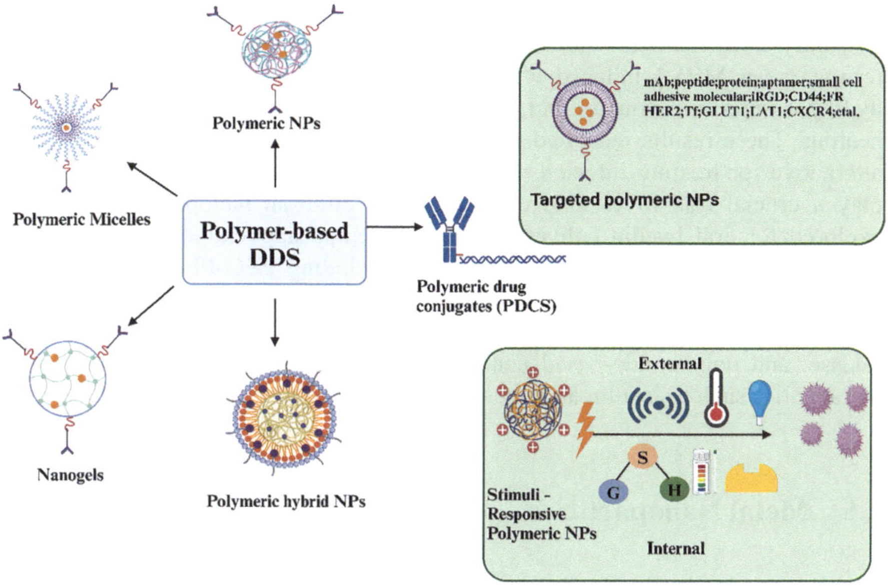


**Fig. 2.3** Polymer-based DDS featuring targeted polymeric NPs and stimuli-responsive NPs for controlled, site-specific drug release via external and internal triggers

cancer stem cells. Mice with RFP-labelled triple-negative breast cancer were separated into four groups: control, PEG-PLGA NPs, HA-PEG-PLGA NPs, and free DOX. By day 25, the mice treated with HA-PEG-PLGA NPs showed significantly reduced tumour growth (118%) compared to the PEG-PLGA group (376%). Furthermore, microscopy revealed a decrease in metastasis to the liver, spleen, colon, and lungs in the HA-PEG-PLGA NP group. These results indicated that HA-PEG-PLGA NPs effectively delivered DOX while reducing metastasis and tumour growth, suggesting potential for further clinical development (Almoustafa et al. 2023). Another study focused on developing an innovative mRNA vaccine capable of inducing a systemic immune response when administered through the gastrointestinal (GI) tract, overcoming the limitations of current mRNA vaccines that require ultracold storage and professional administration. Researchers synthesized a library of cationic branched poly(β-amino ester) (PBAE) polymers and selected one that demonstrated high efficiency in intracellular mRNA delivery and immune stimulation. The resulting mRNA vaccine successfully elicited robust cellular and humoral immunity in mice. Additionally, the formulation's lyophilization conditions were optimized for refrigeration storage at 4 °C. The findings indicated that PBAE nanoparticles were effective platforms for mRNA delivery, promoting B cell and T cell activation and offering a promising alternative for easily administrable oral mRNA vaccines (Kim et al. 2023). Ullah et al. developed a lipase-sensitive nanoparticle-based hydrogel for targeted fusidic acid (FA) delivery in wound infections. Utilizing poly(ε-caprolactone), the formulation produced nanoparticles with a size of approximately 225 nm, a zeta potential of −16 mV, and an entrapment efficiency of 87%. Incorporating FA-NPs into a carbopol hydrogel resulted in significantly enhanced antibacterial activity compared to standard treatments. In vivo testing in an MRSA-induced wound infection model showed that the FA-NPs hydrogel improved wound contraction, reduced bacterial counts, and accelerated healing. These results indicated that the FA-NPs hydrogel could be a promising alternative for treating infected wounds (Ullah et al. 2023). PLGA nanoparticles play a crucial role in vaccine delivery, while chitosan nanoparticles are being explored for oral insulin delivery and gene therapy applications (Omidian et al. 2024). PEGylated polymeric nanoparticles, including PEG-PLGA formulations, improve drug circulation and enhance tumor-targeting efficiency (Zhang et al. 2022). Additionally, polycaprolactone (PCL) nanoparticles facilitate sustained drug release, and polypeptide-based nanoparticles serve as carriers for peptide-based therapeutics such as insulin and interferon.

## 2.5 Metal Nanoparticles

Metal nanoparticles, made from materials like gold, silver, and iron oxide, offer distinctive optical, magnetic, and chemical properties that make them highly effective in drug delivery and diagnostic imaging (Aydin et al. 2019). Their small size (1–100 nm) allows them to interact closely with biological molecules, enabling them to penetrate cell membranes and deliver therapies directly to targeted sites. Gold nanoparticles can carry chemotherapy drugs to tumours, sparing healthy tissue and minimizing side effects. These nanoparticles can be engineered in different shapes and functionalized with therapeutic agents or ligands to target specific tissues or cells, enhancing drug delivery precision and therapeutic efficacy (Yaqoob et al. 2020). In diagnostic imaging, iron oxide nanoparticles serve as MRI contrast agents by accumulating in tumour tissues, thus enhancing image contrast and aiding in early detection. Gold and silver nanoparticles due to their unique light-scattering properties are also used in optical imaging and biosensors. Additionally, metal nanoparticles are highly suited for theranostic, combining drug delivery with imaging in a single system; this allows real-time tracking of treatment progress while reducing potential side effects (Burlec et al. 2023). Their controlled release capabilities, triggered by factors like pH or temperature, enable drugs to be released gradually or at targeted locations, providing an efficient, safer approach to therapy—especially for conditions requiring precision, such as cancer. This dual functionality of metal nanoparticles—targeted therapy and imaging—represents a major step forward in personalized medicine and treatment monitoring (Adepu and Ramakrishna 2021). Gracias et al. explored the biosynthesis of zinc oxide nanoparticles using Cissus quadrangularis (CQ-ZnONPs) and the formulation of chitosan-conjugated nanocomposites (CQ-CS/ZnONCs). Characterization confirmed the formation of the nanoparticles, which exhibited average sizes of 243.3 nm for CQ-ZnONPs and 176.6 nm for CQ-CS/ZnONCs. Both formulations demonstrated strong antioxidant and antibacterial activities. CQ-CS/ZnONCs showed significant inhibition of α-amylase and α-glucosidase, with IC50 values of 73.66 μg/mL and 87.59 μg/mL, respectively. Notably, CQ-CS/ZnONCs achieved impressive wound contraction rates of nearly 99%, indicating their potential as effective drug delivery systems for infection control and tissue repair (Gracias et al. 2023). Another study focused on the green synthesis of zinc oxide nanoparticles (Zn-NPs) and their encapsulation within niosomes to enhance drug delivery. The Zn-NPs, averaging 27.60 nm in size, were characterized using SEM, DLS, FTIR, UV-Vis, and XRD techniques. The optimal niosomal formulation demonstrated an entrapment efficiency of 31.26% and a size of 256.6 nm. The release profile indicated a slow release of Zn-NPs from niosomes compared to free Zn-NPs. Furthermore, niosomes exhibited superior antimicrobial and anti-biofilm effects, along with increased anticancer activity against breast cancer cells, making them promising carriers for targeted Zn-NP delivery (Rezaei et al. 2024). Lastly, AuNPs are widely used for targeted drug delivery and photothermal therapy, particularly in cancer treatment (Riley and Day 2017). AgNPs exhibit strong antimicrobial properties, making them useful in wound dressings and coatings (Zahran et al. 2014). Fe3O4 NPs including Ferumoxytol, function as MRI contrast agents and targeted drug carriers (Li et al. 2020). Additionally, ZnO and TiO2 nanoparticles possess antimicrobial properties and are employed in photodynamic therapy, enhancing their role in cancer treatment and biomedical applications.

## 2.6 Dendrimers

Dendrimers are intricately branched macromolecules that exhibit a unique, tree-like structure, characterized by a central core with multiple branching units. This architecture enables dendrimers to possess a high drug-loading capacity and a precise, uniform shape, making them ideal for a variety of biomedical applications, particularly in drug delivery (Pérez-Ferreiro et al. 2023). Their numerous surface functional groups can be easily modified to attach targeting ligands, imaging agents, or multiple drugs, allowing for personalized and versatile drug delivery systems. Dendrimers enhance the solubility of poorly soluble drugs and facilitate controlled drug release, ensuring a steady therapeutic effect while minimizing side effects (Chis et al. 2020). Additionally, they play a critical role in gene therapy by enabling the effective delivery of genetic material into cells, which is essential for treating genetic disorders. Overall, dendrimers hold great promise for advancing targeted therapies and personalized medicine, improving treatment efficacy across various health conditions (Fateh et al. 2024). Kumar et al. aimed to enhance the brain uptake and pharmacokinetics of galantamine hydrobromide (GH) using a bi-functional surface-engineered dendrimeric formulation. By conjugating vitamin C and polyethylene glycol (PEG) to PAMAM dendrimers via carbodiimide chemistry. GH-loaded formulations (GH@VC@PEG@PM) were developed using solvent injection. Characterized by a particle size of 150.32 nm, zeta potential of 1.60 mV, and PDI of 0.245, these formulations demonstrated a 96-h controlled GH release and improved biocompatibility. *In vitro* studies showed reduced AChE activity and inhibition of amyloid-beta fibril formation. *In vivo*, pharmacokinetics in Sprague Dawley rats indicated a 2.3-fold increase in brain uptake and a four-fold reduction in clearance rate compared to GH alone (V. Kumar et al. 2023). In another study, N-acetyl-d-glucosamine-labelled dendrimers (NAG-Dend) were developed to enhance the targeted delivery of camptothecin (CPT) to A549 lung adenocarcinoma cells, which are known to overexpress glucose transporters and lectin receptors. The NAG-Dend loaded with CPT (NAG-Dend-CPT) demonstrated notably higher cellular uptake and cytotoxicity compared to unlabelled dendrimers (Dend-CPT), showing 4.5 times more toxicity than native CPT in A549 cells. The efficacy of NAG-Dend-CPT was dose- and time-dependent, with increased reactive oxygen species production, elevated apoptosis, and greater inhibition of cell migration relative to Dend-CPT. Additionally, in tumour-bearing mice, NAG-Dend preferentially accumulated in lung tissues, indicating its potential as a targeted delivery system for lung metastasis (Pooja et al. 2020). AMAM (polyamidoamine) dendrimers are commonly used in gene and cancer drug delivery, including dendrimer-paclitaxel conjugates, while PPI (polypropylene imine) dendrimers have shown antimicrobial and anticancer properties (Dey et al. 2022). PEGylated dendrimers improve biocompatibility and circulation, while peptide-dendrimers are being explored for vaccine applications (Sapra et al. 2019). Carbosilane dendrimers also hold potential for antiviral drug delivery, demonstrating their versatility in nanomedicine.

## 2.7 Quantum Dots

Quantum dots (QDs) are semiconductor nanoparticles that stand out for their unique fluorescent properties, making them highly effective for bioimaging and diagnostics (Gil et al. 2021). Ranging from 2 to 10 nanometers in size, they emit light in vibrant, stable colours depending on their dimensions, allowing for precise control over the emitted wavelength. This tunable fluorescence is significantly brighter and more stable than traditional fluorescent dyes, enabling QDs to be observed over long periods without fading, which is essential in detailed, high-resolution imaging applications (Nabil and Megahed 2023). In the field of cancer diagnostics, QDs can be conjugated with targeting molecules that bind to specific cancer cells, illuminating these cells for accurate imaging and improving the precision of early cancer detection. Furthermore, QDs can be linked to therapeutic drugs, acting as markers to track drug distribution and concentration within the body in real time. This monitoring capability allows researchers and clinicians to follow drug pathways, assess tissue targeting, and evaluate treatment efficacy, which is especially advantageous for complex diseases where precise drug localization is critical (Guo et al. 2024). QDs thus provide a versatile platform that merges diagnostic imaging with therapeutic assessment, supporting more effective tracking of therapeutic outcomes and advancing personalized medicine approaches in fields such as oncology (Agarwal et al. 2023). Najafi et al. developed a nanocomposite hydrogel for enhanced drug delivery for cancer treatment, using a water/oil/water method. The nanocomposite achieved high drug loading and encapsulation efficiencies (47.00% and 86.25%, respectively), ranking among the top in similar drug delivery systems (Bhardwaj and Jangde 2023; Chaurasia et al. 2024; Dhiman et al. 2015, 2016a, b; Garg et al. 2022; Hamid et al. 2024; Kamal et al. 2024; Kumar et al. 2015, 2017a, b, 2021, 2022a, b, c, d, e, 2023, 2024a, b; Milan Singh et al. 2025; Murjani et al. 2022; Pandey et al. 2018, 2019; Rai et al. 2021; Rajesh Kumar et al. 2021; Singh et al. 2023a, b, 2024a, b; Suthar and Sharma 2015). Characterized through FTIR and XRD, the hydrogel showed improved QC solubility and optimized particle size (151.57 nm), favouring cancer cell targeting with minimal healthy cell interaction. Morphological analysis and a + 38.8 mV zeta potential demonstrated the nanocarrier's stability, while cell viability assays revealed significant cancer cell cytotoxicity (48%) and high safety for normal cells (92%), positioning this nanocomposite as a promising anticancer delivery platform (Najafi et al. 2024). In another study, Salman et al. developed a novel fluorescent probe using polyamine quantum dots (PA@CQDs) from apricots in a one-step process to measure molnupiravir (MOL) levels. The probe demonstrated a sensitive, cost-effective method for analyzing MOL in plasma and pharmaceutical tablets, with the relative fluorescence intensity significantly quenched by MOL. It provided a linear detection range of 2–70 ng/mL and a limit of quantitation at 1.61 ng/mL. The probe successfully quantified MOL in pharmacokinetic studies, showing a maximum plasma concentration of 920.2 ng/mL, indicating its potential for clinical applications (Salman et al. 2022). Besides these, Cadmium selenide (CdSe) quantum dots are commonly used in cancer imaging (Rahman et al. 2021), while graphene quantum dots (GQDs) show promise for cancer treatment and biosensing applications (Iannazzo et al. 2021). Silicon quantum dots (SiQDs) and carbon quantum dots (CQDs) provide biocompatible options for bioimaging and theranostics, allowing real-time drug monitoring (Priya et al. 2024). Zinc sulfide (ZnS) quantum dots are also being investigated for imaging-assisted drug delivery, highlighting their growing role in nanomedicine.

## 2.8 Mechanism of Drug Delivery

Nanocarriers utilize various mechanisms for cellular penetration, enhancing their efficacy in targeted cancer therapy. Passive diffusion allows nanocarriers to move along concentration gradients, passing through intercellular spaces of cancer cells. Endocytosis, a key internalization pathway, includes processes like phagocytosis, where larger particles form vesicles (phagosomes), and pinocytosis, involving smaller vesicle formation. Clathrin-mediated endocytosis targets receptors on clathrin-coated cell surface areas, while caveolae-mediated uptake involves small, membrane-bound invaginations. Receptor-mediated uptake is facilitated by ligands or antibodies that bind selectively to overexpressed receptors on cancer cells, enhancing absorption and reducing healthy cell interaction. In pH-responsive uptake, nanocarriers respond to the acidic tumour microenvironment, triggering the

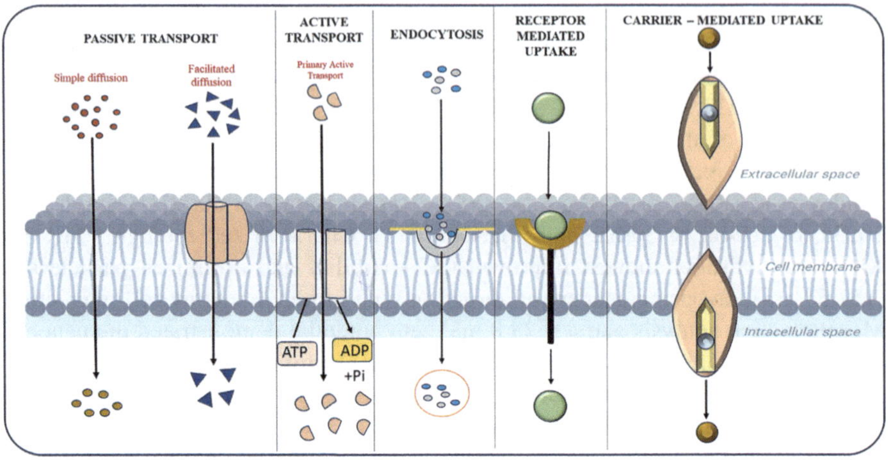


**Fig. 2.4** Representing the various mechanisms of drug transport

release of their therapeutic payload. The enhanced permeability and retention (EPR) effect allows nanocarriers to accumulate in tumour tissues due to leaky vasculature, maximizing therapeutic concentration within tumours and limiting healthy tissue exposure. Together, these mechanisms improve treatment specificity and efficacy by focusing drug action at tumour sites (Fig. 2.4) (Kamal et al. 2024).

## 2.9 Challenges and Limitations

Nanomaterials hold significant promise for enhancing drug delivery, improving drug stability, and increasing bioavailability, yet several critical challenges hinder their clinical application. The complexity involved in designing and manufacturing these materials requires precise control over their size, shape, and surface characteristics, as even slight variations can lead to significant differences in behaviour within biological systems (Kladko et al. 2021). Scaling up production from laboratory synthesis to industrial manufacturing often poses difficulties, leading to inconsistencies in quality and increased costs. Furthermore, concerns regarding biocompatibility arise, as some nanomaterials may induce toxicity in cells or provoke immune responses, which can reduce their effectiveness and raise safety issues (Baer et al. 2013). Biological barriers, such as the blood-brain barrier and the intricate tumour microenvironment, complicate the targeted delivery of drugs, while the body's natural defence mechanisms can clear nanoparticles before they reach their intended targets. Additionally, achieving controlled release of therapeutic agents remains a challenge, as premature release can lead to reduced efficacy and increased toxicity. Regulatory and ethical hurdles also exist, as extensive data on safety and efficacy are required for approval, and the lack of standardized guidelines complicates this process. Lastly, the high production costs and specialized technical expertise needed for nanomaterials can limit accessibility, particularly in low-resource settings (Joudeh and Linke 2022).

## 2.10 Future Perspective and Conclusion

The future of nanomaterials in drug delivery is promising, with ongoing research focused on overcoming current challenges to enhance their clinical applications. Advances in nanotechnology are expected to produce sophisticated nanocarriers for improved targeting and controlled release of therapeutic agents. Innovations like stimuli-responsive nanomaterials, which release drugs in response to specific triggers, could significantly improve treatment efficacy while minimizing side effects. Integrating artificial intelligence and machine learning into nanomaterial design and synthesis will likely speed up discovery, optimize formulations, and predict biological interactions. Personalized medicine will benefit from nanomaterials tailored to individual patient needs, potentially leading to more effective treatments for various diseases, including cancer and neurodegenerative disorders. Collaboration among academia, industry, and regulatory bodies will be essential to establish standardized guidelines for nanomaterial development and testing, ensuring safety and efficacy. Addressing economic barriers to production and accessibility will also be crucial, especially in underserved populations.

## References


Adepu S, Ramakrishna S (2021) Controlled drug delivery systems: current status and future directions. Molecules 26(19). https://doi.org/10.3390/molecules26195905

Agarwal K, Rai H, Mondal S (2023) Quantum dots: an overview of synthesis, properties, and applications. Mater Res Express 10(6):062001. https://doi.org/10.1088/2053-1591/acda17

Ajith S, Almomani F, Elhissi A, Husseini GA (2023) Nanoparticle-based materials in anticancer drug delivery: current and future prospects. Heliyon 9(11):e21227. https://doi.org/10.1016/j.heliyon.2023.e21227

Almoustafa HA, Alshawsh MA, Al-Suede FSR, Alshehade SA, Abdul Majid AM, Chik Z (2023) The chemotherapeutic efficacy of hyaluronic acid coated polymeric nanoparticles against breast cancer metastasis in female NCr-Nu/Nu Nude Mice. Polymers 15(2). https://doi.org/10.3390/polym15020284

Angst MS, Drover DR (2006) Pharmacology of drugs formulated with DepoFoam™: a sustained release drug delivery system for parenteral administration using multivesicular liposome technology. Clin Pharmacokinet 45:1153–1176

Aydin EB, Aydin M, Sezginturk MK (2019) Chapter 2–Immobilization techniques of nanomaterials. In: Ozkan SA, Shah A (eds) New developments in nanosensors for pharmaceutical analysis. Academic, pp 47–78

Badivi S, Kazemi S, Eskandarisani M, Moghaddam NA, Mesbahian G, Karimifard S, Afzali E (2024) Targeted delivery of bee venom to A549 lung cancer cells by PEGylate liposomal formulation: an apoptotic investigation. Sci Rep 14(1):17302. https://doi.org/10.1038/s41598-024-68156-6

Baer DR, Engelhard MH, Johnson GE, Laskin J, Lai J, Mueller K et al (2013) Surface characterization of nanomaterials and nanoparticles: important needs and challenging opportunities. J Vac Sci Technol A 31(5):50820. https://doi.org/10.1116/1.4818423

Bhardwaj H, Jangde RK (2023) Current updated review on preparation of polymeric nanoparticles for drug delivery and biomedical applications. Next Nanotechnol 2:100013. https://doi.org/10.1016/j.nxnano.2023.100013

Burlec AF, Corciova A, Boev M, Batir-Marin D, Mircea C, Cioanca O et al (2023) Current overview of metal nanoparticles' synthesis, characterization, and biomedical applications, with a focus on silver and gold nanoparticles. Pharmaceuticals 16(10):1410. https://doi.org/10.3390/ph16101410

Chaurasia PK, Bharati SL, Singh S (2024) The chemistry inside spices & herbs: research and development: volume 4. Bentham Science Publishers

Chis AA, Dobrea C, Morgovan C, Arseniu AM, Rus LL, Butuca A et al (2020) Applications and limitations of dendrimers in biomedicine. Molecules 25(17):3982. https://doi.org/10.3390/molecules25173982

Dey AD, Bigham A, Esmaeili Y, Ashrafizadeh M, Moghaddam FD, Tan SC et al (2022) Dendrimers as nanoscale vectors: unlocking the bars of cancer therapy. Semin Cancer Biol 86:396–419

Dhiman R, Kumar D, Kumar B, Pandey B, Jayasankar K, Lohadas P et al (2015) Novel LC-MS/MS bioanalytical method validation for gabapentin in human plasma in accordance with current regulatory guidances. Institute of Medical Sciences

Dhiman R, Kumar B, Kumar D, Kumar A (2016a) Liquid-liquid extraction method for Ziprasidone (ZRS) bioanalysis by using ZRS-D8 (stable isotope) as internal standard. J Chem Pharm Res 8(3):722–727

Dhiman R, Kumar D, Kumar B, Pandey B (2016b) Quantitative determination of atorvastatin, ortho-hydroxy atorvastatin, para-hydroxy atorvastatin in human plasma using rosuvastatin as internal standard by LC-MS/MS. Int J Pharm Chem Sci 4:487–500

Elmowafy M, Shalaby K, Elkomy MH, Alsaidan OA, Gomaa HAM, Abdelgawad MA, Mostafa EM (2023) Polymeric nanoparticles for delivery of natural bioactive agents: recent advances and challenges. Polymers 15(5):1123. https://doi.org/10.3390/polym15051123

Fateh ST, Aghaii AH, Aminzade Z, Shahriari E, Roohpour N, Koosha F, Dezfuli AS (2024) Inorganic nanoparticle-cored dendrimers for biomedical applications: a review. Heliyon 10(9):e29726. https://doi.org/10.1016/j.heliyon.2024.e29726

Gagliardi A, Giuliano E, Venkateswararao E, Fresta M, Bulotta S, Awasthi V, Cosco D (2021) Biodegradable polymeric nanoparticles for drug delivery to solid tumors. Front Pharmacol 12:601626. https://doi.org/10.3389/fphar.2021.601626

Garg R, Dobhal K, Saklani T, Gupta AK, Kumar B, Kumar R et al (2022) COVID-19: A challenge towards the sustainability of health in platform era. J Pharm Negat Results 13:1666–1678

Ghezzi M, Pescina S, Padula C, Santi P, Del Favero E, Cantù L, Nicoli S (2021) Polymeric micelles in drug delivery: an insight of the techniques for their characterization and assessment in biorelevant conditions. J Control Release 332:312–336. https://doi.org/10.1016/j.jconrel.2021.02.031

Gil HM, Price TW, Chelani K, Bouillard J-SG, Calaminus SDJ, Stasiuk GJ (2021) NIR-quantum dots in biomedical imaging and their future. iScience 24(3):102189. https://doi.org/10.1016/j.isci.2021.102189

Gracias S, Ayyanar M, Peramaiyan G, Kalaskar M, Redasani V, Gurav N et al (2023) Fabrication of chitosan nanocomposites loaded with biosynthetic metallic nanoparticles and their therapeutic investigation. Environ Res 234:116609. https://doi.org/10.1016/j.envres.2023.116609

Gu W, Andrews GP, Tian Y (2023) Recent clinical successes in liposomal nanomedicines. Int J Drug Discov Pharmacol 2(1):52–59

Guo W, Song X, Liu J, Liu W, Chu X, Lei Z (2024) Quantum dots as a potential multifunctional material for the enhancement of clinical diagnosis strategies and cancer treatments. Nano 14(13):1088. https://doi.org/10.3390/nano14131088

Hamid A, Kumar R, Chaturvedi VK, Dutt S, Sulabh G, Kumar V, Yadav D (2024) Artificial intelligence (AI) based colonoscopy. In: Nanobiotechnology and artificial intelligence in gastrointestinal Diseases. IOP Publishing

Hawthorne D, Pannala A, Sandeman S, Lloyd A (2022) Sustained and targeted delivery of hydrophilic drug compounds: A review of existing and novel technologies from bench to bedside. J Drug Deliv Sci Technol 78:103936. https://doi.org/10.1016/j.jddst.2022.103936

Huang H, Liu R, Yang J, Dai J, Fan S, Pi J et al (2023) Gold nanoparticles: construction for drug delivery and application in cancer immunotherapy. Pharmaceutics 15(7):1868. https://doi.org/10.3390/pharmaceutics15071868

Iannazzo D, Espro C, Celesti C, Ferlazzo A, Neri G (2021) Smart biosensors for cancer diagnosis based on graphene quantum dots. Cancer 13(13):3194

Joudeh N, Linke D (2022) Nanoparticle classification, physicochemical properties, characterization, and applications: a comprehensive review for biologists. J Nanobiotechnol 20(1):262. https://doi.org/10.1186/s12951-022-01477-8

Kamal R, Awasthi A, Paul P, Mir MS, Singh SK, Dua K (2024) Novel drug delivery systems in colorectal cancer: advances and future prospects. Pathol Res Pract 262:155546. https://doi.org/10.1016/j.prp.2024.155546

Kaurav M, Ruhi S, Al-Goshae HA, Jeppu AK, Ramachandran D, Sahu RK et al (2023) Dendrimer: an update on recent developments and future opportunities for the brain tumors diagnosis and treatment. Front Pharmacol 14:1159131. https://doi.org/10.3389/fphar.2023.1159131

Khurana A, Allawadhi P, Khurana I, Allwadhi S, Weiskirchen R, Banothu AK et al (2021) Role of nanotechnology behind the success of mRNA vaccines for COVID-19. Nano Today 38:101142

Kim H, Kirtane AR, Kim NY, Rajesh NU, Tang C, Ishida K et al (2023) Gastrointestinal delivery of an mRNA vaccine using Immunostimulatory polymeric nanoparticles. AAPS J 25(5):81. https://doi.org/10.1208/s12248-023-00844-z

Kladko DV, Falchevskaya AS, Serov NS, Prilepskii AY (2021) Nanomaterial shape influence on cell behavior. Int J Mol Sci 22(10):5266. https://doi.org/10.3390/ijms22105266

Kumar R, Parashar B, Sharma P (2015) World Journal of Pharmacological Research and Technology

Kumar R, Kumar B, Kumar A (2017a) Pharmacokinetic profile of Antitubercular drugs and their clinical applications in therapeutic drug monitoring. Asian J Pharm Pharmacol 3(5):145–154

Kumar A, Kumar B, Kumar R, Kumar A, Singh M, Singh P et al (2017b) Assessment of pain-ameliorative potential of Ethanolic extract of Calotropis proceraand its active components characterization by HPTLC and GC-MS analysis. Drugs

Kumar A, Nandi MK, Kumar B, Kumar A, Kumar R, Singh AK, Singh M (2021) Antiurolithiasis, antioxidant, anti-inflammatory, analgesic, and diuretic activity of Ethanolic extract of seeds of Caesalpinia bonducella. Int J Pharm Investig 11(3):306–311

Kumar A, Kumar B, Kumar R, Kumar A, Singh M, Singh P et al (2022a) Assessment of pain-ameliorative potential of Ethanolic extract of Calotropis procera and its active components characterization by HPTLC and GC-MS analysis. Sys Rev Pharm 13(6):581–586

Kumar A, Kumar B, Kumar R, Kumar A, Singh M, Tiwari V et al (2022b) Acute and subacute toxicity study of ethanolic extract of Calotropis procera (Aiton) Dryand flower in Swiss albino mice. Phytomed Plus 2(2):100224

Kumar A, Nandi MK, Kumar B, Kumar A, Kumar R, Kailashiya V, Singh AK (2022c) Toxicity (acute and subacute) assessment and in-vivo antiurolithiatic activity of ethanolic extract of Caesalpinia bonducella seed in albino Wistar rat. J Appl Pharm Sci 12(2):187–197

Kumar R, Kumar B, Kumar A, Kumar A, Singh M (2022d) GC-MS analysis of Phytocomponents in the methanol extract of Premna latifolia Roxb. Pharm Res 14(1):19–23

Kumar R, Kumar B, Kumar A, Kumar A, Singh M (2022e) HPTLC and GC-MS analyses of biologically active extracts and fractions from Premna latifolia Roxb leaves reveal anti-inflammatory and analgesic effects. Sys Rev Pharm 13(5):579–586

Kumar V, Rana M, Sharma AK, Sinha S, Ajazuddin, Gupta U (2023) Vitamin-C decorated PEGylated PAMAM dendrimers for effective inhibition of amyloid aggregation and improved in vivo pharmacokinetics. J Drug Deliv Sci Technol 89:105058. https://doi.org/10.1016/j.jddst.2023.105058

Kumar R, Dutt S, Goyal A, Kumar A, Kumar B (2024a) Nanotechnology in gastrointestinal endoscopy. IOP

Kumar A, Singh P, Kumar R, Dutt S (2024b) Novel drug delivery systems for inflammatory bowel disease. In: Nanobiotechnology and artificial intelligence in gastrointestinal diseases. IOP, pp 4-1–4-24. https://doi.org/10.1088/978-0-7503-6134-7ch4

Lee D, Shen AM, Garbuzenko OB, Minko T (2024) Liposomal formulations of anti-Alzheimer drugs and siRNA for nose-to-brain delivery: design, safety and efficacy in vitro. AAPS J 26(5):99. https://doi.org/10.1208/s12248-024-00967-x

Li H, Yang S, Hui D, Hong R (2020) Progress in magnetic Fe3O4 nanomaterials in magnetic resonance imaging. Nanotechnol Rev 9(1):1265–1283

Liu P, Chen G, Zhang J (2022) A review of liposomes as a drug delivery system: current status of approved products, regulatory environments, and future perspectives. Molecules 27(4). https://doi.org/10.3390/molecules27041372

Maritim S, Boulas P, Lin Y (2021) Comprehensive analysis of liposome formulation parameters and their influence on encapsulation, stability and drug release in glibenclamide liposomes. Int J Pharm 592:120051. https://doi.org/10.1016/j.ijpharm.2020.120051

Mekuye B, Abera B (2023) Nanomaterials: an overview of synthesis, classification, characterization, and applications. Nano Select 4(8):486–501. https://doi.org/10.1002/nano.202300038

Milan Singh K, Raj K, Amit K, Ankit A, Manish K (2025) Emerging nanotechnology involved in skin cancer: pathogenesis, biomarkers, Ethosomal formulation and future perspective. Curr Pharm Biotechnol 26:1–16. https://doi.org/10.2174/0113892010311407240902050401

Murjani BO, Kadu PS, Bansod M, Vaidya SS, Yadav MD (2022) Carbon nanotubes in biomedical applications: current status, promises, and challenges. Carbon Lett 32(5):1207–1226. https://doi.org/10.1007/s42823-022-00364-4

Nabil M, Megahed F (2023) Quantum dot nanomaterials: preparation, characterization, advanced bio-imaging and therapeutic applications. J Fluoresc 34:2467–2484. https://doi.org/10.1007/s10895-023-03472-0

Najafi M, Pourmadadi M, Abdous M, Rahdar A, Pandey S (2024) Formulation of double nanoemulsions based on pH-sensitive carboxymethyl cellulose/starch copper doped carbon quantum dots for quercetin controlled release. J Mol Liq 400:124543. https://doi.org/10.1016/j.molliq.2024.124543

Nsairat H, Khater D, Sayed U, Odeh F, Al Bawab A, Alshaer W (2022) Liposomes: structure, composition, types, and clinical applications. Heliyon 8(5):e09394. https://doi.org/10.1016/j.heliyon.2022.e09394

Omidian H, Gill EJ, Dey Chowdhury S, Cubeddu LX (2024) Chitosan nanoparticles for intranasal drug delivery. Pharmaceutics 16(6):746

Pandey H, Srivastava S, Kumar R, Tripathi YB (2018) Preclinical acute and repeated dose toxicity of Pueraria tuberosa (PTWE) on charles foster rats. Int J Pharm Sci Res 9:4572–4581

Pandey H, Srivastava S, Kumar A, Kumar R, BhusanTripathi Y (2019) Pharmacokinetics study of puerarin absorption in blood after consumption of Pueraria tuberosa water extract (PTWE) by rats. Explor Anim Med Res 9(2)

Pérez-Ferreiro M, Abelairas AM, Criado A, Gómez IJ, Mosquera J (2023) Dendrimers: exploring their wide structural variety and applications. Polymers (Basel) 15(22):4369. https://doi.org/10.3390/polym15224369

Pham SH, Choi Y, Choi J (2020) Stimuli-responsive nanomaterials for application in antitumor therapy and drug delivery. Pharmaceutics 12(7):630. https://doi.org/10.3390/pharmaceutics12070630

Pooja D, Srinivasa Reddy T, Kulhari H, Kadari A, Adams DJ, Bansal V, Sistla R (2020) N-acetyl-d-glucosamine-conjugated PAMAM dendrimers as dual receptor-targeting nanocarriers for anticancer drug delivery. Eur J Pharm Biopharm 154:377–386. https://doi.org/10.1016/j.ejpb.2020.07.020

Priya L, Mehta S, Gevariya D, Sharma R, Panjwani D, Patel S et al (2024) Quantum dot-based bio-conjugates as an emerging bioimaging tool for cancer Theranostic-A review. Curr Drug Targets 25(4):241–260

Rahman MM, Opo FA, Asiri AM (2021) Cytotoxicity study of cadmium-selenium quantum dots (cdse QDs) for destroying the human HepG2 liver cancer cell. J Biomed Nanotechnol 17(11):2153–2164

Rai SN, Tiwari N, Singh P, Mishra D, Singh AK, Hooshmandi E et al (2021) Therapeutic potential of vital transcription factors in Alzheimer's and Parkinson's disease with particular emphasis on transcription factor EB mediated autophagy. Front Neurosci 15:777347. https://doi.org/10.3389/fnins.2021.777347

Rajesh Kumar BK, Kumar A, Kumar A, Singh M (2021) HPTLC and GC-MS analyses of biologically active extracts and fractions from Premna Latifolia Roxb leaves. J Neurol Disord 9(9):1

Reshma VG, Mohanan PV (2019) Quantum dots: applications and safety consequences. J Lumin 205:287–298. https://doi.org/10.1016/j.jlumin.2018.09.015

Rezaei H, Iranbakhsh A, Sepahi AA, Mirzaie A, Larijani K (2024) Formulation, preparation of niosome loaded zinc oxide nanoparticles and biological activities. Sci Rep 14(1):16692. https://doi.org/10.1038/s41598-024-67509-5

Riley RS, Day ES (2017) Gold nanoparticle-mediated photothermal therapy: applications and opportunities for multimodal cancer treatment. Wiley Interdiscip Rev Nanomed Nanobiotechnol 9(4):e1449

Sajid M (2022) Nanomaterials: types, properties, recent advances, and toxicity concerns. Curr Opi Environ Sci Health 25:100319. https://doi.org/10.1016/j.coesh.2021.100319

Salman BI, Ibrahim AE, El Deeb S, Saraya RE (2022) Fabrication of novel quantum dots for the estimation of COVID-19 antiviral drug using green chemistry: application to real human plasma. RSC Adv 12(26):16624–16631. https://doi.org/10.1039/D2RA02241A

Sapra R, Verma RP, Maurya GP, Dhawan S, Babu J, Haridas V (2019) Designer peptide and protein dendrimers: a cross-sectional analysis. Chem Rev 119(21):11391–11441

Shafiee A, Iravani S, Varma RS (2022) Graphene and graphene oxide with anticancer applications: challenges and future perspectives. MedComm *3*(1):e118. https://doi.org/10.1002/mco2.118

Shekho D, Mishra R, Kamal R, Bhatia R, Awasthi A (2024) Breaking barriers in Alzheimer's disease: the role of advanced drug delivery systems. AAPS PharmSciTech 25(7):207. https://doi.org/10.1208/s12249-024-02923-6

Singh V, Yadav P, Mishra V (2020) Recent advances on classification, properties, synthesis, and characterization of nanomaterials. In: *Green synthesis of nanomaterials for bioenergy applications*, pp 83–97

Singh M, Kumar B, Singh AK, Azad CS, Yadav MK, Kumar A et al (2023a) Pharmacokinetic screening to estimate the drug likeliness characteristics of selected herbal anticancer drugs. Res J Pharm Technol 16(7):3422–3426

Singh A, Prasad LB, Shiv K, Kumar R, Garai S (2023b) Synthesis, characterization, and in vitro antibacterial and cytotoxic study of Co (II), Ni (II), Cu (II), and Zn (II) complexes of N-(4-methoxybenzyl) N-(phenylethyl) dithiocarbamate ligand. J Mol Struct 1288:135835

Singh A, Kumar R, Patel R, Jaiswal T, Gautam R, Bharty MK, Prasad LB (2024a) Dithiocarbamate based novel anti-histaminic agents: synthesis, characterization, crystal structure and thermal study. Dalton Trans 53:14077–14088

Singh A, Kumar R, Shiv K, Pandey SK, Bharty M, Butcher R, Prasad LB (2024b) Synthesis, crystal structure and screening for anticonvulsant and antianxiety activities of three new Ni (II), Cu (II), and Zn (II) dithiocarbamate complexes. J Mol Struct 1298:137052

Suthar SK, Sharma M (2015) Recent developments in chimeric NSAIDs as safer anti-inflammatory agents. Med Res Rev 35(2):341–407. https://doi.org/10.1002/med.21331

Ullah N, Khan D, Ahmed N, Zafar A, Shah KU, ur Rehman, A. (2023) Lipase-sensitive fusidic acid polymeric nanoparticles based hydrogel for on-demand delivery against MRSA-infected burn wounds. J Drug Deliv Sci Technol 80:104110. https://doi.org/10.1016/j.jddst.2022.104110

Yaqoob AA, Ahmad H, Parveen T, Ahmad A, Oves M, Ismail IMI et al (2020) Recent advances in metal decorated nanomaterials and their various biological applications: a review. Front Chem 8:341. https://doi.org/10.3389/fchem.2020.00341

Yin Z, Zheng T, Ho C-T, Huang Q, Wu Q, Zhang M (2022) Improving the stability and bioavailability of tea polyphenols by encapsulations: a review. Food Sci Human Wellness 11(3):537–556. https://doi.org/10.1016/j.fshw.2021.12.011

Zahran M, Ahmed HB, El-Rafie M (2014) Surface modification of cotton fabrics for antibacterial application by coating with AgNPs–alginate composite. Carbohydr Polym 108:145–152

Zhang D, Liu L, Wang J, Zhang H, Zhang Z, Xing G et al (2022) Drug-loaded PEG-PLGA nanoparticles for cancer treatment. Front Pharmacol 13:990505

Zielińska A, Carreiró F, Oliveira AM, Neves A, Pires B, Venkatesh DN et al (2020) Polymeric nanoparticles: production, characterization, toxicology and ecotoxicology. Molecules 25(16). https://doi.org/10.3390/molecules25163731

# Chapter 3
# Biocompatibility and Safety Assessment of Nanomaterials

**Ashutosh Kumar, Pratistha Singh, Devinder Kumar, Sunil Dutt, Pankaj Kalia, and Rajesh Kumar**

**Abstract** Due to the huge surface area to volume ratio and other unique properties, nanomaterials can produce novel mode of actions that could be beneficiary or deleterious on interacting with biological systems. Biocompatibility is the ability of the nanomaterials to perform their designed tasks without causing the body to have additional adverse reactions. On the other hand, safety evaluation examines possible toxicological impact and adverse outcomes that such compounds may have on biological tissues, organs and cells. Some of the physical properties such as surface area, charge, shape and chemical composition of nanomaterials have a marked influence on the toxicity and biocompatibility of the material. For example, high surface reactivity nanoparticles can trigger cytotoxicity, inflammation or oxidative stress reactions. The therapeutic behavior, including safety, of the nanomaterial may be affected by immunological events arising from its likely differential engagement with immune cells. Thus, to assess possible cytotoxicity, genotoxicity, and immunogenicity, extensive assessment involves cell culture studies, as well as animal studies. The regulatory guidelines put forward the principles of standardization and further application of complex analytical and quantification processes in order to ensure the accuracy in biocompatibility and safety assessments. This involves molecular analysis to quantify DNA or protein injury; imaging methods; and assays of cell survival. These evaluations are necessary for formulating policies that eliminate risk factors for health concerns together with harnessing nanoparticles' healing benefits. The fact the nanotechnology is progressing rapidly also poses a risk to environment and health and therefore the measures aimed at minimizing the risks have to be modified very frequently.

---

A. Kumar (✉) · P. Singh
Department of Ophthalmology, University of California, Los Angeles, Los Angeles, CA, USA

D. Kumar · S. Dutt · P. Kalia · R. Kumar (✉)
Department of Pharmacology, Maharaja Agrasen School of Pharmacy, Maharaja Agrasen University, Solan, Himachal Pradesh, India



V. Khanna et al. (eds.), *Nanotechnologies for Drug Delivery and Biopharmaceutical Development*, Progress in Drug Research 77,
https://doi.org/10.1007/978-3-032-00796-4_3

**Keywords** Biocompatibility · Nanomaterials · Safety assessment · Biological tissues · Cytotoxicity · Environmental impact

## 3.1 Introduction

Formulated materials that have dimensions of nanoscale are called nanomaterials and these show variance in properties than their microscale larger counterparts (Gomez-Villalba et al. 2023). Due to the possibility of their interface with biological systems on the nanoscale, these materials are gradually gaining increasing relevance across sectors, especially in the realms of medicine (Barhoum et al. 2022). Since nanomaterials consist of particles at the nanoscale their reactivity, strength, conductivity may be amplified, they have a high surface area to volume ratio, and they possess some flexibility of form. Further, they may be intended too to release therapeutic chemicals that act on particular cells or tissues with improved treatment efficacy and negligible side effects (Harun-Ur-Rashid et al. 2023). For further categorization of nanomaterials, we can categorize them based on the chemical composition, dimensionality or structure. Some of these NM are fullerenes, carbon nanotubes CNTs, and graphene are among the carbon-based nanomaterials (Shao et al. 2021). Due to their unique mechanical, electrical and thermal properties, carbon nanoparticles are particularly suitable for tissue engineering and drug delivery (Fig. 3.1). In view of its ability to penetrate cell membranes and transport drugs to target sites, fullerenes and carbon nanotubes have been discussed for drug carriers. Metallo nanoparticles especially gold, silver and platinum are characterized by unique optic, electric and antibacterial properties. For instance, gold nanoparticles

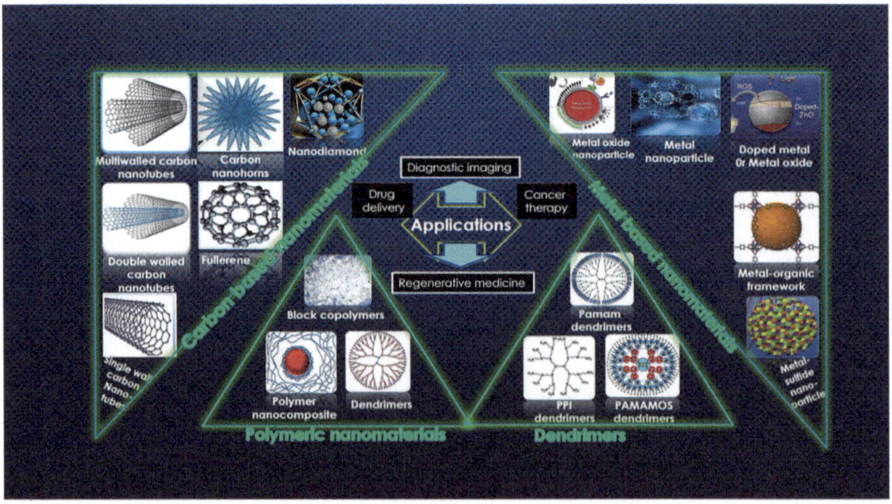


**Fig. 3.1** Expanding the nanomaterials in the diverse area with their types

are used for imaging and photothermal therapy. Silver nanoparticles are often used in medical coatings on its devices and in wound dressing due to its antibacterial capability (Anžlovar and Žagar 2022).

Polymeric nanomaterials refer to organic nanoparticles that are derived from bio compatible and bio degradable polymers. Polymeric nanoparticles have gained tremendous importance due to their remarkably versatile nature and their usability to encapsulate proteins, medicines, or nucleic acids can be released in a controlled and sustained manner (Ma et al. 2024). On account of enhancing biocompatibility and reducing toxicity, they find numerous applications in drug delivery systems for enhancing the therapeutic outcome. Dendrimers are tree shaped, multiterative, branching polymers with highly surface area that can be conjugated with medications or imaging agents. Dendrimers are useful in cancer therapy, gene delivery and imaging because they may be designed for controlled release and site-specific delivery (Wojnarowicz et al. 2020). As the nanoparticles find their use in environmental and medicinal applications, biocompatibility and the safety profile are important aspects. There is a highly likely possibility for nanostructures to exhibit whole range of unique physical, chemical, and biological properties than bulk materials because of their small size and large surface area to volume ratio (Parvin et al. 2024). This is the ability of an implanted substance to perform a certain function in the human body and leak no poisonous or destructive components into the system. Thus, biocompatibility assessments for these materials are more challenging because nanoparticles are known to penetrate cells and tissues in ways that bulk materials do not (Dediu et al. 2023). The safety aspects cover the potential toxicity, inflammatory and carcinogenic effects of nanomaterials. This involves tests for immunogenicity, cytotoxicity, and genotoxicity, that is, indicative or the immune response, injury to cells, or damage to DNA. To address the cellular and systemic level of nanomaterials, other traditional method such as in vitro and in vivo testing are used most of the time (Ferrone et al. 2019). However, due to the unique properties of nanomaterials, these techniques need to be further developed ad adapted. Moreover, advanced imaging approaches and computational modelling depicting the behavior of nanomaterials in biologic systems are emerging (Zhang et al. 2024).

## 3.2 Biomedical Applications of Nanomaterials

Nanomedical applications have been realized in the field that is explored in this paper, some of which includes drug delivery, regenerative medicine, and diagnostic. Medicine delivery is perhaps one of the most revolutionary biomedical applications of nanomaterial (Hekmat et al. 2022). Importantly, drugs may be localized to injured cells or tissues via nanoparticle design, decreasing toxicity and enhancing the benefit of treatments. Liposomes and polymeric nanoparticles have been used in cancer treatment to deliver chemotherapeutic chemicals to tumour cells and sparing normal cells. Fluorescent quantum dots and magnetic nanoparticles are other classes of nanomaterials that are commonly used in imaging to enhance the contrast of procedures such as fluorescence microscopy and Magnetic Resonance Imaging (MRI) (Ranjha et al. 2022). By increasing effectiveness and specificity of imaging procedures, these materials contribute to the development of early diagnostics of diseases. Nanomaterials are applied for fabrication of scaffolds relating to the form of the extracellular matrix in tissue engineering and regenerative medicine that aids tissue regeneration and cell proliferation (Anastasiadis et al. 2022). Carbon nanotubes and hydroxyapatite nanoparticles have been employed to design bone grafts as well as scaffolds for nerve regenerative applications. Cancer treatments involve the use of gold and silver nanoparticles; for example, photothermal therapy where the nanoparticles are used to identify the tumour cells then exposed to light that will cause death of the cancer cells but without affects the other tissues (Hu et al. 2021).

## 3.3 Biocompatibility of Nanomaterials

The ability of a substance to deliver its desired operation in the body without eliciting negative biological responses like toxicity or sharp inflammatory or immunologic responses is referred to as biocompatibility (Boros and Ostafe 2020). Biomaterials cannot be used for such functions in medical programs like tissue engineering and medication delivery, therapeutic reinforcement and diagnosing illnesses and health conditions without being biocompatible. Owing to the small size and heterogeneous morphological and topographical characteristics of the nanomaterials, they have different actions on the biological systems than the bulk materials (Zhu et al. 2021). In order to avoid the aforementioned ill effects on health and enhance the performance of nanomaterials in biomedical sciences, it becomes imperative that the nanomaterials be biocompatible. It was seen that the size and structure topographical factors of nanomaterials greatly influenced their responses on tissue and cell levels (Lebre et al. 2022). Better interaction with deeper tissues is another factor brought by the enhanced ability of small nanoparticles to pass through biological membranes. Although rod like or tubular nanoparticles such as CNs could mechanically injure cells or induce inflammation, spherical nanoparticles tend to engage with the cells in a more orderly manner (Fig. 3.2). The properties which influence the behavior of nanomaterials when in contact with proteins, cells and the immune system are the surface chemistries of the nanomaterials (Kumar et al. 2023). A hydrophobic surface can enhance the abilities of nanoparticles to adsorb proteins; the immune system may then recognize or aggregate the target particles and induce inflammation. Another factor is the surface charge; where compared to the negatively charged or non-charged nanoparticles; the positively charged one may have considerably high cellular applications (Dastgheib et al. 2024). Such factors as toxicity, and biodegradation and solubility also depend with the nature of the material used in the synthesis of the nanoparticles.

For instance, biodegradable polymeric nanoparticles usually exhibit bibliography higher biocompatibility than metal nanoparticles, which includes silver releasing toxic ions (Harish et al. 2022). Inert inorganic nanomaterials such as gold

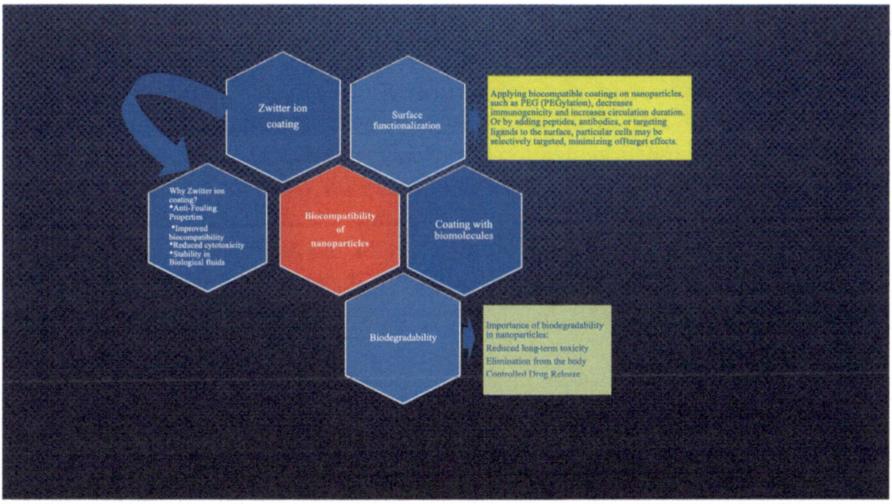


**Fig. 3.2** Biocompatibility of nanomaterials greatly influenced their responses on tissue and cell levels

nanoparticles are generally considered biocompatible due to their non-reactive and non-degradable nature of inorganic materials; however, the long-term residents of inorganic nanoparticles in tissues should be monitored carefully (Butler et al. 2023). High surface area increases the tendency of the nanomaterials to agglomerate, which affects the relation between the material and cells/tissues, and also stability in biological system. Since it is harder to wash out the body with aggregated nanoparticles, they may not be as effective as the well-dispersed ones and may in fact be more toxic (Alshammari et al. 2023).

## 3.4 Biological Interactions of Nanomaterials

Thus, the main forms of interaction between nanomaterials and biological systems can be classified into the following categories: immune system interactions, cellular uptake and adsorption and interaction with proteins (Singh et al. 2023a). Nanoparticles rapidly form a layer of proteins adsorbed from the biological fluid upon their entry into the body known as the protein corona (Kumar et al. 2022a). This corona influences cell recognition built in the nanomaterials and determines whether the cells will absorb the nanomaterials or not. With regard to the properties of nanoparticles, they can exhibit a number of biological effects after entering cells (Singh et al. 2024a). Some nanoparticles are accumulative and get trapped into cells and are resistant to lysosomal degradation while some others may cause oxidative stress and cell injury (Dhiman et al. 2016a). In some cases, immune-interacting Nanoparticles provoke immunogenic reactions that cause inflammatory processes and or hypersensitivity reactions, while in other cases they can be recognized and rejected (Vijaya Durga 2015).

## 3.5 Biocompatibility Enhancement of Nanomaterials

Many strategies have been depicted for enhancing the nanoparticles bio-compatibility (Table 3.1). The most efficient way of enhancing the biocompatibility of nanomaterials is through the functionalization of the nanoparticles by integrating appropriate chemical substances into the otherwise inactive core of the nanomaterials, such as polyethylene glycol (PEG). PEGylation prolongs the time nanoparticles dwell in the bloodstream by enhancing their stability, reducing protein adherence to the particle surface, and reducing the immunological identification of the particles (McInnes and Voelcker 2009). As for example, there are often cases where proteins,

**Table 3.1** Biocompatibility enhancement techniques for nanomaterials, advantages and challenges

| S. no. | Biocompatibility enhancement techniques | Examples | Advantages | Challenges |
|---|---|---|---|---|
| 1 | Surface coating (PEGylation, polymers, lipids) | PEGylated gold nanoparticles, lipid-coated quantum dots | Reduces immune response, increases circulation time | Potential loss of functionality, stability concerns |
| 2 | Biomimetic coating (cell membrane camouflage, protein coating) | Cell membrane-coated nanoparticles, albumin-coated nanocarriers | Improves immune evasion, enhances targeting | Complex fabrication, possible immune recognition over time |
| 3 | Functionalization with biopolymers | Chitosan-functionalized nanoparticles, hyaluronic acid-coated nano systems | Enhances biocompatibility and targeting | Stability and reproducibility issues |
| 4 | Inorganic shell coating (silica, carbon, calcium phosphate) | Silica-coated quantum dots, calcium phosphate shells on drug carriers | Improves stability, reduces toxicity | Potential alteration of physicochemical properties |
| 5 | Controlled release systems (hydrogel/nanogel encapsulation) | Hydrogel-loaded nanoparticles for drug delivery | Sustained release, reduced toxicity | Complexity in tuning release profiles |
| 6 | Charge modification (zwitterionic or neutral coatings) | Zwitterionic polymer-coated nanoparticles | Reduces protein adsorption and immune clearance | May affect cellular uptake and targeting efficiency |
| 7 | Biodegradable nanomaterials (PLGA, PCL, natural polymers) | PLGA nanoparticles for drug delivery, starch-based nanocarriers | Safe degradation into non-toxic byproducts | Limited mechanical stability, controlled degradation issues |
| 8 | Endotoxin removal (detoxification techniques, surface modifications) | Endotoxin-free carbon nanotubes, ultrafiltration for nanoparticle purification | Reduces inflammation and toxicity | Additional processing steps, cost implications |

peptides or lipids can be used to provide specific coatings on nanoparticles reducing their cytotoxic effect and improving their biocompatibility. For example, phospholipid bilayer coating forms lipid nanoparticles like liposomes which are used widely in drug delivery and are relatively very stable (Govindan et al. 2023). The potential hazard of long-term buildup in the body and all the related risks may be significantly lowered by building up biodegradable nanomaterials that break down in the environment into harmless particles. Biocompatible wearable has been prepared from biodegradable materials including polymers like polylactic acid (PLA) and polyglycolic acid (PGA) (Kyriakides et al. 2021). Another approach is employing zwitterionic (neutral) coatings that eradicate most protein adhesion and immune system identification due to rejecting both the positive and negative charges. These coatings minimize the likelihood of the immune response and prolong a blood's circulating life (Safarkhani et al. 2024).

## 3.6 Toxicity and Safety Concerns

Possible toxicity of nanomaterials is a very critical area of concern with respect to the application of these materials environmentally and for biological purposes (Kumar et al. 2021a). Therefore, special chemical and physical characteristics including size, surface area and surface reactivity designated nanomaterials as hazardous because they affect cells, proteins and organs in manners that cannot be compared with that of bulk materials (Pandey et al. 2019). Nanomaterial is toxic in a number of ways, and its toxic effects are discussed below. ROS generation is well known as one of the primary pathways through which nanomaterials can elicit toxic effects (Garg et al. 2022). They again from nanoparticles especially metal based such as titanium dioxide and silver are known to facilitate generation of ROS and cause oxidative stress (Kumar et al. 2022b). This can cause apoptosis, inflammation, and cell dysfunction through direct damage to lipids, proteins, and DNA. As for immune cells, the nanoparticles by virtue of their interactions with leukocytes can induce inflammation (Dhiman et al. 2015a). There are possibilities that have been following activation of neutrophils together with macrophages, the inflammatory cytokines and chemokines are released (Singh et al. 2024b). This inflammation leads to the oxidative stress of tissue, which can worsen tissue damage and result in diseases such as cancer and fibrosis (Kumar et al. 2024a). A number of the nanomaterials are capable of causing either direct or indirect interference to DNA, due to reactions with reactive oxygen species (ROS) (Tsukanov et al. 2022). Some of the nanoparticles can penetrate nucleus and alter DNA, and this leads to formation of chromosomal abnormalities, mutations and most probably cancer. As has been evidenced, metal nanoparticles like cobalt and cadmium are shown to have genotoxic potential. As a result of their small size and high surface energy, nanoparticles are able to mechanically disrupt or harm biological membranes (Wan et al. 2023). The integrity of the membrane may be impaired; the end result could be apoptosis, lack of ability to live or leakage of the cellular content (Mosleh-Shirazi et al. 2022).

The immune system may get activated in response to some nanoparticles because it perceives them as alien substances. This may result in hypersensitivity, allergic responses, or even an autoimmune reaction. For instance, overactivation of the immune system has been linked to carbon nanotubes, which causes fibrotic and inflammatory reactions in tissues (Petrucci et al. 2022).

## 3.7 *In Vitro* Models for Nanomaterial Toxicity Studies

Often in vitro models are used under partly controlled laboratory conditions to study the toxicity of nanomaterials (Table 3.2). These models normally consist of cultured cells exposed to nanomaterials, from which researchers may study biological reactions such as cytotoxicity, oxidative stress, and inflammation (Xie et al. 2023). Cytotoxicity is measured as to how exposed cells interact with nanomaterials and this is determined by cytotoxicity assays (Qamar et al. 2024). Some examples of these indicators are a lactate dehydrogenase (LDH) release test which estimates the membrane integrity and MTT assay that reflects the mitochondrial activity. EROID, AC480, and MitoSOX assays provide an indication of how much ROS was produced in cells after treatment with nanoparticles. For example, the DCFH-DA test recognizes the fluorescence caused by ROS providing insight into the likelihood of oxidation stress of nanoparticles (Rehman et al. 2022). The For the identification of chromosomal abnormalities and DNA damage, other techniques such as the micronucleus tests and DNA damage assays are used. Genomic unstable and bystander effects help in evaluating the genotoxic capability of a nanomaterial (Lebre et al. 2022). The internalization of nanomaterials with cells can be demonstrated by electron microcopy or fluorescent method. The fact that whole living cells are seen in the pictures simplifies the understanding of nanoparticle behavior and more specifically, how they penetrate, degrade, or deposit within cells. Based on immunological elements of inflammation and immune activation, immunoassays evaluate the generation of cytokines or chemokines in response to nanomaterials (Soman et al. 2022).

## 3.8 *In Vivo* Models for Nanomaterial Toxicity Studies

Live things' experience is a more detailed one when it comes to the toxicity of nanomaterials through in vivo models. To evaluate the implication of nanomaterials and the effects on biological system as well as their biodistribution, metabolism, clearance and toxicity, mouse, rat, zebrafish, and rabbit are used (Allan et al. 2021). The biodistribution of nanomaterials in liver, spleen, lungs and brain is assessed in vivo as well as the metabolism and clearance. This provides information on potential organ toxicity data and also on long term accumulation of the compound. Meanwhile, acute and chronic post routes can also be measured in vivo describing both short

**Table 3.2** *In vitro* models for nanomaterial toxicity studies, including their biological reactions, indicators, insights, and applications

| S. no. | In vitro model | Biological reaction studied | Indicators | Insight gained | Applications |
|---|---|---|---|---|---|
| 1 | MTT/XTT/ resazurin (alamar blue) assay | Cytotoxicity & cell viability | Mitochondrial activity, metabolic reduction | Cell viability and proliferation | General cytotoxicity screening |
| 2 | LDH (lactate dehydrogenase) assay | Membrane integrity | LDH leakage into culture medium | Membrane damage and necrosis | Assessing nanoparticle-induced cell damage |
| 3 | Reactive oxygen species (ROS) assay | Oxidative stress | DCFH-DA fluorescence, ROS generation | Oxidative stress levels and inflammation | Nanomaterial-induced oxidative damage |
| 4 | Comet assay (single-cell gel electrophoresis) | DNA damage | DNA strand breaks, tail length in assay | Genotoxicity and DNA fragmentation | Evaluating nanoparticle-induced genetic damage |
| 5 | Micronucleus assay | Chromosomal damage | Formation of micronuclei in cells | Mutagenicity and chromosomal instability | Genetic toxicity screening |
| 6 | Caspase activation assay | Apoptosis induction | Caspase-3/7 activity | Programmed cell death pathway activation | Understanding nanomaterial-induced apoptosis |
| 7 | Annexin V/PI staining | Apoptosis vs. necrosis | Phosphatidylserine exposure, membrane integrity | Differentiates apoptotic vs. necrotic cell death | Toxicity mechanism studies |
| 8 | Cytokine release assay (ELISA, multiplexing) | Inflammatory response | TNF-α, IL-6, IL-1β levels | Inflammatory cytokine secretion profile | Immunotoxicity evaluation |
| 9 | Hemolysis assay | Blood compatibility | Hemoglobin release from RBCs | Nanoparticle-induced hemolysis and RBC damage | Evaluating intravenous nanoparticle safety |
| 10 | Blood coagulation assay | Clotting and Thrombogenicity | Prothrombin time, fibrin clot formation | Effects of nanomaterials on coagulation pathways | Safety assessment for blood-contacting devices |
| 11 | TEER (transepithelial electrical resistance) assay | Barrier integrity (gut, lung, BBB) | Resistance across cell monolayer | Nanoparticle effects on tight junction permeability | Studying nanoparticle transport across barriers |

(continued)

**Table 3.2** (continued)

| S. no. | In vitro model | Biological reaction studied | Indicators | Insight gained | Applications |
|---|---|---|---|---|---|
| 12 | Organoids & 3D cell culture models | Tissue-specific toxicity | Viability, differentiation markers | More physiologically relevant toxicity assessment | Advanced predictive toxicology studies |

term and long-term toxicity of a substance (Hotowy et al. 2024). Protracted low-dose-wellness studies assess the effects of health risks resulting from long-term exposures and are midway tests that mimic realistic scenarios, unlike acute toxicity studies, which encompass massive dose exposures but for a short period (Thangam et al. 2022). On a microscale, the tissues from nanomaterial exposed animals can be examined for fibrosis, inflammation, or tissue injury. For example, carbon nanotubes have been proven to cause lung fibrosis in mouse experiments. The immunosuppressive or allergic effects of NTs nanostructures are investigated using animal models (Pan et al. 2019).

## 3.9 Toxicity Studies of Nanomaterials

The target of extensive toxicity assessments of NM is to determine their efficacy on the toxicity profiles such as systemic, cytotoxic, and genotoxicity. These investigations are important for nanoparticles used in biological applications to be approved by the regulators (Zhang et al. 2021; Kumar et al. 2021b). Some metal nanoparticles that have been shown to cause oxidative stress, genotoxicity and inflammation include: silver, titanium dioxide, and zinc oxide (Fig. 3.3). For instance, high concentration of silver nanoparticles can harm the kidney and liver of a human being. Single-walled and multi-walled carbon nanotubes have fibre-like formations, which has raised alarms because this is similar to asbestos (Kim et al. 2024; Dhiman et al. 2016b).

Research shows that once they are inhaled, the said nanotubes can cause chronic inflammation of the lungs and fibrosis. Made up of semiconductor materials which may include cadmium selenide, quantum dots are mostly used in imaging. The hazardous ions, cadmium, which is released by quantum dots, has potential to induce oxidative stress and DNA damage (Chehelgerdi et al. 2023; Hamid et al. 2024). Long-term application and especially high doses can be dangerous, research indicated. Lipid based nanocarriers, namely liposomes and solid lipid nanoparticles are generally known to be harmless; nevertheless, the safety of these vehicles depends on a number of factors such as their composition and size and the charge on their surface. They are widely used in vaccination as well as medication delivery systems (Balas et al. 2018; Kumar et al. 2024b).

**Fig. 3.3** Toxicity and safety concerns of nanomaterials

## 3.10 Nanomaterial Evaluation

The adoption of nanomaterials in modern biopharmaceuticals makes possible effective drug targeting and enhances bioavailability together with decreased side effects. The safety risks from using nanomaterials emerge from their tiny size as well as their individual properties and biological system compatibility (Table 3.3). A complete evaluation of nanomaterial safety in biopharmaceuticals demands physicochemical analysis as well as laboratory and animal trials along with regulatory evaluations (Bera et al. 2010). The biological activities of nanomaterials become regulated by multiple physicochemical elements involving size, shape and surface charge characteristics and composition content. Nanoparticles smaller than 100 nm possess the capability to cross biological barriers thus risking non-target tissue accumulation due to biodistribution issues through the blood-brain barrier. PEG-coated surfaces enhance biocompatibility among nanomaterials yet provoke immune reactions in particular patients. The effect of nanomaterials on cells and tissues gets evaluated through techniques combining cell viability tests with ROS-producing measurements and hemolysis evaluation methods. In vitro tests have certain boundaries when it comes to accurately representing biological conditions so scientists need to validate their results by using animal test systems (Kelly et al. 2020). Nanomaterial assessment in animals determines how materials distribute in the body and how they break down and leave the body and their potential effects that extend beyond short periods. Safety information becomes crucial through measuring toxicity markers that include organ retention levels and activated immune responses together with changes in tissue histology (Ji et al. 2021).

**Table 3.3** Nanomaterial evaluation regarding safety for some biopharmaceuticals

| S. no. | Evaluation criteria | Examples of nanomaterials | Applications in biopharmaceuticals |
|---|---|---|---|
| 1 | Cytotoxicity assessment | Gold nanoparticles, quantum dots, carbon nanotubes | Ensuring cell viability in drug delivery and imaging |
| 2 | Genotoxicity & mutagenicity | Silver nanoparticles, titanium dioxide ($TiO_2$) | Evaluating DNA damage risks in gene therapy |
| 3 | Immunogenicity & Inflammation | Lipid nanoparticles (LNPs), polymer-based nanocarriers | Assessing immune response in vaccine formulations |
| 4 | Hemocompatibility & coagulation | Iron oxide nanoparticles, PEGylated nanoparticles | Evaluating blood compatibility for IV applications |
| 5 | Biodistribution & pharmacokinetics | Dendrimers, polymeric micelles | Understanding nanodrug metabolism and clearance |
| 6 | Oxidative stress & ROS generation | Silver nanoparticles, zinc oxide nanoparticles | Evaluating oxidative damage in cellular systems |
| 7 | Nanomaterial aggregation & stability | Carbon dots, liposomes | Ensuring formulation stability for biopharma use |
| 8 | Endotoxin contamination & sterility | PLGA nanoparticles, chitosan-based carriers | Preventing immune reactions in biologic therapies |
| 9 | Organ-specific toxicity | Silica nanoparticles, cerium oxide nanoparticles | Assessing nanoparticle toxicity in liver, kidney, brain |
| 10 | Long-term biodegradability & clearance | PLGA nanoparticles, albumin nanoparticles | Ensuring safe degradation and elimination |

The evolving nature of nanotechnology makes it difficult to create standardized testing procedures and regulatory standards. The pharmaceutical industry investigates multiple safety assessment techniques which combine predictive modeling solutions with new testing approaches to improve evaluation processes. Post-market surveillance maintains its importance in monitoring how nanomaterials affect patients in the long run and for protecting patient safety (Kim et al. 2018). The safety concerns surrounding nanomaterials have not impeded their transformative power in drug delivery methods of oncology treatments as well as vaccine development and immunotherapy progress. LNP-based mRNA vaccines demonstrate how nanotechnology handles the improvement of therapeutic effects alongside sufficient patient security. The development of secure biopharmaceuticals depends on sustained nanomaterial-biological interaction investigations along with better toxicology and regulatory science methodologies (Mamun and Sabantina 2023).

## 3.11 Regulatory Guidelines for Nanomaterial Safety

Global nanotechnology growth has expanded numerous industries such as electronics, environmental engineering, cosmetics, and medical industries. But depending on these characteristics such as small size and high surface reactivity the environment or human health may be threatened by dangers from nanomaterials. These concerns have been addressed by different global regulatory bodies developing guidelines for safe manufacture, processing, application and management of nanomaterials (Wei et al. 2023). Risk reduction based on this recommendation ensures that no matter the actual ionization from nanomaterials, adequate measures have been taken towards controlling them. With reference to various approaches to risk assessment and management, a growing number of countries and regions has adopted their specific frameworks for nanomaterials. Even though, we have no common rules and regulations to control it, numbers of important organizations and nations have set some rules and regulations in this regard (Wang et al. 2021). Nanomaterials aspect includes food and medicine packaging, medications, and cosmetics, and medical products applied in healthcare products are under the regulation of the U.S Food and Drug Administration (FDA). Despite emphasizing that the products containing nanomaterials should be in accordance with the existing rules and restrictions more examination may be needed in regards to the specific properties of the nanomaterial (Pei et al. 2024). The FDA makes recommendations that business performs the risk evaluations including, but not limited to environmental, pharmacokinetic and toxicity assessments on nanomaterials (Popescu et al. 2019). The larger particles are covered by the Toxic Substances Control Act commonly referred to as TSCA and are controlled by the Environmental Protection Agency EPA. Under the PMN of nanomaterials, the EPA set some specific guidelines requiring producers to submit information on the chemical name, hazards, and pathways via which the material might cause harm (Díez-Pascual 2022).

Like any other forms of materials, nanomaterials are governed by the Registration, Evaluation, Authorisation, and Restriction of Chemicals (REACH) legislation within the European Chemicals Agency. The producers or importers of nanoparticles have a responsibility under REACH to provide information and register the nanoparticles based on safety, application and effects on the environment (Fytianos et al. 2020). To target nanoforms of chemicals, ECHA developed specific annexes in 2018. These impose on producers' information declarations on the size, surface area, and chemistry of their particles. Cosmetics that contain one or more nanomaterials, and any other goods containing nanomaterials, must be labelled appropriately within the EU to inform customers of their existence (Bobrinetskiy et al. 2021). The specific focus is given to a study of toxicological impacts of nanoparticles utilized as food additives or in packaging materials while the EFSA also provides guidelines for risk assessment of nanomaterials in food and feed. Some of the organizations that are responsible of regulating nanomaterials in Canada include Environment and Climate Change Canada (ECCC) and Health Canada. Although nanomaterials are a comparatively new technology, under the CEPA of Canada, the regulatory treatment of nanomaterials is the same as conventional substances (Wang et al. 2023). To whom new nanomaterials are imported or manufactured have to notify the Ministers of the characteristics of the new nanomaterial and of risks that can be associated with its usage for health and environmental impacts. While Japan has comparatively fewer formal guidelines about nanomaterials, for its management the Ministry of the Environment and Ministry of Health, Labour, and Welfare has issued guidelines concerning the safety of nanomaterials. Such recommendations readily promote sacrificing personal exposures and possible risk to minimize the adverse impacts on the environment and the workers (Sharma et al. 2021).

## 3.12 Key Safety Considerations

The regulatory bodies have required risk assessment of nanomaterials, and the studies of toxicity, bioaccumulation, and the persistence of nanomaterials. As a result, understanding the relationship of nanomaterials and biological systems and ecosystems, both the in vitro and in vivo studies are employed. The potential to aggregate or alter properties of nanoparticles in different environments additionally should be considered. This means that while dealing with nanomaterials, the disposal might not have the same effects of bulk disposal. A lot of times regulatory bodies require testing for attributes such as degradation, ability to move in soil and water, and possibility of accumulating in the tissues of organisms (Guazzo et al. 2018; Kumar et al. 2024c). Employers working with nanomaterials risk exposure of their skin or respiration system. All these regulatory guidelines recommend the provision of adequate fresh air and proper wearing of PPE and training in correct handling methods. Representative example is the OSHA, which provides directions on nanomaterial handling precautions and exposure thresholds. Specifically, the implications of approach means that the customer should be protected through the means of transparency (Krishnan et al. 2020). There are often special rules which require labelling of products containing nanomaterials particularly food and cosmetics. While other regions for example the U.S encourages producer to inform consumers of the use of nanomaterials on the product, it is law in EU that cosmetics containing nanomaterials must be labeled as such. Increasing use of nanotechnology in various industries results in the shift of regulatory norms of nanomaterials safety (Zheng et al. 2023; Singh et al. 2023b). As the regulation strategies of different areas are different, they have the same common goal, which is the environmental and health safety of nanomaterials. All these regulatory strategies target at shielding the community though also creation further expansions of nanotechnology by demanding valuation of danger, obedience to cataloging regulations, and endorsing suitable treatment practices. It is likely that regulatory control will remain or become even more relevant as nanomaterial usage increases in order for all industries to ensure safety (Fang et al. 2023).

## 3.13 Biocompatibility and Toxicity Evaluation

To meet the safe use of nanomaterials in implants, medical devices, as well as treatments, toxicity and biocompatibility must be first defined. The ability of a material to perform its function within a biological environment with some level of acceptance without causing harm to the surrounding tissue in the form reaction such as inflammation, toxicity, or immunogenicity is called biocompatibility. Such assessment is challenging for nanomaterials because of their unique properties, such as particle size, surface area, and bioreactivity (Cafarelli et al. 2021; Kumar et al. 2022c). Classification of a nanomaterial liability to adverse biological effects like oxidative stress, inflammation, cytotoxicity or genotoxicity is called toxicity assessment. To judge biocompatibility and toxicity *in vitro* based on cell culture tests along with *in vivo* tests based on animals are performed. The assays used herein consider aspects such as cytotoxicity, cellular accumulation, generation of reactive oxygen species, and pro-inflammatory effects, as well as effects on immune cells. Long term toxicity studies also evaluate the propensity of the nanomaterials to accumulate, be eliminated, and to exert effects on individual organs (Siraj et al. 2023; Kumar et al. 2021c).

## 3.14 Major Studies on Nanomaterial Biocompatibility and Toxicity

A number of research studies have been conducted to understand the toxicity and biocompatibilities of different nanomaterials. Among all the nanomaterials, the silver nanoparticles (AgNPs) findings are more frequent because they exhibit antibacterial property. Even though AgNPs exhibit outstanding bactericidal properties, it was found that in a significant study they induce both oxidative stress and cytotoxic effects in human cells, mainly due to the release of silver ions (De Armentia et al. 2020). Possible risks include organ deposition and liver toxicity were learned from in-vivo studies to animal subjects which explain why the use of these materials should be controlled in medical applications (Liu et al. 2022). Research has been made to show how single-walled as well as, multi-walled carbon nanotubes can be used in tissue engineering and medication transport. However, their biocompatibility has been questioned in a number of studies (Buitrago et al. 2021; Kumar et al. 2021d). Different in vivo experiments performed on rats have shown that CNTs cause lung inflammation or fibrosis and in certain circumstances provide genotoxicity. Due to the structural properties of CNTs these have been associated with the effects of exposure to asbestos hence there have been attempts to establish the effects of long-term exposure to CNTs. PLGA nanoparticles have garnered considerable interest in drug delivery mainly because the polymer is biocompatible and biodegradable. In vitro and in vivo experiments have revealed that the presence of PLGA nanoparticles does not have many negative impacts because it is a biocompatible material (Mamun and Sabantina 2023; Kumar et al. 2022c). Long term usage is safer since the chemical's by products is naturally metabolized within the body, the byproducts being lactic and glycolic acids. Gold nanoparticles find diverse application in cancer therapy, diagnosis and drug delivery in the contemporary world. Based on a number of in vitro studies, it has been reported that they exhibit low level of cytotoxicity, and are bio compatible. However, experimental animal studies have shown that concentration of AuNPs could rise in some of the target organs comprising spleen and liver at higher concentrations of the nanoparticles. Long-term studies on their ability to bioaccumulate & be cleared are still under relay (Garriga et al. 2020).

## 3.15 Applications of Nanomaterials in Medical Devices, Implants, and Therapeutics

The technology of nanomaterials can revolutionize implants, medical instruments, and treatment methodologies across the world. To boost the efficacy and selectivity of medicinal medicines, nanoparticles have been effectively used as delivery systems (Tang et al. 2022; Kumar et al. 2017). The surface properties of nanoparticles are modifiable to animate the nanoparticles to adhere to specific tissues or cells, reducing side effects and increasing drug uptake for instance in cancer cells (Curulli 2023). Among the most commonly used nanoparticles in delivering medication for chronic diseases including cancer are polymeric nanoparticles like the PLGA and liposomes. Photothermal therapy is a less invasive therapeutic technique for cancer treatment; it employs AuNPs to accumulate in oxidative tumour cells and destroy the cells using laser light to heat the AuNPs (Martínez-Carmona and Vallet-Regí 2020).

Some of the areas in which nanomaterials have been applied include; tissue engineering and biocompatible implants. Better integration with biological tissues is enabled by the ability to mimic the extracellular matrix which promotes cell growth, tissue healing and remodeling (Krasley et al. 2024). Current orthopaedic implants such as tooth and bone implants are being developed from nanocomposite materials because they have better compatibility and mechanical characteristics than the traditional materials. Carbon nanotubes (CNTs) and nanofibers are used to fabricate scaffolds to support regeneration of bone, cartilage and neural tissues (Kashyap et al. 2023; Dhiman et al. 2015b).

Due to the fact that medical gadgets have a higher likelihood of infection or inflammation, functional enhancements are now incorporating nanomaterials into the products. Due to their action against bacteria, they reduce risks of bacterial infections and are applied in catheters, and dressings for open wounds. One of the advantages of implants with nanostructured coatings is improved osseointegration, and reduced inflammation (Soman et al. 2022).

Also, nanoparticles are enhancing application and diagnosing tools in the medical field. Fluorescence imaging and magnetic resonance imaging both make use of iron oxide nanoparticles and quantum dots as contrasting agent. These nanoparticles make it possible to diagnose an illness at an early stage, including cancer, and sharpens the picture resolution. Thus, biosensors with a nanomaterial base are also being created to diagnose diseases through biomarkers that are faster and more accurate diagnostic tools (Ji et al. 2021; Dhiman et al. 2015c).

## 3.16 Long-Term Safety Assessments

Consequently, to comprehend the prospective risks of nanomaterials during long-term exposure that is associated with their application in manufactured products within various industries and medicine as well as their environmental use, long-term effects research must be considered (Hekmat et al. 2022). Owing to the reactivity, large surface area, and nanoscale size of the nanomaterials, one can expect that the material's biological interfaces could be different from those of the corresponding bulk material, thus, the analysis. These assessments include biodistribution, bioaccumulation, persistence and chronic toxicity (Witika et al. 2020).

Some of the nanoparticles can cross the blood-brain barrier and reach the brain, spleen, liver, kidneys and many other parts of the body. Studies using imaging procedures and animals observe the distributions and deposition of these materials in tissues over time. For instance, while gold NPs are reported to be biocompatible, studies have evidenced their use in behaving pecuniarily in the body parts including liver and spleen hence raising a concern of clearance of such particles and determines their impacts in long-run treatments (Vagena et al. 2024).

Some nanoparticles can enter the organs because the body cannot break them down or expel them easily. The slow rate of clearance of metal nanoparticles such as titanium dioxide or silver has been examined because of toxicity and long-term effects (Chen et al. 2024).

To be more precise, oxidative stress, fibrosis, inflammation, and amazingly, carcinogenicity might be outcomes of nanomaterial exposure. These concerns are tested in in vivo animal experiments whereby a test food can take months or years to demonstrate delayed effects. For instance, two long-term studies have bound carbon nanotubes to pulmonary fibrosis or inflammation, like asbestos material (Phua et al. 2022).

## 3.17 Environmental Impact

Nanomaterials pose unique environmentally demanding challenges because they are used in many industries like electronics industry, health industry, agriculture and energy industries (Peng et al. 2022). There is debate over the extensive impact they may have on biological and environmental systems that current courses fail to capture because of their nature, size small and highly reactive. The degree, persistence, toxicity and accumulation of nanomaterials determine the impact on the environment (Guinart et al. 2021) (Fig. 3.4).

One can also consider several ways of the nanomaterials' release into the environment: in emissions, through consumers' products disposal, and through wastewaters with industrial production (Wu et al. 2024). Due to their small size, they are capable of extending through soil water and air and once out never disappear. For instance, metal nanoparticles if not biodegradable like oxides of titanium or silver

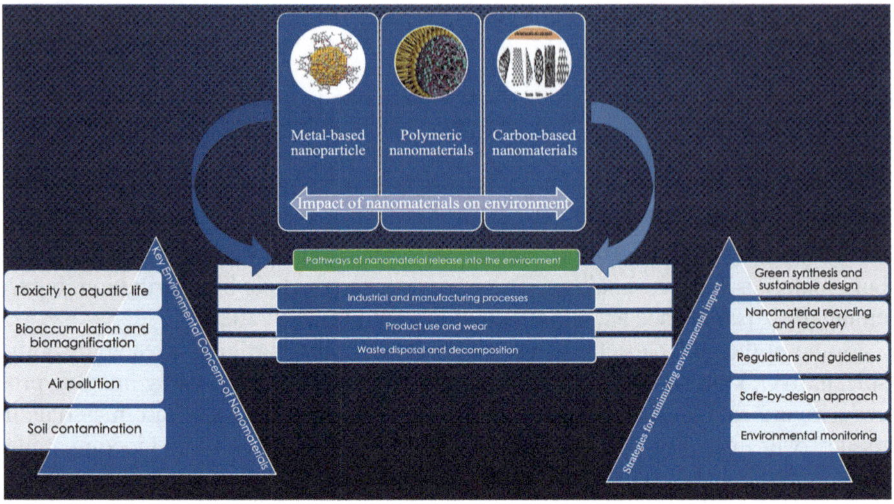


**Fig. 3.4** Impact of nanomaterials on environment

in this case, will hang around in ecosystems with chain effects on the food chain (Bhatt et al. 2023).

As mentioned in recent works, more than 30 types of nanomaterials are toxic to both water and land inhabitants. Used extensively due to their antimicrobial properties, these silver nanoparticles pollute water sources and disorganize water life forms; fish, bacteria and algae inclusive (Paramo et al. 2023). Contact with nanomaterials in water environment has been associated with cellular changes, increased stress rate as well as behavior changes in animals. Besides altering microbial consortiums involved in nutrient cycling, metal oxides and carbon nanotubes may also affect soil health as well (Fritea et al. 2021).

Nanomaterials are cumulative in tissues of living organisms and can actually move up the food chain. Another disadvantage is that nano materials are bioaccumulable and have long lasting ecological impact Once tiny creatures circulate these nano materials, it getting to the next level in the food chain. What makes their impacts on biodiversity, particularly where other organisms are highly sensitive, worrying (Mod et al. 2024). The nanomaterials' ecological performance is being studied by researchers and regulatory bodies, and product alternatives friendly to the environment are being encouraged and new products with minimal impact on the environment are being developed to address these concerns (Tian et al. 2022).

## 3.18 Disposal Concerns of Nanomaterials

Because of certain factors in the nanoscale minuteness of particles, their chemical activity, and potential long lifespan in the environment, nanomaterial disposal threatens the environment and human health (Liao et al. 2020). Application of nanomaterials are gaining popularity in electronics, consumer products, in health and medical sector and so on including the industrial production. Therefore, discharge of such products into the regular waste disposal systems such as the landfill, incineration and wastewater system raise challenges that require special management and consideration (Dimitriou et al. 2023).

Some nanomaterials are virtually stable and refuse to degrade… Examples of such nanomaterials include metal-based nanomaterials particularly those made of titanium dioxide, silver, and gold among others because they are notorious for their long half-life cycle which may take years in natural environments or decomposing in landfills (Petrovic et al. 2024). Since these nanomaterials are fairly persistent in the ecosystems, they could cause research and damage ecosystems in courses that accumulate slowly over time to the environment in various states such as soil, water and air. For example, there is a potential of dissemination of discarded product-derived silver nanoparticles into the groundwater and alter the microbial community, which is vital to the ecosystem (Kang et al. 2021).

Hence, their emissions in waste streams pose specific challenges that are occasioned by some of the nanomaterials that have either dangerous constituents or chemically active surfaces. Some nanomaterials can only be partially decomposed by incineration, which is a common method for the disposal of dangerous waste, and in doing so may emit toxic compounds or disperse nanoparticles in the air. In the same way, the end disposal of nanomaterials may pollute water and soil when they are disposed on landfills. Nanoparticles being smaller than micrometers in water treatment plants reach filtration processes, pose a threat to fish and other aquatic animals, and enter water sources (Paduraru et al. 2021).

The processes of nanomaterial recovery and recycling are technically challenging. Sometimes, nanoparticles cannot be removed or separated from the main bulk of the material using standard recycling methods. Second of all, due to potential capability of nanoparticles to deteriorate the quality of resumed products, or generate new safety risks, mixed waste containing nanomaterials complicates recycling (Xie and Xu 2024). Hence, the new rules of Nanomaterials disposal are needed to address these issues, addressing safe waste management and improving filtering systems for the segregation of non-biodegradable nanomaterials. But to avert dangers to human lives and the environment, nanoparticles must be secured; this needs regulation instruments (Silva et al. 2023).

## 3.19 Conclusion

Considering that the nanoparticles are to be used in various industries, medicine, and environmental enhancement, besides medical uses, the biocompatibility and safety aspects form a crucial part of these nanoparticles. The behavior of nanomaterials in biological systems could be significantly different from that of the bulk materials because of these unique attributes of the nanomaterials, namely size, surface area and reactivity. To better understand risks that range from cytotoxicity, oxidative stress, inflammation to long-term biomagnification, essential safety screenings such as in vitro and in vivo experiments are therefore necessary. Besides, biocompatibility of nanomaterials-referred to as the ability of the nanotechnology to perform its functions within medical systems without calling for side effects-also influences the healthcare innovation success rate. This situation is seen most in cases of medical devices, implants and treatment. Since nanomaterials can remain in place and cause harm to ecosystems through their disposal and usage discharge, the safety assessments also consider broader surroundings. Public policies and standards are emerging to abate the synthesis, use, and removal of nanomaterials so as to address disputed risks associated with nanotechnology. New technologies and safety measures for the explained nanomaterials will continue to be explored so that their usage in other industries would be responsible.

## References

Allan J et al (2021) Regulatory landscape of nanotechnology and nanoplastics from a global perspective. Regul Toxicol Pharmacol 122:104885. https://doi.org/10.1016/j.yrtph.2021.104885

Alshammari BH et al (2023) Organic and inorganic nanomaterials: fabrication, properties and applications. RSC Adv 13(20):13735–13785. https://doi.org/10.1039/d3ra01421e

Anastasiadis SH, Chrissopoulou K, Stratakis E, Kavatzikidou P, Kaklamani G, Ranella A (2022) How the physicochemical properties of manufactured nanomaterials affect their performance in dispersion and their applications in biomedicine: a review. Nanomaterials 12(3):552. https://doi.org/10.3390/nano12030552

Anžlovar A, Žagar E (2022) Cellulose structures as a support or template for inorganic nanostructures and their assemblies. Nanomaterials 12(11):1837. https://doi.org/10.3390/nano12111837

Balas M et al (2018) Coating dependent in vitro biocompatibility of new Fe-Si nanoparticles. Nanomaterials 8(7):1–26. https://doi.org/10.3390/nano8070495

Barhoum A et al (2022) Review on natural, incidental, bioinspired, and engineered nanomaterials: history, definitions, classifications, synthesis, properties, market, toxicities, risks, and regulations. Nanomaterials 12(2):177. https://doi.org/10.3390/nano12020177

Bera D, Qian L, Tseng TK, Holloway PH (2010) Quantum dots and their multimodal applications: a review. Materials 3(4):2260–2345. https://doi.org/10.3390/MA3042260

Bhatt HN et al (2023) Potential and progress of 2D materials in photomedicine for cancer treatment. ACS Appl Bio Mater 6(2):365–383. https://doi.org/10.1021/acsabm.2c00981

Bobrinetskiy I et al (2021) Advances in nanomaterials-based electrochemical biosensors for foodborne pathogen detection. Nanomaterials 11(10):1–26. https://doi.org/10.3390/nano11102700

Boros BV, Ostafe V (2020) Evaluation of ecotoxicology assessment methods of nanomaterials and their effects. Nanomaterials 10(4):610. https://doi.org/10.3390/nano10040610

Buitrago E, Novello AM, Fink A, Riediker M, Rothen-Rutishauser B, Meyer T (2021) Nanosafe III: a user friendly safety management system for nanomaterials in laboratories and small facilities. Nanomaterials 11(10):1–22. https://doi.org/10.3390/nano11102768

Butler J, Handy RD, Upton M, Besinis A (2023) Review of antimicrobial nanocoatings in medicine and dentistry: mechanisms of action, biocompatibility performance, safety, and benefits compared to antibiotics. ACS Nano 17(8):7064–7092. https://doi.org/10.1021/acsnano.2c12488

Cafarelli A et al (2021) Piezoelectric nanomaterials activated by ultrasound: the pathway from discovery to future clinical adoption. ACS Nano 15(7):11066–11086. https://doi.org/10.1021/acsnano.1c03087

Chehelgerdi M et al (2023) Progressing nanotechnology to improve targeted cancer treatment: overcoming hurdles in its clinical implementation. Mol Cancer 22(1):169. https://doi.org/10.1186/s12943-023-01865-0

Chen S et al (2024) Electrospun nanofibrous membranes meet antibacterial nanomaterials: from preparation strategies to biomedical applications. Bioact Mater 42:478–518. https://doi.org/10.1016/j.bioactmat.2024.09.003

Curulli A (2023) Functional nanomaterials enhancing electrochemical biosensors as smart tools for detecting infectious viral diseases. Molecules 28(9):3777. https://doi.org/10.3390/molecules28093777

Dastgheib ZS, Abolmaali SS, Farahavar G, Salmanpour M, Tamaddon AM (2024) Gold nanostructures in melanoma: advances in treatment, diagnosis, and theranostic applications. Heliyon 10(15):e35655. https://doi.org/10.1016/j.heliyon.2024.e35655

De Armentia SL, Del Real JC, Paz E, Dunne N (2020) Advances in biodegradable 3D printed scaffolds with carbon-based nanomaterials for bone regeneration. Materials 13(22):1–49. https://doi.org/10.3390/ma13225083

Dediu V, Ghitman J, Gradisteanu Pircalabioru G, Chan KH, Iliescu FS, Iliescu C (2023) Trends in photothermal nanostructures for antimicrobial applications. Int J Mol Sci 24(11):1–41. https://doi.org/10.3390/ijms24119375

Dhiman R, Durga V, Rio AJ (2015a) Bioanalytical method for lornoxicam determination in human plasma by using piroxicam as internal standard by LC-MS/MS. J Chem Pharm Res 7(8):54–60

Dhiman R, Durga V, Jayasankar K, Raj S (2015b) Solid phase extraction: potential carbocisteine bioanalysis using carbocisteine 13C3 stable isotope: LC-MS/MS technique with pharmacokinetics on healthy human subjects. Der Pharm Lett 7(8):126–134

Dhiman R et al (2015c) Novel LC-MS/MS bioanalytical method validation for gabapentin in human plasma in accordance with current regulatory guidances. Int J Pharm Chem Sci 4(4):462–468. Available: http://www.ijpcsonline.com/files/08-10-15/07-1035.pdf

Dhiman R, Kumar D, Kumar B, Pandey BL (2016a) Quantitative determination of atorvastatin, ortho-hydroxy atorvastatin, para-hydroxy atorvastatin in human plasma using rosuvastatin as internal standard by LC-MS/MS. Int J Pharm Chem Sci 4(4):487–500

Dhiman R, Kumar B, Kumar D, Kumar A (2016b) Liquid-liquid extraction method for Ziprasidone (ZRS) bioanalysis by using ZRS-D 8 (stable isotope) as internal standard. J Chem Pharm Res 8(3):722–727

Díez-Pascual AM (2022) Surface engineering of nanomaterials with polymers, biomolecules, and small ligands for nanomedicine. Materials 15(9):3251. https://doi.org/10.3390/ma15093251

Dimitriou C, Psathas P, Solakidou M, Deligiannakis Y (2023) Advanced flame spray pyrolysis (FSP) technologies for engineering multifunctional nanostructures and nanodevices. Nanomaterials 13(23):3006. https://doi.org/10.3390/nano13233006

Fang L, Shi C, Wang Y, Xiong Z, Wang Y (2023) Exploring the diverse biomedical applications of programmable and multifunctional DNA nanomaterials. J Nanobiotechnology 21(1):1–11. https://doi.org/10.1186/s12951-023-02071-2

Ferrone E, Araneo R, Notargiacomo A, Pea M, Rinaldi A (2019) ZnO nanostructures and electrospun ZnO–polymeric hybrid nanomaterials in biomedical, health, and sustainability applications. Nanomaterials 9(10):1–33. https://doi.org/10.3390/nano9101449

Fritea L et al (2021) Metal nanoparticles and carbon-based nanomaterials for improved performances of electrochemical (bio)sensors with biomedical applications. Materials 14(21):6319. https://doi.org/10.3390/ma14216319

Fytianos G, Rahdar A, Kyzas GZ (2020) Nanomaterials in cosmetics: recent updates. Nanomaterials 10(5):1–16. https://doi.org/10.3390/nano10050979

Garg R et al (2022) COVID-19: a challenge towards the sustainability of health in platform era. Neuroquantology 20(11):2503–2519. https://doi.org/10.47750/pnr.2022.13

Garriga R et al (2020) Toxicity of carbon nanomaterials and their potential application as drug delivery systems: in vitro studies in Caco-2 and MCF-7 cell lines. Nanomaterials 10(8):1–21. https://doi.org/10.3390/nano10081617

Gomez-Villalba LS, Salcines C, Fort R (2023) Application of inorganic nanomaterials in cultural heritage conservation, risk of toxicity, and preventive measures. Nanomaterials 13(9):1454. https://doi.org/10.3390/nano13091454

Govindan B, Sabri MA, Hai A, Banat F, Haija MA (2023) A review of advanced multifunctional magnetic nanostructures for cancer diagnosis and therapy integrated into an artificial intelligence approach. Pharmaceutics 15(3):868. https://doi.org/10.3390/pharmaceutics15030868

Guazzo R et al (2018) Graphene-based nanomaterials for tissue engineering in the dental field. Nanomaterials 8(5):1–31. https://doi.org/10.3390/nano8050349

Guinart A, Perry HL, Wilton-Ely JDET, Tetley TD (2021) Gold nanomaterials in the management of lung cancer. Emerg Top Life Sci 4(6):627–643. https://doi.org/10.1042/ETLS20200332

Hamid A et al (2024) Artificial intelligence (AI) based colonoscopy. In: Nanobiotechnology and artificial intelligence in gastrointestinal diseases. IOP Publishing. https://doi.org/10.1088/978-0-7503-6134-7CH12

Harish V et al (2022) Review on nanoparticles and nanostructured materials: bioimaging, biosensing, drug delivery, tissue engineering, antimicrobial, and agro-food applications. Nanomaterials 12(3):457. https://doi.org/10.3390/nano12030457

Harun-Ur-Rashid M, Jahan I, Foyez T, Bin Imran A (2023) Bio-inspired nanomaterials for micro/nanodevices: a new era in biomedical applications. Micromachines 14(9):1786. https://doi.org/10.3390/mi14091786

Hekmat A, Saso L, Lather V, Pandita D, Kostova I, Saboury AA (2022) Recent advances in nanomaterials of group XIV elements of periodic table in breast cancer treatment. Pharmaceutics 14(12):2640. https://doi.org/10.3390/pharmaceutics14122640

Hotowy A et al (2024) Silver and carbon nanomaterials/nanocomplexes as safe and effective ACE2-S binding blockers on human skin cell lines. Molecules 29(15):3581. https://doi.org/10.3390/molecules29153581

Hu X et al (2021) Tailor-made nanomaterials for diagnosis and therapy of pancreatic ductal adenocarcinoma. Adv Sci 8(7):2002545. https://doi.org/10.1002/advs.202002545

Ji Z, Guo W, Sakkiah S, Liu J, Patterson TA, Hong H (2021) Nanomaterial databases: data sources for promoting design and risk assessment of nanomaterials. Nanomaterials 11(6):1599. https://doi.org/10.3390/nano11061599

Kang HJ, Chen N, Dash BC, Hsia HC, Berthiaume F (2021) Self-assembled nanomaterials for chronic skin wound healing. Adv Wound Care 10(5):221–233. https://doi.org/10.1089/wound.2019.1077

Kashyap BK, Singh VV, Solanki MK, Kumar A, Ruokolainen J, Kesari KK (2023) Smart nanomaterials in cancer theranostics: challenges and opportunities. ACS Omega 8(16):14290–14320. https://doi.org/10.1021/acsomega.2c07840

Kelly IB, Fletcher RB, McBride JR, Weiss SM, Duvall CL (2020) Tuning composition of polymer and porous silicon composite nanoparticles for early endosome escape of anti-microRNA peptide nucleic acids. ACS Appl Mater Interfaces 12(35):39602–39611. https://doi.org/10.1021/acsami.0c05827

Kim Y, Park EJ, Na DH (2018) Recent progress in dendrimer-based nanomedicine development. Arch Pharm Res 41(6):571–582. https://doi.org/10.1007/s12272-018-1008-4

Kim CD, Koo KM, Kim HJ, Kim TH (2024) Recent advances in nanomaterials for modulation of stem cell differentiation and its therapeutic applications. Biosensors 14(8):407. https://doi.org/10.3390/bios14080407

Krasley AT, Li E, Galeana JM, Bulumulla C, Beyene AG, Demirer GS (2024) Carbon nanomaterial fluorescent probes and their biological applications. Chem Rev 124(6):3085–3185. https://doi.org/10.1021/acs.chemrev.3c00581

Krishnan PD et al (2020) Silver nanomaterials for wound dressing applications. Pharmaceutics 12(9):1–27. https://doi.org/10.3390/PHARMACEUTICS12090821

Kumar R, Kumar B, Kumar A (2017) Pharmacokinetic profile of Antitubercular drugs and their clinical applications in therapeutic drug monitoring. Asian J Pharm Pharmacol 3(5):145–154. Available: http://ajpp.in/uploaded/p71.pdf

Kumar A et al (2021a) Antiurolithiasis, antioxidant, anti-inflammatory, analgesic, and diuretic activity of ethanolic extract of seeds of Caesalpinia bonducella. Int J Pharm Investig 11(3):306–311. https://doi.org/10.5530/ijpi.2021.3.54
Kumar R, Kumar B, Kumar A, Kumar A, Singh M (2021b) GC-MS analysis of phytocomponents in the methanol extract of Premna latifolia Roxb. Pharm Res 14(1):19–23. https://doi.org/10.5530/pres.14.1.4
Kumar R et al (2021c) HPTLC and GC-MS analyses of biologically active extracts and fractions from Premna latifolia Roxb leaves reveal anti-inflammatory and analgesic effects. Syst Rev Pharm 12(10):811–818
Kumar, Rajesh, Kumar B, Kumar A and Kumar A, et al. "HPTLC and GC-MS Analyses of Biologically Active Extracts and Fractions from Premna Latifolia Roxb Leaves" J Neurol Disord 9(2021d): 454
Kumar A et al (2022a) Acute and subacute toxicity study of ethanolic extract of Calotropis procera (Aiton) Dryand flower in Swiss albino mice. Phytomed Plus 2:100224. https://doi.org/10.1016/j.phyplu.2022.100224
Kumar A et al (2022b) Toxicity (acute and subacute) assessment and in-vivo antiurolithiatic activity of ethanolic extract of Caesalpinia bonducella seed in albino Wistar rat. J Appl Pharm Sci 12(2):187–197. https://doi.org/10.7324/JAPS.2022.120220
Kumar A et al (2022c) Assessment of pain-ameliorative potential of ethanolic extract of Calotropis procera and its active components characterization by HPTLC and GC-MS analysis. Syst Rev Pharm 13(9):581–586. https://doi.org/10.31858/0975-8453.13.6.659-664
Kumar S et al (2023) Optically active nanomaterials and its biosensing applications—a review. Biosensors 13(1):1–38. https://doi.org/10.3390/bios13010085
Kumar R, Singh Y, Piplani M, Bhateja P, Garg R, Kumar B (2024a) The chemistry and pharmacological study of Copaifera spp. oils. In: The chemistry inside spices & herbs: research and development, vol 4. Bentham Science Publishers, pp 79–116. https://doi.org/10.2174/9789815196832124040006
Kumar R, Dutt S, Goyal A, Kumar A, Kumar B (2024b) Nanotechnology in gastrointestinal endoscopy. In: Nanobiotechnology and artificial intelligence in gastrointestinal diseases. IOP Publishing. https://doi.org/10.1088/978-0-7503-6134-7CH5
Kumar A, Singh P, Kumar R, Dutt S (2024c) Novel drug delivery systems for inflammatory bowel disease. In: Nanobiotechnology and artificial intelligence in gastrointestinal diseases. IOP Publishing. https://doi.org/10.1088/978-0-7503-6134-7CH4
Kyriakides TR et al (2021) Biocompatibility of nanomaterials and their immunological properties. Biomed Mater 16(4):1–49. https://doi.org/10.1088/1748-605X/abe5fa
Lebre F et al (2022) Nanosafety: an evolving concept to bring the safest possible nanomaterials to society and environment. Nanomaterials 12(11):1810. https://doi.org/10.3390/nano12111810
Liao C, Li Y, Tjong SC (2020) Visible-light active titanium dioxide nanomaterials with bactericidal properties. Nanomaterials 10(1):124. https://doi.org/10.3390/nano10010124
Liu Y et al (2022) Emerging theranostic nanomaterials in diabetes and its complications. Adv Sci 9(3):1–54. https://doi.org/10.1002/advs.202102466
Ma C et al (2024) Nanomaterials: leading immunogenic cell death-based cancer therapies. Front Immunol 15:1–31. https://doi.org/10.3389/fimmu.2024.1447817
Mamun A, Sabantina L (2023) Electrospun magnetic nanofiber mats for magnetic hyperthermia in cancer treatment applications—technology, mechanism, and materials. Polymers 15(8):1902. https://doi.org/10.3390/polym15081902
Martínez-Carmona M, Vallet-Regí M (2020) Advances in laser ablation synthesized silicon-based nanomaterials for the prevention of bacterial infection. Nanomaterials 10(8):1–18. https://doi.org/10.3390/nano10081443
McInnes SJP, Voelcker NH (2009) Silicon-polymer hybrid materials for drug delivery. Future Med Chem 1(6):1051–1074. https://doi.org/10.4155/fmc.09.90

Mod B et al (2024) From cane to nano: advanced nanomaterials derived from sugarcane products with insights into their synthesis and applications. Sci Technol Adv Mater 25(1):1–31. https://doi.org/10.1080/14686996.2024.2393568

Mosleh-Shirazi S et al (2022) Nanotechnology advances in the detection and treatment of cancer: an overview. Nanotheranostics 6(4):400–423. https://doi.org/10.7150/ntno.74613

Paduraru AV, Oprea O, Musuc AM, Vasile BS, Iordache F, Andronescu E (2021) Influence of terbium ions and their concentration on the photoluminescence properties of hydroxyapatite for biomedical applications. Nanomaterials 11(9):2442. https://doi.org/10.3390/nano11092442

Pan M, Yin Z, Liu K, Du X, Liu H, Wang S (2019) Carbon-based nanomaterials in sensors for food safety. Nanomaterials 9(9):1–23. https://doi.org/10.3390/nano9091330

Pandey H, Srivastava S, Kumar A, Kumar R, Bhusan Tripathi Y (2019) Pharmacokinetics study of Puerarin absorption in blood after consumption of Pueraria tuberosa water extract (PTWE) by rats. Explor Anim Med Res 9(2):188–196

Paramo L, Jim A, Medina-Ramirez IE, Böhnel HN, Escobar-Alarcón L, Esquivel K (2023) Biocompatibility evaluation of TiO2, Fe3O4, and TiO2/Fe3O4 nanomaterials: insights into potential toxic effects in erythrocytes and HepG2 cells. Nanomaterials 13:2824

Parvin N, Kumar V, Joo SW, Mandal TK (2024) Emerging trends in nanomedicine: carbon-based nanomaterials for healthcare. Nanomaterials 14(13):1085. https://doi.org/10.3390/nano14131085

Pei J et al (2024) Advancements in the synthesis and functionalization of zinc oxide-based nanomaterials for enhanced oral cancer therapy. Molecules 29(11):2706. https://doi.org/10.3390/molecules29112706

Peng T, Xu T, Liu X (2022) Research progress of the engagement of inorganic nanomaterials in cancer immunotherapy. Drug Deliv 29(1):1914–1932. https://doi.org/10.1080/10717544.2022.2086940

Petrovic S, Bita B, Barbinta-Patrascu ME (2024) Nanoformulations in pharmaceutical and biomedical applications: green perspectives. Int J Mol Sci 25(11):5842. https://doi.org/10.3390/ijms25115842

Petrucci R, Bortolami M, Di Matteo P, Curulli A (2022) Gold nanomaterials-based electrochemical sensors and biosensors for phenolic antioxidants detection: recent advances. Nanomaterials 12(6):959. https://doi.org/10.3390/nano12060959

Phua VJX et al (2022) Nanomaterial probes for nuclear imaging. Nanomaterials 12(4):1–23. https://doi.org/10.3390/nano12040582

Popescu RC, Andronescu E, Vasile BS (2019) Recent advances in magnetite nanoparticle functionalization for nanomedicine. Nanomaterials 9(12):1–31. https://doi.org/10.3390/nano9121791

Qamar W et al (2024) An insight into impact of nanomaterials toxicity on human health. PeerJ 12(9):1–45. https://doi.org/10.7717/peerj.17807

Ranjha MMAN et al (2022) Biocompatible nanomaterials in food science, technology, and nutrient drug delivery: recent developments and applications. Front Nutr 8:1–15. https://doi.org/10.3389/fnut.2021.778155

Rehman Y, Qutaish H, Kim JH, Huang XF, Alvi S, Konstantinov K (2022) Microenvironmental behaviour of nanotheranostic systems for controlled oxidative stress and cancer treatment. Nanomaterials 12(14):2462. https://doi.org/10.3390/nano12142462

Safarkhani M et al (2024) Advancements in aptamer-driven DNA nanostructures for precision drug delivery. Adv Sci 11(26):1–30. https://doi.org/10.1002/advs.202401617

Shao Y et al (2021) Shining light on chiral inorganic nanomaterials for biological issues. Theranostics 11(19):9262–9295. https://doi.org/10.7150/thno.64511

Sharma R, Sharma D, Hazlett LD, Singh NK (2021) Nano-biomaterials for retinal regeneration. Nanomaterials 11(8):1–25. https://doi.org/10.3390/nano11081880

Silva DF et al (2023) Biomedical approach of nanotechnology and biological risks: a mini-review. Int J Mol Sci 24(23):16719. https://doi.org/10.3390/ijms242316719

Singh A, Prasad LB, Shiv K, Kumar R, Garai S (2023a) Synthesis, characterization, and in vitro antibacterial and cytotoxic study of Co(II), Ni(II), Cu(II), and Zn(II) complexes of N-(4-

methoxybenzyl) N-(phenylethyl) dithiocarbamate ligand. J Mol Struct 1288:135835. https://doi.org/10.1016/j.molstruc.2023.135835

Singh M et al (2023b) Pharmacokinetic screening to Estimate the drug likeliness characteristics of selected Herbal Anticancer Drugs. Res J Pharm Technol 16(7):3422–3426. https://doi.org/10.52711/0974-360X.2023.00566

Singh A et al (2024a) Synthesis, crystal structure and screening for anticonvulsant and antianxiety activities of three new Ni(II), Cu(II), and Zn(II) dithiocarbamate complexes. J Mol Struct 1298(P2):137052. https://doi.org/10.1016/j.molstruc.2023.137052

Singh A et al (2024b) Dithiocarbamate-based novel anti-histaminic agents: synthesis, characterization, crystal structure and thermal study. Dalton Trans 53:14077–14088. https://doi.org/10.1039/d4dt01777c

Siraj EA, Yayehrad AT, Belete A (2023) How combined macrolide nanomaterials are effective against resistant pathogens? A comprehensive review of the literature. Int J Nanomedicine 18:5289–5307. https://doi.org/10.2147/IJN.S418588

Soman S et al (2022) 2D hetero-nanoconstructs of black phosphorus for breast cancer theragnosis: technological advancements. Biosensors 12(11):1009. https://doi.org/10.3390/bios12111009

Tang L et al (2022) Versatile carbon nanoplatforms for cancer treatment and diagnosis: strategies, applications and future perspectives. Theranostics 12(5):2290–2321. https://doi.org/10.7150/thno.69628

Thangam R, Paulmurugan R, Kang H (2022) Functionalized nanomaterials as tailored theranostic agents in brain imaging. Nanomaterials 12(1):18. https://doi.org/10.3390/nano12010018

Tian T, Li Y, Lin Y (2022) Prospects and challenges of dynamic DNA nanostructures in biomedical applications. Bone Res 10(1):1–15. https://doi.org/10.1038/s41413-022-00212-1

Tsukanov AA, Turk B, Vasiljeva O, Psakhie SG (2022) Computational indicator approach for assessment of nanotoxicity of two-dimensional nanomaterials. Nanomaterials 12(4):650. https://doi.org/10.3390/nano12040650

Vagena IA et al (2024) Functionalized ZnO-based nanocomposites for diverse biological applications: current trends and future perspectives. Nanomaterials 14(5):1–40. https://doi.org/10.3390/nano14050397

Vijaya Durga RD (2015) Determination of S-amlodipine in human plasma using Tizanidine as internal standard by Lc/Ms/Ms method. J Bioanal Biomed 7(3):91–96. https://doi.org/10.4172/1948-593x.1000130

Wan T et al (2023) Tissue-engineered nanomaterials play diverse roles in bone injury repair. Nanomaterials 13(9):1449. https://doi.org/10.3390/nano13091449

Wang J, Li Y, Nie G (2021) Multifunctional biomolecule nanostructures for cancer therapy. Nat Rev Mater 6(9):766–783. https://doi.org/10.1038/s41578-021-00315-x

Wang H, Wang X, Lai K, Yan J (2023) Stimulus-responsive DNA hydrogel biosensors for food safety detection. Biosensors 13(3):320. https://doi.org/10.3390/bios13030320

Wei J, Mu J, Tang Y, Qin D, Duan J, Wu A (2023) Next-generation nanomaterials: advancing ocular anti-inflammatory drug therapy. J Nanobiotechnology 21(1):282. https://doi.org/10.1186/s12951-023-01974-4

Witika BA et al (2020) Biocompatibility of biomaterials for nanoencapsulation: current approaches. Nanomaterials 10(9):1–40. https://doi.org/10.3390/nano10091649

Wojnarowicz J, Chudoba T, Lojkowski W (2020) A review of microwave synthesis of zinc oxide nanomaterials: reactants, process parameters and morphoslogies. Nanomaterials 10(6):1086. https://doi.org/10.3390/nano10061086

Wu X, Zhou Z, Li K, Liu S (2024) Nanomaterials-induced redox imbalance: challenged and opportunities for nanomaterials in cancer therapy. Adv Sci 11(16):e2308632. https://doi.org/10.1002/advs.202308632

Xie W, Xu Z (2024) (Nano)biotechnological approaches in the treatment of cervical cancer: integration of engineering and biology. Front Immunol 15:1461894. https://doi.org/10.3389/fimmu.2024.1461894

Xie J, Li H, Zhang T, Song B, Wang X, Gu Z (2023) Recent advances in ZnO nanomaterial-mediated biological applications and action mechanisms. Nanomaterials 13(9):1500. https://doi.org/10.3390/nano13091500

Zhang C, Wang X, Du J, Gu Z, Zhao Y (2021) Reactive oxygen species-regulating strategies based on nanomaterials for disease treatment. Adv Sci 8(3):1–34. https://doi.org/10.1002/advs.202002797

Zhang D, Chen Y, Hao M, Xia Y (2024) Putting hybrid nanomaterials to work for biomedical applications. Angew Chem Int Ed Engl 63(16):e202319567. https://doi.org/10.1002/anie.202319567

Zheng K et al (2023) Nanomaterial-assisted theranosis of bone diseases. Bioact Mater 24(2022):263–312. https://doi.org/10.1016/j.bioactmat.2022.12.014

Zhu L et al (2021) Applications of aptamer-bound nanomaterials in cancer therapy. Biosensors 11(9):344. https://doi.org/10.3390/bios11090344

# Part II
# Nanotechnology-Based Drug Delivery Systems

# Chapter 4
# Lipid Based Nanocarriers for Drug Delivery

**Girish Kumar, Tarun Virmani, Reshu Virmani, and Kamla Pathak**

**Abstract** The medicinal prospect of nanotechnology has been widely recognized during the past 10 years as a cutting-edge medication delivery method that supplements conventional pharmacology. Lipid nanocarriers (LNPs) have demonstrated noteworthy pharmacological concert and hopeful therapeutic effects, generating significant attention in research. LNPs include liposomes, ethosomes, Niosomes, nanoemulsions, solid lipid nanoparticles, transferosomes, nanostructured lipid carriers, and polymer hybrid nanoparticles. Their nano-size permits the drug transport to tissues or cells with more precision, longer circulation times, and better cellular absorption. Moreover, these can enable controlled, continuous drug release while lowering their toxicity, which is very advantageous for the treatment of chronic illness. This chapter provides exhaustive information about the various types of LNPs, concentrating on their therapeutic potential in treating multiple diseases. Herein, this chapter also explores the recent research for LNPs to alleviate various diseases, clinical trials, and marketed products.

**Keywords** Lipid nanocarriers · Liposomes · Solid lipid nanoparticles · Nanostructured lipid carriers · Nanoemulsions · Transferosomes · Clinical trials

G. Kumar · T. Virmani
Amity Institute of Pharmacy, Amity University, Greater Noida Campus, Uttar Pradesh, India

R. Virmani
School of Pharmaceutical Sciences, MVN University, Palwal, Haryana, India

K. Pathak (✉)
Faculty of Pharmacy, Uttar Pradesh, University of Medical Sciences, Etawah, Uttar Pradesh, India

 
V. Khanna et al. (eds.), *Nanotechnologies for Drug Delivery and Biopharmaceutical Development*, Progress in Drug Research 77,
https://doi.org/10.1007/978-3-032-00796-4_4

## 4.1 Introduction

The involvement and advancement of nanotechnology in the pharmaceutical field have made it possible to develop nanomedicines, signaling the beginning of the "nano age" of medicines. The advent of the nano period officially began in the late 1970s with the development of several nanotools known as nano-based medicine systems (Jain et al. 2021). In recent years, most drugs with promising *in vitro* research results have failed *in vivo*, which has restricted the innovation of new therapeutic molecules. It is credited to poor aqueous solubility, high toxicity, low absorption, slow dissolution, drug distribution to non-diseased organs, and substantial alteration in blood levels due to unexpected bioavailability (Kumar 2019; Kumar et al. 2018). One important way to get around these constraints is to use appropriate drug delivery which allows for regulated and localized drug release (Kumar et al. 2024a). Hence drug formulators are seeking out innovative techniques and formulation strategies like lipid nanocarriers, polymeric nanocarriers, and inorganic nanocarriers to solve these problems and guarantee that patients at risk receive effective care (Virmani et al. 2023a).

Amid various kinds of nanoformulations, LNPs represent one of the most versatile categories that have acquired prevalent practice in drug delivery and medical research (Kumar et al. 2022a). Lipid nanocarriers present numerous advantages like loading both hydrophobic and hydrophilic cargos, boosting solubility, increasing bioavailability, providing GIT stability, drug targeting to desired sites, regulating drug release, reducing toxic effects, and altering biodistribution of drugs (Giordano et al. 2024) (Fig. 4.1). In addition, these nanocarriers can be functionalized with

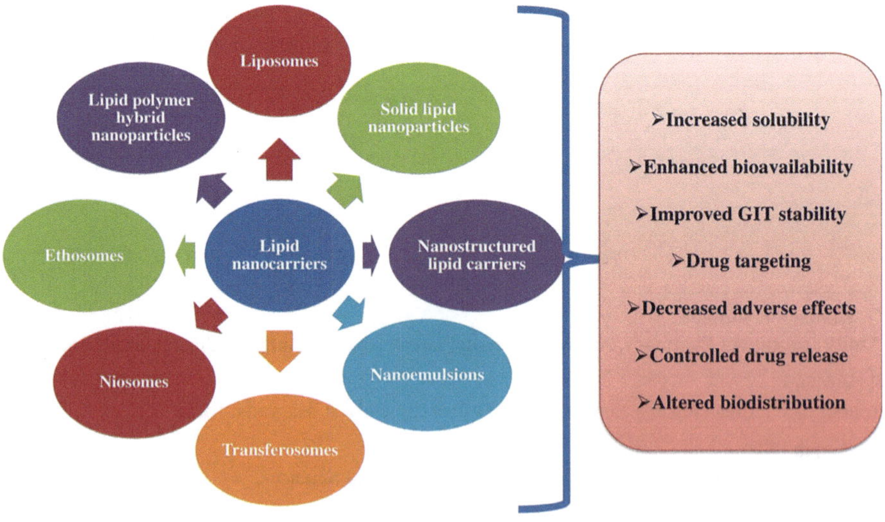


**Fig. 4.1** Pictorial representation of various lipid nanocarriers along with their benefits for drug delivery

certain ligands to target the overexpressed receptors on the cancer cell surface, leading to the accumulation of drugs within cancer cells and avoiding off-target effects (Kumar et al. 2024b). The industrial and scientific sectors have witnessed a substantial increase in recognition of the potential of lipid nanocarriers-based drug delivery in recent times (Dubey et al. 2024). According to estimates, the market value of liposomal medicines for cancer, including Doxil, DepoCyt, Marqibo, Myocet, DaunoXome, and Onivyde, was $3.72 billion in 2021. This figure is projected to get close to $7 billion by 2027. Lipid nanocarrier-based gene therapy was projected to have a $3.5 billion market in 2021, and it is anticipated to develop at a compound annual growth rate of about 18.5% between 2021 and 2026 (Mehta et al. 2023). Apart from treating cancer, lipid nanocarrier-based drugs have received FDA authorization for the treatment of several diseases, such as amyloidosis (Onpattro) and the COVID-19 vaccine (Spikevax, Comirnaty), by delivering mRNA and siRNA, respectively (Wilson and Geetha 2022). This chapter explores recent updates about various types of lipid nanocarriers in treating different ailments.

## 4.2 Types of Lipids Nanocarriers

### *4.2.1 Liposomes*

Liposomes are widely utilized and extensively studied nanocarriers to deliver drugs efficiently. These are artificial circular vesicles (0.025 to 2.5 μm) comprising mainly phospholipids, cholesterol, and emulsifiers (Kumar et al. 2023a). These can load both water-soluble and lipid-soluble drugs (Kumar et al. 2024c). Liposomes have become an auspicious vehicle owing to their many qualities, comprising the capability to load great doses, the ability to extend the drug's circulation time, biocompatibility, biodegradability, enhanced durability, minimum adverse effects, regulated drug release, increased dissolution rate, targeting ability, versatility, and ease of manufacturing (Virmani et al. 2022). They might have their structural modifications to produce the intended therapeutic outcomes. The anticancer activity of liposome-loaded drugs can be boosted by conjugating various liposome ligands to target the escalating receptors on the cancer cells (Wang et al. 2023). In addition, the release of the drugs in a specific tumor environment can be controlled by various stimuli like temperature, redox, pH, and ultrasound (Kumar et al. 2023b). Increased drug circulation and reduced adverse effects can be achieved by coating liposomes with polyethylene glycol (PEG) (Taher et al. 2023). Several liposomes for different ailments have been commercialized, showing the potential of liposomes for efficient drug delivery to treat various ailments. Various researchers have recently assessed new horizons in liposome-based drug delivery in various life-threatening diseases. For instance, As1411 conjugated doxorubicin liposomes were developed in 2024 to target RENCA tumors (kidney cancer). The study's findings demonstrated boosted cellular uptake efficiency and decreased IC50 value. The decreased IC50 value might be due to increased cellular uptake of the doxorubicin by the presence of As1411. In addition, As1411 conjugated liposomes displayed an increased concentration of doxorubicin at the tumor site than plain doxorubicin and un-conjugated doxorubicin-loaded liposomes. This could be attributed to the prolonged circulation time and the targeting of As1411 to receptors on tumor cells. The *in vivo* outcomes assessed the reduced tumor mass in the case of As1411 conjugated liposomes than un-conjugated liposomes and plain doxorubicin, demonstrating the perspective of functionalized liposomes to treat cancer (Zhang et al. 2024). Chen et al. (2024) prepared glucose-modified liposomes of quercetin (QT) to provide increased neuroprotection. The liposomes had a size range of 116–124 nm and a charge of $-1.363 \pm 0.437$ mV. The findings demonstrated that the glucose-modified liposomes of QT could cross the BBB easily through the receptor GLUT1, followed by higher uptake by neuronal cells than unmodified liposomes of QT. Compared to QT, glucose-modified liposomes of QT reduced intracellular ROS and increased cell survival (88.42%) in PC12 cells, attenuating H2O2-induced oxidative damage. The study revealed the potential of modified liposomes to target QT in the brain tissues (Chen et al. 2024a).

### 4.2.2 Solid Lipid Nanoparticles

Using SLNs as a carrier system for various therapeutic purposes has garnered significant interest. SLNs presented an alternative to colloidal drug carriers in the early 1990s (Bharti Sharma et al. 2023). These colloidal carriers have a size range of 50 to 1000 nm (Nemati et al. 2024). The primary benefits of SLNs include their vast surface area, increased stability, high drug-loading capabilities, and drug targeting, all of which enhance pharmacological efficacy (Kumar et al. 2024d). SLNs are more advantageous for intravenous applications because they perform better at managing release kinetics than liquid lipids (Mohammed et al. 2023). SLNs' solid matrix entraps the lipid-soluble drugs or a mixture of hydrophobic drugs, protecting them from chemical degradation and promoting physical stability. These improve the therapeutic value of medications by extending the half-lives in blood by changing their release patterns (Kumar et al. 2023a). The benefits of SLNs, including their biodegradability, stability, and small size, indicate that they may transport drugs to different diseased sites. SLNs can expedite the initial phases of pharmacological action due to their high saturation solubility and quick rate of dissolution (Ghasemiyeh and Mohammadi-Samani 2018). QT-loaded SLNs were obtained to treat colorectal cancer. The optimized SLNs had a particle size of $132.16 \pm 4.1$ nm and an entrapment efficiency of $97.8 \pm 1.16\%$. The findings of the drug release study provided a release of $41.12 \pm 1.6\%$, suggesting a sustained release pattern. The optimized formulation demonstrated an IC50 value of 49 μM/mL for SLNs of QT and 128 μM/mL for free QT at 24 h against Caco-2 cell lines, revealing better anticancer potential in colorectal cancer (Mohamed et al. 2024). Alcantara et al. fabricated cannabidiol-encapsulated SLNs for efficacious treatment of chronic inflammation. The optimized formulation had particle size, a charge, drug loading, and encapsulation efficiency of 123 nm, −32.1 mV, 2.36%, and 95.16%, respectively. It also displayed sustained drug release, ensuring controlled drug release for prolonged periods, which boosts its therapeutic efficiency. Furthermore, optimized formulation dramatically decreased reactive oxygen, proinflammatory cytokines, and nitrogen species in chondrocyte and macrophage cell lines. These repressing outcomes were more significant than coarse cannabidiol (Alcantara et al. 2024). However, SLNs provide several benefits as drug carriers in various ailments; there are certain drawbacks as well, such as drug ejection, little drug loading, arbitrary agglomeration, and polymorphic transitions, which need to be attended (Akanda et al. 2023).

### *4.2.3 Nanostructured Lipid Carriers*

To conquer the constraints of SLNs, NLCs have proven effective as state-of-the-art drug carriers in treating various ailments. Solid and liquid lipids comprise these, causing distinctive nanostructures with increased drug loading, stability, and modified drug release patterns (Mall et al. 2024). These can protect the drugs from oxidation, hydrolysis, and light degradation (Ahalwat et al. 2023). NLCs improve drug bioavailability and shield them from the hostile system of the gastrointestinal tract (Kumar et al. 2024e). Additionally, NLCs enable lymphatic system-based drug targeting, resulting in anticipation of hepatic degradation, diminished hepatotoxicity, and greater absorption (Mahor et al. 2023). Amiri et al. developed NLCs to decrease side effects and dose dumping associated with remdesivir. The optimized NLCs had size, charge, entrapment efficiency, and polydispersity index of 151.0 ± 1.7 nm, −43.8 ± 1.2 mV, 81.34 ± 1.57%, 0.4 ± 0.1 respectively. The optimized NLCs provided sustained release of remdesivir with reduced chances of dose dumping and decreased side effects. Furthermore, remdesivir in NLCs was less cytotoxic to normal, healthy kidney cells, which could lessen renal impairment in COVID-19 patients (Amiri et al. 2024). Chitosan-coated NLCs were developed to simultaneously deliver paclitaxel and PD-L1 siRNA for breast cancer. Chitosan-coated NLCs of paclitaxel and siRNA had particle size and zeta potential of 181.97 ± 0.67 nm and 18.66 ± 0.23 mV. The findings demonstrated increased cytotoxicity and intracellular uptake of the drugs in cell lines. Reduced tumor size was observed in experimental mice on subcutaneous formulation administration. Furthermore, an *in vivo* investigation showed that after administration of chitosan-coated NLCs of paclitaxel and siRNA, the tumor tissue exhibited tumor-specific CD4 + and CD8 + T cell responses, revealing the potential of NLCs to alleviate the cancer by simultaneous delivery of both drugs (Kim et al. 2024).

### 4.2.4 Nanoemulsions

Nanoemulsions (NEs) and colloidal particle-based dispersions are heterogeneous systems comprising lipid and aqueous phases used as drug carriers for efficient drug delivery (Kumar et al. 2022b). These are transparent, thermodynamically unstable, typically spherical systems with a 5–200 nm size range and are produced easily and economically (Kumar et al. 2024f). NEs offer increased aqueous solubility and bioavailability, biodegradability, physical stability, optical clarity, controlled drug release, and enhanced pharmacodynamic and pharmacokinetic profile (Virmani et al. 2022). Drug toxicity can be reduced, and delivery load can be enhanced by employing nanoemulsions that have the drug within their core. NEs can be conjugated with antibodies (Abs) and their fragments to facilitate drug targeting (Virmani et al. 2023b). Numerous studies have demonstrated that these conjugations cause internalization by malignant cells, which facilitates the efficient delivery of drugs via NEs. Additionally, drug-Ab conjugation can be made approachable to cues to increase specificity against malignant cells (Chehelgerdi et al. 2023). The perspective of NEs in delivering the drugs can be assessed by research in which sorafenib was delivered efficiently via NE in hepatocellular carcinoma (HCC). The prepared NE was observed with size, PDI, charge, pH, and viscosity of 121.75 ± 12 nm, 0.310, 12.33 ± 1.34 mV, 4.38 ± 0.3, and 34,776 ± 3276 CPs, respectively. NE exhibited prolonged drug release of sorafenib, demonstrating decreased dosing frequency. The findings of the MTT assay displayed potent cytotoxic properties by NE of sorafenib than suspension of sorafenib as evidenced by reduced IC50 value of 4.73 µg/ml and 11.14 µg/ml for NE and suspension of sorafenib, respectively. The pharmacokinetic results demonstrated an increased area under the curve, Cmax, and half-life for NE than suspension. It showed a better safety profile than its suspension, especially in the liver, brain, and behavior. These results highlight sorafenib-NE's improved pharmacological and toxicological characteristics, indicating its possible use in treating HCC (Zaafar et al. 2024). For another instance, Asiatic acid (AA) loaded spray-dried NE was fabricated to alleviate the hypertension. Maltodextrin was used as a spray-dry carrier to physically encapsulate the AA-loaded NE, which instantly produced nanoparticles with a size of 216.40 ± 2.67 nm after being re-dispersed in an aqueous solvent. Additionally, utilizing Caco-2 cells showed that the development procedure and the matrix were safe to cells. The molecular docking data demonstrated that AA exhibited the ability to inhibit ACE. The spray-dried AA-loaded NE reduced elevated blood pressure more quickly than unformulated AA (raw powder) in hypertension-induced rats (Pradana et al. 2024).

### 4.2.5 Transferosomes

A second class of liposomes, transferosomes, is distinguished by their soft nature, enhanced membrane flexibility, and ultra-deformability (Kumar et al. 2024b). These comprise lipids, alcohols, hydrating media like water, edge activators, and penetration enhancers like spans, Tweens, and sodium deoxycholate (Das et al. 2022). Several therapeutic agents, such as water-soluble actives, bigger molecules, proteins, nucleic acids, and peptides, can be delivered transdermally via transferosomes. Due to their flexible construction and higher hydrophilicity, these are efficient in delivering drugs and other molecules via the skin, employing hydration gradients (Matharoo et al. 2024). This causes improved permeation and regulated release in the skin layers. Because of their compact size, lower polydispersity index, improved trapping, prolonged release, and safety, they are a promising option to deliver the drugs through the skin (Chamel et al. 2024). Furthermore, their physical-chemical characteristics boost stability by halting drug deterioration due to oxidation, light, and temperature (Matharoo et al. 2024). For example, transferosomes were fabricated to improve skin permeation of rutin and alleviate skin cancer. The optimized transferosomes of rutin exhibited particle size of 170.1 ± 1.27 nm, encapsulation efficiency of 87.5 ± 0.58%, and drug loading of 8.119 ± 0 .09% respectively. It exhibited the release of rutin up to 98 ± 0.87% in 24 h, revealing a sustained release pattern. It showed greater inhibition of melanoma cell lines than free rutin, as evidenced by the reduced IC50 value of transferosomes of rutin (Wadher et al. 2024). Glucosamine-encapsulated transferases were developed to treat arthritis efficiently. The optimized formulation had 5% glucosamine loaded in a 90:10 phospholipid: Tween 80 ratio, hydrated with Phosphate buffer saline (PBS) solution. It exhibited a particle size of 126 nm, zeta potential of −6.96 mV, and PDI of 0.337. The poor charge value shows the low stability of developed transferosomes, so these were formulated as transferosomal gel using carbopol 940. The gel formulation provided stable preparation with neutral pH and better viscosity to permit greater remedy to the skin. The findings of *in vivo* studies demonstrated greater cartilage regeneration in 2 rabbits with osteoarthritis without any noteworthy change in complete blood count. At the same time, another rabbit treated with liposomal formulation showed no improvement (Fig. 4.2). It could be due to liposomes' inability to penetrate the skin (Rasheed et al. 2022).

### *4.2.6 Ethosomes*

The third class of liposomes, ethosomes (ETs), are vesicular carriers that boost the bioavailability of drugs via the skin because they contain significant levels of ethanol, phosphatidylcholine, and water (Ahad et al. 2022). Phosphatidylcholines and ethanol in ETs interfere with the stratum corneum's barrier function. Ethanol causes the stratum corneum (SC)'s lipid bilayer to be rearranged, which makes it easier for the pliable and soft ET vesicles to extract and merge with the lipid bilayer, leading to enhanced skin permeation (Nayak et al. 2022). Furthermore, the high ethanol concentration of ETs facilitates the entrapment of hydrophilic or lipophilic active agents, which is especially helpful when working with natural extracts, a class of substances with a wide range of chemical properties (Emanet and Ciofani 2023). Their well-known qualities are their enhanced softness, extended stability, improved

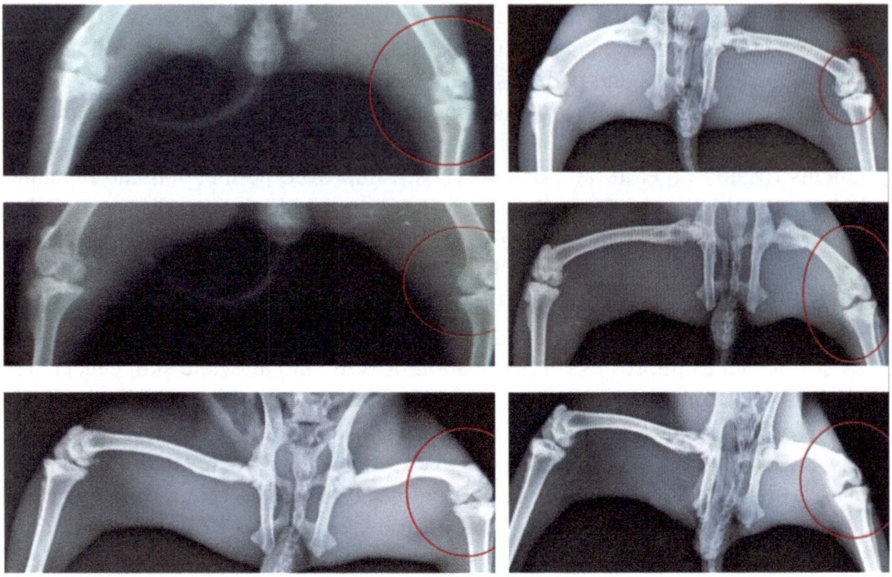

A. After Induction of OA B. After treatmnet

**Fig. 4.2** Pictorial representation of improvement in cartilage regeneration in rabbits with osteoarthritis on the application of drug-loaded transferosomes. (Reprinted with permission of (Rasheed et al. 2022))

solubility of lipophilic drugs, and ability to transport drugs through the deeper layers of the skin, improving drug availability (Kumar et al. 2024b). Mehmood et al. fabricated Vitamin D3 encapsulated ethosomal gel for psoriasis. The optimized ETs of vitamin D3 had a size of 148 nm, a charge of −14 ± 3 mV, and an encapsulation efficiency of 96.25% ± 0.3%. The results demonstrated a greater permeability of ETs compared to free vitamin D3. In 12 h of study, the optimized ethosomal formulation showed permeation of 95.34 ± 3% drug, which was very high than free vitamin D3. The poor penetration of vitamin D3 might be due to the larger size of the pure vitamin D3, while ethosomal formulation has nano size and a high quantity of ethanol, which facilitated greater penetration (Mehmood et al. 2024). Wang et al. developed ETs for efficient delivery of tryptanthrin in the treatment of psoriasis. ETs loaded with tryptanthrin demonstrated good stability and uniform particle size. The fluidic characteristics of ETs made it easier for them to pierce epidermal cells and even reach the skin's deep layers, which improved the administration of tryptanthrin for more effective therapeutic results within skin tissues. It showed remarkable effectiveness in triggering apoptosis and suppressing angiogenesis in aberrantly growing cells in cellular experiments. Additionally, in a mouse model of psoriasis, ETs of tryptanthrin showed promise in reducing inflammation and scaling (Wang et al. 2024).

### 4.2.7 Niosomes

Niosomes are bilayer structures of lipid components (mainly cholesterol) and amphiphilic non-ionic surfactants. These are the most recent vesicular systems to be produced and have an amazing array of applications (Liga et al. 2024). As drug carriers, niosomes considerably improve therapeutic effects while lowering toxicity. Niosomes can improve the stability, solubility, and bioavailability of drugs, decrease their toxicity in the body, and enable lower drug doses. Furthermore, niosomes have better stability, are less expensive, and can carry hydrophilic and hydrophobic medications than nanocarriers like liposomes (Rezaei et al. 2024). Niosomes are carriers in topical vaccine delivery because the aqueous core preserves the antigen, improving skin penetration and triggering an immunological response. These are less harmful and can increase penetration, hence also being researched for the delivery of drugs into the ocular tissues (Moammeri et al. 2023). Niosomes can more accurately administer anticancer drugs while lessening their toxicity and the intensity of their adverse effects. Because proteasomes can accumulate drug-loaded carriers profound into the lung and enhance clinical efficacy, they are particularly interesting for use in nebulizer drug administration (Chen et al. 2019). Soni et al. developed methotrexate (MTX) encapsulated niosomes for skin cancer. The optimized niosomes of MTX exhibited a size of 597.2 nm, a charge of −23.3 mV, PDI of 0.49, and drug encapsulation of 55.05%. Drug release study provided sustained release of MTX with burst release initially. The optimized niosomes of MTX demonstrated potent anticancer activity against melanoma cells, as evidenced by reduced IC50 value. *Ex vivo* experiments validated the entry of niosomes into epidermal layers; their penetration was deeper for smaller particles. There was some modest epidermal erosion *in vivo*, but there was no significant inflammation, indicating satisfactory safety (Soni et al. 2024). Magnetic niosomes were fabricated by Amandi et al. to deliver resveratrol in pancreatic cancer. These exhibited a particle size of 95 nm with a charge and drug encapsulation of −14.0 mV and 85%. These nanoparticles showed sustained release of resveratrol. The study's results demonstrated increased cytotoxicity against Caplan-1 cells for magnetic niosomes than plain resveratrol, revealing greater uptake in cancer cell lines. The study suggested that magnetic niosomes are a suitable carrier with significant promise for use in cancer treatment in the future (Firouzi Amandi et al. 2024).

### 4.2.8 Lipid Polymer Hybrid Conjugates

Lipid polymer hybrid nanoparticles (LPHNPs) are realistic core-shell nanostructures that are thought to be a combination of polymeric nano units and liposomes (Parveen et al. 2023). The dual-structured nature of this novel idea makes it crucial as a combinational drug delivery program for cancer. These have been engineered to maximize the therapeutic efficacy of liposomes and polymeric NPs while dismissing their disadvantages, exploding delivery and RES uptake (Gajbhiye et al. 2023). Their valuable attributes include high cargo loading, stability, enhanced biocompatibility, controlled release, and longer drug half-life (Soomherun et al. 2024). LPHNPs can be decorated in several ways using stimuli-responsive agents (Shah et al. 2022). LPHNPs of fisetin (FST) were developed to alleviate acute pancreatitis. The optimized LPHNPs of FST exhibited a reduced size of 125.39 ± 0.924 nm, high positive charge of + 30.16 ± 1.416 mV, reduced polydispersity index of 0.357 ± 0.012, greater drug entrapment of 61.76 ± 1.254%, high drug loading of 32.18 ± 0.734%, and greater mucoadhesion of 35.64 ± 0.548%. Optimized formulation showed sustained drug release of FST. The findings of in vivo assessment demonstrated that FST-encapsulated LPHNPs inhibited acute pancreatitis induced by L-arginine and various organ destruction as evidenced by biochemical studies and histochemical assessment (Awadeen et al. 2023). Rizwan Ullah et al. fabricated LPHNPs of exemestane (EXE) to enhance oral bioavailability and efficacious management of breast cancer. The EXE-loaded TPGS-LPHNPs exhibited a reduced particle size of 136.37 ± 3.27 nm than EXE-loaded LPHNPs with a particle size of 177.56 ± 5.42 nm. The prepared LPHNPs showed sustained drug release behavior along with enhanced stability in the gastrointestinal tract. The findings *ex vivo* demonstrated increased intestinal permeation from EXE-loaded TPGS-LPHNPs than EXE-loaded LPHPNs without TPGS. The results of the pharmacokinetic assessment displayed 4.69 and 3.58 folds increased oral bioavailability for EXE-loaded LPHNPs without TPGS and EXE-loaded TPGS-LPHNPs, respectively than free EXE. In addition, EXE-loaded TPGS-LPHNPs provided tumor inhibition of 72.72% against MCF-7 cell lines, whereas EXE-loaded LPHNPs without TPGS exhibited tumor inhibition of 61.94%, which was only 30.79% for EXE suspension. Therefore, these findings demonstrate the perspective of LPHNPs to deliver anticancer moieties to improve their therapeutic potential (Rizwanullah et al. 2023). A list of recently studied lipid nanocarriers-based drug delivery for various ailments has been reviewed in Table 4.1.

## 4.3 Clinical Trials for Lipid-Based Nanocarriers

Earlier studies demonstrated the potential of LNPs. Many lipid-based nanocarriers have been created recently for effective drug delivery in various medical ailments. Many studies have so far advanced to the point of clinical trials evaluating the efficacy of lipid-based nanocarriers as drug carriers. Some of these recent studies have been presented in Table 4.2.

**Table 4.1** List of recent research for lipid nanocarriers-based drug delivery in various ailments

| Type of nanocarrier | Therapeutic moiety | Disease | Key outcomes | Year of publication | References |
|---|---|---|---|---|---|
| Liposomes | Doxorubicin | Cancer | As1411 conjugated liposomes exhibited enhanced antitumor activity, which was revealed by reduced IC50 value and tumor mass. | 2024 | Zhang et al. (2024) |
| Liposomes | Quercetin | Neurodegenerative disease | Glucose-modified liposomes of quercetin demonstrated improved penetration of the blood-brain barrier. | 2024 | Chen et al. (2024a) |
| Liposomes | Doxorubicin | Breast cancer | Doxorubicin-loaded stealth liposomes exhibited reduced tumor development in the mice. | 2024 | Paolino et al. (2024) |
| Liposomes | Eremantholide C, goyazensolide | Chronic gout | Liposomes encapsulated with eremantholide C and goyazensolide demonstrated decreased uric acid levels *in vivo* assessment. | 2024 | Cunha Matosinhos et al. (2024) |
| Liposomes | Brilliant blue G-250 | Lung cancer | Liposomal formulation exhibited a reduced IC50 value, revealing enhanced anticancer potential. | 2024 | Gupta et al. (2024) |
| Liposomes | Silybin | Liver disease | Targeted liposomes demonstrated enhanced accumulation and prolonged stay of silybin in the liver and a higher area under the curve than the free drug. | 2024 | Cui et al. (2024) |
| Liposomes | Naringenin | Triple-negative breast cancer | Naringenin-loaded liposomes exhibited reduced IC50 value, cell viability, and cell proliferation than radiation alone. | 2024 | Pearce et al. (2024) |
| Liposomes | Paclitaxel | Breast cancer | Paclitaxel-loaded palmitic acid liposomes demonstrated prolonged circulation and the greatest deposition of the drug in the tumor. | 2024 | Wu et al. (2024) |

(continued)

**Table 4.1** (continued)

| Type of nanocarrier | Therapeutic moiety | Disease | Key outcomes | Year of publication | References |
|---|---|---|---|---|---|
| Liposomes | Paclitaxel with immunotherapy | Non-small cell lung cancer | Liposomes loaded with paclitaxel and immunotherapy provided greater anticancer activity. | 2024 | Li et al. (2024a) |
| Liposomes | Paclitaxel, doxorubicin | Breast cancer | Paclitaxel and doxorubicin co-loaded liposomes exhibited improved inhibition of cell lines than free drugs. | 2024 | Mota Díaz et al. (2024) |
| Liposomes | Nebivolol hydrochloride | Hypertension | Polymer-coated liposomes exhibited improved drug release than uncoated liposomes and noteworthy interaction with mucin. | 2024 | Sessevmez et al. (2024) |
| SLNs | Quercetin | Colorectal cancer | Quercetin-loaded SLNs showed better anti-numerable potential than free quercetin, as evidenced by reduced IC50 value. | 2024 | Mohamed et al. (2024) |
| SLNs | Cannabidiol | Inflammation | The optimized formulation dramatically decreased reactive oxygen, proinflammatory cytokines, and nitrogen species in macrophage cell lines and chondrocytes than free cannabidiol. | 2024 | Alcantara et al. (2024) |
| SLNs | 4-hexylresorcinol | Cancer (Not specific) | 4-hexylresorcinol-loaded SLNs support enhanced anticancer potential than free 4-hexylresorcinol. | 2024 | Yeo et al. (2024) |
| SLNs | Miconazole nitrate | Fungal biofilms | Miconazole-loaded SLN patches demonstrated healing of *Candida albicans* infected wounds after 10 days. | 2024 | Arshad et al. (2024) |
| SLNs | Vildagliptin | Diabetic retinopathy | Ocuserts loaded with vildagliptin-SLNs showed prolonged vildagliptin release and reduced ocular irritation. | 2024 | Ramadan et al. (2024) |

| | | | | | |
|---|---|---|---|---|---|
| SLNs | Bimatoprost | Glaucoma | The optimized formulation displayed a sustained release of bimatoprostup of 12 h compared to a free drug with reduced ocular irritancy. | 2023 | Satyanarayana et al. (2023) |
| SLNs | Insulin | – | Chitosan-coated SLNs of insulin demonstrated sustained insulin release and enhanced retention of their initial loading. | 2024 | Raghunath et al. (2024) |
| SLNs | Curcumin, paclitaxel | Lung cancer | The SLNs co-loaded with curcumin and paclitaxel demonstrated 12 times decreased tumor volume in tumor-induced animals, revealing improved anticancer potential. | 2024 | Li et al. (2024b) |
| NLCs | Remdesivir | Covid-19 | The NLCs provided reduced chances of dose dumping and decreased side effects, along with less cytotoxicity to healthy kidney cells. | 2024 | Amiri et al. (2024) |
| NLCs | Paclitaxel, siRNA | Breast cancer | The NLCs exhibited increased cytotoxicity and intracellular uptake in the cancer cell lines along with reduced tumor size on subcutaneous administration. | 2024 | Kim et al. (2024) |
| NLCs | *Zingiber officinale* | – | NLCs provided a stable and well-prepared formulation having uniform size distribution. | 2024 | Shazwani et al. (2024) |
| NLCs | Ezetimibe | Hyperlipidemia | Drug-loaded NLCs markedly corrected serum HDL-C levels in rats while dramatically lowering increased serum lipid markers. | 2024 | Elkhayat et al. (2024) |
| NLCs | Donepezil hydrochloride | Alzheimer disease | Drug-loaded NLCs demonstrated a nine-fold enhanced drug concentration in the brain on intranasal administration than intravenous administration. | 2024 | Tekade et al. (2024) |

(continued)

**Table 4.1** (continued)

| Type of nanocarrier | Therapeutic moiety | Disease | Key outcomes | Year of publication | References |
|---|---|---|---|---|---|
| NLCs | Silibinin | Alzheimer disease | The discrimination rate was meaningfully excellent in the silibinin-NLCs treated group than in the aβ group, along with significant residence time in the target area than in the silibinin and Aβ groups. | 2024 | Khodabandelou et al. (2024) |
| NLCs | Indomethacin | Breast cancer | Indomethacin-loaded NLCs exhibited superior inhibition of MDA-MB-468 cancer cells, revealing enhanced anticancer potential against breast cancer. | 2024 | Thiruchenthooran et al. (2024) |
| NLCs based gels | Paclitaxel and 5-fluorouracil | Cancer (not specific) | Drugs-loaded NLCs displayed improved cytotoxicity along with prolonged residence time for intraductal administration. | 2024 | Passos et al. (2024) |
| NLCs | Propolis | Breast cancer | Encapsulation of propolis in NLCs enhanced anticancer activity and increased the effectiveness of 5-fluorouracil on co-administration. | 2023 | Shaker et al. (2023) |
| NLCs | Metformin and thymoquinone | Breast cancer | NLCs co-loaded with metformin and thymoquinone demonstrated improved cytotoxicity against MCF-7 cell lines and increased cellular internalization of the drugs. | 2023 | Kumari and Dang (2023) |
| NLCs | Roflumilast | Parkinson | Drug-loaded NLCs dramatically improved antioxidant parameters, decreased TNF-α and IL-6 levels in the striatum, and alleviated behavioral impairments more than pure drugs. | 2024 | Roy et al. (2024) |
| NLCs | Tanshinone IIA | Parkinson | Chitosan-coated NLCs of tanshinone IIA displayed enhanced antiparkinsonian and antidepressant activity in diseased animals. | 2024 | Hassan et al. (2024) |

| | | | | | |
|---|---|---|---|---|---|
| NLCs | Dopamine hydrochloride | Parkinson | Drug-loaded NLCs demonstrated 2 times improved drug permeation than free drugs and showed significant effects on locomotion activity. | 2023 | Neha et al. (2023) |
| NEs | Sorafenib | Hepatocellular carcinoma | Drug-loaded NE demonstrated potent cytotoxicity as evidenced by reduced IC50 value and increased area under the curve and Cmax than drug suspension. | 2024 | Zaafar et al. (2024) |
| NEs | Asiatic acid | Hypertension | The spray-dried Asiatic acid-loaded NE reduced elevated blood pressure more quickly than an unformulated drug in hypertension-induced rats. | 2024 | Pradana et al. (2024) |
| NEs | Letrozole | Breast cancer | Magnetic NE is a better option for delivering letrozole in a controlled and sustained way to alleviate breast cancer. | 2024 | Karthikeyan et al. (2024) |
| NEs | Fisetin | Alzheimer disease | NEs demonstrated 8.81 times increased concentration of fisetin in the brain than plain fisetin, and it also showed noteworthy improvements in cognition function in diseased animals. | 2024 | Vishwas et al. (2024) |
| NEs | Pterostilbene | Alzheimer disease | Drug-loaded NEs demonstrated an improved influence on memory and learning function compared to pure drugs, enhanced protection of hippocampal neurons, and prevention of oxidative stress and apoptosis compared to pure drugs. | 2024 | Liu et al. (2024) |
| NEs | 2-care | Breast cancer | Drug-loaded NEs exhibited increased cytotoxicity against cancer cells and illustrated considerable apoptotic and anticancer activity in breast cancer-induced animals. | 2024 | Nayila et al. (2024) |

(continued)

**Table 4.1** (continued)

| Type of nanocarrier | Therapeutic moiety | Disease | Key outcomes | Year of publication | References |
|---|---|---|---|---|---|
| NEs | Eugenol | Breast cancer | Eugenol-encapsulated NEs showed a considerable anticancer effect against MCF-7 cells. | 2024 | Velho et al. (2024) |
| NEs | *Baccharisdracunculifolia*extract | Hypertension | The encapsulation of *Baccharis dracunculifolia* extract in NE increased antihypertensive activity in diseased animals. | 2024 | Batista et al. (2024) |
| NEs | Curcumin | Dry eye disease | The encapsulation of curcumin in NE demonstrated improved efficacy, with the greatest dose revealing the most resemblance to the control group (healthy group). | 2024 | Hadipour Jahromy et al. (2024) |
| Transferosomes | Rutin | Melanoma | Transferosomal formulation of rutin showed greater inhibition of melanoma cell lines than free rutin as evidenced by reduced IC50 value. | 2024 | Wadher et al. (2024) |
| Transferosomes | Glucosamine | Arthritis | Glucosamine-loaded transferases showed greater cartilage regeneration without any noteworthy change in complete blood count. | 2022 | Rasheed et al. (2022) |
| Transferosomes | Genistein | Skin cancer | Genistein-loaded transferosomes displayed strong anticancer properties, enhancing permeation via the skin cancer spheroids. | 2024 | Motawea et al. (2024) |
| Transferosomes | Lycorine | Cutaneous cell carcinoma | Cell-penetrating peptides modified transfer of some of the lycorine provided improved skin penetration for better management of skin cancer. | 2023 | Li et al. (2023) |
| Transferosomes | Epigallocatechin-3-gallate and ascorbic acid | Alzheimer disease | Co-loaded transferosomes displayed 5 times greater memory enhancement. | 2024 | Mishra et al. (2024) |

| | | | | | |
|---|---|---|---|---|---|
| Transferosomes | Itraconazole | Skin cancer | Itraconazole-loaded transferases demonstrated two-fold more potency against A375 cells than free drugs, as evidenced by cell viability assessment. | 2024 | Kudi et al. (2024) |
| Transferosomes | Purpurin-18 sodium salt and doxorubicin hydrochloride | Cancer (Not specific) | Transferosomes showed effective delivery of drugs in cancer cells under the influence of pH and increased cytotoxicity against HeLa and A549 cell lines. | (2023 | Yeo et al. (2023) |
| Transferosomes | R-carvedilol | Skin cancer | R-carvedilol-loaded transferosomes displayed better skin permeation and prolonged retention than coarse drugs. | 2023 | Shamim et al. (2023) |
| Transferosomes | Fluconazole | Fungal infections | Fluconazole-loaded transferases demonstrated improved antifungal activity. | 2024 | Cheng et al. (2024) |
| Ethosomes | Vitamin D3 | Psoriasis | Vitamin D3-loaded ETs showed permeation of 95.34 ± 3% drug, which was much higher than free vitamin D3. | 2024 | Mehmood et al. (2024) |
| Ethosomes | Tryptanthrin | Psoriasis | ETs of tryptanthrin showed remarkable effectiveness in triggering apoptosis and suppressing angiogenesis in aberrantly growing cells in cellular experiments. | 2024 | Wang et al. (2024) |
| Ethosomes | Curcumin and piperine | Skin damage | ETs provided skin penetration of curcumin and increased antioxidant properties on co-delivery of curcumin and piperine in ETs. | 2024 | Ferrara et al. (2024) |
| Transethosomal gel | Apigenin | Skin cancer | Presented upgraded apigenin permeation via skin along with an improved reduction in cancer cell viability compared to conventional gels. | 2023 | Adnan et al. (2023) |
| Ethosomes | Insulin | – | Ethosomes of insulin demonstrated prolonged stability, higher encapsulation, and enhanced penetration through the skin. | 2024 | Nabila et al. (2024) |

(continued)

**Table 4.1** (continued)

| Type of nanocarrier | Therapeutic moiety | Disease | Key outcomes | Year of publication | References |
|---|---|---|---|---|---|
| Niosomes | Methotrexate | Skin cancer | Drug-loaded niosomes demonstrated potent anticancer activity against melanoma cells. | 2024 | Soni et al. (2024) |
| Niosomes | Resveratrol | Pancreatic cancer | Resveratrol-loaded niosomes showed increased cytotoxicity against Caplan-1 cells under the influence of a magnetic field than plain resveratrol. | 2024 | Firouzi Amandi et al. (2024) |
| Niosomes | Letrozole | Breast cancer | The addition of letrozole-loaded niosomes into the scaffold showed improved anticancer activity. | 2024 | Mahdizadeh et al. (2024) |
| Niosomes | Bee Venom | Cancer (not specific) | Bee venom-encapsulated niosomes showed improved therapeutic efficiency against various cancer cell lines and increased anti-inflammatory activity. | 2024 | Pinto et al. (2024) |
| Niosomes | Paclitaxel and sodium oxamate | Breast cancer | Hyaluronic acid conjugated niosomes of drugs exhibited improved cytotoxicity against MCF-7 cells while non-toxic against normal cells. | 2024 | Kaveh Zenjanab et al. (2024) |
| Niosomes | 5-fluorouracil | Skin cancer | Hyaluronic acid conjugated niosomes demonstrated enhanced retention of 5-fluorouracil in tumor tissues. | 2023 | Khalid et al. (2023) |
| LPHNPs | Fisetin | Pancreatitis | Fisetin-encapsulated LPHNPs inhibited acute pancreatitis induced by L-arginine and various organ destruction. | 2023 | Awadeen et al. (2023) |

| | | | | | |
|---|---|---|---|---|---|
| LPHNPs | Exemestane | Breast cancer | The pharmacokinetic assessment displayed 4.69 and 3.58 folds increased oral bioavailability for exemestane-loaded LPHNPs without TPGS and exemestane-loaded TPGS-LPHNPs, respectively than the free drug. In addition, exemestane-loaded TPGS-LPHNPs provided increased tumor inhibition against MCF-7 cell lines than suspension. | 2023 | Rizwanullah et al. (2023) |
| LPHNPs | siRNA | Triple-negative breast cancer | EGFR-conjugated LPHNPs of siRNA demonstrated improved targeting of cancer cells than unconjugated nanoparticles. | 2024 | Mehta et al. (2024) |
| LPHNPs | Doxorubicin | – | Doxorubicin-loaded LPHNPs displayed a 2 times greater area under the curve than free doxorubicin. | 2023 | Shafique et al. (2023) |

**Table 4.2** A list of clinical trials recently conducted for lipid-based nanocarriers in drug delivery

| Type of nanocarrier | Therapeutic moiety | Indication | Sponsor | Phase | Recruitment status | Estimated completion year | Clinical trial no. |
|---|---|---|---|---|---|---|---|
| Liposomes | Irinotecan | Neuroendocrine carcinoma | First Affiliated Hospital Xi'an Jiaotong University | Phase 4 | Not yet recruiting | 2028 | NCT06485739 |
| Liposomes | Irinotecan | Small cell lung cancer | Tang-Du Hospital | Phase IV | Recruiting | 2028 | NCT06467786 |
| Liposomes | Irinotecan | Ewing sarcoma | Peking University People's Hospital | Phase I | Recruiting | 2026 | NCT06340204 |
| Liposomes | Irinotecan | Metastatic colorectal cancer | West China Hospital | | Not yet recruiting | 2026 | NCT06341309 |
| Cationic liposomes | Paclitaxel | Advanced solid tumors | CSPC ZhongQi Pharmaceutical Technology Co., Ltd. | Phase I | Recruiting | 2025 | NCT06632574 |
| Liposomes | Doxorubicin | Advanced malignant tumors | Fudan University | Phase II | Unknown | 2022 | NCT05354076 |
| Liposomes | Doxorubicin | Desmoidtumor | Sun Yat-sen University | Phase III | Unknown | 2023 | NCT05561036 |
| Liposomes | Mitoxantrone hydrochloride | Pediatric tumors | Sun Yat-sen University | | Active, not recruiting | 2025 | NCT05620862 |
| Liposomes | Mitoxantrone hydrochloride | Advanced malignant tumors | CSPC ZhongQi Pharmaceutical Technology Co., Ltd | Phase II | Suspended | 2022 | NCT05089461 |
| Liposomes | Irinotecan | Rectal cancer | Zhejiang Cancer Hospital | Phase I | Recruiting | 2025 | NCT06300489 |
| Liposomes | Bupivacaine | Gynecologic cancer | M.D. Anderson Cancer Center | Phase III | Terminated | 2021 | NCT02740114 |

| | | | | | | | |
|---|---|---|---|---|---|---|---|
| SLNs | miR-193a-3p | Solid tumors | InteRNA | Phase I | Terminated | 2023 | NCT04675996 |
| SLNs | Akt-1 Antisense Oligonucleotide | Advanced solid tumors | Zhejiang Haichang Biotech Co., Ltd | Phase I | Recruiting | 2024 | NCT05267899 |
| SLNs | mRNAs Encoding Human OX40L, IL-23, and IL-36γ | Lymphoma | ModernaTX, Inc. | Phase I | Active, not recruiting | 2026 | NCT03739931 |
| Lipid nanoparticles | H3 mRNA vaccine | Influenza immunization | Sanofi Pasteur | Phase I | Complete | 2024 | NCT05829356 |
| Niosomes | Dexmedetomidine | Postoperative pain | Assiut University | Phase II, III | Recruiting | 2023 | NCT05340725 |

## 4.4 Commercialized Lipid Nanocarriers

Commercially available lipid nanocarriers have provided a revolutionary development in drug delivery, offering sophisticated approaches for safer and tailored therapeutic delivery (Plaza-Oliver et al. 2021). Among the well-known lipid nanocarriers, liposomes, SLNs, and NLCs have all seen successful commercialization for a range of medicinal uses. One of the first lipid-based nanocarriers to hit the market was liposomes, used for cancer. Drugs such as Doxil® (doxorubicin liposome) reduce toxic effects and improve patient consequences by delivering chemotherapy directly to cancer cells (Chen et al. 2024b). Likewise, lipid nanoparticles have been used in mRNA vaccine technology. Notably, the vaccines (COVID-19) from Moderna and Pfizer-BioNTech employ lipid nanoparticles to safely transport and preserve the mRNA into cells, showcasing lipid-based nanocarriers efficiency in the delivery of nucleic acids (Wilson and Geetha 2022). A list of commercialized LNPs for drug delivery has been illustrated in Table 4.3.

## 4.5 Conclusion

Lipid nanocarriers have ushered in a new age for the pharmaceutical industry. Drugs and lipid nanocarriers may be combined to create clinical applications that target specific cells and tissues to maximize therapeutic effects while minimizing adverse effects. Lipid nanocarriers can overcome biological barriers and enhance cellular uptake of encapsulated drugs by modifying active, passive, and co-transport mechanisms. Thermolabile drugs that have poor pharmacokinetic profiles or are physicochemically incompatible can all be delivered to the target site using lipid nanocarriers. Additionally, proteins and peptides can be transported more effectively and with less toxicity. Many lipid nanocarriers for drug delivery have been advanced to the point of clinical trials, and some have been commercialized in recent years, revealing the popularity of lipid nanocarriers to deliver drugs effectively in various kinds of diseases. One of the biggest difficulties in the creation of drug-loaded LNPs is maintaining the stability of the drug nanoparticles. The structure and potency of nanocarriers at the cellular level, both *in vitro* and *in vivo*, require more investigation. Furthermore, utilizing personalized medicine with LNPs customized to each patient's profile, particularly in gene therapy and cancer, may greatly improve treatment results. To facilitate clinical translation, regulatory standards must also change to guarantee the efficacy, safety, and consistency of LNPs. LNPs have the potential to revolutionize drug delivery and provide more potent treatments in several medical specialties with further development.

**Table 4.3** A list of commercialized LNPs in drug delivery

| Type of nanocarrier | Active moiety | Indication | Name of product | Manufacturer |
|---|---|---|---|---|
| SLNs | Black currant seed oil and manuka oil | Regeneration of dry, rough, and ageing skin | NanoLipid Restore CLR | ChemischesLaboratium Dr. Kurt Richter, CLR-Berlin, Germany |
| Lipid nanoparticles | BNT162b2 | Covid-19 | Comirnaty® | Pfizer-BioNTech |
| Lipid nanoparticles | mRNA | Covid-19 | SPIKEVAX™ | Moderna Inc. |
| Lipid nanoparticles | Recombinant varicella-zoster virus glycoprotein E | Shingles and port-herpetic neuralgia | Shingrix® | GSK Pharmaceuticals |
| Lipid nanoparticles | Patisiran (siRNA) | Hereditary transthyretin amyloidosis | Onpattro | Alnylam Pharmaceuticals |
| Lipid nanoparticles | Givosiran (siRNA) | Acute hepatic porphyria | Givlaari | Alnylam Pharmaceuticals |
| Liposomes | Amphotericin B | Fungal infections | Ambisome® | AstellasPharma |
| Liposomes | Cytarabine:Daunorubicin | Cancer | Vyxeos® | Jazz Pharmaceuticals |
| Liposomes | Doxorubicin | Cancer | Lipodox® | Sun Pharma |
| Liposomes | Doxorubicin | Cancer | Doxil® | Sequus Pharmaceuticals |
| Liposomes | Cytarabine | Cancer | Depocyt® | SkyPharma Inc. |
| Liposomes | Vincristine | Cancer | Marqibo® | Talon Therapeutics |
| Liposomes | Paclitaxel | Cancer | Lipusu® | Luye Pharma Group |
| Liposomes | Irinotecan | Cancer | Onivyde™ | Merrimack Pharmaceuticals |
| Liposomes | Morphine sulfate | Analgesic | DepoDur™ | SkyPharma |
| Liposomes | Verteporphin | Photodynamic therapy | Visudyne® | Novartis AG |
| Liposomes | Bupivacaine | Analgesic | Exparel® | Pacira pharmaceuticals |
| Liposomes | Doxorubicin | Cancer | Zolsketil® | Accord Healthcare |
| Liposomes | Amikacin | Lung infections | Arikayce® | Insmed |
| Liposomes | Bupivacaine | Anesthesia | Exparel® | Pacira Pharmaceuticals |
| Nanoemulsion | Cyclosporine | Immunosuppressant | Neoral® | Novartis |
| Nanoemulsion | Cyclosporine | Immunosuppressant | Gengraf® | Concord Biotech ltd |

(continued)

**Table 4.3** (continued)

| Type of nanocarrier | Active moiety | Indication | Name of product | Manufacturer |
|---|---|---|---|---|
| Nanoemulsion | Saquinavir | HIV infections | Fortovase® | F. Hoffmann-La Roche AG |
| Nanoemulsion | Ritonavir | HIV infections | Norvir® | AbbVie, Inc |
| Nanoemulsion | Propofol | Anesthetics | Diprivan® | Fresenius Kabi |
| Nanoemulsion | Dexamethasone | Steroid | Limethasone® | Mitsubishi Pharmaceutical |
| Nanoemulsion | Flurbiprofen | NSAIDs | Ropivon® | Neon Laboratories |
| Nanoemulsion | Vitamins A, D,E, and K | Nutrition | Vitalipid® | Fresenius Kabi |

## References

Adnan M, Afzal O, Altamimi ASA, Alamri MA, Haider T, Faheem Haider M (2023) Development and optimization of transethosomal gel of apigenin for topical delivery: in-vitro, ex-vivo and cell line assessment. Int J Pharm 631:122506. https://doi.org/10.1016/j.ijpharm.2022.122506

Ahad A et al (2022) Chapter 14–Ethosomes: a potential vesicular carrier for drug delivery. In: Nayak AK, Hasnain MS, Aminabhavi TM, Torchilin VP (eds) Systems of nanovesicular drug delivery. Academic, pp 221–237. https://doi.org/10.1016/B978-0-323-91864-0.00003-6

Ahalwat S et al (2023) Mannose-functionalized isoniazid-loaded nanostructured lipid carriers for pulmonary delivery: in vitro prospects and in vivo therapeutic efficacy assessment. *Pharmaceuticals* 16(8):1108. https://doi.org/10.3390/ph16081108

Akanda M, Mithu MSH, Douroumis D (2023) Solid lipid nanoparticles: an effective lipid-based technology for cancer treatment. J Drug Deliv Sci Technol 86:104709. https://doi.org/10.1016/j.jddst.2023.104709

Alcantara KP, Malabanan JWT, Nalinratana N, Thitikornpong W, Rojsitthisak P, Rojsitthisak P (2024) Cannabidiol-loaded solid lipid nanoparticles ameliorate the inhibition of Proinflammatory cytokines and free radicals in an in vitro inflammation-induced cell model. Int J Mol Sci 25(9):4744. https://doi.org/10.3390/ijms25094744

Amiri F, Ziaei Chamgordani S, Ghourchian H (2024) An in vitro study for reducing the cytotoxicity and dose dumping risk of remdesivir via entrapment in nanostructured lipid carriers. Sci Rep 14(1):19360. https://doi.org/10.1038/s41598-024-70003-7

Arshad MS et al (2024) Fabrication of miconazole nitrate solid lipid nanoparticle loaded microneedle patches for the treatment of Candida albicans biofilms. RSC Pharm 1(3):458–471. https://doi.org/10.1039/D4PM00042K

Awadeen RH, Boughdady MF, Zaghloul RA, Elsaed WM, Abu Hashim II, Meshali MM (2023) Formulation of lipid polymer hybrid nanoparticles of the phytochemical Fisetin and its in vivo assessment against severe acute pancreatitis. Sci Rep 13(1):19110. https://doi.org/10.1038/s41598-023-46215-8

Batista MAC et al., (2024) Antihypertensive effect of a nanoemulsion of baccharis dracunculifolia leaves extract in sodium-dependent hypertensive rats. Nat Prod Res. Accessed 25 Oct 2024. [Online]. Available: https://www.tandfonline.com/doi/full/10.1080/14786419.2024.2397724

Bharti Sharma J et al (2023) Statistical optimization of tetrahydrocurcumin loaded solid lipid nanoparticles using box Behnken design in the management of streptozotocin-induced diabetes mellitus. Saudi Pharm J 31(9):101727. https://doi.org/10.1016/j.jsps.2023.101727

Chamel S, Mishra A, Gull A (2024) Transferosomes as innovative drug delivery systems for enhanced antifungal therapy: a comprehensive review. J Drug Deliv Sci Technol 95:105545. https://doi.org/10.1016/j.jddst.2024.105545

Chehelgerdi M et al (2023) Progressing nanotechnology to improve targeted cancer treatment: overcoming hurdles in its clinical implementation. Mol Cancer 22:169. https://doi.org/10.1186/s12943-023-01865-0

Chen S, Hanning S, Falconer J, Locke M, Wen J (2019) Recent advances in non-ionic surfactant vesicles (niosomes): fabrication, characterization, pharmaceutical and cosmetic applications. Eur J Pharm Biopharm 144:18–39. https://doi.org/10.1016/j.ejpb.2019.08.015

Chen J et al (2024a) A novel quercetin encapsulated glucose modified liposome and its brain-target Antioxidative neuroprotection effects. Molecules 29(3):607. https://doi.org/10.3390/molecules29030607

Chen J et al (2024b) Recent advances and clinical translation of liposomal delivery systems in cancer therapy. Eur J Pharm Sci 193:106688. https://doi.org/10.1016/j.ejps.2023.106688

Cheng Z et al (2024) Optimizing fluconazole-embedded transfersomal gel for enhanced antifungal activity and compatibility studies. Front Pharmacol 15:1353791. https://doi.org/10.3389/fphar.2024.1353791

Cui J et al (2024) The pharmacokinetics and liver-targeting evaluation of Silybin liposomes mediated by the NTCP/OCTN2 dual receptors. Mol Pharm 21(10):4912–4923. https://doi.org/10.1021/acs.molpharmaceut.3c01245

Cunha Matosinhos R et al (2024) Development and characterization of liposomal formulations containing sesquiterpene lactones for the treatment of chronic gout. Sci Rep 14(1):6991. https://doi.org/10.1038/s41598-024-57663-1

Das B, Nayak AK, Mallick S (2022) Chapter 7 – Transferosomes: a novel nanovesicular approach for drug delivery. In: Nayak AK, Hasnain MS, Aminabhavi TM, Torchilin VP (eds) Systems of nanovesicular drug delivery. Academic, pp 103–114. https://doi.org/10.1016/B978-0-323-91864-0.00022-X

Dubey S et al (2024) Breaking barriers in eco-friendly synthesis of plant-mediated metal/metal oxide/bimetallic nanoparticles: antibacterial, anticancer, mechanism elucidation, and versatile utilizations. J Nanomater 2024. https://doi.org/10.1155/2024/9914079

Elkhayat D, Abdelmalak NS, Amer R, Awad HH (2024) Ezetimibe-loaded nanostructured lipid carrier for Oral delivery: response surface methodology; in vitro characterization and assessing the Antihyperlipidemic effect in rats. ACS Omega 9(7):8103–8116. https://doi.org/10.1021/acsomega.3c08428

Emanet M, Ciofani G (2023) Ethosomes as promising transdermal delivery systems of natural-derived active compounds. Adv NanoBiomed Res 3(10):2300020. https://doi.org/10.1002/anbr.202300020

Ferrara F et al (2024) Ethosomes for curcumin and Piperine cutaneous delivery to prevent environmental-stressor-induced skin damage. Antioxidants (Basel) 13(1):91. https://doi.org/10.3390/antiox13010091

Firouzi Amandi A et al (2024) Fabrication of magnetic niosomal platform for delivery of resveratrol: potential anticancer activity against human pancreatic cancer Capan-1 cell. Cancer Cell Int 24(1):46. https://doi.org/10.1186/s12935-024-03219-2

Gajbhiye KR, Salve R, Narwade M, Sheikh A, Kesharwani P, Gajbhiye V (2023) Lipid polymer hybrid nanoparticles: a custom-tailored next-generation approach for cancer therapeutics. Mol Cancer 22(1):160. https://doi.org/10.1186/s12943-023-01849-0

Ghasemiyeh P, Mohammadi-Samani S (2018) Solid lipid nanoparticles and nanostructured lipid carriers as novel drug delivery systems: applications, advantages and disadvantages. Res Pharm Sci 13(4):288–303. https://doi.org/10.4103/1735-5362.235156

Giordano A, Provenza AC, Reverchon G, Baldino L, Reverchon E (2024) Lipid-based Nanocarriers: bridging diagnosis and cancer therapy. Pharmaceutics 16(9):1158. https://doi.org/10.3390/pharmaceutics16091158

Gupta T et al (2024) Preparation and in vitro evaluation of BBG-250 loaded liposomal formulation for anticancer potential. Futur J Pharm Sci 10(1):19. https://doi.org/10.1186/s43094-024-00581-w

Hadipour Jahromy M et al (2024) Evaluation of curcumin-based ophthalmic nano-emulsion on atropine-induced dry eye in mice. Heliyon 10(7):e29009. https://doi.org/10.1016/j.heliyon.2024.e29009

Hassan DM, El-Kamel AH, Allam EA, Bakr BA, Ashour AA (2024) Chitosan-coated nanostructured lipid carriers for effective brain delivery of Tanshinone IIA in Parkinson's disease: interplay between nuclear factor-kappa β and cathepsin B. Drug Deliv Transl. Res. 14(2):400–417. https://doi.org/10.1007/s13346-023-01407-7

Jain V, Kumar H, Chand P, Jain S, Preethi S (2021) Lipid-based Nanocarriers as drug delivery system and its applications. In: Nanopharmaceutical advanced delivery systems. Wiley, pp 1–29. https://doi.org/10.1002/9781119711698.ch1

Karthikeyan D, Subbiah L, Palanisamy S (2024) Development of Letrozole-Loaded Magnetic Nanoemulsion Used for Breast Cancer Treatment. Proceedings 100(1):7. https://doi.org/10.3390/proceedings2024100007

Kaveh Zenjanab M et al (2024) Hyaluronic Acid-Targeted Niosomes for Effective Breast Cancer Chemostarvation Therapy. ACS Omega 9(9):10875–10885. https://doi.org/10.1021/acsomega.3c09782

Khalid W et al (2023) 5-fluorouracil-loaded hyaluronic acid-coated Niosomal vesicles: fabrication and ex vivo evaluation for skin drug delivery. ACS Omega 8(48):45405–45413. https://doi.org/10.1021/acsomega.3c04457

Khodabandelou S et al (2024) Development of silibinin-loaded nanostructured lipid carriers for Alzheimer's disease induced by amyloid beta in Wistar rats. J Mater Chem B 12:11426–11443. https://doi.org/10.1039/d4tb00775a

Kim H-E et al (2024) Fabrication and evaluation of chitosan-coated nanostructured lipid carriers for co-delivery of paclitaxel and PD-L1 siRNA. Int J Pharm 666:124835. https://doi.org/10.1016/j.ijpharm.2024.124835

Kudi P, Sen S, Murkute S, Mohapatra P, Ranjan OP (2024) Quality by design (QbD) based approach for development of itraconazole-loaded transferosomes for skin cancer: in vitro, ex vivo and cell line studies. Drug Dev Ind Pharm 0(0):1–14. https://doi.org/10.1080/03639045.2024.2400203

Kumar R (2019) Lipid-based nanoparticles for drug-delivery systems. In: Nanocarriers for drug delivery. Elsevier, pp 249–284. https://doi.org/10.1016/B978-0-12-814033-8.00008-4

Kumar G, Khar RK, Virmani T, Jogpal V, Virmani R (2018) Comparative evaluation of fast dissolving tablet of atorvastatin calcium using natural and synthetic super disintegrating agents. Res J Pharm Technol 11(11):5001–5007. https://doi.org/10.5958/0974-360X.2018.00912.5

Kumar G, Virmani T, Pathak K, Alhalmi A (2022a) A revolutionary blueprint for mitigation of hypertension via Nanoemulsion. Biomed Res Int 2022:4109874. https://doi.org/10.1155/2022/4109874

Kumar G, Virmani T, Pathak K, Kamaly OA, Saleh A (2022b) Central composite design implemented Azilsartan Medoxomil loaded Nanoemulsion to improve its aqueous solubility and intestinal permeability: In Vitro and Ex Vivo Evaluation. Pharmaceuticals (Basel) 15(11):1343. https://doi.org/10.3390/ph15111343

Kumar G, Virmani T, Sharma A, Pathak K (2023a) Codelivery of phytochemicals with conventional anticancer drugs in form of Nanocarriers. Pharmaceutics 15(3):889. https://doi.org/10.3390/pharmaceutics15030889

Kumar M et al (2023b) Nanocarriers in tuberculosis treatment: challenges and delivery strategies. Pharmaceuticals (Basel) 16(10):1360. https://doi.org/10.3390/ph16101360

Kumar G, Virmani T, Misra SK, Sharma A, Pathak K (2024a) Exploration of chitosan and its modified derivatives as vaccine adjuvant: a review. Carbohydr Polym Technol Appl 8:100537. https://doi.org/10.1016/j.carpta.2024.100537

Kumar G et al (2024b) Enhancing therapy with nano-based delivery systems: exploring the bioactive properties and effects of apigenin. Ther Deliv 15(9):717–735. https://doi.org/10.1080/20415990.2024.2386928

Kumar G, Virmani T, Sharma A, Virmani R, Pathak K (2024c) Chapter 14–Nanoparticulate drug delivery systems for colonic disorders. In: Dureja H, Loebenberg R, Singh SK, Dhanasekaran M, Dua K (eds) Advanced drug delivery Systems for Colonic Disorders. Academic, pp 317–344. https://doi.org/10.1016/B978-0-443-14044-0.00009-0

Kumar G, Virmani R, Sharma A, Virmani T, Pathak K (2024d) Solid lipid nanoparticles for the management of allergic airway diseases. In: Prasher P, Sharma M, Singh SK, MacLoughlin R, Pabreja K, Dua K (eds) Understanding allergic airway diseases: contemporary treatment paradigm. Springer Nature, Singapore, pp 183–201. https://doi.org/10.1007/978-981-97-1953-2_11

Kumar G, Virmani T, Sharma A, Virmani R, Pathak K (2024e) Synbiotics in Oral drug delivery. In: Dua K (ed) Synbiotics in human health: biology to drug delivery. Springer Nature, Singapore, pp 413–433. https://doi.org/10.1007/978-981-99-5575-6_20

Kumar A et al (2024f) Unraveling the multifaceted role of nanoemulsions as drug delivery system for the management of cancer. J Drug Deliv Sci Technol 100:106056. https://doi.org/10.1016/j.jddst.2024.106056

Kumari P, Dang S (2023) Dual drug loaded nanostructured lipid carrier for cytotoxic effect against breast cancer – a drug repurposing approach. Surf Interfaces 40:103138. https://doi.org/10.1016/j.surfin.2023.103138

Li Y et al (2023) Lycorine transfersomes modified with cell-penetrating peptides for topical treatment of cutaneous squamous cell carcinoma. J Nanobiotechnol 21(1):139. https://doi.org/10.1186/s12951-023-01877-4

Li R et al (2024a) Paclitaxel liposome (Lipusu) based chemotherapy combined with immunotherapy for advanced non-small cell lung cancer: a multicenter, retrospective real-world study. BMC Cancer 24(1):107. https://doi.org/10.1186/s12885-024-11860-3

Li M et al (2024b) Co-delivery of paclitaxel and curcumin loaded solid lipid nanoparticles for improved targeting of lung cancer: *in vitro* and *in vivo* investigation. Heliyon 10(9):e30290. https://doi.org/10.1016/j.heliyon.2024.e30290

Liga S, Paul C, Moacă E-A, Péter F (2024) Niosomes: composition, formulation techniques, and recent Progress as delivery systems in cancer therapy. Pharmaceutics 16(2):223. https://doi.org/10.3390/pharmaceutics16020223

Liu J et al (2024) Pterostilbene nanoemulsion promotes Nrf2 signaling pathway to downregulate oxidative stress for treating Alzheimer's disease. Int J Pharm 655:124002. https://doi.org/10.1016/j.ijpharm.2024.124002

Mahdizadeh N, Khorshid Shabestari M, Tafvizi F, Khodarahmi P (2024) Delivery of letrozole-encapsulated niosomes via a 3D bioprinting gelatin–alginate scaffold for potential breast cancer treatment. Cancer Nanotechnol 15(1):33. https://doi.org/10.1186/s12645-024-00271-5

Mahor AK et al (2023) Nanostructured lipid carriers for improved delivery of therapeutics via the Oral route. J Nanotechnol 2023(1):4687959. https://doi.org/10.1155/2023/4687959

Mall J, Naseem N, Haider MF, Rahman MA, Khan S, Siddiqui SN (2024) Nanostructured lipid carriers as a drug delivery system: a comprehensive review with therapeutic applications. Int Pharm. https://doi.org/10.1016/j.ipha.2024.09.005

Matharoo N, Mohd H, Michniak-Kohn B (2024) Transferosomes as a transdermal drug delivery system: dermal kinetics and recent developments. Wiley Interdiscip Rev Nanomed Nanobiotechnol 16(1):e1918. https://doi.org/10.1002/wnan.1918

Mehmood Y et al (2024) Synthesis of vitamin D3 loaded ethosomes gel to cure chronic immune-mediated inflammatory skin disease: physical characterization, in vitro and ex vivo studies. Sci Rep 14(1):23866. https://doi.org/10.1038/s41598-024-72951-6

Mehta M, Bui TA, Yang X, Aksoy Y, Goldys EM, Deng W (2023) Lipid-based nanoparticles for drug/gene delivery: an overview of the production techniques and difficulties encountered in their industrial development. ACS Mater Au 3(6):600–619. https://doi.org/10.1021/acsmaterialsau.3c00032

Mehta M, Bui TA, Care A, Deng W (2024) Targeted polymer lipid hybrid nanoparticles for in-vitro siRNA therapy in triple-negative breast cancer. J Drug Deliv Sci Technol 98:105911. https://doi.org/10.1016/j.jddst.2024.105911

Mishra G et al (2024) Development of epigallocatechin and ascorbic acid dual delivery transferosomes for managing Alzheimer's disease: in vitro and in vivo studies. ACS Omega 9(33):35463–35474. https://doi.org/10.1021/acsomega.4c02140

Moammeri A et al (2023) Current advances in niosomes applications for drug delivery and cancer treatment. Mater Today Bio 23:100837. https://doi.org/10.1016/j.mtbio.2023.100837

Mohamed JMM et al (2024) Optimization and characterization of quercetin-loaded solid lipid nanoparticles for biomedical application in colorectal cancer. Cancer Nanotechnol 15(1):16. https://doi.org/10.1186/s12645-024-00249-3

Mohammed HA et al (2023) Solid lipid nanoparticles for targeted natural and synthetic drugs delivery in high-incidence cancers, and other diseases: roles of preparation methods, lipid composition, transitional stability, and release profiles in nanocarriers' development. Nanotechnol Rev 12(1):1–30. https://doi.org/10.1515/ntrev-2022-0517

Mota Díaz II et al (2024) Co-encapsulation of paclitaxel and doxorubicin in liposomes layer by layer. Colloids Interface 8(4):42. https://doi.org/10.3390/colloids8040042

Motawea A, Maria SN, Maria DN, Jablonski MM, Ibrahim MM (2024) Genistein transfersome-embedded topical delivery system for skin melanoma treatment: in vitro and ex vivo evaluations. Drug Deliv 31(1):2372277. https://doi.org/10.1080/10717544.2024.2372277

Nabila FH et al (2024) Ionic liquid-mediated ethosome for transdermal delivery of insulin. Chem Commun (Camb) 60(30):4036–4039. https://doi.org/10.1039/d3cc06130b

Nayak D, Mohanta YK, Rauta PR, Chakrabartty I, Saravanan M (2022) Ethosomes for dermal and transdermal drug delivery systems. In: Barabadi H, Mostafavi E, Saravanan M (eds) Pharmaceutical Nanobiotechnology for targeted therapy. Springer, Cham, pp 313–336. https://doi.org/10.1007/978-3-031-12658-1_11

Nayila I et al (2024) Formulation, characterization and evaluation of anti-breast cancer activity of 2-carene nanoemulsion; *in silico*, *in vitro* and *in vivo* study. Arab J Chem 17(9):105937. https://doi.org/10.1016/j.arabjc.2024.105937

Neha SL, Mishra AK, Rani L, Paroha S, Dewangan HK, Sahoo PK (2023) Design and evaluations of a nanostructured lipid carrier loaded with dopamine hydrochloride for intranasal bypass drug delivery in Parkinson's disease. J Microencapsul 40(8):599–612. https://doi.org/10.1080/02652048.2023.2264386

Nemati S, Mottaghi M, Karami P, Mirjalali H (2024) Development of solid lipid nanoparticles-loaded drugs in parasitic diseases. Discov Nano 19(1):7. https://doi.org/10.1186/s11671-023-03955-w

Paolino D et al (2024) Improved anti-breast cancer activity by doxorubicin-loaded super stealth liposomes. Biomater Sci 12(15):3933–3946. https://doi.org/10.1039/D4BM00478G

Parveen S, Gupta P, Kumar S, Banerjee M (2023) Lipid polymer hybrid nanoparticles as potent vehicles for drug delivery in cancer therapeutics. Med Drug Discov 20:100165. https://doi.org/10.1016/j.medidd.2023.100165

Passos JS, Apolinario AC, Ishida K, Martins TS, Lopes LB (2024) Nanostructured lipid carriers loaded into in situ gels for breast cancer local treatment. Eur J Pharm Sci 192:106638. https://doi.org/10.1016/j.ejps.2023.106638

Pearce K, Cairncross SI, Benjeddou M (2024) Liposomal-Naringenin Radiosensitizes triple-negative breast cancer MDA-MB-231 cells in vitro. IET Nanobiotechnol 2024(1):3786627. https://doi.org/10.1049/2024/3786627

Pinto MB et al (2024) Bee venom-loaded Niosomes as innovative platforms for cancer treatment: development and Therapeutical efficacy and safety evaluation. Pharmaceuticals 17(5):572. https://doi.org/10.3390/ph17050572

Plaza-Oliver M, Santander-Ortega MJ, Lozano MV (2021) Current approaches in lipid-based nanocarriers for oral drug delivery. Drug Deliv Transl Res 11(2):471–497. https://doi.org/10.1007/s13346-021-00908-7

Pradana A, Ritthidej G, Limprasutr V, Wongtayan A, Lipipun V, Iksen I (2024) Antihypertensive activity of spray-dried nanoemulsion containing Asiatic acid-palm oil in high salt diet-fed rats. Pharmacia 71:1–10. https://doi.org/10.3897/pharmacia.71.e115091

Raghunath I, Koland M, Sarathchandran C, Saoji S, Rarokar N (2024) Design and optimization of chitosan-coated solid lipid nanoparticles containing insulin for improved intestinal permeability using piperine. Int J Biol Macromol 280:135849. https://doi.org/10.1016/j.ijbiomac.2024.135849

Ramadan AEH et al (2024) Development and optimization of vildagliptin solid lipid nanoparticles loaded ocuserts for controlled ocular delivery: a promising approach towards treating diabetic retinopathy. Int J Pharm: X 7:100232. https://doi.org/10.1016/j.ijpx.2024.100232

Rasheed MS, Ansari SF, Shahzadi I (2022) Formulation, characterization of glucosamine loaded transfersomes and in vivo evaluation using papain induced arthritis model. Sci Rep 12(1):19813. https://doi.org/10.1038/s41598-022-23103-1

Rezaei H, Iranbakhsh A, Sepahi AA, Mirzaie A, Larijani K (2024) Formulation, preparation of niosome loaded zinc oxide nanoparticles and biological activities. Sci Rep 14(1):16692. https://doi.org/10.1038/s41598-024-67509-5

Rizwanullah M et al (2023) Polymer-lipid hybrid nanoparticles of exemestane for improved oral bioavailability and anti-tumor efficacy: an extensive preclinical investigation. Int J Pharm 642:123136. https://doi.org/10.1016/j.ijpharm.2023.123136

Roy D et al (2024) Roflumilast-loaded nanostructured lipid carriers attenuate oxidative stress and neuroinflammation in Parkinson's disease model. J Drug Target 33:1–16. https://doi.org/10.1080/1061186X.2024.2408724

Satyanarayana SD et al (2023) Ocular delivery of Bimatoprost-loaded solid lipid nanoparticles for effective Management of Glaucoma. Pharmaceuticals 16(7):1001. https://doi.org/10.3390/ph16071001

Sessevmez M, Sinani G, Çevikelli T (2024) Formulation, optimization and in vitro evaluation of polymer-coated liposomes encapsulating nebivolol hydrochloride. Int J Polym Mater Polym Biomater 73(10):897–907. https://doi.org/10.1080/00914037.2023.2225117

Shafique M et al (2023) Formulation development of lipid polymer hybrid nanoparticles of doxorubicin and its in-vitro, in-vivo and computational evaluation. Front Pharmacol 14:1025013. https://doi.org/10.3389/fphar.2023.1025013

Shah S, Famta P, Raghuvanshi RS, Singh SB, Srivastava S (2022) Lipid polymer hybrid nanocarriers: insights into synthesis aspects, characterization, release mechanisms, surface functionalization and potential implications. Colloid Interfac Sci Commun 46:100570. https://doi.org/10.1016/j.colcom.2021.100570

Shaker SA et al (2023) Propolis-loaded nanostructured lipid carriers halt breast cancer progression through miRNA-223 related pathways: an in-vitro/in-vivo experiment. Sci Rep 13(1):15752. https://doi.org/10.1038/s41598-023-42709-7

Shamim MA et al (2023) Transfersome encapsulated with the R-carvedilol enantiomer for skin cancer chemoprevention. Nano 13(5):929. https://doi.org/10.3390/nano13050929

Shazwani SS, Marlina A, Misran M (2024) Development of nanostructured lipid carrier-loaded flavonoid-enriched Zingiber officinale. ACS Omega 9(15):17379–17388. https://doi.org/10.1021/acsomega.4c00091

Soni S, Baghel K, Soni ML, Kashaw SK, Soni V (2024) Size-dependent effects of niosomes on the penetration of methotrexate in skin layers. Futur J Pharm Sci 10(1):48. https://doi.org/10.1186/s43094-024-00624-2

Soomherun N, Kreua-ongarjnukool N, Niyomthai ST, Chumnanvej S (2024) Lipid-polymer hybrid nanoparticles synthesized via lipid-based surface engineering for a robust drug delivery platform. Colloids Surf B: Biointerfaces 237:113858. https://doi.org/10.1016/j.colsurfb.2024.113858

Taher M et al (2023) PEGylated liposomes enhance the effect of cytotoxic drug: a review. Heliyon 9(3):e13823. https://doi.org/10.1016/j.heliyon.2023.e13823

Tekade A, Susar R, Kulkarni G, Surwade S, Gaikwad A (2024) Nanostructured lipid carriers of donepezil hydrochloride for the treatment of Alzheimer's disease. Curr Alzheimer Res 21:710–722. https://doi.org/10.2174/0115672050288659240229080535

Thiruchenthooran V et al (2024) Combination of indomethacin with nanostructured lipid carriers for effective anticancer therapy. *IJN* 19:7033–7048. https://doi.org/10.2147/IJN.S464239

Velho MC et al (2024) Eugenol-loaded nanoemulsions: Antiproliferative activity against breast cancer cells and hemocompatibility. J Drug Deliv Sci Technol 101:106248. https://doi.org/10.1016/j.jddst.2024.106248

Virmani T, Kumar G, Virmani R, Sharma A, Pathak K (2022) Nanocarrier-based approaches to combat chronic obstructive pulmonary disease. Nanomedicine 17(24):1833–1854. https://doi.org/10.2217/nnm-2021-0403

Virmani T et al (2023a) Amelioration of cancer employing chitosan, its derivatives, and chitosan-based nanoparticles: recent updates. Polymers 15(13):2928. https://doi.org/10.3390/polym15132928

Virmani T, Kumar G, Sharma A, Pathak K (2023b) Chapter 2– An overview of ocular drug delivery systems—conventional and novel drug delivery systems. In: Rai M, Occhiutto ML, Talegaonkar

S (eds) Nanotechnology in ophthalmology. Academic, pp 23–48. https://doi.org/10.1016/B978-0-443-15264-1.00007-5

Vishwas S et al (2024) Formulation and optimization of fisetin nanoemulsion for the treatment of Alzheimer's disease in rats: pharmacokinetic and pharmacodynamic assessment. J Drug Deliv Sci Technol 101:106269. https://doi.org/10.1016/j.jddst.2024.106269

Wadher K, Trivedi S, Rarokar N, Umekar M (2024) Development and assessment of rutin loaded transfersomes to improve *ex vivo* membrane permeability and *in vitro* efficacy. Hybrid Adv 5:100144. https://doi.org/10.1016/j.hybadv.2024.100144

Wang S, Chen Y, Guo J, Huang Q (2023) Liposomes for tumor targeted therapy: a review. Int J Mol Sci 24(3):2643. https://doi.org/10.3390/ijms24032643

Wang P et al (2024) Ethosomes-mediated tryptanthrin delivery as efficient anti-psoriatic nano therapy by enhancing topical drug absorption and lipid homeostasis. J Nanobiotechnol 22(1):584. https://doi.org/10.1186/s12951-024-02860-3

Wilson B, Geetha KM (2022) Lipid nanoparticles in the development of mRNA vaccines for COVID-19. J Drug Deliv Sci Technol 74:103553. https://doi.org/10.1016/j.jddst.2022.103553

Wu X, Chen X, Wang X, He H, Chen J, Wu W (2024) Paclitaxel-lipid prodrug liposomes for improved drug delivery and breast carcinoma therapy. Chin Chem Lett 35(2):108756. https://doi.org/10.1016/j.cclet.2023.108756

Yeo S, Kim MJ, Yoon I, Lee WK (2023) pH-responsive Nano-transferosomes of Purpurin-18 sodium salt and doxorubicin for enhanced anticancer efficiency by photodynamic and chemo combination therapy. ACS Omega 8(18):16479–16490. https://doi.org/10.1021/acsomega.3c01654

Yeo S, Jung S, Kim H, Ahn J-H, Hwang S-J (2024) 4-hexylresorcinol loaded solid lipid nanoparticles for enhancing anticancer activity. Pharmaceuticals 17(10):1296. https://doi.org/10.3390/ph17101296

Zaafar D et al (2024) Preparation and characterization of Sorafenib nano-emulsion: impact on pharmacokinetics and toxicity; an in vitro and in vivo study. Drug Deliv Transl Res 14(11):3089–3111. https://doi.org/10.1007/s13346-024-01530-z

Zhang D et al (2024) As1411-modified liposomes to enhance drug utilization and augment the antitumor efficacy. Cancer Nanotechnol 15(1):25. https://doi.org/10.1186/s12645-024-00262-6

# Chapter 5
# Polymeric Nanoparticles in Biopharmaceutical Development

**Jibrin Muhammad Yelwa, Haruna Musa, Alhagie Drammeh, Kalu M. Kalu, Philbert Nzeyimana, Hareg Zewdu Alehegn, Fatima B. Tanimu, and Husaina G. Anchau**

**Abstract** The emergence of polymeric nanoparticles (PNPs) as a promising platform for drug delivery, vaccine development, and genetic therapy encompasses superior bioavailability, targeted delivery, and controlled release. Most importantly, PNP has the potential to facilitate the delivery of small molecules, peptides, proteins, and nucleic acids against the special barriers of poor solubility, immune clearance, and endosomal escape. Advances in multi-stimuli responsive systems (MRS) and artificial intelligence (AI)-assisted drug design have further enhanced their potential. Despite such great potential, manufacturing challenges still exist on a large scale, through stability, and regulatory landscapes that need to be negotiated. The case studies of approved PNP-based therapeutics illustrate both potential opportunities and complexities for the development and marketability of such systems. This chapter discusses the basic principles, fabrication methods, encapsulation

J. M. Yelwa (✉) · F. B. Tanimu · H. G. Anchau
Department of Scientific and Industrial Research, National Research Institute for Chemical Technology, Zaria, Nigeria
e-mail: jibrin.yelwa@narict.gov.ng; husaina.garba@narict.gov.ng

H. Musa
Department of Pure and Industrial Chemistry, Bayero University Kano, Kano, Nigeria
e-mail: h.musa@chm.buk.edu.ng

A. Drammeh
Chemistry Unit, Division of Physical and Natural Sciences, School of Arts and Sciences, University of The Gambia, Serrekunda, Gambia

Pan African University Life and Earth Sciences Institute (Including Health and Agriculture), University of Ibadan, Ibadan, Oyo, Nigeria

K. M. Kalu
Department of Chemical Sciences, Gombe State University, Gombe, Nigeria
e-mail: kaymichael@gsu.edu.ng

P. Nzeyimana · H. Z. Alehegn
Pan African University Life and Earth Sciences Institute (Including Health and Agriculture), University of Ibadan, Ibadan, Oyo, Nigeria

 
V. Khanna et al. (eds.), *Nanotechnologies for Drug Delivery and Biopharmaceutical Development*, Progress in Drug Research 77,
https://doi.org/10.1007/978-3-032-00796-4_5

strategies, and pharmacokinetic profiles that make PNPs effective in biopharmaceutical applications. Further research efforts will focus on biodegradable polymers (BDPs) for tailored medicine applications, smart drug delivery systems, and regulatory standardization to maximize clinical utility while ensuring their safety and efficacy.

**Keywords** AI-assisted design · Batch reproducibility · Biodegradable polymers · Controlled release · Drug delivery · Endosomal escape · Gene therapy · Immune clearance · Immunogenicity · Multi-stimuli responsive · Nanotechnology · Pharmacokinetics · Pharmacodynamics · Personalized medicine · Peptides · Polymeric nanoparticles · Regulatory frameworks · Scalability · Stability · Vaccine delivery

## 5.1 Introduction

Nanotechnology has indeed changed the face of drug delivery by rendering therapeutic interventions more specific, controlled, and site-specific. All types of nanoparticles, ranging from liposomes to polymer carriers, have the advantages of improved drug solubility and enhancement of bioavailability, apart from reduced toxicity because of drug delivery at targeted tissue and organs. (Hussen and Yadav 2023). Advances in nanotechnology have taken further steps toward the fields of oncology, neurology, cardiology, and immunology by overcoming biological barriers and by achieving optimal therapeutic performance (Nguyen et al. 2021). Advances in vaccine delivery, nanocarriers loaded with drugs, and gene therapy have brought nanoparticles to the doorstep of a new revolution in drug interaction with biological environments (Chatterjee and Dhibar 2023).

This is because PNPs has come into limelight among different nanocarriers due to its biocompatibility, biodegradability, and shielding of encapsulated drugs against enzymatic degradation (Jhaveri et al. 2021). The, PNP can provide sustainable and controlled release of the drug, thereby enhancing its therapeutic efficiency while minimizing the adverse side effects. The delivery of small molecules, peptides, proteins, and nucleic acids using these nanoparticles has turned out to be highly promising in diversified applications, which range from cancer therapy to vaccine development (Patole et al. 2021). Active treatments via targeting with PNPs allow site-specific delivery without systemic exposure and toxicity for diseases involving the central nervous system (Ayub and Wettig 2022).

This chapter discusses the development, fabrication, and applications of polymeric nanoparticles in biopharmaceutical. This review discusses the method of PNP synthesis, surface modifications toward targeting, and drug-release mechanisms toward optimization of therapeutic outcome (Barik et al. 2022). The study outlines the pharmacokinetics of PNPs, dealing with absorption, distribution, and elimination pathways and discusses toxicity, stability, and regulatory problems that face the development and commercialization processes (Ndebele et al. 2021). These includes the development of trends, such as smart nanoparticles that respond to environmental stimuli (Jahan 2022), and the implementation of AI in drug delivery design to give a comprehensive overview of the future landscape of PNPs (Huang and Ding 2022).

This chapter seeks to broaden and focus the reader's perspective on exactly how PNP are changing the dimensions of drug delivery and biopharmaceutical innovation and allow for further discussion in subsequent sections regarding fabrication techniques, the mechanism of drug release, and clinical applications of PNPs.

## 5.2 Fundamentals of Polymeric Nanoparticles

### *5.2.1 Definition and Classification of Polymeric Nanoparticles*

Polymeric nanoparticles are generally nanostructures of biodegradable polymers with variable sizes ranging from 10 to 100 nm. The roles nanoparticles are executing in today's drug delivery systems include the encapsulation of therapeutic agents, bioavailability enhancement of drugs, and active drug targeting. PNPs have plays a very important role in pharmaceutical development due to controlled release of hydrophilic and hydrophobic drugs over extended periods. The biocompatibility and biodegradability of the polymers improve the safety profile by minimizing the possibility of toxicity among other adverse events in treatment (Amin and Boateng 2022).

Figure 5.1 illustrates some different types of polymeric nanoparticles used for drug delivery. Synthetic polymeric nanoparticles have well-defined control over drug release; polymeric nanogels and natural polymeric nanoparticles exhibit biocompatibility while sustaining release in some cases. The complexes of lipid-polymer can combine with stability and enhancement of absorption. Polymeric micelles improve the delivery in hydrophobic drugs. Lastly, dendrimers propose targeted delivery due to their branched structure, so the PNP may be versatile enough for biopharmaceutical applications.

#### 5.2.1.1 Nanospheres vs. Nanocapsules

PNP can be divided into two major subtypes namely, nanospheres and nanocapsules.

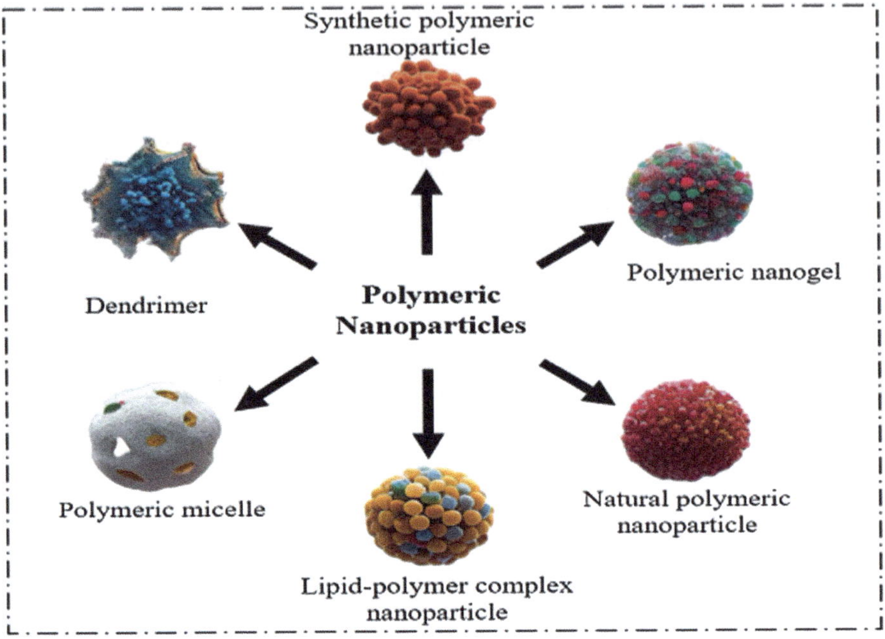


**Fig. 5.1** Schematic representation of polymeric nanoparticles (PNPs) types

Nanospheres

Nanospheres are solid, homogeneous particles in which the drug is dispersed throughout the polymer matrix. Such structures provide gradual and sustained drug release depending on the degradation of the matrix. Nanospheres can offer a lot of advantages, such as the encapsulation of hydrophobic drugs, improving their solubility and bioavailability (Malabanan et al. 2023).

Nanocapsules

Nanocapsules have a core-shell structure in which the drug is confined in a cavity surrounded by a polymeric membrane. This design allows for precise control of the release kinetics due to diffusive drug transport across the outer shell, or by degradation of the outer shell (Melo et al. 2021).

#### 5.2.1.2 Natural Polymers vs. Synthetic Polymers

PNPs can be further classified based on the origin of polymers used.

Polymeric nanoparticles may be developed using natural or synthetic polymers, each of which has its own unique set of benefits for drug delivery. Natural polymers such as chitosan and alginate are highly biocompatible, whereas synthetic

**Table 5.1** Comparison of natural vs. synthetic polymers for PNPs

| Feature | Natural polymers (e.g., Chitosan, Alginate) | Synthetic polymers (e.g., PLGA, PEG) | References |
|---|---|---|---|
| Biodegradability | High (enzymatically degraded) | Variable (controlled by composition) | Amin and Boateng (2022) |
| Biocompatibility | Excellent | Moderate to excellent (depends on polymer) | Ryu et al. (2021) |
| Stability | Sensitive to pH and enzyme content | High chemical and mechanical stability | Sheir et al. (2021) |
| Drug release | Sustained but influenced by the environment | Controlled and programable | Jusu et al. (2021) |
| Typical applications | Wound healing, mucosal delivery, and vaccine carriers | Cancer therapy, gene delivery, and long-circulating drugs | Jhaveri et al. (2021) |

polymers such as PLGA and PEG offer benefits in terms of controlling drug release and stability. A comparison of the various types of polymers are compared in Table 5.1 below.

Natural Polymers

Examples of natural-occurring polymers for synthesizing nanoparticles include chitosan and alginate. Chitosan is a derivative of chitin and thus possesses very good biocompatibility and antibacterial properties, hence suitable for wound healing applications (Ryu et al. 2021). Alginate generally originates from brown seaweed and possesses good mucoadhesive properties, which enable this polymer to attach to biological membranes and enhance drug absorption (Sheir et al. 2021).

Synthetic Polymers

The properties of some synthetic polymers such as poly(lactic-co-glycolic acid) (PLGA) and polyethylene glycol (PEG) can be control in nanoparticles with higher precision. PLGA is a very valuable polymer for its controlled degradation which enables sustained release of drugs. When used as a surface modifier, PEG enhances nanoparticle stability and reduces protein adsorption, extending circulation time in the bloodstream (Jusu et al. 2021).

### 5.2.2 *Physicochemical Properties of Polymeric Nanoparticles*

The performance of polymeric nanoparticles as drug vectors is governed by their properties, such as size, surface charge, morphology, biodegradability, biocompatibility, and stability. These properties must be optimized according to the intended therapeutic application to achieve optimal drug delivery.

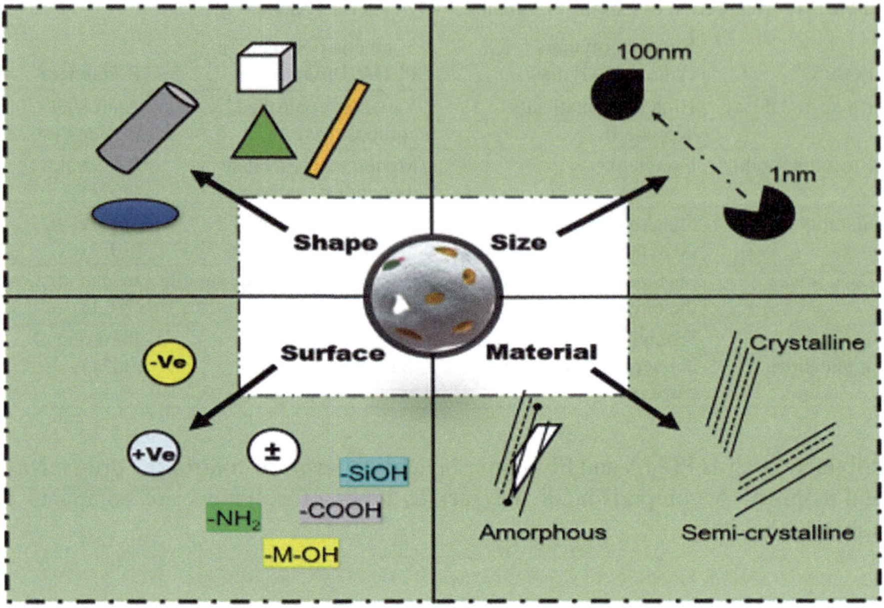


**Fig. 5.2** Physicochemical properties of polymeric nanoparticles

Figure 5.2 highlights some important physicochemical properties of polymeric nanoparticles, such as size, morphology, biodegradability, surface charge, and stability, which are affecting performances in drug delivery.

#### 5.2.2.1 Size, Surface Charge, and Morphology

The efficacy of polymeric nanoparticles as drug delivery carriers is greatly dependent on their physicochemical properties, including size, surface charge, and morphology. These properties affect the cellular uptake, circulation time, and targeting efficiency. Table 5.2 below summarizes the primary correlations between the properties of nanoparticles and their influence on drug delivery performance.

Size

The size of nanoparticles affects their distribution, cellular uptake, and biodistribution. Particles with smaller sizes (<100 nm) enhanced cellular uptake and thus can permeate the biological barriers better. In contrast, larger particles (>200 nm) allow for longer circulation times and, hence, are better candidates for controlled release of drugs (Alkholief et al. 2022).

**Table 5.2** Effects of PNP properties on drug delivery

| Property | Effects of drug delivery | References |
|---|---|---|
| Size | Smaller (<100 nm) enhances cellular uptake, larger (>200 nm) particles allow longer circulation. | Alkholief et al. (2022) |
| Surface charge | Positive charges promote cell uptake but may cause toxicity; negative charges extend the circulation time. | Chaghervand et al. (2021) |
| Morphology | Spherical nanoparticles improve stability; irregular shapes alter biodistribution. | Zegarra-Urquia et al. (2022) |
| Biodegradability | Faster degradation improves drug release, slower degradation allows sustained delivery. | Gaur et al. (2023) |

Surface Charge

A surface charge or zeta potential defines the stability of nanoparticles in suspension, as well as their interaction with biological membranes. Positively charged particles exhibit increased cellular uptake due to the interaction with negatively charged cell membranes, while negatively charged particles have longer circulation times (Chaghervand et al. 2021).

Morphology

The shape and surface are critical features determining the capability of nanoparticles for immune evasion and adhesion to target tissues. Spherical nanoparticles, due to their dispersal homogeneity and ease of preparation, are the most commonly used (Zegarra-Urquia et al. 2022).

Biodegradability and Biocompatibility

Biodegradable polymers such as PLGA and chitosan degrade into non-toxic byproducts makes them very useful materials in biomedical applications. Moreover, the rate of degradation can be modified by changing the composition of the polymer; therefore, sustained drug release occurs over a specific period of time (Amin and Boateng 2022). Biocompatibility should, in general, not lead to adverse immune responses or toxicities. The toxicity expressed by both natural and synthetic polymers used in the nanoparticle is minimal. Normally, PEGylation is performed to increase biocompatibility and low immunogenicity (Gaur et al. 2023).

Stability Under Physiological Conditions

The stability of polymeric nanoparticles under physiological conditions is important for ensuring drug efficacy. Temperature, pH, and ionic strength are some of the factors responsible for the stability of polymeric nanoparticles. For example, chitosan nanoparticles can provide good stability in gastric fluids, thus offering a potential candidate for oral drug carriage (Sharma et al. 2022). PEG-coated nanoparticles remain stable in blood circulation due to inhibition of protein adsorption and aggregation (Ramnandan et al. 2021). Maintaining the stability of nanoparticles provides assurance that the drug in their core will not degenerate prior to reaching the target site.

The properties of polymeric nanoparticles mostly refer to their classification and physicochemical features, which are crucial in the design of effective drug delivery systems. The proper selection and optimization of polymers, size, surface charge, and stability enable the development of nanoparticles that enhance bioavailability, target specific tissues, and minimize adverse effects, all of which make PNPs a promising candidate for future development in biopharmaceutical.

## 5.3 Methods of Fabrication and Preparation

Polymeric nanoparticles have become an essential tool in drug delivery because of their versatility and accuracy throughout the drug-delivery process. The performance and therapeutic potential of PNPs strongly depend on the preparation technique and subsequent post-surface modifications. Various fabrication methods are tailored to create particles with optimal size, surface properties, and release characteristics, while surface modifications improve the circulation time and targeting ability. The following section describes several fabrication methods into polymeric nanoparticles in Table 5.3, including the advantages of each method, typical applications, and limitations: emulsion-solvent evaporation, nanoprecipitation, spray drying, and ionic gelation.

**Table 5.3** PNP fabrication methods, advantages, applications and limitations

| Method | Advantages | Applications | Limitations | References |
|---|---|---|---|---|
| Emulsion-solvent evaporation | Suitable for hydrophilic and hydrophobic drugs | Oncology, chronic diseases | Organic solvents | Vallorz et al. (2022) |
| Nanoprecipitation | Precise control over particle size; easy scalability | Gene therapy, chemotherapy | Limited to low polymer concentration | Chen et al. (2022) |
| Spray drying | Excellent for large-scale production; high drug loading efficiency | Vaccine formulations | High-energy input required | Strojewski and Krupa (2022) |
| Ionic gelation | Ideal for sensitive biomolecules; simple process | Protein and peptide delivery | Limited drug-loading capacity | Grune et al. (2021) |

## 5.3.1 PNP Synthesis Techniques

### 5.3.1.1 Emulsion-Solvent Evaporation

The most commonly used technique for synthesizing polymeric nanoparticles is the emulsion-solvent evaporation method. This technique involves dissolving a polymer and drug in an organic solvent, which is further emulsified into an aqueous phase using surfactants. After evaporation of organic solvent, the produced nanoparticles are solid. Typically, evaporation is performed under reduced pressure to enhance the efficiency. This gives critical stability to the emulsion, one of which is one the main determining factors for the size and homogeneity of the particles produced (Vallorz et al. 2022). This approach is very useful in the case of hydrophobic drugs, making it the most commonly used in oncology and chronic disease treatment. Lyophilization techniques are also employed to enhance the stability of nanoparticles during long-term storage.

This method provides the benefits of nanoparticles with controlled release, which is useful for sustained drug delivery. Solvent evaporation limits the encapsulation of hydrophilic drugs and may require modification of the emulsion system in order to overcome this issue (Grune et al. 2021).

Figure 5.3 shows polymeric nanoparticles synthesized by the emulsion-solvent technique, where one of the fabrication methods followed step by step.

### 5.3.1.2 Nanoprecipitation

Nanoprecipitation, otherwise known as solvent displacement, rapidly mixes a polymer solution with a nonsolvent which precipitates nanoparticles. This technique has excellent control over particle size and distribution without high-energy input and hence is suitable for scale-up (Chen et al. 2022).

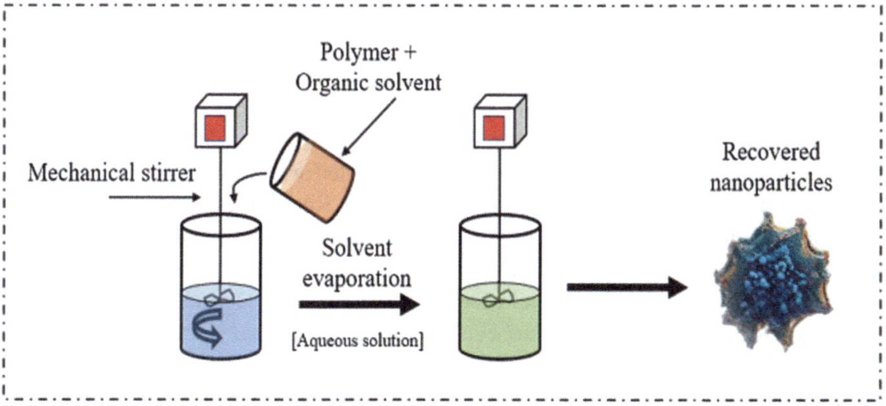


**Fig. 5.3** Emulsion-solvent synthesis of nanoparticles

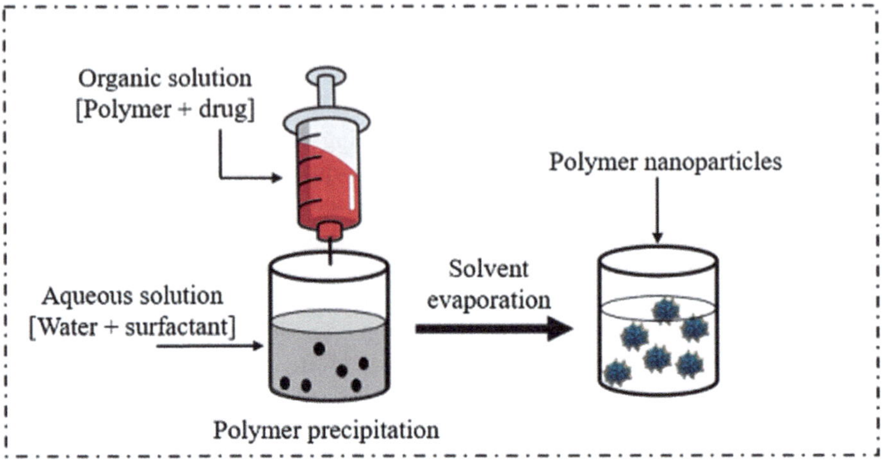


**Fig. 5.4** Nanoparticle synthesis technique

Nanoprecipitation is suitable for the encapsulation of hydrophobic drugs within biodegradable polymers, including poly(lactic-co-glycolic acid) (PLGA). Scalable and reproducible, this method is favoured for the production of nanoparticles useful in gene delivery and chemotherapy. However, the success of nanoprecipitation depends on parameters such as solvent polarity, temperature, and polymer concentration, which have to be carefully optimized in nanoprecipitation.

Figure 5.4 illustrates the nanoprecipitation technique: a rapid mixing of solvent-nonsolvent resulted in rapid precipitation and consequent formation of nanoparticles.

#### 5.3.1.3 Spray Drying

Spray drying converts liquid solutions or suspensions into solid particles. It involves the atomization of a solution in a chamber that is hot enough to evaporate the solvent, leaving the solids as nanoparticles. Spray drying is easily scalable and hence suitable for commercial production (Strojewski and Krupa 2022). A variation of this technique-nano spray drying—can realize the precision required in controlling particle size at the earliest stages of drug development.

Amongst all, spray drying is especially applicable for the preparation of nanoparticles with high drug loading efficiency. In addition, this technique is generally used to encapsulate poorly water-soluble drugs with improved bioavailability. Moreover, during the process, changes in parameters such as airflow, temperature, and solvent concentration can be made to achieve the desired particle morphology and the release profile accordingly (Heidari et al. 2022).

Figure 5.5 illustrates the spray drying process, which is efficient at high drug loading and can be scaled up for vaccine formulations and other large-scale applications.

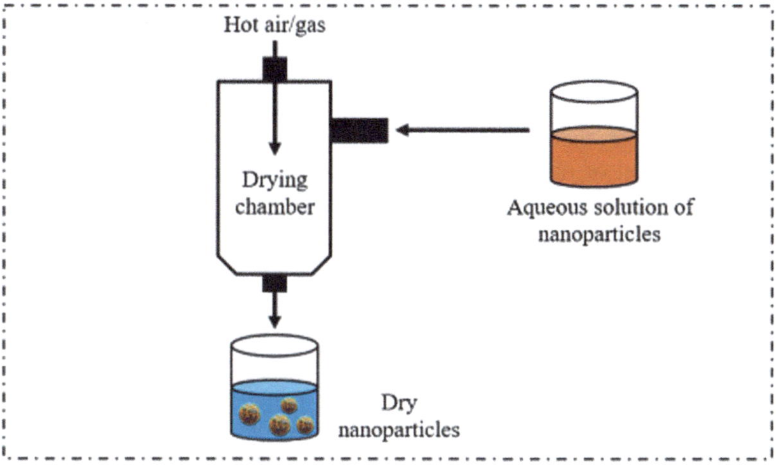


**Fig. 5.5** Spray drying of nanoparticles

#### 5.3.1.4 Ionic Gelation

Ionic gelation relies on electrostatic interactions between two oppositely charged polymers, which result in the formation of nanoparticles. A well-known example is the mixing of chitosan, in its positively charged polymeric form, being mixed with negatively charged sodium tripolyphosphate to yield stable nanoparticles. This technique finds fertile ground among sensitive bioactive encapsulation proteins, peptides, and nucleic acids (Potdar and Jain 2023).

Ionic gelation can be performed under mild conditions, making it suitable for drug delivery systems requiring biocompatibility and minimal toxicity. On the other hand, the process is quite simple and effective, making it cost effective for the development of nanoparticles applied in vaccines and gene therapy applications.

### 5.3.2 *Surface Modification and Functionalization*

Surface modification is critical for nanoparticles to enhance their circulating time, stability, and targeting ability. Functionalization with molecules such as polyethylene glycol (PEG) and targeting ligands can therefore offer site-specific drug delivery and enhance therapeutic efficacy.

#### 5.3.2.1 PEGylation for Enhanced Circulation Time

PEGylation is a process in which PEG chains are attached to the surface of a nanoparticle. PEGylation potentially enhances the stability of nanoparticles through the suppression of protein adsorption and its poor recognition by the immune system. This stealth effect prolongs the circulation time of nanoparticles, especially in applications that require extended drug exposure (Park et al. 2021).

PEGylation of nanoparticles enables the accumulation of drugs at the site of action through the enhanced permeability and retention (EPR) effect and is hence widely used in cancer therapy and for other chronic diseases. In addition to preventing the aggregation of nanoparticles, PEGylation enhances the solubility of NPs making it valuable for intravenously administered NPs.

#### 5.3.2.2 Targeted Drug Delivery Using Ligands (Antibodies, Peptides)

Functionalizing nanoparticles with targeting ligands such as antibodies, peptides, or aptamers enables site-specific drug delivery to specific tissues or cells. This is a widely used approach in cancer therapy, where nanoparticles are functionalized to recognize overexpressed receptors on tumor cells, such as HER2 or EGFR (Leach et al. 2023).

Because of the nature of functionalization, antibody-functionalized nanoparticles can bind to certain receptors in tumor cells and induce receptor-mediated endocytosis in the cells. The encapsulated drug is released into the cell upon endocytosis. Another promising strategy involves peptide-based targeting, which typically exhibits high specificity with minimal immunogenicity. Targeting of drug delivery improves the therapeutic outcome by minimizing off-target effects and reducing systemic toxicity.

The functionalized nanoparticles also have applications in imaging, carrying diagnostic agents for real-time tracking of drug delivery. This capability is important in personalized medicine because the effectiveness of a treatment can be monitored and adjusted based on the response by the patient (Ma et al. 2021).

The methods of fabrication and surface modification of polymeric nanoparticles are important approaches that will tune the optimal drug delivery performance. It includes several techniques such as emulsion-solvent evaporation, nanoprecipitation, spray drying, and ionic gelation, allowing flexibility in the production of nanoparticles with tailored properties. Techniques such as PEGylation and ligand functionalization enhance the circulation time and targeting ability while improving therapeutic efficacy and patient outcomes. These modifications and fabrication techniques, together with continuous research, will ensure that a bright future is on the horizon while developing next-generation drug delivery systems capable of more efficient treatments for diseases.

## 5.4 Mechanisms of Drug Encapsulation and Release

In the design of effective polymeric nanoparticle (PNP) drug delivery systems, the mechanisms of encapsulation and release are crucial. Strategies toward encapsulation determine how well the drug is guarded in the carrier, while controlled mechanisms for release ensure proper delivery at the site and at a therapeutic concentration versus time.

## 5.4.1 Encapsulation Strategies

### 5.4.1.1 Covalent Attachment vs. Physical Encapsulation

Drug encapsulation can be performed via covalent attachment or physical encapsulation. Covalent attachment is achieved through chemical attachment of the drug molecule to a polymer backbone or matrix. This provides stable and, controlled release via cleavage of chemical bonds. This is an effective drug delivery technique for pharmaceuticals that must be delivered with precision dosing and in certain specific environments, such as inside the cells of cancer (Kanwal et al. 2021).

In contrast, physical encapsulation depends on non-covalent interactions of van der Waals, hydrogen bonding, and hydrophobic interactions to trap the drug inside the polymer matrix. This is easier to execute and is pretty useful for hydrophobic drugs that must be protected from aqueous surroundings (Villamizar-Sarmiento et al. 2021). Physical encapsulation therefore offers a higher loading capacity but may have problems of drug leakage during transit.

### 5.4.1.2 Hydrophobic vs. Hydrophilic Drug Loading

Loading hydrophobic drugs should be performed by polymers containing hydrophobic segments able to encapsulate such molecules, protecting them from the aqueous environment. This approach will prevent a premature degradation and allows targeted delivery only in hydrophobic tissues, such as tumors (Biswas et al. 2021). Hydrophilic drugs can be encapsulated within hydrophilic polymer layers or cavities to protect them from rapid dissolution in physiological fluids. It is worth noting that hydrophilic drugs may diffuse out of the carrier prematurely requiring further modification, such as PEGylation, to enhance stability (Marinelli et al. 2022).

## 5.4.2 Controlled Release Mechanisms

Table 5.4 compares different mechanisms of controlled drug release, including diffusion-controlled, pH-response, enzyme-sensitive, and stimuli-sensitive among others, with their corresponding stimulus triggers, representative release profile, and applications. Figure 5.6 shows some of the controlled-release mechanisms of polymeric nanoparticles targeted for specific stimuli responses like pH, temperature, and enzymes active delivery.

### 5.4.2.1 Diffusion-Controlled Release

In diffusion-controlled systems, the drug is released through diffusion through the polymer matrix and/or the nanoparticle nanoparticle the carriers, the process of diffusion is related to several factors, such as the gradient of drug concentration,

**Table 5.4** Comparison of drug release mechanisms

| Mechanism | Stimulus | Release profile | Applications | References |
|---|---|---|---|---|
| Diffusion-controlled | Concentration gradient | Sustained release | Chronic disease management | Wang et al. (2021a, 2021b) |
| pH-responsive | pH variation | Triggered in acidic environments | Cancer therapy | Mondal et al. (2023) |
| Enzyme-sensitive | Enzyme presence | Activated by specific enzymes | Intracellular delivery | Naeem et al. (2014) |
| Stimuli-responsive | Temperature or light | External stimuli trigger release | Remote drug release | Huang et al. (2021) |

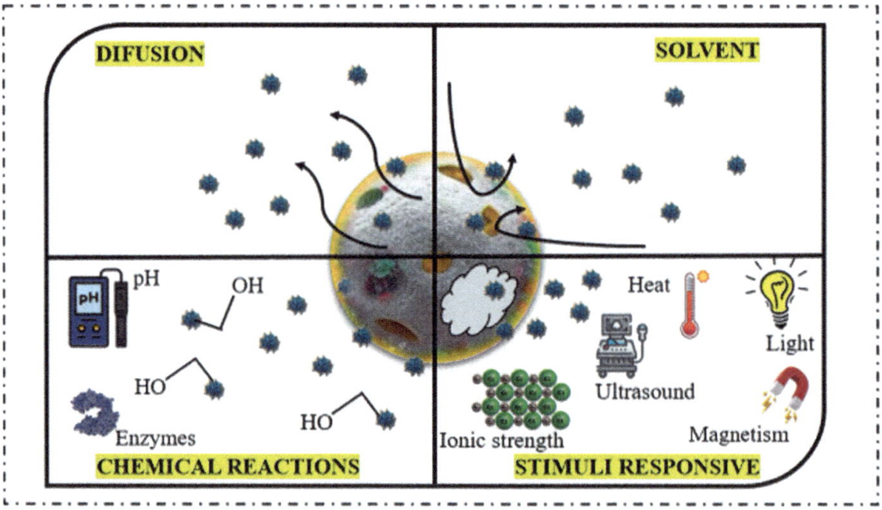


**Fig. 5.6** Controlled release mechanisms

porosity of the polymer, and the rate of matrix degradation (Wang et al. 2021a, b). For example, ferulic acid-loaded nanoparticles rely on quasi-Fickian diffusion to release drugs, ensuring sustained delivery over time (García et al. 2023). Diffusion-controlled release systems can provide consistency in drug levels over long and very extended periods.

#### 5.4.2.2 pH-Responsive and Enzyme-Responsive Nanoparticles

pH-responsive systems take advantage of the pH difference between healthy and diseased tissues. Such nanoparticles are stable at physiological pH 7.4 but degrade into acidic environments, such as tumor tissues or intracellular lysosomes, thereby initiating drug release (Mondal et al. 2023), In contrast, enzyme enzyme-sensitive nanoparticles release drugs in response to enzymes overexpressed in certain diseases. Certain polymersomes, for example, contain enzyme-sensitive linkages and selectively release their payload in specific cancerous environments where certain enzymes are more abundant Naeem et al. (2014).

#### 5.4.2.3 Stimuli-Responsive Polymers

Stimuli-responsive polymers release drugs upon the application of any form of external stimulus, such as temperature or light. Elevated temperature induces changes in the structure of temperature-responsive systems, facilitating the release of drugs at targeted diseased sites, such as inflamed or cancerous tissues (Yang et al. 2021). Likewise, photo-sensitive systems take advantage of UV or infrared light to change the structure of the polymer and the release of the encapsulated drug. For example, molecular motor-doped nanoparticles can release their payload upon exposure to UV light, making them suitable for remote-controlled therapies (Huang et al. 2021).

Stimuli-responsive polymers offer precise control over drug release, enhancing therapeutic outcomes by minimizing off-target effects. In fact, Ferrero et al. (2022) reported that a new dual-responsive systems that combine multiple stimuli, such as pH and redox sensitivity, have shown promising applications in more effective and finely controlled drug delivery.

Figure 5.6 schematically shows the different mechanisms of controlled drug release in polymeric nanoparticles. Release can occur either by a diffusional process, where the drug slowly diffuses out of the nanoparticle, or because of some solvent interaction that dissolves the drug out. Other chemical reactions include change in pH or enzyme activity changes that trigger release mechanisms through degradation of the structural features of the nanoparticles. With external stimuli, the triggers are heat, light, ultrasound, magnetism, or ionic strength for stimuli-responsive release. It will enable to precise control of location of drug delivery.

The mechanisms for the encapsulation and release of bioactive agents are the basis on which polymeric nanoparticles can be developed into potent drug delivery systems. Covalent attachment and physical encapsulation would offer different advantages in stabilizing drugs. Loading strategies for hydrophilic and hydrophobic drugs could be designed for tailored delivery. This also would involve controlled release mechanisms such as diffusion, pH-responsive, enzyme-sensitive, and stimuli-responsive systems to result in the correct release of a drug at the correct site to enhance its therapeutic efficacy with reduced side effects. Continued innovation in the field of stimulus-responsive technologies will improve the capability of polymeric nanoparticles in precision medicine.

## 5.5 Applications in Biopharmaceutical Development

Over the years, Polymeric nanoparticles (PNPs) have also become versatile carriers in biopharmaceutical settings and have enabled a wide array of therapeutic agents such as small molecules, proteins, nucleic acids, and vaccines to reach their target. This section provides an overview of recent the major applications of nanoparticles in drug delivery, vaccine development, and gene therapy.

### *5.5.1 Enhanced Delivery in Biopharmaceuticals Through Polymeric Nanoparticles (PNPs)*

Polymeric nanoparticles (PNPs) represent a new drug delivery system for the administration of biologics, including peptides, proteins, and nucleic acids. These macromolecules are often plagued by a lack of stability, rapid clearance, and poor bioavailability, but PNPs offer solutions by enhancing the delivery efficiency, reducing immunogenicity, and promoting controlled release.

#### 5.5.1.1 Case Studies on Improved Biologic Delivery

Protein and Peptide Delivery

Ma et al. (2021) demonstrated that PEGylated lipid-polymer hybrid nanoparticles enabled the delivery of hydroxycamptothecin, a biological drug, via improved solubility and stability in the bloodstream. The encapsulated drug exhibited 2.5-fold better bioavailability than free hydroxycamptothecin, demonstrating the potential of PNPs in stabilizing susceptible biologics.

Similarly, insulin was encapsulated in PLGA nanoparticles coated with chitosan, which protected it against enzymatic degradation and prolonged its release. In vivo testing showed that nanoparticles loaded with insulin had hypoglycemic effect for up to 24 h, which was significantly longer in comparison to traditional insulin injections (Maher et al. 2023).

Nucleic Acid Delivery

One of the most promising biopharmaceutical developments is the use of PNPs for RNA-based treatment. Lysebetten et al. (2021) reported that lipid-polyglutamate nanoparticle carriers enhanced intracellular delivery of mRNA vaccines by promoting endosomal escape and protecting the mRNA from degradation. The technology was instrumental in the development of the Pfizer-BioNTech and Moderna COVID-19 vaccines, which demonstrated high efficacy due to optimized nanoparticle formulations.

Mechanisms for Biologic Delivery

PNPs enhance biological delivery by several mechanisms:

**Enhanced Stability**
Protects peptides and proteins from enzymatic degradation in biological fluids.
**Improved Targeting**
PNPs that are functionalized with ligands such as antibodies or peptides selectively bind to the receptors of target cells and improve uptake.
**Controlled Release**
PNPs prolong drug release through diffusion or polymer degradation, reducing the need for repeated dosing.
**Reduced Immunogenicity**
Surface modifications, such as PEGylation, help PNPs avoid immune detection, increasing circulation time.

## *5.5.2 Polymeric Nanoparticles for Drug Delivery*

### 5.5.2.1 Delivery of Small Molecules, Peptides, and Proteins

Poor aqueous solubility, low bioavailability, and rapid body clearance are among the problems that PNP overcomes to facilitate the superior delivery of small molecules, peptides, and proteins. The encapsulation of these agents within PNP generally enhances their pharmacokinetic profile, leading increased efficacy and increased compliance among patients. For instance, encapsulated small molecules are protected from enzymatic degradation, and PNPs definitely facilitate targeted delivery toward specific tissues (Steen et al. 2021). PNP can be further functionalized with some ligands, which may lead to improved cellular uptakes to enhanced higher drug delivery efficiency.

### 5.5.2.2 Improving Drug Solubility and Bioavailability

Most therapeutic agents are poorly water-soluble and their clinical applications are highly limited. PNP help in the solubilization of these therapeutic agents by encapsulation in a hydrophobic core, thus making it possible to deliver them in an aqueous environment. This results not only enhanced solubility but also sustained drug release at the target site. For instance, Tenchov et al. (2021) showed that hybrids of polymer-lipid nanoparticles can improve the solubility and delivery efficiency of hydrophobic drugs like paclitaxel.

### 5.5.3 Role in Vaccine Delivery

#### 5.5.3.1 Antigen Encapsulation for Sustained Immune Response

PNP has more scope in vaccine development by encapsulating antigens, maintaining its stability, and providing controlled release. The controlled release stimulates immunity for a longer period without repeated administration of booster doses. Polymer-lipid hybrid nanoparticles have efficiently delivered protein-and mRNA-based vaccines (Lysebetten et al. 2021). These PNPs ensure that antigens remain stable during storage and are delivered effectively to antigen-presenting cells. PNPs protect vaccine antigens from premature degradation so that they successfully reach antigen-presenting cells (APCs). It has been shown that PLGA-based nanoparticles encapsulating viral proteins lead to prolonged antigen presentation, which induces a stronger and more lasting immune response compared to traditional vaccines.

For example, Ma et al. (2021) demonstrated that polymeric nanoparticles encapsulating SARS-CoV-2 spike protein fragments induced higher titers of neutralizing antibodies compared to free antigen administration. The sustained release of antigens over the course of weeks obviated the need for multiple booster dosing.

#### 5.5.3.2 Adjuvant Properties of PNPs

Apart from antigen delivery, PNPs can also be used as adjuvants to enhance immune responses against the encapsulated antigen. For example, lipid-polymer hybrid nanoparticles have been used to co-deliver antigens and immune-stimulating molecules, boosting the production of antigen-specific T cells (Grego et al. 2021). Such multifunctionality simplifies vaccine formulation and improves immunogenicity. This dual role makes PNPs an effective platform for vaccines in the future.

#### 5.5.3.3 PNP-Based Vaccine Delivery Mechanisms

APC Endocytosis

PNPs facilitate dendritic cell uptake, enhancing antigen processing and presentation by major histocompatibility complex (MHC) molecules for robust immune stimulation.

Slow Antigen Release

PNPs protect encapsulated antigens from enzymatic degradation, facilitating prolonged antigen release over weeks, which results in persistent immune stimulation and enhanced memory response.

Enhanced Cellular Immunity

Co-delivery of antigen and immune-stimulatory molecules such as Toll-like receptor (TLR) agonists promote dendritic cell maturation, leading to increased T-cell proliferation and stronger cytotoxic responses.

Mucosal Immunity Enhancement

Certain PNP preparations, particularly those designed for nasal or oral administration, facilitate mucosal immune responses via augmented uptake through Peyer's patches and M cells.

## *5.5.4 PNPs in Gene Therapy*

### 5.5.4.1 Delivery of DNA, RNA, and CRISPR Components

Gene therapy relies on the effective delivery of nucleic acids, including DNA, RNA, and CRISPR components, into cells. PNPs protect these fragile biomolecules from enzymatic degradation and deliver them into target cells. Stimuli-responsive nanoparticles deliver CRISPR-Cas9 complexes and mRNA vaccines effectively with a high degree of gene editing accuracy (Wang et al. 2021a, b).

### 5.5.4.2 Overcoming Barriers to Nucleic Acid Delivery

In gene therapy, nucleic acid delivery into cells faces numerous obstacles and must overrcome barriers like endosomal entrapment. The PNPs designed with pH-sensitive polymers support endosomal escape, thus ensuring that the genetic cargo reaches the cytoplasm for effective therapy (Álvarez-Benedicto et al. 2021). Similarly, nanoparticles modified with coiled-coil peptides enhance membrane fusion, bypassing the endocytosis pathway and improving cytosolic delivery (Zeng et al. 2023).

PNPs have immense potential for enhancing drug solubility, targeting its delivery, improving vaccine efficacy, and enabling gene therapy from various aspects of biopharmaceutical development. The versatility and characteristics of biological barrier crossing make them a highly valued tool in modern medicine. With continued advancement, the role of PNPs is likely to further increase in personalized medicine and next-generation therapies. PNP facilitate gene therapy by:

Shielding Nucleic Acids

Protects against premature degradation before delivery to target cells via ncapsulation in polymeric matrices.

Enhancing Endosomal Escape

The majority of nucleic acids are trapped in lysosomes following cellular uptake; PNPs incorporating pH-responsive polymers facilitate their escape into the cytosol for effective gene translation or silencing.

Targeted Delivery

Nanoparticles that are functionalized with targeting ligands (aptamers, antibodies, etc.) enhance the specificity of delivery to targeted tissues or cell types, resulting in better therapeutic effects and fewer off-target effects.

#### 5.5.4.3 Case Study: CRISPR-Cas9 Delivery Using PNPs

Wang et al. (2021a, b) demonstrated that CRISPR-Cas9 complexes can be efficiently delivered to cancer cells by lipid-polymer nanoparticles, with precise gene editing and minimal off-target effects. The authors demonstrated 70% improvement in gene correction efficiency compared with conventional lipid-based transfection methods. In addition, the use of biodegradable PLGA nanoparticles for the delivery of CRISPR alleviated the cytotoxicity and improved intracellular uptake.

#### 5.5.4.4 Mechanisms of Gene Therapy Enhancement Using PNPs

pH-Sensitive Release

PNP degrade under acidic conditions, i.e., in endosomes, to facilitate the release of nucleic acids into the cytoplasm for efficient translation or genome editing.

Receptor-Mediated Targeting

Nanoparticles functionalized with ligands bind to certain cell surface receptors, enabling specific delivery of genetic material to cells relevant to the disease, for example, hepatocytes for liver disease or neurons for neurodegenerative disorders.

Stimuli-Responsive Activation

PNPs engineered with light, temperature, or ultrasound-responsive components facilitate controlled release of genes at specific anatomical sites, enhancing therapeutic specificity.

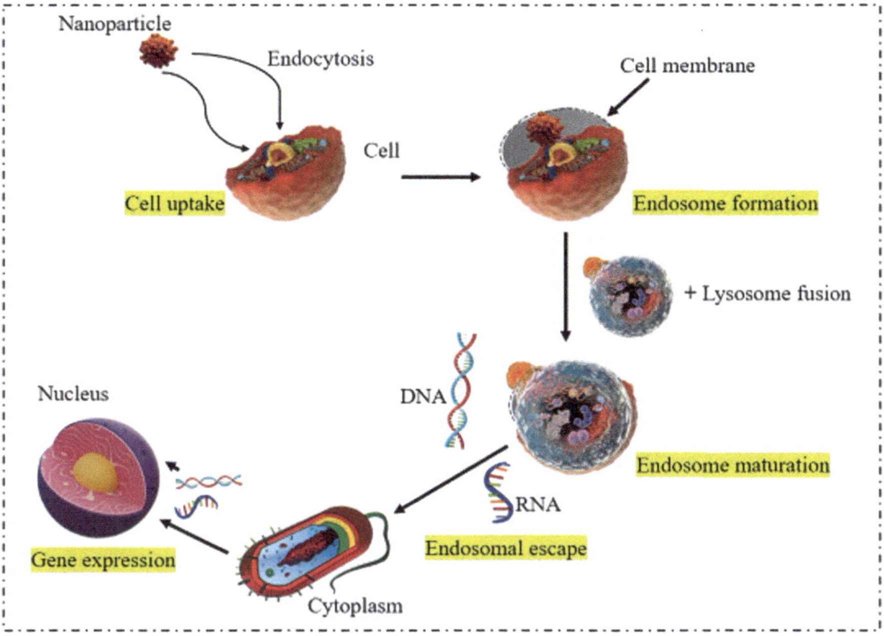


**Fig. 5.7** Mechanisms of endosomal escape and gene delivery

Immune Evasion Strategies

Surface modification like PEGylation or biomimetic coatings derived from cell membranes (e.g., red blood cell membrane-coated PNPs) allows nanoparticles to escape immune clearance, thereby increasing their circulation half-life and efficacy.

Figure 5.7 shows some mechanisms of endosomal escape used by polymeric nanoparticles to ensure efficient delivery of the genetic material, which is a crucial theme in gene therapy applications.

## 5.6 Pharmacokinetics and Pharmacodynamics of Polymeric Nanoparticles

Pharmacokinetics (PK) and pharmacodynamics (PD) are important for defining the efficacy and safety profiles of the PNPs by drug delivery. PK investigations generally focus on the absorption, distribution, metabolism, and elimination (ADME) of drugs while PD covers the biochemical and physiological profilings of drug action. The unique opportunities to enhance drug delivery with PNPs have been demonstrated, from biodistribution modulation; poor solubility of drugs; extended circulation time; to targeted therapy.

## 5.6.1 *Drug Absorption and Distribution*

### 5.6.1.1 Role of the Enhanced Permeability and Retention (EPR) Effect

The EPR effect effectively contributes to the passive targeting of PNPs to tumors and inflamed tissues. Many times, vasculature in tumors and other diseased tissues tends to be leaky, while lymphatic drainage is impaired. Such factors allow nanoparticles to passively accumulate selectively in these tissues, thus enhancing therapeutic efficacy by minimizing systemic toxicity (Chong et al. 2021). In particular, the EPR effect is highly advantageous for polymeric drug carriers, like PEG-coated nanoparticles, which evade immediate immune system clearance and, with some time, end up getting accumulated in target tissues.

### 5.6.1.2 Impact on Biodistribution and Plasma Half-Life

PNP alters the biodistribution of drugs, targeting specific tissues that must achieve drug accumulation, such as tumors or sites of inflammation. Nanoparticles modified with PEG provide evidence of extended plasma half-life due to their ability to avoid reticuloendothelial system (RES) clearance. This extended circulation time therefore leads to increased drug accumulation at target sites but reduces off-target effects (Xenaki et al. 2021). Moreover, biodegradable polymeric NPs enhance the bioavailability of hydrophobic drugs in such a way that they serve well for many diseases requiring long-duration controlled drug release (Ernst et al. 2023).

## 5.6.2 *Clearance and Elimination Pathways*

### 5.6.2.1 Renal vs. Hepatic Clearance

The main factors regulating the clearance of PNPs include their size, surface chemistry, and composition. Particles less than 5 nm in diameter are primarily eliminated through renal excretion, as they easily diffuse through the glomerular filtration barrier and are thus excreted into the urine (Laurent et al. 2021). Larger nanoparticles are more likely to be sequestered in the liver and spleen and eliminated via the hepatobiliary pathway. Consequently, PEGylated nanoparticles showed slower clearance, which improved drug retention and therapeutic efficiency (Xu et al. 2023).

Hepatic removal mechanisms involve the uptake of liver sinusoidal endothelial cells and Kupffer cells, followed by excretion into the bile. PEGylation or albumin coatings are other surface modifications that can delay hepatic excretion, a premise to increase the systemic circulation of nanoparticles (Mills et al. 2022).

#### 5.6.2.2 Role of the Reticuloendothelial System (RES)

The RES consists of macrophages in the liver, spleen, and lymph nodes and is considered the most important barrier to the clearance of nanoparticles. Nanoparticles that are not sufficiently stealthy are rapidly taken up by macrophages and removed from circulation, which limits the therapeutic window. PEGylation methods or albumin-binding approaches were developed for evading RES uptake and allowing longer plasma half-life (Takagi et al. 2021). Apart from this, the transient suppression of RES by small-molecule drugs like chloroquine can also alter the retention and biodistribution of nanoparticles (Glass et al. 2022).

The pharmacokinetics and pharmacodynamics of polymeric nanoparticles hold many advantages of drug delivery, such as improved absorption, distribution, and retention within tissues. The EPR effect facilitates selective accumulation in diseased tissues, while surface modifications such as PEGylation extend plasma half-life. Pathways through which nanoparticles would be cleared includes renal and hepatic clearing and are influenced by nanoparticle size and surface chemistry. Besides, bypassing the reticuloendothelial system is another important strategy that contributes to enhancing the efficacy of nanoparticles. As the field advances, optimization of PK/PD profiles will further unlock the full potential of polymeric nanoparticles in personalized medicine.

Figure 5.8 shows a diagram of the pharmacokinetic pathways of polymeric nanoparticles, routes of absorption, distribution, and clearance in the body.

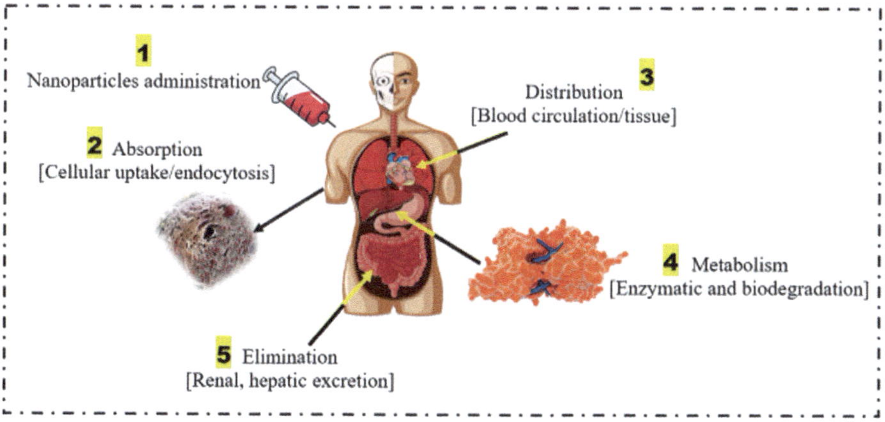


**Fig. 5.8** Pharmacokinetic pathways of PNPs

## 5.7 Challenges in the Development and Commercialization of Polymeric Nanoparticles

The manufacturing, stability, and safety of polymeric nanoparticles pose unique challenges that burden development and commercialization. These challenges impact scalability, storage, and eventual regulatory approval that affect translation into market-ready therapeutics. This section discusses some important challenges and arising strategies to address them.

### *5.7.1 Manufacturing and Scalability Issues*

#### 5.7.1.1 Batch-to-Batch Reproducibility

Essentially, among the major challenges of PNP commercialization, batch-to-batch consistency is one of the most significant. Variability in manufacturing processes might alter particle size, drug encapsulation efficiency, and release profile, affecting the safety and efficacy of the final product. Microfluidic devices, which can easily ensure batch-to-batch reproducibility, enable accurate control of particle size and surface properties (Truong et al. 2014). However, making microfluidic processes scale up in operation is still a problem, with innovation at least in the techniques of parallelization for industrial demands to be met (Shepherd et al. 2021).

#### 5.7.1.2 Regulatory Constraints and Quality Control

The complex nature of therapies based on PNP dictates that they must sail through tough regulatory requirements. In that respect, regulatory agencies demand physicochemical characterizations covering broad stability studies, toxicity profiles, and reproducibility assessments. However, because of the lack of uniform guidelines with respect to nanoparticle-based drugs, regulatory compliance is challenging (Ramos et al. 2022). The use of Quality by Design (QbD) makes it possible for developers to anticipate risks and manufacture robustly concerning consistent product quality (Sipos et al. 2021).

### *5.7.2 Stability and Storage Considerations*

#### 5.7.2.1 Aggregation and Degradation Issues

Poor storage conditions cause aggregation and degradation of PNPs that might alter the particle size, profile of drug release, and bioactivity. Surface chemistry of the nanoparticles, very important for their stability. The wrong formulation might result in particle aggregation or leakage of drugs from the particles (Malhotra et al. 2022). For maintaining colloid stability, a wide range of stabilizers and surfactants is used. However, such type of additives may introduce new problems connected, for example, with phase separation during long-term storage.

#### 5.7.2.2 Lyophilization Strategies

Freeze-drying removes water through lyophilization and has become one of the most prevalent strategies leading to improvements in stability and increasing the shelf life of PNPs. However, it is important not to apply inappropriate conditions of lyophilization, as this can lead to aggregation or loss of activities of the particles. Cryoprotectants such as sucrose are added in order not to collapse the nanoparticles during freeze-drying (Trenkenschuh et al. 2021). Advancement in controlled ice nucleation and annealing processes confer further protection against structural failure in lyophilized products, therefore assuring their efficacy upon reconstitution.

### *5.7.3 Safety and Toxicity Concerns*

#### 5.7.3.1 Immunogenicity of Nanoparticles

Concerns regarding the immunotoxicity of PNP exposure are increasing. This is one of the major concerns regarding PNP exposure: that PNP has the potential to induce immune responses. Although it has advantages in vaccine delivery, the unexpected induction of immune responses may trigger several adverse effects. To minimize these immunogenic properties, nanoparticles are generally coated with PEG or another biocompatible material (Nam et al. 2021). However, repeated treatment with PEGylated nanoparticles sometimes elicits the production of anti-PEG antibodies, compromising their therapeutic efficacy.

#### 5.7.3.2 Preclinical Toxicity Studies

The identification of risks from PNP led to such conclusions, because preclinical toxicity studies are prerequisite. This will evaluate the biodistribution, metabolism, and elimination of nanoparticles to predict their safety profiles in humans. Polymeric nanoparticles exhibit reduced systemic toxicity compared with free drugs. However, several issues related to accumulation in the liver or spleen remain important concerns (Colby et al. 2021). Other innovations currently being explored in the design of nanoparticles, including stimulus-responsive systems and targeted delivery, are envisioned to minimize off-target effects and enhance safety.

Table 5.5 summarizes some of the challenges in the commercialization of polymeric nanoparticles, along with some proposed solutions, including microfluidic

**Table 5.5** Challenges in PNP commercialization and solutions

| Challenge | Impact | Mitigation strategy | Solutions | References |
|---|---|---|---|---|
| Batch-to-batch variability | Inconsistent product performance | Adopt microfluidic processes | Parallelization of microfluidics | Truong et al. (2014) |
| Regulatory compliance | Delays in approval | Implement Quality by Design (QbD) | Regulatory harmonization | Ramos et al. (2022) |
| Aggregation during storage | Reduced therapeutic efficacy | Use stabilizers and surfactants | Lyophilization techniques | Malhotra et al. (2022) |
| Immunogenicity concerns | Adverse immune responses | PEGylation of nanoparticles | Preclinical immunogenicity testing | Nam et al. (2021) |

processes to provide consistency between batches and lyophilization to improve stability.

Polymeric nanoparticles face numerous challenges in their development and commercialization related to manufacturing variability, regulatory barriers, stability problems, and safety issues. Strategies include microfluidic manufacturing, Quality by Design frameworks, advanced lyophilization techniques, and detailed preclinical toxicity assessments. Overcoming these challenges is expected to successfully unlock the comprehensive functionalities of PNP for drug delivery translation, which will be very crucial from the laboratory to the clinic.

## 5.8 Regulatory Landscape for Polymeric Nanoparticles

The opportunities and challenges go hand in hand within a regulatory landscape for polymeric nanoparticles. So far, regulatory agencies like the Food and Drug Administration (FDA) and European Medicines Agency (EMA) have started to issue guidelines with regard to nanotechnology-based pharmaceuticals. The route to approval of these advanced drug delivery systems faces complex issues associated with manufacturing scalability, quality control, and product safety.

### *5.8.1 Regulatory Requirements (FDA, EMA)*

The FDA and EMA have introduced specific frameworks to emphasize the safety, efficacy, and quality control of nanomedicines. The FDA has shown keen interests in studies related to biocompatibility and physicochemical characterization. In addition, product developers must establish batch consistency and prove their hypothesis that a nanoparticle does not generate unexpected biological interactions.

**Table 5.6** Regulatory guidelines for nanotechnology-based pharmaceuticals

| Regulatory body | Requirements | Focus | Documentation | References |
|---|---|---|---|---|
| FDA | Comprehensive physicochemical characterization and toxicity testing | Batch-to-batch reproducibility, product safety | IND and NDA filings | Singh et al. (2021) |
| EMA | Risk-based assessment, focusing on early risk identification | Quality by Design (QbD), early-stage risk assessment | Marketing Authorization Application (MAA) | Csóka et al. (2021) |

Contrarily, the EMA supports a risk-based approach where more profound risk assessments are expected in the early stages of development. The adoption of Quality by Design (QbD) principles applied in this respect ensure greater predictability in manufacturing outcomes and regulatory compliance (Csóka et al. 2021). Both agencies emphasize the importance of conducting preclinical and clinical studies under strict regulatory protocols to ensure product safety and efficacy.

Table 5.6 above illustrated the regulatory guidelines by FDA and EMA, putting in a nutshell the requirements for the approval of drugs based on PNP, the focus areas of the regulation, and documentation.

### *5.8.2 Current Guidelines for Nanotechnology-Based Pharmaceuticals*

Current regulatory guidelines require pharmaceutical developers to provide comprehensive documentation on physicochemical properties, toxicity, pharmacokinetics, and pharmacodynamics of PNPs. In vivo toxicity studies are compulsory by the agencies for the establishment of safe dosing ranges. The regulators also mandate the submission of dossiers concerning the methods involved in the manufacturing, stability testing, and risk assessment of nanoparticle-based pharmaceutical drugs (Sipos et al. 2021).

Where the FDA-approved nanotechnology products are concerned, the most critical issues revolve around scalability with quality control in manufacturing. Stability testing protocols ensure that nanoparticles maintain their properties under variation in environmental conditions (Singh et al. 2021).

### *5.8.3 Case Studies of Approved PNP-Based Products*

Various products based on PNP have already gained approval by either FDA or EMA to show the potential of such systems. The most striking products are PLGA-based formulations, as can be noted with products like Lupron Depot having achieved regulatory approval in the treatment of cancer. Lipid-polymer hybrid

**Table 5.7** Case studies of approved PNP-based products

| Product name | Company | Polymer used | Indication | Approval year | References |
|---|---|---|---|---|---|
| Lupron Depot® | AbbVie | PLGA | Prostate cancer | 1985 | Namiot et al. (2023) |
| Pfizer-BioNTech COVID-19 Vaccine | Pfizer | Lipid-polymer hybrid | COVID-19 | 2020 | Ramos et al. (2022) |
| Moderna COVID-19 vaccines | Moderna | Lipid-polymer hybrid | COVID-19 | 2020 | Tenchov et al. (2021) |

nanoparticles played a key role in mRNA COVID-19 vaccines developed by Pfizer-BioNTech and Moderna (Namiot et al. 2023).

These few examples underpin the wide range of applications going through oncology, vaccination, and gene therapy, among others. However, because of the strict requirements for product characterization and validation, more complex nanoparticle systems-like multi-drug formulation-are receiving very limited regulatory approval (Operti et al. 2021).

Table 5.7 showed some of FDA-or EMA-approved PNP-based products along with associated companies, types of polymers used, involved therapeutic indications, and year of approval.

## 5.9 Future Perspectives and Emerging Trends

In this regard, the development of polymeric nanoparticles is rapid, releasing a number of key trends, such as the development of multi-stimuli responsive systems, integration of artificial intelligence into drug design, and application of biodegradable polymers for personalized medicine.

### *5.9.1 Advancements in Smart Polymeric Nanoparticles (Multi-stimuli Responsive)*

Smart polymeric nanoparticles are presenting dynamic drug delivery mechanisms, multiple response being produced by pH, temperature, and redox conditions. Drug release can thus be well controlled for the least off-target effect in such systems. Multi-stimuli-responsive PNPs often provide great benefit in cancer therapy since tumors commonly exhibit a special microenvironment compared with the normal tissues (Yu et al. 2023). Such nanoparticles can sense both pH and temperature changes simultaneously and release their drug payload at conditions that will correspond to the target tissue, therefore minimizing systemic toxicity and increasing therapeutic effectiveness (Schoppa et al. 2021).

### 5.9.2 Integration with Artificial Intelligence for Drug Design

The new era of Artificial Intelligence-based technologies has changed the field of drug development by enabling predictive modeling of drug interactions with polymeric carriers. Machine learning algorithms allow one to optimize nanoparticle formulations capable of predicating the best combinations of drugs and polymers for improved efficacy. AI is accelerating designing multifunctional nanoparticles for a range of applications in therapy by optimizing many factors that may further hasten the process of development (Nikkhah and Thompson 2021).

The integration of AI in molecular modeling brings a profound understanding of the behaviour of smart polymers in biological media, thereby allowing more personalized and very specific treatments.

### 5.9.3 Use of Biodegradable Polymers in Personalized Medicine

Biodegradable polymers such as PLGA constitute the mainstay in personalized medicine due to their excellent control-release features and minimal toxicity. The degradation products emitted by these polymers are nontoxic, and are ideal for long-term therapy treatment. Personalized polymeric formulations made according to individual needs will assure better treatment results, thereby reducing the toxicity of therapy (Sonawane et al. 2023). It has also recently been applied in increasing volumes in gene therapy and drug targeting-now part of the domain of personalized medicine-pursuing precision in treatment.

The regulatory landscape is constantly changing for polymeric nanoparticles. Regulatory understanding by FDA and EMA provides a framework that ensures product quality, safety, and efficacy. Challenges remain, however, with regard to scaling up the production process as well as ensuring regulatory compliance of final products. Future areas that could be interesting to explore are multistimuli-responsive systems, AI-based drug design, and biodegradable polymers that further open a new window in using polymeric nanoparticles for personalized and precision medicine.

## 5.10 Conclusion

Polymeric nanoparticles, representing a game-changing approach in the development of biopharmaceuticals, may address challenges in drug delivery, vaccine carriage, and even gene therapy developments. These nanoparticles enhance the bioavailability of drugs-especially poorly soluble drugs-and allow targeted delivery of small molecules, peptides, proteins, and nucleic acids. With the encapsulation of drugs with control release profiles, therapeutic efficacy is enhanced with minimal systemic toxicity due to the PNPs. In vaccine preparation, PNPs hold an equally important position because they act as a highly effective antigen carrier and adjuvant. This dual role has been highly instrumental in the recent mRNA-based vaccines for COVID-19—a role that overtly shows how nanoparticles are contributing toward the advancement of immunotherapy.

By applying gene therapy, it becomes possible for the PNPs to deliver DNA, RNA, and CRISPR components while overcoming major biological obstacles that demand improving endosomal escape to better their precision and efficacy. Their role in pharmacokinetic and pharmacodynamic enhancement is noteworthy because they facilitate the circulation time through the enhanced permeation and retention effect. In addition, it could further be altered by surface modification like PEGylation to fine-tune their distribution versus elimination, although this remains a challenge in seeking an optimal balance between the two. Factors such as scalability, stability, and regulatory issues usually impede commercialization because it has to be ensured that the product will be safe, batches will be consistent, and compatible with the immune system.

Moving ahead, the future of PNP will lie in modern advancements related to multi-stimuli responsive technologies, artificial intelligence, and degradable polymers for advanced personalized medicine and controlled drug delivery. Results from such research on PNP, responding to environmental variables of pH, temperature, or redox conditions, make treatments more specific and potent for such complicated diseases as cancer. AI tools will be harnessed in simplifying nanoparticle design by making predictions of therapeutic outcomes and optimization of the choice of polymers to speed up drug development. Besides, there are biodegradable polymers that include PLGA, offering controlled release and reduced toxicity. As PNP technology continues to improve, harmonization between different regions' regulatory frameworks and standard protocols for testing will support rapid approvals and commercialization. Beyond traditional drug delivery, their versatility in the encapsulation of such biologics and gene therapies positions them as potentially key tools in next-generation healthcare. With ongoing research effort and collaboration between industry and regulator, PNPs could be capable, in a not-so-distant future, of driving huge innovation into a wide range of medial fields by successfully using the power of improved therapeutic precision and improved patient outcomes.

## References

Alkholief M, Kalam MA, Anwer MK, Alshamsan A (2022) Effect of solvents, stabilizers and the concentration of stabilizers on the physical properties of poly(d,l-lactide-*co*-glycolide) nanoparticles: encapsulation, in vitro release of indomethacin and cytotoxicity against HepG2-Cell. Pharmaceutics 14(4):870. https://doi.org/10.3390/pharmaceutics14040870

Álvarez-Benedicto E, Farbiak L, Márquez Ramírez M, Wang X, Johnson LT, Siegwart DJ (2021) Optimization of phospholipid chemistry for improved lipid nanoparticle (LNP) delivery of messenger RNA (mRNA). Biomater Sci 10(2):549–559. https://doi.org/10.1039/d1bm01454d

Amin MK, Boateng JS (2022) Enhancing stability and mucoadhesive properties of chitosan nanoparticles by surface modification with sodium alginate and polyethylene glycol for potential oral mucosa vaccine delivery. Mar Drugs 20(3):156. https://doi.org/10.3390/md20030156

Ayub A, Wettig S (2022) An overview of nanotechnologies for drug delivery to the brain. Pharmaceutics 14(2):224. https://doi.org/10.3390/pharmaceutics14020224

Barik SR, Mohapatra RK, Mohapatra PK, Mahal A, El-ajaily MM (2022) Recent developments on biopolymeric nanoparticles for drug delivery systems: an overview. Micro Nanosyst 14(2):92–100. https://doi.org/10.2174/1876402913666210405155127

Biswas G, Jena BC, Samanta P, Mandal M, Dhara D (2021) Synthesis, self-assembly, and drug release study of a new dual-responsive biocompatible block copolymer containing phenylalanine derivative. J Macromol Sci A 58(9):792–803. https://doi.org/10.1080/10601325.2021.1947748

Chaghervand M, Torbati M, Shaabanzadeh M, Ahmadi A, Tafvizi F (2021) Hydroxyurea-loaded Fe3O4/SiO2/chitosan-g-mPEG2000 nanoparticles; pH-dependent drug release and evaluation of cell cycle arrest and altering p53 and lincRNA-p21 genes expression. Arch Pharmacol 395:51–63. https://doi.org/10.1007/s00210-021-02168-1

Chatterjee P, Dhibar S (2023) Nanomaterial marvels: pioneering applications and cutting-edge advancements in drug delivery. Nano Med Mater 3(1):220. https://doi.org/10.59400/nmm.v3i2.220

Chen H, Celik AE, Mutschler A, Combes A, Runser A, Klymchenko AS, Lecommandoux S, Serra CA, Reisch A (2022) Assembly of fluorescent polymer nanoparticles using different microfluidic mixers. Langmuir 38(26):7945–7955. https://doi.org/10.1021/acs.langmuir.2c00534

Chong S, Agarwal S, Agarwal S, Aluri K, Arciprete M, Brown C, Charissé K, Cichocki J, Fitzgerald K, Goel V, Gu Y, Guenther D, Habtemariam B, Jadhav V, Janas M, Jayaraman M, Kurz J, Li J, Liou S, Liu J, Liu X, Maclauchlin C, Maier M, Manoharan M, McDougall R, Nair J, Ramsden D, Robbie G, Schmidt K, Smith P, Theile C, Vaishnaw A, Waldron S, Wu J, Xu Y, Zhang X, Zlatev I, Castellanos-Rizaldos E (2021) The nonclinical disposition and pharmacokinetic/pharmacodynamic properties of N-Acetylgalactosamine–conjugated small interfering RNA are highly predictable and build confidence in translation to human. Drug Metab Dispos 50:781–797. https://doi.org/10.1124/dmd.121.000428

Colby AH, Liu R, Doyle RP, Merting AD, Zhang H, Savage N, Chu NQ, Hollister B, McCulloch W, Burdette J, Pearce C, Liu K, Oberlies N, Colson Y, Grinstaff M (2021) Pilot-scale production of expansile nanoparticles: practical methods for clinical scale-up. J Control Release 337:144–154. https://doi.org/10.1016/j.jconrel.2021.07.012

Csóka I, Ismail R, Jójárt-Laczkovich O, Pallagi E (2021) Regulatory considerations, challenges and risk-based approach in nanomedicine development. Curr Med Chem 28(36):7461–7476. https://doi.org/10.2174/0929867328666210406115529

Ernst LM, Mondragón L, Ramis J, Gustà MF, Yudina T, Casals E, Bastús NG, Fernández-Varo G, Casals G, Jiménez W, Puntes V (2023) Exploring the long-term tissue accumulation and excretion of 3 nm cerium oxide nanoparticles after single dose administration. Antioxidants 12(3):765. https://doi.org/10.3390/antiox12030765

Ferrero C, Casas M, Caraballo I (2022) Redox-responsive polymersomes as smart doxorubicin delivery systems. Pharmaceutics 14(8):1724. https://doi.org/10.3390/pharmaceutics14081724

García L, Palma-Florez S, Espinosa V, Rokni F, Lagunas A, Mir M, García-Celma M, Samitier J, Rodríguez-Abreu C, Grijalvo S (2023) Ferulic acid-loaded polymeric nanoparticles prepared from nano-emulsion templates facilitate internalisation across the blood-brain barrier in model membranes. Nanoscale 15(17):7929–7944. https://doi.org/10.1039/d2nr07256d

Gaur M, Maurya S, Akhtar M, Yadav AB (2023) Synthesis and evaluation of BSA-loaded PLGA–chitosan composite nanoparticles for the protein-based drug delivery system. ACS Omega 8(21):18751–18759. https://doi.org/10.1021/acsomega.3c00738

Glass EM, Kulkarni S, Eng C, Feng S, Malaviya A, Radhakrishnan R (2022) Multiphysics pharmacokinetic model for targeted nanoparticles. Front Med Technol 4:934015. https://doi.org/10.3389/fmedt.2022.934015

Grego EA, Siddoway AC, Uz M, Liu L, Christiansen JC, Ross KA, Kelly SM, Mallapragada SK, Wannemuehler MJ, Narasimhan B (2021) Polymeric nanoparticle-based vaccine adjuvants and delivery vehicles. Curr Top Microbiol Immunol 433:29–76. https://doi.org/10.1007/82_2020_226

Grune C, Zens C, Czapka A, Scheuer K, Thamm J, Hoeppener S, Jandt K, Werz O, Neugebauer U, Fischer D (2021) Sustainable preparation of anti-inflammatory atorvastatin PLGA nanoparticles. Int J Pharm 599:120404. https://doi.org/10.1016/j.ijpharm.2021.120404
Heidari M, Golenser J, Greiner A (2022) Meeting the needs of a potent carrier for malaria treatment: encapsulation of Artemisone in poly(lactide-co-glycolide) micro- and nanoparticles. Part Part Syst Charact 39(5). https://doi.org/10.1002/ppsc.202100152
Huang S, Ding X (2022) Precise design strategies of nanotechnologies for controlled drug delivery. J Funct Biomater 13(4):188. https://doi.org/10.3390/jfb13040188
Huang R, Lan R, Shen C, Zhang Z, Wang Z, Bao J, Wang Z, Zhang L, Hu W, Yu Z, Zhu S, Wang L, Yang H (2021) Remotely controlling drug release by light-responsive microcapsules. ACS Appl Mater Interfaces 13(49):56315–56327. https://doi.org/10.1021/acsami.1c16367
Hussen MA, Yadav S (2023) Nanotechnology: reviewing the new concept of drug delivery. Int J Med Phar Sci 13(01):01–08. https://doi.org/10.31782/ijmps.2023.13101
Jahan I (2022) Nanotechnology for drug delivery and cancer therapy. In: Garg R, Garg R, Eddy N (eds) Handbook of research on green synthesis and applications of nanomaterials. IGI Global Scientific Publishing, pp 338–362. https://doi.org/10.4018/978-1-7998-8936-6.ch015
Jhaveri J, Raichura Z, Khan T, Momin M, Omri A (2021) Chitosan nanoparticles-insight into properties, functionalization and applications in drug delivery and theranostics. Molecules 26(2):272. https://doi.org/10.3390/molecules26020272
Jusu SM, Obayemi JD, Salifu A, Nwazojie C, Uzonwanne V, Odusanya O, Soboyejo WO (2021) PLGA-CS-PEG microparticles for controlled drug delivery in the treatment of triple-negative breast cancer cells. Appl Sci 11(15):7112. https://doi.org/10.3390/app11157112
Kanwal S, Naveed M, Arshad A, Arshad A, Firdous F, Faisal A, Yameen B (2021) Reduction-sensitive dextran-paclitaxel polymer-drug conjugate: synthesis, self-assembly, and anticancer efficacy. Bioconjug Chem 32(12):2516–2529. https://doi.org/10.1021/acs.bioconjchem.1c00492
Laurent G, Benbalit C, Chrétien C, Dupuis C, Pellequer Y, Bazzi R, Thakare V, Denat F, Roux S, Béduneau A (2021) Characterization and biodistribution of Au nanoparticles loaded in PLGA nanocarriers using an original encapsulation process. Colloids Surf B: Biointerfaces 205:111875. https://doi.org/10.1016/j.colsurfb.2021.111875
Leach A, Finnegan M, Machado MS, Ferguson L, Steven J, Smyth P, Porter A, Barelle C, Themistou E, Scott C (2023) Functionalization of polymeric nanoparticles with targeting VNAR ligands using vinyl sulfone conjugation. J Mater Chem B 11(19):4181–4190. https://doi.org/10.1039/d2tb01985j
Lysebetten VD, Malfanti A, Deswarte K, Koynov K, Golba B, Ye T, Zhong Z, Kasmi S, Lamoot A, Chen Y, Van Herck S, Lambrecht B, Sanders N, Lienenklaus S, David S, Vicent M, De Koker S, De Geest B (2021) Lipid-polyglutamate nanoparticle vaccine platform. ACS Appl Mater Interfaces 13(5):6011–6022. https://doi.org/10.1021/acsami.0c20607
Ma Z, Liu J, Li X, Xu Y, Liu D, He H, Wang Y, Tang X (2021) Hydroxycamptothecin (HCPT)-loaded PEGylated lipid–polymer hybrid nanoparticles for effective delivery. Drug Deliv Transl Res 12(2):306–324. https://doi.org/10.1007/s13346-021-00939-0
Maher R, Moreno-Borrallo A, Jindal D, Mai BT, Ruiz-Hernández E, Harkin A (2023) Intranasal polymeric and lipid-based nanocarriers for CNS drug delivery. Pharmaceutics 15(3):746. https://doi.org/10.3390/pharmaceutics15030746
Malabanan J, Alcantara K, Jantaratana P, Pan Y, Nalinratana N, Vajragupta O, Rojsitthisak P, Rojsitthisak P (2023) Enhancing physicochemical properties and biocompatibility of hollow porous iron oxide nanoparticles through polymer-based surface modifications. ACS Appl Bio Mater 6(12):5426–5441. https://doi.org/10.1021/acsabm.3c00657
Malhotra K, Fuku R, Kumar B, Hrovat D, Van Houten J, Piunno P, Gunning P, Krull U (2022) Unlocking long-term stability of upconversion nanoparticles with biocompatible phosphonate-based polymer coatings. Nano Lett 22(18):7285–7293. https://doi.org/10.1021/acs.nanolett.2c00437
Marinelli L, Ciulla M, Ritsema JAS, van Nostrum CF, Cacciatore I, Dimmito MP, Palmerio F, Orlando G, Robuffo I, Grande R, Puca V, Di Stefano A (2022) Preparation, characterization,

and biological evaluation of a hydrophilic peptide loaded on PEG-PLGA nanoparticles. Pharmaceutics 14(9):1821. https://doi.org/10.3390/pharmaceutics14091821

Melo M, Pereira FM, Rocha MA, Ribeiro JG, Junges A, Monteiro W, Diz FM, Ligabue R, Morrone FB, Severino P, Fricks A (2021) Chitosan and chitosan/PEG nanoparticles loaded with indole-3-carbinol: characterization, computational study, and potential effect on human bladder cancer cells. Mater Sci Eng C 124:112089. https://doi.org/10.1016/j.msec.2021.112089

Mills J, Humphries J, Simpson J, Sonderegger S, Thurecht K, Fletcher N (2022) Modulating macrophage clearance of nanoparticles: comparison of small-molecule and biologic drugs as pharmacokinetic modifiers of soft nanomaterials. Mol Pharm 19(11):4080–4097. https://doi.org/10.1021/acs.molpharmaceut.2c00528

Mondal A, Das S, Ali SM, Kolay S, Sengupta A, Molla MR (2023) Bioderived lipoic acid-based dynamic covalent nanonetworks of poly(disulfide)s: enhanced encapsulation stability and cancer cell-selective delivery of drugs. Bioconjug Chem 34(3):489–500. https://doi.org/10.1021/acs.bioconjchem.2c00493

Nam J, Son S, Park KS, Moon J (2021) Modularly programmable nanoparticle vaccine based on polyethyleneimine for personalized cancer immunotherapy. Adv Sci 8(5):2002577. https://doi.org/10.1002/advs.202002577

Namiot ED, Sokolov AV, Chubarev VN, Tarasov VV, Schiöth HB (2023) Nanoparticles in clinical trials: analysis of clinical trials, FDA approvals and use for COVID-19 vaccines. Int J Mol Sci 24(1):787. https://doi.org/10.3390/ijms24010787

Naeem M, Kim W, Cao J, Jung Y, & Yoo J-W (2014) Enzyme/pH dual sensitive polymeric nanoparticles for targeted drug delivery to the inflamed colon. Colloids SurfS B: Biointerfaces 123:271–278. https://doi.org/10.1016/j.colsurfb.2014.09.026

Ndebele RT, Yao Q, Shi Y, Zhai Y-Y, Xu H, Zhao Y-Z (2021) Progress in the application of nano- and micro-based drug delivery systems in pulmonary drug delivery. BIO Integr 3(2):71–83. https://doi.org/10.15212/bioi-2021-0028

Nguyen T, Vo TK, Nguyen TN, Nguyen MK, Vo TV, Vo GV (2021) Nanotechnology-based drug delivery for central nervous system disorders. Biomed Pharmacother 143:112117. https://doi.org/10.1016/j.biopha.2021.112117

Nikkhah SJ, Thompson D (2021) Molecular modelling guided modulation of molecular shape and charge for design of smart self-assembled polymeric drug transporters. Pharmaceutics 13(2):141. https://doi.org/10.3390/pharmaceutics13020141

Operti MC, Bernhardt A, Grimm S, Engel A, Figdor CG, Tagit O (2021) PLGA-based nanomedicines manufacturing: technologies overview and challenges. Int J Pharm 605:120807. https://doi.org/10.1016/j.ijpharm.2021.120807

Park H, Kim J-S, Kim S, Ha E-S, Kim M-S, Hwang S (2021) Pharmaceutical applications of supercritical fluid extraction of emulsions for micro−/nanoparticle formation. Pharmaceutics 13(11):1928. https://doi.org/10.3390/pharmaceutics13111928

Patole KP, Danane AA, Nikam A, Patil A (2021) Review on nanotechnology and its utilization in pharmaceuticals. Asian J Res Pharm Sci 11(4):319–312. https://doi.org/10.52711/2231-5659.2021.00050

Potdar MB, Jain NK (2023) Formulation and in vitro cytotoxicity studies of Avastin nanoparticle-mediated targeted nose-to-brain delivery system. EAS J Pharm Pharmacol 5(5):122–131. https://doi.org/10.36349/easjpp.2023.v05i05.001

Ramnandan D, Mokhosi S, Daniels A, Singh M (2021) Chitosan, polyethylene glycol and polyvinyl alcohol modified $MgFe_2O_4$ ferrite magnetic nanoparticles in doxorubicin delivery: a comparative study in vitro. Molecules 26(13):3893. https://doi.org/10.3390/molecules26133893

Ramos T, Villacis-Aguirre CA, López-Aguilar KV, Padilla LS, Altamirano C, Toledo JR, Vispo NS (2022) The hitchhiker's guide to human therapeutic nanoparticle development. Pharmaceutics 14(2):247. https://doi.org/10.3390/pharmaceutics14020247

Ryu S, Park S, Lee HY, Lee H, Cho C, Baek J-S (2021) Biodegradable nanoparticles-loaded PLGA microcapsules for enhanced encapsulation efficiency and controlled release of hydrophilic drugs. Int J Mol Sci 22(6):2792. https://doi.org/10.3390/ijms22062792

Schoppa T, Jung D, Rust T, Mulac D, Kuckling D, Langer K (2021) Light-responsive polymeric nanoparticles for photosensitizer delivery. Int J Pharm 579:120326. https://doi.org/10.1016/j.ijpharm.2021.120326
Sharma DK, Pattnaik G, Behera A (2022) Preparation and in vitro/in vivo characterisation of pioglitazone-loaded chitosan/PEG blended PLGA nanoparticles. J Biomater Sci Polym Ed 33(13):1623–1643. https://doi.org/10.1080/09205063.2022.2068947
Sheir MM, Nasra MMA, Abdallah OY (2021) Chitosan alginate nanoparticles as a platform for the treatment of diabetic and non-diabetic pressure ulcers: formulation and in vitro/in vivo evaluation. Int J Pharm 607:120963. https://doi.org/10.1016/j.ijpharm.2021.120963
Shepherd SJ, Issadore D, Mitchell MJ (2021) Microfluidic formulation of nanoparticles for biomedical applications. Biomaterials 274:120826. https://doi.org/10.1016/j.biomaterials.2021.120826
Singh S, Molugulu N, Devi L, Kesharwani P (2021) Regulatory pathway to introduce a nanomedicine product in the market at international level. In: Theory and applications of nonparenteral nanomedicine. Elsevier, pp 489–499. https://doi.org/10.1016/B978-0-12-820466-5.00020-X
Sipos B, Katona G, Csóka I (2021) A systematic, knowledge space-based proposal on quality by design-driven polymeric micelle development. Pharmaceutics 13(5):702. https://doi.org/10.3390/pharmaceutics13050702
Sonawane SS, Pingale P, Amrutkar S (2023) PLGA: a wow smart biodegradable polymer in drug delivery system. Indian J Pharm Educ Res 57(2):s189–s197. https://doi.org/10.5530/ijper.57.2s.23
Steen AVD, Khalife R, Colant N, Khan HM, Deveikis M, Charalambous S (2021) Bioengineering bacterial encapsulin nanocompartments as targeted drug delivery system. Synth Syst Biotechnol 6(3):231–241. https://doi.org/10.1016/j.synbio.2021.09.001
Strojewski D, Krupa A (2022) Spray drying and nano spray drying as manufacturing methods of drug-loaded polymeric particles. Polym Med 54(2):101–111. https://doi.org/10.17219/pim/152230
Takagi K, Somiya M, Jung J, Iijima M, Kuroda S (2021) Polymerized albumin receptor of hepatitis B virus for evading the reticuloendothelial system. Pharmaceuticals 14(5):408. https://doi.org/10.3390/ph14050408
Tenchov R, Bird R, Curtze AE, Zhou Q (2021) Lipid nanoparticles—from liposomes to mRNA vaccine delivery, a landscape of research diversity and advancement. ACS Nano 15(11):16982–17015. https://doi.org/10.1021/acsnano.1c04996
Truong N P, Whittaker MR, Mak CW, & Davis TP (2014) The importance of nanoparticle shape in cancer drug delivery. Expert Opin Drug Deliv 12(1):129–142. https://doi.org/10.1517/17425247.2014.950564
Trenkenschuh E, Savšek U, Friess W (2021) Formulation, process, and storage strategies for lyophilizates of lipophilic nanoparticulate systems. Int J Pharm 606:120929. https://doi.org/10.1016/j.ijpharm.2021.120929
Vallorz EL, Encinas-Basurto D, Schnellmann RG, Mansour HM (2022) Design, development, physicochemical characterization, and in vitro drug release of formoterol PEGylated PLGA polymeric nanoparticles. Pharmaceutics 14(3):638. https://doi.org/10.3390/pharmaceutics14030638
Villamizar-Sarmiento MG, Guerrero J, Moreno-Villoslada I, Oyarzun-Ampuero F (2021) The key role of the drug self-aggregation ability to obtain optimal nanocarriers. Eur J Pharm Biopharm 166:19–29. https://doi.org/10.1016/j.ejpb.2021.05.023
Wang Y, Shahi P, Wang X, Xie R, Zhao Y, Wu M, Roge S, Pattnaik B, Gong S (2021a) In vivo targeted delivery of nucleic acids and CRISPR genome editors enabled by GSH-responsive silica nanoparticles. J Control Release 336:296–309. https://doi.org/10.1016/j.jconrel.2021.06.030
Wang X, Yang Y, Liu C, Guo H, Chen Z, Xia J, Liao Y, Tang C, Law W (2021b) Photo- and pH-responsive drug delivery nanocomposite based on o-nitrobenzyl functionalized upconversion nanoparticles. Polymer 229:123961. https://doi.org/10.1016/j.polymer.2021.123961
Xenaki KT, Dorresteijn B, Muns J, Adamzek K, Doulkeridou S, Houthoff H, Oliveira S, Van Bergen En Henegouwen P (2021) Homogeneous tumor targeting with a single dose of HER2-targeted

albumin-binding domain-fused nanobody-drug conjugates results in long-lasting tumor remission in mice. Theranostics 11(11):5525–5538. https://doi.org/10.7150/thno.57510

Xu L, Cao Y, Xu Y, Li R, & Xu X (2023) Redox-responsive polymeric nanoparticle for nucleic acid delivery and cancer therapy: Progress, opportunities and challenges. Macromol Biosci 24(3):2300238. https://doi.org/10.1002/mabi.202300238

Yang Z, Zhao H, Wang D, Yin L, Cai K, Lin Z et al (2021) DPD simulations on mixed polymeric DOX-loaded micelles for dual pH/reduction-responsive release. Phys Chem Chem Phys 23(34):19011–19021. https://doi.org/10.1039/d1cp02750f

Yu Z, Shen X, Tu H, Chittasupho C, Zhao Y (2023) Smart polymeric nanoparticles in cancer immunotherapy. Pharmaceutics 15(3):775. https://doi.org/10.3390/pharmaceutics15030775

Zegarra-Urquia C, Santiago J, Bumgardner J, Vega-Baudrit J, Hernández-Escobar C, Zaragoza-Contreras E (2022) Synthesis of nanoparticles of the chitosan-poly((α,β)-DL-aspartic acid) polyelectrolyte complex as hydrophilic drug carriers. Int J Polym Mater Polym Biomater 72(6):497–506. https://doi.org/10.1080/00914037.2022.2029440

Zeng Y, Shen M, Pattipeiluhu R, Zhou X, Zhang Y, Kros A (2023) Efficient mRNA delivery using lipid nanoparticles modified with fusogenic coiled-coil peptides. Nanoscale 15(37):15206–15218. https://doi.org/10.1039/d3nr02175k

# Chapter 6
# Dendrimers and Nanogels: Emerging Trends in Drug Delivery

**Aishat Adejoke Obalola, Heidi Abrahamse, and Sathish Sundar Dhilip Kumar**

**Abstract** Nano-delivery systems address many challenges of traditional drug administration by enhancing aqueous solubility, stability, drug efficiency, and controlled release in chemotherapy. Dendrimers and nanogels offer unique platforms for drug attachment, enabling both passive and active drug administration. Dendrimers are highly branched polymeric structures with a well-defined nanosize, low polydispersity, and a numerous active amine groups on their surface. Their high drug encapsulation capacity and passive targeting ability make them effective drug carriers. On the other hand, nanogels are a flexible drug delivery strategy since they combine the qualities of hydrogels and nanoparticles. When it comes to the distribution of bioactive mediators at specified sites or under temporal control, nanogels have garnered a lot of interest as tiny drug delivery systems. Nanogels have shown tremendous potential for a variety of uses, including gene transfer, medication administration during chemotherapy, organ targeting, diagnostics, and many more. This chapter covers dendrimers and nanogels drug loading capacity, drug release mechanism, and the most significant current usage in various drug delivery domains. Additionally, we address the difficulties, options, and future of dendrimer- and nanogel-based drug delivery technologies.

**Keywords** Dendrimer · Nanogel · Drug delivery system · Nanocarriers · Targeted drug delivery

## Abbreviations

| | |
|---|---|
| DDS | drug delivery system |
| DOX | doxorubicin |
| NG | nanogels |

A. A. Obalola · H. Abrahamse · S. S. Dhilip Kumar (✉)
Laser Research Centre, University of Johannesburg, Johannesburg, South Africa
e-mail: sathishd@uj.ac.za



V. Khanna et al. (eds.), *Nanotechnologies for Drug Delivery and Biopharmaceutical Development*, Progress in Drug Research 77,
https://doi.org/10.1007/978-3-032-00796-4_6

| | |
|---|---|
| NPs | nanoparticles(NPs) |
| PAMAM | polyamidoamine |
| siRNA | small interfering RNAs |

## 6.1 Introduction

Notwithstanding the wealth of research on therapeutic delivery, there are still many unanswered questions about the creation of an appropriate platform, or carrier, for the transport of medications or nucleic acids to certain bodily locations and target cells to cure illnesses (Bildstein et al. 2011; Reischl and Zimmer 2009; Torchilin 2008). Nevertheless, nanotechnology has overcome this obstacle and is predicted to continue producing innovative nanomaterials for treating a wide range of illnesses (Couvreur 2013). Particles with at least one nanoscale dimension are called nanoparticles (NPs) (Chen et al. 2014). NPs can overcome various challenges faced by drugs (Astashkina et al. 2012; Mastrobattista et al. 2006; Tan et al. 2008). To achieve maximal effectiveness while minimizing or eliminating undesirable side effects, the perfect delivery system would convey therapies through the vasculature without drug leakage (Bourzac 2012). The development of dendrimer and nanogel systems was made necessary by nanotechnology, which confirms the materials' potential to deliver medications in a regulated, continuous, and targeted manner. Dendrimers and nanogels (NG) offer a consistent surface for attaching drugs, enabling drug delivery methods that include both passive and active drug delivery to be effectively achieved.

Hydrogel nanoparticles (NPs) with a controllable size range of 1–1000 nm that are created by chemical or physical cross-linking networks are referred to as "nanogels." Drugs and DNA may all be incorporated into nanogel networks, which also ensure that multivalent bioconjugation has a lot of surface area. They could have a chemical structure that can be adjusted to influence mechanical strength, biocompatibility, and water absorption. By altering their volume, water absorption, and refractive index, they may respond to external stimuli more quickly (Bawa et al. 2009; De las Heras Alarcón et al. 2005; Onaca et al. 2009). They are also known as smart materials as they react to various stimuli. To functionalize the nanogel and make it responsive to the intended stimuli, additional molecules may be added to its surface. These stimuli may be used to create nanogels that work well as carriers in tissue engineering, biosensing bioimaging, and medical diagnostics (Hayashi et al. 2009; Oishi et al. 2007; Peng et al. 2010).

Some essential characteristics of a flexible drug administration mechanism include an optimal and adjustable particulate matter for enhanced permeability and holding power; basic non-covalent drug imprisonment; preventing the release of drugs too soon; release has to be brought about by outside factors; absence of inherent harmfulness; and a dependable and repeatable delivery process. Regaining the characteristics of stimulus-responsive nanogels involves removing the stimulus. Additionally, they exhibit unique qualities such as the capacity to pass through physiological obstacles, a hydrophilic internal network for the loading and releasing of drugs, increased steadiness for extended blood movement, and the ability to precisely aim the intended spot. Since they are soft nanocarriers, they may flatten in the vascular plane and simultaneously cling to many locations (Motornov et al. 2010). Polyethyleneimine and poly (ethylene glycol) (PEG) were the first DDSs of polynucleotides that were documented. Moreover, it has been shown that gold nanorods for photodynamic treatment and magnetic nanoparticles may be included for optical and magnetic imaging of live cells. Hydrophilic ionizable carriers, which are also gaining interest as DDS, combine with biomolecules with the opposite charge to produce polyionic complexes (Das et al. 2007; Gupta and Wells 2004; Zhang et al. 2004). NGs are useful nano-transporters for the administration of drugs in vivo for the treatment of various illnesses, including cancer, neurological disorders, bone deterioration, etc., because of their cell-mediated characteristics and intricately controlled architecture (Vinogradov 2007).

Dendrimer molecules are branching structures resembling tree limbs. In order to efficiently transport medicines and nucleic acids to target areas without harming healthy cells, they may attach to these substances and be changed by ligands unique to those targets (Rajani et al. 2020). Conversion of synthons to the central core covalently forms the overall molecular structures, and a variety of medications with distinct operational divisions may be bonded to their exterior as sealing substances (Choudhary et al. 2017; Santos et al. 2019). A wide range of functionalized and ligand-anchored dendrimers, including citric acid dendrimers, polyamidoamine dendrimers, polypropyleneimine peptide dendrimers, poly(L-lysine) dendrimers, and carbohydrate-based dendrimers, are categorized according to their functional moiety in both the central and auxiliary domains (Chis et al. 2020; Mignani et al. 2018) Vaccine distribution, enzyme, antibiotic, gene delivery, and anticancer medication carriers are just a few of the many uses for dendrimers (Chenthamara et al. 2019; Scioli Montoto et al. 2020).

One important class of nanostructured carriers that is being exploited to treat many diseases and enhance nanomedicine is dendrimers. Because of their structural diversity and adaptability, dendrimers have been used in a number of ways to carry DNA and medications (Fig. 6.1). For instance, dendrimers with a hydrophilic outer layer and a hydrophobic core may function as unimolecular micelles. They have been used to solubilise hydrophobic medications by enclosing them inside the intramolecular cavity (Menjoge et al. 2010; Rao et al. 2015). As non-viral gene carriers, cationic dendrimers have been widely used (Parekh 2007). Drugs and other functional moieties may be coupled with dendrimer surface groups (Gupta and Nayak 2015; Kurtoglu et al. 2010). Conjugates of dendrimers and polysaccharides are often used to provide nanomaterials with better compatibility and appealing binding qualities (Agrawal et al. 2007; Pooresmaeil and Namazi 2021). A hyaluronic acid-associated PAMAM dendrimer revealed an improved tumor penetrating potential because hyaluronic acid has a significant affinity for CD44 receptors, which cancer stem cells and tumor cells overexpress, because mannose may bind selectively to mannose reporter expressed on the surface of macrophages (He et al. 2018).

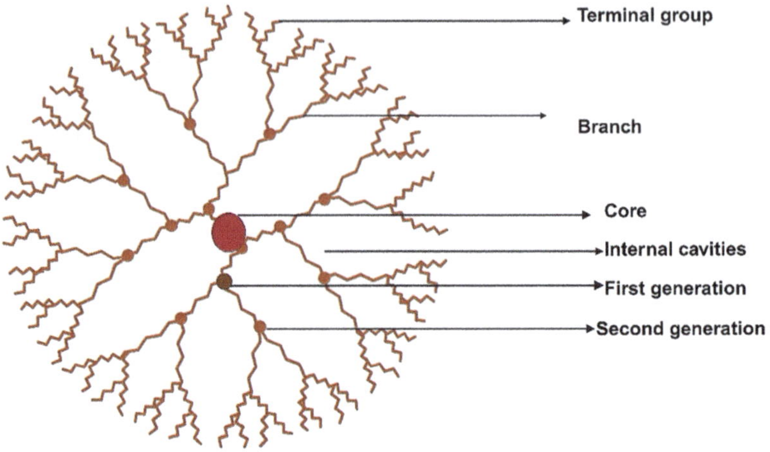


**Fig. 6.1** Structure of dendrimer- created in BioRender. Dhilip kumar, S. (2025) https://BioRender.com/q08r390

## 6.2 Drug Delivery System (DDS)

Finding effective ways to combat diseases and continuously improving treatment is one of the biggest difficulties in nanomedicine. The drugs and active compounds that are now on the market are often useful instruments for treating certain illnesses; nonetheless, difficulties in their delivery significantly limit their effectiveness; as a result, Many efforts have gone into creating what is known as "drug delivery systems," which are gadgets that are capable of moving medications and active ingredients to the locations where their pharmacological effects are felt (Tiwari et al. 2012). Researchers centered their attention on developing drug administration methods to overcome barriers to the clinical use of pharmacological substances, such as toxicity, and adverse effects, before using them (Sherje et al. 2018). Generating carriers with regulated kinetics of release has been the main goal of drug delivery vector research throughout the last 10 years in an effort to enhance medication pharmacokinetics and biodistribution (Li et al. 2018). An ideal therapeutic vector should have tissue-specific targeting, minimal toxicity, optimal release kinetics, biocompatibility, and excellent drug-loading capacity (Zhu and Shi 2013). Numerous drug delivery methods have been developed to date, such as liposomes, polymers, nanoparticles, nanotubes, quantum dots, and nanogels (Ghaffari et al. 2018). In the realm of medication administration, nanotechnologies and nanoparticles are unquestionably significant: appropriately high amounts of drug loading and regulated drug release may be achieved using well-designed nanocarriers, and their small size and Functionality may lessen toxicity and adverse effects by increasing the delivery process's selectivity (Ramasamy et al. 2017).

## 6.3 Dendrimers in Drug Delivery

Exact control over medication distribution is crucial to eliminating the common problems associated with conventional treatment. As a result, a medication delivery system needs to be created. Dendrimer drug delivery has advanced considerably in the last several years in terms of enhanced biodistribution, and controlled release of medication to the intended area (Allen and Cullis 2004). Dendrimers' high water solubility, polyvalency, and molecular weight have sparked a lot of interest in biological applications (Duncan and Izzo 2005; Patton et al. 2006); These characteristics make them the ideal vehicle for targeted applications and the administration of medications (Kitchens et al. 2005; Tomalia 2005).

Various medication classes have been integrated into these adaptable nanoscale carriers. The highly functional lauryl-modified G3 PAMAM dendrimer-paclitaxel conjugates showed a 12-fold increase in permeability across the monolayers of pig brain endothelial cells and Caco-2 cells compared to paclitaxel alone, as well as excellent stability under physiological circumstances. Prior research by Ooya et al. found that paclitaxel's solubility increased 400 times when added to polyglycerol dendrimers (Ooya et al. 2004).

Chemotherapeutics must be delivered with precision to minimize the deleterious consequences of traditional treatment, which may cause healthy tissues to amass toxic drug levels. By altering dendrimers' surfaces with various targeting moieties, the drug might be delivered to particular sites. Several effective active and passive targeting efforts were made by manipulating the dendrimer surface groups and division units. To administer methotrexate specifically, Patri et al. coupled FA to G5 PAMAM dendrimer. They then observed receptor-mediated drug delivery, which exhibited delayed drug release and great selectivity for the cells (Patri et al. 2005). Recently, it was reported that FA-attached G5 PAMAM dendrimers were used to specifically transport 2-methoxy estradiol to cancerous cells, as verified by the MTT experiment (Wang et al. 2011). By selectively targeting MCF-7 and C26 cells, the aptamer-modified dendrimers increase the cytotoxicity of epirubicin while preventing internalization by healthy cells, reducing side effects. Apart from in vitro investigations, an in vivo investigation conducted demonstrated the capability of dendrimers modified with aptamers and loaded with epirubicin to inhibit the formation of tumors (Taghdisi et al. 2016). In vitro and in vivo administration of camptothecin for the treatment of colon cancer was made possible by an experiment using PEGylated polyamidoamine dendrimers. The internalization of camptothecin in colon cancer cells was improved by AS1411-modified polyamidoamine dendrimers, resulting in a 93.67% encapsulation efficiency and a reduction in viability (Alibolandi et al. 2017). Lung cancer cells undergo 14% late apoptosis when their surface is modified with the AS1411 aptamer, which also increases the ability of short hairpin RNA in gene silencing. Notably, AS1411-modified dendrimers loaded with short hairpin RNA indicate their stability and ability to internalize into tumor cells (Ayatollahi et al. 2017). Using folic acid to modify the surface of polyamidoamine dendrimers, cervical and ovarian cancer cells that overexpress the folate receptor are specifically targeted (Luong et al. 2016). The results of dendrimer as a drug delivery system are discussed in Table 6.1.

**Table 6.1** Dendrimer as a drug delivery system

| Dendrimer Type | Drug | Cell line | Application |
|---|---|---|---|
| Polyamidoamine (PAMAM) dendrimer | Curcumin and small interfering RNA | HeLa cells | Owing to the combined impact of co-delivering Cur and Bcl-2 siRNA, the nano bioconjugate demonstrated better suppression of the development of malignant cells and improved cellular absorption (Ghaffari et al. 2020) |
| Polyether-copolyester PEPE dendrimers | Methotrexate (MTX) | U87 MG and U 343 MGa cells. | These MTX-loaded dendrimers demonstrated their capacity to overcome MTX resistance by killing even MTX-resistant cells (Dhanikula et al. 2008). |
| PAMAM dendrimers | Doxorubicine (DOX) | Glioma Cell | With a transporting ratio of 6.06% in 3 h, the in vitro test demonstrated that G4-DOX-PEG-Tf-TAM had a greater capacity for BBB transportation. (Li et al. 2012) |
| Poly(propyleneimine) (PPI) dendritic | Paclitaxel (PTX) | U373MG human astrocytoma cells | Rats' in vivo pharmacokinetics and biodistribution investigations revealed a noticeably greater build-up of PTX in the brain than in free PTX (Patel et al. 2016). |
| Carboxylated PAMAM dendrimer | Cisplatin | Lung cancer cell, NCI-H460 | An in vitro investigation revealed that the drug-nanocarrier combination reduced the cytotoxicity of cisplatin. (Nguyen et al. 2015). |
| Polyamidoamine (PAMAM) dendrimer and carboxylated PAMAM dendrimer | 5-fluorouracil (5-FU). | Breast cancer | The drug-nanocarrier system demonstrated antiproliferative effects on MCF-7 cells. (Tran et al. 2013). |
| Poly(amidoamine) dendrimer | Carboplatin | HeLa, A549, and MCF-7. | The outcome showed that G4-PEG@CAR was able to kill three representative cells. (Ho et al. 2019). |
| Poly(amidoamine) Dendrimers | Paclitaxel | Human cervical epithelial cells | G4-TOS-PEG-Tf-PTX greatly reduced the development of human cervical epithelial cell. (Bhatt et al. 2019). |
| Polyamidoamine (PAMAM) dendrimers | Oxaliplatin | HeLa, A549, MCF-7 cells | The result showed decreased cell viability (Nguyen et al. 2019). |

## 6.4 Dendrimer Drug-Loading Mechanisms

Using hydrogen bonds or hydrophobic linkages, drug molecules can be covalently conjugated to a dendrimer's end clusters or confined to the central (Hughes 2005). The amount of generations influence the drug's ability to load because a larger generation number results in more surface functional groups for drug conjugation and greater space for guest medications. Due to its surface's high concentration of functional groups, dendrimers have been thoroughly investigated as drug transporters and dendrimers. There are three ways that anti-cancer drugs interact with dendrimers: covalent conjugation, electrostatic interaction, and physical encapsulation (Howell and Fan 2010).

Dendrimers' open design, hollow interior voids, and ellipsoidal or spheroidal form enable the simple encapsulation of visitor molecules inside the macromolecule. These interior spaces are vacant because they are hydrophobic and may interact with medications that are poorly soluble using hydrophobic interactions. Additionally, the oxygen atoms may use hydrogen to interact with the medicinal molecules to create bonds. Given these particular characteristics, the connection between dendrimers' internal chambers and such supramolecular interactions may be involved in medicinal compounds, such as hydrophobic contact and physical encapsulation. (Brownlie et al. 2004; Liu et al. 1999).

The surface of dendrimers such as carboxyl and amine groups, possesses potential uses for improving the solubility of hydrophobic medicines by electrostatic contact. The G3 Ammonia-cored PAMAM dendrimer is regarded as a for instance. Its density of amino groups is much greater than that of traditional linear polymers. Ibuprofen and other medications with carboxyl groups Indomethacin, naproxen, and ketoprofen have been extensively complexed via electrostatic means with dendrimers exchanges. Certain antimicrobial and anti-cancer medications have also been documented to be integrated by this kind of communication. These medicinal compounds all have the characteristic of being weak medicines containing carboxyl groups in their compounds that are acidic (Neerman et al. 2004).

A variety of drugs with relevant functional divisions may be covalently conjugated with dendrimers because of their abundant surface functional groups. Here, the medication is bonded covalently to dendrimers, and their enzymatic or chemical release hydrolytically labile bond cleavage. The summary of medication molecules within hydrophobic spaces or the assimilation of medications to dendrimer surfaces by electrostatic relationships maintains the chemical purity and drug compounds' pharmacological characteristics, but covalent medication binding to dendrimer surface groups via chemical connections provides greater drug control release, enabling the regulated medication and tissue targeting delivery (Tripathy and Das 2013). Figure 6.2 illustrates the Dendimer drug loading mechanism.

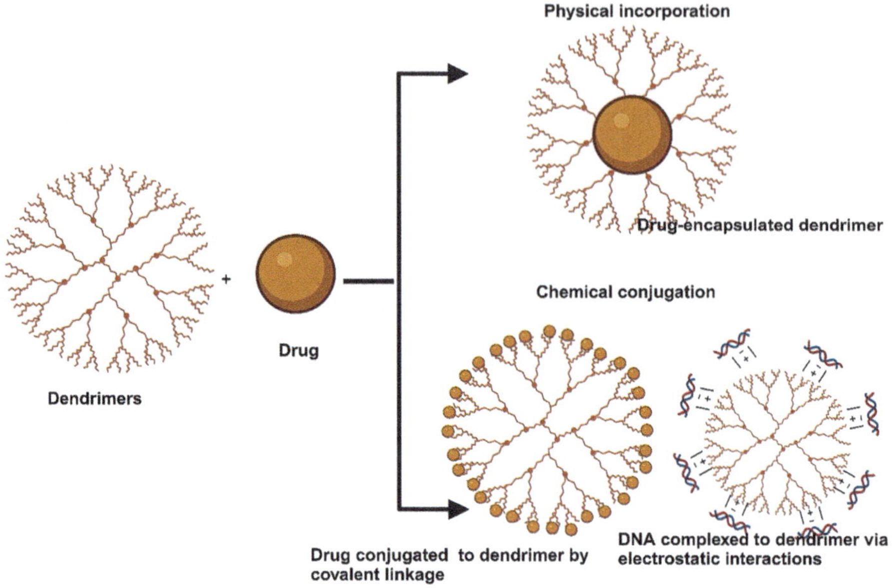


**Fig. 6.2** Dendimer drug loading mechanism- created in BioRender. Dhilip kumar, S. (2025) https://BioRender.com/x58u870

## 6.5 Dendrimer Controlled Release

Drug molecules may be contained both within the dendrimers and on the surface groups, thanks to the dendrimers' compact and globular shape and many surface functional groups. Medications may be loaded into dendrimers via their structure, or they can interact with medications in their terminal functional groups through covalent or electrostatic interactions to serve as drug carriers (Nanjwade et al. 2009). The drug and the outer surface of dendrimers establish a covalent connection. It has been noted that dendrimers' external surfaces may interact with medications. The structural layout of a dendrimer may influence how many drug molecules are integrated into it. Drug delivery strategies include the covalent link between the drug and its dendrimer being broken down in vivo by appropriate enzymes and the drug being released from the dendrimer due to external stimuli like temperature or pH fluctuation.

Encapsulation-based drug dendrimer complexes are simpler to create than conjugation-based drug dendrimer complexes. Furthermore, the researchers have additional flexibility in quickly delivering the medication to the desired spot because of the "pH switch" and the drug's simple release caused by noncovalent interactions. Drugs used to treat brain illnesses are often delivered using dendrimers, particularly PAMAM dendrimers (Palmerston Mendes et al. 2017). Carbomazepine is incorporated into G4.5 dendrimers with carboxyl terminus with the purpose of treating neurodegenerative illnesses linked to protein aggregation toxicity, such as

Huntington's disease (Igartúa et al. 2018). He et al. combined the targeted ligands transferrin and wheat germ agglutinin with doxorubicin by encapsulating it in PEGylated PAMAM G4 dendrimers. According to their results, doxorubicin was more effectively administered to brain tumor areas by this combination than by a free medication (He et al. 2011). Swami et al. encapsulated Docetaxel into PAMAM G4 dendrimers with p-hydroxyl benzoic acid covalently linked to the surface (Swami et al. 2015). Compared to the medication alone, the combination kills glioblastoma cells efficiently and increases the amount of docetaxel reaching the target region.

Drugs covalently bound to dendrimers or other macromolecules need to have their connections engineered to break chemically, releasing the active ingredient. In an ideal world, these connections may be made to specifically cleave at the therapeutic target in response to certain biological circumstances. Doxorubicin has been conjugated to polyester dendrimer-poly(ethylene oxide) (PEO) hybrids, antibodies via hydrazone linkages (Lee et al. 2006; Ulbrich et al. 2003).

## 6.6 The Use of Dendrimers for the Delivery of Biopharmaceuticals

Dendrimers are hyperbranched with sizes ranging from 1 to 10 nm. Dendrimers' large hydrophobic cavity, well-defined molecular weight, variable surface functionality, are some of their unique advantages for controlled and prolonged drug delivery (Mignani et al. 2013).

For the oral administration of camptothecin, Ghandehari and colleagues studied cationic amine- PAMAM dendrimer and anionic PAMAM. The increased integration of camptothecin in the G4.0 formulation compared to the G3.5 formulation was most likely caused by the anionic drug's electrostatic contact with the G4.0 formulation's surface. Oral camptothecin absorption increased two to three times with both formulations, although none was superior to the other (Sadekar et al. 2013). Another research examined the delivery of simvastatin using PAMAM dendrimers, Rat PK tests showed that formulations containing NH2, OH, and PEG functions improved oral bioavailability by 5.6, 3.3, and 7.5 times and increased Cmax by 1.3, 1.4, and 1.7 times, respectively, in comparison to the drug solution. It was proposed that the decreased Cmax and AUC values for the G4-PAMAM-OH formulation were caused by loose drug-dendrimer interaction and reduced drug encapsulation (Kulhari et al. 2013). In contrast to the findings of this study, Qi et al. found that G5-NH2 significantly improved simvastatin's aqueous solubility and in vitro transepithelial transport when compared to G4.5-COOH (Qi et al. 2015). In recent work, chitosan was anchored to poly (propylene imine) dendrimers to create a new mucoadhesive dendrimer-based complex. A popular anthelmintic medication, albendazole, was added to the nanostructure, lyophilised, and then formulated into a matrix tablet. Compared to ordinary albendazole, drug release from mucoadhesive tablets was shown to be successfully regulated. This suggests that chitosan anchoring to PPI dendrimers significantly reduces haemolysis. Albendazole's Cmax and AUC were raised by 12 and 24 times, respectively, using this method (Mansuri et al. 2016).

PAMAM dendrimer covalently bound to ibuprofen was created by Kolhe et al. The findings indicated that, in comparison to pure ibuprofen, the dendrimer–ibuprofen combination caused a faster pharmacological response and increased medication effectiveness via improving cellular transport (Kolhe et al. 2006). Dendrimer was conjugated with cisplatin, a strong anti-cancer medication with low water solubility and some toxicity. In addition to its ability to preferentially accumulate cisplatin in solid tumours, this compound demonstrated improved permeability, lower systemic toxicity, and greater solubility in vivo (Malik et al. 1999). Doxorubicin was coupled with a polyester dendrimer and then added to an injectable formulation. In mice, this mixture demonstrated the ability to treat C-26 colon carcinomas (Lee et al. 2006).

In order to treat vascular thromboembolism, investigated the PAMAM dendrimer, The positively charged PAMAM dendrimers were determined to be a suitable nanocarrier for the delivery of enoxaparin into the lungs without resulting in lung damage (Bai et al. 2007). Luo et al. investigated PEG-modified PAMAM dendrimers in delivering DNA to cells. In both basic research laboratories and possible therapeutic contexts, They proved that dendrimer is a very effective and reasonably priced method of delivering DNA. (Luo et al. 2002).

Chauhan and associates looked at the PAMAM dendrimer's increased bioavailability in a transdermal medication application. By boosting the indomethacin flow through the skin in both in vitro and in vivo settings, they demonstrated the efficacy of the PAMAM dendrimer as a drug delivery method (Chauhan et al. 2003). Dendrimer coupled with doxorubicin was administered pulmonary to rats, which resulted in increased drug retention in their lungs (Kaminskas et al. 2014).

Folic acid-conjugated dendrimers were used by Konda et al. as a diagnostic contrast agent in magnetic resonance imaging. Compared to cells that were not treated, cells expressing the folate receptor showed a longitudinal relaxation rate of almost 100% (Konda et al. 2001). Trivedi et al. found that the usage of PAMAM dendrimers with hydroxyl or carboxylic groups increased the drug's residence time and enhanced the bioavailability of pilocarpine in the eye because of the dendrimers (Trivedi et al. 2012).

Using the PAMAM dendrimer, Katare et al. examined the effectiveness of antipsychotic medication when administered intranasally. In comparison to a control formulation of the medication, they revealed a much greater distribution in the brain and plasma and showed that the dendrimer-based formulation enhanced the drug's water solubility. This research showed how dendrimers might enhance the brain's absorption of water-insoluble medications (Katare et al. 2015). The Biopharmaceutical application of dendrimer is shown in Table 6.2.

**Table 6.2** Biopharmaceutical application of dendrimer

| Biopharmaceutical application | Dendrimer | Drug loaded | Application |
|---|---|---|---|
| Gene delivery | PAMAM | microRNA-22 (miR-22) | Studies using preclinical animal models have shown excellent therapeutic effectiveness (Chen and Jiang 2019) |
| CNS drug delivery | PAMAM | Prion protein (PrP) | In Alzheimer's disease, the complex prevents the production of β-amyloid plaque since they are strong neurotoxins (Nieznański et al. 2020) |
| Cancer targeting | Peptide-dendrimer | Docetaxel | Prostate cancer was targeted, imaged, and treated by conjugating the peptide with the dendrimer (Wu and Chen-Yun 2022). |
| Drug-delivery system | Asymmetric dendrimer | Paclitaxel | When compared to free paclitaxel, the anti-tumor treatment is improved (Suzuki et al. 2021). |
| Drug-delivery system | Camptothecin | G4 PAMAM | Low quantities were hemocompatible for intravenous delivery, and it reduced the viability of cancer cells (Oledzka et al. 2023) |
| Drug-delivery system | Lactoferrin | G3 DAB dendrimer | Enhancement of anti-proliferative properties (Almowalad et al. 2021) |

## 6.7 Nanogels in Drug Delivery

Hydrogels with three dimensions that are nanosized and exhibit properties common to both hydrogels and nanoparticles are called nanogels (Hamidi et al. 2008). Cross-linking hydrophilic polymers in aqueous environments is a physical or chemical synthesis method that may manufacture nanogels. With a structure akin to a sponge that has a high water retention rate and swells when in contact with aqueous fluids, The ratio of surface area to volume is high in nanogels that enhance their drug loading capacity without being dissolved in the watery medium (Anooj et al. 2021). Furthermore, because of their innate ability to expand, nanogels may deliver their payload to specific tissues or cells in reaction to inputs from the outside or the inside (Chang et al. 2021). According to studies, chemotherapeutic medications may be selectively released by nanogels into the tumor environment to eradicate cancer cells without harming nearby healthy cells (Faraji et al. 2021; Zhang et al. 2020). They may be given multiple ways to target different organs since their surface and size characteristics can be modulated to adapt to the varying pH levels within the human body. By shielding the encapsulated pharmaceuticals from enzymatic degradation, nanogels may extend their half-life, which is another significant benefit in drug delivery (Anooj et al. 2021). Nanogel is starting to be heralded as Paul Ehrlich's "magical remedy" for the delivery of drugs due to its capacity to increase medicinal effectiveness while decreasing related side effects (Strebhardt and Ullrich 2008).

Nanogels are able to penetrate biological barriers with ease since their sizes vary from 1 to 1000 nm, with many of them being between 20 and 200 nm (Jain et al. 2019). This also increases the efficiency of cellular absorption and penetration. Additionally, cargo protection and enhanced stability are provided by nanogels, which are operationalized to lengthen the cargo circulation period in vivo and reduce the necessary medication dosage and dosing frequency. Nanogels have a unique design that permits loading of drugs that are both hydrophilic and hydrophobic, expanding their potential uses for a variety of therapeutic agents (Raemdonck et al. 2009). Because of their hydrophilic nature, hydrophilic pharmaceuticals can be incorporated into them, and because of their versatile design, hydrophobic medications may be loaded into nanogel networks by adding surfactants or hydrophobic chains. Moreover, as high biocompatibility and biodegradability materials are often utilized to produce nanogels, this might improve the long-term safety profile of nanogel-based treatment and reduce interference with the body's biological processes (Yallapu et al. 2011). Above all, nanogels are a desirable option for targeted DDS because to their chemical and size tunability.

They may be given multiple ways to target different organs since their surface and size characteristics can be modulated to adapt to the varying pH levels within the human body. The fact that nanogels may react to outside stimuli like ionic strength has also drawn a lot of interest in them (Kazakov et al. 2006). In response to these (the temperature (Cao et al. 2015), pH (He et al. 2015) as well as electromagnetic fields (Koc and Alveroglu 2016)) stimuli, these nanocarriers alter their volume, flexibility, and refractive index, among other characteristics, by swelling or contracting. Because nanogels are readily absorbed by cells, they enhance these treatments' in vivo bioavailability and safety (Ahmad et al. 2006). Biodegradable nanogels have been shown by Akiyoshi et al. to be an intracellular protein delivery platform. These materials' potential for tissue engineering makes them very interesting in the aspect of nanomedicine cancer treatment scaffolding uses picturing theranostics gene delivery as well as pH-responsive actions (Chan and Almutairi 2016; Hashimoto et al. 2015; Kar et al. 2017; Nagahama et al. 2015; Thompson and Dua 2017). The results of of nanogel as a drug delivery system are discussed in Table 6.3.

## 6.8 Nanogel Drug Loading Mechanism

Therapeutic agent delivery is commonly used for nanogels. A large drug loading capacity reduces the transferor required for an effective nano-delivery platform. There are many ways to load up medicines onto nanogels. One is the Covalent Conjugation technique, where the functional groups of the nanogel and the appropriate drug groups establish a covalent link. Since the most often employed covalent bonds are stimuli-sensitive, the medication may be released into the body in the appropriate area. Recently, solid tumor medication delivery and activation were achieved by the application of a pH-sensitive hydrazone bond connecting

**Table 6.3** Nanogel as a drug delivery system

| Nanogel type | Drug | Cell line | Application |
|---|---|---|---|
| DOX-Fe3O4/ P(NIPAM-AA-MAPEG) Nanogel | Doxorubicin (DOX) | Breast Cancer | The studies exhibited that the nanobioconjugate could inhibit the tumor growth (Chen et al. 2018) |
| Dual-responsive nanogel | Cisplatin | HepG2 cell line | The drug-free nanogel's cytotoxicity test on the HepG2 cell line demonstrated its biocompatibility and suitability as a drug carrier (Salimi et al. 2018) |
| p(DEAEMA-co-HEMA-g-PEGMA) nanogels | Small interfering RNAs (siRNA) | H1299, L929 (mouse fibroblasts) and Cal27 | The nanogels can deliver siRNA with tolerable levels of cytotoxicity and exhibit excellent serum stability. (Notabi et al. 2023) |
| pH-responsive alginate-based nanogels | Etoposide (Et) | Lung cancer (H1299) cells | The ALG-g-PHPMA@Et nanogels had a stronger anti-cancer impact on lung cancer cells than free etoposide. (Geyik et al. 2023) |
| Folic acid-modified biocompatible Pullulan/poly(acrylic acid) nanogels | Doxorubicin (DOX) | L929 and MCF-7 cells. | Studies conducted in vitro revealed that DOX-loaded nanogels were efficient against cancer cells. (Ghaffarlou et al. 2023) |
| pH-responsive nanoscale hydrogels | siRNA | Human epithelial colorectal adenocarcinoma cell line. | These results imply that this platform has qualities that make it a promising platform for siRNA delivery (Liechty et al. 2019) |
| pH-responsive alginate nanogel (pH-AN) | Doxorubicin (DOX) | HeLa cells | Tests for in vitro cytotoxicity showed the substantial suppression of HeLa cell proliferation. (Xue et al. 2015) |
| PEG-PDMAEMA nanogels responsive nanogels | DOX | HeLa,. | The dual stimuli-responsive nano gels generated in this work showed a number of advantageous characteristics that make them a potential option for anti-cancer drug administration. (Kumar et al. 2021) |
| Dual-sensitive fluorescent branched alginate-polyethyleneimine copolymer (bAPSC) nanogels | DOX | HeLa cells | The nanogels were shown to be less cytotoxic and hemocompatible in cellular cytotoxicity and hemolysis tests, indicating they are the best substitute drug carriers for drug delivery systems. (Wu et al. 2017) |
| Redox-responsive nanogels | DOX | A2780CP ovarian cancer cells | Highly cytocompatible and efficiently endocytosed by A2780CP the nanogels are a potential class of nanomaterials for effective intracellular delivery of anti-cancer medicines. (Asadi and Khoee 2016) |

doxorubicin to the polymeric nanogel made of methacrylamide (Chen et al. 2017). Using a direct addition method, the medication and monomer are dissolved in the emulsion's aqueous phase prior to the nanogel's creation. As a result, during the nanogel's formation, the drug is confined inside it by hydrophobic or electrostatic interaction. Photoisomerization was used to create aspirin-containing nanogels using a direct addition technique. The hydrophobic substituent was prepared in the first phase of the synthesis, and its interaction with dextran occurred in the second. Subsequently, photoisomerization-induced non-covalent self-aggregation produced the nanogel (Patnaik et al. 2007). The soaking method is helpful when working with amphiphilic nanogels that include hydrophobic components like cholesterol. The medicine might be administered by simply dipping the nanogels into a supersaturated drug solution. For instance, indomethacin-carrying nanogels were created using this technique (Sahiner et al. 2007).

## 6.9 Nanogel Drug Release Mechanisms

When the aqueous liquid crosses the porous particle surface of the nanogel network, the medication is released. When an aqueous solution is added, the nanogel network swells, which results in the release of the medication. The way that nanogels swell in a given media determines how quickly drugs are released. Passive diffusion is among the greatest widely used methods for hydrogel-based medication release. Three types of drug release mechanisms may be distinguished in relation to nanogels: chemically controlled, swelling-controlled, and diffusion-controlled. The medications' physicochemical characteristics and the manner in which they are placed into the nanogels determine the drug release mechanism. Covalent and noncovalent interactions, as well as physical entrapment, are the main methods used to load drugs into nanogels. When there is physical contact, the laden cargo should come free of the nanogels as soon as they swell. Typically, macromolecular therapies are physically incorporated into nanogels. The loaded cargo will instantly escape from the nanogel matrix upon swelling, provided the mesh size permits it. The mesh size within the nanogel matrix determines how much of the medication is released. Generally, bigger molecules slowly release from the matrix, while tiny drug molecules diffuse out more readily. When drugs are loaded into nanogels via covalent and noncovalent contacts, the medicine's extraction from the matrix of the nanogel is as determined as the nanogel's swelling. In conclusion, The release mechanism is linked to the stimuli that the system encounters and initiates the released medication, regardless of whether the drug is physically loaded or covalently attached to the system. (Suhail et al. 2019; Zhang et al. 2016). Figure 6.3 illustrates the nano gel release mechanism.

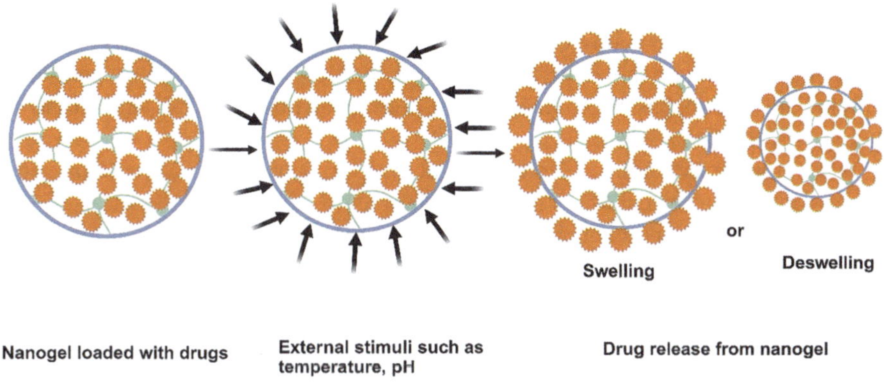


**Fig. 6.3** Nanogel drug release mechanism- created in BioRender. Dhilip kumar, S. (2025) https://BioRender.com/h55q981

## 6.10 Application of Dendrimers and Nanogels

Since they are soluble in water, biocompatible, biodegradable, and have stable encapsulation and intelligent release, nanogels and dendrimers have shown great promise as vehicles for the transportation of drugs. Due to their excellent stability of encapsulation, regulated release of medication characteristics, and great structural flexibility, dendrimers and nanogels have become strong candidates for developing various drug delivery systems. Both dendrimers and nanogel have vast inner dimensions and flexible shapes that allow for loading hydrophilic and hydrophobic medicines. Most significantly, the NGs may stop the medications from degrading quickly while in circulation (Yin et al. 2020). The anti-cancer asparaginase enzyme was coprecipitated in different research by Steinhiber et al. (Steinhilber et al. 2013) to produce biodegradable polyglycerol nanogels (NGs) loaded with enzymes that have an encapsulation efficacy of over 99%. The breakdown of NGs in an acidic environment caused the asparaginase enzyme to be released, and the liberated enzyme maintained its inherent activity.

Because photothermal treatment is extremely targeted and less invasive, It is among the most successful forms of therapy. (Mendes et al. 2017). These days, a variety of polymer-based NPs may be employed to thermally ablate tumors while irradiating them with NIR lasers without harming surrounding normal tissues (Molina et al. 2016; Theune et al. 2019). Pyrole monomers were loaded into thermoresponsive dendrimer-based NGs for in situ polymerization, serving as a nanoreactor. (Theune et al. 2019). The resultant nanobioconjugate had high water dispersibility and outstanding photostability; utilizing an A459 tumor model, the created PPY-Co-dPG NGs showed outstanding tumor PTT effectiveness. Using PAMAM dendrimer–porphyrin conjugates, Peng et al. increased the photodynamic effectiveness of hydrophobic porphyrin while minimizing adverse effects (Peng et al. 2006). According to Taratula et al. (Taratula et al. 2013), phthalocyanine-dendritic complexes modified with PEG that target the LHRH moiety have plenty of potential for photodynamic treatment and NIR fluorescence image-guided drug delivery. Fourth-generation PAMAM dendrimers were coupled with a peptide by Narsireddy et al.(Narsireddy et al. 2015) for targeted in vivo photodynamic treatment. A hydrophilic nanoconjugate that improves Ce6's water solubility and intracellular absorption has been reported by Lee and Kim (Lee and Kim 2018) to increase PDT effectiveness.

A developing technique called biological imaging may identify a live organism's tissues' molecular or cellular functions (Chen and Chen 2010; Pysz et al. 2010). Imaging agents based on nanotechnology have drawn a lot of interest because they can diagnose malignancies with excellent selectivity and specificity. In particular, therapeutic and imaging molecules may be combined with various nanoplatforms to provide imaging-guided cancer treatment. Dendrimer-based NGs have also been used in cancer treatment with imaging guidance. In vitro fluorescent imaging and treatment of cancer cells were made possible by integrating FITC and DOX into AG NGs generated using the G5 dendrimers, as described by Gonçalves et al. (Gonçalves et al. 2014).

Gene therapy is used to identify and cure human illnesses caused by genetic material transferred to certain cells (Mulligan 1993). The monodispersity, functional groups, and multivalence architectures of dendrimers make them useful in gene transfer (Tomalia 2005). Dendrimers were used for gene therapy by Haensler et al. (Haensler and Szoka Jr. 1993). Using a fifth-generation PAMAM dendrimer, Li et al. (Li et al. 2017) modified gold nanoparticles to provide a safe delivery method for regulated gene delivery in the treatment of breast cancer. Dendrimer increased the efficacy of gene transfection and improved cellular absorption in the A549 cell line in the study of Aptamer-conjugated PAMAM dendrimer nanoparticles created by Wang et al. to transport specific genes (Wang et al. 2015). Nanogels have also demonstrated their potential in gene delivery; they have the capacity to expand and release their enclosed payload in an intelligent manner. Since they are pH-responsive polymers, dendritic polyglycerol (dPG) and low molecular weight PEI acrylamide have been developed for targeted siRNA delivery. This has led to improved gene silencing results as well as decreased cytotoxicity (Dimde et al. 2017). Table 6.4 below shows the Clinical trial status of dendrimers and nanogels.

## 6.11 Challenges Associated with Dendrimers and Nanogels for Drug Delivery Systems

The objective of delivering medications derived from diverse plant sources to the appropriate locations in the body to treat patients has come a long way, and a range of delivery techniques have been used successfully recently. However, the sort of treatment that these systems may provide has several drawbacks and difficulties.

The paucity and diversity of available materials are major barriers to advancing drug delivery methods. The development of any research is contingent upon the availability of critical data in the literature. A primary hindrance to the progress of nanotechnology in medicine is the discrepancy in published studies about the

**Table 6.4** Clinical trial status of dendrimers and nanogels

| ClinicalTrials. Gov Identifier | Composition | Application | Phase | Status |
|---|---|---|---|---|
| NCT05771077 | Poly(Amidoamine) dendrimer | White spot lesions | Phase 1 | Completed |
| NCT04262076 | Pulpine with Polyamidoamine dendrimer | Deep caries | Not applicable | Completed |
| NCT06358911 | Poly amido amine dendrimer (PAMAM) | Tooth hypersensitivity | Phase 4 | Completed |
| NCT06358924 | PAMAM-carboxylic acid dendrimer | Tooth hypersensitivity | Phase 4 | Completed |
| NCT03500627 | OP-101 (Dendrimer N-Acetyl-Cysteine) | | Phase 1 | Completed |
| NCT03255343 | IMDENDRIM | Liver cancer | Phase 1 | Unknown status |
| NCT05105607 | D-4517.2 (Hydroxyl Dendrimer VEGFR Tyrosine Kinase Inhibitor) | Neovascular age-related macular degeneration Diabetic macular edema | Phase 1 | Completed |
| NCT05783206 | MIR 19 ® contains anti-SARS-CoV-2 siRNAs/KK-46 (peptide dendrimer) complexes: | COVID-19 | Phase 2 and 3 | Completed |
| NCT05268718 | Nano-shell hydrogel | Resistance, microbial Photothermal therapy | Early phase 1 | Unknown status |

recorded description of reported experimental data. (Youn and Bae 2018). Future advances in nanomedicines may be hampered by the incomplete and inconsistent data that ought to guide industry, delaying the transition from laboratory and research to clinical application. Researchers widely acknowledge that nanoparticles can have positive or negative effects (Pasut 2019). On the other hand, the safety of these particles, their degree of interaction with non-target proteins, and how they move through and affect organs other than those they are intended to target are all poorly understood. (Sharma et al. 2021).

One way to solve the issues associated with employing much bigger particles is to utilize much smaller particles to be administered to the human body. Large particles, which some of these delivery systems employ as carriers, are not very helpful for treatment because they can present issues like inadequate absorption and solubility, instability in vivo, low bioavailability, difficulties with target-specific distribution, and a number of adverse side effects upon treatment (Patra et al. 2018).

Target-specific delivery complexity is one issue that all delivery systems face. Target-specific delivery is more successful in lowering toxicity, but its effectiveness cannot be assured until it reaches the intended location in large enough quantities. This is seen when siRNA is administered systemically; at high concentrations, its negative charge prevents the cells from absorbing them, which leads to little or no body absorption. Because bodily enzymes readily break them down, they seldom reach the intended cell or organ (Ezike et al. 2023). Targeted delivery has many challenges, including the inability to provide a dosage to an unconscious patient, limited solubility and permeability at the target region, potential for dietary interactions, and potential degradation by the stomach flora (Ezike et al. 2023).

Toxicology of the particles used for distribution is another significant problem for medication delivery systems; some nanomaterials may harm the environment and human health. (Singh et al. 2017). When PAMAM was tested on embryonic zebrafish models, Pryor et al. discovered that, at the same dose, compared to neutral PAMAM generation 6 and anionic PAMAM generation 6, cationic PAMAM generation 6 was statistically more detrimental. (Pryor et al. 2014).

Drug delivery systems face several difficulties, including acceptability (the medication's capacity to enter the body and be absorbed without triggering the immune system) and biocompatibility (The drug's capacity to interact with the body in particular circumstances) due to the body's differing reactions to biological and synthetic materials (Singh et al. 2017).

The body's natural detoxification processes include the kidney and liver, which may handle nanoparticles as possible waste products. Their actions may hinder the delivery of drugs and cause nanoparticles to accumulate in these organs. Nanomaterials mostly gather in the liver. After entering the renal system, the kidneys' size, charge, and shape dictate what happens to the nanomaterials (Bartucci et al. 2020).

## 6.12 Future Perspectives and Conclusions

One of the most fascinating fields of modern science is drug delivery systems, which have drawn much attention for investigation, testing, and clinical trials in recent years. Although there have been barriers to clinical adoption, the new medication delivery technology has a lot of promise. To attain this efficiency, we must work across academic theory, laboratory experiments, medical knowledge, pharmaceutical knowledge, and superior research [148]. Cell treatments may greatly mitigate the issues with drug delivery systems' bio-acceptability. Additionally, they assert that it will provide an efficient single dose, preventing a significant accumulation of medications in the system. Cell therapies have the potential to break down innate biological barriers, trigger spontaneous responses inside the body, and provide an endless supply of complex biologics.

One way to increase medication delivery effectiveness is to utilize prime agents with the ability to alter the biological environment in which they are administered. These agents should be able to change tissue structure and function so that the drug is favorable without posing a risk to the patient. Furthermore, as cells are an inherent human body component, cell-based medication systems—including nanomaterials with cells—should be considered in the biomaterials industry. While still, in theory, this innovative strategy seems to be very inventive and encourages the use of various drug delivery techniques to produce the most effective drug delivery pattern. Further investigation and clinical studies are still required to prove both the effectiveness and drawbacks of these contemporary drug delivery methods.

## References


Agrawal P, Gupta U, Jain N (2007) Glycoconjugated peptide dendrimers-based nanoparticulate system for the delivery of chloroquine phosphate. *Biomaterials 28*(22):3349–3359

Ahmad Z, Pandey R, Sharma S, Khuller GK (2006) Pharmacokinetic and pharmacodynamic behaviour of antitubercular drugs encapsulated in alginate nanoparticles at two doses. *Int J Antimicrob Agents 27*(5):409–416

Alibolandi M, Taghdisi SM, Ramezani P, Shamili FH, Farzad SA, Abnous K, Ramezani M (2017) Smart AS1411-aptamer conjugated pegylated PAMAM dendrimer for the superior delivery of camptothecin to colon adenocarcinoma in vitro and in vivo. *Int J Pharm 519*(1–2):352–364

Allen TM, Cullis PR (2004) Drug delivery systems: entering the mainstream. *Science 303*(5665):1818–1822

Almowalad J, Somani S, Laskar P, Meewan J, Tate RJ, Mullin M, Dufès C (2021) Lactoferrin-bearing gold nanocages for gene delivery in prostate cancer cells in vitro. *Int J Nanomed* Volume 16:4391–4407

Anooj E, Charumathy M, Sharma V, Vibala B, Gopukumar S, Jainab SB, Vallinayagam S (2021) Nanogels: An overview of properties, biomedical applications, future research trends and developments. *J Mol Struct 1239*(130):446

Asadi H, Khoee S (2016) Dual responsive nanogels for intracellular doxorubicin delivery. *Int J Pharm 511*(1):424–435

Astashkina A, Mann B, Grainger DW (2012) A critical evaluation of in vitro cell culture models for high-throughput drug screening and toxicity. *Pharmacol Ther 134*(1):82–106

Ayatollahi S, Salmasi Z, Hashemi M, Askarian S, Oskuee RK, Abnous K, Ramezani M (2017) Aptamer-targeted delivery of Bcl-xL shRNA using alkyl modified PAMAM dendrimers into lung cancer cells. *Int J Biochem Cell Biol 92*:210–217

Bai S, Thomas C, Ahsan F (2007) Dendrimers as a carrier for pulmonary delivery of enoxaparin, a low-molecular weight heparin. *J Pharm Sci 96*(8):2090–2106

Bartucci R, Paramanandana A, Boersma YL, Olinga P, Salvati A (2020) Comparative study of nanoparticle uptake and impact in murine lung, liver and kidney tissue slices. *Nanotoxicology 14*(6):847–865

Bawa P, Pillay V, Choonara YE, Du Toit LC (2009) Stimuli-responsive polymers and their applications in drug delivery. *Biomed Mater 4*(2):022001

Bhatt H, Kiran Rompicharla SV, Ghosh B, Torchilin V, Biswas S (2019) Transferrin/α-tocopherol modified poly (amidoamine) dendrimers for improved tumor targeting and anticancer activity of paclitaxel. *Nanomedicine 14*(24):3159–3176

Bildstein L, Dubernet C, Couvreur P (2011) Prodrug-based intracellular delivery of anticancer agents. *Adv Drug Deliv Rev 63*(1–2):3–23

Bourzac K (2012) Nanotechnology: carrying drugs. *Nature 491*(7425):S58–S60

Brownlie A, Uchegbu I, Schätzlein A (2004) PEI-based vesicle-polymer hybrid gene delivery system with improved biocompatibility. *Int J Pharm 274*(1–2):41–52

Cao H, Wang Q, Li M, Chen Z (2015) Synthesis of stimuli-responsive poly (ethylene glycol) diacrylate/methacrylic acid-based nanogels and their application as drug delivery vehicle. *Colloid Polym Sci 293*:441–451

Chan M, Almutairi A (2016) Nanogels as imaging agents for modalities spanning the electromagnetic spectrum. *Materials horizons 3*(1):21–40

Chang D, Ma Y, Xu X, Xie J, Ju S (2021) Stimuli-responsive polymeric nanoplatforms for cancer therapy. Front Bioeng Biotechnol 9(707):319

Chauhan AS, Sridevi S, Chalasani KB, Jain AK, Jain SK, Jain N, Diwan PV (2003) Dendrimer-mediated transdermal delivery: enhanced bioavailability of indomethacin. *J Contr Release 90*(3):335–343

Chen K, Chen X (2010) Design and development of molecular imaging probes. *Curr Top Med Chem 10*(12):1227–1236

Chen J, Jiang X (2019) Methods for treating acute myeloid leukemia and nanoparticle complexes of miR-22 utilized therein. Google Patents

Chen G, Qiu H, Prasad PN, Chen X (2014) Upconversion nanoparticles: design, nanochemistry, and applications in theranostics. *Chem Rev 114*(10):5161–5214

Chen Y, Tezcan O, Li D, Beztsinna N, Lou B, Etrych T, Ulbrich K, Metselaar J, Lammers T, Hennink WE (2017) Overcoming multidrug resistance using folate receptor-targeted and pH-responsive polymeric nanogels containing covalently entrapped doxorubicin. *Nanoscale 9*(29):10404–10419

Chen Y, Cheng Y, Zhao P, Zhang S, Li M, He C, Zhang X, Yang T, Yan R, Ye P, Ma X, Xiang G (2018) Co-delivery of doxorubicin and imatinib by pH sensitive cleavable PEGylated nanoliposomes with folate-mediated targeting to overcome multidrug resistance. *Int J Pharm 542*(1–2):266–279. https://doi.org/10.1016/j.ijpharm.2018.03.024

Chenthamara D, Subramaniam S, Ramakrishnan SG, Krishnaswamy S, Essa MM, Lin F-H, Qoronfleh MW (2019) Therapeutic efficacy of nanoparticles and routes of administration. *Biomater Res 23*(1):20

Chis AA, Dobrea C, Morgovan C, Arseniu AM, Rus LL, Butuca A, Juncan AM, Totan M, Vonica-Tincu AL, Cormos G (2020) Applications and limitations of dendrimers in biomedicine. *Molecules 25*(17):3982

Choudhary S, Gupta L, Rani S, Dave K, Gupta U (2017) Impact of dendrimers on solubility of hydrophobic drug molecules. *Front Pharmacol 8*:261

Couvreur P (2013) Nanoparticles in drug delivery: past, present and future. *Adv Drug Deliv Rev 65*(1):21–23

Das M, Sanson N, Fava D, Kumacheva E (2007) Microgels loaded with gold nanorods: photothermally triggered volume transitions under physiological conditions. *Langmuir 23*(1):196–201

De las Heras Alarcón C, Pennadam S, Alexander C (2005) Stimuli responsive polymers for biomedical applications. *Chem Soc Rev 34*(3):276–285

Dhanikula RS, Argaw A, Bouchard J-F, Hildgen P (2008) Methotrexate loaded polyether-copolyester dendrimers for the treatment of gliomas: enhanced efficacy and intratumoral transport capability. *Mol Pharm 5*(1):105–116

Dimde M, Neumann F, Reisbeck F, Ehrmann S, Cuellar-Camacho JL, Steinhilber D, Ma N, Haag R (2017) Defined pH-sensitive nanogels as gene delivery platform for siRNA mediated in vitro gene silencing. *Biomater Sci 5*(11):2328–2336

Duncan R, Izzo L (2005) Dendrimer biocompatibility and toxicity. *Adv Drug Deliv Rev 57*(15):2215–2237

Ezike TC, Okpala US, Onoja UL, Nwike CP, Ezeako EC, Okpara OJ, Okoroafor CC, Eze SC, Kalu OL, Odoh EC (2023) Advances in drug delivery systems, challenges and future directions. *Heliyon 9*(6):e17488

Faraji N, Esrafili A, Esfandiari B, Abednezhad A, Naghizadeh M, Arasteh J (2021) Synthesis of pH-sensitive hyaluronic acid nanogels loaded with paclitaxel and interferon gamma: Characterization and effect on the A549 lung carcinoma cell line. *Colloids Surf B: Biointerfaces 205*(111):845

Geyik G, Güncüm E, Işıklan N (2023) Design and development of pH-responsive alginate-based nanogel carriers for etoposide delivery. *Int J Biol Macromol 250*(126):242

Ghaffari M, Dehghan G, Abedi-Gaballu F, Kashanian S, Baradaran B, Dolatabadi JEN, Losic D (2018) Surface functionalized dendrimers as controlled-release delivery nanosystems for tumor targeting. *Eur J Pharm Sci 122*:311–330

Ghaffari M, Dehghan G, Baradaran B, Zarebkohan A, Mansoori B, Soleymani J, Dolatabadi JEN, Hamblin MR (2020) Co-delivery of curcumin and Bcl-2 siRNA by PAMAM dendrimers for enhancement of the therapeutic efficacy in HeLa cancer cells. *Colloids Surf B: Biointerfaces 188*(110):762

Ghaffarlou M, Sütekin SD, Karacaoğlu E, Turan SK, İnci ÖG, Güven O, Barsbay M (2023) Folic acid-modified biocompatible pullulan/poly (acrylic acid) nanogels for targeted delivery to MCF-7 cancer cells. *Eur J Pharm Biopharm 184*:189–201

Gonçalves M, Maciel D, Capelo D, Xiao S, Sun W, Shi X, Rodrigues J, Tomás H, Li Y (2014) Dendrimer-assisted formation of fluorescent nanogels for drug delivery and intracellular imaging. *Biomacromolecules 15*(2):492–499

Gupta V, Nayak SK (2015) Dendrimers: a review on synthetic approaches. *J Appl Pharm Sci 5*(3):117–122

Gupta AK, Wells S (2004) Surface-modified superparamagnetic nanoparticles for drug delivery: preparation, characterization, and cytotoxicity studies. *IEEE Trans Nanobiosci 3*(1):66–73

Haensler J, Szoka FC Jr (1993) Polyamidoamine cascade polymers mediate efficient transfection of cells in culture. *Bioconjug Chem 4*(5):372–379

Hamidi M, Azadi A, Rafiei P (2008) Hydrogel nanoparticles in drug delivery. *Adv Drug Deliv Rev 60*(15):1638–1649

Hashimoto Y, Mukai S-A, Sawada S-I, Sasaki Y, Akiyoshi K (2015) Nanogel tectonic porous gel loading biologics, nanocarriers, and cells for advanced scaffold. *Biomaterials 37*:107–115

Hayashi C, Hasegawa U, Saita Y, Hemmi H, Hayata T, Nakashima K, Ezura Y, Amagasa T, Akiyoshi K, Noda M (2009) Osteoblastic bone formation is induced by using nanogel-crosslinking hydrogel as novel scaffold for bone growth factor. *J Cell Physiol 220*(1):1–7

He H, Li Y, Jia X-R, Du J, Ying X, Lu W-L, Lou J-N, Wei Y (2011) PEGylated Poly (amidoamine) dendrimer-based dual-targeting carrier for treating brain tumors. *Biomaterials 32*(2):478–487

He L, Liang H, Lin L, Shah BR, Li Y, Chen Y, Li B (2015) Green-step assembly of low density lipoprotein/sodium carboxymethyl cellulose nanogels for facile loading and pH-dependent release of doxorubicin. *Colloids Surf B: Biointerfaces 126*:288–296

He H, Yuan Q, Bie J, Wallace RL, Yannie PJ, Wang J, Lancina MG III, Zolotarskaya OY, Korzun W, Yang H (2018) Development of mannose functionalized dendrimeric nanoparticles for targeted delivery to macrophages: use of this platform to modulate atherosclerosis. *Transl Res 193*:13–30

Ho MN, Bach LG, Nguyen TH, Ho MH, Nguyen DH, Nguyen CK, Nguyen CH, Nguyen NV, Hoang Thi TT (2019) PEGylated poly (amidoamine) dendrimers-based drug loading vehicles for delivering carboplatin in treatment of various cancerous cells. *J Nanopart Res 21*:1–12

Howell B, Fan D (2010) Poly (amidoamine) dendrimer-supported organoplatinum antitumour agents. *Proc R Soc A 466*(2117):1515–1526

Hughes GA (2005) Nanostructure-mediated drug delivery. *Nanomedicine 1*(1):22–30

Igartúa DE, Martinez CS, Temprana CF, Alonso SDV, Prieto MJ (2018) PAMAM dendrimers as a carbamazepine delivery system for neurodegenerative diseases: a biophysical and nanotoxicological characterization. *Int J Pharm 544*(1):191–202

Jain S, Ancheria RK, Shrivastava S, Soni SL, Sharma M (2019) An overview of nanogel–novel drug delivery system. *Asian J Pharm Res Dev 7*(2):47–55

Kaminskas LM, McLeod VM, Ryan GM, Kelly BD, Haynes JM, Williamson M, Thienthong N, Owen DJ, Porter CJ (2014) Pulmonary administration of a doxorubicin-conjugated dendrimer enhances drug exposure to lung metastases and improves cancer therapy. *J Contr Release 183*:18–26

Kar M, Fechner L, Nagel G, Glitscher E, Noe Rimondino G, Calderón M (2017) Responsive nanogels for anti-cancer Therapy. The Royal Society of Chemistry

Katare YK, Daya RP, Sookram Gray C, Luckham RE, Bhandari J, Chauhan AS, Mishra RK (2015) Brain targeting of a water insoluble antipsychotic drug haloperidol via the intranasal route using PAMAM dendrimer. *Mol Pharm 12*(9):3380–3388

Kazakov S, Kaholek M, Gazaryan I, Krasnikov B, Miller K, Levon K (2006) Ion concentration of external solution as a characteristic of micro-and nanogel ionic reservoirs. *J Phys Chem B 110*(31):15107–15116

Kitchens KM, El-Sayed ME, Ghandehari H (2005) Transepithelial and endothelial transport of poly (amidoamine) dendrimers. *Adv Drug Deliv Rev 57*(15):2163–2176

Koc K, Alveroglu E (2016) Tuning the gel size and LCST of N-isopropylacrylamide nanogels by anionic fluoroprobe. *Colloid Polym Sci 294*:285–290

Kolhe P, Khandare J, Pillai O, Kannan S, Lieh-Lai M, Kannan RM (2006) Preparation, cellular transport, and activity of polyamidoamine-based dendritic nanodevices with a high drug payload. *Biomaterials 27*(4):660–669

Konda SD, Aref M, Wang S, Brechbiel M, Wiener EC (2001) Specific targeting of folate–dendrimer MRI contrast agents to the high affinity folate receptor expressed in ovarian tumor xenografts. *MAGMA 12*:104–113

Kulhari H, Kulhari DP, Prajapati SK, Chauhan AS (2013) Pharmacokinetic and pharmacodynamic studies of poly (amidoamine) dendrimer based simvastatin oral formulations for the treatment of hypercholesterolemia. *Mol Pharm 10*(7):2528–2533

Kumar P, Behl G, Kaur S, Yadav N, Liu B, Chhikara A (2021) Tumor microenvironment responsive nanogels as a smart triggered release platform for enhanced intracellular delivery of doxorubicin. *J Biomater Sci Polym Ed 32*(3):385–404

Kurtoglu YE, Mishra MK, Kannan S, Kannan RM (2010) Drug release characteristics of PAMAM dendrimer–drug conjugates with different linkers. *Int J Pharm 384*(1–2):189–194

Lee S-R, Kim Y-J (2018) Hydrophilic chlorin e6-poly (amidoamine) dendrimer nanoconjugates for enhanced photodynamic therapy. *Nanomaterials 8*(6):445

Lee CC, Gillies ER, Fox ME, Guillaudeu SJ, Fréchet JM, Dy EE, Szoka FC (2006) A single dose of doxorubicin-functionalized bow-tie dendrimer cures mice bearing C-26 colon carcinomas. *Proc Natl Acad Sci 103*(45):16649–16654

Li Y, He H, Jia X, Lu W-L, Lou J, Wei Y (2012) A dual-targeting nanocarrier based on poly (amidoamine) dendrimers conjugated with transferrin and tamoxifen for treating brain gliomas. *Biomaterials 33*(15):3899–3908

Li C, Xia F, Wang K, Wang C, Xu P, Zhang H, Wang J, Toru A, Ni J, Cui D (2017) Dendrimer-modified gold nanorods as high efficient controlled gene delivery release system under near-infrared light irradiation. *Nano Biomed Eng 9*(1)

Li J, Liang H, Liu J, Wang Z (2018) Poly (amidoamine)(PAMAM) dendrimer mediated delivery of drug and pDNA/siRNA for cancer therapy. *Int J Pharm 546*(1–2):215–225

Liechty WB, Scheuerle RL, Ramirez JEV, Peppas NA (2019) Cytoplasmic delivery of functional siRNA using pH-responsive nanoscale hydrogels. *Int J Pharm 562*:249–257

Liu M, Kono K, Fréchet JM (1999) Water-soluble dendrimer–poly (ethylene glycol) starlike conjugates as potential drug carriers. *J Polym Sci A Polym Chem 37*(17):3492–3503

Luo D, Haverstick K, Belcheva N, Han E, Saltzman WM (2002) Poly (ethylene glycol)-conjugated PAMAM dendrimer for biocompatible, high-efficiency DNA delivery. *Macromolecules 35*(9):3456–3462

Luong D, Kesharwani P, Deshmukh R, Amin MCIM, Gupta U, Greish K, Iyer AK (2016) PEGylated PAMAM dendrimers: enhancing efficacy and mitigating toxicity for effective anticancer drug and gene delivery. *Acta Biomater 43*:14–29

Malik N, Evagorou EG, Duncan R (1999) Dendrimer-platinate: a novel approach to cancer chemotherapy. *Anti-Cancer Drugs 10*(8):767–776

Mansuri S, Kesharwani P, Tekade RK, Jain NK (2016) Lyophilized mucoadhesive-dendrimer enclosed matrix tablet for extended oral delivery of albendazole. *Eur J Pharm Biopharm 102*:202–213

Mastrobattista E, van der Aa MA, Hennink WE, Crommelin DJ (2006) Artificial viruses: a nanotechnological approach to gene delivery. *Nat Rev Drug Discov 5*(2):115–121

Mendes R, Pedrosa P, Lima JC, Fernandes AR, Baptista PV (2017) Photothermal enhancement of chemotherapy in breast cancer by visible irradiation of gold nanoparticles. *Sci Rep 7*(1):10,872

Menjoge AR, Kannan RM, Tomalia DA (2010) Dendrimer-based drug and imaging conjugates: design considerations for nanomedical applications. *Drug Discov Today 15*(5–6):171–185

Mignani S, El Kazzouli S, Bousmina M, Majoral J-P (2013) Expand classical drug administration ways by emerging routes using dendrimer drug delivery systems: a concise overview. *Adv Drug Deliv Rev 65*(10):1316–1330

Mignani S, Rodrigues J, Tomas H, Zablocka M, Shi X, Caminade A-M, Majoral J-P (2018) Dendrimers in combination with natural products and analogues as anti-cancer agents. *Chem Soc Rev 47*(2):514–532

Molina M, Wedepohl S, Calderón M (2016) Polymeric near-infrared absorbing dendritic nanogels for efficient in vivo photothermal cancer therapy. *Nanoscale 8*(11):5852–5856

Motornov M, Roiter Y, Tokarev I, Minko S (2010) Stimuli-responsive nanoparticles, nanogels and capsules for integrated multifunctional intelligent systems. *Prog Polym Sci 35*(1–2):174–211

Mulligan RC (1993) The basic science of gene therapy. *Science 260*(5110):926–932

Nagahama K, Sano Y, Kumano T (2015) Anticancer drug-based multifunctional nanogels through self-assembly of dextran–curcumin conjugates toward cancer theranostics. *Bioorg Med Chem Lett 25*(12):2519–2522

Nanjwade BK, Bechra HM, Derkar GK, Manvi F, Nanjwade VK (2009) Dendrimers: emerging polymers for drug-delivery systems. *Eur J Pharm Sci 38*(3):185–196

Narsireddy A, Vijayashree K, Adimoolam MG, Manorama SV, Rao NM (2015) Photosensitizer and peptide-conjugated PAMAM dendrimer for targeted in vivo photodynamic therapy. *Int J Nanomed* 10:6865–6878

Neerman MF, Zhang W, Parrish AR, Simanek EE (2004) In vitro and in vivo evaluation of a melamine dendrimer as a vehicle for drug delivery. *Int J Pharm 281*(1–2):129–132

Nguyen H, Nguyen NH, Tran NQ, Nguyen CK (2015) Improved method for preparing cisplatin-dendrimer nanocomplex and its behavior against NCI-H460 lung cancer cell. *J Nanosci Nanotechnol 15*(6):4106–4110

Nguyen DTD, Bach LG, Nguyen TH, Ho MH, Ho MN, Nguyen DH, Nguyen CK, Hoang Thi TT (2019) Preparation and characterization of oxaliplatin drug delivery vehicle based on PEGylated half-generation PAMAM dendrimer. *J Polym Res 26*:1–14

Nieznański K, Nieznańska H, Surewicz WK, Surewicz K, Bandyszewska M (2020) Prion protein-dendrimer conjugates for use in treatment of Alzheimer disease. Google Patents

Notabi MK, Arnspang EC, Peppas NA, Andersen MØ (2023) siRNA delivery mediated by pH and redox responsive p (DEAEMA-co-HEMA-g-PEGMA) nanogels. *J Drug Deliv Sci Technol 86*(104):510

Oishi M, Sumitani S, Nagasaki Y (2007) On– off regulation of 19F magnetic resonance signals based on pH-sensitive PEGylated nanogels for potential tumor-specific smart 19F MRI probes. *Bioconjug Chem 18*(5):1379–1382

Oledzka E, Paśnik K, Domańska I, Zielińska-Pisklak M, Piotrowska U, Sobczak M, Szeleszczuk Ł, Laskowska A (2023) Poly (amidoamine) dendrimer/camptothecin complex: from synthesis to in vitro cancer cell line studies. *Molecules 28*(6):2696

Onaca O, Enea R, Hughes DW, Meier W (2009) Stimuli-responsive polymersomes as nanocarriers for drug and gene delivery. *Macromol Biosci 9*(2):129–139

Ooya T, Lee J, Park K (2004) Hydrotropic dendrimers of generations 4 and 5: synthesis, characterization, and hydrotropic solubilization of paclitaxel. *Bioconjug Chem 15*(6):1221–1229

Palmerston Mendes L, Pan J, Torchilin VP (2017) Dendrimers as nanocarriers for nucleic acid and drug delivery in cancer therapy. *Molecules 22*(9):1401

Parekh H (2007) The advance of dendrimers-a versatile targeting platform for gene/drug delivery. *Curr Pharm Des 13*(27):2837–2850

Pasut G (2019) Grand challenges in nano-based drug delivery, vol 1. Frontiers Media SA, p 1

Patel HK, Gajbhiye V, Kesharwani P, Jain NK (2016) Ligand anchored poly (propyleneimine) dendrimers for brain targeting: comparative in vitro and in vivo assessment. *J Colloid Interface Sci 482*:142–150

Patnaik S, Sharma AK, Garg B, Gandhi R, Gupta K (2007) Photoregulation of drug release in azo-dextran nanogels. *Int J Pharm 342*(1–2):184–193

Patra JK, Das G, Fraceto LF, Campos EVR, Rodriguez-Torres MDP, Acosta-Torres LS, Diaz-Torres LA, Grillo R, Swamy MK, Sharma S (2018) Nano based drug delivery systems: recent developments and future prospects. *J Nanobiotechnol 16*:1–33

Patri AK, Kukowska-Latallo JF, Baker JR Jr (2005) Targeted drug delivery with dendrimers: comparison of the release kinetics of covalently conjugated drug and non-covalent drug inclusion complex. *Adv Drug Deliv Rev 57*(15):2203–2214

Patton D, Cosgrove Sweeney Y, McCarthy T, Hillier S (2006) Preclinical safety and efficacy assessments of dendrimer-based (SPL7013) microbicide gel formulations in a nonhuman primate model. *Antimicrob Agents Chemother 50*(5):1696–1700

Peng C, Lai P, Chiu C, Chang C, Lai T, Shieh M (2006) Synthesis and Photodynamic Efficacy of PAMAM-Photosensitiser Conjugates in Vitro. NSTI-Nanotech

Peng H, Stolwijk JA, Sun L-N, Wegener J, Wolfbeis OS (2010) A nanogel for ratiometric fluorescent sensing of intracellular pH values. *Angew Chem 122*(25):4342e4345

Pooresmaeil M, Namazi H (2021) Advances in development of the dendrimers having natural saccharides in their structure for efficient and controlled drug delivery applications. *Eur Polym J 148*(110):356

Pryor JB, Harper BJ, Harper SL (2014) Comparative toxicological assessment of PAMAM and thiophosphoryl dendrimers using embryonic zebrafish. *Int J Nanomed* 9:1947–1956

Pysz MA, Gambhir SS, Willmann JK (2010) Molecular imaging: current status and emerging strategies. *Clin Radiol 65*(7):500–516

Qi R, Zhang H, Xu L, Shen W, Chen C, Wang C, Cao Y, Wang Y, van Dongen MA, He B (2015) G5 PAMAM dendrimer versus liposome: a comparison study on the in vitro transepithelial transport and in vivo oral absorption of simvastatin. *Nanomedicine 11*(5):1141–1151

Raemdonck K, Demeester J, De Smedt S (2009) Advanced nanogel engineering for drug delivery. *Soft Matter 5*(4):707–715

Rajani C, Borisa P, Karanwad T, Borade Y, Patel V, Rajpoot K, Tekade RK (2020) Cancer-targeted chemotherapy: emerging role of the folate anchored dendrimer as drug delivery nanocarrier. In: *Pharmaceutical applications of dendrimers*. Elsevier, pp 151–198

Ramasamy T, Ruttala HB, Gupta B, Poudel BK, Choi H-G, Yong CS, Kim JO (2017) Smart chemistry-based nanosized drug delivery systems for systemic applications: a comprehensive review. *J Contr Release 258*:226–253

Rao B, Viswanath V, Reddy K, Fathima S, Surekha P, Bhuvaneswari S (2015) Dendrimers–structure, synthesis, encapsulation, characterization and application. *J Global Trends Pharm Sci 6*:2860–2866

Reischl D, Zimmer A (2009) Drug delivery of siRNA therapeutics: potentials and limits of nanosystems. *Nanomedicine 5*(1):8–20

Sadekar S, Thiagarajan G, Bartlett K, Hubbard D, Ray A, McGill L, Ghandehari H (2013) Poly (amido amine) dendrimers as absorption enhancers for oral delivery of camptothecin. *Int J Pharm 456*(1):175–185

Sahiner N, Alb AM, Graves R, Mandal T, McPherson GL, Reed WF, John VT (2007) Core–shell nanohydrogel structures as tunable delivery systems. *Polymer 48*(3):704–711

Salimi F, Dilmaghani KA, Alizadeh E, Akbarzadeh A, Davaran S (2018) Enhancing cisplatin delivery to hepatocellular carcinoma HepG2 cells using dual sensitive smart nanocomposite. *Artif Cells Nanomed Biotechnol 46*(5):949–958

Santos A, Veiga F, Figueiras A (2019) Dendrimers as pharmaceutical excipients: synthesis, properties, toxicity and biomedical applications. *Materials 13*(1):65

Scioli Montoto S, Muraca G, Ruiz ME (2020) Solid lipid nanoparticles for drug delivery: pharmacological and biopharmaceutical aspects. *Front Mol Biosci 7*(587):997

Sharma S, Parveen R, Chatterji BP (2021) Toxicology of nanoparticles in drug delivery. *Curr Pathobiol Rep* 9:1–12

Sherje AP, Jadhav M, Dravyakar BR, Kadam D (2018) Dendrimers: a versatile nanocarrier for drug delivery and targeting. *Int J Pharm 548*(1):707–720

Singh N, Joshi A, Toor AP, Verma G (2017) Drug delivery: advancements and challenges. In: *Nanostructures for drug delivery*. Elsevier, pp 865–886

Steinhilber D, Witting M, Zhang X, Staegemann M, Paulus F, Friess W, Küchler S, Haag R (2013) Surfactant free preparation of biodegradable dendritic polyglycerol nanogels by inverse nanoprecipitation for encapsulation and release of pharmaceutical biomacromolecules. *J Contr Release 169*(3):289–295

Strebhardt K, Ullrich A (2008) Paul Ehrlich's magic bullet concept: 100 years of progress. *Nat Rev Cancer 8*(6):473–480

Suhail M, Rosenholm JM, Minhas MU, Badshah SF, Naeem A, Khan KU, Fahad M (2019) Nanogels as drug-delivery systems: a comprehensive overview. *Ther Deliv 10*(11):697–717

Suzuki M, Hatano K, Yoshida S, Yamashita Y (2021) Carbosilane dendrimer and aggregatable carrier obtained using said dendrimer for drug delivery system. Google Patents

Swami R, Singh I, Kulhari H, Jeengar MK, Khan W, Sistla R (2015) p-Hydroxy benzoic acid-conjugated dendrimer nanotherapeutics as potential carriers for targeted drug delivery to brain: an in vitro and in vivo evaluation. *J Nanopart Res 17*:1–11

Taghdisi SM, Danesh NM, Ramezani M, Lavaee P, Jalalian SH, Robati RY, Abnous K (2016) Double targeting and aptamer-assisted controlled release delivery of epirubicin to cancer cells by aptamers-based dendrimer in vitro and in vivo. *Eur J Pharm Biopharm 102*:152–158

Tan JP, Wang Q, Tam KC (2008) Control of burst release from nanogels via layer by layer assembly. *J Contr Release 128*(3):248–254

Taratula O, Schumann C, Naleway MA, Pang AJ, Chon KJ, Taratula O (2013) A multifunctional theranostic platform based on phthalocyanine-loaded dendrimer for image-guided drug delivery and photodynamic therapy. *Mol Pharm 10*(10):3946–3958

Theune LE, Buchmann J, Wedepohl S, Molina M, Laufer J, Calderón M (2019) NIR-and thermo-responsive semi-interpenetrated polypyrrole nanogels for imaging guided combinational photothermal and chemotherapy. *J Contr Release 311*:147–161

Thompson J, Dua R (2017) Nanogels for tissue engineering. The Royal Society of Chemistry

Tiwari G, Tiwari R, Sriwastawa B, Bhati L, Pandey S, Pandey P, Bannerjee SK (2012) Drug delivery systems: an updated review. *Int J Pharm Investig 2*(1):2–11

Tomalia DA (2005) Birth of a new macromolecular architecture: dendrimers as quantized building blocks for nanoscale synthetic polymer chemistry. *Prog Polym Sci 30*(3–4):294–324

Torchilin VP (2008) Cell penetrating peptide-modified pharmaceutical nanocarriers for intracellular drug and gene delivery. *Pept Sci 90*(5):604–610

Tran NQ, Nguyen CK, Nguyen TP (2013) Dendrimer-based nanocarriers demonstrating a high efficiency for loading and releasing anticancer drugs against cancer cells in vitro and in vivo. *Adv Nat Sci Nanosci Nanotechnol 4*(4):045013

Tripathy S, Das MK (2013) Dendrimers and their applications as novel drug delivery carriers. *J Appl Pharm Sci 3*(9):142–149

Trivedi V, Patel U, Bhimani B, Daslaniya D, Patel G, Vyas B (2012) Dendrimer: polymer of twenty-first century. *IJPRBS J 1*(2):1–21

Ulbrich K, Etrych T, Chytil P, Jelınková M, Říhová, B. (2003) HPMA copolymers with pH-controlled release of doxorubicin: in vitro cytotoxicity and in vivo antitumor activity. *J Contr Release 87*(1–3):33–47

Vinogradov SV (2007) Polymeric nanogel formulations of nucleoside analogs. *Expert Opin Drug Deliv 4*(1):5–17

Wang Y, Guo R, Cao X, Shen M, Shi X (2011) Encapsulation of 2-methoxyestradiol within multifunctional poly (amidoamine) dendrimers for targeted cancer therapy. *Biomaterials 32*(12):3322–3329

Wang H, Zhao X, Guo C, Ren D, Zhao Y, Xiao W, Jiao W (2015) Aptamer-dendrimer bioconjugates for targeted delivery of miR-34a expressing plasmid and antitumor effects in non-small cell lung cancer cells. *PLoS One 10*(9):e0139136

Wu H-C, Chen-Yun Y (2022) Peptide-conjugated nanoparticles for targeting, imaging, and treatment of prostate cancer. Google Patents

Wu S-Y, Debele TA, Kao Y-C, Tsai H-C (2017) Synthesis and characterization of dual-sensitive fluorescent nanogels for enhancing drug delivery and tracking intracellular drug delivery. *Int J Mol Sci 18*(5):1090

Xue Y, Xia X, Yu B, Luo X, Cai N, Long S, Yu F (2015) A green and facile method for the preparation of a pH-responsive alginate nanogel for subcellular delivery of doxorubicin. *RSC Adv 5*(90):73416–73423

Yallapu MM, Jaggi M, Chauhan SC (2011) Design and engineering of nanogels for cancer treatment. *Drug Discov Today 16*(9–10):457–463

Yin Y, Hu B, Yuan X, Cai L, Gao H, Yang Q (2020) Nanogel: a versatile nano-delivery system for biomedical applications. *Pharmaceutics 12*(3):290

Youn YS, Bae YH (2018) Perspectives on the past, present, and future of cancer nanomedicine. *Adv Drug Deliv Rev 130*:3–11

Zhang S, Xu Y, Wang B, Qiao W, Liu D, Li Z (2004) Cationic compounds used in lipoplexes and polyplexes for gene delivery. *J Contr Release 100*(2):165–180

Zhang H, Zhai Y, Wang J, Zhai G (2016) New progress and prospects: the application of nanogel in drug delivery. *Mater Sci Eng C 60*:560–568

Zhang Y, Dosta P, Conde J, Oliva N, Wang M, Artzi N (2020) Prolonged local in vivo delivery of stimuli-responsive nanogels that rapidly release doxorubicin in triple-negative breast cancer cells. *Adv Healthc Mater 9*(4):e1901101

Zhu J, Shi X (2013) Dendrimer-based nanodevices for targeted drug delivery applications. *J Mater Chem B 1*(34):4199–4211

# Chapter 7
# Nanocapsules in Dermal Drug Delivery

**Bancha Yingngam**

**Abstract** The skin acts as a great barrier against dermal drug delivery, posing a major problem associated with the low bioavailability of topically applied drugs. Submicron-sized carriers, termed nanocapsules, with a core–shell structure, appear to represent a promising approach to ameliorate these limitations and enhance drug delivery through the skin. In this chapter, the current state of the art in the application of nanocapsule-based dermal drug delivery is described briefly. This chapter aims to explain the main advantages and mechanisms by which nanocapsules increase dermal drug absorption through improved drug solubility, controlled release, and penetration through the skin. This chapter discusses different types of nanocapsules, including polymeric, lipidic, and hybrid systems; fabrication methods; and factors affecting their performance. Additionally, recent developments in the design of nanocapsules, including the use of stimuli-responsive and targeted delivery systems, are discussed. Overall, the platform of nanocapsules appears to be easily tunable for optimizing dermal drug delivery. Rationally designing nanocapsules and utilizing their unique properties allows researchers to devise new strategies for addressing the obstacles involving dermal drug delivery and treating skin diseases. However, ultimately, the long-term safety and scalability of these systems need to be addressed for their potential for successful clinical translation.

**Keywords** Controlled release · Lipid-based nanoparticles · Nanocarriers · Polymeric nanoparticles · Skin penetration enhancement · Stimuli-responsive nanomaterials · Targeted drug delivery · Transdermal drug delivery

B. Yingngam (✉)
Department of Pharmaceutical Chemistry and Technology, Faculty of Pharmaceutical Sciences, Ubon Ratchathani University, Ubon Ratchathani, Thailand
e-mail: bancha.y@ubu.ac.th



V. Khanna et al. (eds.), *Nanotechnologies for Drug Delivery and Biopharmaceutical Development*, Progress in Drug Research 77,
https://doi.org/10.1007/978-3-032-00796-4_7

## Abbreviations

| | |
|---|---|
| EGFR | Epidermal growth factor receptor |
| HA | Hyaluronic acid |
| LbL | layer-by-layer |
| PDGF | Platelet-derived growth factor |
| PLA | Poly(lactic acid) |
| PLGA | Poly(lactic-co-glycolic acid) |
| VEGF | Vascular endothelial growth factor |

## 7.1 Introduction

The outer covering or skin is the largest organ in the human body and acts as the first line of defense against any substance, including a drug. One of the greatest challenges for dermal drug delivery is barrier function, which, according to most of the literature, is assigned to the stratum corneum (Farago et al. 2024). The traditional oil-in-water emulsion systems used in topical preparations have difficulty overcoming this barrier, thus resulting in poor drug uptake and hence bioavailability (Gomes et al. 2024). Indeed, the physicochemical characteristics of most drugs, such as hydrophilicity, high molecular weight, and charged nature, make permeation across the skin barrier even more challenging (Abdel-Emam et al. 2024). This makes it appropriate to search for new approaches for skin targeting and optimizing the therapeutic effects of dermatological drugs.

Dermal drug delivery systems have limitations in that nanotechnology is a promising approach. The limitations of traditional dermal drug delivery systems have prompted the emergence of nanotechnology as a promising approach to overcome these limitations. Among all the nanocarriers investigated, nanocapsules have received much interest because of their special structure and various properties (Dahmash et al. 2024). The core–shell structure of the nanocapsule systems consists of a submicron-sized core, which is the drug-containing core, and a polymeric or lipid shell (Aboomeirah et al. 2024). The advantages of such an architecture include the protection of the incorporated active substance from degradation, controlling release kinetics, and the ability to guide the deposition to targeted skin strata or appendages (Karami et al. 2024). Furthermore, owing to their small dimensions, nanocapsules improve skin penetration and permeation efficiency and even increase the site bioavailability of drugs (Do et al. 2024). These properties make nanocapsules attractive new platforms for dermal drug delivery that may revolutionize therapy for skin disease.

This chapter describes the use of nanocapsules for dermal drug delivery. This review highlights different types of nanocapsules, primarily polymeric, lipidic and hybrid systems, along with their properties and potential applications. Second, different strategies for the synthesis of nanocapsules are discussed in terms of interfacial polymerization, nanoprecipitation, and emulsion-based approaches. The mechanisms of dermal drug delivery, which are mediated by controlled release, enhanced penetration and targeted delivery of more soluble compounds, are discussed through the lens of the use of nanocapsules. Moreover, this chapter evaluates how size, surface properties, drug loading, and skin conditions affect nanocapsule performance. Advanced nanocapsule designs, including stimuli-responsive and functionalized systems along with their potential applications in combination with therapy and diagnostics, will also be presented in this chapter. In conclusion, this chapter discusses how nanocapsules can be applied, including industrial cases, the formulation process, the optimization of formulations, the safety and biocompatibility of nanocapsules, and the path from the bench for biomedical applications. At the end of this chapter, readers understand state-of-the-art nanocapsule-based dermal drug delivery and the future prospects of this interesting field.

## 7.2 Types of Nanocapsules for Dermal Drug Delivery

Dermal drug delivery nanocapsules fall into three main types on the basis of what they are made of: those built from lipids, those made from polymers, and hybrid versions that combine different materials. Each type of nanocapsule has unique benefits and is appropriate for loading dissimilar classes of drugs.

### *7.2.1 Lipid-Based Nanocapsules*

Lipid-based nanocapsules, known as lipid nanocapsules in the present chapter, exhibit an oily core confined to a solid or semisolid lipid shell (Fig. 7.1a) (Aparicio-Blanco et al. 2024). An overview of their strengths and weaknesses is provided in Table 7.1. Core materials, including triglycerides, fatty acids, and waxes, can be employed, while phospholipids or surfactants mainly construct the shell of the particle (Yingngam et al. 2019). However, the described approach is particularly suitable for the formation of lipid-based nanocapsules for lipophilic drugs since the lipid core helps in drug solubilization and stabilization (Ibrahim et al. 2024). Compared with conventional systems, lipid-based nanocapsules are small in size and biocompatible, thereby resulting in enhanced percutaneous penetration while minimizing skin redness or irritation, even at higher concentrations of the formulation (Otavi et al. 2024). Furthermore, the solid lipid shell offers advantages of occlusive properties since it is capable of holding drugs within the skin, resulting in prolonged drug treatment (Alam et al. 2024). The dermal use of lipid composite nanocapsules has been reported for various drugs, including antioxidants, anti-inflammatory agents, and anticancer drugs, which have been successfully used (Feng et al. 2025; Tiatragoon et al. 2024).

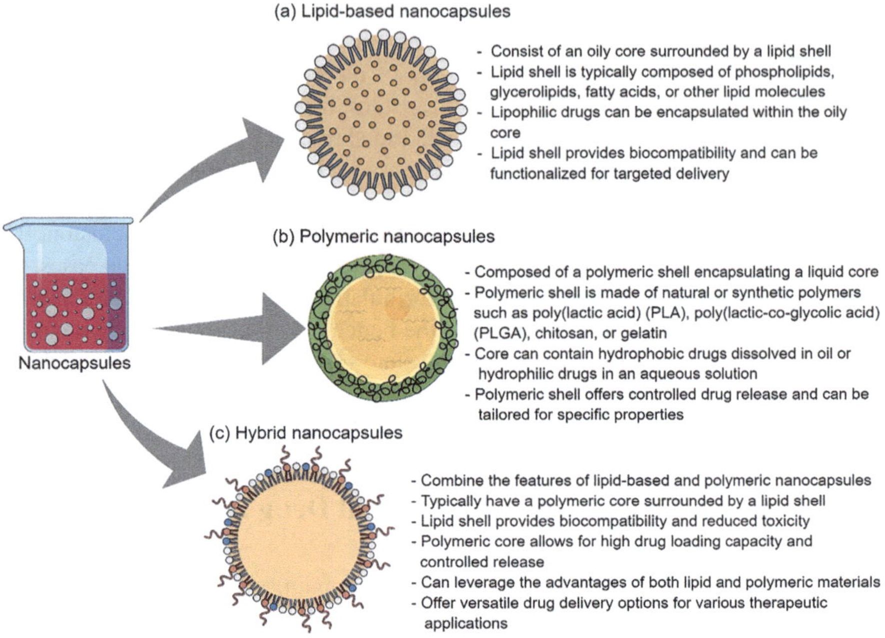


**Fig. 7.1** Schematic representation of the structures of (**a**) lipid-based nanocapsules, (**b**) polymeric nanocapsules, and (**c**) hybrid nanocapsules (generated via the licensed Mind the Graph software, with no permission needed)

## 7.2.2 Polymeric Nanocapsules

Polymeric nanocapsules are polymeric shells surrounding a liquid or semisolid core containing a drug (Fig. 7.1b) (Gomes et al. 2024). Typically, the polymeric shell is made of biodegradable and biocompatible polymers, such as poly(ε-caprolactone), poly(lactic acid) (PLA), poly(lactic-co-glycolic acid) (PLGA), and chitosan (Farago et al. 2024; Naser et al. 2024). The mechanical stability of these polymers is excellent; they can be used for controlled drug release and surface property modification for targeted delivery (Naser et al. 2024). Polymeric nanocapsules are capable of containing drugs with variable sizes and solubilities (lipophilic and hydrophilic) and macromolecules such as proteins and peptides (Mei et al. 2024). Administration is potentially adaptable to other types of cancer and can be combined with the use of analgesic drugs (Oliveira Santos et al. 2024). Drug release from polymeric nanocapsules is affected by the polymer composition, molecular weight, and crosslink drugs or nanocapsules (Can Karaca et al. 2025). Moreover, the surface of polymeric nanocapsules can be functionalized with ligands or antibodies for targeted delivery to specific skin receptors or diseased sites (Karati et al. 2024). Because of these positive attributes, polymeric nanocapsules are promising vehicles for controlled and targeted dermal drug release (Sethi et al. 2024).

**Table 7.1** Advantages and limitations of polymeric, lipid-based, and hybrid nanocapsules for dermal drug delivery

| Nanocapsule types | Advantages | Limitations | References |
|---|---|---|---|
| Lipid-based nanocapsules | Excellent biocompatibility | Lower stability compared to polymeric systems | Desai et al. (2024) |
| | Enhanced skin penetration | Potential drug leakage during storage | |
| | High drug loading capacity for lipophilic drugs | Temperature sensitivity | |
| | Natural composition reduces toxicity concerns | Limited mechanical strength | |
| | Easier scale-up process | Variable drug release rates | |
| | Cost-effective production | Possible oxidation of lipid components | |
| | Self-assembling properties | Limited long-term stability | |
| | Good skin adhesion properties | Batch-to-batch reproducibility challenges | |
| Polymeric nanocapsules | High stability during storage and application | Potential toxicity of some synthetic polymers | Gomes et al. (2024) |
| | Controlled and sustained drug release | Limited drug loading capacity for hydrophilic drugs | |
| | Tunable degradation rates | Sometimes requires organic solvents in preparation | |
| | Versatile surface modification options | Batch-to-batch variability | |
| | Protection of sensitive drug molecules | Potentially high production costs | |
| | Good mechanical strength | Complex scale-up process | |
| | Biodegradability (with appropriate polymers) | Possible inflammatory responses | |
| | Customizable size and surface properties | Limited penetration of larger particles | |
| Hybrid nanocapsules | Combines advantages of both systems | Complex manufacturing process | Yingngam et al. (2024) |
| | Enhanced stability compared to lipid-only systems | Higher production costs | |
| | Improved drug loading capacity | Challenging quality control | |
| | Better control over drug release | Complex characterization requirements | |
| | Versatile functionality | Limited standardization | |
| | Both hydrophilic and lipophilic drug delivery | Regulatory complexity | |
| | Tunable surface properties | Scale-up challenges | |
| | Enhanced penetration capabilities | Limited long-term stability data | |

### 7.2.3 *Hybrid Nanocapsules*

Novel hybrid nanocapsules based on both polymeric and lipidic systems have emerged as versatile platforms for dermal drug delivery that can be achieved by complementing the best properties of the systems. Typically, such nanocapsules possess some material that encircles the core, which could likewise be lipid or the other way around (Fig. 7.1c) (Yadava et al. 2024). The lipid aspect helps with solubilization of the drug and compatibility with skin, and the polymeric aspect provides mechanical stability and controlled-release capabilities (Santhanakrishnan et al. 2024). Hybrid nanocapsules are fabricated via different techniques, such as nanoprecipitation, emulsion–solvent evaporation, and self-assembly methods (Sengel-Turk et al. 2024). Both lipids and polymers can combine, allowing the encapsulation of various drugs possessing different physicochemical properties (Iglicki et al. 2024). Additionally, hybrid nanocapsules can be customized to react to particular stimuli, such as pH or temperature changes, to trigger drug delivery to the area of interest (Farnoudian-Habibi et al. 2024). This dual system in which polymers and lipids are synergistically incorporated to prepare hybrid nanocapsules is an attractive opportunity to increase dermal drug delivery safety and efficacy.

## 7.3 Fabrication Methods for Nanocapsules

Notably, nanocapsules that perform dermal drug delivery have been prepared, and numerous groups of researchers have produced nanocapsules. To achieve the desired characteristics, such as size, morphology and drug entrapment efficiency, of nanocapsules, many strategies with different approaches have been utilized. This section describes four common fabrication approaches: interfacial polymerization, nanoprecipitation, emulsion-based methods, and layer-by-layer assembly (Fig. 7.2).

### 7.3.1 *Interfacial Polymerization*

Another well-established process to obtain polymeric nanocapsules is based on polymerization under interfacial conditions. The approach used most widely to create emulsions is to combine two nonmiscible phases, such as aqueous and organic phases (Farago et al. 2024). The emulsion is stabilized by the surfactant aqueous phase, while the drug and the monomers are in the organic phase (Amantino et al. 2024). The process involves polymerization at the interface of the two phases, resulting in the formation of a polymeric shell around the drug (Wang et al. 2024a). The polymer concentration and reaction parameters govern the thickness and permeability of the shell (Dahmash et al. 2024). Various drugs (retinoids, antibiotics and anticancer agents) have been loaded into and successfully prepared through the interfacial polymerization of nanocapsules.

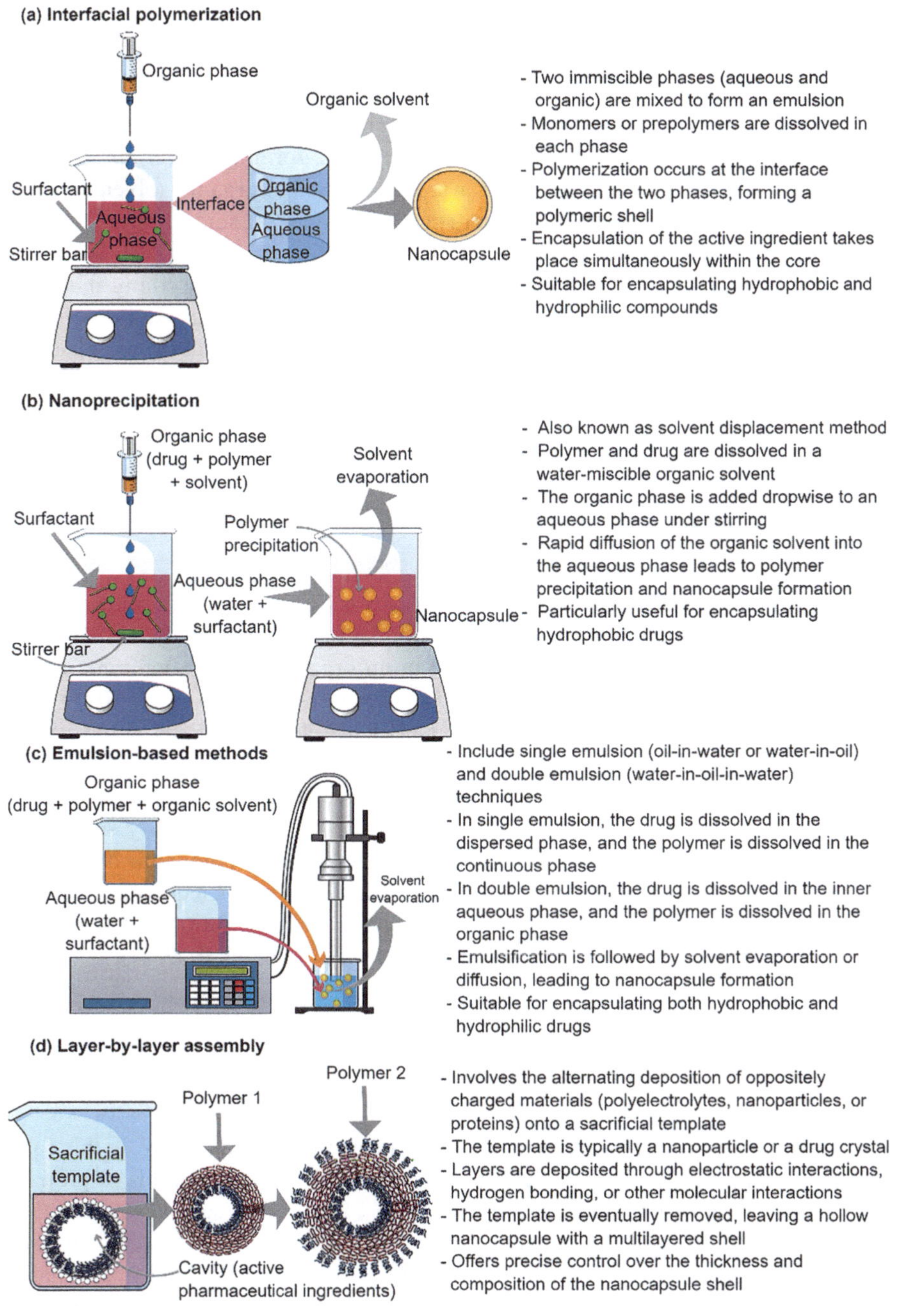


**Fig. 7.2** Schematic diagram of the fabrication methods for nanocapsules: (**a**) interfacial polymerization, (**b**) nanoprecipitation, (**c**) emulsion-based methods, and (**d**) layer-by-layer assembly (generated from the licensed Mind the Graph software, with no permission required)

### 7.3.2 Nanoprecipitation

A simple and fast method, 'nanoprecipitation' (or 'solvent displacement' or 'interfacial deposition' methods) (Chai et al. 2024), is used to prepare polymeric or lipid-based nanocapsules. The drug and the nanocapsule-forming materials, i.e., the polymers or the lipids, are solubilized in a water-miscible organic solvent via this method. Nanocapsule formation is spontaneous when an organic solution is added dropwise into a solution of the surfactant in water (Pereira et al. 2024). When the organic solvent dissolves in aqueous solution, the nanocapsule-forming materials precipitate and self-assemble into nanocapsules (sealing the drug within the nanocapsule) (Chiarentin et al. 2024). For encapsulating hydrophobic drugs, nanoprecipitation is particularly suitable and has been widely applied for the preparation of polymeric nanocapsules made from PLA, PLGA, and PLC (Kumar et al. 2024b).

### 7.3.3 Emulsion-Based Methods

The most common method to obtain nanocapsules is by means of different emulsion methods (single emulsification and double emulsification). Using the single emulsion approach, the drug is then suspended in an organic phase containing a high concentration of the nanocapsule-creating materials (Ahmadian et al. 2024). The previous organic phase is subsequently emulsified into an aqueous phase with a surfactant using high-energy techniques (ultrasonication or high-pressure homogenization) (Cerro et al. 2024). Nanocapsules are obtained via evaporation/extraction of the organic solvent (Saleh et al. 2023). This method is particularly suitable to load hydrophilic drugs using a water-in-oil-in-water (w/o/w) emulsion- based double emulsion method (Ahmadian et al. 2024). Emulsion-based techniques support the formation of nanocapsules with desired payload encapsulation efficiency and adjustable nanocapsule size (Ahmadian et al. 2024; Bufalini et al. 2024).

### 7.3.4 Layer-by-Layer Assembly

Such types of nanocapsules are obtained using layer-by-layer assembly. In this case, the use of charges is applied to the drugcore envelop by different lipids or proteins or the application of opposite charge (Xie et al. 2024b). The second layer is then added to result in a multilayered nanocapsule which can be continued to the required number of layers. (Gao et al. 2024). The thickness, composition, and functionality of the shell assembling are controlled with high precision through layer-by-layer (LbL) assembly (Ramamoorthy et al. 2024). Moreover, it is possible to achieve specific molecular answer to a stimulus, thereby allowing the controllable delivery of a drug (Motta et al. 2024). More recently, LbL-assembled nanocapsuls have been used as vesicular systems for dermal delivery of active principles like antioxidants or growth factors or also DNA (Iqbal et al. 2024; Narayanan et al. 2024).

In Table 7.1, the four main methods of nanocapsule fabrication are compared: interfacial polymerization, nanoprecipitation, emulsion-based methods and layer-by-layer assembly methods. Nanocapsules can be widely produced through different methods, each with its own advantages and disadvantages. Interfacial polymerization can have precise control over the process of shell formation, but there are toxicity concerns associated with it; nanoprecipitation is simple and mild, but hydrophilic drug loading can be poor; emulsion and emulsion/solvent evaporation techniques are versatile and can be scaled up but require organic solvents and multiple steps; and LbL assembly provides precise control of the final shell properties but is time consuming and labor intensive. Low-energy solvent evaporation methods include, but are not limited to, various types of solvent evaporation methods that adjust the environmental conditions, resulting in a low-energy environment for the produced substance and allowing proper conformational interactions without breaking bonds too quickly, resulting in particles that match the desired characteristics for the active pharmaceutical ingredient in the drug. Table 7.2, given below shows a Comparison of various methods for fabricating nanocapsules

**Table 7.2** Comparison of various methods for fabricating nanocapsules

| Fabrication methods | Principles | Advantages | Disadvantages | References |
|---|---|---|---|---|
| Interfacial polymerization | Polymerization of monomers at the interface of two immiscible phases | One-step process | Potential residual monomer toxicity | Farago et al. (2024) |
| | Formation of polymer membrane at the interface | High encapsulation efficiency | Complex reaction conditions | |
| | *In situ* shell formation through chemical reaction | Controlled shell thickness | Limited polymer selection | |
| | Cross-linking at the droplet surface | Good mechanical strength | Difficult to control reaction kinetics | |
| | | Suitable for various core materials | Possible cross-reactions | |
| | | Tunable membrane properties | Scale-up challenges | |

(continued)

**Table 7.2** (continued)

| Fabrication methods | Principles | Advantages | Disadvantages | References |
|---|---|---|---|---|
| Nanoprecipitation | Spontaneous emulsification through solvent displacement | Simple and rapid process | Limited to specific polymer-solvent pairs | Amantino et al. (2024) |
| | Diffusion of organic solvent into aqueous phase | Mild preparation conditions | Poor loading of hydrophilic drugs | |
| | Polymer precipitation at the interface | No high-energy input required | Residual organic solvent concerns | |
| | Self-assembly of polymer chains | Small particle size | Batch-to-batch variability | |
| | | Narrow size distribution | Limited drug loading capacity | |
| | | Good reproducibility | Scale-up limitations | |
| Emulsion-based methods | Formation of oil-in-water (O/W) or water-in-oil (W/O) emulsions | Versatile process | Use of organic solvents | Ahmadian et al. (2024) |
| | Solvent evaporation/ extraction | High drug loading capacity | Multiple processing steps | |
| | Polymer precipitation and shell formation | Well-established methodology | Energy-intensive process | |
| | Stabilization by surfactants | Scalable production | Broad size distribution | |
| | | Suitable for various materials | Stability challenges | |
| | | Good batch reproducibility | Surfactant removal needed | |
| Layer-by-layer assembly | Sequential deposition of oppositely charged layers | Precise control over shell properties | Time-consuming process | Gao et al. (2024) |
| | Electrostatic interactions between layers | No organic solvents needed | Multiple washing steps | |
| | Surface charge alternation | Room temperature process | Limited to charged materials | |
| | Stepwise shell construction | Multiple functionality possible | Labor-intensive | |
| | | High reproducibility | Complex optimization | |
| | | Fine control over thickness | Difficult scale-up | |

## 7.4 Mechanisms of Nanocapsule-Mediated Dermal Drug Delivery

Compared with conventional formulations, nanocapsules have many advantages as dermal drug carriers. The enhancement of drug delivery across the skin barrier through nanocapsules occurs predominantly through improved drug solubility and stability, controlled and sustained release, increased skin penetration and permeation, and targeted delivery to specific skin layers or appendages (Fig. 7.3). This section describes these mechanisms in detail.

### *7.4.1 Enhanced Drug Solubility and Stability*

Poor aqueous solubility is a common problem with many drugs that could have therapeutic benefits in skin diseases but are ineffective in conventional topical formulations (Pan et al. 2024). In fact, approximately 40% of approved drugs and close to 90% of drug candidates in the development pipeline are estimated to suffer from poor water solubility, which can significantly impact their bioavailability and

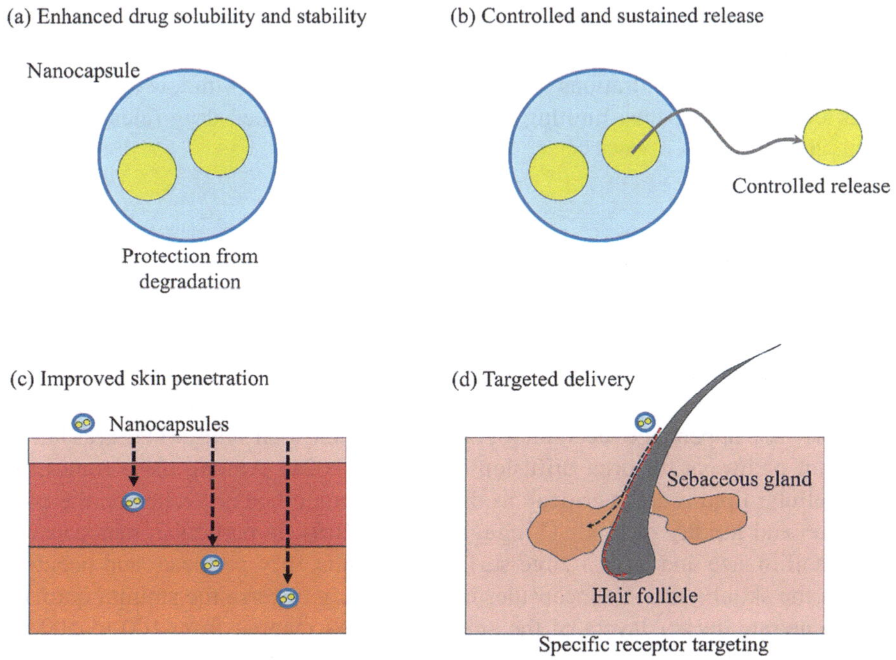


**Fig. 7.3** Mechanisms of nanocapsule-mediated dermal drug delivery: (**a**) enhanced drug solubility and stability, (**b**) controlled and sustained release, (**c**) improved skin penetration, and (**d**) targeted delivery. (created by the author, with no permission required)

efficacy (Xie et al. 2024a). The problem is that hydrophobic drugs are poorly soluble when dissolved, so nanocapsules with a core can encapsulate the drug to increase its solubility and bioavailability (Yang et al. 2024b). The lipids or polymers in the nanocapsule core are termed the microenvironment in which hydrophobic drugs can be solubilized (Kabil et al. 2024). They also protect drugs from degradation caused by external factors such as light and oxygen, as well as the drug's own enzymes once they are encapsulated in nanocapsules (da Costa Rodrigues et al. 2024). The improved solubility and stability of drugs in nanocapsules confer improved dermal delivery and therapeutic efficacy (Dahmash et al. 2024).

### *7.4.2 Controlled and Sustained Drug Release*

Nanocapsules can be designed to release a drug in a controlled, sustained manner suitable for drugs intended for dermal delivery. As a result, various elements, including release modifiers, can affect the release kinetics of drugs from nanocapsules as well as the composition and thickness of the shell particles along with the characteristics of the drug (Yu et al. 2024). For example, drug release from polymeric nanocapsules with thicker or cross-linked shells is slower than that from nanocapsules with thinner or noncross-linked shells (Foroutani et al. 2024). Nanocapsules enable sustained drug release, which promotes prolonged drug retention in the skin, thereby reducing dosing frequencies and increasing patient compliance (Gomes et al. 2024). Moreover, drug concentrations remain within the therapeutic window with a limited peak concentration, thus limiting side effects via controlled drug release (Heikal et al. 2024).

### *7.4.3 Enhanced Skin Penetration and Permeation*

Specifically, the skin provides three potential pathways for drug permeation, namely, the intercellular lipid pathway, the transcellular pathway and the appendageal pathway via hair follicles and sweat glands (Busch et al. 2024; Khan et al. 2024). Notably, skin appendages account for only 0.1% of the total skin surface area but act as shunt pathways for drug diffusion (Akl et al. 2024). Among these routes, the intercellular lipid route is critical, as drug absorption efficiency relies on the composition and fluidity of lipids (Yingngam et al. 2019). In particular, nanocapsules are small in size and have a large surface area; thus, they permeate and penetrate across the skin barrier. Nanocapsules can effectively traverse the stratum corneum and penetrate deeper layers of the skin, with sizes ranging from 100 to 500 nm (Yingngam et al. 2019).

Figure 7.4 shows the influence of nanocapsule size on the penetration ability of the skin in three size ranges: small nanocapsules (<100 nm) possess better penetration ability because their small size allows them to reach the deep dermis layer; the

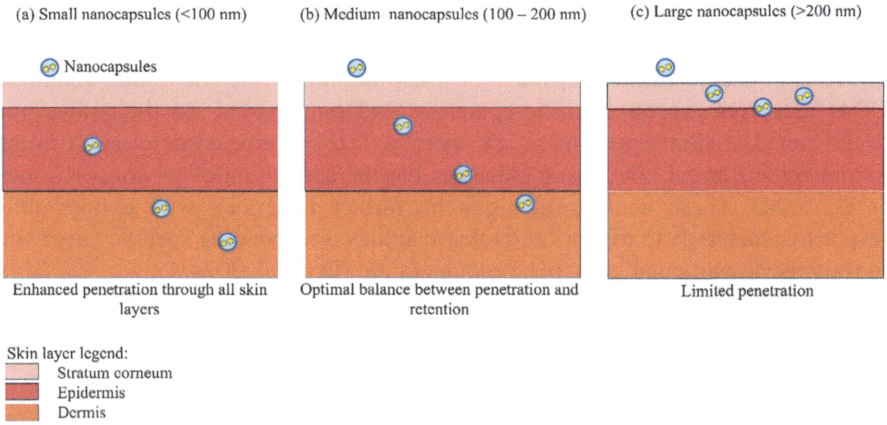


**Fig. 7.4** Effect of nanocapsule size on skin penetration. The different categories in terms of size for nanocapsules are as follows: (**a**) small nanocapsules, which are <100 nm in size and exhibit high penetration; (**b**) medium nanocapsules, which are in the size range of 100–200 nm, demonstrating optimal penetration and retention; and (**c**) large nanocapsules greater than 200 nm in size and showing poor penetration. (Created by the author, with no permission required)

next size increase (100–200 nm) provides an optimal balance of good penetration with good retention time in the target skin layers; and large nanocapsules (>200 nm) show limited penetration ability, being unable to pass through skin barriers and remaining primarily at the stratum corneum and upper epidermis owing to their size. Given that penetration behavior plays an important role in determining the depth of drug delivery and how long it lasts, this work will inform the design of dermal drug delivery systems for the treatment of a wide variety of diseases. Nanocapsules are usually absorbed through several routes—intercellular and follicular (Yingngam et al. 2019). After nanocapsules penetrate the stratum corneum, they can allow low inflammatory drug release in a localized manner (An et al. 2024). Nanocapsules can also be surface modified with penetration enhancers or cell-penetrating peptides for better permeation on the skin (Bhavsar et al. 2024; Toussaint et al. 2024). Hence, the perforation and penetration of drugs through nanocapsules increases drug bioavailability and therapeutic efficacy (Dahmash et al. 2024).

### *7.4.4 Targeted Delivery to Specific Skin Layers or Applications*

The fact that their formulation is clear makes it easier to design these nanocapsules to deliver drugs to certain layers of the skin or appendages, such as hair follicles or sebaceous glands, for more targeted drug delivery (Desai et al. 2024). Ligands or antibodies on the surface of nanocapsules are being modified to allow delivery to the targeted site (Lei et al. 2024). For example, folate-functionalized nanocapsules can be used for targeted delivery of anticancer drugs to skin cancer cells where surface folate receptors are overexpressed (Di Filippo et al. 2023). Likewise, hyaluronic acid-decorated nanocapsules may also be directed to the highly abundant CD44 receptors expressed in the epidermis and can be used for the delivery of drugs aimed at skin regeneration (Desai et al. 2024). Targeted delivery of drugs minimizes off-target effects and enhances the therapeutic index of drugs (Gomes et al. 2024). These nanocapsules can be further integrated with extrastimuli-responsive materials to trigger drug release at the corresponding specific target site in response to particular cues, pH, or temperature (Do et al. 2024).

### *7.4.5 Nanocapsules for Dermal Delivery of Biopharmaceuticals*

Nanocapsules are now gaining importance in the dermal delivery of biopharmaceuticals, as proteins, peptides, antibodies and nucleic acids fall within the realm of biopharmaceuticals, in which nanocapsules can easily be used as alternatives to existing formulations. Biopharmaceuticals are usually limited in topical applications owing to their considerable molecular weight, high hydrophilicity, and vulnerability to enzymatic degradation in the skin (Petrovic et al. 2024). Nanocapsules protect the environment, protecting sensitive biomolecules from degradation and promoting their penetration through the skin barrier.

#### 7.4.5.1 Protein and Peptide Delivery

Polymeric and lipid nanocapsules have been used as dermal carriers of several therapeutic proteins and peptides. In particular, chitosan and phospholipid-based hybrid nanocapsules have been fabricated that deliver insulin for the treatment of microorganism-aggravated diabetic wounds, demonstrating a superior ability to promote wound healing than free insulin formulations (Farasati Far et al. 2023). These nanocapsules protected insulin from proteolytic degradation and led to sustained insulin release at the wound site. Tang et al. (2024) also prepared platelet-derived growth factor (PDGF)-loaded PLGA-based nanocapsules for chronic wound healing. These nanocapsules exhibited enhanced protein stability, controllable release kinetics, and improved dermal penetration, leading to accelerated re-epithelialization in animal wound models. Importantly, this nanocapsule shell-core configuration was effective at maintaining the secondary structure of PDGF and thus maintaining its biological activity.

#### 7.4.5.2 Nucleic Acid Delivery

Nanocapsules also have potential applications in the area of biopharmaceuticals, particularly for the transdermal delivery of nucleic acids such as small interfering RNAs (siRNAs), microRNAs (miRNAs), and plasmid DNA. Sallam et al. (2021) also covered cationic lipid nanocapsules for inflammatory cytokine targeting through the delivery of siRNA in psoriasis. The data showed that anocapsules can protect siRNAs from degradation by nucleases and significantly increase their uptake by keratinocytes, which in turn represses markers of inflammation in a psoriasis skin model. Gene therapy applications do not suffer from either of these issues, as described by Marinho et al. (2021), who explored the polyepoxydation of polymeric nanocapsules coated with hyaluronic acid for the delivery of plasmid DNA encoding vascular endothelial growth factor (VEGF) for skin regeneration. These nanocapsules successfully transfect skin cells to produce local VEGF and have superior angiogenesis-inducing capabilities compared with other formulations in a skin wound healing model.

#### 7.4.5.3 Antibody Delivery

Monoclonal antibodies constitute an additional class of biopharmaceuticals that could benefit from nanocapsule-mediated skin delivery. Kang et al. (2022) reviewed lipid–polymer hybrid nanocapsules for the topical delivery of anti-tumor necrosis factor-alpha (TNF-$\alpha$) antibodies for treating inflammatory skin conditions. The nanocapsules maintained the antibody structure and function while allowing penetration into deeper layers of skin, thus demonstrating improved efficacy relative to standard formulations. The mechanisms that allow transdermal delivery of biopharmaceuticals by nanocapsules are as follows:

- Enzymatic degradation is prevented because of encapsulation in the core–shell structure
- Enhanced skin penetration via hair follicle portals and skin lipid interactions
- Failure to deliver release profiles that maintain therapeutic levels at the target site
- In addition, stimuli-responsive materials can also be conjugated to nanocapsules to facilitate the tailored release of biopharmaceuticals in diseased skin, for example, altered pH, increased pH, or specific enzyme signaling (Sudharsan et al. 2025).

## 7.5 Factors Influencing Nanocapsule Performance

The four important variables that drive the interaction of capsule particles with the skin and progressively enhance dermal drug delivery—nanocapsule size and surface properties, drug loading and encapsulation efficiency, nanocapsule composition and materials, skin conditions and barrier function—will be emphasized in this section, guided by the principles of dermal host defense. Understanding these parameters is critical to help tailor nanocapsules to achieve the desired treatment effects. These factors impacting nanocapsule performance are described in detail below.

### *7.5.1 Nanocapsule Size and Surface Properties*

The crucial properties influencing skin permeability, drug delivery, and cell uptake include the size and surface of such nanocapsules. Smaller nanocapsules are more permeable to the skin than larger nanocapsules are because smaller carriers have greater potential to penetrate the intercellular lipid spaces of the stratum corneum, usually 200 nm or less (Yingngam et al. 2019). However, smaller nanocapsules (< 10 nm) can be rapidly cleared from the skin. Therefore, the potential size range generally required for the dermal delivery of drugs is 100–200 nm (Yingngam et al. 2019). Likewise, the interactions of the skin are also affected by the surface charge of the nanocapsules, as the modification of the positive charge was observed to result in greater adhesion and penetration capacity through the skin than negatively charged or uncharged nanocapsules (Bhavsar et al. 2024). In fact, the stability and prevention of nanocapsule aggregation for improved performance are further enhanced by the use of functional groups (e.g., polyethylene glycol) on the surface of nanocapsules (Aboomeirah et al. 2024).

### *7.5.2 Drug Loading and Encapsulation Efficiency*

The loading capacity and encapsulation efficiency of nanocapsules are important parameters for determining their therapeutic efficacy. On the other hand, drug loading is defined as the amount of drug per gram of the nanocapsules, whereas encapsulation efficiency is the percentage of drug that is to be encapsulated according to the total amount of drug used (Yingngam et al. 2019). Therefore, higher drug loading and encapsulation efficiency would be ideal to deliver the lowest possible doses of the nanocapsules (Karami et al. 2024). Drug–nanocapsule interactions, the viability of water-insoluble drugs in core formulations, and fabrication techniques all influence nanocapsule drug loading and encapsulation efficiency (Yingngam et al. 2019). These results show that such parameters should not be neglected in future studies aiming to ensure high biomolecule loading together, alongside high encapsulation efficiency, which would enable dermal drug delivery to be established (Karami et al. 2024).

### 7.5.3 Nanocapsule Composition and Materials

These parameters are also important for dermal drug delivery of such nanocapsule formulations. The drug loading and release capabilities as well as the compatibility with biological systems and stability of nanocapsules are determined by the choice of the core as well as that of the surrounding shell materials (Yingngam et al. 2019). Compared with lipid-based materials, polymer-based materials can encapsulate different classes of drugs regardless of their solubility and are better suited for the encapsulation of lipophilic drugs (Abdel-Mottaleb et al. 2011; Kamel et al. 2024; Simon et al. 2020). The biodegradability and biocompatibility of compounds such as chitosan and PLGA allow highly effective and safe drug delivery to deeper layers of the skin (Abdel-Emam et al. 2024; Gomes et al. 2024). In addition, pH- or temperature-sensitive polymers enable target-specific absorption and release of therapeutic cargo when exposed to stimuli associated with skin diseases (Dahmash et al. 2024).

### 7.5.4 Skin Conditions and Barrier Function

The performance of nanocapsules in dermal drug delivery is significantly influenced by skin conditions and barrier function. The stratum corneum (the outermost layer of the skin) is an important barrier for exogenous substances, including nanocapsules, but is permeable for delivery. The thickness and integrity of the stratum corneum layers differ according to age, body site and skin disease (Yingngam et al. 2019). Indeed, in patients suffering from inflammatory skin diseases, such as atopic dermatitis and psoriasis, the stratum corneum barrier is often impaired and therefore leads to increased absorption of nanocapsules (Gomes et al. 2024). In contrast, in aged or dry skin, the stratum corneum barrier may become thickened and unable to absorb nanocapsules (Jose et al. 2024; Sethi et al. 2024). Therefore, the natural history of skin disorders and the barrier functionality of the target population are key to designing nanocapsule systems that can overcome the skin barrier and deliver drugs to the correct target site of action.

## 7.6 Advanced Nanocapsule Designs for Dermal Drug Delivery

Nanocapsules with improved functionality and performance for dermal drug delivery have been developed on the basis of advances in nanotechnology. Nanocapsules are categorized into stimuli-responsive nanocapsules, functionalized nanocapsules (targeted delivery), a wide range of nanocapsules for combination therapy, and diagnostic and theranostic applications. The following section presents these drug release systems and their potential applications in dermal drug delivery.

### 7.6.1 Stimuli-responsive Nanocapsules

In this method, the drug cargo of nature is liberated when stimulated by particular internal or external stimuli, for example, pH, temperature, light and compounds (Fig. 7.5) (Dahmash et al. 2024). These can also be fashioned to respond to the native microenvironment of the skin, which could activate and modulate their delivery (Do et al. 2024). For example, pH-sensitive nanocapsules can be covalently linked to therapeutic agents for release at the skin surface in situations with high matching pH in the source skin tissue or low pH in the skin tumor environment (Toussaint et al. 2024). Temperature-responsive nanocapsules can also release drugs in a controllable manner under psoriasis or dermatitis conditions, taking advantage of the increased temperature of inflamed skin (Yadav et al. 2024). Light-responsive nanocapsules can release drugs in response to specific wavelengths of light, enabling spatiotemporal control of drug delivery (Liu et al. 2024). Enzyme-responsive nanocapsules with skin-specific lipases or proteases allow the release of drugs specifically anywhere (Pawar et al. 2023). Stimuli-responsive nanocapsules could help prevent premature drug release and improve the therapeutic effect of dermal drug delivery systems (Do et al. 2024).

Table 7.3 presents a comprehensive overview of stimuli-responsive nanocapsules designed for dermal drug delivery, revealing six distinct triggering mechanisms: pH-responsive systems (using chitosan and Eudragit), temperature-sensitive formulations (utilizing PNIPAM and lipid-based systems), light-activated carriers (incorporating gold and upconverting nanoparticles), ultrasound-responsive systems (based on perfluorocarbons and polymer-shelled nanobubbles), magnetic field-responsive carriers (using iron oxide and lipid-based magnetic nanocapsules), and enzymatic-responsive systems (employing peptide and polysaccharide-based nanocapsules), each designed to achieve specific therapeutic objectives ranging from targeted delivery to controlled release in response to various physiological and external stimuli.

### 7.6.2 Functionalized Nanocapsules for Targeted Delivery

A better delivery of the drug to the target site is achieved through targeted drug delivery, which is designed using functionalized nanocapsules that target specific skin cells, receptors, or pathways (Fig. 7.6). This approach demonstrates that, in theory, one can functionalize a nanocapsule with a targeting ligand that has other molecules, such as antibodies, peptides, or aptamers, to bind to particular molecules in the skin (Hussain et al. 2024). For example, if we functionalize an antibody against the epidermal growth factor receptor (EGFR) on the surface of nanocapsules, these nanocapsules selectively target and deliver anticancer drugs to skin cancer cells overexpressing EGFR (Desai et al. 2024). Thus, anti-inflammatory agents can be conjugated to the core or shell of nanocapsules developed from hyaluronic acid, which target CD44 receptors that are overexpressed in psoriatic skin (Adamus-Grabicka et al. 2024; Fernández-Mariño et al. 2023). Additionally, suitably chosen

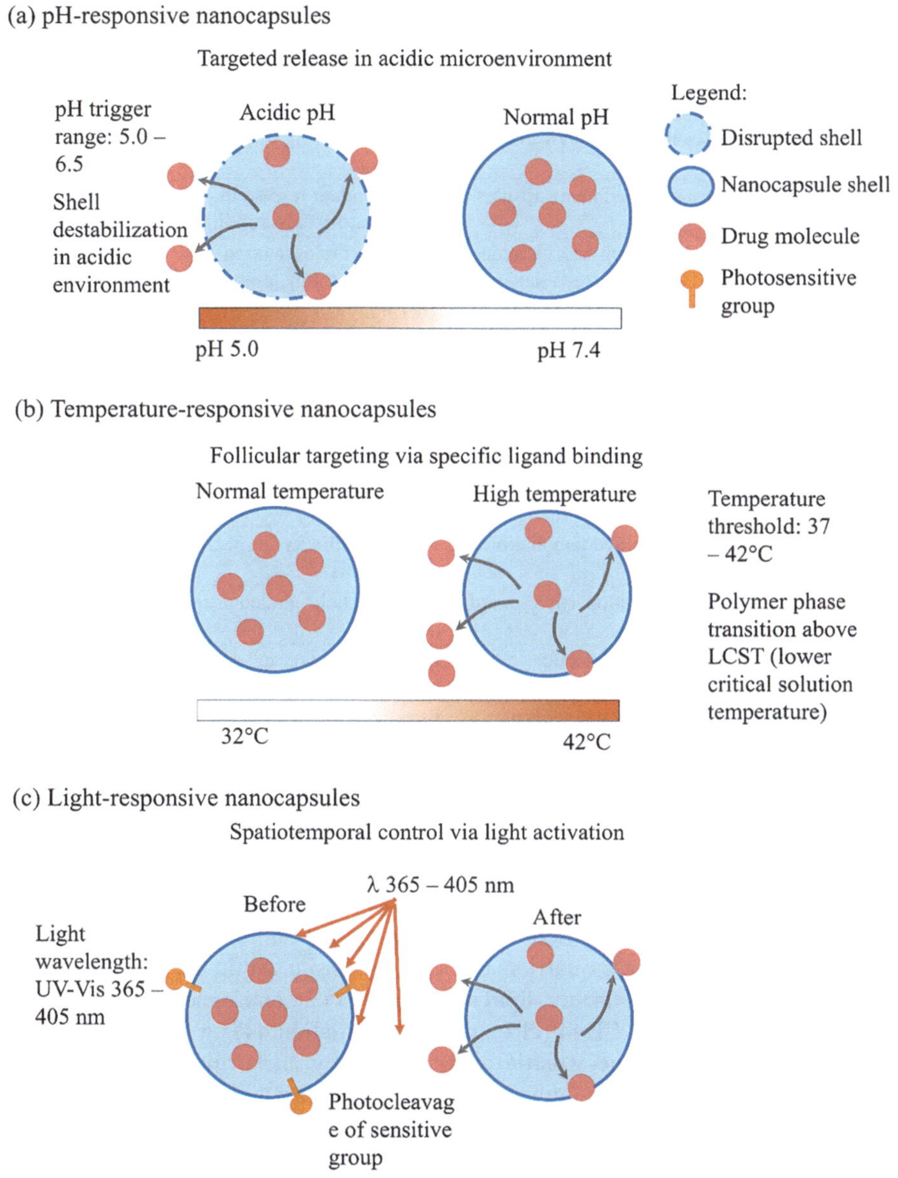


**Fig. 7.5** Stimuli-responsive nanocapsules. The smart nanoveehicles include (**a**) pH-sensitive nanocarriers for targeted medication delivery in an acidic skin environment, (**b**) thermosensitive nanocarriers for medication delivery in inflamed skin, and (**c**) photosensitizer-sensitive nanocarriers for precise space-time drug delivery. (Created by the author, with no permission required)

ligands can bind to specific skin layers (e.g., epidermis and dermis); thus, nanocapsules can be functionalized for layered skin (Huang et al. 2023). Drug delivery on the basis of nanocapsule functionalization has shown that the concentration of drugs at the site of action is increased and that fewer side effects are observed, thus enhancing dermal drug therapeutic efficiency (Fan et al. 2023).

**Table 7.3** Stimuli-responsive nanocapsules for dermal drug delivery, triggering stimuli, and potential applications

| Triggering stimuli | Nanocapsule systems | Potential applications | References |
|---|---|---|---|
| pH | Chitosan-based nanocapsules | Targeted delivery to specific skin layer | Naghib et al. (2024) |
| | Eudragit-based nanocapsules | pH-dependent release in inflammatory skin conditions | Lei et al. (2024) |
| Temperature | Poly(N-isopropylacrylamide) (PNIPAM) nanocapsules | Controlled release upon exposure to body temperature | Jafarzadeh et al. (2024) |
| | Lipid-based thermosensitive nanocapsules | Thermally triggered drug delivery for localized therapy | Yang et al. (2024a) |
| Light | Gold nanoparticle-based nanocapsules | Spatiotemporal control of drug release | Desai et al. (2024) |
| | Upconverting nanoparticle-based nanocapsules | Photodynamic therapy for skin diseases | Luo et al. (2024) |
| Ultrasound | Perfluorocarbon-based nanocapsules | Enhanced skin penetration and drug delivery | De et al. (2024) |
| | Polymer-shelled nanobubbles | Targeted delivery to specific skin regions | Zeng et al. (2024) |
| Magnetic field | Iron oxide nanoparticle-based nanocapsules | Magnetic field-guided drug delivery | Sardar et al. (2024) |
| | Lipid-based magnetic nanocapsules | Combination therapy with hyperthermia | Sardar et al. (2024) |
| Enzymatic | Peptide-based nanocapsules | Enzyme-triggered release in response to skin microenvironment | Bhavsar et al. (2024) |
| | Polysaccharide-based nanocapsules | Targeted delivery to enzymatically active skin sites | Bhavsar et al. (2024) |

Various forms of functionalized nanocapsules for dermal targeted drug delivery are presented in Table 7.4, where emphasis is placed on how the functionalizing ligands assist in directing the carriers to the intended site within the skin. For example, via a hyaluronic acid anchor, lipid nanocapsules can reach and interact with the hyaluronic acid receptor CD44 (cluster of differentiation 44) on keratinocytes and the RGD (arginine-glycine-aspartic acid) peptide by anchoring and attaching to $\alpha5\beta1$ integrin-containing nanofibers. The table also highlights several examples, such as folic acid-modified chitosan nanocapsules for targeting melanoma cells, TAT (transactivator of transcription) peptide PCL nanocapsules for better cell membrane penetration, mannose-modified hybrid nanocapsules for targeting Langerhans cells, and antikeratin antibody-functionalized PLA nanocapsules for targeting hair follicles. This rational functionalization facilitates focused and targeted drug delivery to certain structures of the skin or certain skin cell populations.

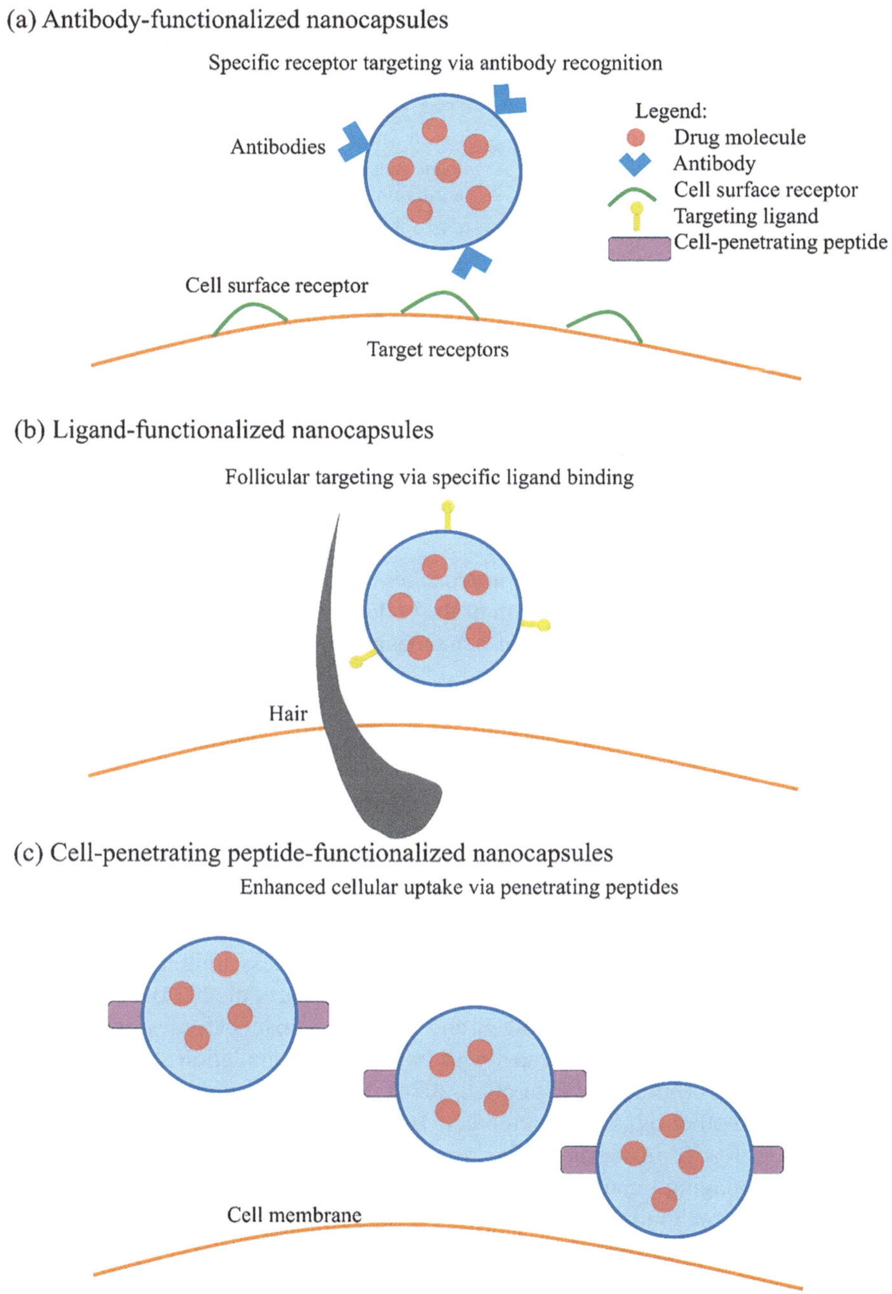


**Fig. 7.6** Functionalized nanocapsules for targeted dermal drug delivery. Nanocapsules with different functional ends include (a) antibody end-functionalized nanocapsules, which have high skin receptor selectivity; (b) ligand end-functionalized nanocapsules, which selectively target the hair follicle; and (c) cell penetrating peptide end-functionalized nanocapsules, which increase intracellular uptake (created by the author, with no permission required)

**Table 7.4** Functionalized nanocapsules for targeted dermal drug delivery, their ligands, and specific skin targets

| Ligand | Nanocapsule system | Specific skin target | References |
|---|---|---|---|
| Hyaluronic acid (HA) | HA-coated polymeric nanocapsules | CD44 receptors overexpressed in skin cancer cells | Hussain et al. (2024) |
| Folic acid | Folate-conjugated lipid nanocapsules | Folate receptors overexpressed in inflammatory skin conditions | Pahwa et al. (2024) |
| Peptides (e.g., RGD) | Peptide-functionalized lipid nanocapsules | $\alpha5\beta1$ integrins on fibroblasts | Farasatkia et al. (2024) |
| TAT peptide | Polycaprolactone nanocapsules | Cell membrane penetration | Wu et al. (2024) |
| Mannose | Hybrid lipid-polymer nanocapsules | Langerhans cells | Mochida and Uchida (2024) |
| Anti-keratin antibody | Polylactic acid nanocapsules | Keratin in hair follicles | Hansen et al. (2024) |

### 7.6.3 Combination Therapy with Nanocapsules

Figure 7.7 illustrates three strategic approaches to combination therapy using nanocapsules. The threefold classification of the codelivery strategies includes (i) bringing together two or more drugs in a nanocapsule system with the intention of effective combined action once released; (ii) a system where certain drugs are released immediately while others are released in the system over an extended period of time; and (iii) a combination of therapeutic drugs and adjuvants, whereby the adjuvants act as assistants to boost the efficacy.

This form of therapy occurs when a drug combination with synergistic properties is delivered via nanocapsules. By using a suitable choice of biopolymer and core material and a core–wall design, nanocapsules can be developed to encapsulate and deliver two or more drugs with different physicochemical characteristics and targets (Ataide et al. 2023). Nanocapsules, in combination with other drugs, have certain benefits, including denser and more effective treatment, lessening the development of drug resistance, and reducing side effects (dos Santos et al. 2023). For example, simultaneous delivery of granzyme B and perforin in nanocapsules enhances the therapeutic impact of tumor therapy, with improved accumulation at the tumor site over traditional formulations (Shi et al. 2024). Similarly, a system that codelivers flavonoids and other bioactive compounds from pitaya peel as a sunscreen agent in nanocapsules loaded in dermal applications will offer better efficacy than photoprotective and antiaging treatments (Xue et al. 2025). Nanocapsules can thus be engineered for dual or multiple drugs with fluctuating release rates and profiles. The short release rate can reflect the rapid action of the drug to curb the immediate impact caused by a disease, whereas a long release rate aims to address the progressive treatment for chronic ailments (Shi et al. 2024). Combination therapy using nanocapsules is an efficient strategy for addressing the multifactorial features of skin diseases and improving treatment effectiveness.

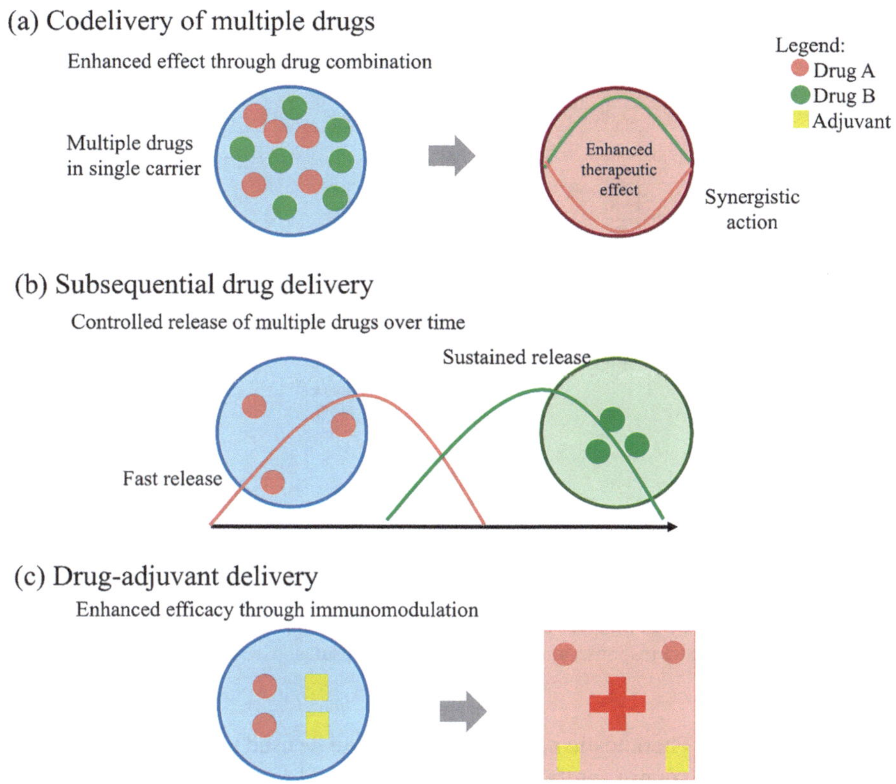


**Fig. 7.7** Nanocapsule combination therapy. This system can enable (**a**) simultaneous release of multiple therapeutics with complementary activity profiles, (**b**) sustained/progressive release of two or more drugs with dissimilar release profiles and (**c**) dual/triple combination therapy of therapeutics and adjuvants (generated by the author without permission)

### 7.6.4 *Nanocapsules for Diagnostic and Theranostic Applications*

Nanocapsules are versatile platforms used in dermatology for both diagnostic and theranostic purposes. For example, as shown in Fig. 7.8, nanocapsules can take many forms, with diagnostic agents that encapsulate imaging agents such as fluorescent dyes and contrast agents being mentioned to monitor and visualize skin conditions without invasive detection (Wang et al. 2024b). For example, in skin cancer, near-infrared fluorescent dye-loaded nanocapsules were employed to identify skin cancer early and examine the demarcation of skin cancer margins (Li and Li 2020). Moreover, theranostic nanocapsules are biocompatible agents that can image skin diseases with a simultaneous treatment function (Sarbadhikary et al. 2021). For example, theranostic nanocapsules coloaded with a photosensitizer and an anticancer drug can be applied as image-guided photodynamic therapy against skin tumors

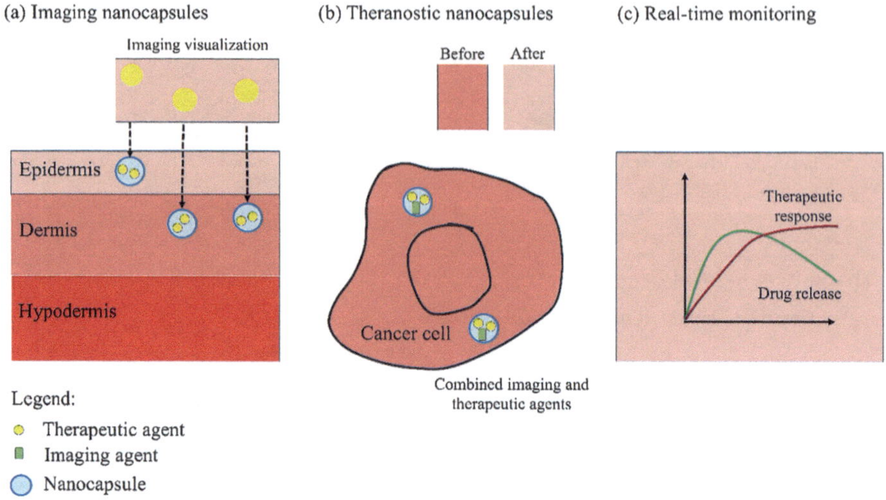


**Fig. 7.8** Nanocapsules for diagnostic and theranostic applications in dermatology. This is based on the following potential applications of nanocapsules: (**a**) imaging agents confined to nanocapsules for noninvasive imaging of skin tissue; (**b**) theranostic agents, which are nanocapsules that function simultaneously as imaging and treatment agents for skin cancer; and (**c**) real-time tracking of drug delivery and response to treatment via nanocapsules. (Created by the author, with no permission required)

(Rata et al. 2021). Theranostic nanocapsules that can be used to monitor drug release and therapeutic response, preferably in real time, and to inform personalized and adaptive treatment are also possible (Sun et al. 2020). A more accurate diagnosis, low invasiveness, improved patient prognosis, and improved quality of life are needed for targeted therapy, early diagnosis and noninvasive detection of skin disease via diagnostic and theranostic nanocapsules (Wang et al. 2024b).

## 7.7 Safety, Biocompatibility, and Clinical Translation

For successful bench-to-bedside translation of nanocapsule-based dermal drug delivery systems, their safety, biocompatibility, and clinical efficacy need to be evaluated. This section discusses important areas of safety assessment, such as *in vitro* and *in vivo* toxicity studies, skin irritation and sensitization tests, preclinical and clinical studies, market approval considerations, and commercialization challenges. Figure 7.9 shows that the clinical translation pathway for nanocapsule-based dermal drug delivery systems follows a systematic progression through four major stages: preclinical development, which includes initial formulation design, comprehensive physicochemical characterization, *in vitro* testing, and animal studies to establish safety and efficacy profiles; regulatory approval, involving pre-IND

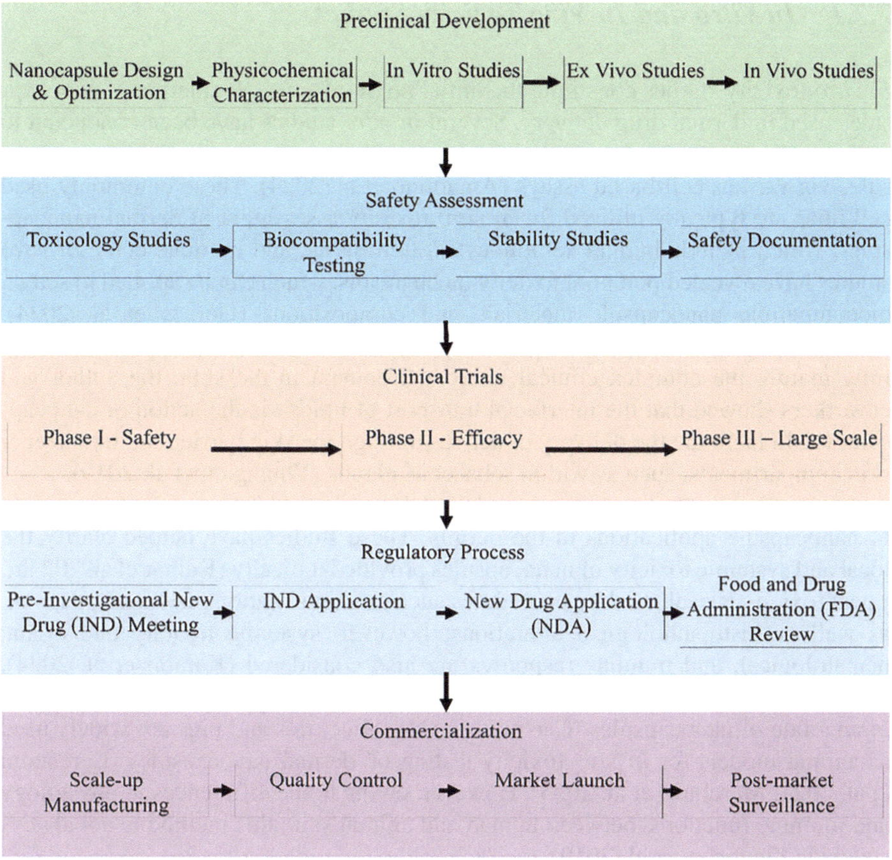


**Fig. 7.9** Schematic overview of the clinical translation pathway for nanocapsule-based dermal drug delivery systems, highlighting the key stages from preclinical development to commercialization. (Created by the author, with no permission required)

(investigational new drug) consultations with regulatory authorities, IND application submission, and regulatory review; clinical trials, progressing through phase I (safety assessment in small human populations), phase II (efficacy evaluation in target patient groups), and phase III (large-scale confirmation of safety and efficacy); and finally, commercialization, encompassing new drug application (NDA) or marketing authorization submission, manufacturing scale-up, market launch, and ongoing postmarket surveillance. At each stage, developers must meet stringent regulatory requirements, ensure product quality and consistency, and demonstrate both safety and efficacy before advancing to the next phase, ultimately working toward successful market introduction and patient benefit. Furthermore, some details are discussed in the following subsections.

### 7.7.1 In Vitro and In Vivo Toxicity Studies

*In vitro* toxicity studies constitute the initial phase of safety evaluation of nanocapsules used in dermal drug delivery. Several *in vitro* studies have been conducted to evaluate the cytotoxicity, genotoxicity, and oxidative stress potential of nanocapsules via various cell-based assays (Amantino et al. 2024). These commonly used cell lines are typically utilized for *in vitro* toxicity assessment of dermal nanocapsules, which include human keratinocytes, fibroblasts, and immune cells. *In vitro* studies have revealed potential toxicity mechanisms, which can be applied to screen biocompatible nanocapsule materials and compositions (Gomes et al. 2024). Although baked *in vitro* as a comparative simplification that cannot be considered to fully mature the complex clinical microenvironment in the skin, the author and coworkers showed that the interfacial transport of lipids via the action of nanocapsules could facilitate the delivery of actives through the skin barrier and into deeper skin compartments, such as within sebaceous glands (Yingngam et al. 2019).

*In vivo* toxicity studies using animal models would help better evaluate the safety of nanocapsule applications in the dermis. These studies have helped clarify the local and systemic toxicity of nanocapsules provided topically (Kumar et al. 2024a). Local toxicity is evaluated by procedures such as skin irritation (erythema, edema), as well as histopathological alterations; however, systemic toxicity and organ, hematological, and immune responses are also considered (Karami et al. 2024). *In vivo* studies may also be conducted to assess the biodistribution, clearance, and *in vivo* fate of nanocapsules (Cao et al. 2024). Mice, rats and pigs are widely used as animal models for *in vivo* toxicity testing of dermal nanocapsules (Lorenzoni et al. 2021; Marchiori et al. 2023). However, owing to the differences in physiology and immune functions between human and animal skin, this method is not always suitable (Yingngam et al. 2019).

### 7.7.2 Skin Irritation and Sensitization

The safety issues of skin irritation and sensitization involving dermal drug delivery systems (nanocapsules) are discussed. When applied to the skin, nanocapsules cause a local inflammatory reaction, which is a phenomenon known as skin irritation, and sensitization is the process that results in the emergence of an allergic reaction (Potter et al. 2024). *In vitro* skin irritation tests (e.g., the reconstructed human epidermis model and the human R repeat insult patch test) can be used for initial screening of nanocarrier irritation potential (McLean et al. 2024; Mendoza-Muñoz et al. 2023). However, skin irritation and sensitization studies using animal models, the Draize test and the local lymph node assay are generally more definitive (Ngoc et al. 2023). The knowledge gained from these studies aids in the selection of nanocapsule formulations with low potential for such irritating properties and in the proper selection of skin models for future safety screening (Masri et al. 2024).

### 7.7.3 Preclinical and Clinical Studies

Animal models of skin diseases (psoriasis, atopic dermatitis, and skin cancer) are used in preclinical studies of nanocapsule-based dermal drug delivery systems: criteria for testing the effect and safety of the system. The application of these studies allows the optimization of nanocapsule formulations, dosing regimens, and methods of application for various skin diseases (Anwar et al. 2023). Preclinical studies provide important pharmacokinetic and pharmacodynamic data about drug absorption, distribution, metabolism, and elimination (Gnanakani et al. 2024). Successful preclinical studies are needed to advance nanocapsule formulations to clinical trials (Kant et al. 2024).

Dermal drug delivery with drug-loaded nanocapsules is in its final stages, and human clinical trials must be carried out to establish the viability of the drug formulation. Phase I studies of nanocapsules in healthy adults have evaluated their toxicity, whereas phase II and phase III studies have investigated their efficacy in patients whose specific skin conditions are targeted by the nanocapsules (da Silva et al. 2024). Randomized, double-blind, placebo-controlled designs have been commonly used in clinical trials of dermal nanocapsules, as bias and confounding factors are minimized (Gobbato et al. 2019). Clinical trials can have endpoint measures, including skin barrier function, inflammation, histological changes, and patient-reported outcomes (Ramezani et al. 2023). Regulatory approval and commercialization of dermal drug delivery systems based on nanocapsule technology depend on the successful completion of clinical trials (Sucheta et al. 2024).

### 7.7.4 Regulatory Considerations and Commercial Challenges

Delivery via dermal nanocapsules is a demanding and evolving area of regulation. Hence, on the basis of the regulations of policymakers, including the United States Food and Drug Administration (US FDA) and the European Medicines Agency (EMA) (Sucheta et al. 2024), the biophysical characterization of nanocapsules in nanomedicines is necessary. Major emphasis has been placed on characterizing nanocapsule physicochemical properties, quality control, and safety assessment (Lunter et al. 2024). The regulatory requirements of the nanocapsule-based dermal drug delivery system depend on the type of drug, indications, and route of administration (Das et al. 2024). The successful translation of nanocapsule-based products through the regulatory landscape requires effective communication and collaboration between researchers, regulators, and manufacturers (Mehta et al. 2024).

However, the commercialization of nanocapsule-based dermal drug delivery systems is not without challenges: scalability, manufacturing consistency, and cost-effectiveness (Yingngam 2024a). Currently, precise control of the size, composition, and drug loading of nanocapsules is challenging at the industrial scale (Yingngam et al. 2024). Commercial success also depends critically on guaranteeing batch-to-batch consistency and the long-term stability of the nanocapsule formulations (Yingngam 2024b). Moreover, compared with traditional formulations, nanocapsule-based products are likely to be more expensive to produce and develop (Yingngam 2024b). These challenges can be overcome only with collaboration among academia, industry, and regulatory agencies to facilitate the translation of nanocapsule-based dermal drug delivery systems from the laboratory to the clinic [30].

Table 7.5 outlines the regulatory requirements for nanocapsule-based dermal drug delivery systems across six major global markets: the United States (FDA), the European Union (EMA), Japan (PMDA), India (CDSCO), China (NMPA), and Thailand (Thai FDA). The table is structured into six key regulatory aspects: primary guidelines, characterization requirements, safety requirements, quality control, labeling requirements, and postmarket requirements. While there are some regional variations, common themes emerge across all jurisdictions: all require thorough particle characterization, comprehensive safety assessments, strict quality control measures, and postmarket surveillance. The EU and US generally have the most stringent requirements, with specific emphasis on nanospecific characterization and safety assessments, whereas Asian markets often require additional local testing and language-specific documentation. However, some differences must be noted, including the obligatory mention of "(nano)" in EU labeling, Japanese requirements for environmental conditions during stability testing, the Indian focus on localized clinical trials, China's emphasis on population-specific data, and Thailand's alignment with ASEAN requirements. While each region has established its own set of overarching guidelines and technical requirements, the key objective remains the same: to establish the safety, quality, and efficacy of products through robust regulatory oversight.

## 7.8 Market Analysis of the Effects of the Nanocapsules on Drug Delivery

The market for dermal drug delivery via nanocapsules is growing exponentially and is supported by various macro- and microscale factors. Macrolevel influences (e.g., economic forecasts, regulatory environment changes) should be examined in conjunction with microlevel factors (e.g., new technologies, changing consumer preferences) to understand the market dynamics of the period across geographies.

**Table 7.5** Regulatory considerations for nanocapsule-based dermal drug delivery systems on key markets

| Regulatory aspect | US (FDA) | EU (EMA) | Japan (PMDA) | India (CDSCO) | China (NMPA) | Thailand (Thai FDA) |
|---|---|---|---|---|---|---|
| Primary guidelines | 21 CFR part 58<br>FDA guidance for industry on nanomaterials | Regulation (EC) No 1223/2009<br>EMA/CHMP/79769/2006 | Guidelines for Nanomedicine products<br>Safety guidelines for nanomaterials | Guidelines for Nanopharmaceuticals<br>Schedule M requirements | Technical guidelines for nanomaterials<br>Nanobased drug delivery guidelines | Guidelines for Nanobased products<br>ASEAN cosmetic directive |
| Characterization requirements | Complete particle characterization<br>Size distribution<br>Surface properties<br>Stability data | Physicochemical characterization<br>Stability studies<br>Surface modification data | Particle characterization<br>Stability under local conditions<br>Quality parameters | Basic characterization<br>Stability data<br>Process validation | Full characterization<br>Environmental stability<br>Manufacturing validation | Basic characterization<br>ASEAN stability requirements<br>Quality specifications |
| Safety requirements | Toxicological evaluation<br>Dermal absorption studies<br>Systemic exposure assessment | Enhanced safety assessment<br>Environmental risk assessment<br>Postmarket surveillance | Local tolerance studies<br>Photostability testing<br>Long-term monitoring | Local clinical trials<br>Bioequivalence studies<br>Safety monitoring | Local safety studies<br>Population-specific data<br>Environmental assessment | Safety assessment report<br>Local tolerance studies<br>Postmarket surveillance |
| Quality control | Validated methods<br>Batch consistency<br>Process validation | GMP compliance<br>Process validation<br>Stability testing | Quality parameters<br>Stability testing<br>Process controls | GMP compliance<br>In-process controls<br>Validation requirements | Process validation<br>Quality controls<br>Stability testing | Quality specifications<br>Process controls<br>Stability data |
| Labeling requirements | Nanospecific warnings<br>Complete ingredients list<br>Usage instructions | "(nano)" designation<br>Full ingredient list<br>Special precautions | Japanese language<br>Nano declaration<br>Safety information | Local language<br>Nano identification<br>Usage guidelines | Chinese language<br>Nano declaration<br>Safety warnings | Thai language<br>ASEAN requirements<br>Safety information |

(continued)

**Table 7.5** (continued)

| Regulatory aspect | US (FDA) | EU (EMA) | Japan (PMDA) | India (CDSCO) | China (NMPA) | Thailand (Thai FDA) |
|---|---|---|---|---|---|---|
| Postmarket requirements | Safety monitoring<br>Periodic reports<br>Change control | Pharmacovigilance<br>Safety updates<br>Risk management | Re-examination period<br>Safety monitoring<br>Change notification | PSURs<br>Safety monitoring<br>Change approvals | Annual reports<br>Safety updates<br>Change control | Safety monitoring<br>Periodic updates<br>Change notification |

Note: Requirements may vary on the basis of specific product classification and intended use. Manufacturers should consult current regional guidelines and engage with regulatory authorities during development

Abbreviations: *ASEAN* Association of Southeast Asian Nations, *CDSCO* Central Drugs Standard Control Organization, *CFR* Code of Federal Regulations, *CHMP* Committee for Medicinal Products for Human Use, *EMA* European Medicines Agency, *FDA* Food and Drug Administration, *GMP* Good Manufacturing Practice, *NMPA* National Medical Products Administration, *PMDA* Pharmaceuticals and Medical Devices Agency, *PSUR* Periodic Safety Update Report, *SCCS* Scientific Committee on Consumer Safety

### 7.8.1 Current Market Trends

The burden of skin-related healthcare has been growing worldwide, and the pharmaceutical industry is progressively leading toward the promising potential of nanocapsuls to improve the bioavailability of drugs with reduced adverse effects. These findings open the door to nanomedicine, which is becoming fundamental to therapeutic management. Nanocapsules are anticipated to grow to USD 3.9 billion in the dermal drug delivery market in 2023, with an estimated compound annual growth rate (CAGR) of 6.2% from 2024–2034. The growth of the market can be attributed to such technological developments, the growing consumer demand for effective skincare solutions, and the rising awareness of personal grooming among various demographics. Not surprisingly, key market segments such as creams and other topical formulations stand an even better chance of accessibility through retail attunes (The Niche Research 2024; Transparency Market Research 2024). The market analysis employs a bottom-up approach, focusing on individual countries to estimate regional market sizes. This includes evaluating growth drivers, technological innovations, and constraints specific to each region, culminating in accurate regional growth projections. The Asia–Pacific region, in particular, is showing promising growth potential due to rising consumer awareness and increased investment in skincare innovations (GrandViewResearch 2023).

### 7.8.2 Key Players

Some notable market players in the global nanocapsule market are included in the dermal drug delivery segment. Noteworthy organizations include BioDelivery Sciences, Camurus, Carlina Technologies, Cerulean Pharma, Inc., Nano Green Science, Inc., and NanoNutra. These companies have used advanced technologies and strategic initiatives (through mergers, acquisitions, and partnerships) to bolster their product offerings and to solidify their market position. One area is nanotechnology, which has been employed by BioDelivery Sciences, a neuroscience drug delivery company that can develop novel drug delivery systems to enhance therapeutic effects. They are in the pole position on the market, with a focus on personalized medicine.

### 7.8.3 Future Projections

Some of the significant factors that are anticipated to impact the market during the next 8 years include the increasing adoption of transdermal drug delivery systems owing to their noninvasive features and the increasing prevalence of chronic diseases that require long-term medication management. Moreover, new collaborations and partnerships in the industry are expected to affect the market and competitive actions (Liu et al. 2023).

## 7.9 Conclusion and Future Perspectives

Recent studies using nanocapsules highlight their potential as a platform technology for enhancing dermal drug delivery while addressing the weaknesses inherent in conventional topical formulations. This unique core–shell structure and flexible properties of nanocapsules enable enhanced solubility of drugs, controlled release of drugs, and enhanced skin penetration followed by targeted delivery to specific skin layers or appendages. The excellent drug loading and release performance of the nanocapsules could be attributed to their ability to have controlled sizes, surface properties, compositions, and drug loadings. Furthermore, recent developments in stimuli-responsive nanocapsules, functionalized nanocapsules for targeted delivery, and combination therapy approaches have expanded the potential applications of nanocapsules for dermal drug delivery. These innovative formulations provide a means of not only delivering effective treatments for disease with enhanced targeted delivery to the area, increasing the efficacy and reducing side effects but also ensuring that the disease is targeted at the cellular level. Moreover, this section outlines the utilization of nanocapsules in noninvasive diagnostic and theranostic approaches and the image-guided treatment of dermatoses. However, there are still challenges in the translation of nanocapsule-based dermal drug delivery systems from the laboratory to the clinic. The complete safety, biocompatibility, and toxicity of nanocapsule-based dermal drug delivery systems have been thoroughly evaluated through *in vitro* and *in vivo* studies and clinical trials before their use in human subjects to confirm the safe and effective use of nanocapsules in humans. The successful development and market entry of nanocapsule-based products must also address regulatory considerations and commercialization challenges, including scalability, manufacturing consistency, and cost-effectiveness.

There are several key areas worthy of future research in nanocapsule-based dermal drug delivery. The first category of challenges relates to the design of smarter nanocapsules that are equipped to communicate with the exotic skin microenvironment. Second, the integration of nanocapsules with other new techniques, such as microneedles, iontophoresis, and 3D bioprinting, could improve nanocapsule efficiency and increase the feasibility of translating dermal drug delivery into the clinic. Third, appropriate approaches for characterizing nanocapsules, assessing their safety, and controlling their quality are yet to be standardized to replicate research studies, compare outcomes with others, and obtain approval from regulatory authorities. Finally, advanced imaging and analytical techniques could be applied to investigate the fate of nanocapsules and drug release in the skin noninvasively, which is crucial for the further improvement of nanocapsule architecture and function.

In conclusion, nanoparticle approaches, particularly nanocapsules, represent a promising and versatile type of dermal drug delivery platform that will transform the treatment of skin diseases. However, there is much work to be done in research and innovation to overcome the difficulties associated with nanocapsule utilization in various applications in different areas of biomedicine. Through continued multidisciplinary collaboration and research, nanocapsule-based dermal drug delivery systems can be translated from the benchtop to the bedside to improve the quality of life for patients around the world with a number of skin-related conditions.

## References

Abdel-Emam RA, Ali MF, Hassan AS, Abd-Ellatief RB (2024) Development and evaluation of dexamethasone-loaded bioadhesive polymeric nanocapsules for mitigating cardiac and gastric adverse effects of free dexamethasone. J Pharm Investig 54:825–844. https://doi.org/10.1007/s40005-024-00686-7

Abdel-Mottaleb MMA, Neumann D, Lamprecht A (2011) Lipid nanocapsules for dermal application: a comparative study of lipid-based versus polymer-based nanocarriers. Eur J Pharm Biopharm 79(1):36–42. https://doi.org/10.1016/j.ejpb.2011.04.009

Aboomeirah AA, Kabil MF, Azzazy HME-S (2024) Polyvinyl alcohol-chitosan-polyethylene glycol-glycerol incorporated with Peganum harmala loaded in lipid nanocapsules as an elastic nanocomposite surgical sealant to control bleeding. Int J Biol Macromol 280:135987. https://doi.org/10.1016/j.ijbiomac.2024.135987

Adamus-Grabicka AA, Hikisz P, Sikora J (2024) Nanotechnology as a promising method in the treatment of skin cancer. Int J Mol Sci 25(4):2165. https://doi.org/10.3390/ijms25042165

Ahmadian S, Kenari RE, Amiri ZR, Sohbatzadeh F, Khodaparast MHH (2024) Fabrication of double nanoemulsions loaded with hyssop (*Hyssopus officinalis* L.) extract stabilized with soy protein isolate alone and combined with chia seed gum in controlling the oxidative stability of canola oil. Food Chem 430:137093. https://doi.org/10.1016/j.foodchem.2023.137093

Akl MA, Eldeen MA, Kassem AM (2024) Beyond skin deep: phospholipid-based nanovesicles as game-changers in transdermal drug delivery. AAPS PharmSciTech 25(6):184. https://doi.org/10.1208/s12249-024-02896-6

Alam A, Imran M, Ansari MD, Elkirdasy AF (2024) Preparation of *Aloysia citriodora* (*lemon verbena*) extract loaded lipid-nano capsules for topical antibacterial treatment. J Drug Deliv Sci Technol 101:106164. https://doi.org/10.1016/j.jddst.2024.106164

Amantino CF, do Amaral SR, Aires-Fernandes M, Oliani SM, Tedesco AC, Primo FL (2024) Development of 3D skin equivalents for application in photodynamic biostimulation therapy assays using curcumin nanocapsules. Heliyon 10(12):e32808. https://doi.org/10.1016/j.heliyon.2024.e32808

An X, Yang J, Cui X, Zhao J, Jiang C, Tang M et al (2024) Advances in local drug delivery technologies for improved rheumatoid arthritis therapy. Adv Drug Deliv Rev 209:115325. https://doi.org/10.1016/j.addr.2024.115325

Anwar M, Muhammad F, Akhtar B, Fatima S, Khan H, Chou C-C (2023) Nanomedicines in the treatment of skin diseases. In: Akhtar B, Muhammad F, Sharif A (eds) Nanomedicine in treatment of diseases. Springer, Singapore, pp 285–306

Aparicio-Blanco J, Pucci C, De Pasquale D, Marino A, Debellis D, Ciofani G (2024) Development and characterization of lipid nanocapsules loaded with iron oxide nanoparticles for magnetic targeting to the blood–brain barrier. Drug Deliv Transl Res 14(12):3494–3511. https://doi.org/10.1007/s13346-024-01587-w

Ataide JA, Coco JC, dos Santos ÉM, Beraldo-Araujo V, Silva JRA, de Castro KC et al (2023) Coencapsulation of drugs for topical application–a review. Molecules 28(3):1449. https://doi.org/10.3390/molecules28031449

Bhavsar J, Kasture K, Salvi BV, Shende P (2024) Strategies for transportation of peptides across the skin for treatment of multiple diseases. Ther Deliv 16:63–86. https://doi.org/10.1080/20415990.2024.2411943

Bufalini C, Palazzo I, Casazza AA, Ferrari PF, Campardelli R, Firpo G et al (2024) Encapsulation of Arthrospira platensis polyphenolic extract using supercritical emulsion-based process. J Supercrit Fluids 212:106335. https://doi.org/10.1016/j.supflu.2024.106335

Busch L, Asadzadeh D, Klein AL, Suriyaamporn P, Vollrath MK, Keck CM, Meinke MC (2024) The penetration efficiency of a dissolved model drug into hair follicles depends on the concentration of added nanoparticles. Drug Deliv Transl Res 15(4):1444–1452. https://doi.org/10.1007/s13346-024-01718-3

Can Karaca A, Rezaei A, Qamar M, Assadpour E, Esatbeyoglu T, Jafari SM (2025) Lipid-based nanodelivery systems of curcumin: recent advances, approaches, and applications. Food Chem 463:141193. https://doi.org/10.1016/j.foodchem.2024.141193

Cao X, Wu X, Zhang Y, Qian X, Sun W, Zhao Y (2024) Emerging biomedical technologies for scarless wound healing. Bioact Mater 42:449–477. https://doi.org/10.1016/j.bioactmat.2024.09.001

Cerro D, Torres A, Romero J, Streitt C, Rojas A, Matiacevich S, Machuca S (2024) Supercritical fluid extraction of emulsion-assisted encapsulation of hypocholesterolemic bioactive compounds. J Supercrit Fluids 211:106306. https://doi.org/10.1016/j.supflu.2024.106306

Chai L, Chen Y, Yan X, Alcouffe P, Ganachaud F, Fleury E, Bernard J (2024) Poly(vinyl alcohol)s and their glycoderivatives as efficient shell-builders of nanocapsules by nanoprecipitation. Biomacromolecules 25(6):3596–3606. https://doi.org/10.1021/acs.biomac.4c00213

Chiarentin R, Pereira Bottcher D, Zeni B, Grave C, Neutzling Kaufmann F, Emmanoella Sebulsqui Saraiva T et al (2024) Development and pharmacological evaluation of liposomes and nanocapsules containing paroxetine hydrochloride. Int J Pharm 660:124304. https://doi.org/10.1016/j.ijpharm.2024.124304

da Costa Rodrigues K, da Silva Neto MR, dos Santos Barboza V, Hass SE, de Almeida Vaucher R, Giongo JL et al (2024) New curcumin-loaded nanocapsules as a therapeutic alternative in an amnesia model. Metab Brain Dis 39(4):589–609. https://doi.org/10.1007/s11011-023-01329-w

da Silva EL, Pedraça ES, Salgueiro AP, Gazzi RP, Nunes JS, Cavagni J et al (2024) Efficacy and safety of a 0.05% nanoencapsulated imiquimod hydrogel for the treatment of actinic cheilitis: drug release analysis and clinical study. Nanomedicine 62:102779. https://doi.org/10.1016/j.nano.2024.102779

Dahmash EZ, Attiany LM, Ali D, Assaf SM, Alkrad J, Alyami H (2024) Development and characterization of transdermal patches using novel thymoquinone-l-arginine-based polyamide nanocapsules for potential use in the management of psoriasis. AAPS PharmSciTech 25(4):69. https://doi.org/10.1208/s12249-024-02781-2

Das U, Bala A, Molla R, Mandal S (2024) Chapter 11 – Nanomedicine in infectious disease challenges and regulatory concerns. In: Beg S, Shukla R, Handa M, Rahman M, Dhir A (eds) Nanostructured drug delivery systems in infectious disease treatment. Academic, pp 237–259

De A, Jee J-P, Park Y-J (2024) Why perfluorocarbon nanoparticles encounter bottlenecks in clinical translation despite promising oxygen carriers? Eur J Pharm Biopharm 199:114292. https://doi.org/10.1016/j.ejpb.2024.114292

Desai VM, Kumbhar P, Kadam AY, Swarup J, Priya S, Jain A, Singhvi G (2024) Exploring the therapeutic modalities of targeted treatment approach for skin carcinoma: cutting-edge strategies and key insights. Expert Opin Drug Deliv 21(8):1213–1233. https://doi.org/10.1080/17425247.2024.2392799

Di Filippo LD, de Paula MC, Duarte JL, de Paula GA, Frigieri I, Chorilli M (2023) Nanocarrier functionalization strategies for targeted drug delivery in skin cancer therapy: current progress and upcoming challenges. Expert Opin Drug Deliv 20(6):831–849. https://doi.org/10.1080/17425247.2023.2221026

Do UT, Nguyen QT, Kim J, Luu QS, Park Y, Song M et al (2024) Tailored synthesis of pH-responsive biodegradable microcapsules incorporating gelatin, alginate, and hyaluronic acid for effective-controlled release. Int J Biol Macromol 270:132178. https://doi.org/10.1016/j.ijbiomac.2024.132178

dos Santos FL, Zilles JC, Machado AU, Marques MS, da Costa BS, Kulkamp Guerreiro IC et al (2023) Polymeric nanocapsules containing ozonated oil and terbinafine hydrochloride as a potential treatment against dermatophytes. AAPS PharmSciTech 24(7):198. https://doi.org/10.1208/s12249-023-02657-x

Fan L, Wang Z, Shi D (2023) Targeted nanoscale drug delivery systems for melanoma therapy. J Drug Deliv Sci Technol 86:104724. https://doi.org/10.1016/j.jddst.2023.104724

Farago PV, Camargo GDA, Mendes MB, Semianko BC, Camilo Junior A, Dias DT et al (2024) Computational simulation on the study of tacrolimus and its improved dermal retention using poly(ε-caprolactone) nanocapsules. J Mol Graph Model 126:108625. https://doi.org/10.1016/j.jmgm.2023.108625

Farasati Far B, Naimi-Jamal MR, Sedaghat M, Hoseini A, Mohammadi N, Bodaghi M (2023) Combinational system of lipid-based nanocarriers and biodegradable polymers for wound healing: an updated review. J Funct Biomater 14(2):115. https://doi.org/10.3390/jfb14020115

Farasatkia A, Maeso L, Gharibi H, Dolatshahi-Pirouz A, Stojanovic GM, Edmundo Antezana P et al (2024) Design of nanosystems for melanoma treatment. Int J Pharm 665:124701. https://doi.org/10.1016/j.ijpharm.2024.124701

Farnoudian-Habibi A, Aliebrahimi S, Sehati F, Nabavizadeh F, Asadi H, Montazeri V et al (2024) Development a novel nanoplatform for thrombolysis acceleration by thrombin sensitive polymer-peptide hybrid nancapsules. Int J Pharm 663:124561. https://doi.org/10.1016/j.ijpharm.2024.124561

Feng R, Wang M, Song Z, Jia Y, Li Y, Liu Q et al (2025) pH-responsive hydrogel containing curcumin-loaded lipid nanocapsules for oral curcumin's stability, bioaccessibility and intestinal absorption improvement. Food Hydrocoll 158:110488. https://doi.org/10.1016/j.foodhyd.2024.110488

Fernández-Mariño I, Anfray C, Crecente-Campo J, Maeda A, Ummarino A, Teijeiro-Valiño C et al (2023) Mannose-modified hyaluronic acid nanocapsules for the targeting of tumor-associated macrophages. Drug Deliv Transl Res 13(7):1896–1911. https://doi.org/10.1007/s13346-022-01265-9

Foroutani M, Jahanmardi R, Atai M, Nodehi A (2024) Temperature-sensitive fragrance-releasing polymeric microcapsules. Eur Polym J 216:113256. https://doi.org/10.1016/j.eurpolymj.2024.113256

Gao Y, Xiao W, Gu C, Yu S, Li S, Zou A (2024) Layer-by-layer assembled decomposable nanocapsules for light-responsive release of pesticide imidacloprid on *Aphis craccivora* Koch. Pest Manag Sci 80(7):3207–3214. https://doi.org/10.1002/ps.8023

Gnanakani SPE, Kirubakaran JJ, Rama P, Saritha M, Jayavarapu KR, Sathish A et al (2024) Harnessing the targeting potential of Nanobiomaterials to treat autoimmune skin disorders. In: Pradhan M, Yadav K, Singh Chauhan N (eds) Biomaterial-inspired Nanomedicines for targeted therapies. Springer, Singapore, pp 183–208

Gobbato C, Gobbato A, Magalhães TB, Mendes GD, Ilha JO, Moreno RA et al (2019) Randomized, double-blind, phase III clinical study of a novel nanotechnological topical anesthetic formulation containing lidocaine 25 mg/g and prilocaine 25 mg/g (nanorap) in skin phototypes I–III patients with ablative fractional CO2 laser treatment indication in the forehead. Lasers Surg Med 51(7):609–615. https://doi.org/10.1002/lsm.23071

Gomes GS, Benin T, Ferreira ECJ, Palma VM, Alves ACS, Visioli F et al (2024) Pectin-based hydrogel loaded with polymeric nanocapsules for the cutaneous delivery of tacrolimus: a targeted approach for the treatment of psoriasis. J Drug Deliv Sci Technol 99:105967. https://doi.org/10.1016/j.jddst.2024.105967

GrandViewResearch (2023) Nanocapsules market size, share & growth report, 2030 – Grand View Research. Retrieved from https://www.grandviewresearch.com/industry-analysis/global-nanocapsules-market

Hansen ME, Ibrahim Y, Desai TA, Koval M (2024) Nanostructure-mediated transport of therapeutics through epithelial barriers. Int J Mol Sci 25(13):7098. https://doi.org/10.3390/ijms25137098

Heikal LA, El-Habashy SE, El-Kamel AH, Mehanna RA, Ashour AA (2024) Bioactive baicalin rhamno-nanocapsules as phytotherapeutic platform for treatment of acute myeloid leukemia. Int J Pharm 661:124458. https://doi.org/10.1016/j.ijpharm.2024.124458

Huang M, Zhai B-T, Fan Y, Sun J, Shi Y-J, Zhang X-F et al (2023) Targeted drug delivery systems for curcumin in breast cancer therapy. Int J Nanomedicine 18(null):4275–4311. https://doi.org/10.2147/IJN.S410688

Hussain Z, Akbari AH, Barbuor SH, Dawood Alshetiwi DS, Ahmed IS, Rawas-Qalaji M (2024) Hyaluronic acid based functionalization of nanodelivery systems: a promising strategy for CD44-receptors-mediated targeted therapy of lung cancer. J Drug Deliv Sci Technol 101:106183. https://doi.org/10.1016/j.jddst.2024.106183

Ibrahim MM, Basalious EB, El-Nabarawi MA, Makhlouf AIA, Sayyed ME, Ibrahim IT (2024) Nose to brain delivery of mirtazapine via lipid nanocapsules: preparation, statistical optimization, radiolabeling, *in vivo* biodistribution and pharmacokinetic study. Drug Deliv Transl Res 14(9):2539–2557. https://doi.org/10.1007/s13346-024-01528-7

Iglicki D, Kahn ML, Goubault C, Blot M, Jarry U, Pedeux R et al (2024) Simple elaboration of drug-SPION nanocapsules (hybridosomes®) by solvent shifting: effect of the drug molecular structure and concentration. Int J Pharm 649:123645. https://doi.org/10.1016/j.ijpharm.2023.123645

Iqbal MH, Kerdjoudj H, Boulmedais F (2024) Protein-based layer-by-layer films for biomedical applications. Chem Sci 15(25):9408–9437. https://doi.org/10.1039/d3sc06549a

Jafarzadeh F, Peyman H, Roshanfekr H, Azizi S, Idris AO, Maaza M (2024) Fabrication of a nano-magnetic smart polymer carrier as a potential candidate for a drug delivery system. Arab J Sci Eng 49(7):9381–9394. https://doi.org/10.1007/s13369-024-08724-0

Jose J, Shenoy S, Shetty S, Zulfa S, Haseena T (2024) Chapter 18 – Nanomaterials in cosmetics. In: Malik MI, Hussain D, Shah MR, Guo D-S (eds) Handbook of nanomaterials, vol Volume 2. Elsevier, pp 497–522

Kabil MF, Gaber SAA, Hamzawy MA, El-Sherbiny IM, Nasr M (2024) Folic/lactobionic acid dual-targeted polymeric nanocapsules for potential treatment of hepatocellular carcinoma. Drug Deliv Transl Res 14(5):1338–1351. https://doi.org/10.1007/s13346-023-01467-9

Kamel YN, El-Marakby EM, Gad HA (2024) Intravenous delivery of furosemide using lipid-based versus polymer-based nanocapsules: in vitro and in vivo studies. Pharm Dev Technol 29(7):738–750. https://doi.org/10.1080/10837450.2024.2389855

Kang H, Zuo Z, Lin R, Yao M, Han Y, Han J (2022) The most promising microneedle device: present and future of hyaluronic acid microneedle patch. Drug Deliv 29(1):3087–3110. https://doi.org/10.1080/10717544.2022.2125600

Kant S, Kaur H, Mishra AK, Challa RR, Vallamkonda B, Popli P, Swami R (2024) Harmonizing nature and technology: lipidic nanophytotherapeutics in the management of atopic dermatitis. Rev Bras 34(5):948–968. https://doi.org/10.1007/s43450-024-00570-3

Karami H, Niavand MR, Haddadi R, Noriyan A, Vafaei SY (2024) Development of a hydrogel containing bisabolol-loaded nanocapsules for the treatment of atopic dermatitis in a BALB/c mice model. Int J Pharm 656:124029. https://doi.org/10.1016/j.ijpharm.2024.124029

Karati D, Mukherjee S, Prajapati B, Bose A, Paul S, Elossaily GM, Roy S (2024) A review on lipid-polymer hybrid nanocarriers in cancer. J Drug Deliv Sci Technol 97:105827. https://doi.org/10.1016/j.jddst.2024.105827

Khan SU, Ullah M, Saeed S, Saleh EAM, Kassem AF, Arbi FM et al (2024) Nanotherapeutic approaches for transdermal drug delivery systems and their biomedical applications. Eur Polym J 207:112819. https://doi.org/10.1016/j.eurpolymj.2024.112819

Kumar M, Mahmood S, Chopra S, Bhatia A (2024a) Biopolymer based nanoparticles and their therapeutic potential in wound healing–A review. Int J Biol Macromol 267:131335. https://doi.org/10.1016/j.ijbiomac.2024.131335

Kumar S, Singh S, Quadir SS, Joshi G, Chouhan M, Puri D, Choudhary D (2024b) Chapter 13 – Polymeric (PLGA-based) nanocomposites for application in drug delivery: current state of the art and forthcoming perspectives. In: Verma D, Okhawilai M, Goh KL, Ramakrishna S, Pasbakhsh P, Sharma M (eds) Bioresorbable polymers and their composites. Woodhead Publishing, pp 277–324

Lei Y, Jiang W, Peng C, Wu D, Wu J, Xu Y et al (2024) Advances in polymeric nanodelivery systems targeting hair follicles for the treatment of acne. Drug Deliv 31(1):2372269. https://doi.org/10.1080/10717544.2024.2372269

Li X, Li C (2020) Tumor diagnosis patterns. In: Huang R, Wang Y (eds) New nanomaterials and techniques for tumor-targeted systems. Springer, Singapore, pp 87–133

Liu L, Zhao W, Ma Q, Gao Y, Wang W, Zhang X et al (2023) Functional nanosystems for transdermal drug delivery and skin therapy. Nanoscale Adv 5(6):1527–1558. https://doi.org/10.1039/d2na00530a

Liu Y, Zhang S, Jiang K, Ding S, Xu L, Liu Y et al (2024) Near-infrared light responsive upconversion-DNA nanocapsules for remote-controlled CRISPR-Cas9 genome editing. Chin Chem Lett 110282:110282. https://doi.org/10.1016/j.cclet.2024.110282

Lorenzoni R, Contri RV, Lima CKFD, Barreto F, Araujo BVD, Pohlmann AR et al (2021) Dermatopharmacokinetic and pharmacodynamic evaluation of a novel nanostructured formulation containing capsaicinoids for treating neuropathic pain. Int J Pharm 596:120294. https://doi.org/10.1016/j.ijpharm.2021.120294

Lunter D, Klang V, Eichner A, Savic SM, Savic S, Lian G, Erdő F (2024) Progress in topical and transdermal drug delivery research–focus on nanoformulations. Pharmaceutics 16(6):817. https://doi.org/10.3390/pharmaceutics16060817

Luo Z, Mao D, Li X, Luo J, Gong C, Liu X (2024) Lanthanide-based nanoparticles for cancer phototherapy. Coord Chem Rev 508:215773. https://doi.org/10.1016/j.ccr.2024.215773

Marchiori MCL, Rigon C, Jardim F d S, Jardim F, Giuliani LM, Copetti PM, Sagrillo MR et al (2023) Hydrogel containing silibinin-loaded pomegranate oil-based nanocapsules for cutaneous application: *in vitro* safety investigation and human skin biometry and permeation studies. AAPS PharmSciTech 24(6):138. https://doi.org/10.1208/s12249-023-02606-8

Marinho A, Nunes C, Reis S (2021) Hyaluronic acid: a key ingredient in the therapy of inflammation. Biomolecules 11(10):1518. https://doi.org/10.3390/biom11101518

Masri S, Fauzi MB, Rajab NF, Lee W-H, Zainal Abidin DA, Siew EL (2024) *In vitro* 3D skin culture and its sustainability in toxicology: a narrative review. Artif Cells Nanomed Biotechnol 52(1):476–499. https://doi.org/10.1080/21691401.2024.2407617

McLean P, Marshall J, García-Bilbao A, Beal D, Katsumiti A, Carrière M, Boyles MSP (2024) A comparison of dermal toxicity models; assessing suitability for safe(r)-by-design decision-making and for screening nanomaterial hazards. Toxicol In Vitro 97:105792. https://doi.org/10.1016/j.tiv.2024.105792

Mehta N, Shetty S, Prajapati BG, Shetty S (2024) Chapter 12 – Regulatory and ethical concerns in the use of nanomaterials. In: Prajapati BG, Chellappan DK, Kendre PN (eds) Alzheimer's disease and advanced drug delivery strategies. Academic, London, pp 197–212

Mei H, Sha C, Lv Q, Liu H, Jiang L, Song Q et al (2024) Multifunctional polymeric nanocapsules with enhanced cartilage penetration and retention for osteoarthritis treatment. J Control Release 374:466–477. https://doi.org/10.1016/j.jconrel.2024.08.031

Mendoza-Muñoz N, Leyva-Gómez G, Piñón-Segundo E, Zambrano-Zaragoza ML, Quintanar-Guerrero D, Del Prado Audelo ML, Urbán-Morlán Z (2023) Trends in biopolymer science applied to cosmetics. Int J Cosmet Sci 45(6):699–724. https://doi.org/10.1111/ics.12880

Mochida Y, Uchida S (2024) mRNA vaccine designs for optimal adjuvanticity and delivery. RNA Biol 21(1):1–27. https://doi.org/10.1080/15476286.2024.2333123

Motta MA, Mulko L, Marin E, Larrañaga A, Calderón M (2024) Polypeptide-based multilayer nanoarchitectures: controlled assembly on planar and colloidal substrates for biomedical applications. Adv Colloid Interf Sci 331:103248. https://doi.org/10.1016/j.cis.2024.103248

Naghib SM, Ahmadi B, Mikaeeli Kangarshahi B, Mozafari MR (2024) Chitosan-based smart stimuli-responsive nanoparticles for gene delivery and gene therapy: recent progresses on cancer therapy. Int J Biol Macromol 278:134542. https://doi.org/10.1016/j.ijbiomac.2024.134542

Narayanan M, Gnanasekaran C, Balamurugan P, Govindan R, Chelliah CK, Govindan R, Natesan M (2024) Role of polymer-based Nanocarriers in drug delivery system. In: Nanoscience and nanotechnology for smart prevention, diagnostics and therapeutics: fundamentals to applications, pp 311–337

Naser SS, Gupta A, Choudhury A, Yadav A, Sinha A, Kirti A et al (2024) Biophysical translational paradigm of polymeric nanoparticle: embarked advancement to brain tumor therapy. Biomed Pharmacother 179:117372. https://doi.org/10.1016/j.biopha.2024.117372

Ngoc LTN, Moon JY, Lee YC (2023) Plant extract-derived carbon dots as cosmetic ingredients. Nano 13(19):2654. https://doi.org/10.3390/nano13192654

Oliveira Santos Md J, Teles-Souza J, de Araújo-Calumby RF, Copeland RL, Marcelino HR, Vilas-Bôas DS (2024) Advances, limitations and perspectives in the use of celecoxib-loaded nanocarriers in therapeutics of cancer. Discov Nano 19(1):142. https://doi.org/10.1186/s11671-024-04070-0

Otavi S, Lad N, Shah S, Navale A, Acharya S, Kaur G et al (2024) Lipidic nanosystem as state-of-the-art nanovehicle for biomedical applications. Indian J Microbiol 64(2):429–444. https://doi.org/10.1007/s12088-024-01298-3

Pahwa R, Saini S, Chhabra J, Goyal R, Kumar S, Awasthi R, Dureja H (2024) Harnessing nanotechnology for enhanced delivery of erlotinib: a dynamic duo in cancer treatment. Beni-Suef Univ J Basic Appl Sci 13(1):69. https://doi.org/10.1186/s43088-024-00528-3

Pan X, Xiao S, Wang B, Cai Y, Chen X, Wang J (2024) Curcumin Nanocapsules based on carbon dots for photodynamic sterilization toward *Listeria monocytogenes* and *Staphylococcus aureus*. Food Chem 458:140295. https://doi.org/10.1016/j.foodchem.2024.140295

Pawar V, Maske P, Khan A, Ghosh A, Keshari R, Bhatt M, Srivastava R (2023) Responsive nanostructure for targeted drug delivery. J Nanotheranostics 4(1):55–85. https://doi.org/10.3390/jnt4010004

Pereira DGS, Torquato LC, Marques MB d F, Senna CA, Menezes RP, Mosqueira VCF, Araújo RS (2024) Effects of continuous millifluidic process on the physicochemical properties and freeze-drying of cloxacillin benzathine-loaded nanocapsules. J Drug Deliv Sci Technol 99:105969. https://doi.org/10.1016/j.jddst.2024.105969

Petrovic S, Bita B, Barbinta-Patrascu M-E (2024) Nanoformulations in pharmaceutical and biomedical applications: green perspectives. Int J Mol Sci 25(11):5842. https://doi.org/10.3390/ijms25115842

Potter TM, Neun BW, Dobrovolskaia MA (2024) *In vitro* and *in vivo* methods for analysis of nanoparticles' potential to induce delayed-type hypersensitivity reactions. In: Clogston JD, Crist RM, Dobrovolskaia MA, Stern ST (eds) Characterization of nanoparticles intended for drug delivery. Springer, New York, NY, pp 193–207

Ramamoorthy S, Prasanen P, Nehru S, Sundaramurthy A (2024) Incorporation of CuS nanorods in polyelectrolyte multilayer microcapsules improved cancer cell cytotoxicity and signal intensity in ultrasound imaging. Int J Pharm 664:124638. https://doi.org/10.1016/j.ijpharm.2024.124638

Ramezani V, Ghadirian S, Shabani M, Boroumand MA, Daneshvar R, Saghafi F (2023) Efficacy of curcumin for amelioration of radiotherapy-induced oral mucositis: a preliminary randomized controlled clinical trial. BMC Cancer 23(1):354. https://doi.org/10.1186/s12885-023-10730-8

Rata DM, Cadinoiu AN, Atanase LI, Popa M, Mihai C-T, Solcan C et al (2021) Topical formulations containing aptamer-functionalized nanocapsules loaded with 5-fluorouracil – an innovative concept for the skin cancer therapy. Mater Sci Eng C 119:111591. https://doi.org/10.1016/j.msec.2020.111591

Saleh MN, Salam MA, Capanoglu E (2023) Encapsulation of black rice bran extract in a stable nanoemulsion: effects of thermal treatment, storage conditions, and *in vitro* digestion. ACS Omega 9(11):12585–12595. https://doi.org/10.1021/acsomega.3c07060

Sallam MA, Prakash S, Kumbhojkar N, Shields CW IV, Mitragotri S (2021) Formulation-based approaches for dermal delivery of vaccines and therapeutic nucleic acids: recent advances and future perspectives. Bioeng Transl Med 6(3):e10215. https://doi.org/10.1002/btm2.10215

Santhanakrishnan KR, Koilpillai J, Narayanasamy D (2024) PEGylation in pharmaceutical development: current status and emerging trends in macromolecular and immunotherapeutic drugs. Cureus 16(8):e66669. https://doi.org/10.7759/cureus.66669

Sarbadhikary P, George BP, Abrahamse H (2021) Recent advances in photosensitizers as multifunctional theranostic agents for imaging-guided photodynamic therapy of cancer. Theranostics 11(18):9054–9088. https://doi.org/10.7150/thno.62479

Sardar MS, Kashinath KP, Gupta U, Roy S, Kaity S (2024) Polymeric nanotheranostics for solid tumor management: recent developments and global regulatory landscape. Polym Adv Technol 35(6):e6461. https://doi.org/10.1002/pat.6461

Sengel-Turk CT, Paksoy AO, Alpturk O (2024) The state of the art in core–shell-type lipid–polymer hybrid nanocarriers and beyond. Polym Bull 81(6):4771–4800. https://doi.org/10.1007/s00289-023-04951-x

Sethi M, Rana R, Sambhakar S, Chourasia MK (2024) Nanocosmeceuticals: trends and recent advancements in self care. AAPS PharmSciTech 25(3):51. https://doi.org/10.1208/s12249-024-02761-6

Shi Z, Yan J, Zhao M, Li S, She T, Qian X (2024) Coencapsulation of granzyme B and perforin in nanocapsules for tumor therapy: biomimicking immune cells. J Contr Release 369:658–667. https://doi.org/10.1016/j.jconrel.2024.04.017

Simon L, Lapinte V, Lionnard L, Marcotte N, Morille M, Aouacheria A et al (2020) Polyoxazolines based lipid nanocapsules for topical delivery of antioxidants. Int J Pharm 579:119126. https://doi.org/10.1016/j.ijpharm.2020.119126

Sucheta, Britya NV, Jha M, Yadav R, Chahal K, Nagori K et al (2024) Clinical and regulatory consideration for Nanobiomaterials. In: Pradhan M, Yadav K, Singh Chauhan N (eds) Biomaterial-inspired Nanomedicines for targeted therapies. Springer, Singapore, pp 451–476

Sudharsan MS, Praveen Kumar M, Gokul V, Pazhamalai V, Hariharan N, Saravanan M (2025) Stimuli-responsive hybrid microgels for a sustainable future: a comprehensive review of classification, synthesis, and multifaceted applications. Polym Adv Technol 36(1):e70087. https://doi.org/10.1002/pat.70087

Sun H, Erdman W, Yuan Y, Mohamed MA, Xie R, Wang Y et al (2020) Crosslinked polymer nanocapsules for therapeutic, diagnostic, and theranostic applications. Wiley Interdiscip Rev Nanomed Nanobiotechnol 12(6):e1653. https://doi.org/10.1002/wnan.1653

Tang L, Cai S, Lu X, Wu D, Zhang Y, Li X et al (2024) Platelet-derived growth factor nanocapsules with tunable controlled release for chronic wound healing. Small 20(27):2310743. https://doi.org/10.1002/smll.202310743

The Niche Research (2024) Nanocapsules market size, share, growth report, 2034. Retrieved from https://thenicheresearch.com/reports/global-nanocapsules-market-forecast-2024-2034/

Tiatragoon W, Laffleur F, Netsomboon K (2024) Development of curcumin-loaded lipid nanocapsules for drug delivery across mucus. J Drug Deliv Sci Technol 99:106014. https://doi.org/10.1016/j.jddst.2024.106014

Toussaint F, Lepeltier E, Franconi F, Pautu V, Jérôme C, Passirani C, Debuigne A (2024) Diversely substituted poly(N-vinyl amide) derivatives toward nontoxic, stealth and pH-responsive lipid nanocapsules. Colloids Surf B: Biointerfaces 235:113788. https://doi.org/10.1016/j.colsurfb.2024.113788

Transparency Market Research (2024) Nanocapsules market for cosmetics size, share, trend 2034. Retrieved from https://www.transparencymarketresearch.com/nanocapsules-market-for-cosmetics.html

Wang T, Liu X, Luo J, Sun G, Liu R (2024a) Interfacial polymerization for preparation of capsaicin microcapsules: achieving sustained release in UV-curable coatings on plastic substrates. Colloids Surf A Physicochem Eng Asp 696:134379. https://doi.org/10.1016/j.colsurfa.2024.134379

Wang M, Skripka A, Zhang Y, Cheng T, Ng M, Wong SY et al (2024b) Theranostic nanocapsules: heating, imaging, and luminescence nanothermometry. Chem Mater 36(7):3285–3295. https://doi.org/10.1021/acs.chemmater.3c03216

Wu H, Wang Y, Ren Z, Cong H, Shen Y, Yu B (2024) The nanocarrier strategy for crossing the blood–brain barrier in glioma therapy. Chin Chem Lett 36:109996. https://doi.org/10.1016/j.cclet.2024.109996

Xie B, Liu Y, Li X, Yang P, He W (2024a) Solubilization techniques used for poorly water-soluble drugs. Acta Pharm Sin B 14:4683–4716. https://doi.org/10.1016/j.apsb.2024.08.027

Xie L, Zhang Y, Chen L, Wang T, Zhang S, Li X (2024b) Structural changes of layer-by-layer self-assembled starch-based nanocapsules in the gastrointestinal tract: implications for their M cell-targeting delivery and transport efficiency. Int J Biol Macromol 261:129786. https://doi.org/10.1016/j.ijbiomac.2024.129786

Xue L, Xin X, Hui Z, Junchi Y, Chao Y, Caisheng W (2025) Preparation of novel biosunscreen using nanocapsules encapsulating pitaya peel flavonoids. Waste Biomass Valori 16:701–712. https://doi.org/10.1007/s12649-024-02692-0

Yadav K, Kumar Sahu K, Sucheta, Yadav R, Raza W, Minz S et al (2024) A complex molecular landscape to drug delivery concept for achieving precise therapy in psoriasis. Med Drug Discov 22:100183. https://doi.org/10.1016/j.medidd.2024.100183

Yadava SK, Reddy BPK, Prausnitz MR, Cicerone MT (2024) Hybrid lipid nanocapsules: a robust platform for mRNA delivery. ACS Appl Mater Interfaces 16(13):15981–15992. https://doi.org/10.1021/acsami.4c00992

Yang Q, Luo S, Tong S, Zhao H, Chen R, Song B, Wu J (2024a) Advances in the study of liposomes gel with stimulus responsiveness in disease treatment. J Clust Sci 35(3):701–714. https://doi.org/10.1007/s10876-023-02510-y

Yang Z, Xie Y, Song J, Liu R, Chen J, Weitz DA et al (2024b) Self-assembly of biocompatible core-shell nanocapsules with tunable surface functionality by microfluidics for enhanced drug delivery. Adv Funct Mater 34(44):2407112. https://doi.org/10.1002/adfm.202407112

Yingngam B (2024a) Advances in nanomaterials for drug delivery. In: Prabhakar PK, Prakash A (eds) Cutting-edge applications of nanomaterials in biomedical sciences. IGI Global, Hershey, pp 22–85

Yingngam B (2024b) Nanocoatings for Cosmetotextiles. In: Garg R, Anjum A (eds) Sustainable approach to protective Nanocoatings. IGI Global, Hershey, pp 173–205

Yingngam B, Chiangsom A, Pharikarn P, Vonganakasame K, Kanoknitthiran V, Rungseevijitprapa W, Prasitpuriprecha C (2019) Optimization of menthol-loaded nanocapsules for skin application using the response surface methodology. J Drug Deliv Sci Technol 53:101138. https://doi.org/10.1016/j.jddst.2019.101138

Yingngam B, Sethabouppha B, Yingngam K, Netthong R, Khumsikiew J, Donsamak S, Navabhatra A (2024) Biomedical applications of hybrid nanomaterials. In: Khanna V, Sharma P, Mahajan P (eds) Technological applications of Nano-hybrid composites. IGI Global, Hershey, pp 170–225

Yu X, Zhao Y, Chen G, Yu J, Wang R, Deng Y et al (2024) Preparation and characterization of titanium oxide nanocapsules containing retinoic acid for sustained release. J Clust Sci 35(2):419–428. https://doi.org/10.1007/s10876-023-02496-7

Zeng H, Zhang Y, Liu N, Wei Q, Yang F, Li J (2024) Stimulus-responsive nanodelivery and release systems for cancer gene therapy: efficacy improvement strategies. Int J Nanomed 19(null):7099–7121. https://doi.org/10.2147/IJN.S470637

# Part III
# Nanotechnology in Biopharmaceutical Formulation

# Chapter 8
# Nanoscale Formulation Techniques for Biotherapeutics

**Fezile Motsoene, Sivakumar Singaravelu, Heidi Abrahamse, and Sathish Sundar Dhilip Kumar**

**Abstract** The development of nanoscale formulations (NFs) in biotherapeutics has revolutionised drug delivery, diagnostics, and treatment strategies, resulting in a new era of precision medicine. NFs were developed to overcome the limitations of conventional drug delivery systems and navigate biological barriers effectively. This chapter highlights the crucial role of NFs in developing personalised and effective therapies. Additionally, the chapter provides a thorough understanding of the synthetic methods, characterisations, and the application of various NFS in biotherapeutics. Moreover, the integration of NFs within various therapeutic platforms, including antibody conjugation, protein fusion, gene expression and monoclonal-antibody delivery, are discussed within this chapter. Subsequently, the benefits, limitations and enhancement strategies associated with NFs drug bioavailability, targeting specificity, stability, and the potential for controlled release will be discussed The final section discusses emerging trends, future perspectives, and key recommendations for advancing the field of nanomedicine, spotlighting its potential to address limitations in current therapeutic approaches.

**Keywords** Nanoscale formulations · Nanomedicine · Biotherapeutics · Bioavailability · Drug delivery

## 8.1 Introduction

Nanotechnology refers to the process of modifying various materials into small particles at the nanoscale, typically ranging from 1 to 1000 nm. This innovative field finds applications across a diverse array of disciplines, including biotherapeutics,

F. Motsoene · S. Singaravelu · H. Abrahamse · S. S. Dhilip Kumar (✉)
Laser Research Centre, University of Johannesburg, Johannesburg, South Africa
e-mail: sathishd@uj.ac.za



V. Khanna et al. (eds.), *Nanotechnologies for Drug Delivery and Biopharmaceutical Development*, Progress in Drug Research 77,
https://doi.org/10.1007/978-3-032-00796-4_8

nanomaterials, nanochemistry, and nanoscience (Harish et al. 2022). Nanoscaled formulations are division of drug delivery systems with various physical, chemical and biological properties and are differentiated based on composition, dimensionality, dispersion state, functionality, morphology, origin phase variation, and porosity (Khan et al. 2019 & El-Khawaga et al. 2023). NFs have a broad spectrum of distinct benefits including enhanced drug solubility and encapsulation, targeted drug delivery and reduced side effects (Jampílek and Kráľová 2015; Malik et al. 2023). Moreover, specialised formulations such as liposomes and dendrimers enable precisely targeted drug delivery to target organs like the brain, liver and kidneys, while conjugated nanoparticles with monoclonal antibodies stem cells facilitate highly specific targeting of cellular receptors (Rizvi and Saleh 2018; Hami 2021; Manimaran et al. 2023).

The small size of NFs facilitates the encapsulation of hydrophilic, hydrophobic, and lipophilic drugs, allowing the therapeutic drug molecules to cross biological barriers while enhancing cellular interaction and leading to potential advancements in personalised medicine and disease treatment (Amiri et al. 2023; Manimaran et al. 2023). Additionally, NFs have reduced side effects of medicinal drugs, minimising damage to non-target tissues and improving the therapeutic index. Moreover, their customisable surface properties enable precise targeting and improved circulation time while inhibiting premature degradation and drug stability (Hami 2021). Nanofibers (NFs) exhibit multifunctionality and possess the capability to encapsulate various therapeutic agents, enabling combination therapies. Additionally, the diversity in NF scaffold design and dosage formulations enhances bioavailability and facilitates drug delivery via multiple administration routes, thereby optimising effectiveness for topical, oral, and injectable applications. Furthermore, NFs play a crucial role in mitigating drug resistance, especially in the context of antimicrobial, cancer, and wound healing therapies, by circumventing traditional resistance mechanisms. (Vega-Vásquez et al. 2020; Hsu et al. 2023).

Nanoscale formulations represent a diverse range of innovative materials, including nanocrystals, nanoparticles (i.e. lipid, polymeric or metallic materials), and self-assembling nanoscale carriers intended to deliver drugs, vaccines, or genetic material (Ladj et al. 2013). An overview of the diverse therapeutic modalities investigated for the development of current and next-generation biopharmaceuticals are shown in Fig. 8.1. Overall, this chapter provides a comprehensive overview of the types of NFs currently available, characterisation, synthesis methods, and application of nanoscaled formulations in biotherapeutics. It includes a broad spectrum of therapeutic modalities, including monoclonal antibodies, antibody-drug conjugates, bispecific antibodies, fusion proteins, gene therapies, cell therapies, oligonucleotides, and peptides. This chapter concludes with a discussion of the future prospects of NFs and offers several recommendations.

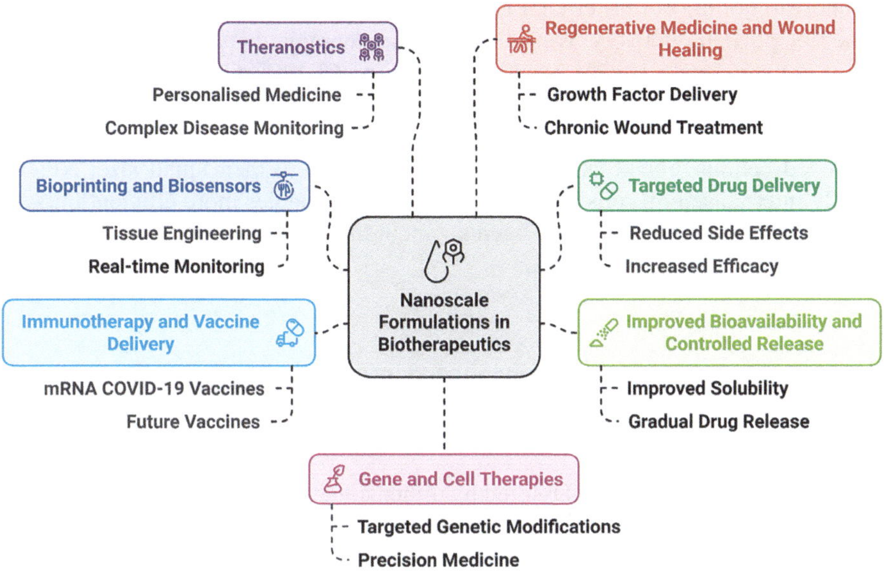


**Fig. 8.1** The therapeutic modalities used in the development of current and next-generation biopharmaceuticals

## 8.2 Fundamentals of Nanoscale Formulation

The development and application of NFs in biotherapeutics has transformed.

Nanoscale formulations in biotherapeutics represent a significant advancement in drug delivery and targeting, enhancing the efficacy of biologically derived treatments like proteins, peptides, and nucleic acids (Jeevanandam et al. 2016). Their nanometric size provides distinct benefits, including increased surface area, thermal stability, and modified optical properties which enable the formulation to effectively address the challenges posed by traditional therapeutic methods (Baig et al. 2021). NFs, which can exhibit dimensions ranging from zero to three dimensions, significantly impact the behaviours and interaction of encapsulated therapeutic agents. The shape and chemical composition of these materials play a crucial role in determining various properties, such as the distribution of nanoscale carriers, mechanisms of drug release, and the protection of the biological payload. Ultimately, these factors contribute to enhancing the therapeutic potential of the drug (Byakodi et al. 2022).

Nanoscale formulations offer a key advantage through their multi-layered structure, which facilitates controlled and targeted drug delivery. These systems have a functionalized surface layer that incorporates molecules like fatty acids or polymers, enhancing stability and even dispersion of therapeutic agents (Patra et al. 2018; Elmowafy and Al-Sanea 2021). Below this is a shell made of a different material from the core, which affects drug-loading capacity and release profile. The composition of the core is crucial for determining drug release rates, enabling sustained or pulsatile delivery as needed. In gene therapy, this layered design effectively delivers nucleic acids, such as siRNA or CRISPR/Cas9, while protecting them from enzymatic degradation (Shao et al. 2016; Jacob et al. 2022). NFs are further distinguished by their ability to enhance therapeutic efficacy through improved cellular uptake and targeted delivery. Due to their small size, NFs are readily internalised by cells via endocytosis, which enables more efficient intracellular delivery of therapeutic agents, including drugs, genes, and proteins (Chenthamara et al. 2019).

### *8.2.1 Classification of Nanoscaled Formulations Based on Dimensionality*

Nanoscale formulations are classified by dimensionality which refers to the number of dimensions exhibiting nanoscale properties impacting their interactions, functionality, and applications (Harish et al. 2022). They are categorised as zero-dimensional (0D), one-dimensional (1D), two-dimensional (2D), and three-dimensional (3D), each offering unique advantages for biotherapeutics such as drug delivery, tissue engineering, diagnostics, and regenerative medicine (Soares et al. 2018; Joudeh and Linke 2022). This dimensionality enables tailored approaches for specific therapeutic needs, making nanoscale formulations versatile tools in modern medicine (Mansoori et al. 2022).

Zero-dimensional (0D) nanoscale formulations, like quantum dots and nanospheres, are particles under 100 nm that interact with biological systems due to their size and properties (Saleh 2020). Their biocompatibility, chemical inertness, cell permeability, and optical stability make them valuable for drug delivery and imaging. They have a high surface area-to-volume ratio, improving drug loading and interaction with target tissues (Haider et al. 2023; Mekuye and Abera 2023). In contrast, one-dimensional (1D) nanoscale formulations, such as nanofibers, nanorods, nanotubes, and nanowires, are characterised by linear dimensions and can exceed 100 nm (Samykano 2021). Composed of carbon, metals, and polymers, 1D NFs facilitate improved drug loading, controlled release, and targeted delivery (Haider et al. 2023). Their unique properties, including directional conductivity and enhanced mechanical strength, allow them to filter contaminants and serve as scaffolds for wound dressing and tissue engineering (Harish et al. 2022).

Nanoscaled formulations with two-dimensional (2D) structures are characterised by plate-like particles that have measurable length and width but negligible thickness (Rizwan et al. 2021). These formulations possess unique physicochemical properties, making them highly advantageous in biotherapeutics. With a high surface area-to-volume ratio and exceptional flexibility, 2D nanostructures are ideal for applications in drug delivery, diagnostics, and tissue engineering. Their enhanced mechanical strength, biocompatibility, and customizable features make them suitable for chronic disease treatments, precise drug delivery, and imaging (Joudeh and Linke 2022). Examples include nanofilms, nanosheets, and nanoplates, used in wound healing, cancer therapy, sensors, and optical devices (Haider et al. 2023).

Nanoscale formulations with three-dimensional characteristics exhibit higher complexity and functionality compared to lower-dimensional counterparts (Paras et al. 2022). These 3D nanomaterials, including dendrimers, nanocages, and mesoporous nanoparticles, serve as biotherapeutic scaffolds and super-absorbents for applications such as antibacterial therapy, cancer treatment, and drug delivery (Wang et al. 2012; Paramasivam et al. 2021). Each type offers unique advantages; for example, dendrimers feature controlled architectures, while mesoporous nanoparticles allow for effective drug encapsulation (Mittal et al. 2021). Moreover, 3D NFs can be customised to offer biodegradable or biocompatible systems, thus rendering them applicable for long-term therapeutic use. Furthermore, they composed of multiple layers of polymers or metals, 3D nanoscale formulations can be functionalized to enhance targeting and stability. Gold and mesoporous silica nanoparticles have internal cavities for transporting drugs, while surfacing modifications, like PEG attachment, improve stability and circulation time (Dristant et al. 2023; Saleh and Hassan 2023; Patra et al. 2018; Wang et al. 2021). These formulations usually have diameters ranging from 10 to 500 nanometers and demonstrate high chemical, mechanical, and thermal stability, with possibilities for controlled degradation and sustained therapeutic release (Shao et al. 2023).

## *8.2.2 Classification of Nanoscaled Formulations Based on Composition*

The classification of nanofibers (NFs) based on their composition is essential for understanding their structure, functionality, and applications in biotherapeutics. Variations in design have integrated biological and physical sciences by using nanostructures across medicinal fields (Patra et al. 2018). NFs are classified into components such as lipid, metal, polymeric, hybridised, and protein-based types., as discussed below.

### 8.2.2.1 Lipid-Based Nanoscaled Formulations

Lipid-Based Nanoscale Formulations (L-NFs) are organic nanocarriers made primarily of lipids, designed to mimic biological membranes (Puri et al. 2009). They consist of lipid cores, surfactants, and lipid bilayers, offering high biocompatibility and biodegradability for delivering diverse therapeutics, especially hydrophobic drugs. L-NFs, including liposomes, solid lipid nanoparticles (SLNs), nanostructured lipid carriers (NLCs), lipid-polymer hybrid nanoparticles (LPNs), and cubosomes, are known for their stability and targeted drug delivery through surface modifications while protecting encapsulated agents (Mohite et al. 2023; Sherif et al. 2022; Syama et al. 2022).

Cubosomes are L-NFs with a cubic liquid crystalline structure formed from amphiphilic lipids that self-assemble in water (Rarokar and Khedekar 2017). They create internal nanostructures with large aqueous channels, enabling efficient loading of both hydrophilic and hydrophobic drugs (Umar et al. 2022). Additionally, surfactants like Pluronic and targeting ligands enhance their stability and site-specific delivery (Rao et al. 2018; Fornasier et al. 2020). While they show promise in drug delivery and vaccine applications, challenges in physiological conditions and scalability remain. Ongoing research is focused on overcoming these challenges, particularly for gene therapy and cancer treatment (Abourehab et al. 2022; Sivadasan et al. 2023).

Liposomes are spherical nano-vesicles formed by hydrating dried phospholipids, characterized by a central aqueous core surrounded by one or more phospholipid bilayers. They effectively encapsulate hydrophobic drugs and can vary in size, composition, and structure (Abdellatif and Alsowinea 2021; Patra et al. 2018). The main types include: (i) Conventional liposomes that are composed of phospholipids, these surround drug molecules and use cholesterol to enhance stability and regulate drug release. Additionally, PEGylated liposomes (ii) are coated with polyethylene glycol for steric stability, they are used in chronic wounds and cancer therapy due to their extended circulation time. Moreover, Immunoliposomes (iii) are composed of ligands like antibodies attached to their surface for targeted delivery. (iv) Furthermore, thermosensitive liposomes (iv) release drugs at elevated temperatures (39–45 °C) and are often combined with nanoparticles for cancer treatment (Sercombe et al. 2015; Patra et al. 2018; Tseu and Kamaruzaman 2023). Although liposomes enhance drug absorption, biodistribution, and biocompatibility, challenges such as complex manufacturing and immunogenicity remain (Sercombe et al. 2015). Despite these issues, their ability to encapsulate diverse drugs makes them a potent platform in modern drug delivery systems, including cancer therapy and vaccine delivery (Robson et al. 2018).

Solid Lipid Nanoparticles (SLNs) are lipid-based nanoscale formulations that efficiently transport drugs. Composed of a biocompatible and biodegradable lipid matrix stabilised with surfactants, SLNs remain solid at body temperature (Duan et al. 2020). Unlike traditional colloidal carriers like liposomes and emulsions, SLNs encapsulate both lipophilic and hydrophilic drugs due to their solid core, made from various lipids, including fatty acids and triglycerides (Viegas et al. 2023). They can incorporate components like antibiotics and polymers for controlled and targeted drug release (Ahmed et al. 2021). Benefits of SLNs include increased drug stability and reduced toxicity (Kishore et al. 2014), though challenges like particle agglomeration and drug expulsion during storage persist. Overall, SLNs demonstrate versatility for biotherapeutic applications, particularly in dermal regeneration, diabetic foot ulcers, cancer therapy, and vaccine development (Dhiman et al. 2021).

Nanostructure lipid carriers (NLCs) are a class of biodegradable and biocompatible lipid-based nanoscaled formulations designed for delivering various biotherapeutic compounds, including antibiotics, antibodies, and vaccines (Elkhateeb et al. 2023). NLCs enhance the stability of lipophilic drugs by encapsulating them, addressing the limitations of solid lipid nanoparticles (SLNs) using liquid lipids and emulsifiers. This results in improved drug loading capacity, stability, and a controlled release profile, reducing systemic toxins and minimising off-target effects for safer long-term use (Fitriani et al. 2024). NLCs exhibit three distinct morphologies: amorphous, imperfect crystals, and multiple-type lipid carriers. Amorphous NLCs are formulated to minimize drug leakage during crystallization, while multiple-type NLCs feature oil-in-fat-in-water delivery systems that prevent drug leakage and allow for controlled release (Obeid et al. 2018; Khosa et al. 2019; Attama et al. 2012). Imperfect crystals comprise a mix of lipids, enabling high drug loading, although encapsulation efficacy is hindered by low solubility (Shah et al. 2015; Dhiman et al. 2021). However, NLCs face challenges due to complex formulation processes and high production costs.

Lipid-Polymer Hybrid Nanoparticles (LPHNs) are innovative nanoscaled formulations that merge the advantageous properties of lipid-based and polymer-based nanoparticles (Xu et al. 2022). They consist of a core-shell structure, with a polymer core and a lipid outer shell, enhancing stability, biocompatibility, and drug loading efficiency (Gajbhiye et al. 2023). The lipid layer, typically made from phospholipids and cholesterols, improves biological interactions and minimizes toxicity and side effects (Mukherjee et al. 2019). LPHNs offer several benefits for drug delivery, such as extended circulation time, controlled drug release, and versatile encapsulation of various drug types, although their complex formulations can limit scalability (Mandal et al. 2013; Bou et al. 2020).

#### 8.2.2.2 Polymeric Nanoscale Formulations

Polymeric nanoparticles are another class of nanoscale formulations derived from natural and synthetic biomaterials such as polyethylene glycol (PEG) alginate, hyaluronic acid and poly (lactic-co-glycolic acid) (PLGA) (Oliveira et al. 2023). They are easily modified through advanced chemical processes that allow them to form 3D structures with precisely engineered properties. Additionally, the flexibility of PNFs allows them to be incorporated into various scaffolds and matrices, enhancing their potential in adjuvant therapy, targeted drug delivery and tissue engineering and remodelling (Bovone et al. 2019). Moreover, these nanoparticles can be functionalised with specific ligands or molecules, such as antibodies or peptides, to achieve tailor-made surface functionality, improving the specificity of drug-targeting diseased tissues or cells (Zielińska et al. 2020).

Designed to encapsulate and protect therapeutic agents from degradation, P-NFs offer a controlled and sustained release profile, which is particularly beneficial for chronic conditions requiring long-term treatment. This sustained release capability minimises the need for frequent drug administration, thus reducing patient burden and improving compliance (Jiang et al. 2017). The protective encapsulation provided by these nanoparticles also increases the bioavailability of sensitive therapeutic agents, such as proteins, peptides, or nucleic acids, ensuring that a higher concentration of the drug reaches its target site intact. Additionally, the biocompatibility of these polymeric formulations reduces the risk of toxicity or adverse immune responses, making them ideal candidates for long-term therapeutic applications. Furthermore, polymeric nanoparticles are not limited to conventional drug delivery (Li et al. 2024). They are being increasingly explored for gene delivery, immunotherapy, and the delivery of complex biologics such as monoclonal antibodies or vaccines. Their ability to be engineered with multiple functional layers opens possibilities for combination therapies, where a single nanoparticle system can deliver both therapeutic agents and diagnostic tools, thus enabling more personalised and efficient treatments (Levit and Tang 2021). With their scalable manufacturing processes and adaptability to a wide range of therapeutic modalities, polymeric nanoparticle formulations are poised to play a crucial role in the future of precision medicine and biotherapeutics.

Polymeric nanoparticles (PNFs) composed of natural polymers are derived from renewable sources, including animal biomass, microbial by-products, and plant extracts (Negut et al. 2010). These natural polymers, including alginate chitosan, collagen, gelatine, honey, hyaluronic acid and silk, offer significant advantages in biotherapeutic applications due to their inherent biocompatibility, biodegradability, and bioactivity (Oliveira et al. 2023). Natural PNFs are well-suited for various health-related applications, including drug delivery, tissue engineering, extracellular matrix replacement, and wound healing. Their biological origin renders them highly compatible with living tissues, making them particularly advantageous for such medical uses. Upon undergoing enzymatic degradation within the body, these natural polymers decompose into by-products that are generally non-toxic and well-tolerated (Bao et al. 2009). This process reduces the likelihood of adverse reactions and enhances patient safety. However, a significant limitation associated with their biological degradation is the limited ability to control the enzymatic degradation process. These natural PNFs are influenced by various physiological factors, making it difficult to accurately regulate the efficacy of different encapsulation kinetics, reducing the nanoparticles' shelf-life during tissue engineering and the circulation period within the body for some therapeutic applications (Elmowafy et al. 2023).

In contrast, synthetic polymer-based nanoparticles are engineered from chemically synthesised materials such as polyvinyl pyrrolidone (PVP), polyethylene glycol (PEG), polycaprolactone (PCL) and Polyethylene oxide (PEO) (Ansari and Darvishi 2024). Synthetic PNFs provide enhanced regulation over their physicochemical characteristics, such as molecular weight, hydrophobicity, and degradation rates. This enables more foreseeable and uniform drug release patterns. In addition, synthetic PNFs exhibit precise control over their solvent absorption and thermal stability due to their custom-designed degradation rates, which enables accurate regulation of drug release duration and maintenance of the structural integrity of the nanoparticle or scaffold ((Shahravi et al. 2020). As a result, the properties displayed by synthetic PNFs make them well-suited for applications that require precise therapeutic delivery, such as sustained-release drug formulations and long-term implants. Furthermore, synthetic polymers can be tailored to include functional groups that improve their mechanical properties, enhance their stability in biological environments, and facilitate surface functionalization for targeted drug delivery (Elmowafy et al. 2023).

Hybrid polymer-based PNFs integrate the advantageous attributes of both natural and synthetic polymers, uniting biocompatibility and biodegradability with enhanced manipulation of physical and chemical characteristics (Huang et al. 2024). Hybrid PNFs comprise a synthetic polymer core for structural integrity and controlled release, enclosed in a natural polymeric shell to ensure biocompatibility and minimise immune reactions. These versatile systems can be tailored for targeted delivery and are increasingly utilised to encapsulate various therapeutic agents such as drugs, genes, and proteins (Lavrador et al. 2021; Gajbhiye et al. 2023). Hybrid PNFs are engineered to address the limitations of natural and synthetic systems by amalgamating their advantages, achieving a harmonious blend of biodegradability, biocompatibility, and precise control over degradation and drug release (Shukla and Maiti 2019). This adaptability renders them well-suited for complex therapeutic applications, including cancer treatment, regenerative medicine, and gene therapy, where both biological compatibility and precise control are indispensable (Kamaly et al. 2016).

#### 8.2.2.3 Peptide and Protein-Based Nanoparticles

Peptide and protein-based nanoscale formulations have become increasingly important in biotherapeutics due to their biocompatibility and ability to mimic biological processes (Katyal et al. 2019). These formulations use peptides and proteins as carriers for drug delivery, leveraging their natural functions to enhance therapy effectiveness. Peptides, short chains of amino acids, can be engineered for targeted therapies (Li et al. 2022), while protein-based formulations consist of larger, more complex molecules that carry greater drug payloads and interact intricately with biological systems.

Despite similarities in amino acid composition, peptide-based NFs are smaller and more flexible, enabling precise targeting and reduced off-target effects for applications like antimicrobial treatment, cancer therapy, and gene delivery (Zhang et al. 2021). In contrast, protein-based formulations, such as albumin and ferritin, are larger and more complex, making them suitable for enzyme replacement therapies and as carriers for monoclonal antibodies in cancer treatment (Verma et al. 2018). Both types are widely used in biotherapeutics. Peptide-based systems effectively target tumour markers in cancer therapy and enhance antibiotic delivery in infections, while protein formulations aid in restoring deficient enzymes in genetic disorders (Verma et al. 2018). Additionally, both are being explored for gene therapy applications, leveraging their biocompatibility to deliver genetic material to target cells (Boisguérin et al. 2021).

#### 8.2.2.4 Metal and Metal Oxide Nanoscale Formulations

Metal and metal oxide-based nanoscale formulations are effective in biotherapeutics due to their unique physicochemical properties, including large surface area, conductivity, optical attributes, and reactivity (Pan et al. 2021; Shabatina et al. 2023). Metal nanoparticles like copper, gold, silver, and platinum are prized for their ability to functionalise with biomolecules, making them suitable for drug delivery, medical imaging, and diagnostics. They ensure targeted delivery and controlled release of therapeutic agents (Nikolova et al. 2023). In contrast, metal oxide nanoparticles such as zinc oxide ($ZnO_2$), iron oxide ($Fe_3O_4$), and titanium dioxide ($TiO_2$) possess magnetic and photocatalytic properties, which enhance their use in imaging, hyperthermia treatments, and antimicrobial applications (Shabatina et al. 2023; Subhan 2022).

Furthermore, metal nanoparticles, especially gold and silver, are well-regarded for their biocompatibility and their capacity for targeted drug delivery modifications (Siddique and Chow 2020). In contrast, metal oxide nanoparticles offer distinct advantages, such as magnetic properties suitable for MRI and photocatalytic capabilities beneficial for photodynamic therapy (Khalil et al. 2017; Yi et al. 2020). Despite their merits, both types of nanoparticles come with certain drawbacks; metal nanoparticles can raise concerns regarding long-term toxicity due to their non-biodegradable nature, while metal oxides may cause oxidative stress if not carefully engineered (Manuja et al. 2021). Nonetheless, their potential to enhance drug targeting, improve diagnostic processes, and integrate therapeutic functions positions them as crucial elements in the future of personalized medicine (Manuja et al. 2021; Rathi et al. 2024).

#### 8.2.2.5 Carbon-Based Nanoscale Formulations

Carbon-based nanoscale formulations are a versatile class of nanomaterials in biotherapeutics due to their unique structural properties, large surface area, and ability to interact with biological molecules (Maiti et al. 2018). They include structures like graphene, graphene oxide, carbon nanotubes, fullerenes, and nanodiamonds, which are known for their electrical conductivity, chemical stability, and mechanical strength. These properties make them suitable for drug delivery, gene therapy, tissue engineering, and biosensing (Mostofizadeh et al. 2011; Bhattacharya et al. 2016). Their biocompatibility and ease of functionalization enhance their effectiveness in biomedical applications, allowing for customization in solubility, targeting, and drug loading (Riley and Narayan 2021).

Carbon-based NFs include various nanostructures, each with distinct characteristics influencing their behaviour and applications (Maiti et al. 2018). Graphene, a two-dimensional (2D) structure of carbon atoms in a honeycomb lattice, is celebrated for its high electrical conductivity and large surface area. Graphene oxide

(GO), a functionalised derivative, improves dispersibility in biological systems, making it suitable for biomedical applications (Pratibha et al. 2019; Ghulam et al. 2022). Carbon nanotubes (CNTs), one-dimensional (1D) cylindrical structures made of rolled graphene, come in single-walled (SWCNTs) and multi-walled (MWCNTs) forms, each with unique properties (He et al. 2013; Thiruvengadam et al. 2021). Fullerenes, like C60, are zero-dimensional (0D) carbon molecules capable of encapsulating drugs (Paukov et al. 2023). Nanodiamonds, also zero-dimensional, are highly biocompatible and ideal for targeted drug delivery and imaging. These various nanomaterials can be tailored for specific therapeutic applications (Chauhan et al. 2020).

Carbon-based nanoscale formulations are highly versatile for various biotherapeutic applications. Graphene and graphene oxide are notable for their large surface area, making them effective in drug delivery systems due to their capacity for high drug loading and controlled release (Pratibha et al. 2019). Graphene oxide is particularly researched for gene delivery, enabling the transport of nucleic acids to the targeted cells (Shafiee et al. 2022). These materials also serve in tissue engineering, promoting cell proliferation and differentiation, especially in bone, nerve, and cardiac repairs (Maleki et al. 2020; Guo et al. 2023).

Carbon nanotubes (CNTs) are used in cancer therapies for their ability to enter cells and transport drugs to nuclei, as well as in photothermal therapy to eliminate cancer cells with heat. Their conductive properties also enhance biosensor development for detecting biological molecules (Tang et al. 2021; Deng et al. 2022). Fullerenes show promise in antioxidant therapies by crossing the blood-brain barrier to scavenge free radicals (Schuhmann and Fluri 2017; Grebowski et al. 2022). Nanodiamonds contribute to drug delivery, antibacterial therapy, and imaging due to their unique properties, facilitating targeted therapy and diagnostic imaging (Perevedentseva et al. 2013; Chauhan et al. 2020), while also playing a significant role in tissue engineering due to their mechanical strength and biocompatibility (Curbelo et al. 2020).

Carbon-based nanoscale formulations offer significant benefits but also have limitations. Graphene and CNTs provide excellent conductivity and mechanical strength but can pose biocompatibility challenges, including cytotoxicity and inflammatory responses, especially in their non-functionalized forms. Their size, shape, and surface properties must be carefully managed for safe biological interaction. Fullerenes face solubility issues due to their hydrophobic nature, limiting their use in aqueous environments unless functionalized. Concerns about the long-term toxicity of carbon-based NFs due to potential accumulation in the body exist, but when properly engineered, they provide superior drug loading capacity, targeting precision, and multi-functionality. Their diverse structures, ranging from 0D fullerenes to 2D graphene, support various biomedical applications, making them adaptable for modern biotherapeutics.

#### 8.2.2.6 Mechanisms of Action for Nanoscale Formulations

Understanding nanoparticle nucleation and growth mechanisms is crucial for developing nanoscale formulations with precise control over size, composition, and stability (Harish et al. 2022). These mechanisms allow for flexibility in controlling parameters such as nanoparticle diameter, shape, and molecular orientation. A thorough understanding of nucleation is essential for managing nanoparticle morphology and achieving specific functionalities, particularly in drug delivery, imaging, and diagnostics (Patra et al. 2018). Nucleation marks the initial phase of forming nanostructures from a reactive phase, with atoms or molecules arranging in a pattern that determines the material's crystal structure. This nucleation site serves as a foundation for further particle growth.

Numerous variables influence the nucleation and growth stages of nanoparticles, which are crucial for customising them for specific applications. Temperature significantly affects these processes; higher temperatures reduce the critical level of supersaturation and increase the nucleation rate, allowing for better control of particle size and uniformity. Maintaining a consistent reaction volume is also essential for uniform nucleation rates. Additionally, Ionic strength impacts the activity of reaction species and the growth rate by affecting ion adsorption on nuclei, thus influencing nanoparticle size. Moreover, factors like agitation and the presence of foreign particles can lead to heterogeneous nucleation, enhancing nucleation rates. By carefully controlling these variables, one can design nanostructures with tailored sizes, shapes, and functionalities for various biotherapeutic applications.

## 8.3 Techniques for Nanoscale Formulation

Based on the preceding sections, it is evident that nanoscale formulations play a pivotal role across various domains such as drug delivery, nanomedicine, material science, cell therapy, and regeneration therapy due to their unique properties and high surface-to-volume ratio. The choice of suitable preparation methods for nanofibers (NFs) depends on their physical and chemical properties, structural composition, intended functional use, and applications (Pal et al. 2011; Jeevanandam et al. 2016). Various techniques have been employed in the production of NFs, with the most common methods falling into distinct categories, including the Bottom-up approach and Top-down approach, Biological or Green Synthesis, Physicochemical Methods, Hybrid Methods and Specialised Techniques for Nanocarrier Formulations, as illustrated in Table 8.1. Moreover, these methods can be further categorised into distinct subclasses based on the specific operation, reaction conditions, and implemented protocol (Khan et al. 2019).

**Table 8.1** The synthetic methods used for NFs formulation

| Preparation technique | Approach | Mechanism | NFs Produced | Benefits | Limitations | Application | References |
|---|---|---|---|---|---|---|---|
| Top-down method | Grinding | Mechanical reduction of bulk materials into smaller particles through shear and compressive forces. | Metallic and metal oxides<br>Polymeric NFs | Straightforward<br>Scalability<br>Adaptability<br>Enhanced reactivity and solubility<br>Bioavailability | Contamination from grinding medium<br>Irregular particle shapes<br>Uniformity<br>Stability | Catalysis drug delivery | Loh et al. (2015) and Lepore (2024) |
| | High-pressure homogenisation | High pressure is used to force materials through a narrow gap, creating shear forces, cavitation, and impact that break down materials into nanoscale particles | Liposomes<br>SLNs<br>NLCS<br>Exosomes | Structure<br>Functionality<br>Controlled drug release<br>Large scale production<br>Uniformity | Degradation of protein and enzyme | Drug delivery<br>Dispersion | Naseri et al. (2015), Mourdikoudis et al. (2018), Ferreira et al. (2021), Shi et al. (2022), and Karle et al. (2022) |
| | Ultrasonication | This technique uses high-frequency sound waves to generate cavitation bubbles in liquids. To produce intense shear forces that break down particles into nanoscale dimensions | Liposomes<br>Polymeric NFs<br>Metal and metal oxide NFs | Particle dispersion and uniformity<br>Low shear stress<br>High energy input | Degradation of heat sensitive materials<br>Scalability<br>Metal contamination | Drug delivery,<br>Catalysis<br>Dispersion | Iqbal et al. (2012), Gheisari et al. (2019), Glaubitz et al. (2022), Liu and Ge (2022), and Shahhosseinlou et al. (2023) |

(continued)

**Table 8.1** (continued)

| Preparation technique | Approach | Mechanism | NFs Produced | Benefits | Limitations | Application | References |
|---|---|---|---|---|---|---|---|
| Bottom-up method | Solvent evaporation | The drug and polymer dissolve in an organic solvent, then emulsify into an aqueous phase. As the solvent evaporates, NFs are formed | Polymeric NFS<br>Lipid-polymer NFs | Particle size and dispersion | Solvent toxicity<br>Ineffectiveness for water-soluble drugs | Drug delivery<br>Controlled release | Hadinoto et al. (2013), Davoodabadi and Ghasemi (2021), and Procopio et al. (2022) |
| | Solvent displacement | The rapid diffusion of polymer in a water-miscible organic solvent are injected into aqueous phases leading to NFs formation | Polymeric NFs<br>Lipid NFs | Compatible for temperature-sensitive drugs and hydrophobic drugs | Scalability<br>Limited solvent selection | Nanoencapsulation<br>Drug delivery | Shepherd et al. (2021), Pulingam et al. (2022), and Piñón-Segundo et al. (2006) |
| | Self-assembly | NFs spontaneously form nanoscale structures through non-covalent interactions like hydrogen bonding and hydrophobic forces. | Polymeric NFs<br>Lipid NFs<br>Metalic NFs<br>Micelles, liposomes | Controlled size and of intricate NFs<br>Simple and effective | Template removal in needed<br>Limited to certain materials | Drug delivery in nanodevices<br>NFs coating | Thiruvengadathan et al. (2013), Zhu et al. (2012), and Yadav et al. (2020) |
| | Chemical reduction | NFs are formed by the reduction of metal ions in a solution by a reducing agent | Metal NFs<br>Carbon nanotubes graphene NFs | Uniform distribution<br>Desirable\optical and catalytic properties<br>Controlled drug release<br>Scalability | Potential toxicity of residual chemicals<br>Agglomeration<br>Stability | Drug delivery<br>Drug and NFs coating<br>Catalysis | Beyene et al. (2017), Nikam et al. (2018), Hachem et al. (2022), and Nyabadza et al. (2023) |

| | | | | | | | |
|---|---|---|---|---|---|---|---|
| | Coacervation | NFs are formed through phase separation in a polymer-rich liquid as the solvent is removed | Ionic NFs<br>Protein based NFs | Controlled drug release<br>Protein and nucleic acid stabilisation<br>High surface area | Incomplete drug encapsulation<br>Polydisperse particles | Catalysis<br>Drug delivery<br>Protein stabilisation | Sing and Perry, (2020), Zhou et al. (2020), Moulik et al. (2022), and Napiórkowska and Kurek (2022) |
| Hybrid method | Supercritical fluid technology | Lipid, polymeric, metallic and organic NFs | Metallic NFs | Controlled size and shape<br>Environmentally friendly<br>Uniformity<br>High conservation rate | Limited scalability | Drug delivery | Adschiri and Yoko (2018), Franco and De Marco (2021) |
| | Plasma assisted synthesis | Plasma energy enhances nanoparticle formation | Plasma grown and metallic NFs | Fast process | Expensive to conduct<br>High energy cost | Drug delivery | Sohn and Murali (2021), Shah et al. (2024), and Messai et al. (2024) |
| Green-synthesis | Microbial synthesis | Bacteria, fungi and other microorganisms are used as reducing agents to form metallic NFs | Metalic and | Biocompatible | Difficult to scale<br>Lengthy synthesis process | Drug delivery and coating | Das et al. (2017), Kumar et al. (2021), Alsaiari et al. (2023), and Singh et al. (2023) |

(continued)

**Table 8.1** (continued)

| Preparation technique | Approach | Mechanism | NFs Produced | Benefits | Limitations | Application | References |
|---|---|---|---|---|---|---|---|
| | Plant extract synthesis | Plant extracts with reducing agents are used to form metallic NFs | Metallic and hybridised NFs | Biocompatible<br>Cost effective suitable for large-scale production<br>Strong antimicrobial and anticancer properties | Particle size and shape optimisation | Drug delivery and coating | Anandalakshmi et al. (2016), Amale et al. (2021), and Kazemi et al. (2023) |
| Physicochemical method | Ionic gelation | NFs are formed through cross-linking polymers through ionic interactions | Polymeric NFs | Enhance encapsulation efficiency<br>Controlled size and particle distribution<br>Enhanced protein stability Maqbool and Noreen 2019 | High sensitivity to environmental factors | Drug delivery and functionality<br>Gene delivery | Giri (2016), Sacco et al. (2021), Pedroso-Santana and Fleitas-Salazar (2020), and Fernández-Quiroz et al. (2023) |
| | Spray drying | NFs are formed by spraying a solution or suspension into a stream of hot air, causing rapid evaporation of the solvent | Polymer- or protein-based NFs | Cost-effective<br>Effective for thermosensitive drugs<br>Enhance delivery of hydrophobic drugs | Agglomeration<br>High energy consumption | Drug delivery<br>Thermal stability | Maqbool and Noreen (2019), Kumar et al. (2020), Marante et al. (2020), and Malamatari et al. (2020) |
| | Freeze drying | Eliminates water from NFs suspensions through sublimation | Lipid NFs<br>Metallic NFs<br>Polymeric NFs | Preserves and stabilised NFS<br>Extended shelf-life<br>Biocompatibility | Time-consuming<br>Expensive | NFs preservation | Abdelwahed et al. (2006), Fonte et al. (2016), and Gatto and Najahi-Missaoui (2023) |

## 8.4 Characterisation of Nanoscale Formulations

Nanoscale materials exhibit distinct properties compared to their bulk counterparts, primarily due to their high surface-to-volume ratio, significantly enhancing molecular reactivity (Thanh et al. 2014). These unique characteristics span electronic, optical, chemical, and mechanical domains, making nanoparticles (NPs) a subject of intense study for academic exploration and technological applications (Koczkur et al. 2015). Nanomaterials can be synthesised via various methods, including mechanical and chemical pathways, and recent advancements have led to the development of a broader range of nanostructures in larger quantities. However, precise characterisation of these materials remains challenging due to their small size and multidisciplinary nature, often requiring access to various analytical techniques (Wang et al. 2013).

Comprehensive characterisation of nanoparticles is crucial, and no single technique is universally applicable for analysing all properties. It is essential to combine different methods for a thorough understanding of NPs. Techniques such as transmission electron microscopy (TEM), high-resolution TEM (HRTEM), and atomic force microscopy (AFM) provide insights into size, morphology, and crystal structure. Other specialised methods, like superconducting quantum interference device (SQUID), vibrating sample magnetometry (VSM), ferromagnetic resonance (FMR), and X-ray magnetic circular dichroism (XMCD), are used for specific material groups, such as magnetic nanomaterials. Additionally, X-ray diffraction, spectroscopy, and scattering offer detailed information on structure, elemental composition, and optical properties (Kim et al. 2004).

The hydrodynamic size and surface charge of nanoparticle dispersions significantly influence an organism's biological response upon exposure, necessitating thorough characterization of these properties. Nanoparticle size is critical in governing interactions with biological systems, affecting absorption, distribution, metabolism, and excretion (Renwick et al. 2001; Borm et al. 2006.; Choi et al. 2007). Surface characteristics, including charge, directly impact nanoparticle agglomeration in dispersions, as even slight changes in surface charge can markedly alter hydrodynamic size distributions. The uptake and translocation of nanoparticles by organisms are similarly influenced by surface charge (Hoshino et al. 2004). For instance, changes in nanoparticle surface charge have been shown to affect the integrity and permeability of the blood-brain barrier (Lockman et al. 2004).

The characterization of hydrodynamic size and surface charge has been highlighted as critical in several nanotoxicology review articles (Powers et al. 2007; Oberdörster et al. 2005). Despite this, there is a scarcity of systematic studies examining these parameters. Murdock et al. (2024) characterised the hydrodynamic size and surface charge of various nanoparticles (e.g., Ag, $TiO_2$, and carbon nanotubes) in biologically relevant solutions. However, their study primarily reported observational results and did not discuss key factors influencing nanoparticle dispersion, such as solution ionic strength (IS), pH, surface charge, and surface coatings. These variables are critical for understanding and controlling nanoparticle behaviour in biological environments.

Many therapeutic drugs exhibit poor solubility and low absorption rates and are prone to degradation, chemical reactions, or rapid clearance when administered directly. Encapsulating these drugs within nanoparticles can significantly improve their solubility, stability, and bioavailability, while also enhancing therapeutic efficacy. Nanoparticle encapsulation offers protection to the drug and can be engineered to enable controlled drug release, as well as selective targeting of specific cells, thus maintaining therapeutic levels and minimising toxicity (Cheng et al. 2023).

Encapsulation assessment can be conducted for various nanoparticle systems, including:

(a) Virus-like particles (VLPs), such as adeno-associated virus (AAV), capsids
(b) Liposomal drug formulations
(c) Lipid nanoparticle (LNP) payloads

Despite these advantages, the manufacturing and long-term stability of therapeutic formulations poses significant challenges. Suboptimal production conditions-such as the introduction of contaminants, exposure to shear stress, freeze-thaw cycles, or light can negatively affect product quality attributes. These adverse conditions may lead to the formation of impurities, misassembled particles, loss of nanoparticle integrity, or aggregation. Furthermore, such conditions can result in the production of nanoparticles lacking payload, which diminishes the effective concentration of the therapeutic agent and compromises both efficacy and safety. Therefore, precise characterization of the encapsulation efficiency or drug loading fraction of nanoparticles within a formulation is crucial. This assessment ensures that the correct therapeutic dose is delivered, while minimising the presence of empty delivery vehicles, ultimately optimising both the efficacy and safety of nanoparticle-based drug delivery systems (Jain et al. 2021).

The release of drugs from pharmaceutical nanoparticles is a critical factor influencing their biological effects and evaluating drug release kinetics is essential in this field. Kinetic models are often employed to elucidate the mechanisms of drug release, which in turn can aid in the controlled release of therapeutic agents. A key advantage of using kinetic models is their ability to represent complex release data with one or two parameters, providing a simplified yet informative description of the release process. To date, a comprehensive kinetic model for drug release from nanoparticles that could serve as a general framework has not been extensively reported in the literature. Parameters derived from such a generalised model would be valuable for comparing different drug delivery systems and correlating these findings with bioavailability data. Without a standardised model, it is difficult to make meaningful comparisons or correlations between parameters obtained from fitting release data to various kinetic models (Costa and Sousa Lobo 2001).

Several mathematical models have been developed to understand drug release kinetics better. For example, zero-order kinetics describe a constant drug release rate, independent of the drug concentration in the system, which is ideal for maintaining stable therapeutic levels over time. The development of more robust models could further enhance the understanding and control of nanoparticle-based drug delivery systems (Ahmed et al. 2019).

$$Qt - Q_0 = K_0 t \tag{8.1}$$

Where $K_0$ is the zero-order model constant, $Q_0$ is the total fraction of drug released, $Q_t$ is the fraction released in time $t$. First-order kinetics represent concentration-dependent release systems and are given by the expression below (Ahmed et al. 2019; Dash et al. 2010).

$$Q_t = Q_0\left(1 - e - k1t\right) \tag{8.2}$$

Where $K1$ is the first-order rate constant. The drug release from polymer matrix systems based solely on diffusion where the drug concentration is directly proportional to the square root of time ($\propto \sqrt{t}$) is described in the Higuchi model (Rabha et al. 2021). In its simplified form, it is represented as,

$$Qt = K_{\mathrm{Ht}}^{1/2} \tag{8.3}$$

Here, $K_H$ is the Higuchi dissolution constant. The Peppas-Sahlin model considers the influences of both diffusional and relaxational mechanisms in systems exhibiting anomalous drug release (Sera et al. 2006).

Nanoencapsulation techniques play a crucial role in protecting drugs from degradation in biological fluids and enhancing their cellular penetration. These methods are particularly advantageous for hydrophobic molecules, as the ultra-dispersed pharmaceutical dosage forms provided by nanoencapsulation facilitate rapid drug dissolution (Brigger et al. 2002). The encapsulation of DMEO within an appropriate polymer is an effective strategy to overcome its poor solubility, as the polymer functions as a rate-controlling membrane, enabling controlled drug release. However, the physical stability of pure compounds remains a significant challenge for pharmaceutical scientists aiming to improve solubility. According to guidelines established by the International Council for Harmonization (ICH), physical stability tests should be conducted under accelerated (6 months) and/or long-term (12 months) storage conditions. In this context, physical stability is typically assessed using powder X-ray diffraction (PXRD) analysis (Huynh-Ba and Zahn 2009).

The unique properties of drug nanoparticles have facilitated their widespread application across various dosage forms, including oral, parenteral, ocular, pulmonary, dermal, and specialised delivery systems (Gao et al. 2008). While different dosage forms may share common stability issues such as sedimentation, particle agglomeration, or crystal growth their effects on drug products can vary significantly. For example, particle agglomeration poses a major concern in pulmonary drug delivery, as it can affect the deposition site and dosage, thereby impacting drug efficacy. In intravenous formulations, agglomeration can lead to blood capillary blockage, obstructing blood flow and posing serious safety risks. Moreover, the selection of stabilizers is highly dependent on the dispersion medium, and dosage form, and is strictly regulated by the FDA. While there is a broad range of FDA-approved stabilizers for oral dosage forms, the options for inhalation formulations remain limited. This highlights the importance of carefully selecting stabilizers based on the specific characteristics and regulatory requirements of the intended dosage form (Rogueda 2005).

## 8.5 Applications in Biotherapeutics

### *8.5.1 Antimicrobial Therapy*

Nanoparticle-based antimicrobial therapies have been extensively studied for combating microbial infections, including multidrug-resistant (MDR) bacterial strains. Various types of nanoparticles (NPs), such as polymer-based NPs, immune cell-based nanoformulations, and liposomal systems, have proven effective in delivering conventional antibiotics through sustained release mechanisms without the need for increased concentrations (Yan et al. 2022). A prime example is ciprofloxacin-loaded liposomes (Lipoquin), developed for respiratory infections, which enable extended antibiotic release without adverse effects, reducing the need for drug repurposing, higher doses, or combination therapies.

Liposomal formulations, due to their unique structure and characteristics, have also shown promise as nanocarriers for antifungal drugs like amphotericin B, significantly reducing cytotoxicity. These formulations have been used to treat neutropenia, histoplasmosis, and even viral infections. Organic nanoparticles, such as chitosan nanoparticles (ChNPs), prepared via ionic gelation (193–530 nm) (Alqahtani et al. 2020) and electrospray methods (average size ~200 nm), have demonstrated efficacy against MDR pathogens, including *Neisseria gonorrhoeae* and biofilm-forming oral microbes such as *Staphylococcus spp*, *Enterococcus spp*, and *Candida spp* (Ibrahim et al. 2021).

Silver nanoparticles (AgNPs) have emerged as potent antimicrobial agents against MDR microorganisms and have been incorporated into implant materials to enhance antimicrobial (Dutt et al. 2022a, b), antifungal (Kim et al. 2009), anti-inflammatory (Nadworny et al. 2010), and antiviral properties (Lara et al. 2010). AgNPs have also been widely utilised in wound infection management, with products such as Silverline, Verigene, Acticoat, and Endorem used in clinical applications. Nanocomposites of silver, fluoride, and chitosan, synthesised through chemical methods, have demonstrated antimicrobial activity against pathogenic *Enterococcus spp* and *Candida spp*; however, nanocomposites smaller than 10 nm were found to be more toxic to mouse macrophages than larger particles (Freire et al. 2016). Additionally, polymeric nanoparticles conjugated with microcin J25 antimicrobial peptides (AMPs) have shown dose-dependent bactericidal effects against tetracycline-resistant *E. coli* K88 and methicillin-resistant *Staphylococcus aureus* (MRSA), without exhibiting cytotoxicity to *Caenorhabditis elegans*. These advancements underscore the potential of nanoparticle-based systems in the treatment of microbial infections, particularly those caused by resistant pathogens (Haitao et al. 2022).

#### 8.5.1.1 Tissue Regeneration

Delayed wound healing, often complicated by multidrug-resistant (MDR) microbial infections and immunocompromised states, has driven significant interest in the fields of tissue engineering (TE), tissue regeneration (TR), transplantation, and alternative biological approaches aimed at repairing or reconstructing damaged or diseased tissues, skin, or organs (Dutt et al. 2023). Unlike traditional bandages or drug-based treatments, functionalised nanoparticles (NPs) exhibit superior interaction with underlying cells, the wound environment, and microbes (Maurer et al. 2016). NPs-based biomaterials and TR techniques create an environment that accelerates tissue regeneration by engaging with cells and addressing oxygen-deficient microenvironments (Mihai et al. 2019).

Stem cells and enzymes within the tissue respond to the microenvironment, triggering crucial processes for tissue regeneration and remodelling (Yan et al. 2022). Various fabrication techniques, such as electrospinning and coaxial electrospinning, along with advanced biomaterials, have been explored for tissue regeneration applications (Yan et al. 2021). These techniques produce nanofibrous scaffolds, nanogels, hydrogels, thread-based patches, and sponge scaffolds, all enhancing the healing process by acting as functional barriers that promote faster wound closure (Dutt et al. 2023; Xue et al. 2019). Electrospinning applies an electrical potential to a biopolymer solution,

Forming threads or filaments, while coaxial electrospinning enables the simultaneous incorporation of drugs, enzymes, or cells into the core of the nanofibers, with a biopolymer forming the outer shell (Xue et al. 2019). Nanofiber-based fabrications have been extensively applied to wound healing, ulcers, and lesions, with promising results (Pal et al. 2017; Wu et al. 2018). These skin substitutes, patches, or scaffolds not only act as physical barriers but also facilitate accelerated healing by incorporating therapeutic agents such as drugs, growth factors, enzymes, or cells (e.g., fibroblasts, keratinocytes, or osteoblasts) (Dutt et al. 2022a, 2022b).

However, scaffolds designed for soft tissue regeneration are unsuitable for hard tissue applications, such as bone and dentine repair. For hard tissues, biomaterials must mimic the mechanical properties, including strength, porosity, and texture, of natural tissues. Calcium phosphate-based hydroxyapatite (Hap) is widely used as a bone substitute due to its similarity to the bone matrix and its ability to induce osteoblast activity when incorporated into nanofibers (Zhou and Lee 2011; Wepener et al. 2012). Cartilage regeneration poses a greater challenge due to limited physiological remodelling and the difficulty of repairing damaged tissues post-inflammation. This is often attributed to the improper presentation of progenitor cells. Incorporating either undifferentiated mesenchymal stem cells (MSCs) or differentiated chondrocytes into tissue regeneration methods has shown potential for enhancing cartilage repair. Nanofibrous scaffolds or cartilage matrices embedded with these cells have demonstrated improved outcomes in tissue regeneration, offering promising solutions for regenerating complex tissues like cartilage (Shafiee et al. 2011; Planka et al. 2012).

#### 8.5.1.2 Drugs and Medications

Nano-sized inorganic particles, whether of simple or complex composition, exhibit unique physical and chemical properties, making them increasingly valuable in the development of advanced nanodevices for various applications across physical, biological, biomedical, and pharmaceutical fields (Loureiro et al. 2016; Nikalje 2015; Martis et al. 2012). Nanoparticles (NPs) have garnered significant attention in medicine due to their ability to deliver drugs in optimised dosages, resulting in enhanced therapeutic efficacy, reduced side effects, and improved patient compliance (Alexis et al. 2008). Iron oxide particles, such as magnetite ($Fe_3O_4$) and its oxidised form maghemite ($Fe_2O_3$), are among the most widely used nanoparticles for biomedical applications (Ali et al. 2016). The choice of NPs for applications in contrast imaging, biological and cell imaging, and photothermal therapy is primarily based on their optical properties. Calculations of absorption, scattering efficiencies, and optical resonance wavelengths for common nanoparticles such as gold (Au) NPs, silica-Au NPs, and Au nanorods can be performed using Mie theory and discrete dipole approximation methods (Jain et al. 2006).

In recent years, the development of hydrophilic nanoparticles as drug carriers has presented a significant challenge. Among various approaches, polyethylene oxide (PEO) and polylactic acid (PLA) nanoparticles have demonstrated great promise for the intravenous delivery of drugs (Calvo et al. 1997). Superparamagnetic iron oxide nanoparticles, when functionalised with appropriate surface chemistries, can be utilised in numerous in vivo applications, including MRI contrast enhancement, tissue repair, immunoassays, detoxification of biological fluids, hyperthermia, drug delivery, and cell separation. For these applications, NPs must exhibit high magnetization, particle sizes below 100 nm, and a narrow size distribution (Laurent et al. 2008). In targeted drug delivery, biodegradable nanoparticles have gained considerable interest as effective vehicles for therapeutic agents. Polymers used in drug delivery can enhance therapeutic outcomes by delivering drugs precisely to target sites, thereby increasing efficacy while minimising side effects. Controlled drug release at therapeutically optimal doses remains a key objective in designing such systems (Khlebtsov and Dykman 2010; Zhang and Saltzman 2013).

Liposomes once considered a potential drug carrier due to their ability to protect drugs from degradation, target specific sites, and reduce toxicity, have faced challenges such as low encapsulation efficiency, rapid leakage in biological fluids, and poor storage stability. In contrast, polymeric nanoparticles offer distinct advantages over liposomes, including improved drug stability, enhanced solubility, and controlled drug release properties. Semiconductor and metallic nanoparticles also hold significant potential in cancer diagnosis and therapy due to their surface plasmon resonance (SPR) properties, which enhance light scattering and absorption. Gold nanoparticles, for instance, can convert absorbed light into localised heat, enabling their use in selective laser photothermal cancer therapy (Jain et al. 2007). Additionally, some nanoparticles, such as multihydroxylated $[Gd@C_{82}(OH)_{22}]n$ nanoparticles, exhibit antineoplastic activity, inhibiting tumour growth with lower toxicity (Chen et al. 2005). Nanoparticles are increasingly utilised in wound dressings, catheters, and household products due to their antimicrobial properties. Antimicrobial agents are crucial in textiles, medicine, water disinfection, and food packaging (Hajipour et al. 2012).

## 8.6 Challenges and Limitations

Nanoparticles (NPs) hold promise for biomedical applications, but their adoption faces significant challenges. A key issue is navigating biological barriers and clearance mechanisms effectively (Ezike et al. 2023). NPs must cross mucosal layers, epithelial cells, and the blood-brain barrier (BBB) to reach targets (Wu et al. 2023; Vashist et al. 2023). Factors such as size and hydrophobicity can hinder penetration. Challenges for intravascular delivery include immune clearance in the liver and spleen, endothelium permeation, tissue interstitium penetration, endocytosis, cytoplasmic diffusion, and potential nuclear entry (Barua and Mitragotri 2014). The body's defence mechanisms often recognize NPs as foreign, leading to quick removal and limited circulation time, which diminishes their therapeutic efficacy. Addressing these hurdles to create formulations that evade immune recognition, extend circulation, and target specific tissues is a crucial challenge in the field (Hoshyar et al. 2016; Rosenblum et al. 2018). The challenges in the development and commercialisation of nanoscale formulations are shown in Fig. 8.2.

A significant hurdle in the development of nanoscale formulations for biological treatments involves regulatory and safety concerns. The current regulatory framework is still evolving, as nanoparticles possess unique properties, such as small size and the ability to cross biological barriers, which raise new safety issues (Ramos

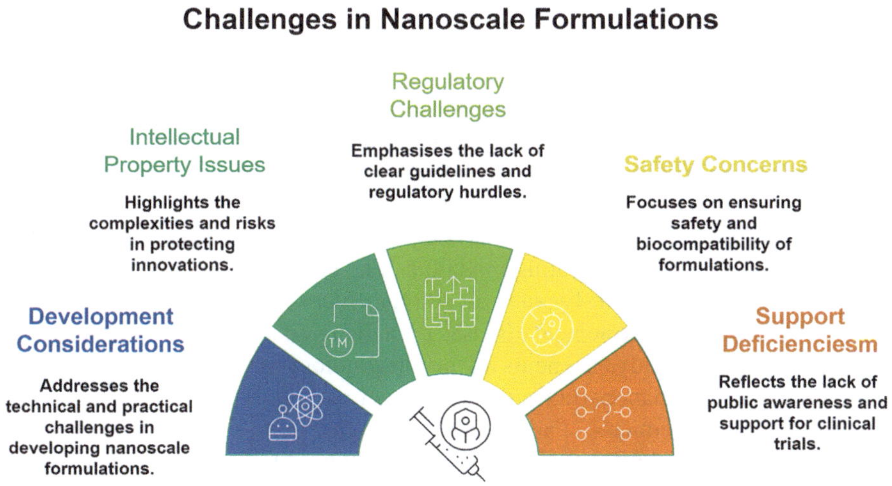


**Fig. 8.2** Challenges in the development and commercialisation of nanoscale formulations

et al. 2022). These nanoparticles can accumulate in vital organs, leading to potential long-term toxicity and complicating safety assessments. The lack of standardized safety protocols and extensive toxicity data poses challenges for regulatory agencies like the FDA in establishing guidelines for nanoparticle-based drugs (Souto et al. 2024). Additionally, clinical trials often face delays due to the need for thorough safety evaluations, which prolongs the timeline for bringing these treatments to market (Thapa and Kim 2023).

Nanoscale formulations face significant challenges related to storage and stability, particularly for nanoparticles from biological or organic materials, which are sensitive to factors like temperature, pH, and humidity (Andra et al. 2022). This sensitivity can lead to degradation or aggregation, compromising efficacy and safety. Improper storage conditions may also alter release profiles and reduce drug bioavailability, raising concerns for both commercial scalability and clinical applications (Malik et al. 2023). Ensuring stability often requires complex storage methods, increasing production and distribution costs.

Additionally, the adoption of nanoscale formulations is hindered by ethical issues and limitations in clinical studies. Concerns about the long-term environmental and health impacts of nanoparticles, especially their behaviour in the human body, highlight the need for more transparency in clinical trials (Ezike et al. 2023). The lack of clear guidelines for ethical trials using nanotechnology poses risks to patient safety. Consequently, delays in FDA approval occur due to the need for rigorous studies. Ensuring informed consent and understanding of risks in nanotechnology treatments further complicates ethical considerations in nanomedicine trials (Malik et al. 2023).

## 8.7 Future Perspectives

Advances in nanotechnology are revolutionising drug delivery and personalised medicine through nanoscale formulations that enhance the bioavailability and effectiveness of biotherapeutics while minimising side effects and improving stability (Ezike et al. 2023). These formulations enable precise drug delivery with targeted release triggered by changes in pH, temperature, or light, which is especially useful in acidic environments like bacterial infections and tumours (Pei et al. 2024). This targeted approach reduces potential harm to healthy tissue and enhances cancer and bacterial treatments by integrating modalities like chemotherapy and immunotherapy into a unified platform (Egwu et al. 2024; Oehler et al. 2024).

The treatment of neurological diseases is challenging due to the blood-brain barrier (BBB). Recent advancements in nanotechnology are addressing this issue through engineered nanoparticles that can target specific brain regions, enhancing therapies for Alzheimer's, Parkinson's, and other neurodegenerative diseases (Vashist et al. 2023; Mittal et al. 2020; Ekhator et al. 2023). Additionally, nanorobotics offers potential for disease detection and targeted therapy through nanoscale robots capable of navigating the body (Elnaggar et al. 2024). In regenerative medicine, incorporating nanoparticles into scaffolds aids tissue growth and repair, with applications in wound healing and organ repair (Carton and Malatesta 2024). Nanoparticles also support stem cell differentiation, providing new treatment avenues for injuries and degenerative diseases. Furthermore, green nanotechnology focuses on sustainable practices by using eco-friendly materials for synthesis and aiding environmental cleanup.

As nanomedicine evolves, there will be increased need to update ethical and regulatory frameworks to ensure long-term safety, environmental impact, and ethical use in human health. Comprehensive safety testing will be vital for public trust and regulatory approval. Governments and organisations like the FDA must adapt to balance the benefits and risks of nanomedicine. Additionally, advancements in manufacturing will aim to improve the scalability and accessibility of nanoscale formulations, lowering costs and simplifying processes to make nanomedicine widely available for therapeutic use and across various industries.

## 8.8 Conclusion

Nanoscale formulations (NFs) in biotherapeutics are rapidly evolving, offering transformative potential for drug delivery and personalized medicine. Ongoing research is refining these formulations to overcome challenges like degradation, low bioavailability, and poor targeting. Advances in nanoparticle engineering have led to multifunctional nanocarriers capable of delivering multiple therapeutic agents, enhancing treatment outcomes for complex diseases while minimising adverse effects. The development of biomimetic nanoparticles improves cellular uptake and reduces immune detection, making treatments safer. These innovations hold promise for tackling complex health issues such as cancer, neurodegenerative disorders, and autoimmune diseases, where conventional therapies may fall short.

As this technology matures, it aims to provide personalized treatment plans tailored to individual patient needs. By combining advanced engineering with biocompatible materials, researchers are working towards safer, more efficient, and accessible therapeutic options. Ultimately, nanoscale formulations have the potential to revolutionize treatment precision and safety across various medical conditions, significantly enhancing patient outcomes worldwide.

**Acknowledgments** This work is based on the research supported by the South African Research Chairs Initiative of the Department of Science and Technology and National Research Foundation of South Africa (Grant No 98337), as well as grants received from the University of Johannesburg (URC), the National Research Foundation (NRF), and the CSIR (Council for Scientific and Industrial Research)—NLC (National Laser Centre) Laser Rental Pool Programme.

**Consent for Publication** All authors agree for publication.

**Competing Interests** The authors affirm that they have no conflicts of interest to disclose.

## References


Abdellatif AAH, Alsowinea AF (2021) Approved and marketed nanoparticles for disease targeting and applications in COVID-19. Nanotechnol Rev 10(1):1941–1977. https://doi.org/10.1515/ntrev-2021-0115

Abdelwahed W, Degobert G, Stainmesse S, Fessi H (2006) Freeze-drying of nanoparticles: formulation, process and storage considerations. Adv Drug Deliv Rev 58(15):1688–1713. https://doi.org/10.1016/j.addr.2006.09.017

Adschiri T, Yoko A (2018) Supercritical fluids for nanotechnology. J Supercrit Fluids 134:167–175. https://doi.org/10.1016/j.supflu.2017.12.033

Ahmed L, Atif R, Eldeen TS, Yahya I, Omara A, Eltayeb M (2019) Study the using of nanoparticles as drug delivery system based on mathematical models for controlled release. Int J Latest Technol Eng Manag Appl Sci 8(5):52–56

Alexis F, Pridgen E, Molnar LK, Farokhzad OC (2008) Factors affecting the clearance and biodistribution of polymeric nanoparticles. Mol Pharm 5(4):505–515. https://doi.org/10.1021/mp800051m

Ali A, Zafar H, Zia M, Ul Haq I, Phull AR, Ali JS, Hussain A (2016) Synthesis, characterization, applications, and challenges of iron oxide nanoparticles. Nanotechnol Sci Appl 9:49–67. https://doi.org/10.2147/NSA.S99986

Alsaiari NS, Alzahrani FM, Amari A, Osman H, Harharah HN, Elboughdiri N, Tahoon MA (2023) Plant and microbial approaches as green methods for the synthesis of nanomaterials: synthesis, applications, and future perspectives. Molecules (Basel, Switzerland). 28(1):463. https://doi.org/10.3390/molecules28010463

Amiri S, Pashizeh F, Moeinabadi-Bidgoli K, Eyvazi Y, Akbari T, Salehi Moghaddam Z, Eskandarisani M, Farahmand F et al (2023) Co-encapsulation of hydrophilic and hydrophobic drugs into niosomal nanocarrier for enhanced breast cancer therapy: in silico and in vitro studies. Environ Res 239(117292):117292. https://doi.org/10.1016/j.envres.2023.117292

Anandalakshmi K, Venugobal J, Ramasamy V (2016) Characterization of silver nanoparticles by green synthesis method using Pedalium murex leaf extract and their antibacterial activity. Appl Nanosci 6(3):399–408. https://doi.org/10.1007/s13204-015-0449-z

Andra VVSNL, Pammi SVN, Bhatraju LVKP, Ruddaraju LK (2022) A comprehensive review on novel liposomal methodologies, commercial formulations, clinical trials and patents. BioNanoScience 12(1):274–291. https://doi.org/10.1007/s12668-022-00941-x

Ansari M, Darvishi A (2024) A review of the current state of natural biomaterials in wound healing applications. Front Bioeng Biotechnol 12:1309541. https://doi.org/10.3389/fbioe.2024.1309541

Attama AA, Momoh MA, Builders PF (2012) Lipid nanoparticulate drug delivery systems: a revolution in dosage form design and development. In: Recent advances in novel drug carrier systems. InTech

Bao P, Kodra A, Tomic-Canic M, Golinko MS, Ehrlich HP, Brem H (2009) The role of vascular endothelial growth factor in wound healing. J Surg Res 153(2):347–358. https://doi.org/10.1016/j.jss.2008.04.023

Barua S, Mitragotri S (2014) Challenges associated with penetration of nanoparticles across cell and tissue barriers: a review of current status and future prospects. Nano Today 9(2):223–243. https://doi.org/10.1016/j.nantod.2014.04.008

Beyene HD, Werkneh AA, Bezabh HK, Ambaye TG (2017) Synthesis paradigm and applications of silver nanoparticles (AgNPs), a review. Sustain Mater Technol 13:18–23. https://doi.org/10.1016/j.susmat.2017.08.001

Bhattacharya K, Mukherjee SP, Gallud A, Burkert SC, Bistarelli S, Bellucci S, Bottini M, Star A et al (2016) Biological interactions of carbon-based nanomaterials: from coronation to degradation. Nanomedicine 12(2):333–351. https://doi.org/10.1016/j.nano.2015.11.011

Boisguérin P, Konate K, Josse E, Vivès E, Deshayes S (2021) Peptide-based nanoparticles for therapeutic nucleic acid delivery. Biomedicine 9(5):583. https://doi.org/10.3390/biomedicines9050583

Borm PJA, Robbins D, Haubold S, Kuhlbusch T, Fissan H, Donaldson K, Schins R, Stone V et al (2006) The potential risks of nanomaterials: a review carried out for ECETOC. Part Fibre Toxicol 3(1):11. https://doi.org/10.1186/1743-8977-3-11

Bou S, Wang X, Anton N, Bouchaala R, Klymchenko AS, Collot M (2020) Lipid-core/polymer-shell hybrid nanoparticles: synthesis and characterization by fluorescence labeling and electrophoresis. Soft Matter 16(17):4173–4181. https://doi.org/10.1039/d0sm00077a

Bovone G, Steiner F, Guzzi EA, Tibbitt MW (2019) Automated and continuous production of polymeric nanoparticles. Front Bioeng Biotechnol 7:423. https://doi.org/10.3389/fbioe.2019.00423

Brigger I, Dubernet C, Couvreur P (2002) Nanoparticles in cancer therapy and diagnosis. Adv Drug Deliv Rev 54(5):631–651. https://doi.org/10.1016/s0169-409x(02)00044-3

Byakodi M, Shrikrishna NS, Sharma R, Bhansali S, Mishra Y, Kaushik A, Gandhi S (2022) Emerging 0D, 1D, 2D, and 3D nanostructures for efficient point-of-care biosensing. Biosens Bioelectron X 12(100284):100284. https://doi.org/10.1016/j.biosx.2022.100284

Calvo P, Remuñán-López C, Vila-Jato JL, Alonso MJ (1997) Novel hydrophilic chitosan-polyethylene oxide nanoparticles as protein carriers. J Appl Polym Sci 63(1):125–132. https://doi.org/10.1002/(sici)1097-4628(19970103)63:1<125::aid-app13>3.0.co;2-4

Carton F, Malatesta M (2024) Nanotechnological research for regenerative medicine: the role of hyaluronic acid. Int J Mol Sci 25(7):3975. https://doi.org/10.3390/ijms25073975

Chauhan S, Jain N, Nagaich U (2020) Nanodiamonds with powerful ability for drug delivery and biomedical applications: recent updates on in vivo study and patents. J Pharm Anal 10(1):1–12. https://doi.org/10.1016/j.jpha.2019.09.003

Chen C, Xing G, Wang J, Zhao Y, Li B, Tang J, Jia G, Wang T et al (2005) Multihydroxylated [Gd@C82(OH)22]n nanoparticles: antineoplastic activity of high efficiency and low toxicity. Nano Lett 5(10):2050–2057. https://doi.org/10.1021/nl051624b

Cheng X, Xie Q, Sun Y (2023) Advances in nanomaterial-based targeted drug delivery systems. Front Bioeng Biotechnol 11:1177151. https://doi.org/10.3389/fbioe.2023.1177151

Choi HS, Liu W, Misra P, Tanaka E, Zimmer JP, Itty Ipe B, Bawendi MG, Frangioni JV (2007) Renal clearance of quantum dots. Nat Biotechnol 25(10):1165–1170. https://doi.org/10.1038/nbt1340

Costa P, Sousa Lobo JM (2001) Modeling and comparison of dissolution profiles. Eur J Pharm Sci 13(2):123–133. https://doi.org/10.1016/s0928-0987(01)00095-1

Curbelo S, Meneses R, Pereira-Prado V, Tapia G (2020) Bone regeneration as an example of tissue engineering in dentistry, with an emphasis on the development of scaffolds. Odontoestomatologia 22(36):74–86. https://doi.org/10.22592/ode2020n36a9

Das RK, Pachapur VL, Lonappan L, Naghdi M, Pulicharla R, Maiti S, Cledon M, Dalila LMA et al (2017) Biological synthesis of metallic nanoparticles: plants, animals and microbial aspects. Nanotechnol Environ Eng 2(1):18. https://doi.org/10.1007/s41204-017-0029-4

Dash S, Murthy PN, Nath L, Chowdhury P (2010) Kinetic modeling on drug release from controlled drug delivery systems. Acta Pol Pharm 67(3):217–223

Dhiman N, Awasthi R, Sharma B, Kharkwal H, Kulkarni GT (2021) Lipid nanoparticles as carriers for bioactive delivery. Front Chem 9:580118. https://doi.org/10.3389/fchem.2021.580118

Dristant U, Mukherjee K, Saha S, Maity D (2023) An overview of polymeric nanoparticles-based drug delivery system in cancer treatment. Technol Cancer Res Treat 22:15330338231152083. https://doi.org/10.1177/15330338231152083

Dutt Y, Dhiman R, Singh T, Vibhuti A, Gupta A, Pandey RP, Raj VS, Chang C-M et al (2022a) The association between biofilm formation and antimicrobial resistance with possible ingenious bio-remedial approaches. Antibiotics (Basel, Switzerland) 11(7):930. https://doi.org/10.3390/antibiotics11070930

Dutt Y, Pandey RP, Dutt M, Gupta A, Vibhuti A, Samuel Raj V, Chang C-M, Priyadarshini A (2022b) Synthesis and biological characterization of Phyto-fabricated silver nanoparticles from *Azadirachta indica*. J Biomed Nanotechnol 18(8):2022–2057. https://doi.org/10.1166/jbn.2022.3402

Dutt Y, Pandey RP, Dutt M, Gupta A, Vibhuti A, Raj VS, Chang C-M, Priyadarshini A (2023) Silver nanoparticles phytofabricated through Azadirachta indica: anticancer, apoptotic, and wound-healing properties. Antibiotics (Basel, Switzerland). 12(1):121. https://doi.org/10.3390/antibiotics12010121

Egwu CO, Aloke C, Onwe KT, Umoke CI, Nwafor J, Eyo RA, Chukwu JA, Ufebe GO et al (2024) Nanomaterials in drug delivery: strengths and opportunities in medicine. Molecules (Basel, Switzerland). 29(11):2584. https://doi.org/10.3390/molecules29112584

Ekhator C, Qureshi MQ, Zuberi AW, Hussain M, Sangroula N, Yerra S, Devi M, Naseem MA et al (2023) Advances and opportunities in nanoparticle drug delivery for central nervous system disorders: a review of current advances. Cureus 15(8):e44302. https://doi.org/10.7759/cureus.44302

Elkhateeb O, Badawy MEI, Tohamy HG, Abou-Ahmed H, El-Kammar M, Elkhenany H (2023) Curcumin-infused nanostructured lipid carriers: a promising strategy for enhancing skin regeneration and combating microbial infection. BMC Vet Res 19(1):206. https://doi.org/10.1186/s12917-023-03774-2

El-Khawaga AM, Zidan A, El-Mageed AIAA (2023) Preparation methods of different nanomaterials for various potential applications: a review. J Mol Struct 1281(135148):135148. https://doi.org/10.1016/j.molstruc.2023.135148

Elmowafy M, Al-Sanea MM (2021) Nanostructured lipid carriers (NLCs) as drug delivery platform: advances in formulation and delivery strategies. Saudi Pharm J 29(9):999–1012. https://doi.org/10.1016/j.jsps.2021.07.015

Elmowafy M, Shalaby K, Elkomy MH, Alsaidan OA, Gomaa HAM, Abdelgawad MA, Mostafa EM (2023) Polymeric nanoparticles for delivery of natural bioactive agents: recent advances and challenges. Polymers 15(5):1123. https://doi.org/10.3390/polym15051123

Elnaggar A, Kang S, Tian M, Han B, Keshavarz M (2024) State of the art in actuation of micro/nanorobots for biomedical applications. Small Sci 4(3):2300211. https://doi.org/10.1002/smsc.202300211

Fernández-Quiroz D, Tohidi MM, Paymard B, Lucero-Acuña A (2023) Immobilization of essential oils in biopolymeric matrices: recent approaches for controlled delivery systems. In: Studies in natural products chemistry. Elsevier, pp 365–401

Fitriani EW, Avanti C, Rosana Y, Surini S (2024) Nanostructured lipid carriers: a prospective dermal drug delivery system for natural active ingredients. Farmatsiia 71:1–15. https://doi.org/10.3897/pharmacia.71.e115849

Fonte P, Reis S, Sarmento B (2016) Facts and evidences on the lyophilization of polymeric nanoparticles for drug delivery. J Contr Release 225:75–86. https://doi.org/10.1016/j.jconrel.2016.01.034

Fornasier M, Biffi S, Bortot B, Macor P, Manhart A, Wurm FR, Murgia S (2020) Cubosomes stabilized by a polyphosphoester-analog of Pluronic F127 with reduced cytotoxicity. J Colloid Interface Sci 580:286–297. https://doi.org/10.1016/j.jcis.2020.07.038

Franco P, De Marco I (2021) Nanoparticles and nanocrystals by supercritical CO2-assisted techniques for pharmaceutical applications: a review. Appl Sci (Basel, Switzerland) 11(4):1476. https://doi.org/10.3390/app11041476

Freire PLL, Albuquerque AJR, Farias IAP, da Silva TG, Aguiar JS, Galembeck A, Flores MAP, Sampaio FC et al (2016) Antimicrobial and cytotoxicity evaluation of colloidal chitosan—silver nanoparticles–fluoride nanocomposites. Int J Biol Macromol 93:896–903. https://doi.org/10.1016/j.ijbiomac.2016.09.052

Gajbhiye KR, Salve R, Narwade M, Sheikh A, Kesharwani P, Gajbhiye V (2023) Lipid polymer hybrid nanoparticles: a custom-tailored next-generation approach for cancer therapeutics. Mol Cancer 22(1):160. https://doi.org/10.1186/s12943-023-01849-0

Gao L, Zhang D, Chen M (2008) Drug nanocrystals for the formulation of poorly soluble drugs and its application as a potential drug delivery system. J Nanopart Res 10(5):845–862. https://doi.org/10.1007/s11051-008-9357-4

Gatto MS, Najahi-Missaoui W (2023) Lyophilization of nanoparticles, does it really work? Overview of the current status and challenges. Int J Mol Sci 24(18):14041. https://doi.org/10.3390/ijms241814041

Ghulam AN, Dos Santos OAL, Hazeem L, Pizzorno Backx B, Bououdina M, Bellucci S (2022) Graphene oxide (GO) materials-applications and toxicity on living organisms and environment. J Funct Biomater 13(2):77. https://doi.org/10.3390/jfb13020077

Giri TK (2016) Nanoarchitectured polysaccharide-based drug carrier for ocular therapeutics. In: *Nanoarchitectonics for smart delivery and drug targeting*. Elsevier, pp 119–141

Grebowski J, Kazmierska-Grebowska P, Cichon N, Konarska A, Wolszczak M, Litwinienko G (2022) Fullerenol C60(OH)36 protects the antioxidant enzymes in human erythrocytes against oxidative damage induced by high-energy electrons. Int J Mol Sci 23(18):10939. https://doi.org/10.3390/ijms231810939

Guo J, Cao G, Wei S, Han Y, Xu P (2023) Progress in the application of graphene and its derivatives to osteogenesis. Heliyon 9(11):e21872. https://doi.org/10.1016/j.heliyon.2023.e21872

Hachem K, Ansari MJ, Saleh RO, Kzar HH, Al-Gazally ME, Altimari US, Hussein SA, Mohammed HT et al (2022) Methods of chemical synthesis in the synthesis of nanomaterial and nanoparticles by the chemical deposition method: a review. BioNanoScience 12(3):1032–1057. https://doi.org/10.1007/s12668-022-00996-w

Hadinoto K, Sundaresan A, Cheow WS (2013) Lipid-polymer hybrid nanoparticles as a new generation therapeutic delivery platform: a review. Eur J Pharm Biopharm 85(3 Pt A):427–443. https://doi.org/10.1016/j.ejpb.2013.07.002

Haider A, Ikram M, Rafiq A (2023) Introduction. In: Green nanomaterials as potential antimicrobials. Springer, Cham, pp 1–23

Haitao Y, Yifan C, Mingchao S, Shuaijuan H (2022) A novel polymeric nanohybrid antimicrobial engineered by antimicrobial peptide MccJ25 and chitosan nanoparticles exerts strong antibacterial and anti-inflammatory activities. Front Immunol 12:811381. https://doi.org/10.3389/fimmu.2021.811381

Hajipour MJ, Fromm KM, Ashkarran AA, Jimenez de Aberasturi D, de Larramendi IR, Rojo T, Serpooshan V, Parak WJ et al (2012) Antibacterial properties of nanoparticles. Trends Biotechnol 30(10):499–511. https://doi.org/10.1016/j.tibtech.2012.06.004

Hami Z (2021) A brief review on advantages of nano-based drug delivery systems. Ann Military Health Sci Res 19(1):e112274. https://doi.org/10.5812/amh.112274

Harish V, Ansari MM, Tewari D, Gaur M, Yadav AB, García-Betancourt M-L, Abdel-Haleem FM, Bechelany M et al (2022) Nanoparticle and nanostructure synthesis and controlled growth methods. Nanomaterials (Basel, Switzerland). 12(18):3226. https://doi.org/10.3390/nano12183226

He H, Pham-Huy LA, Dramou P, Xiao D, Zuo P, Pham-Huy C (2013) Carbon nanotubes: applications in pharmacy and medicine. Biomed Res Int 2013:1–12. https://doi.org/10.1155/2013/578290

Hoshino A, Fujioka K, Oku T, Suga M, Sasaki YF, Ohta T, Yasuhara M, Suzuki K et al (2004) Physicochemical properties and cellular toxicity of nanocrystal quantum dots depend on their surface modification. Nano Lett 4(11):2163–2169. https://doi.org/10.1021/nl048715d

Hoshyar N, Gray S, Han H, Bao G (2016) The effect of nanoparticle size on in vivo pharmacokinetics and cellular interaction. Nanomedicine (Lond) 11(6):673–692. https://doi.org/10.2217/nnm.16.5

Hsu C-Y, Rheima AM, Kadhim MM, Ahmed NN, Mohammed SH, Abbas FH, Abed ZT, Mahdi ZM et al (2023) An overview of nanoparticles in drug delivery: properties and applications. S Afr J Chem Eng 46:233–270. https://doi.org/10.1016/j.sajce.2023.08.009

Huang X, He T, Liang X, Xiang Z, Liu C, Zhou S, Luo R, Bai L et al (2024) Advances and applications of nanoparticles in cancer therapy. MedComm – Oncol 3(1):e67. https://doi.org/10.1002/mog2.67

Huynh-Ba K, Zahn M (2009) Understanding ICH guidelines applicable to stability testing. In: *Handbook of stability testing in pharmaceutical development*. Springer, New York, NY, pp 21–41

Ibrahim A, Moodley D, Uche C, Maboza E, Olivier A, Petrik L (2021) Antimicrobial and cytotoxic activity of electrosprayed chitosan nanoparticles against endodontic pathogens and Balb/c 3T3 fibroblast cells. Sci Rep 11(1):24487. https://doi.org/10.1038/s41598-021-04322-4

Iqbal MA, Md S, Sahni JK, Baboota S, Dang S, Ali J (2012) Nanostructured lipid carriers system: recent advances in drug delivery. J Drug Target 20(10):813–830. https://doi.org/10.3109/1061186X.2012.716845

Jacob EM, Borah A, Sakthi Kumar D (2022) CRISPR/Cas9 nano-delivery approaches for targeted gene therapy. In: Nanotechnology in the life sciences. Springer, Cham, pp 27–64

Jain PK, Lee KS, El-Sayed IH, El-Sayed MA (2006) Calculated absorption and scattering properties of gold nanoparticles of different size, shape, and composition: applications in biological imaging and biomedicine. J Phys Chem B 110(14):7238–7248. https://doi.org/10.1021/jp057170o

Jain PK, El-Sayed IH, El-Sayed MA (2007) Au nanoparticles target cancer. *Nano Today* 2(1):18–29. https://doi.org/10.1016/s1748-0132(07)70016-6

Jain K, Salamat-Miller N, Taylor K (2021) Freeze–thaw characterization process to minimize aggregation and enable drug product manufacturing of protein based therapeutics. Sci Rep 11(1):11332. https://doi.org/10.1038/s41598-021-90772-9

Jampílek J, Kráľová K (2015) Applications of Nanoformulations in agricultural production and their impact on food and human health. Proc ECOpole 9(2):465–472

Jeevanandam J, Chan YS, Danquah MK (2016) Nano-formulations of drugs: recent developments, impact and challenges. Biochimie 128–129:99–112. https://doi.org/10.1016/j.biochi.2016.07.008

Jiang T, Jin K, Liu X, Pang Z (2017) Nanoparticles for tumor targeting. In: *Biopolymer-Based Composites*. Elsevier, pp 221–267

Joudeh N, Linke D (2022) Nanoparticle classification, physicochemical properties, characterization, and applications: a comprehensive review for biologists. J Nanobiotechnol 20(1):262. https://doi.org/10.1186/s12951-022-01477-8

Kamaly N, Yameen B, Wu J, Farokhzad OC (2016) Degradable controlled-release polymers and polymeric nanoparticles: mechanisms of controlling drug release. Chem Rev 116(4):2602–2663. https://doi.org/10.1021/acs.chemrev.5b00346

Karle MAA, Yadav MG, Jain DA, Rane DB (2022) Nanosuspension formulation by high pressure homogenization (HPH). Int J Sci Res Sci Technol 9(4):115–122. https://doi.org/10.32628/ijsrst229414

Katyal P, Meleties M, Montclare JK (2019) Self-assembled protein- and peptide-based nanomaterials. ACS Biomater Sci Eng 5(9):4132–4147. https://doi.org/10.1021/acsbiomaterials.9b00408

Kazemi S, Hosseingholian A, Gohari SD, Feirahi F, Moammeri F, Mesbahian G, Moghaddam ZS, Ren Q (2023) Recent advances in green synthesized nanoparticles: from production to application. Mater Today Sustain 24(100500):100500. https://doi.org/10.1016/j.mtsust.2023.100500

Khalil M, Jan BM, Tong CW, Berawi MA (2017) Advanced nanomaterials in oil and gas industry: design, application and challenges. Appl Energy 191:287–310. https://doi.org/10.1016/j.apenergy.2017.01.074

Khan I, Saeed K, Khan I (2019) Nanoparticles: properties, applications and toxicities. Arab J Chem 12(7):908–931. https://doi.org/10.1016/j.arabjc.2017.05.011

Khlebtsov NG, Dykman LA (2010) Optical properties and biomedical applications of plasmonic nanoparticles. J Quant Spectrosc Radiat Transf 111(1):1–35. https://doi.org/10.1016/j.jqsrt.2009.07.012

Kim F, Connor S, Song H, Kuykendall T, Yang P (2004) Platonic gold nanocrystals. Angew Chem 43(28):3673–3677. https://doi.org/10.1002/anie.200454216

Kim K-J, Sung WS, Suh BK, Moon S-K, Choi J-S, Kim JG, Lee DG (2009) Antifungal activity and mode of action of silver nano-particles on Candida albicans. Biometals 22(2):235–242. https://doi.org/10.1007/s10534-008-9159-2
Koczkur KM, Mourdikoudis S, Polavarapu L, Skrabalak SE (2015) Polyvinylpyrrolidone (PVP) in nanoparticle synthesis. Dalton Trans 44(41):17883–17905. https://doi.org/10.1039/c5dt02964c
Kumar N, Gautam V, Kumar V, Maurya PK (2020) Nanoparticle-based macromolecule drug delivery to lungs. In: Targeting chronic inflammatory lung diseases using advanced drug delivery systems. Elsevier, pp 227–259
Kumar JA, Krithiga T, Manigandan S, Sathish S, Renita AA, Prakash P, Prasad BSN, Kumar TRP et al (2021) A focus to green synthesis of metal/metal based oxide nanoparticles: various mechanisms and applications towards ecological approach. J Clean Prod 324(129198):129198. https://doi.org/10.1016/j.jclepro.2021.129198
Ladj R, Bitar A, Eissa MM, Fessi H, Mugnier Y, Le Dantec R, Elaissari A (2013) Polymer encapsulation of inorganic nanoparticles for biomedical applications. Int J Pharm 458(1):230–241. https://doi.org/10.1016/j.ijpharm.2013.09.001
Lara HH, Ayala-Nuñez NV, Ixtepan-Turrent L, Rodriguez-Padilla C (2010) Mode of antiviral action of silver nanoparticles against HIV-1. J Nanobiotechnol 8(1):1. https://doi.org/10.1186/1477-3155-8-1
Laurent S, Forge D, Port M, Roch A, Robic C, Vander Elst L, Muller RN (2008) Magnetic iron oxide nanoparticles: synthesis, stabilization, vectorization, physicochemical characterizations, and biological applications. Chem Rev 108(6):2064–2110. https://doi.org/10.1021/cr068445e
Lavrador P, Esteves MR, Gaspar VM, Mano JF (2021) Stimuli-responsive nanocomposite hydrogels for biomedical applications. Adv Funct Mater 31(8):2005941. https://doi.org/10.1002/adfm.202005941
Levit SL, Tang C (2021) Polymeric nanoparticle delivery of combination therapy with synergistic effects in ovarian cancer. Nanomaterials (Basel, Switzerland). 11(4):1048. https://doi.org/10.3390/nano11041048
Li T, Lu X-M, Zhang M-R, Hu K, Li Z (2022) Peptide-based nanomaterials: self-assembly, properties and applications. Bioact Mater 11:268–282. https://doi.org/10.1016/j.bioactmat.2021.09.029
Li X, Peng X, Zoulikha M, Boafo GF, Magar KT, Ju Y, He W (2024) Multifunctional nanoparticle-mediated combining therapy for human diseases. Signal Transduct Target Ther 9(1):1. https://doi.org/10.1038/s41392-023-01668-1
Liu X, Ge W (2022) The emerging role of ultrasonic nanotechnology for diagnosing and treatment of diseases. Front Med 9:814986. https://doi.org/10.3389/fmed.2022.814986
Liu Y, Liang Y, Yuhong J, Xin P, Han JL, Du Y, Yu X, Zhu R et al (2024b) Advances in nanotechnology for enhancing the solubility and bioavailability of poorly soluble drugs. Drug Des Devel Ther 18:1469–1495. https://doi.org/10.2147/dddt.s447496
Lockman PR, Koziara JM, Mumper RJ, Allen DD (2004) Nanoparticle surface charges alter blood-brain barrier integrity and permeability. J Drug Target 12(9–10):635–641. https://doi.org/10.1080/10611860400015936
Loh ZH, Samanta AK, Sia Heng PW (2015) Overview of milling techniques for improving the solubility of poorly water-soluble drugs. Asian J Pharm Sci 10(4):255–274. https://doi.org/10.1016/j.ajps.2014.12.006
Loureiro A, Azoia NG, Gomes AC, Cavaco-Paulo A (2016) Albumin-based nanodevices as drug carriers. Curr Pharm Des 22(10):1371–1390. https://doi.org/10.2174/13816128226661601251149 00
Maiti D, Tong X, Mou X, Yang K (2018) Carbon-based nanomaterials for biomedical applications: a recent study. Front Pharmacol 9:1401. https://doi.org/10.3389/fphar.2018.01401
Malamatari M, Charisi A, Malamataris S, Kachrimanis K, Nikolakakis I (2020) Spray drying for the preparation of nanoparticle-based drug formulations as dry powders for inhalation. Processes (Basel, Switzerland) 8(7):788. https://doi.org/10.3390/pr8070788

Maleki M, Zarezadeh R, Nouri M, Sadigh AR, Pouremamali F, Asemi Z, Kafil HS, Alemi F et al (2020) Graphene oxide: a promising material for regenerative medicine and tissue engineering. Biomol Concepts 11(1):182–200. https://doi.org/10.1515/bmc-2020-0017

Malik S, Muhammad K, Waheed Y (2023) Emerging applications of nanotechnology in healthcare and medicine. Molecules (Basel, Switzerland) 28(18):6624. https://doi.org/10.3390/molecules28186624

Mandal B, Bhattacharjee H, Mittal N, Sah H, Balabathula P, Thoma LA, Wood GC (2013) Core-shell-type lipid-polymer hybrid nanoparticles as a drug delivery platform. Nanomedicine 9(4):474–491. https://doi.org/10.1016/j.nano.2012.11.010

Manimaran V, Nivetha RP, Tamilanban T, Narayanan J, Vetriselvan S, Fuloria NK, Chinni SV, Sekar M et al (2023) Nanogels as novel drug nanocarriers for CNS drug delivery. Front Mol Biosci 10:1232109. https://doi.org/10.3389/fmolb.2023.1232109

Manuja A, Kumar B, Kumar R, Chhabra D, Ghosh M, Manuja M, Brar B, Pal Y et al (2021) Metal/metal oxide nanoparticles: toxicity concerns associated with their physical state and remediation for biomedical applications. Toxicol Rep 8:1970–1978. https://doi.org/10.1016/j.toxrep.2021.11.020

Maqbool I, Noreen S (2019) A review of novel techniques for nanoparticles preparation. *Global drug design & development.* Review IV(I):41–50. https://doi.org/10.31703/gdddr.2019(iv-i).05

Marante T, Viegas C, Duarte I, Macedo AS, Fonte P (2020) An overview on spray-drying of protein-loaded polymeric nanoparticles for dry powder inhalation. Pharmaceutics. 12(11):1032. https://doi.org/10.3390/pharmaceutics12111032

Martis E, Badve R, Degwekar M (2012) Nanotechnology based devices and applications in medicine: an overview. Chron Young Sci 3(1):68. https://doi.org/10.4103/2229-5186.94320

Maurer LL, Yang X, Schindler AJ, Taggart RK, Jiang C, Hsu-Kim H, Sherwood DR, Meyer JN (2016) Intracellular trafficking pathways in silver nanoparticle uptake and toxicity in Caenorhabditis elegans. Nanotoxicology 10(7):831–835. https://doi.org/10.3109/17435390.2015.1110759

Mekuye B, Abera B (2023) Nanomaterials: an overview of synthesis, classification, characterization, and applications. Nano Select 4(8):486–501. https://doi.org/10.1002/nano.202300038

Messai R, Ferhat MF, Belmekki B, Alam MW, Al-Othoum MAS, Sadaf S (2024) GAD plasma-assisted synthesis of ZnO nanoparticles and their photocatalytic activity. Mater Res Express 11(1):015006. https://doi.org/10.1088/2053-1591/ad1a82

Mihai MM, Dima MB, Dima B, Holban AM (2019) Nanomaterials for wound healing and infection control. Materials 12(13):2176. https://doi.org/10.3390/ma12132176

Mittal P, Saharan A, Verma R, Altalbawy FMA, Alfaidi MA, Batiha GE-S, Akter W, Gautam RK et al (2021) Dendrimers: a new race of pharmaceutical nanocarriers. Biomed Res Int 2021:8844030. https://doi.org/10.1155/2021/8844030

Mohite P, Singh S, Pawar A, Sangale A, Prajapati BG (2023) Lipid-based oral formulation in capsules to improve the delivery of poorly water-soluble drugs. Front Drug Deliv 3:1232012. https://doi.org/10.3389/fddev.2023.1232012

Moulik SP, Rakshit AK, Pan A, Naskar B (2022) An overview of coacervates: the special disperse state of amphiphilic and polymeric materials in solution. Colloid Interfaces 6(3):45. https://doi.org/10.3390/colloids6030045

Mourdikoudis S, Pallares RM, Thanh NTK (2018) Characterization techniques for nanoparticles: comparison and complementarity upon studying nanoparticle properties. Nanoscale 10(27):12871–12934. https://doi.org/10.1039/c8nr02278j

Mukherjee A, Waters AK, Kalyan P, Achrol AS, Kesari S, Yenugonda VM (2019) Lipid-polymer hybrid nanoparticles as a next-generation drug delivery platform: state of the art, emerging technologies, and perspectives. Int J Nanomed 14:1937–1952. https://doi.org/10.2147/IJN.S198353

Nadworny PL, Wang J, Tredget EE, Burrell RE (2010) Anti-inflammatory activity of nanocrystalline silver-derived solutions in porcine contact dermatitis. J Inflamm (London, England) 7(1):13. https://doi.org/10.1186/1476-9255-7-13

Napiórkowska A, Kurek M (2022) Coacervation as a novel method of microencapsulation of essential oils—a review. Molecules (Basel, Switzerland) 27(16):5142. https://doi.org/10.3390/molecules27165142

Naseri N, Valizadeh H, Zakeri-Milani P (2015) Solid lipid nanoparticles and nanostructured lipid carriers: structure, preparation and application. Adv Pharm Bull 5(3):305–313. https://doi.org/10.15171/apb.2015.043

Nikalje AP (2015) Nanotechnology and its applications in medicine. Med Chem 5(2):081–089. https://doi.org/10.4172/2161-0444.1000247

Nikam AV, Prasad BLV, Kulkarni AA (2018) Wet chemical synthesis of metal oxide nanoparticles: a review. CrystEngComm 20(35):5091–5107. https://doi.org/10.1039/c8ce00487k

Obeid MA, Tate RJ, Mullen AB, Ferro VA (2018) Lipid-based nanoparticles for cancer treatment. In: Lipid Nanocarriers for drug targeting. Elsevier, pp 313–359

Oberdörster G, Maynard A, Donaldson K, Castranova V, Fitzpatrick J, Ausman K, Carter J, Karn B et al (2005) Principles for characterizing the potential human health effects from exposure to nanomaterials: elements of a screening strategy. Part Fibre Toxicol 2(1):8. https://doi.org/10.1186/1743-8977-2-8

Oehler JB, Rajapaksha W, Albrecht H (2024) Emerging applications of nanoparticles in the diagnosis and treatment of breast cancer. J Pers Med 14(7):723. https://doi.org/10.3390/jpm14070723

Oliveira C, Sousa D, Teixeira JA, Ferreira-Santos P, Botelho CM (2023) Polymeric biomaterials for wound healing. Front Bioeng Biotechnol 11:1136077. https://doi.org/10.3389/fbioe.2023.1136077

Pal SL, Jana U, Manna PK, Mohanta GP, Manavalan R (2011) Nanoparticle: an overview of preparation and characterization. J Appl Pharm Sci 1(6):228–234

Pal P, Dadhich P, Srivas PK, Das B, Maulik D, Dhara S (2017) Bilayered nanofibrous 3D hierarchy as skin rudiment by emulsion electrospinning for burn wound management. Biomater Sci 5(9):1786–1799. https://doi.org/10.1039/c7bm00174f

Pan S, Goudoulas TB, Jeevanandam J, Tan KX, Chowdhury S, Danquah MK (2021) Therapeutic applications of metal and metal-oxide nanoparticles: Dermato-cosmetic perspectives. Front Bioeng Biotechnol 9:724499. https://doi.org/10.3389/fbioe.2021.724499

Paramasivam G, Palem VV, Sundaram T, Sundaram V, Kishore SC, Bellucci S (2021) Nanomaterials: synthesis and applications in theranostics. Nanomaterials (Basel, Switzerland). 11(12):3228. https://doi.org/10.3390/nano11123228

Paras, Yadav K, Kumar P, Teja DR, Chakraborty S, Chakraborty M, Mohapatra SS, Sahoo A et al (2022) A review on low-dimensional nanomaterials: nanofabrication, characterization and applications. Nanomaterials (Basel, Switzerland). 13(1):160. https://doi.org/10.3390/nano13010160

Patra JK, Das G, Fraceto LF, Campos EVR, Rodriguez-Torres MDP, Acosta-Torres LS, Diaz-Torres LA, Grillo R et al (2018) Nano based drug delivery systems: recent developments and future prospects. J Nanobiotechnol 16(1):71. https://doi.org/10.1186/s12951-018-0392-8

Paukov M, Kramberger C, Begichev I, Kharlamova M, Burdanova M (2023) Functionalized fullerenes and their applications in electrochemistry, solar cells, and nanoelectronics. Materials. 16(3):1276. https://doi.org/10.3390/ma16031276

Pedroso-Santana S, Fleitas-Salazar N (2020) Ionotropic gelation method in the synthesis of nanoparticles/microparticles for biomedical purposes. Polym Int 69(5):443–447. https://doi.org/10.1002/pi.5970

Pei J, Yan Y, Palanisamy CP, Jayaraman S, Natarajan PM, Umapathy VR, Gopathy S, Roy JR et al (2024) Materials-based drug delivery approaches: recent advances and future perspectives. Green Process Synth 13(1):20230094. https://doi.org/10.1515/gps-2023-0094

Piñón-Segundo E, Ganem-Quintanar A, Rafael Garibay-Bermúdez J, Juan Escobar-Chávez J, López-Cervantes M, Quintanar-Guerrero D (2006) Preparation of nanoparticles by solvent displacement using a novel recirculation system. Pharm Dev Technol 11(4):493–501. https://doi.org/10.1080/10837450600940824

Planka L, Srnec R, Rauser P, Stary D, Filova E, Jancar J, Juhasova J, Kren L et al (2012) Nanotechnology and mesenchymal stem cells with chondrocytes in prevention of partial growth plate arrest in pigs. Biomed Pap Med Fac Univ Palacky, Olomouc, Czech 156(2):128–134. https://doi.org/10.5507/bp.2012.041

Powers KW, Palazuelos M, Moudgil BM, Roberts SM (2007) Characterization of the size, shape, and state of dispersion of nanoparticles for toxicological studies. Nanotoxicology 1(1):42–51. https://doi.org/10.1080/17435390701314902

Pratibha, Singh S, Kumar S, Reddy KR, Naveen S, Sadhu V (2019) Hybrid carbon nanostructures for chemical and biological sensors. In: Advanced biosensors for health care applications. Elsevier, pp 357–374

Procopio A, Lagreca E, Jamaledin R, La Manna S, Corrado B, Di Natale C, Onesto V (2022) Recent fabrication methods to produce polymer-based drug delivery matrices (experimental and in silico approaches). Pharmaceutics. 14(4):872. https://doi.org/10.3390/pharmaceutics14040872

Pulingam T, Foroozandeh P, Chuah J-A, Sudesh K (2022) Exploring various techniques for the chemical and biological synthesis of polymeric nanoparticles. Nanomaterials (Basel, Switzerland). 12(3):576. https://doi.org/10.3390/nano12030576

Puri A, Loomis K, Smith B, Lee J-H, Yavlovich A, Heldman E, Blumenthal R (2009) Lipid-based nanoparticles as pharmaceutical drug carriers: from concepts to clinic. Crit Rev Ther Drug Carrier Syst 26(6):523–580. https://doi.org/10.1615/critrevtherdrugcarriersyst.v26.i6.10

Rabha B, Bharadwaj KK, Boro N, Ghosh A, Gogoi SK, Varma RS, Baishya D (2021) Cheilocostus speciosus extract-assisted naringenin-encapsulated poly-ε-caprolactone nanoparticles: evaluation of anti-proliferative activities. Green Chem 23(19):7701–7711. https://doi.org/10.1039/d1gc02260a

Ramos TI, Villacis-Aguirre CA, López-Aguilar KV, Santiago Padilla L, Altamirano C, Toledo JR, Santiago Vispo N (2022) The hitchhiker's guide to human therapeutic nanoparticle development. Pharmaceutics 14(2):247. https://doi.org/10.3390/pharmaceutics14020247

Renwick LC, Donaldson K, Clouter A (2001) Impairment of alveolar macrophage phagocytosis by ultrafine particles. Toxicol Appl Pharmacol 172(2):119–127. https://doi.org/10.1006/taap.2001.9128

Riley PR, Narayan RJ (2021) Recent advances in carbon nanomaterials for biomedical applications: a review. Curr Opin Biomed Eng 17(100262):100262. https://doi.org/10.1016/j.cobme.2021.100262

Rizvi SAA, Saleh AM (2018) Applications of nanoparticle systems in drug delivery technology. Saudi Pharm J 26(1):64–70. https://doi.org/10.1016/j.jsps.2017.10.012

Rizwan M, Shoukat A, Ayub A, Razzaq B, Tahir MB (2021) Types and classification of nanomaterials. In: Nanomaterials: synthesis, characterization, hazards and safety. Elsevier, pp 31–54

Robson A-L, Dastoor PC, Flynn J, Palmer W, Martin A, Smith DW, Woldu A, Hua S (2018) Advantages and limitations of current imaging techniques for characterizing liposome morphology. Front Pharmacol 9:80. https://doi.org/10.3389/fphar.2018.00080

Rogueda P (2005) Novel hydrofluoroalkane suspension formulations for respiratory drug delivery. Expert Opin Drug Deliv 2(4):625–638. https://doi.org/10.1517/17425247.2.4.625

Rosenblum D, Joshi N, Tao W, Karp JM, Peer D (2018) Progress and challenges towards targeted delivery of cancer therapeutics. Nat Commun 9(1):1410. https://doi.org/10.1038/s41467-018-03705-y

Sacco P, Pedroso-Santana S, Kumar Y, Joly N, Martin P, Bocchetta P (2021) Ionotropic gelation of chitosan flat structures and potential applications. Molecules (Basel, Switzerland). 26(3):660. https://doi.org/10.3390/molecules26030660

Saleh TA (2020) Nanomaterials: classification, properties, and environmental toxicities. Environ Technol Inn 20(101067):101067. https://doi.org/10.1016/j.eti.2020.101067

Saleh HM, Hassan AI (2023) Synthesis and characterization of nanomaterials for application in cost-effective electrochemical devices. Sustainability 15(14):10891. https://doi.org/10.3390/su151410891

Samykano M (2021) Progress in one-dimensional nanostructures. Mater Charact 179(111373):111373. https://doi.org/10.1016/j.matchar.2021.111373
Schuhmann MK, Fluri F (2017) Effects of fullerenols on mouse brain microvascular endothelial cells. Int J Mol Sci 18(8):1783. https://doi.org/10.3390/ijms18081783
Sercombe L, Veerati T, Moheimani F, Wu SY, Sood AK, Hua S (2015) Advances and challenges of liposome assisted drug delivery. Front Pharmacol 6:286. https://doi.org/10.3389/fphar.2015.00286
Shabatina TI, Vernaya OI, Shimanovskiy NL, Melnikov MY (2023) Metal and metal oxides nanoparticles and nanosystems in anticancer and antiviral theragnostic agents. Pharmaceutics. 15(4):1181. https://doi.org/10.3390/pharmaceutics15041181
Shafiee A, Soleimani M, Chamheidari GA, Seyedjafari E, Dodel M, Atashi A, Gheisari Y (2011) Electrospun nanofiber-based regeneration of cartilage enhanced by mesenchymal stem cells. J Biomed Mater Res A 99(3):467–478. https://doi.org/10.1002/jbm.a.33206
Shafiee A, Iravani S, Varma RS (2022) Graphene and graphene oxide with anticancer applications: challenges and future perspectives. MedComm 3(1):e118. https://doi.org/10.1002/mco2.118
Shah R, Eldridge D, Palombo E, Harding I (2015) Lipid nanoparticles: production, characterization and stability. Springer, Cham
Shah K, Patel J, Kumar S, Pandey R, Maity G, Dubey S (2024) Perspectives on sustainable and efficient routes of nanoparticle synthesis: an exhaustive review on conventional and microplasma-assisted techniques. Nanoscale 16(44):20374–20404. https://doi.org/10.1039/d4nr02478h
Shahhosseinlou F, Reza Farahpour M, Sonboli A (2023) Fabrication of novel polysaccharide hybrid nanoliposomes containing citral for targeting MRSA-infected wound healing. J Ind Eng Chem 118:187–195. https://doi.org/10.1016/j.jiec.2022.11.004
Shahravi Z, Sisakht MM, Bahrami S, Amirkhani MA (2020) Concise overview on biomaterial based wound dressing. Biomed J Sci Tech Res 24(1):17972–17977. https://doi.org/10.26717/bjstr.2019.24.003997
Shao M, Xu T-R, Chen C-S (2016) The big bang of genome editing technology: development and application of the CRISPR/Cas9 system in disease animal models. Dong wu xue yan jiu [Zool Res] 37(4):191–204. https://doi.org/10.13918/j.issn.2095-8137.2016.4.191
Shao X, Shi C, Wu S, Wang F, Li W (2023) Review of the pharmacokinetics of nanodrugs. Nanotechnol Rev 12(1). https://doi.org/10.1515/ntrev-2022-0525
Shepherd SJ, Issadore D, Mitchell MJ (2021) Microfluidic formulation of nanoparticles for biomedical applications. Biomaterials 274:120826. https://doi.org/10.1016/j.biomaterials.2021.120826
Sherif AY, Harisa GI, Alanazi FK, Nasr FA, Alqahtani AS (2022) Engineered nanoscale lipid-based formulation as potential enhancer of gefitinib lymphatic delivery: cytotoxicity and apoptotic studies against the A549 cell line. AAPS PharmSciTech 23(6):183. https://doi.org/10.1208/s12249-022-02332-7
Shi Y, Zhang M, Chen K, Wang M (2022) Nano-emulsion prepared by high pressure homogenization method as a good carrier for Sichuan pepper essential oil: preparation, stability, and bioactivity. Lebensmittel-Wissenschaft und Technologie [Food Sci Technol] 154(112779):112779. https://doi.org/10.1016/j.lwt.2021.112779
Shukla A, Maiti P (2019) Biodegradable polymer-based nanohybrids for controlled drug delivery and implant applications. In: Materials horizons: from nature to nanomaterials. Springer, Singapore, pp 3–19
Singh NA, Narang J, Garg D, Jain V, Payasi D, Suleman S, Swami RK (2023) Nanoparticles synthesis via microorganisms and their prospective applications in agriculture. Plant Nano Biology 5(100047):100047. https://doi.org/10.1016/j.plana.2023.100047
Sivadasan D, Sultan MH, Alqahtani SS, Javed S (2023) Cubosomes in drug delivery-a comprehensive review on its structural components, preparation techniques and therapeutic applications. Biomedicine 11(4):1114. https://doi.org/10.3390/biomedicines11041114
Sohn HY, Murali A (2021) Plasma synthesis of advanced metal oxide nanoparticles and their applications as transparent conducting oxide thin films. Molecules (Basel, Switzerland). 26(5):1456. https://doi.org/10.3390/molecules26051456

Souto EB, Blanco-Llamero C, Krambeck K, Kiran NS, Yashaswini C, Postwala H, Severino P, Priefer R et al (2024) Regulatory insights into nanomedicine and gene vaccine innovation: safety assessment, challenges, and regulatory perspectives. Acta Biomater 180:1–17. https://doi.org/10.1016/j.actbio.2024.04.010

Subhan MA (2022) Advances with metal oxide-based nanoparticles as MDR metastatic breast cancer therapeutics and diagnostics. RSC Adv 12(51):32956–32978. https://doi.org/10.1039/d2ra02005j

Syama K, Jakubek ZJ, Chen S, Zaifman J, Tam YYC, Zou S (2022) Development of lipid nanoparticles and liposomes reference materials (II): cytotoxic profiles. Sci Rep 12(1):18071. https://doi.org/10.1038/s41598-022-23013-2

Thanh NTK, Maclean N, Mahiddine S (2014) Mechanisms of nucleation and growth of nanoparticles in solution. Chem Rev 114(15):7610–7630. https://doi.org/10.1021/cr400544s

Thapa RK, Kim JO (2023) Nanomedicine-based commercial formulations: current developments and future prospects. J Pharm Investig 53(1):19–33. https://doi.org/10.1007/s40005-022-00607-6

Thiruvengadam M, Rajakumar G, Swetha V, Ansari MA, Alghamdi S, Almehmadi M, Halawi M, Kungumadevi L et al (2021) Recent insights and multifactorial applications of carbon nanotubes. Micromachines 12(12):1502. https://doi.org/10.3390/mi12121502

Thiruvengadathan R, Korampally V, Ghosh A, Chanda N, Gangopadhyay K, Gangopadhyay S (2013) Nanomaterial processing using self-assembly-bottom-up chemical and biological approaches. Rep Prog Phys Phys Soc (Great Britain) 76(6):066501. https://doi.org/10.1088/0034-4885/76/6/066501

Tseu GYW, Kamaruzaman KA (2023) A review of different types of liposomes and their advancements as a form of gene therapy treatment for breast cancer. Molecules (Basel, Switzerland) 28(3):1498. https://doi.org/10.3390/molecules28031498

Umar H, Wahab HA, Gazzali AM, Tahir H, Ahmad W (2022) Cubosomes: design, development, and tumor-targeted drug delivery applications. Polymers 14(15):3118. https://doi.org/10.3390/polym14153118

Vashist A, Manickam P, Raymond AD, Arias AY, Kolishetti N, Vashist A, Arias E, Nair M (2023) Recent advances in nanotherapeutics for neurological disorders. ACS Appl Bio Mater 6(7):2614–2621. https://doi.org/10.1021/acsabm.3c00254

Vega-Vásquez P, Mosier NS, Irudayaraj J (2020) Nanoscale drug delivery systems: from medicine to agriculture. Front Bioeng Biotechnol 8:79. https://doi.org/10.3389/fbioe.2020.00079

Verma D, Gulati N, Kaul S, Mukherjee S, Nagaich U (2018) Protein based nanostructures for drug delivery. J Pharm 2018:9285854. https://doi.org/10.1155/2018/9285854

Viegas C, Patrício AB, Prata JM, Nadhman A, Chintamaneni PK, Fonte P (2023) Solid lipid nanoparticles vs. nanostructured lipid carriers: a comparative review. Pharmaceutics. 15(6):1593. https://doi.org/10.3390/pharmaceutics15061593

Wang Y, Wan D, Xie S, Xia X, Huang CZ, Xia Y (2013) Synthesis of silver octahedra with controlled sizes and optical properties via seed-mediated growth. ACS Nano 7(5):4586–4594. https://doi.org/10.1021/nn401363e

Wang H, Brown PC, Chow ECY, Ewart L, Ferguson SS, Fitzpatrick S, Freedman BS, Guo GL et al (2021) 3D cell culture models: drug pharmacokinetics, safety assessment, and regulatory consideration. Clin Transl Sci 14(5):1659–1680. https://doi.org/10.1111/cts.13066

Wepener I, Richter W, van Papendorp D, Joubert AM (2012) In vitro osteoclast-like and osteoblast cells' response to electrospun calcium phosphate biphasic candidate scaffolds for bone tissue engineering. J Mater Sci Mater Med 23(12):3029–3040. https://doi.org/10.1007/s10856-012-4751-y

Wu T, Xue J, Li H, Zhu C, Mo X, Xia Y (2018) General method for generating circular gradients of active proteins on nanofiber scaffolds sought for wound closure and related applications. ACS Appl Mater Interfaces 10(10):8536–8545. https://doi.org/10.1021/acsami.8b00129

Wu JR, Hernandez Y, Miyasaki KF, Kwon EJ (2023) Engineered nanomaterials that exploit blood-brain barrier dysfunction for delivery to the brain. Adv Drug Deliv Rev 197(114820):114820. https://doi.org/10.1016/j.addr.2023.114820

Xu L, Wang X, Liu Y, Yang G, Falconer RJ, Zhao C-X (2022) Lipid nanoparticles for drug delivery. Adv Nanobiomed Res 2(2):2100109. https://doi.org/10.1002/anbr.202100109

Xue J, Wu T, Dai Y, Xia Y (2019) Electrospinning and electrospun nanofibers: methods, materials, and applications. Chem Rev 119(8):5298–5415. https://doi.org/10.1021/acs.chemrev.8b00593

Yadav S, Sharma AK, Kumar P (2020) Nanoscale self-assembly for therapeutic delivery. Front Bioeng Biotechnol 8:127. https://doi.org/10.3389/fbioe.2020.00127

Yan N, Xu J, Liu G, Ma C, Bao L, Cong Y, Wang Z, Zhao Y et al (2022) Penetrating macrophage-based nanoformulation for periodontitis treatment. ACS Nano 16(11):18253–18265. https://doi.org/10.1021/acsnano.2c05923

Yi C, Yu Z, Ren Q, Liu X, Wang Y, Sun X, Yin S, Pan J et al (2020) Nanoscale ZnO-based photosensitizers for photodynamic therapy. Photodiagn Photodyn Ther 30(101694):101694. https://doi.org/10.1016/j.pdpdt.2020.101694

Zhang J, Saltzman M (2013) Engineering biodegradable nanoparticles for drug and gene delivery. Chem Eng Prog 109(3):25–30

Zhang W, Chen Q, Wu F, Dai J, Ding D, Wu J, Lou X, Xia F (2021) Peptide-based nanomaterials for gene therapy. Nanoscale Adv 3(2):302–310. https://doi.org/10.1039/d0na00899k

Zhou H, Lee J (2011) Nanoscale hydroxyapatite particles for bone tissue engineering. Acta Biomater 7(7):2769–2781. https://doi.org/10.1016/j.actbio.2011.03.019

Zhou L, Shi H, Li Z, He C (2020) Recent advances in complex coacervation design from macromolecular assemblies and emerging applications. Macromol Rapid Commun 41(21):e2000149. https://doi.org/10.1002/marc.202000149

Zielińska A, Carreiró F, Oliveira AM, Neves A, Pires B, Venkatesh DN, Durazzo A, Lucarini M et al (2020) Polymeric nanoparticles: production, characterization, toxicology and ecotoxicology. Molecules (Basel, Switzerland). 25(16):3731. https://doi.org/10.3390/molecules25163731

# Chapter 9
# Nanoencapsulation as a Strategy to Enhance the Pharmacokinetics of Bioactive Compounds

**Dhanalekshmi Unnikrishnan Meenakshi, Gurpreet Kaur Narde, Selvasudha Nandhakumar, and Alka Ahuja**

**Abstract** Natural bioactive compounds/ biopharmaceuticals with medicinal properties have been beneficial in maintaining a healthy lifestyle since the dawn of humankind. Secondary metabolites are the main bioactive components that are found in large quantities in natural sources, especially plants, animals, and minerals. In the previous several decades, research or exploration into health have proven that consuming bioactive compounds could protect against various diseases, and they are considered better therapeutic agents for treating various ailments. These bioactive compounds can be inherently unstable due to their sensitivity to several in-vitro and in-vivo factors. Hence, proper formulation approaches are crucial to maintaining therapeutic efficacy and bioavailability. Pharmaceutical encapsulation techniques play a vital role in protecting bioactive compounds in relevant aspects. Nanoencapsulation can enhance the stability of bioactive molecules by forming a protective shield, thus improving the pharmacokinetics of drugs at physiologically active sites. This chapter overviews nanoencapsulation techniques and their applications in encapsulating pharmaceutically critical bioactive compounds. It delves into the benefits of nanoencapsulation of bioactive compounds, their properties, and challenges in formulation. It also spotlights the mechanism of nanoencapsulation along with the factors influencing the encapsulation efficiency and release kinetics. Furthermore, it discusses the potential benefits of bioactive compounds in modifying the pharmacokinetic pattern, especially the bio-enhancement of combinational drugs. The chapter also focuses on insights into the current challenges, emerging trends, and future directions in nanoencapsulation research. The chapter will summarize the key findings, reinforcing the potential of nanoencapsulation in advancing the delivery and efficacy of pharmaceutically important biopharmaceuticals.

D. Unnikrishnan Meenakshi · G. K. Narde · A. Ahuja (✉)
College of Pharmacy, National University of Science and Technology, Muscat, Oman
e-mail: alkaahuja@nu.edu.om

S. Nandhakumar
Department of Pharmaceutics, Kamarajar College of Pharmacy, Chidambaram, India

 
V. Khanna et al. (eds.), *Nanotechnologies for Drug Delivery and Biopharmaceutical Development*, Progress in Drug Research 77,
https://doi.org/10.1007/978-3-032-00796-4_9

**Keywords** Bioactive compounds · Nanoencapsulation · Bio enhancer · Pharmacokinetics · Formulation · Drug delivery · Biopharmaceuticals

## 9.1 Introduction

Natural bioactive compounds/ biopharmaceuticals with medicinal properties are beneficial in maintaining a healthy lifestyle since the dawn of humankind. They are secondary metabolites of plants and animals that serve as growth stimulants and defense against biotic and abiotic stressors and have nutritional and therapeutic value. Plants are one of the significant sources/leads of therapeutically effective biomolecules. Various bioactive components, such as phenols, alkaloids, essential oils, proteins, carbohydrates, terpenoids, and flavonoids, can be found in different plants (Nogueira et al. 2020). Bioactive molecules have garnered the interest of academics and researchers recently because of their potent therapeutic efficacy, e.g. Morphine. Their consumption has been linked to positive physical and mental health effects. Most of the bioactive constituents possess anti-microbial, antioxidant, anti-inflammatory, and/or neurodegenerative properties, and hence, they are specially categorized as pharmaceutically important bioactive constituents. These compounds can act as inhibitors or modulators of specific biological pathways, offering targeted treatment options that minimize side effects.

Traditional medicine has long utilized bioactive molecules derived from plants, fungi and other natural sources, laying the groundwork for modern drug discovery and development. For example, compounds such as artemisinin from *Artemisia annua* have proven effective in treating malaria, inspiring the creation of synthetic derivatives (Costa et al. 2021). Another notable bioactive molecule is paclitaxel, originally extracted from the Western yew tree and commonly employed in cancer therapy (Ravi et al. 2018). Additionally, curcumin from turmeric has led to investigations in various formulations. The ongoing exploration of these traditional bioactive molecules provides valuable insights for developing novel, effective pharmaceuticals (Chowdhury et al. 2024).

Bioactive compounds are typically formulated into various pharmaceutical dosage forms for human consumption, particularly for therapeutic applications. This formulation strategy is believed to improve patient adherence and increase the steadiness of bioactive compounds. However, several challenges arise in formulating bioactive molecules into pharmaceutical dosage forms. Advances in drug delivery systems, such as nanoparticles, boost the pharmacokinetics and effectiveness of bioactive molecules (Nogueira et al. 2020). Moreover, ongoing research into their mechanism of action is paving the way for novel therapies. Consequently, this chapter focuses on the nanoformulation strategy of pharmaceutically significant bioactive components, which is supported by the subsequent introduction.

Hydrophilic bioactive compounds like polyphenols or lipophilic compounds like carotenoids benefit human health. Still, they are inherently unstable due to their sensitivity to several in-vitro and in-vivo factors. Hence, proper formulation approaches are crucial to maintain bioavailability and therapeutic efficacy. Nanoencapsulation is one of the technologies that show promise for solving the issues. Nanosize formulation approach with suitable carrier encapsulation improves surface area and physiochemical function. This improves solubility, stability, bioavailability of the bioactive constituents and therapeutic efficiency (Chowdhury et al. 2024). Formulating an optimized nano-encapsulated formulation of bioactive compounds with physical and chemical stability and cost-effectiveness entails many challenges.

Though bioactive compounds have versatile, functional properties, their use is restricted by their unpredictable stability during storage, preparation, and in the human system. The unstable compounds produced during storage might lead to toxic products. Exposure to temperature, pH and light during storage, influence of body fluids, enzymes and interaction with foods when in contact with human body are challenging for the drug delivery of bioactive compounds (Pateiro et al. 2021). In general, bioactive molecules occur along with others as complex structures, and their stability is disturbed when it is isolated. Hence, technologies that retain the stability and functionality of biomolecules are emerging. Among all, nanoencapsulation methods have become more popular for maintaining stability and functionality of the bioactive compounds (Chowdhury et al. 2024).

This chapter briefly overviews the various nanoencapsulation techniques and their applications in encapsulating pharmaceutically critical bioactive compounds. A brief discussion was held on successfully developed nanoencapsulation methods for bioactive compounds. Through absorption, incorporation, chemical interaction, or dispersion into nanocarriers, various bioactive compounds can be nanoencapsulated (Nicolescu et al. 2023). Additional characterization techniques and several physicochemical metrics should be used to evaluate the characteristics of encapsulated bioactive. The release pattern and therapeutic results of the bioactive molecules are influenced by the choice of the carriers used in the formulation approach.

Various materials, including lipids, polymers, and inorganic nanoparticles, have been extensively researched as carriers for these bioactive compounds. Each type of carrier presents unique advantages in the pharmacokinetic and pharmacodynamic aspects (Liga et al. 2023). Thus, understanding the interaction between bioactive compounds and their carriers is essential for optimizing nanoformulation strategies in pharmaceutical applications. Important bioactive materials used for nanoencapsulation employing state-of-the-art techniques are highlighted in the current chapter. Naturally occurring biologically active compounds, including flavonoids, alkaloids, terpenoids, etc., exert a crucial function in the physiological functions of biological systems. This chapter identifies several significant bioactive compounds and provides a comprehensive discussion on how nanoencapsulation modifies their pharmacokinetics, thereby enhancing their therapeutic efficacy. Additionally, a section is dedicated to examining the molecular mechanisms and targets that can impede the therapeutic effects of these bioactive compounds. The chapter also focuses on insights into the current challenges, emerging trends and future directions in nanoencapsulation research and thereby summarizes the potential of nanoencapsulation in advancing the delivery and efficacy of pharmaceutically important biopharmaceuticals.

## 9.2 Bioactive Compounds Stability and Bioavailability

Bioactive compounds have complex molecular structures and their conversion into essential components for beneficial health effects in humans is influenced by many factors. The complicated complex absorption mechanism, which impairs the bioavailability of the bioactive substances, is a challenge when compared to the drug molecules which need disintegration and dissolution alone for better absorption (Rein et al. 2013). In contrast to drug molecules, when we take as food these bioactive compounds, they should be released from the matrix of food, be absorbed in the GIT, subjected to metabolism, reach the systemic circulation for better bioavailability, then finally enter the target site to produce their action. Among all, bioaccessibility is a challenging process. The physicochemical and physiological properties like temperature, pH, texture, crystallinity, density, osmotic pressure, presence of bile acid, enzymes at the GIT etc. modify the kinetics of release pattern of natural bioactive molecules (Nicolescu et al. 2023). The pre-requisite ordered path of digestion, such as hydrolysis by gastric acid, emulsification by the bile acids, and action by lipases, is essential for the bioavailability of bioactive compounds, particularly for the lipid bioactives such as polyunsaturated fatty acid, fat-soluble vitamins, and carotenoids, etc. (Luo 2020). Release of monoacylglycerol along with free fatty acid is the crucial step as it forms micelles aiding the absorption of both lipophilic and hydrophilic bioactive compounds, and if the process does not occur as sequence, it also becomes challenging (Acevedo-Fani and Singh 2022).

Apart from this, gut microbiota also influences the bioavailability of bioactive substances. For instance, polysaccharides act as pre-biotics and are degraded by gut microbiomes, producing secondary metabolites like butyrates, and acetate, which exert their pharmacological action. As the colon is the space for millions of versatile microbes, the fate of different bioactive compounds and their role in various diseases couldn't be elucidated (Plamada and Vodnar 2021). Another challenge in the bioavailability of bioactive compounds is inter-subject variability, which in turn depends on genetic makeup, gut microbiota composition, etc. As most of bioactive compounds are BCS IV or II drugs, only 0.5% to 45% of active compounds are absorbed leading to low plasma concentration (Vamanu and Gatea 2020). Due to these factors, bioactive compounds are limited to functional materials. Both lipophilic or hydrophilic biomolecules exhibit different solubility profiles and hence the mechanism of absorption varies between them. There are various types of lipids i.e. glycolipids, phospholipids, sterols, fat soluble vitamins, triglycerides, and free fatty acids. Though the unstirred water layer facilitates the absorption of lipid biomolecules across the intestine, the size becomes a barrier which should be reduced. The absorption of lipids by enterocytes is thought to occur through passive diffusion as well as facilitated diffusion involving specific transporters (Iqbal and Hussain 2009). Triacylglycerols are created when fatty acids and monoacylglycerols esterify inside the enterocytes. Then, via triacylglycerol-rich lipoproteins called chylomicrons, these triacylglycerols are secreted into the lymphatic system. Lipid-soluble substances are harder for the body to get rid of than hydrophilic ones. Rather, they are either reabsorbed into the bloodstream as lipoproteins and then stored in adipose tissue, or they are stored in the liver (Rashidinejad et al. 2022).

In comparison to lipophilic bioactive compounds, the hydrophilic compounds possess simple absorption mechanisms. But they exist along with glycosides, polysaccharide and esters which couldn't be absorbed as such. For instance, polyphenols are absorbed after enzymatic hydrolysis which occurs at brush border of epithelial cells. As the result, aglycone is released which then cross the enterocyte and subjected to phase II reaction, then eliminated after the bioavailability pathway. Here the bioactive compound faces some challenges and acted upon by ABC transporters leading to efflux reaction (Liu and Hu 2007). The escaped bio molecule from the gut excretion, again acted upon by the liver, and excreted, which further reduces the bioavailability. The result of the above-mentioned process, which is well depicted as metabolic detoxication, is almost eliminating aglycones from the circulatory system. Since the conjugated forms of flavonoids are highly efficacious to exhibit a therapeutic efficacy, previous theories hypothesized that aglycones are the therapeutically effective isomer of flavonoids and hence further research is warranted (Liga et al. 2023).

The structure and molecular weight of bio active compounds are another determining factor for the effective absorption. High molecular weight compounds like lipids, glycoproteins, some oligomers like proanthocyanidins should be breakdown before absorption as they couldn't pass through the gut walls. In general b-glucosides and flavonoids co-exist, and for its absorption it should be cleaved by the particular enzymes like hydrolase, b-glucosidase (Nie et al. 2023). Some other bio molecules are cleaved in the large intestine only, as they can be acted upon only by the microbiota like quercetin, which exists in conjugation with rhamnose moiety. Apart from this, the molecular configuration also plays a major role in drug absorption and the stereoisomers have different absorption patterns. The bioavailability of (−)-epicatechin and (+)-catechin and the bioavailability of lycopene's cis and all-trans isomers are all examples of this (Rein et al. 2013). In human blood and urine samples, hesperitin-7-glucoside was discovered to have an R: S ratio of 39:69, suggesting that the S configuration might be more readily absorbed by the body (Rein et al. 2013).

An additional challenge for the bioavailability of biomolecules/bioactive is transport mechanisms. Among the three basic transport mechanisms, for active and facilitated transport, the biomolecules need to diffuse against the concentration gradient through the intestine for better bioavailability. As most biomolecules have either low solubility or poor permeability, they couldn't diffuse passively and need special carrier systems, i.e., transporters, which have two mechanisms in common, influx (OAT1, GLUT-1) or efflux (P-gp, BCRP). Certain nutrients block intestinal transport systems, which have an impact on the bioavailability of biomolecules, as well as there is selective transport of different bioactive compounds. Certain bioactive compounds can be reserved by OATs, which may prolong the biomolecules plasma half-lives. A study proved that ellagic acid exhibits a high affinity for OATs, specifically hOAT1, which may result in drug interactions with some drugs like antivirals and antibiotics (Vaou et al. 2022). In another study it was reported that ferulic acid can prevent nateglinide from being absorbed, which lowers its bioavailability. Similarly, flavonoids are known substrates for ABC transporters, which contributes to their low bioavailability.

In general, the uptake of lipids is passive, some carrier proteins like FATP4, FABPpm and CD36 could play a role in the movement of lipids over the intestinal barrier. Hence the transport of multiple compounds through the same transport system leads to the competitive inhibition of biomolecules. As a result, the decrease in bioavailability is observed. This effect was reported in a study for the sterols which reduces the absorption of cholesterol and vitamin K (Ehehalt et al. 2008).

The interaction of drugs and bioactive compounds with enzymes, particularly cytochrome P450 (CYP) enzymes has major part in the biomolecules' bioavailability. Biomolecules undergo Phase I metabolism via CYP enzymes, which alter the structures of compounds through various reactions like reduction, amination and oxidation. But some bioactive molecules are not CYP substrates like polyphenols but subjected to conjugation with methyl, sulfate, or glucuronyl groups affecting bioavailability (Ref). Few bioactive molecules activate or render CYP enzymes, which alter the drug metabolism. For instance, sesame lignin (sesamin) increases γ-tocopherol concentrations by inhibiting CYP enzymes, and grapefruit juice increases drug bioavailability by inhibiting CYP3A4. The furanocoumarin compounds found in grapefruit juice, such as bergamottin, have the ability to increase a drug's bioavailability and produce toxic plasma concentrations (Lim et al. 2012; Kiani and Imam 2007). Lipids such as EPA and DHA can enhance drug absorption, as seen in increased bioavailability of ciclosporin A due to CYP3A inhibition by DHA. Thus, various types of interactions have significant implications for safety and efficiency of bioactive compounds and drugs, highlighting the demand for careful consideration in drug administration.

One of the lipophilic bioactive molecules is carotenoids which has low solubility and susceptible to oxidation leading to decrease in its functionality (Focsan et al. 2019). Another class of bioactive compound is a phenolic compound which is hydrophilic and has several limitations like low permeability, rapid release, and degradation in environment and unstable in the human system (Grgić et al. 2020). Nanoencapsulation is a technique which is used to address these issues. But it also has some challenges. The encapsulated bioactive compound may not be released due to some physiological factors like metabolism, microbiota and molecular transformation which should be considered during formulation techniques (Chowdhury et al. 2024).

Continued research into innovative nanoencapsulation techniques could enhance the bioavailability of these bioactive molecules, facilitating more effective therapeutic applications. Additionally, the development of smart carriers that respond to physiological conditions may improve the controlled release of encapsulated bioactive compounds, mitigating issues related to metabolism and molecular transformation. Future studies should also focus on the interaction between encapsulated compounds and the gut microbiota to optimize formulations for improved stability and efficacy in the human system. Overall, interdisciplinary approach is essential for advancing nanoencapsulation strategies and enhancing the functionality of bioactive compounds in pharmaceutical formulations.

## 9.3 Nanoencapsulation Techniques

In general, there are two fundamental classes of nanostructures i.e. single-core and multicore. Mechanisms related to nanosphere formulation and nanoencapsulation specifically in relation to nanocarriers are described. In a study the author categorized the various methods used to create nanocapsules into five groups "(1) lipid formulations (nanoemulsions, nanoliposomes, and nanostructured lipid carriers)., (2) natural nanocarriers (caseins, cyclodextrins, nanocrystals, and amylose nanostructures)., (3) specialized equipment (electrospun nanofibers, electrosprayed NPs, and nanospray dried particles)., (4) functional biopolymeric NPs., and (5) disparate methods (protein nanotubes and carbohydrate nanogels)" (Jafari 2017). In all cases the following techniques are common either for nanoparticle synthesis or nanoencapsulation.

### *9.3.1 First-Generation Nanotechniques*

First-generation nanocrystals are classically synthesized as colloidal delivery systems without the use of a carrier. During the initial stages, it was developed using a one-step process of either bottom-up or top-down approaches. This involves concepts like precipitation and high-pressure homogenization. The stability of the formulated drug nanoparticle, defined particle size with the improved pharmacokinetic as well as therapeutic properties not much satisfied through this simple process (Haripriyaa and Suthindhiran 2023).

### *9.3.2 Bottom-Up Approach*

Bottom-up methods are also known as precipitation techniques that operate on the basic principle of precipitating bioactive nanoparticles from a solution which has reached supersaturation. The advantages of bottom-up methods are (1) simple operation, (2) less expensive and (3) more suitable for intravenous administration of anticancer compounds. However, residual organic solvents might lead to instability/toxicity, and thus not suitable for drugs that are insoluble in organic solvents (Tehrani 2019). This is broadly grouped into four major types as a solvent- anti-solvent precipitation using simple mixing or modified methods, precipitation using supercritical fluid (SCF), conventional or special freezing based solvent removal technique and precipitation coupled with high energy process. In this technique, stabilizer selection is considered significant, which is usually based on the hydrophilic-lipophilic balance (HLB). Accordingly, stabilizers like tweens, pluronics with high HLB value have been frequently used (Liu et al. 2021). The three main categories of bottom-up approaches include: solvent-anti-solvent precipitation (regular static mixing or modified mixing), supercritical fluid technology and solvent removal techniques.

The liquid solvent-anti-solvent precipitation method is a simple technique that could be performed either by simple mixing or facilitated mixing through the modified method. This process involves eco-friendly and non-genotoxic organic solvents often selected from Class II and III solvents. Some include acetone, ethanol and isopropanol. Followed by the hydrosol used through this type of precipitation, NanoMorph® technology has been developed and utilized for a variety of molecules (Wu et al. 2019).

Compared to the simple static mixing process, the addition of some external factors to the precipitation or performing the precipitation under some altered experimental conditions resulted in improved size reduction. This includes sonoprecipitation where ultrasonic source fitted with the mixing device resulting in size reduced amorphous particle with almost uniform size distribution. on controlling the process parameters like sonication duration and intensity, synthesis of desired nano sized particles become possible (Sandhya et al. 2021). Some other modified mixing-based methods include high gravity reactive precipitation (HGRP) and evaporative precipitation into an aqueous solution (EPAS). Relatively lower cost and a technique suitable for mass production called HGRP, the one that was previously said to be Higee technology, is primarily used for commercial production (Yu et al. 2011). However, this process needs critical optimization of the process parameters to attain nanosized particles. Next to this modern technique, flash nanoprecipitation is also widely being utilized for industrial-scale production as it can provide long term stability to the final product. This stability is achieved through the minimization of Ostwald ripening and nanoparticle agglomeration. The term flash nanoprecipitation comes from the process that involves drug solution and anti-solvent flash mixing leading to the formation of high supersaturation conditions. The particles are further size reduced by simultaneous stabilization using precipitation. In contrast to the previous technique of electrohydrodynamic atomization, it uses surfactants and polymers in aiding the stabilization of the drug molecules (Tao et al. 2019).

The second approach based on the utilization of supercritical fluid (SCF) helps in preparing the nanonized particles. It is an economically feasible and non-toxic technique for commercial purposes using supercritical solvents like carbon dioxide, ammonia, fluoroform, ethane and ethylene. The method suitable for even thermolabile drug compounds with almost no organic solvent in the final drug nanocrystal product opens a wide application of this technique in pharmaceutical usage. The SCF-based techniques that are widely used for the precipitation of organic compounds comprised of rapid expansion of supercritical solution (RESS), the rapid expansion of a supercritical solution into a liquid solvent (RESOLVE), and supercritical antisolvent (SAS) (Liu et al. 2021).

Regarding the solvent removal process of drug nanoparticle synthesis, conventional spray drying and freeze-drying types have not shown much significant efficiency in particle size reduction. This was modified using some special freezing techniques proposed by researchers called spray freezing into liquid (SFL). Recently, Electrohydrodynamic atomization has become a promising technique for overcoming various physicochemical stability issues during storage. The safety issues related to organic solvents, stabilizers or surfactants are also minimal in this technique as there is minimal or no residual solvent after processing in the final product (Chen et al. 2022).

### 9.3.3 Top-Down Approach

The top-down approach of the nanosizing process to produce various nanomedicines probably incorporates two basic methods of milling and homogenization. This is conventional modern technique, useful for several clinically important bioactives. Based on the results of varied studies in drug nano-encapsulation, it is clearly understood that the top-down methods often provide high drug loading compared to bottom-up techniques. This beneficial advantage impressed most of the pharmaceutical industries in utilizing top-down approaches with minimal rate of drug wastage. This property also aids in reducing the dose and toxicity to the consuming patients (Dahiya 2017). Accordingly, the top-down methods are found to be easier in scaling up and commercialization of the process. All the advantages of top-down methods drive the interest of researchers as well as industrial people in many drugs nanocrystal preparation process. Sometimes, there might be chances of product contamination and instability developed by high shear and temperature. A top-down approach involves milling technique and high-pressure homogenization.

### 9.3.4 Milling

The milling chamber, milling motor, milling pearls with recirculation chamber progress the milling process for the particle size reduction. This technique is an industrially feasible one due to the following valuable merits. Traditionally, jet milling was used for reducing the particle size followed by the wet milling (Loh et al. 2015). This includes ease of large-scale production, relatively lower cost and could be performed as either batch, continuous or recirculation in obtaining particles with narrow particle size distribution. Pharmaceutical manufacturing industries are often using the continuous mode of processing through either multipass continuous or cascade-continuous for large scale production of nanomedicines. Moreover, the recent advancement resulted in Pearl milling called Nanocrystal Technology has been frequently utilized in producing most of the oral formulations either in solid or liquid dosage forms. This technique requires the solidification step that was achieved using spray or free-drying equipment at the industrial level (Joshi et al. 2019). Further, wet ball milling is an advanced and versatile technique reported for treating and formulating poorly water-soluble bioactives. Their efficiency in particle size reduction has been proved by the preparation of various drug nanocrystals with the desired nanonized particles. High energy milling is also one of the commercial level milling processes that could perform particle size reduction in a shorter time. The easier scaling up and developmental process makes it more appreciable in the pharmaceutical industries (Loh et al. 2015).

### *9.3.5 High-Pressure Homogenization*

This technique employs either microfluidization or piston gap homogenization principle. In earlier days, a Canadian company RTP developed microfluidizer based high-pressure homogenization that worked on the principle of collision and cavitation. Due to this working principle, the microfluidization technique is also said to be air-jet milling or jet stream homogenization (Salazar et al. 2014). Then it is tracked by the Piston-gap homogenization method comprises of Dissocubes homogenization followed by the second variant called Nanopure® technique. The contamination problem associated with the bead milling is not observed in this technique that makes it potential in the preparation of parenteral nanosuspensions. Additionally, the low cost and simplicity make this technique approved for the commercial production of pharmaceutical cosmetics and food products. Coarse emulsion, dispersion or suspension generally used as premix to prepare nanoparticles by HPH (Grumbach 2022). This technique could also be easily scale-up from laboratory to industrial-scale production. With the progression and advancement of formulation technology, Nanopure Technology replaced the classical approach of using water with oil, glycerol and polyethylene glycol (PEG). This technique allowed the direct filling of nanoformulations into either capsules or injectables (Suk 2016).

### *9.3.6 Second Generation Nanotechnology*

The recent innovation of precipitation combined with a high energy process has received much research interest in synthesizing various nanomedicines. Precipitation using a high energy process is a coupled technique commonly said to be a combination technique. A carrier-based colloidal delivery system called second-generation drug nanopartciles are being developed using non-ionic surfactants and polymers or stabilizers. Second generation nanocrystals are prepared by a variety of precombinative treatments followed by the process of high-pressure homogenization. This advanced technology has resulted in the preparation of Smart crystals.

Typically, both top-down and bottom-up or combinations of both techniques get involved in the synthesis of these second-generation drug nanocapsules. One such example is the emulsion diffusion-high-pressure homogenization-solvent evaporation technique (EHS). These techniques are developed to overcome the basic and significant barriers like Ostwald ripening and agglomeration forming a physically stable formulation. Some of the commonly used polymers, in this case, are Poloxamer 188, poly (D, L-lactide-co-glycolide) (PLGA), polyethylene glycol, polyvinylpyrrolidone and other cellulose derivatives (Fontana et al. 2018; Panda et al. 2019).

Combination technique is used in the novel preparation of nanosuspension. This is almost like the modified solvent-antisolvent technique indicating that high-pressure homogenization, ultrasonic wave or high energy mixing are utilized as annealing steps after the precipitation in reducing the particle size. Combination techniques have shown promising results in the size reduction process compared to traditional nanonization techniques. Some of the recent combination techniques are Nanoedge, H42, H69, H96. Nanoedge is precipitation followed by high-pressure homogenization, H42 is spray drying followed by high-pressure homogenization, H69 is cavi-precipitation followed by high-pressure homogenization and lyophilization followed by high-pressure homogenization. These methods have both advantages and limitations (Salazar et al. 2014). For instance, bottom-up approaches are cost effective and simple to operate but crystal growth is difficult to control. Similarly topdown approaches are easy to scale up but it is high cost. Table 9.1 shows some of the nanoencapsulation techniques of the pharmaceutically important bioactive molecules.

## 9.4 Encapsulation of Bioactive Compounds- Role of Carriers

Hybrid nanocarriers represent an innovative approach in advanced drug delivery systems, comprising two key components: an internal framework, often based on metals or polymers, and an external layer typically made of single or multi layers lipid fabrication. This lipid fabrication decoration not only protects the bioactive molecule from degradation but also minimizes water permeation, enhancing their stability. These hybrid nanocarriers hold significant role in the delivery of several phytopharmaceuticals and synthetic drugs. Chitosan coated nano liposomes demonstrate remarkable efficacy in encapsulating bioactive components, exhibiting slow -release properties compared to their non coated counterparts. Niosomes, nanoliposomes, phytosomes and particulate carriers are collectively known as vesicular carriers. The self-assembly of surfactant molecules in aqueous solutions leads to the formation of spherical bilayers, which serve as effective encapsulation systems for a variety of bioactive substances, including peptides. Nano-lipid particles are produced by combining liquid and solid lipids, while solid lipid nanoparticles are formed from solid lipid blends.

Table 9.1 Nanoencapsulation Techniques of Bioactive compounds

| S. no | Name of Bioactive compounds | Encapsulation technique | Encapsulating polymers | Reference |
|---|---|---|---|---|
| 1. | Artemisinin | High pressure homogenization<br>Ionic gelation<br>Single emulsion solvent evaporation technique | Bovine serum albumin<br>Chitosan<br>PVA and heparin | Natesan et al. (2017)<br>Odera et al. (2024) |
| 2. | Catechin | Thin flim hydration method.<br>Wet milling and ultrasonication | Span 60 and cholesterol<br>Beta-glucan | Gadapa et al. (2022)<br>Shah et al. (2022) |
| 3. | Caffenic acid | Electrospining method<br>Capsulation | Carob flour, whey protein concentrate (WPC), and polyethylene oxide (PEO)<br>Hypromellose | Zeren et al. (2022)<br>Stanciauskaite et al. (2023) |
| 4. | Camptothecin | Flash Nano precipitation | Poly(lactic acid)-block-poly(ethylene glycol) | Wilson et al. (2021) |
| 5. | Combretastatin | Amphiphilic random copolymerization | Copolymers composed of tri(ethylene glycol) methacrylate (TEGMA) or poly(ethylene glycol) methyl ether methacrylate (PEGMA) and perfluorohexylethyl Acrylate (FA) were | Calosi et al. (2022) |
| 6. | Curcumin | Antisolvent co-precipitation method<br>pH-driven loading method | Zein/HP-β-CD<br>Saponin | Qiu et al. (2023)<br>Peng et al. (2018) |
| 7. | Epigallocatechin-3-gallate | Thin film hydration method<br>Anti-solvent precipitation method | 1,2-Dioleoyl-3-trimethylammonium propane (DOTAP)<br>Zein-gum arabic | Li et al. (2023)<br>Jin et al. (2022) |
| 8. | Epicatechin | Desolvation method<br>Solvent evaporation (double emulsion) method | Bovine serum albumin<br>PLGA | Yadav et al. (2014)<br>Ahmad et al. (2020a, b) |
| 9. | Genistein | Nanoprecipitation technique<br>Wet milling | Poly-ε-caprolactone<br>HPMC (0.6% w/v), Pluronic F-68 | Abriata et al. (2024)<br>Iqbal et al. (2024) |
| 10. | Lycopene | Melt-solvent technique<br>Coupling micro-filtration and diafiltration processes, | Polyvinylpyrrolidone-K30,modified HI-CAP 100 starch and Maltodextrin.<br>Sodium alginate | Mirahmadi et al. (2024)<br>Sampaio et al. (2019) |

(continued)

**Table 9.1** (continued)

| S. no | Name of Bioactive compounds | Encapsulation technique | Encapsulating polymers | Reference |
|---|---|---|---|---|
| 11. | Paclitaxel | Ionic gelation method<br>Conventional electrophilic radioiodination and desolvation method | Chitosan<br>SnBu3-paclitaxel as a precursor, and albumin | Menon et al. (2022)<br>Lee et al. (2024) |
| 12. | Quercetin | Ionic gelation method<br>Ionic gelation process | Chitosan<br>Gelatin | Nalini et al. (2023)<br>Kang et al. (2023) |
| 13. | Resveratrol | Extrusion technique<br>Nanoencapsulation-based edible coating | 1,2-dipalmitoyl-sn-glycero-3-phosphocholine and cholesterol<br>Chitosan | SOO et al. (2016)<br>Hu et al. (2023) |
| 14. | Silibinin | Interfacial deposition of preformed polymer method<br>Interfacial deposition of preformed polymer, | Ethylcelullose, span 80, and medium chain triglycerides<br>Gellan gum | Gehrcke et al. (2021)<br>Marchiori et al. (2023) |
| 15. | Thymoquinone | Solvent dispersion and ultrasonication method<br>Nanoemulsification technique | Glycerol monooleate<br>Labrafil M 2125 CS, Labrafac Lipophile WL 1349, Labrafac PG, Labrafil, and Compritol 888 ATO | Yaghmur et al. (2019)<br>Rathore et al. (2023) |

Lipid-based nanoencapsulation is prominently enhancing the bioavailability and targeted delivery of phytoconstituents/drugs and several research are focused on this encapsulation technique. Polymeric nano-carriers have also emerged as powerful tools for encapsulating and delivering bioactive compounds. Natural materials such as starch, chitosan, casein, albumin, and whey protein are frequently utilized in their design. Chitosan is used to encapsulate the marine carotenoid fucoxanthin and this encapsulation method significantly boosting the fucoxanthin anticancer effects (Ravi et al. 2018). Another noteworthy study successfully encapsulated bioactive extracts from grape pomace using chitosan and alginate nanoparticles. This approach not only enhanced the bioactivity of the extracts but also effectively protected their bioactive components from hydrolysis in the gastrointestinal tract (Costa et al. 2021). These advancements accentuate the potential of hybrid, lipid and polymer nano-carriers in revolutionizing drug delivery, particularly in enhancing the efficacy and bioavailability by improving the pharmacokinetics of bioactive molecules.

## 9.5 Improvement in Pharmacokinetics and Efficacy of Pharmaceutically Important Bioactive Compounds by Nanoencapsulation/Nanoformulation Approaches

The bioactive compounds are finding spectrum of applications as effective pharmaceutical candidates for life-threatening disorders such as cancer and also as food supplements (Asma et al. 2022). However, the susceptibility of bioactive compounds to various physical attributes such as temperature, light, ionic strength, oxygen, pH etc. limits the potential of direct employment of bioactive compounds in different industries such as food or pharma (Dahiya et al. 2023). The pharmacokinetics of bioactive agents can be improved by encapsulation in nanoparticle formulation. Nanoparticles such as liposomes, micelles, polymeric nanoparticles, dendrimers etc. have distinct properties due to the small size ranging between 10–200 nm. Nanoparticles of correct size have prolonged presence in blood circulation as they can evade the renal clearance (Adhipandito et al. 2021). Particularly in cancer, nanoparticles have found an astounding potential because of the enhanced permeability and retention (EPR) environment in tumour cells (Ahuja et al. 2021). Research has shown that the ADME pharmacokinetics of bioactive compounds increased in nanoparticle formulations (Mabrouk et al. 2021). As shown in Fig. 9.1, this chapter emphasizes key bioactive molecules used therapeutically and those being explored in nanoencapsulation research. Various studies indicate that nanoformulation of these phytoconstituents shows promising results in overcoming formulation and therapeutic challenges. The advancements in nanoencapsulation techniques augment the potential bioavailability of these compounds, leading to improved treatment outcomes. The details of pharmaceutically important bioactive molecules are discussed in the following sub-sections.

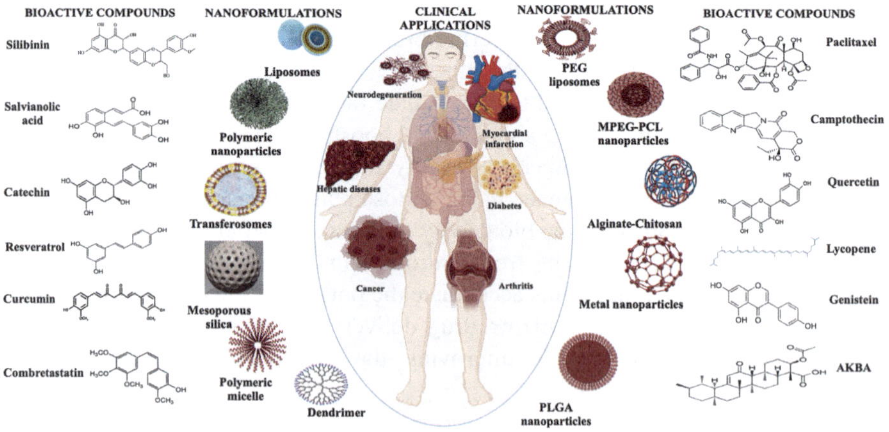


**Fig. 9.1** Important bioactive compounds when formulated in various types of nanoparticles can improve the bioavailability and ADMET profile

### 9.5.1 *Artemisinin*

**Artemisinin** is an antimalarial drug that also shows potential benefits as an anticancer agent. The peroxide bridge in this sesquiterpene plays an important role in its biological mode of action. The major drawback is its pharmacokinetic pattern, especially the amount of drug that reaches circulation. Artemisinin and its derivatives, such as artesunate, artemether, dihydroartemisinin, etc., are used in clinical settings because of their beneficial therapeutic effects. Moreover, the therapeutic efficacy and pharmacokinetic parameters of artemisinin and its derivatives can be enhanced by formulating them into nanoparticles (Devalapally et al. 2007). Various surface modification strategies have been used to increase the plasma half-life of this bioactive molecule with a controlled and well-adjusted pharmacokinetic pattern (Pimentel et al., 2007). For example, pharmacokinetic analysis of artemisinin nanoparticles reported by Yameogo et al. showed $IC_{50}$ of 7.0 ng/mL in *in vitro* antiplasmodial activity. The intravenous injection in healthy rats resulted in plasma half-life by 6.25 folds thereby increasing the bioavailability of artemisinin in amphiphilic γ-cyclodextrin nanoparticles (Gerard Yomeogo et al., 2020). As reported by Isacchi et al., intraperitoneal administration of PEGylated liposomes loaded with artemisinin in a healthy animal model was detectable after 24 h in blood circulation compared to 3 h blood-circulation time for liposomes and 1 h for free artemisinin. Compared to free artemisinin, the nanoliposomes of artemisinin showed a six-fold greater area under the curve (AUC). In addition to that, the nanoliposome formulation of artemisinin showed five-fold enhancement in half-life. These enhanced pharmacokinetic properties of the liposomal formulation imply that they can be used successfully in the management of various ailments including malaria (Isacchi et al. 2011). Various derivatives of Artemisinin have also been reported to have better half-life, $IC_{50}$ and kinetics when encapsulated in polymer based, lipid based or metal-based nanoparticles (Alven and Aderibigbe 2020; Wang et al. 2021; Guo et al. 2021).

### 9.5.2 *Silibinin*

Another unique phytochemical **Silibinin** is used as potent hepatoprotective drug and is also finding applications as anticancer agent. According to the Biopharmaceutics Classification System (BCS) Silibinin is classified as class II drug. It has poor dissolution and poor solubility resulting in limited bioavailability (Sahibzada et al. 2017). The $T_{max}$ of silibinin is 2–4 h and the half-life ($t_{1/2}$) is 6 h implying that it is absorbed rapidly when administered orally (Wu et al. 2009). Absorption of silibinin from the gastrointestinal tract was only 20–50% and is further undergoing major biotransformation in the liver resulting in poor bioavailability (Wu et al. 2007). The pharmacokinetics of silibinin showed considerable improvement with encapsulation in nanoparticles. As reported by Sahebzada et al., the bioavailability of silibinin in rabbit when administered with silibinin nanoparticles encapsulated in propylene glycol was 23 mg/mL compared to 3.45 mg/mL for silibinin alone. Compared to free silibinin, the AUC and $C_{max}$ of silibinin nanoformulation produced by antisolvent precipitation with a syringe pump were found to be 15.56 and 6.88 folds greater (Sahibzada et al. 2020). In the preclinical studies, it was found that the silibinin availability to hepatocytes was improved when silibinin bovine serum albumin nanoparticles were injected in mice with induced liver toxicity (Ding et al. 2022).

### *9.5.3 Salvianolic Acid*

A wonder drug, **Salvianolic acid** extracted from radix of Chinese herb *Salvia miltiorrhiza* has immense anti-inflammatory and antioxidant capabilities resulting in protective effect on heart, lungs, kidneys, bones, intestine, liver and skin. It has been extensively researched as an anticancer agent (He et al. 2023). Salvianolic acid is highly hydrophilic and soluble in water. However, the challenge is the stability of salvianolic acid in an aqueous medium and, hence, its sustained release in vivo. This limitation has been addressed by encapsulating salvianolic acid in various nano formulations. For instance, in a study salvianolic acid was loaded on polydopamine nanoparticles and elastin-mimic peptide hydrogel. Salvianolic acid is released at the site of infarction for up to 4 weeks (Chen et al. 2022). In another study, mice with induced breast cancer were treated with PEGylated salvianolic acid liposomes. The slow drug release pattern of salvianolic acid resulted in an improvement in the tumor immunity (Lu et al. 2022). Other studies in rats have also shown that salvianolic acid bioavailability increased about 2.9 folds when encapsulated in phospholipid complex and the $C_{max}$ increased from 0.9 μg/mL to 3.4 μg /mL in (Shahzadi et al. 2020). Nanoparticles of salvianolic acid phospholipid complex can reduce the oral tongue carcinoma growth rate by 5.5% compared to 12.2% tumor growth reduction with free salvianolic acid. (Liu et al. 2022).

### *9.5.4 Curcumin*

The bioactive phytoconstituent of turmeric, **curcumin** has many potent therapeutic properties and they are reported to be effective against inflammation, neurodegeneration and cardiac complications. It also has therapeutic applications in Alzheimer's disease. The therapeutic use of curcumin is hindered due to its poor water solubility, instability and low bioavailability. Poor absorption, high metabolic rate and rapid systemic clearance by conjugation reactions producing sulfated and glucuronidated curcumin are main reasons for the low bioavailability of curcumin (Anand et al. 2007; Mundekkad and Cho 2023). Nanoformulations help overcome these limitations because it increases the stability, solubility and bioavailability of curcumin. A simple nanosuspension of curcumin prepared using poly vinly alcohol showed 4 times increase in $C_{max}$ from 84.6 μg /mL to 410.2 μg /mL in rat plasma (Ravichandran

2013). In another study by Tanvi et al., the lipid coated nanoparticles of curcumin increased its solubility by $1.4 \times 10^6$ times in aqueous medium the solubility of powder curcumin being 11 ng/mL. The study also showed 69.78 times higher oral bioavailability of curcumin coated in lipidic nanoparticles (Gupta et al. 2020). It has been shown that curcumin nanoparticles are effective against cervical cancer and the results showed a 75% increase in tumor growth inhibition. Several *vivo* experiments suggest that anti-tumor activity increases by 50–70% in nanoformulated form compared to curcumin alone (Silvestre et al. 2023). Nanoformulated curcumin has shown dramatic effects not only in cancer but also in cardiovascular diseases, hepatic diseases and as a potent anti diabetic, anti-inflammatory and anti-microbial agent (Chopra et al. 2021). Commercially available curcumin nanoformulations such as Gelucire® and BioCurc® have higher absorption of >400-fold showing highest bioavailability as compared to unformulated curcumin (Stohs et al. 2018).

### *9.5.5 Combretastatin*

There is other plant derived phytoconstituents such as paclitaxel, camptothecin and combretastatin that have been shown to have potential benefits in diseases such as cancer. The nanoformulation of stilbene combretastatin showed several advantages in the nanoformulated form, such as delivery of high drug load, sustained release, reversal of drug resistance, etc. A successful nanoformulation of combrestatin using poly (d,l-lactic-co-glycolic acid) showed showed 77% bioavailability and 41% tumor inhibition, which was significantly higher than free CA-4P in S180 subcutaneous xenotransplant tumor models. The biggest limitation of solubility is overcome by nanoencapsulation of hydrophobic drugs such as combretastatin. As shown in a recent study, the solubility of combretastatin increased from 11.8 to 240 μg/mL in Polyamidoamine (PAMAM) dendrimers (Karatoprak et al. 2020). Sun et al. have also shown that viability of squamous cell carcinoma cells SCC-4 decreased from 90% to 17% in when treated with 40 μM of combretastatin encapsulated in hyaluronic acid (HA-CA4P-NPs), the cytotoxicity effect of free combretastatin was approximately 4 times lower than that of HA-CA4P-NPs (Sun et al. 2024).

### *9.5.6 Paclitaxel*

**Paclitaxel** is commercially available as Taxol® for chemotherapy. One disadvantage of this formulation that has been commonly reported by the formulation researcher is its poor water solubility (0.1–0.32 μg/ml) (Borga et al., 2019). A recent phase I trial with nanoparticles of paclitaxel, Nanotax® provided higher and prolonged paclitaxel levels and reduced toxicity compared to IV paclitaxel administration (Fu et al. 2016). It was reported that paclitaxel loaded PLGA nanoparticles showed a 1.33 folds increase in $C_{max}$ and 4.6 folds higher $t_{1/2}$ in rabbits compared to paclitaxel drug (Sakhi et al. 2022).

### 9.5.7 Camptothecin (CPT)

A naturally occurring alkaloid belonging to the quinolone class is **Camptothecin (CPT)**, which was first isolated from the Chinese medicinal plant Camptotheca acuminata. CPT has potent anticarcinogenic abilities; however, clinical applications are hindered because of its toxicity. Formulation scientists also reported that physicochemical properties are not good enough to get into different drug delivery approaches. The encapsulation of CPT in Poly (methacrylic acid-co-methyl methacrylate) resulted in improved pharmacokinetics. For instance, when tested in rats, the $T_{max}$ increased from 0.5 to 12 h, the $C_{max}$ increased from 135 to 197 ng/mL and the $t_{1/2}$ increased from 0.8 to 3.2 h. There was also a significant difference in *in vitro* cytotoxicity of pure CPT and nano encapsulated CPT on Caco2 cell lines (Manikandan and Kannan 2017). It was found that the nanoparticles sustained diffusion of CPT at pH 5.0 which could be beneficial in treating cancers owing to the acidic tumor environment (Jackson et al. 2023).

### 9.5.8 Quercetin

**Quercetin,** a flavanol, is one of the important bioactive molecules present in numerous fruits and vegetables. Flavonoids in vegetables and fruits are a major source of antioxidants that can provide protection against deadly diseases. Quercetin is a lipophilic molecule that makes its bioavailability a limitation. Additionally, the quick first-pass metabolism adds on to the limitation. The polymer-based nano formulations can overcome these limitations and improve the pharmacokinetics of quercetin. As shown by Ashil et al., there is a major improvement of 62 folds in bioavailability of quercetin in galactomannan hydrogels compared to free quercetin (Joseph et al. 2022). The $C_{max}$ of quercetin improved from 39 μg/mL to 73 μg/mL when encapsulated in an amphiphilic block copolymers of monomethoxy poly(ethylene glycol)–poly(εcaprolactone) (MPEG–PCL) (Xu et al. 2015). Quercetin is also a very potent bioenhancer because of its CYP3A4 inhibition property. Quercetin nanoparticles can enhance the bioavailability of other phytochemicals by several folds. As reported by Unnikrishnan et al., the half-life of an anticancer drug methotrexate (MTX) encapsulated in a nano emulsion with quercetin enhanced to 8.2 h in comparison to 3.6 h of MTX nanoparticles alone (Unnikrishnan Meenakshi et al. 2024a, b).

### 9.5.9 Lycopene

**Lycopene** is another important flavonoid found abundantly in tomatoes composed of several polyene bonds making it a potent antioxidant. The main limitations of lycopene are its unsatisfactory pharmacokinetic characteristics. Encapsulation of lycopene in lipids or other polymers can significantly improve its pharmacokinetics. As shown by Yusuf, lycopene in phosphatidylserine-chitosan coated with polysorbate-80 resulted in 4.6 times increase in $C_{max}$ and a relative increase of 20.4 in bioavailability of lycopene across the blood brain barrier in albino mice (Yusuf 2021). The relative bioavailability of lycopene increases by 297% when encapsulated in alginate beads. The study also inferred that the nanoparticles resulted in sustained release (Alam et al. 2023).

### *9.5.10 Genistein*

**Genistein** is an isoflavanol found abundantly in soybeans, is labelled as a phytoestrogen and is a potent anticancer agent. The limitation of poor solubility and extensive hepatic metabolism restricts its use. It was shown that genistein phytosomes showed improved pharmacokinetics with an increased $C_{max}$ of 0.888 μg/mL compared to 0.22 μg/mL for free genistein (Komeil et al. 2022). Nanometric porous Metal-Organic Frameworks of genistein resulted in a significant 12 folds increase in $C_{max}$ of genistein in mice (Botet-Carreras et al. 2021).

AKBA is isolated from extracts of the gum resin of Boswellia serrata. It is a traditional ayurvedic medicine and is becoming popular owing to its potent anticancer, antioxidant and anti-inflammatory properties. The deterrent in using AKBA is lipophilicity and rapid clearance resulting in low bioavailability. Various methods of encapsulating AKBA to improve its pharmacokinetics have been tried. In one such study, 11.38 times increase in bioavailability was found when AKBA was encapsulated in chitosan nanoparticles (Gaur et al. 2022; Bairwa and Jachak 2015).

### *9.5.11 Catechin, Epigallocatechin-3-Gallate (EGCG) and Epicatechin*

There are several potent anticancer phytochemicals found in beverages such as green tea, black tea and red wine. For example, **catechin, epigallocatechin-3-gallate (EGCG) and epicatechin** from green tea, **caffeic acid** from coffee and red wine and resveratrol from red wine have all shown very potential anticancer activities. For example, the limitation with resveratrol is bioavailability which is less than 1%. Moreover, the $C_{max}$ and $T_{max}$ are 2.4 μM and 1.5 h respectively. However, in numerous nanoformulation approaches, the pharmacokinetics of resveratrol showed significant improvement. Resveratrol polymeric nanoparticle formulation using mPEG-PCL (poly(ε-caprolactone)–poly (ethylene glycol) showed improved $IC_{50}$ values of 47.85 ± 0.96 μg/mL *in vitro* in HT29 cell line. In the *vivo* study, chitosan coated resveratrol nanoparticles showed $C_{max}$ of 9.7 μg/mL in rats (Unnikrishnan Meenakshi et al. 2024a, b). Similarly, different catechins such as EGCG and epicatechin have found a significant role as anticancer, antidiabetic and antioxidant agents. When administered orally, the peak plasma concentration of catechins was found to be suboptimal for therapeutic levels in humans or animals and hence it needs optimization in the formulation aspects to reach the therapeutic level range (1–100 μmol/L). Poor systemic absorption after oral administration, including low absorption, poor pharmacokinetics and bioavailability, scarce biodistribution, first-pass metabolism, trivial penetration and low accumulation in the body's related tissues, or low targeting efficacy, limits the therapeutic potential of catechins. Catechins' vulnerability to oxygen, pH fluctuations, metal ions, and other stressors are addressed via delivery systems based on nanoparticles (Cai et al. 2018). Encapsulation of catechins in various types of lipid polymers increases the stability significantly. As shown by Niyaz et al., in treatment of epilepsy, the bioavailability of chitosan coated PLGA nanopolymer with catechin increased from 5.38 to 7.23 in blood of mice. The bioavailability of the active constituent catechin increased 4 times from 247.26 to 1063.11 in the brain tissue of mice. The release kinetics followed Higuchi model indicating catechin was released through diffusion and swelling methods in a controlled manner (Ahmad et al. 2020a, b).

#### 9.5.11.1 Molecular Signaling Pathways of Bioactive Compounds for Therapeutic Efficacy

As shown in Fig. 9.2 **Arteminisin** nanoparticles exert the anticancer effect by various mechanisms such as induction of ferroptosis, downregulating the metastatic pathways VEGF, Cox-2, MMP-9, upregulating the metastatic pathway Bax (Ma et al. 2021). **Salvianolic acid** exerts its antitumor effect by inhibiting the AKT/PI3K, MAPK, ERK, Keap, NF-κB, JNK pathways resulting in apoptosis and chemokines. The oxidative stress is relieved by increasing the production of super oxide dismutase (SOD) and decreasing lipid peroxidation by downregulating malondialdehyde (MDA) (He et al. 2023). **Silibinin** nanoparticles play an important role in hepatoprotection by upregulating the antioxidant pathway, Nrf2/ARE pathway, which is a main mechanism for cellular resistance to oxidative stress (Ding et al. 2022). Silibinin nanoparticles also provide protective effects by inhibiting STAT3, tyrosine kinase and NF-κB signaling pathways (Durymanov et al. 2020). **Resveratrol** nanoparticles exert anticancer effect by downregulating various apoptotic pathways such MAPK, p53, NF-κB, β-catenin and Cox 2 (Unnikrishnan Meenakshi et al. 2024a, b). Various studies have shown that catechin nanoparticles exert their anticancer effect by downregulating the NF-κB pathway (Jiang et al. 2021). By lowering Bax expression and restoring the redox status via the Nrf2-ERK pathway, Kumar et al. demonstrated that oral administration of green tea **catechins** loaded on bovine serum albumin greatly enhanced the radioprotective properties of green tea in vivo (Kumar et al. 2016). **Curcumin** nanoparticles have been shown to exert their anticancer effect by various mechanisms including the Jak/STAT3 signaling pathway, by downregulating the Wnt pathway, and suppression of the Akt and mTOR proteins, which are involved in cell cycle regulation. Tyrosine kinase receptors

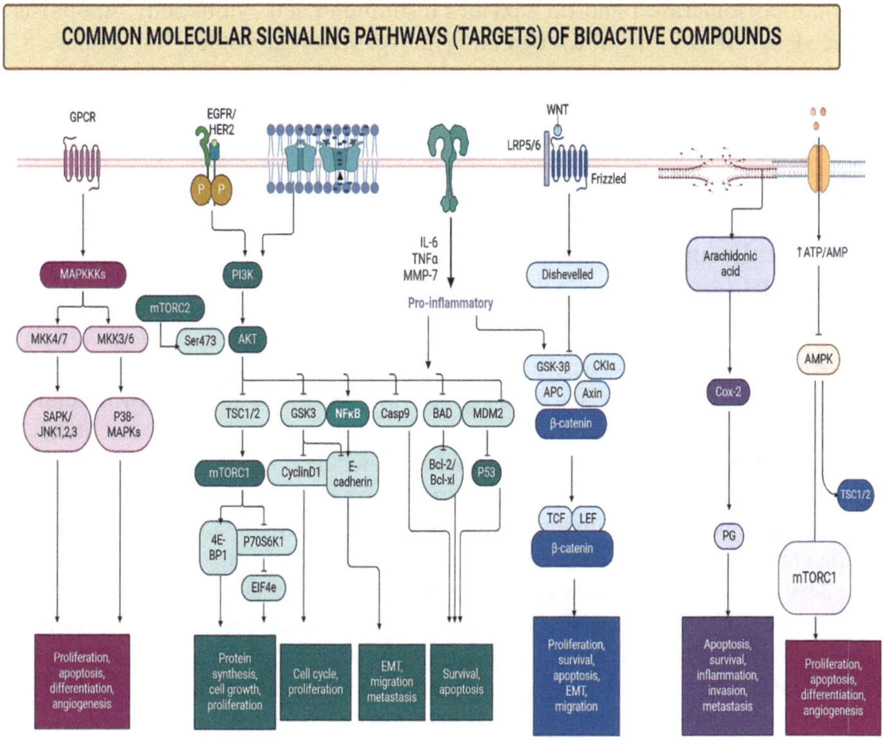


**Fig. 9.2** Different mechanistic pathways targeted by bioactive compounds

regulate a variety of signaling cascade activities involved in cancer, including migration, vascular permeability, survival, and proliferation. It has been discovered that curcumin inhibits nearly every element of the downstream pathway and the tyrosine kinase pathway, including NF-kB, MAPK, PI3K/Akt, and Jak/STAT3 (Mundekkad and Cho 2023). **Combretastain** exerts its anti-inflammatory effect by inhibiting NF-kb, p-AKT, Mcl-1 and Wnt/β-catenin pathways. It is also known to inhibit polymerization of tubulin resulting in apoptosis and thereby acting as an anticancer agent (Karatoprak et al. 2020). Similarly, **paclitaxel** also possesses the ability to attack microtubules resulting in cell cycle arrest and eventually apoptosis (Sharifi-Rad et al. 2021). **Camptothecin** inhibits DNA topoisomerase I, thus interfering with DNA replication resulting in cell death. Camptothecin with a targeted delivery approach using nanotechnology can be a very effective strategy in cancer treatment (Huang et al. 2008). **Quercetin** nanoparticles exert its anti-inflammatory effect by inhibiting the cyclooxygenase (COX) and lipooxygenase (LOX) enzymes. Some studies have also reported that quercetin can inhibit p38, extracellular signal-regulated kinase (ERK) 1/2 mitogen-activated protein kinases (MAPKs), and nuclear factor E2-like 2 (Nrf2) resulting in reduction in glutathione (GSH) production (Tomou et al. 2023; Long et al. 2024). **AKBA** nanoparticles is also known to inhibit topoisomerase I and IIa enzymes resulting in cell cytotoxicity. Apoptosis is induced by AKBA via downregulating the pathway dependent on caspase-8 in human leukemia, hepatoma, colon, and in a wide range of cancer cell lines. Inhibition of the signaling pathway of PI3K/Akt by AKBA nanoparticles is effective in breast cancer cell line (Trivedi et al. 2023).

## 9.6 Challenges in Nanoencapsulation of Bioactive Compounds

Nanoencapsulation of bioactive compounds faces several critical challenges that impact their effectiveness and applicability in industries such as pharmaceuticals, food and cosmetics. Phytoconstituents are complex and sensitive to heat, light, oxygen, and humidity, in contrast to medication molecules, which are distinct chemical entities. They typically have a varied particle size distribution, poor flow, and bulk density. Thus, a thorough understanding of the basic physiochemical characteristics of the various formulation ingredients and excipient types is necessary for the effective development of nano-encapsulated bioactive substances. One significant challenge is the chemical stability of bioactive compounds during the encapsulation process and storage (Puri et al., 2022). Many bioactive agents, such as vitamins and polyphenols, are sensitive to environmental factors, including heat, light, and oxygen, which can lead to degradation and loss of bioactivity. This significantly reduces their bioactive potential (Šeregelj et al. 2020). Hence, a suitable formulation technology/methodology is warranted to prepare a stable formulation that is in demand, and there is a necessity to develop cost-effective technologies (Soukoulis and Bohn 2018). An additional challenge is the selection of appropriate encapsulating materials/carriers. Biodegradable polymers, liposomes, and solid lipid nanoparticles are common choices, but their compatibility with specific bioactive agents can vary widely. Hence, formulating a product that balances stability, compatibility, and bioavailability remains complex. Achieving controlled and sustained release is another hurdle. Many nanoencapsulated formulations do not provide the desired release profiles, resulting in rapid degradation or insufficient delivery of the active ingredients to target sites to maximize their therapeutic efficacy. Lastly, scalability and economic feasibility pose a significant challenge. Overall, overcoming these challenges requires a multidisciplinary approach, integrating insights from materials science, pharmacology, and engineering.

The most widely used approach to addressing formulation challenges is developing isolation protocols for bioactive compounds from sources. Another promising approach is modifying drug delivery systems. The development of a novel drug delivery system with different carriers modifies the properties of various bioactive components positively, especially the pharmacokinetics, as discussed elsewhere in the chapter (Jampilek et al., 2019).

## 9.7 Impact on Pharma Industry

The pharmaceutical industry is fundamental for healthcare units all over the globe, which involves various private and public administrations that invent, discover, progress, produce, and market medications for the sake of animal and human health. Advancements in medicines for protecting the patients in the pandemic condition are carried out within the pharmaceutical industry by medicinal chemists, molecular biologists, and pharmacists, which will result in the increased strength and specificity of drugs. Pharmaceutical firms function in both multinational and national marketplaces which promote the pharmaceutical industry. Hence, the movements in connection to design, formulation, quality control, clinical trials, and marketing should be regulated through legislation and appropriate policies (Spilker 1994). Variation occurs between large scale establishments that involve sets of activities viz. drug discovery and development, production and quality control, sales marketing and smaller-scale establishments which concentrate on a definite objective. Utmost multinational pharmaceutical concerns are involved in all said activities., but, they may focus on a particular feature based upon local market demands.

Legal protections for proprietary medicines i.e. patent rights are maintained by many countries and after the expiry of the same, the pharma companies initiate production and sales of generic medicines. High venture capital is required for the Pharma industry to contribute to manufacturing, quality assurance and control, R&D, legal approval sales. Government rules for marketing the product, made stringent requirements for GMP to guarantee the reliability of pharmaceutical products. Many international regulatory entities Viz. World Health Organization (or WHO), US Food and Drug Administration (or FDA), Medicines and Healthcare Products Regulatory Agency (or MHRA) monitor drug safety (Cole 1990). This legal requirement along with funding, taxes, trading practices affects the pharmaceutical industry growth. Moreover, dissimilarities existing between developed and developing countries, concerning their requirements for medicine also play as an obstacle for the development.

Production of the Pharma industry varies according to region and requirements. For instance, developing countries, are need of vitamins, anti-infective drugs and nutritional supplements while developed countries, demand for disease-oriented medicines. One side is in the necessity to produce bulk drug and another one is dosage form production. Bulk drug production involves biological /natural extraction, fermentation or organic chemical synthesis (Theodore and McGuinn 1992) which may be a pilot plant, huge production or a combination of both. API is combined with pharmaceutical additives based on dosage forms to be designed and desired properties viz. diluents, binders, organoleptic agents, antioxidants, buffers, bases, etc. and processed to final formulation.

The value of the pharmaceutical industry worldwide was assessed $1 trillion in 2014 and generated profits of $980.1 billion during 2013. In 2018, the pharma industry universally raised to $1.2 trillion and expected to be $1.5 trillion by 2023.

Major publicly operated companies include GlaxoSmithKline, Novartis AG, Pfizer Inc. (PFE), Johnson & Johnson,Merck (MRK) and Sanofi, etc. Among all, Pfizer is the largest pharmaceutical company globally with a turnover of $53.7 billion. Numerous FDA approvals were sanctioned for this firm, but now Johnson & Johnson is able to grab Pfizer, which was the most commercial pharma company in 2019, with a turnover of $80bn and above per year. These types of pharmaceutical firms with the biggest revenue spend more on supply chain management. Johnson & Johnson employs around $30 billion, and GlaxoSmithKline expends about $4.5 billion annually for both manufacturing and supplying the medications. Equally, other firms like Pfizer, Teva and Merck allow millions of dollars to warrant the safe supply of their goods. The nanoencapsulation market was valued at USD 7.8 million in 2021 is expected to reach the value of USD 14.76 million by 2029, at a CAGR of 8.3% during the forecast period of 2022–2029′. Nanoencapsulation of bioactive compounds are now followed in the food industry to protect the shelf life of the molecules. The safety of nanomedicines, their commercial success, and patient assurance are the primary factors influencing the market for nanotechnology. More public knowledge of the advantages, dangers, and safety issues associated with nanoformulations is necessary for the future market to overcome this obstacle.

## 9.8 Future Perspectives

Nanoencapsulation of bioactive compounds represents a cutting-edge trend with significant potential in both the food and pharmaceutical sectors, paving the way for new commercialization opportunities. This technology primarily involves creating nanostructured ingredients that enhance the solubility, stability, and texture of products. Additionally, encapsulation effectively improves the bioavailability of bioactive compounds and facilitating controlled release in pharmaceutical applications. It also helps mitigate the volatility of the encapsulated compounds. Looking ahead, the integration of nanotechnology in product development is likely to transform both industries, leading to innovative offerings that cater to evolving consumer demands. Many bioactive substances originating from plants are currently uncharacterized in terms of their toxicity, edibility, and physical-chemical characteristics. This fact suggests that further research is still required to develop novel formulation strategies and optimize the available strategy to improve the stability of bioactive including experiments that evaluate the effectiveness of plant-derived bioactive compounds both in vitro and in vivo.

## 9.9 Conclusion

The exploration of nanoencapsulation techniques presents a promising frontier in the optimization and delivery of bioactive compounds, essential for enhancing their therapeutic efficacy and stability. As we investigate deeper into the diverse methodologies and materials used for encapsulating these pharmaceutically significant biomolecules, it becomes increasingly clear that the interplay between bioactive compounds and their carriers is crucial for achieving targeted and effective drug delivery. Despite the challenges posed by instability and degradation, innovative nanoformulation strategies are paving the way for more reliable therapeutic options, thereby expanding the potential applications of these compounds in modern medicine. This chapter not only highlights successful nanoencapsulation methods but also underscores the necessity for continued research and development in this field. By addressing the existing limitations and leveraging emerging trends, we can significantly improve the pharmacokinetic profiles of bioactive compounds, ultimately enhancing patient outcomes and fostering advancements in biopharmaceuticals.

## References

Abriata JP, Luiz MT, Viegas JSR, Fumagalli F, Melo SMG, Emery F, Marchetti JM, Vicentini FTMdeC (2024) Poly-ε-caprolactone based nanoparticles for delivery of genistein in melanoma treatment. Braz J Pharm Sci 60. https://doi.org/10.1590/s2175-97902024e22542

Acevedo-Fani A, Singh H (2022) Biophysical insights into modulating lipid digestion in food emulsions. Prog Lipid Res 85(101129):101129. https://doi.org/10.1016/j.plipres.2021.101129

Adhipandito CF, Cheung S-H, Lin Y-H, Wu S-H (2021) Atypical renal clearance of nanoparticles larger than the kidney filtration threshold. Int J Mol Sci 22(20):11182. https://doi.org/10.3390/ijms222011182

Ahmad N, Ahmad R, Alrasheed RA, Almatar HMA, Al-Ramadan AS, Amir M, Sarafroz M (2020a) Quantification and evaluations of catechin hydrate polymeric nanoparticles used in brain targeting for the treatment of epilepsy. Pharmaceutics 12(3):203. https://doi.org/10.3390/pharmaceutics12030203

Ahmad N, Ahmad R, Alrasheed RA, Almatar HMA, Al-Ramadan AS, Buheazah TM, AlHomoud HS, Al-Nasif HA, Alam MA (2020b) A Chitosan-PLGA based catechin hydrate nanoparticles used in targeting of lungs and cancer treatment. Saudi J Biolog Sci 27(9):2344–2357

Ahuja A, Narde GK, Wadi NMA, Meenakshi DU (2021) Drug targeting approaches and use of drug delivery systems in management of cancer. Curr Pharm Des 27(45):4593–4609. https://doi.org/10.2174/1381612827666210728105017

Alam MAU, Khatun M, Alam MAU (2023) Recent trend of nanotechnology applications to improve bio-accessibility of lycopene by nanocarrier: a review. Retrieved from http://arxiv.org/abs/2301.10397

Alven S, Aderibigbe BA (2020) Nanoparticles formulations of artemisinin and derivatives as potential therapeutics for the treatment of cancer, leishmaniasis and malaria. Pharmaceutics 12(8):748. https://doi.org/10.3390/pharmaceutics12080748

Anand P, Kunnumakkara AB, Newman RA, Aggarwal BB (2007) Bioavailability of curcumin: problems and promises. Mol Pharm 4(6):807–818. https://doi.org/10.1021/mp700113r

Asma ST, Acaroz U, Imre K, Morar A, Shah SRA, Hussain SZ et al (2022) Natural products/bioactive compounds as a source of anticancer drugs. Cancer 14(24):6203. https://doi.org/10.3390/cancers14246203

Bairwa K, Jachak SM (2015) Development and optimisation of 3-Acetyl-11-keto-β-boswellic acid loaded poly-lactic-co-glycolic acid-nanoparticles with enhanced oral bioavailability and in-vivo anti-inflammatory activity in rats. J Pharm Pharmacol 67(9):1188–1197. https://doi.org/10.1111/jphp.12420

Botet-Carreras A, Tamames-Tabar C, Salles F, Rojas S, Imbuluzqueta E, Lana H, Blanco-Prieto MJ, Horcajada P (2021) Improving the genistein oral bioavailability *via* its formulation into the metal-organic framework MIL-100(Fe). J Mater Chem B 9(9):2233–2239

Cai Z-Y, Li X-M, Liang J-P, Xiang L-P, Wang K-R, Shi Y-L et al (2018) Bioavailability of tea catechins and its improvement. Molecules (Basel, Switzerland) 23(9):2346. https://doi.org/10.3390/molecules23092346

Calosi M, Guazzelli E, Braccini S, Lessi M, Bellina F, Galli G, Martinelli E (2022) Self-assembled amphiphilic fluorinated random copolymers for the encapsulation and release of the hydrophobic Combretastatin A-4 drug. Polymers 14(4):774. https://doi.org/10.3390/polym14040774

Chen L, Ru C, Zhang H, Zhang Y, Wang H, Hu X, Li G (2022) Progress in electrohydrodynamic atomization preparation of energetic materials with controlled microstructures. Molecules (Basel, Switzerland) 27(7):2374. https://doi.org/10.3390/molecules27072374

Chopra H, Dey PS, Das D, Bhattacharya T, Shah M, Mubin S et al (2021) Curcumin nanoparticles as promising therapeutic agents for drug targets. Molecules (Basel, Switzerland) 26(16):4998. https://doi.org/10.3390/molecules26164998

Chowdhury S, Kar K, Mazumder R (2024) Exploration of different strategies of nanoencapsulation of bioactive compounds and their ensuing approaches. Future J Pharm Sci 10(1). https://doi.org/10.1186/s43094-024-00644-y

Cole G (1990) Pharmaceutical production facilities: design and applications. Ellis Horwood Ltd, Chichester, West Sussex

Costa JR, Xavier M, Amado IR, Gonçalves C, Castro PM, Tonon RV et al (2021) Polymeric nanoparticles as oral delivery systems for a grape pomace extract towards the improvement of biological activities. Mater Sci Eng C Mater Biol Appl 119(111551):111551. https://doi.org/10.1016/j.msec.2020.111551

Dahiya S (2017) Drug nanonization: an overview of industrially feasible top-down technologies for nanocrystal production. Bull Pharm Res 7(2). https://doi.org/10.21276/bpr.2017.7.2.2

Dahiya D, Terpou A, Dasenaki M, Nigam PS (2023) Current status and future prospects of bioactive molecules delivered through sustainable encapsulation techniques for food fortification. Sustain Food Technol 1(4):500–510. https://doi.org/10.1039/d3fb00015j

Devalapally H, Chakilam A, Amiji MM (2007) Role of nanotechnology in pharmaceutical product development. J Pharm Sci 96(10):2547–2565. https://doi.org/10.1002/jps.20875

Ding Y, Zhang S, Sun Z, Tong Z, Ge Y, Zhou L et al (2022) Preclinical validation of silibinin/albumin nanoparticles as an applicable system against acute liver injury. Acta Biomater 146:385–395. https://doi.org/10.1016/j.actbio.2022.04.021

Durymanov M, Permyakova A, Reineke J (2020) Pre-treatment with PLGA/silibinin nanoparticles mitigates dacarbazine-induced hepatotoxicity. Front Bioeng Biotechnol 8:495. https://doi.org/10.3389/fbioe.2020.00495

Ehehalt R, Sparla R, Kulaksiz H, Herrmann T, Füllekrug J, Stremmel W (2008) Uptake of long chain fatty acids is regulated by dynamic interaction of FAT/CD36 with cholesterol/sphingolipid enriched microdomains (lipid rafts). BMC Cell Biol 9(1):45. https://doi.org/10.1186/1471-2121-9-45

Focsan AL, Polyakov NE, Kispert LD (2019) Supramolecular carotenoid complexes of enhanced solubility and stability-the way of bioavailability improvement. Molecules (Basel, Switzerland) 24(21):3947. https://doi.org/10.3390/molecules24213947

Fontana F, Figueiredo P, Zhang P, Hirvonen JT, Liu D, Santos HA (2018) Production of pure drug nanocrystals and nano co-crystals by confinement methods. Adv Drug Deliv Rev 131:3–21. https://doi.org/10.1016/j.addr.2018.05.002

Fu Q, Hargrove D, Lu X (2016) Improving paclitaxel pharmacokinetics by using tumor-specific mesoporous silica nanoparticles with intraperitoneal delivery. Nanomedicine 12(7):1951–1959. https://doi.org/10.1016/j.nano.2016.04.013

Gadapa S, Battula SN, Mor S, Pushpadass HA, Naik LN, Emerald ME (2022) Green tea catechin loaded niosomes: formulation and their characterization for food fortification. J Food Sci Technol 59(9):3669–3682. https://doi.org/10.1007/s13197-022-05384-6

Gaur PK, Puri D, Singh AP, Kumar N, Rastogi S (2022) Optimization and pharmacokinetic study of boswellic acid–loaded chitosan-guggul gum nanoparticles using box-Behnken experimental design. J Pharm Innov 17(2):485–500. https://doi.org/10.1007/s12247-020-09527-0
Gehrcke M, de Bastos Brum T, da Rosa LS, Ilha BD, Soares FZM, Cruz L (2021) Incorporation of nanocapsules into gellan gum films: a strategy to improve the stability and prolong the cutaneous release of silibinin. Mater Sci Eng C Mater Biol Appl 119(111624):111624. https://doi.org/10.1016/j.msec.2020.111624
Gérard Yaméogo JB, Mazet R, Wouessidjewe D, Choisnard L, Godin-Ribuot D, Putaux J-L et al (2020) Pharmacokinetic study of intravenously administered artemisinin-loaded surface-decorated amphiphilic γ-cyclodextrin nanoparticles. Mater Sci Eng C Mater Biol Appl 106(110281):110281. https://doi.org/10.1016/j.msec.2019.110281
Gessica Sampaio LA, Sidney P, Ana Paula OR, Melicia CG, Flávia SG, Renata VT (2019) Encapsulation of a lycopene-rich watermelon concentrate in alginate and pectin beads: Characterization and stability. LWT 116:108589. https://doi.org/10.1016/j.lwt.2019.108589
Grgić J, Šelo G, Planinić M, Tišma M, Bucić-Kojić A (2020) Role of the encapsulation in bioavailability of phenolic compounds. Antioxidants (Basel, Switzerland) 9(10):923. https://doi.org/10.3390/antiox9100923
Grumbach C, Czermak P (2022) Process analytical technology for the production of parenteral lipid emulsions according to good manufacturing practices. Processes 10(6):1174. https://doi.org/10.3390/pr10061174
Guo W, Li N, Ren G, Wang R, Chai L, Li Y et al (2021) Murine pharmacokinetics and antimalarial pharmacodynamics of dihydroartemisinin trimer self-assembled nanoparticles. Parasitol Res 120(8):2827–2837. https://doi.org/10.1007/s00436-021-07208-6
Gupta T, Singh J, Kaur S, Sandhu S, Singh G, Kaur IP (2020) Enhancing bioavailability and stability of curcumin using solid lipid nanoparticles (CLEN): a covenant for its effectiveness. Front Bioeng Biotechnol 8:879. https://doi.org/10.3389/fbioe.2020.00879
Haripriyaa M, Suthindhiran K (2023) Pharmacokinetics of nanoparticles: current knowledge, future directions and its implications in drug delivery. Future J Pharm Sci 9(1). https://doi.org/10.1186/s43094-023-00569-y
He G, Chen G, Liu W, Ye D, Liu X, Liang X, Song J (2023) Salvianolic acid B: a review of pharmacological effects, safety, combination therapy, new dosage forms, and novel drug delivery routes. Pharmaceutics 15(9). https://doi.org/10.3390/pharmaceutics15092235
Hu Q, Zhou F, Ly NK, Ordyna J, Peterson T, Fan Z, Wang S (2023) Development of multifunctional Nanoencapsulated *trans*-resveratrol/chitosan nutraceutical edible coating for strawberry preservation. ACS Nano 17(9):8586–8597
Huang Z-R, Hua S-C, Yang Y-L, Fang J-Y (2008) Development and evaluation of lipid nanoparticles for camptothecin delivery: a comparison of solid lipid nanoparticles, nanostructured lipid carriers, and lipid emulsion. Acta Pharmacol Sin 29(9):1094–1102. https://doi.org/10.1111/j.1745-7254.2008.00829.x
Iqbal J, Hussain MM (2009) Intestinal lipid absorption. Am J Physiol Endocrinol Metab 296(6):E1183–E1194. https://doi.org/10.1152/ajpendo.90899.2008
Iqbal FM, Rodríguez-Nogales C, Boulens N, Delie F (2024) Formulation and optimization of transferrin-modified genistein nanocrystals: in vitro anti-cancer assessment and pharmacokinetic evaluation. Int J Pharm 124863. https://doi.org/10.1016/j.ijpharm.2024.124863
Isacchi B, Arrigucci S, la Marca G, Bergonzi MC, Vannucchi MG, Novelli A, Bilia AR (2011) Conventional and long-circulating liposomes of artemisinin: preparation, characterization, and pharmacokinetic profile in mice. J Liposome Res 21(3):237–244. https://doi.org/10.3109/08982104.2010.539185
Jackson N, Ortiz AC, Jerez A, Morales J, Arriagada F (2023) Kinetics and mechanism of camptothecin release from transferrin-gated mesoporous silica nanoparticles through a pH-responsive surface linker. Pharmaceutics 15(6). https://doi.org/10.3390/pharmaceutics15061590

Jafari SM (2017) An overview of nanoencapsulation techniques and their classification. In nanoencapsulation technologies for the food and nutraceutical industries. Academic Press. Elsevier Inc 1–34. https://doi.org/10.1016/B978-0-12-809436-5.00001-X

Jampilek J, Kos J, Kralova K (2019) Potential of nanomaterial applications in dietary supplements and foods for special medical purposes. Nanomaterials 9(2):296. https://doi.org/10.3390/nano9020296

Jiang Y, Jiang Z, Ma L, Huang Q (2021) Advances in nanodelivery of green tea catechins to enhance the anticancer activity. Molecules (Basel, Switzerland) 26(11):3301. https://doi.org/10.3390/molecules26113301

Jin J, Liu C, Tong H, Sun Y, Huang M, Ren G, Xie H (2022) Encapsulation of EGCG by Zein-gum Arabic complex nanoparticles and in vitro simulated digestion of complex nanoparticles. Foods (Basel, Switzerland) 11(14):2131. https://doi.org/10.3390/foods11142131

Joseph A, Shanmughan P, Balakrishnan A, Maliakel B, Krishnakumar IM (2022) Enhanced bioavailability and pharmacokinetics of a natural self-emulsifying reversible hybrid-hydrogel system of quercetin: a randomized double-blinded comparative crossover study. ACS Omega 7(50):46825–46832. https://doi.org/10.1021/acsomega.2c05929

Joshi K, Chandra A, Jain K, Talegaonkar S (2019) Nanocrystalization: an emerging technology to enhance the bioavailability of poorly soluble drugs. Pharma Nanotechnol 7(4):259–278. https://doi.org/10.2174/2211738507666190405182524

Kang MJ, Kim D-Y, Baek Y, Lee HG (2023) Enhancement of antioxidant activities and stabilities of quercetin and isoquercetin through the combination of nanoencapsulation and hydrogel incorporation. Food Biosci 55(102963):102963. https://doi.org/10.1016/j.fbio.2023.102963

Karatoprak GŞ, Küpeli Akkol E, Genç Y, Bardakci H, Yücel Ç, Sobarzo-Sánchez E (2020) Combretastatins: an overview of structure, probable mechanisms of action and potential applications. Molecules (Basel, Switzerland) 25(11):2560. https://doi.org/10.3390/molecules25112560

Kiani J, Imam SZ (2007) Medicinal importance of grapefruit juice and its interaction with various drugs. Nutr J 6(1):33. https://doi.org/10.1186/1475-2891-6-33

Komeil IA, Abdallah OY, El-Refaie WM (2022) Surface modified genistein phytosome for breast cancer treatment: in-vitro appraisal, pharmacokinetics, and in-vivo antitumor efficacy. Eur J Pharm Sci 179(106297):106297. https://doi.org/10.1016/j.ejps.2022.106297

Kumar S, Meena R, Rajamani P (2016) Fabrication of BSA-green tea polyphenols-chitosan nanoparticles and their role in radioprotection: a molecular and biochemical approach. J Agric Food Chem 64(30):6024–6034. https://doi.org/10.1021/acs.jafc.6b02068

Lee Y, Park SM, Song IH, Kim BS, Park HS, Moon BS, Kim H-H (2024) Pharmacokinetic evaluation of paclitaxel, albumin-binding paclitaxel, and liposomal-encapsulated albumin-binding paclitaxel upon gastric subserosal administration. Front Pharmacol 15:1381406. https://doi.org/10.3389/fphar.2024.1381406

Li D, Liu D, Wang Y, Sun Q, Sun R, Zhang J et al (2023) Multifunctional liposomes co-encapsulating epigallocatechin-3-gallate (EGCG) and miRNA for atherosclerosis lesion elimination. Nanoscale Adv 6(1):221–232. https://doi.org/10.1039/d3na00369h

Liga S, Paul C, Péter F (2023) Flavonoids: overview of biosynthesis, biological activity, and current extraction techniques. Plants 12(14). https://doi.org/10.3390/plants12142732

Lim Y-P, Ma C-Y, Liu C-L, Lin Y-H, Hu M-L, Chen J-J et al (2012) Sesamin: a naturally occurring lignan inhibits CYP3A4 by antagonizing the pregnane X receptor activation. Evid Based Complement Alternat Med 2012:242810. https://doi.org/10.1155/2012/242810

Liu Z, Hu M (2007) Natural polyphenol disposition via coupled metabolic pathways. Expert Opin Drug Metab Toxicol 3(3):389–406. https://doi.org/10.1517/17425255.3.3.389

Liu G, Li J, Deng S (2021) Applications of supercritical anti-solvent process in preparation of solid multicomponent systems. Pharmaceutics 13(4):475. https://doi.org/10.3390/pharmaceutics13040475

Liu W, Zhou Z, Zhu L, Li H, Wu L (2022) Chemopreventive efficacy of salvianolic acid B phospholipid complex loaded nanoparticles against experimental oral carcinogenesis: implication of sustained drug release. Ann Transl Med 10(5):244. https://doi.org/10.21037/atm-21-4457

Loh ZH, Samanta AK, Sia Heng PW (2015) Overview of milling techniques for improving the solubility of poorly water-soluble drugs. Asian J Pharm Sci 10(4):255–274. https://doi.org/10.1016/j.ajps.2014.12.006
Long Y, Paengkoum S, Lu S, Niu X, Thongpea S, Taethaisong N et al (2024) Physicochemical properties, mechanism of action of lycopene and its application in poultry and ruminant production. Front Veterin Sci 11:1364589. https://doi.org/10.3389/fvets.2024.1364589
Lu P, Li J, Liu C, Yang J, Peng H, Xue Z, Liu Z (2022) Salvianolic acid B dry powder inhaler for the treatment of idiopathic pulmonary fibrosis. Asian J Pharm Sci 17(3):447–461. https://doi.org/10.1016/j.ajps.2022.04.004
Luo Y (2020) Perspectives on important considerations in designing nanoparticles for oral delivery applications in food. J Agric Food Res 2(100031):100031. https://doi.org/10.1016/j.jafr.2020.100031
Ma Z, Woon CY-N, Liu C-G, Cheng J-T, You M, Sethi G et al (2021) Repurposing artemisinin and its derivatives as anticancer drugs: a chance or challenge? Front Pharmacol 12:828856. https://doi.org/10.3389/fphar.2021.828856
Mabrouk M, Das DB, Salem ZA, Beherei HH (2021) Nanomaterials for biomedical applications: production, characterisations, recent trends and difficulties. Molecules (Basel, Switzerland) 26(4):1077. https://doi.org/10.3390/molecules26041077
Manikandan M, Kannan K (2017) Pharmacokinetic and Pharmacodynamic evaluation of Camptothecin encapsulated poly (methacylic acid-co-methyl methacrylate) nanoparticles. J Appl Pharm Sci. https://doi.org/10.7324/japs.2017.70303
Marchiori MCL, Rigon C, da S Jardim F, Giuliani LM, Copetti PM, Sagrillo MR et al (2023) Hydrogel containing silibinin-loaded pomegranate oil-based nanocapsules for cutaneous application: in vitro safety investigation and human skin biometry and permeation studies. AAPS PharmSciTech 24(6):138. https://doi.org/10.1208/s12249-023-02606-8
Mirahmadi M, Kamali H, Azimi-Hashemi S, Lavaee P, Gharaei S, Sherkatsadi K et al (2024) Evaluation of novel carriers for enhanced dissolution of lycopene. J Food Measur Character 18(6):4718–4732. https://doi.org/10.1007/s11694-024-02526-6
Mundekkad D, Cho WC (2023) Applications of curcumin and its nanoforms in the treatment of cancer. Pharmaceutics 15(9). https://doi.org/10.3390/pharmaceutics15092223
Nalini T, Khaleel Basha S, Mohamed Sadiq A, Sugantha Kumari V (2023) Fabrication and evaluation of nanoencapsulated quercetin for wound healing application. Polymer Bull (Berlin, Germany) 80(1):515–540. https://doi.org/10.1007/s00289-022-04094-5
Natesan S, Ponnusamy C, Sugumaran A, Chelladurai S, Shanmugam Palaniappan S, Palanichamy R (2017) Artemisinin loaded chitosan magnetic nanoparticles for the efficient targeting to the breast cancer. Int J Biol Macromol 104:1853–1859. https://doi.org/10.1016/j.ijbiomac.2017.03.137
Nicolescu A, Babotă M, Barros L, Rocchetti G, Lucini L, Tanase C et al (2023) Bioaccessibility and bioactive potential of different phytochemical classes from nutraceuticals and functional foods. Front Nutr 10:1184535. https://doi.org/10.3389/fnut.2023.1184535
Nie F, Liu L, Cui J, Zhao Y, Zhang D, Zhou D et al (2023) Oligomeric Proanthocyanidins: an updated review of their natural sources, synthesis, and potentials. Antioxidants (Basel, Switzerland) 12(5). https://doi.org/10.3390/antiox12051004
Nogueira GF, de Oliveira RA, Velasco JI, Fakhouri FM (2020) Methods of incorporating plant-derived bioactive compounds into films made with agro-based polymers for application as food packaging: a brief review. Polymers 12(11):2518. https://doi.org/10.3390/polym12112518
Odera PA, Otieno G, Onyango JO, Owuor JJ, Oloo FA, Ongas M et al (2024) NANOPARTICLE-BASED formulation of dihydroartemisinin-lumefantrine duo-drugs: preclinical evaluation and enhanced antimalarial efficacy in a mouse model. Heliyon 10(6):e26868. https://doi.org/10.1016/j.heliyon.2024.e26868
Panda BP, Krishnamoorthy R, Bhattamisra SK, Shivashekaregowda NKH, Seng LB, Patnaik S (2019) Fabrication of second generation smarter PLGA based nanocrystal carriers for improvement of drug delivery and therapeutic efficacy of gliclazide in type-2 diabetes rat model. Sci Rep 9(1):17331. https://doi.org/10.1038/s41598-019-53996-4

Pateiro M, Gómez B, Munekata PES, Barba FJ, Putnik P, Kovačević DB, Lorenzo JM (2021) Nanoencapsulation of promising bioactive compounds to improve their absorption, stability, functionality and the appearance of the final food products. Molecules (Basel, Switzerland) 26(6):1547. https://doi.org/10.3390/molecules26061547

Peng S, Li Z, Zou L, Liu W, Liu C, McClements DJ (2018) Improving curcumin solubility and bioavailability by encapsulation in saponin-coated curcumin nanoparticles prepared using a simple pH-driven loading method. Food Funct 9(3):1829–1839. https://doi.org/10.1039/c7fo01814b

Pimentel LF, Jacome Junior AV, Furtado Mosqueira VC, Santos-Magalhães NS (2007) Application of pharmaceutical nanotechnology to the treatment of malaria. Rev Bras Farm Sci 43(4). https://doi.org/10.1590/S1516-93322007000400003

Plamada D, Vodnar DC (2021) Polyphenols-gut microbiota interrelationship: a transition to a new generation of prebiotics. Nutrients 14(1):137. https://doi.org/10.3390/nu14010137

Puri V, Nagpal M, Singh I, Singh M, Dhingra GA, Huanbutta K, Dheer D, Sharma A, Sangnim TA (2022) Comprehensive Review on Nutraceuticals: Therapy Support and Formulation Challenges. Nutrients 14(21):4637. https://doi.org/10.3390/nu14214637

Qiu C, Zhang Z, Li X, Sang S, McClements DJ, Chen L et al (2023) Co-encapsulation of curcumin and quercetin with zein/HP-β-CD conjugates to enhance environmental resistance and antioxidant activity. Npj Sci Food 7(1):29

Rashidinejad A, Singh H, Marze S, Rashidinejad A, Singh H, Digestion SML, Lipid-Soluble B (2022) Lipid digestion and bioaccessibility of lipid-soluble compounds to cite this version. HAL Id

Rathore C, Hemrajani C, Sharma AK, Gupta PK, Jha NK, Aljabali AAA, Gupta G, Singh SK, Yang JC, Dwivedi RP, Dua K, Chellappan DK, Negi P, Tambuwala MM (2023) Self-nanoemulsifying drug delivery system (SNEDDS) mediated improved oral bioavailability of thymoquinone: optimization, characterization, pharmacokinetic, and hepatotoxicity studies. Drug Deliv Transl Res 13(1):292–307. https://doi.org/10.1007/s13346-022-01193-8

Ravi H, Kurrey N, Manabe Y, Sugawara T, Baskaran V (2018) Polymeric chitosan-glycolipid nanocarriers for an effective delivery of marine carotenoid fucoxanthin for induction of apoptosis in human colon cancer cells (Caco-2 cells). Mater Sci Eng C Mater Biol Appl 91:785–795. https://doi.org/10.1016/j.msec.2018.06.018

Ravichandran R (2013) Pharmacokinetic study of nanoparticulate curcumin: oral formulation for enhanced bioavailability. J Biomater Nanobiotechnol 04(03):291–299. https://doi.org/10.4236/jbnb.2013.43037

Rein MJ, Renouf M, Cruz-Hernandez C, Actis-Goretta L, Thakkar SK, da Silva Pinto M (2013) Bioavailability of bioactive food compounds: a challenging journey to bioefficacy. Br J Clin Pharmacol 75(3):588–602. https://doi.org/10.1111/j.1365-2125.2012.04425.x

Sahibzada MUK, Sadiq A, Khan S, Faidah HS, Ullah N, Khurram M et al (2017) Fabrication, characterization and in vitro evaluation of silibinin nanoparticles: an attempt to enhance its oral bioavailability. Drug Des Devel Ther 11:1453–1464. https://doi.org/10.2147/dddt.s133806

Sahibzada MUK, Sadiq A, Zahoor M, Naz S, Shahid M, Qureshi NA (2020) Enhancement of bioavailability and hepatoprotection by silibinin through conversion to nanoparticles prepared by liquid antisolvent method. Arab J Chem 13(2):3682–3689. https://doi.org/10.1016/j.arabjc.2020.01.002

Sakhi M, Khan A, Iqbal Z, Khan I, Raza A, Ullah A et al (2022) Design and characterization of paclitaxel-loaded polymeric nanoparticles decorated with trastuzumab for the effective treatment of breast cancer. Front Pharmacol 13:855294. https://doi.org/10.3389/fphar.2022.855294

Salazar J, Müller RH, Möschwitzer JP (2014) Combinative particle size reduction technologies for the production of drug nanocrystals. J Pharm 2014:265754. https://doi.org/10.1155/2014/265754

Sandhya M, Ramasamy D, Sudhakar K, Kadirgama K, Harun WSW (2021) Ultrasonication an intensifying tool for preparation of stable nanofluids and study the time influence on distinct properties of graphene nanofluids – a systematic overview. Ultrason Sonochem 73(105479):105479. https://doi.org/10.1016/j.ultsonch.2021.105479

Šeregelj V, Ćetković G, Čanadanović-Brunet J, Tumbas Šaponjac V, Vulić J, Stajčić S (2020) Encapsulation and degradation kinetics of bioactive compounds from sweet potato peel during storage. Food Technol Biotechnol 58(3):314–324. https://doi.org/10.17113/ftb.58.03.20.6557

Shah A, Ashraf Z, Gani A, Jhan F, Gani A, Sidiq M (2022) Encapsulation of Catechin into β-Glucan Matrix Using Wet Milling and Ultrasonication as a Coupled Approach: Characterization and Bioactivity Retention. Foods 11(10):1493. https://doi.org/10.3390/foods11101493

Shahzadi I, Ali Z, Bukhari S, Narula AS, Mirza B, Mohammadinejad R (2020) Possible applications of salvianolic acid B against different cancers. Explor Targeted Anti-Tumor Ther 1(4):218–238. https://doi.org/10.37349/etat.2020.00014

Sharifi-Rad J, Quispe C, Patra JK, Singh YD, Panda MK, Das G et al (2021) Paclitaxel: application in modern oncology and nanomedicine-based cancer therapy. Oxidative Med Cell Longev 2021(1):3687700. https://doi.org/10.1155/2021/3687700

Silvestre F, Santos C, Silva V, Ombredane A, Pinheiro W, Andrade L et al (2023) Pharmacokinetics of curcumin delivered by nanoparticles and the relationship with antitumor efficacy: a systematic review. Pharmaceuticals (Basel, Switzerland) 16(7). https://doi.org/10.3390/ph16070943

Soukoulis C, Bohn T (2018) A comprehensive overview on the micro-and nano-technological encapsulation advances for enhancing the chemical stability and bioavailability of carotenoids. Crit Rev Food Sci Nutr 58:1–36

Spilker B (1994) Multinational pharmaceutical companies: principles and practices. Raven Press, New York

Stanciauskaite M, Poskute M, Kurapkiene V, Marksa M, Jakstas V, Ivanauskas L, Ramanauskiene K (2023) Optimization of delivery and bioavailability of encapsulated caffeic acid. Foods (Basel, Switzerland) 12(10). https://doi.org/10.3390/foods12101993

Stohs SJ, Ji J, Bucci LR, Preuss HG (2018) A comparative pharmacokinetic assessment of a novel highly bioavailable curcumin formulation with 95% curcumin: a randomized, double-blind, crossover study. J Am Coll Nutr 37(1):51–59. https://doi.org/10.1080/07315724.2017.1358118

Suk JS, Xu Q, Kim N, Hanes J, Ensign LM (2016) PEGylation as a strategy for improving nanoparticle-based drug and gene delivery. Adv Drug Deliv Rev 99(Pt A):28–51. https://doi.org/10.1016/j.addr.2015.09.012

Sun C, Zhou Z, Liu F, Li H, Liu Z (2024) Combretastatin A4 phosphate encapsulated in hyaluronic acid nanoparticles is highly cytotoxic to oral squamous cell carcinoma. Arch Med Sci 20(3):1022–1028. https://doi.org/10.5114/aoms/189535

Tao J, Chow SF, Zheng Y (2019) Application of flash nanoprecipitation to fabricate poorly water-soluble drug nanoparticles. Acta Pharm Sin B 9(1):4–18. https://doi.org/10.1016/j.apsb.2018.11.001

Tehrani AA, Omranpoor MM, Vatanara A, Seyedabadi M, Ramezani V (2019) Formation of nanosuspensions in bottom-up approach: theories and optimization. Daru 27(1):451–473. https://doi.org/10.1007/s40199-018-00235-2

Theodore L, Mcguinn YC (1992) Pollution prevention. Van Nostrand Reinhold, New York

Tomou E-M, Papakyriakopoulou P, Saitani E-M, Valsami G, Pippa N, Skaltsa H (2023) Recent advances in nanoformulations for quercetin delivery. Pharmaceutics 15(6). https://doi.org/10.3390/pharmaceutics15061656

Trivedi VL, Soni R, Dhyani P, Sati P, Tejada S, Sureda A et al (2023) Anti-cancer properties of boswellic acids: mechanism of action as anti-cancerous agent. Front Pharmacol 14:1187181. https://doi.org/10.3389/fphar.2023.1187181

Unnikrishnan Meenakshi D, Narde GK, Ahuja A, Akhtar MJ, Khan SA (2024a) Role of natural phytoconstituents as a potential bioenhancer of anti-cancer and anti-microbial agents: spotlight on the mechanism of action, clinical studies and patents. Processes (Basel, Switzerland) 12(10):2060. https://doi.org/10.3390/pr12102060

Unnikrishnan Meenakshi D, Narde GK, Ahuja A, Al Balushi K, Francis AP, Khan SA (2024b) Therapeutic applications of nanoformulated resveratrol and quercetin phytochemicals in colorectal cancer-an updated review. Pharmaceutics 16(6):761. https://doi.org/10.3390/pharmaceutics16060761

Vamanu E, Gatea F (2020) Correlations between microbiota bioactivity and bioavailability of functional compounds: a mini-review. Biomedicine 8(2):39. https://doi.org/10.3390/biomedicines8020039

Vaou N, Stavropoulou E, Voidarou CC, Tsakris Z, Rozos G, Tsigalou C, Bezirtzoglou E (2022) Interactions between medical plant-derived bioactive compounds: focus on antimicrobial combination effects. Antibiotics (Basel, Switzerland) 11(8):1014. https://doi.org/10.3390/antibiotics11081014

Wang R, Ren G, Chai L, Guo W, Li Y, Wang R et al (2021) Pharmacokinetics and antimalarial activities of reduction-responsive releasing dihydroartemisinin prodrug self-assembled nanoparticles in rodents. J Drug Deliv Sci Technol 63(102515):102515. https://doi.org/10.1016/j.jddst.2021.102515

Wilson BK, Sinko PJ, Prud'homme RK (2021) Encapsulation and controlled release of a camptothecin prodrug from nanocarriers and microgels: tuning release rate with nanocarrier excipient composition. Mol Pharm 18(3):1093–1101. https://doi.org/10.1021/acs.molpharmaceut.0c01012

Wu J-W, Lin L-C, Hung S-C, Chi C-W, Tsai T-H (2007) Analysis of silibinin in rat plasma and bile for hepatobiliary excretion and oral bioavailability application. J Pharm Biomed Anal 45(4):635–641. https://doi.org/10.1016/j.jpba.2007.06.026

Wu J-W, Lin L-C, Tsai T-H (2009) Drug-drug interactions of silymarin on the perspective of pharmacokinetics. J Ethnopharmacol 121(2):185–193. https://doi.org/10.1016/j.jep.2008.10.036

Wu M, Feng Z, Deng Y, Zhong C, Liu Y, Liu J, Fu Y (2019) Liquid antisolvent precipitation: an effective method for ocular targeting of lutein esters. Int J Nanomedicine 14:2667–2681. https://doi.org/10.2147/IJN.S194068

Xu Q, Ensign LM, Boylan NJ, Schön A, Gong X, Yang J-C, Hanes J (2015) Impact of surface polyethylene glycol (PEG) density on biodegradable nanoparticle transport in mucus ex vivo and distribution in vivo. ACS Nano 9(9):9217–9227. https://doi.org/10.1021/acsnano.5b03876

Yadav R, Kumar D, Kumari A, Yadav SK (2014) Encapsulation of catechin and epicatechin on BSA NPS improved their stability and antioxidant potential. EXCLI J 13:331–346. Retrieved from https://www.ncbi.nlm.nih.gov/pubmed/26417264

Yaghmur A, Tran BV, Moghimi SM (2019) Non-lamellar liquid crystalline nanocarriers for thymoquinone encapsulation. Molecules (Basel, Switzerland) 25(1):16. https://doi.org/10.3390/molecules25010016

Yu L, Li C, Le Y, Chen J-F, Zou H (2011) Stabilized amorphous glibenclamide nanoparticles by high-gravity technique. Mater Chem Phys 130(1–2):361–366. https://doi.org/10.1016/j.matchemphys.2011.06.049

Yusuf M (2021) Formulation and cognitive evaluation of self-assembled phosphatidylserine-chitosan nanoparticles of lycopene, an innovative technique to lessen STZ-induced oxidative stress: a vital persuader of major neurological diseases. J Drug Deliv Sci Technol 63(102534):102534. https://doi.org/10.1016/j.jddst.2021.102534

Zeren S, Sahin S, Sumnu G (2022) Encapsulation of Caffeic acid in carob bean flour and whey protein-based nanofibers via electrospinning. Foods (Basel, Switzerland) 11(13):1860

# Chapter 10
# Nanoparticle Based Vaccines and Immunotherapies

**Pallavi Kumari, Sandini Garg, Harmik Sohi, Jasjeet Kaur Narang, Alka Ahuja, and Shweta Dang**

**Abstract** Nanoparticles offer unique advantages such as enhanced delivery efficiency, targeted release, and improved stability of antigens and therapeutic agents. This chapter tries to explore the use of various types of nanoparticles in the development of novel vaccines and immunotherapies. Nanotechnology provides a revolutionary approach in contrast to traditional immunotherapies, which frequently result in detrimental side effects because of hyperactivity, lengthy treatment periods, and inflated prices. Nanoparticles produce highly focused tissue delivery systems by utilising the better physicochemical characteristics of nanocarriers, such as a significant specific surface area that improves interactions with the immune system. There exists many promising nanoadjuvants and nanoparticulate delivery systems that demonstrate that the vaccine developers can now select specific adjuvants and delivery systems tailored to combat various infectious diseases. There are challenges associated with the development of these nanoparticle-based vaccine and therapy development, including manufacturing, scalability, and regulatory hurdles. The chapter concludes by envisioning future directions and the transformative potential of nanoparticle-based approaches in revolutionizing healthcare and disease management.

**Keywords** Nanoparticles · Nanotechnology · Vaccines · Immunotherapy

---

P. Kumari · S. Garg · S. Dang (✉)
Department of Biotechnology, Jaypee Institute of Information Technology, Noida, Uttar Pradesh, India
e-mail: shweta.dang@jiit.ac.in

H. Sohi
Manufacturing Science and Technology: Small Molecules and Biologics, Civica, Petersburg, VA, USA

J. K. Narang
Khalsa College of Pharmacy, Amritsar, Punjab, India

A. Ahuja
College of Pharmacy, National University of Science and Technology, Muscat, Oman

 
V. Khanna et al. (eds.), *Nanotechnologies for Drug Delivery and Biopharmaceutical Development*, Progress in Drug Research 77,
https://doi.org/10.1007/978-3-032-00796-4_10

## 10.1 Introduction

The medical world has been more interested in nanomedicine recently since it has made it possible to modify engineering devices for delivery to and interaction with cell environments. Nanotechnology has led to radical changes in numerous scientific and medical fields, providing an opportunity to increase the precision, efficacy, as well as safety of therapeutic interventions. Nanocarriers, meticulously engineered on the nanoscale to deliver therapeutic agents, combine unique precision with a specific efficiency in delivering the right drug to the right site. They have been applied years ago for targeted immune responses, improved therapeutic delivery and less systemic toxicity (Haleem et al. 2023; Malik et al. 2023) but recently they were extended to the field of immunotherapies and vaccine development. Nanoparticle-based systems are known to be a powerful solution for a variety of immune complex related disorders and infections, infectious diseases in which conventional strategies have failed in their intervention. Nanocarriers have witnessed some interesting advances in the area of immunotherapy, an emerging field that harnesses the power of the body to fend off diseases like cancer. The drug delivery vehicle, which can be specifically programmed to target immune cells and release immune-crowning agents, provides an intensifying focused immune response. This targeted delivery is pivotal to make sure that our treatment works and not lead to uncontrolled side effects (Farooq et al. 2024; Rana et al. 2023). As for vaccination, nanocarrier is the central idea to deliver efficiently to immune cells. This would improve upon their recognition and subsequently response to infections, in return producing more robust and enduring immunity. This novel method not only increases vaccine potency but also the dose level that is protective might be decreased (Bezbaruah et al. 2022).

In general, the nanocarriers integration will be disrupting immunotheraphies and vaccines providing smarter agents that will be more precise, specific and efficient. With improved delivery of therapeutic compounds, researchers are working towards safer and more efficacious therapies that can better prevent infections as well as fight diseases. The chapter summarizes the radical potential of nanoparticles for transformation in a comprehensive way Under the umbrella of vaccines and immunotherapies to underscore the differing form of nanocarriers, their mode of action and therapeutic uses to prevent or treat several diseases. We: examine these novel technologies from inception to practical application in a clinical setting, highlight others with the potential for promise for future innovation. Looking into the extensive influence of nanotechnology on vaccine and immunotherapy landscapes, aims at understanding how these innovations are transforming healthcare, fostered new more effective patient-centric and accessible treatments that encapsulate diagnostics. in these state-of-the-art Such emerging strategies aim to address several of the shortcomings in conventional therapies and vaccines across the board in order for safer, more efficacious treatment modalities which appears to offer significant promise in next-generation healing.

## 10.2 Nanocarriers: An Overview

Advanced, nanometer-sized particles called nanocarriers for high-efficient delivering of drugs, antigens or therapeutic molecules to checkpoint cells/organisms. Their size allows them to extravasate into the complex biological environment of the body, improving the distribution, uptake by cells and biodistribution of these agents. This is a feature that makes nanocarriers crucial within modern medicine and especially to immunotherapy and vaccine aiming to improve outcomes of health treatment as they route directed toward their target (Yetisgin et al. 2020; Edis et al. 2021). Vaccination is one of the most successful tools to prevent infectious diseases, but many vaccines are plagued by limitations such as antigen instability, subpar immune responses and reliance on adjuvants to improve their effectiveness. Nanocarrier–based vaccines: promising for improving vaccine stability and enhancing the therapeutic efficiency. Nanocarriers can protect the antigens from degradation and release them as a sustained-release agent to drive a robust immune response when encapsulating these into nanocarriers. Moreover, nanocarriers can launch immunostimulatory agents or adjuvants to permit vaccine tailoring. The incorporation of nanocarriers in vaccine technology is a tremendous progress that is tightening the grip on infectious diseases, changing strategies for immunization. Integrating nanocarriers into the vaccine suite allows for a scaling down of traditional vaccination, which improves stability and immune activation of antigen. Nanocarriers can encapsulate or load with antigens, preventing them from being degraded and allowing specific delivery to antigen presenting cells (APCs) uses like dendritic cells. Once inside the body, lymph nodes are the final destination for nanocarriers that process and present antigens by APCs to T cells for a robust and specific immune response. This mechanism is not only improving on the immunogenicity of vaccines (Moni et al. 2023), but also contributes to longer lasting immunity, resulting in lower dosage required or even complete lack thereof (Facciolà et al. 2022).

### *10.2.1 Types of Nanoparticles Used to Deliver Vaccine*

A variety of nanoparticles have been designed to tailor the interaction with immune, each having their own merits depending on the materials they are composed of, size, shape and surface functionalities. Nanoparticles based on gold, dendrimers and carbon or polymers (Liposomes) were employed most effectively for vaccine delivery (Aljabali et al. 2023; Joudeh and Linke 2022; Pati et al. 2018; Ray et al. 2021). All of the nanoparticle types can induce cytokine production and provoke powerful antibody responses which enhance vaccine efficiency the overall. This article will be divided into the types of nanoparticles used in vaccines and immunotherapies and will describe more detailed information highlighting the features and applications.

#### 10.2.1.1 Lipid Based Nanoparticles

Lipid-based nanoparticles (LNPs, named so due to their composition of lipids are bilayers or multilayers of lipids that form a membrane simulating that within the cell) consisting predominantly of lipids: a disc (bilayer/multilayer) demonstrating a hydrophilic and hydrophobic head as their structure. This composition feature it to be a staple for all kind of antigens and adjuvants. There are many benefits with these features like well-defined formulations, self-assembly, biocompatible, high bioavailability and big cargo payload (Albertsen et al. 2022). They are also called liposomes. One of the first nanomedicine vectors that successfully went from an idea to clinic. LNP being the talk of town for playing a very significant role in the COVID-19 mRNA vaccines, where they invariably maintain and deliver an mRNA molecule to the cells (Tenchov et al. 2021). LNPs have further been developed into the use for other applications such as agriculture, nutrition, medicine imaging, cosmetics and other mesmerizing sectors/n-fields of nanoreactor.

The method used to synthesize LNPs is crucial, as it directly influences both the size of the LNPs and the efficiency of mRNA encapsulation (Swetha et al. 2023). Typically, LNPs are created by precipitating lipids from an ethanol solution into water. The mRNA can either be dissolved in the aqueous phase or encapsulated during the lipid condensation process or be complexed with the LNPs in a subsequent step (Tenchov et al. 2021). LNPs are designed to encapsulate mRNA encoding the viral protein of interest. After the vaccine is injected, muscle cells absorb the LNPs, allowing the mRNA to be released into the cytosol, where the host cell machinery translates it into the target protein. Simultaneously, the LNPs trigger danger-associated signals that attract innate immune cells, such as neutrophils, monocytes, macrophages, dendritic cells, and others. Antigen-presenting cells then process and present the antigen to T cells, which differentiate into effector T cells and support B cell-mediated responses. Activated cytotoxic T cells target and destroy infected cells, while antibodies produced by B cells or plasma cells neutralize the virus (Fig. 10.1) (Hou et al. 2021).

#### 10.2.1.2 Polymer-Based Nanoparticles

Polymer-based nanoparticles are composed of some biodegradable and biocompatible polymers like poly (lactic-co-glycolic acid) (PLGA), polyethylene glycol (PEG), and chitosan. The US Food and Drug Administration (FDA) and the European Medicines Agency (EMA) have approved PLGA for use in humans, and PLGA has long been used to deliver medications. It is widely used in drug delivery applications due to their ability of controlled sustained release of the encapsulated material. This is particularly valuable in vaccine delivery, where a steady and sustained release of antigens can enhance immune activation and lead to more robust and long-lasting immune responses. It offers great benefits like inducing immunity quickly and persistently, having a longer shelf life at higher temperatures, and improving patient compliance (Grego et al. 2020). Figure 10.2 represents the polymer based nanoparticles for delivering the vaccines.

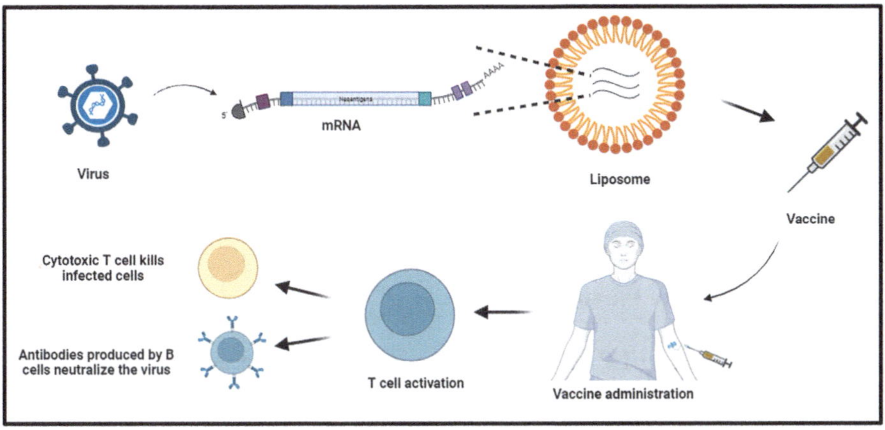


**Fig. 10.1** Immune response by mRNA vaccines. Schematic representation of a lipid-based nanoparticle (liposome) encapsulating mRNA for vaccine delivery

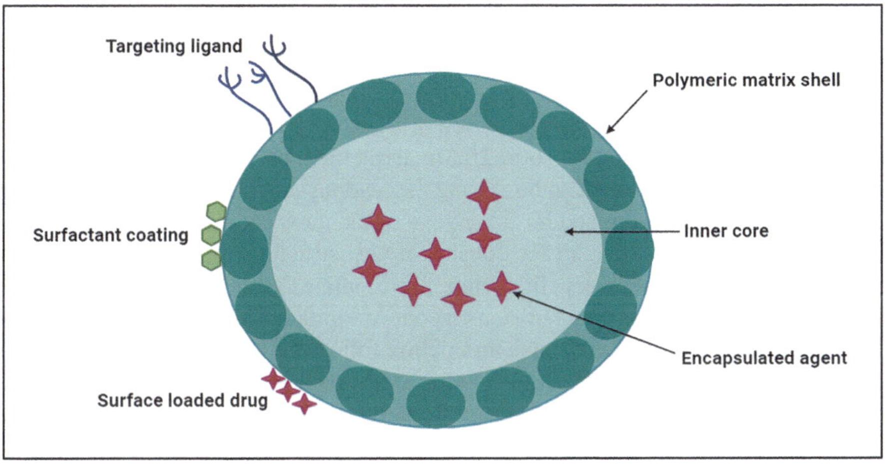


**Fig. 10.2** Basic structure of a polymeric nanoparticle. (Adapted from: https://link.springer.com/article/10.1007/s11481-016-9692-7)

In an investigational study by Hardenberg et al. (2024), who studied the poly (amido) amine (PAA)-based polymeric nanoparticles (PAAQ NP) as an mRNA delivery system for an influenza vaccine. A strong humoral and cellular immune response was observed in ferrets without causing local or systemic reactogenicity. Vaccinated animals showed reduced clinical signs and lower virus loads compared to controls after virus exposure. These promising results from the study highlights the advantages of using polymeric NPs like their ability to carry diverse types of nucleic acids, their flexibility and simplicity in formulation, and the capacity to be lyophilized which makes them a compelling option for biomedical applications.

Polymeric NPs have been additionally exploited with co-delivery of more than one therapeutic agents to tumor cells, which results benefit in that will polymer-based immunotherapeutic (Hardenberg et al. 2024) employs. Polymeric nanoparticles in general represent a strong and versatile tool for vaccine delivery/immunotherapy with enhancement of antigen stability, enabling controlled release and preferential targeting of various immune cells which makes them one of the most interesting approaches in the field of next generation immunotherapeutic.

#### 10.2.1.3 Virus Like Particles

Virus-like particles (VLPs) are non-infectious protein structures derived from viruses and typically consist of one or more polymers that can self-assemble. They are made in the shape and size on what appears to be a viral particle, however they don't contain any genetic material so it cannot infect cells. Due to their high immunogenicity, VLPs can elicit both antibody-mediated and cell-mediated immune responses. Different mechanisms induce the responses as compared (Tariq et al. 2022) that are activated by inactivated viral vaccines. Before now, the VLP-based vaccines have been developed for Hepatitis B virus and the research in this area has progressed a long way. Numerous VLP-based vaccines have been developed and now in the clinics for the targeting of human papilloma virus (HPV), Hepatitis E virus (Donaldson et al. 2018). VLPs: role of VLPs in the design of HCMV (human cytomegalovirus) vaccine (Perotti 2020), explained previously in acute injury (Gupta et al. 2023) continues to be among the leading causes of morbidity and mortality in immunocompromised hosts, andone of the most common viral lesions during neonatal examination. VLPs consist of the structural genes of target viral structural proteins—the very first step for cloning. Vectors then express self-assembling viral proteins in both eukaryotic or prokaryotic expression systems. They are removing cell aggregates and debris following cell harvest and lysis in a clarification phase. Via further purification steps such ion-exchange chromatography and ultracentrifugation one can improve the whole process, which ultimately ends up with polishing out any remaining host proteins and nucleic acids. Solely by sterilized filtering and formulation this ends up at the end of a safe, successful and efficient vaccine based on VLP (Gupta et al. 2023). Virus like particles (VLPs) have been constructed from components of multiple virus families illustrating the potential of vaccine development using these particles. These families include Parvoviridae, such as adeno-associated virus; Retroviridae, which includes HIV; Flaviviridae, represented by the Hepatitis C virus; Paramyxoviridae, such as the Nipah virus; and bacteriophages, including Qβ and AP205. By utilizing the structural proteins of these viruses, researchers can create VLPs that mimic the native virus's architecture without containing infectious material, making them promising candidates for safe and effective vaccines (Tornesello et al. 2022; Hess et al. 2019). Figure 10.3 represents the process of manufacturing virus-like particles (VLPs).

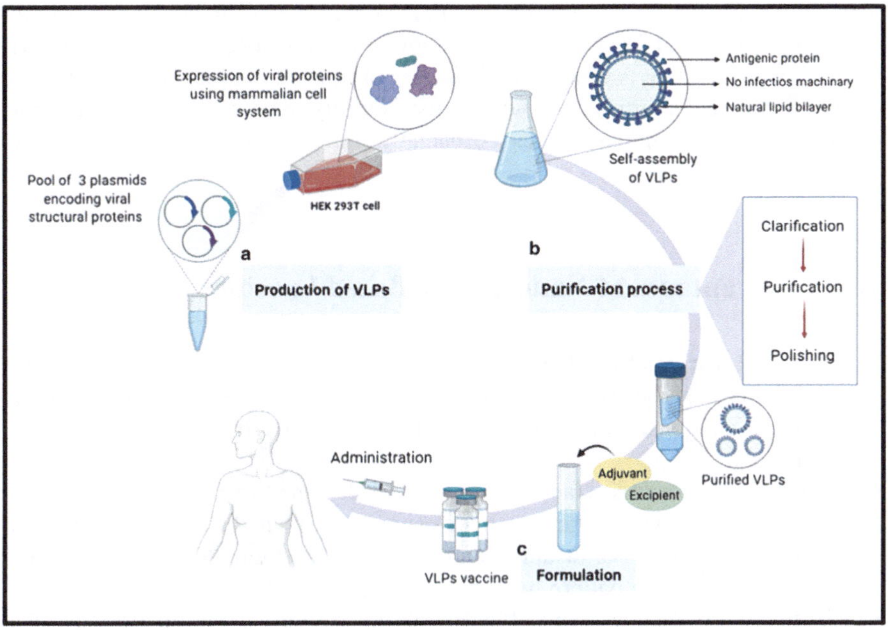


**Fig. 10.3** Process of preparation virus-like particles (VLPs)

#### 10.2.1.4 Inorganic Nanoparticles

Inorganic NPs are principally made of the materials gold, silica, iron oxide, and calcium phosphate. Most inorganic materials are perfect for delivering antigens in vaccines because they have reduced particle sizes, improved stability, precise tunability, greater permeability, high drug-loading capacities, and controlled release profiles. Other unique properties like optical, electric, and magnetic characteristics, well-defined chemistry, high surface-to-volume ratio, and high avidity display make them even more useful for the development of vaccines, cancer immunotherapy, and autoimmune therapies. These cutting-edge materials are frequently created as hybrid inorganic nanomaterials, which have an organic outer shell and an inorganic core (Niżnik et al. 2024).

Gold nanoparticles (AuNPs) have diverse applications in fields like computing, catalysis, sensing, and drug delivery, and are particularly well-suited for vaccine development due to their low toxicity and chemical versatility (Huang et al. 2023). Tao et al. (2017) formulated a peptide-based AuNP vaccine bearing the M2e peptide that efficiently elicited protective immunity towards influenza A. The vaccine adjuvanted with CpG oligodeoxynucleotide to induce TLR-9 stimulation yielded total protection in H1N1 influenza challenged mice as determined through high M2e-specific IgG antibodies levels (Tao et al. 2017). Iron oxide NPs (IONPs) have been explored in recent times as a promising vaccine adjuvant because they can manipulate the immune responses. First, they were identified because of their function as

MRI contrast agents and consist of Fe2O3 and $Fe_3O_4$ magnetite (Shah and Dobrovolskaia 2018). Pusic et al. (2013) showed that an inorganic nanoparticle with an iron oxide core that is water-soluble is very promising for vaccine delivery based on its adjuvant capabilities and low toxicity. Stability in both solution and dry form of the IO-vaccine conjugate makes it especially convenient to employ in field conditions (Pusic et al. 2013).

## 10.3 Advantages of Nanocarrier-Based Vaccines

- **Improved Antigen Stability**: Antigens are sensitive molecules that can degrade quickly in the bloodstream or during storage. Nanocarriers encapsulate and protect these antigens, enhancing their stability and ensuring they reach their target intact.
- **Adjuvant Properties**: Nanocarriers can be designed to function as adjuvants themselves, enhancing the immune response without the need for additional adjuvant molecules. For instance, nanoparticles can stimulate the innate immune system by activating pattern recognition receptors (PRRs) on APCs.
- **Multivalency**: Nanocarriers offer the ability to present multiple copies of an antigen on their surface, increasing the likelihood that the immune system will recognize and respond to the antigen. This multivalent presentation mimics the structure of viruses, making nanocarrier-based vaccines more effective in eliciting a strong immune response.

### *10.3.1 Application in mRNA Vaccines: The COVID-19 Case*

The success of the mRNA vaccines, during the coronavirus pandemic, pointed out the importance of some of the basic components, in particular, the nanocarriers-lipid nanoparticles (LNPs)-in the further advancement of vaccine technology. By introducing a molecule of messenger RNA that contains instructions telling the cell to produce viral proteins, such as the spike protein of the SARS-CoV-2, these vaccines causes cells to produce viral proteins, solving the immunological responses. Nevertheless, mRNA is generally regarded as an unstable molecule subject to degradation in the biological environment. LNPs protect this fragile mRNA, keeping it intact for delivery. In addition, LNPs serve to facilitate cell uptake of the mRNA, allowing for efficient translation into the viral protein. This is then recognized by the immune system to prepare for future encounters with the virus. The trial successes of lipid nanoparticles (LNPs) in the form of the Pfizer-BioNTech and Moderna COVID-19 vaccines demonstrated their immense applicability, opened the doors for mRNA vaccines to be utilized for treating and preventing other infectious diseases, cancers, and even genetic disorders (Wilson and Geetha 2022; Swetha et al. 2023; Kutikuppala et al. 2024).

## 10.4 Nanocarriers in Immunotherapies

The field of both cancer and non-cancer disease is now revolutionized by Immunotherapies, a therapy which takes advantage of the ability of our immune system to recognize & eliminate cells that are abnormal. Although traditional immunotherapies have been extremely successful (a few examples are monoclonal antibodies and cytokines as well as immune checkpoint inhibitors) Use falls short often due to suboptimal drug delivery, from bioavailability to non-specificity and systemic toxicity. Nanocarriers promise an alternative to these problems as they facilitate the delivery, stability and therapeutic potency of immunotherapeutic formulations. Utilizing the nanocarriers properties optimize immunotherapy treatments to be highly accurate and effective in the human body by reasoned (Sanmamed and Chen 2018; Lôbo et al. 2021).

### *10.4.1 Role of Nanocarriers in Cancer Immunotherapy*

A wide range of interests have been focussed on Nanocarriers for designing a more efficient cancer immunotherapy. Provide several advantages such as solubilization of drugs, targeting the delivery, tumor penetration enhancement and stir up an abundance of immune activation. Moreover, nanocarriers may also remodel the whole tumor microenvironment (TME) and circumvent immunosuppressive barriers that prevent immune system access to cancer cells (Huang et al. 2024; Chen et al. 2023). Nanocarriers based immunotherapies have the following advantages:

#### 10.4.1.1 Improved Drug Solubility

Immunotherapeutic agents such as peptides, nucleic acids and small-molecule inhibitors which are generally very poor water soluble undergo numerous limitations in terms of failing to release bioavailable & efficacious to the body. When administered in their native form most of these molecules are rapidly catabolized by the biological system or only poorly absorbed leading to suboptimal concentrations on target sites for diseases such as cancer (Elumalai et al. 2024). Nanocarriers as a unique approach can be employed for overcoming this and can encapsulate hydrophobic therapeutic agents in order to optimize their solubility in physiological fluids. The solubility is improved and the drug is also protected from immediate degradation by this protecting agent of encapsulation, to prolong the stability and efficiency of drugs in biological systems (Mitchell et al. 2021). Different kind of nanocarriers (liposomes, polymeric nanoparticles and lipid-based nanoparticles) are designed to accomplish this. These nanocarriers would be designed to release the entrapped drug under either a controlled or sustained manner ensuring that therapeutic agent is released over a long period at the target site and for this purpose it increase not only its bioavailability but also its therapeutic action. This adds a bonus to immune checkpoint inhibitors (key players in cancer immunotherapy) which are poorly solubilized and rapidly cleared by systemic circulation (Lu et al. 2021). Tornesello et al. showed in 2020 that the encapsulation peptide antigens in liposome-based nanocarriers almost massively enhanced the solubility, stability and abrload to APCs (antigen presenting cells) of these, eliciting healthier immune response with increased T-cell activation and longer remission from tumor (Tornesello et al. 2020). Another, Lee et al. (2022) also indicated that nanoparticle improved their pharmacokinetics in the anti-PD-1 antibody (nearly similar with anti-PD-1 monoclonal antibody)like immune checkpoint inhibitors efficacy and it improved both bioavailability and sustained concentration of drugs which were closely associated with an enhanced anti-tumor immune response (Lee et al. 2022). Solubility issues and poor pharmacokinetic properties of these therapeutic agents are resolved with nanocarriers that result in improved, targeted and sustained immunotherapy providing a more effective treatment in patients.

#### 10.4.1.2 Targeted Delivery

One of the most important applications of nanocarrier systems in precision immunotherapy is targeted drug delivery where drugs are delivered to tumor cells or immune cells by a nanocarrier which can only have minimal effect on normal tissues. Through the functionalization of the nanocarrier surface with ligands or antibodies that can target either cancer cells or immune cells, systems in this family allow for specific and localized drug delivery. The resulting targeted mode of action leads to a higher therapeutic index, i.e., increased therapeutic efficacy matched with strong reduction of the systemic toxicity ensuring scarcely collateral damage to healthy tissue. At a mechanistic level it occurs due to the functionalization of nanocarriers with these ligands or antibodies that specifically recognize overexpressed receptors, such as PD-L1, HER2 or any cancer/cell surface antigen. The nanocarriers would then target the therapeutics directly to the tumor microenvironment for efficient therapeutic drug delivery upon binding (Yao et al. 2020). Experiment have shown this selective targeting in a study conducted by Venditti et al. (2022) where AuNPs functionalized anti-PDL-I antibodies were synthesized, and then exploited for specific delivery of targeted PD-L1-expressing melanoma cells. This has the potential, by massively increasing the amount of the therapeutic agent that would be taken up by the tumor to facilitate superior tumor suppression and lower systemic toxicity compared to non-targeted therapy. Actually, nanocarriers based on antibodies raised against specific tumor antigens as PD-L1 or HER2 can be used to deliver immunotherapeutic agents in an tumour microenvironments. This does not only increase drug potency, but also forces direct action of immune-modulating agents in the tumor microenvironment and improves therapeutic effectively while preventing collateral damage of healthy tissues (Venditti et al. 2022).

#### 10.4.1.3 Enhanced Tumor Penetration

One of the biggest problems in cancer care is that drugs rarely penetrate deeply into tumors, which have characteristic dense and stenographic vasculature which actually prohibits drug from entering surrounding tissue nodes and results in incomplete treatment. The size and tumor-tropic nature of nanocarriers allow the delivery of a higher amount of drug deep into the tumor tissue to achieve better homogeneity. This is based on the enhanced permeability and retention (EPR) effect, by which tumors form new vasculature that leak nanoparticles. Once within these nanocarriers, they can be engineered to release their contents at controlled or even triggered manner, thus leading to higher drug concentration at the tumoral site. For example, Silva et al. showed in 2015 that polymeric nanoparticles loaded with toll-like receptor (TLR) agonists could be delivered to distant from the periphery tumor tissue, providing activation of immune system that gave remarkable regression of tumor. The use of NSs promisingly increased the probability of complete tumor eradication and decreased the risk for development of recurrence or resistance due to the cascade effects (Silva et al. 2015).

#### 10.4.1.4 Immune Activation

Therapeutic immunotherapy requires full activation of the immune system, especially by cytotoxic T lymphocytes that are capable to kill cancer cells. Nanocarriers might be engineered to specifically deliver immunostimulatory molecules [e.g. For example tumor antigens, cytokines or toll-like receptor (TLR) agonists directly to APCs such as dendritic cells by means of perfusion through the nanoparticles, therefore extending and/or bypassing the immunosuppressive environment imposed by a tumour. These nanocarriers deliver tumor antigens and adjuvants to dendritic cells (key for T-cell priming) in a unique platform (Irvine and Dane 2020), which specifically targets dendritic cells. Bhagchandani et al. (2021) created a nanoscale system through which the co-delivery of tumor antigens and TLR9 agonists resulted in maximum dendritic cell maturation along with significant augmentation of T-cell activation, giving rise to improved anti-tumor immune responses associated with prolonged regression of tumors in preclinical models. Co-delivery system allows for the simultaneous presentation of necessary immune signals to induce potent tumor-specific immune response while maximally activating and minimizing off-targets enhancing the efficacy and safety of cancer immunotherapies (Bhagchandani et al. 2021).

## *10.4.2 Overcoming Immunosuppression in the Tumor Microenvironment (TME)*

The tumor microenvironment (TME) is usually immunosuppressive, allowing cancer cells to circumvent immune recognition. Tumor suppressor cells such as regulatory T-cells (Tregs) and myeloid-derived suppressor cells are recruited by the tumors to suppress immune response. This could be a unique opportunity provided by nanocarriers to manipulate the TME to nullify immunosuppression and increase immunotherapy effectiveness (Tie et al. 2022).

### 10.4.2.1 Reversing Immune Checkpoints

Immune checkpoint inhibitors such as anti-PD-1 and anti-CTLA-4 antibodies have become central pillar of cancer immunotherapy as these drugs enable the patient immune system to recognize and destroy tumors. Nevertheless, clinical administration of these inhibitors is often associated with major systemic immune-related adverse effects (irAEs) forfeiting therapeutic advantages. Nanocarriers may alleviate this by offering a targeted delivery system for checkpoint inhibitors with low systemic delivery and less side effects. Engineered nanoparticle nanocarriers are designed to deliver checkpoint inhibitors towards tumor microenvironment (TME) (Shiravand et al. 2022) These nanoparticles can be engineered as to encapsulate checkpoint molecules and specifically deliver them to the tumor microenvironment: Jia et al. Targeted release of these agents specifically at the tumor site allows inhibition of PD-1/PD-L1 interaction between T-cells and cancer cells, hence, the mechanism of action which localized blockade allows reactivating T-cells at site of tumor inhibiting their ability to be activated systemically and at the same time counteracts systemic immune activation that could result in irAEs (Han et al. 2020). For example, nanoparticles containing anti-PD-1 antibodies conjugated to lung cancer cell-targeting peptides were constructed by Wang et al. (2020) which show improved anti-tumor immunity and drastically decreased systemic toxicity in several animal models. This novel method led to a beneficial effect on anti-tumor immunity and notably decreased systemic toxicity in preclinical models with respect to the immunotherapy-induced local blockade of immune checkpoints in the TME (Wang et al. 2020).

### 10.4.2.2 Modulating the Tumor Microenvironment

The tumor microenvironment (TME) is highly immunosuppressive and underlies many of major barriers towards cancer control as it suppresses immune responses and masks cancer cells from detection (Munn and Bronte 2016). Nanocarriers provide a smart strategy for the administration of theranostic nanocarriers, which aim to reprogram TME by targeting immunosuppressive pathways while promoting a pro-inflammatory environment that can increase immune cell can be established. The nanocarriers are able to load agents like TLR agonists (or inhibitors), indoleamine 2,3-dioxygenase (IDO) inhibitors. IDO inhibitors. The de novo synthesis of tryptophan metabolite IDO also suppresses immune function as it depletes T-cell functional essential amino acid. Through nanocarriers to deliver IDO inhibitors can silence this escape pathway and re-invigorate T-cell function, anti-tumor immunity (Zhu et al. 2020). Despite the fact that the mechanism of action was very promising in vitro 24, a recent study by Smith et al. (2020) demonstrated a clear in vivo activity of nanoparticles co-delivering TLR7/8 agonists with IDO inhibitors directly to tumors. It effectively reversed immunosuppression and greatly amplified T cell activity in tumor-draining lymph nodes of animals with this combination therapy. Nanocarrier-mediated utilization of agents that target immunosuppressive mechanisms provides a new paradigm in which the TME can be modified from a hospitable space for cancer cells to one where vigorous immune activation is enhanced. Not only does this strategy significantly raise the microtome specificity of already proven immunotherapies; it also suggests the possibility of provoking more robust responses that extend beyond local tumors (Smith et al. 2020).

## 10.5 Clinical Trials with Vaccine and Immunotherapies Delivery by NPs

The quick FDA clearance of RNA and LNP-based vaccines in response to the COVID-19 pandemic demonstrated the efficacy of this strategy in treating complicated diseases. The NP-based vaccines are not only being developed for infectious diseases, but they are also making progress against cancer and hyperlipidaemia. Many clinical trials are still in the early phases, and many are almost finished, indicating that a growing number of products will probably hit the market soon (Zhang et al. 2023). Table 10.1 shows a list of ongoing or completed clinical trials for nanoparticle-based vaccines and immunotherapies (Clinicaltrials.gov 2024).

## 10.6 Challenges Associated with NP Vaccines and Immunotherapies

The intricate design and manufacturing procedures needed to create reliable and repeatable NP vaccinations are among the primary challenges. On a large scale, it is still challenging to provide homogeneous size, shape, and surface qualities, despite the fact that this is essential for the best immunological response. A further difficulty in gaining regulatory approval for NP-based vaccinations is that the unique characteristics of nanoparticles may not be adequately covered by current rules, necessitating the development of new standards and thorough safety analyses.

**Table 10.1** List of clinical trials with vaccine delivery by nanoparticles

| Title | Phase | Clinical status | Disease/ condition | Trial number |
|---|---|---|---|---|
| NoveRNA-nanoparticle Vaccine for the Treatment of Early Melanoma Recurrence Following Adjuvant Anti-PD-1 Antibody Therapy | Phase 1 | Recruiting | Melanoma | NCT05264974 |
| Phase 3 Lot Consistency Study of Trivalent Hemagglutinin Nanoparticle Influenza Vaccine | Phase 3 | Not yet recruiting | COVID-19 | NCT06485752 |
| A Study to Evaluate the Safety and Immunogenicity of COVID-19 and Influenza Combination Vaccine (COVID-19) | Phase 2 | Completed | COVID-19, influenza | NCT05519839 |
| A Phase-I Study of a Nanoparticle-based Peptide Vaccine Against Dengue Virus | Phase 1 | Completed | Dengue | NCT04935801 |
| A Phase-I Study of a Nanoparticle-based Peptide Vaccine Against SARS-CoV-2 | Phase 1 | Completed | COVID-19 | NCT05113862 |
| Safety and Immunogenicity of ChulaCov19 BNA159 mRNA Vaccine | Phase 1 | Recruiting | COVID-19 | NCT05231369 |
| Safety and Immunogenicity of an Epstein-Barr Virus (EBV) gp350-Ferritin Nanoparticle Vaccine in Healthy Adults with or without EBV Infection | Phase 1 | Active | EBV infection | NCT04645147 |
| Study of a Respiratory Syncytial Virus Candidate Encapsulated in a Lipid Nanoparticle Based Formulation in Adults Aged 18 to 50 Years and 60 Years and Older | Phase 2 | Active | Respiratory syncytial virus immunization | NCT05639894 |
| Evaluation of the Safety and Immunogenicity of a Recombinant Trivalent Nanoparticle Influenza Vaccine with Matrix M-1 Adjuvant (NanoFlu) | Phase 2 | Completed | Influenza | NCT03293498 |
| Safety, Reactogenicity, and Immunogenicity Study of Heterologous Booster Vaccination of a SARS-CoV-2 Recombinant Protein Nanoparticle Vaccine (GBP510) | Phase 2 | Active | COVID-19 (healthy volunteers) | NCT05175950 |
| Safety and Immunogenicity Study of SARS-CoV-2 Nanoparticle Vaccine (GBP510) Adjuvanted with Aluminum Hydroxide (COVID-19) | Phase 2 | Completed | COVID-19 | NCT04742738 |
| Immunogenicity and Safety Study of SK SARS-CoV-2 Recombinant Nanoparticle Vaccine (GBP510) Adjuvanted with AS03 (COVID-19) | Phase 3 | Completed | COVID-19 | NCT05007951 |

(continued)

**Table 10.1** (continued)

| Title | Phase | Clinical status | Disease/ condition | Trial number |
|---|---|---|---|---|
| Evaluation of the Safety and Immunogenicity of a SARS-CoV-2 rS Nanoparticle Vaccine With/Without Matrix-M Adjuvant | Phase 2 | Completed | COVID-19 | NCT04368988 |
| Immunogenicity and Safety Study of Heterologous Booster Vaccination of a SARS-CoV-2 Recombinant Protein Nanoparticle Vaccine (GBP510) Adjuvanted with AS03 in Adults Aged 18 Years and Older | Phase 3 | Active | COVID-19 | NCT05501522 |

Lipid-based nanoparticle technologies have demonstrated great potential in treating a range of ailments through their clinical advancement by using nucleic acid and chemotherapeutics. However, in comparison to the large number of formulations presently undergoing (pre)clinical trials, the number of successful products that have reached the market is disproportionately low. This disparity underscores the persistent obstacles and complexities involved in transferring these nanoparticles from animal models to human uses. Furthermore, NP vaccines and immunotherapies are more expensive to develop and produce than conventional vaccinations, which could restrict accessibility, particularly in environments with limited resources as represented in Fig. 10.4. For NP vaccinations to be used widely, safely, and effectively in the future, these obstacles must be overcome (Desai 2012; Nandedkar 2012; Thi et al. 2021; Bezbaruah et al. 2022; Zielińska et al. 2020; Sharma et al. 2021).

In biomedical applications, the potential toxicity of nanoparticles has raised concerns since normally benign compounds might turn dangerous when reduced to the size of nanoparticles. Nanoparticles are easily able to enter many tissues and organs because of their small size, yet this can have advantages as well as disadvantages. For example, oral ingestion could result in gastrointestinal side effects, nasal administration could create respiratory complications, and injecting nanovaccines into the skin could induce dermatological concerns. Parenteral administration has the potential to cause cardiovascular problems and brain damage if nanoparticles manage to get beyond the blood-brain barrier (Bezbaruah et al. 2022). The interplay of complexity, and multicomponent nanomedicines is what makes forecasting their behavior in biological systems rather hard. Besides, Nanotechnological demands are specific that the regulatory authorities has to align itself to meet the goal of nanomedicine and convey sharp guidelines and standards for clinical testing & development. Nanomedicine with its great expectation, especially in oncology and the prospect of personalized therapy shows a lot of promise despite the hurdles. Some of the challenges are highlighted as below:

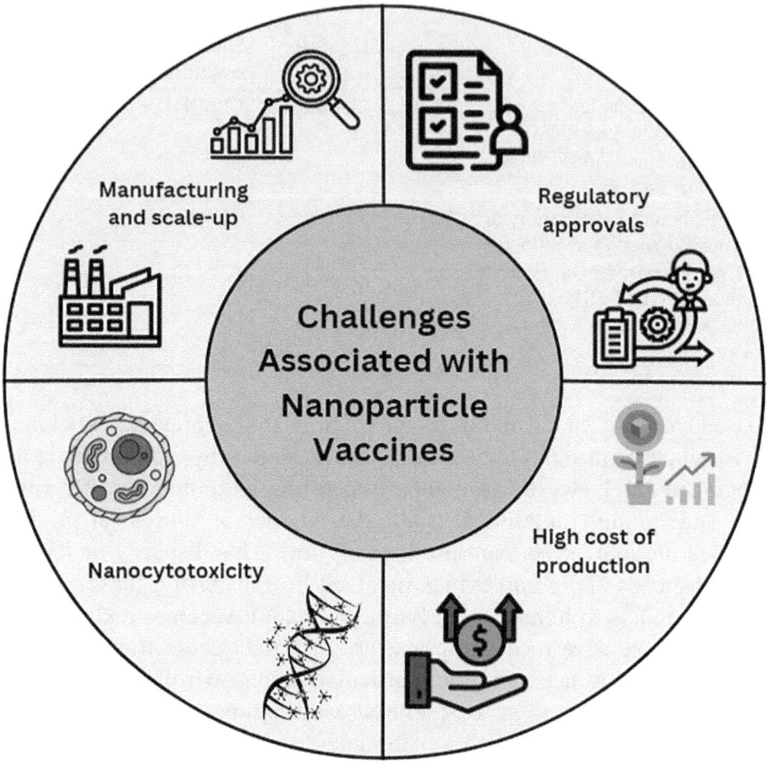


**Fig. 10.4** Some of the challenges associated with nanoparticle based formulations

1. **Safety and Toxicity**: The most important issue is safety related to nanoparticles (mainly long term release and systemic effects on human). The size and surface properties being unique physicochemical properties of nanoparticles often cause the unexpected interactions with biological systems. Nanoparticles can also improve immune responses, but at that they could be toxic if they got stuck in organs like liver, splenic or renal. Because of the longer term exposure or higher concentration of nanoparticles, unexpected side effects such as organ damage or systemic inflammation, oxidative stress can be associated. So proper assessment of biocompatibility and long term safety is fundamental before any clinical applications of such materials. Figure 10.5 shows the cytotoxic effect nanoparticles.
2. **Immune Overactivation and Inflammatory Brigades**: Nanoparticles are engineered to increase immune recognition, but not the operative immune cascade is fully understood. Overactivation of the immune system develop inflammation and may cause autoimmune diseases or hypersensitivity. Some nanoparticles in particular non-biodegradable nanoparticles or those that are foreign-like, could trigger a wrong body immune response causing harmful effects such as cytokine storms and other immune-mediated toxicities.

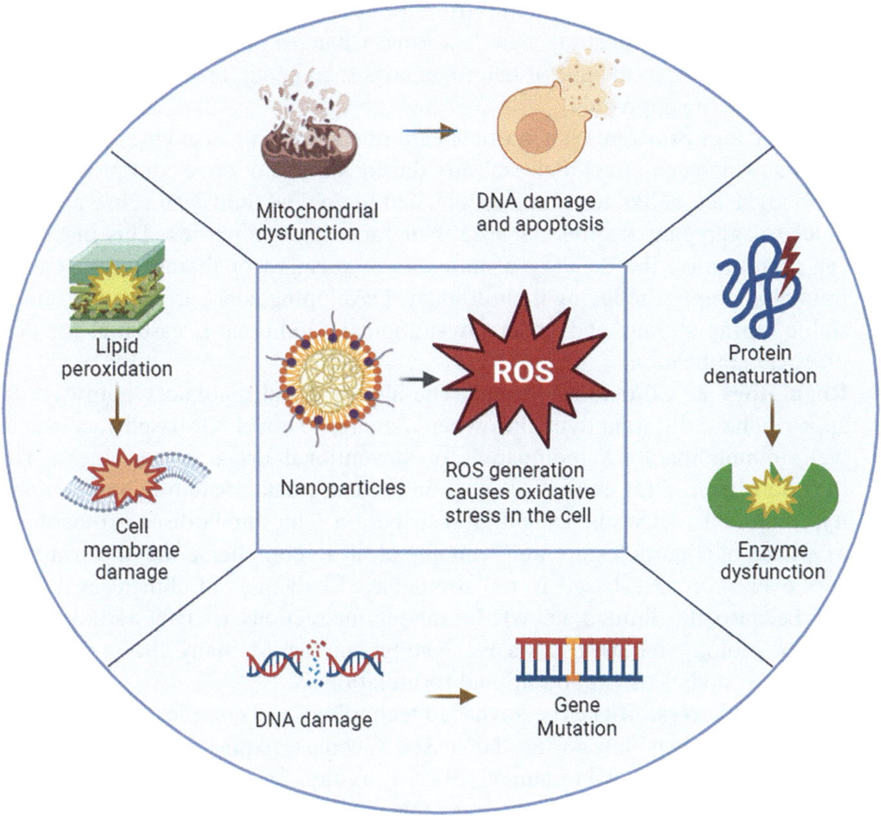


**Fig. 10.5** Cytotoxicity by Nanoparticles. Nanoparticles can act as catalysts to produce reactive oxygen species (ROS), which can interfere with cellular processes. This interference may result in decreased cell viability and proliferation, initiate inflammatory reactions, cause oxidative stress, and lead to cell death through apoptosis

3. **Nanoparticles are biocompatible and cleared naturally from the body**: Material composition of nanoparticle has an impact on how the nanoparticles are recognized, processed and eliminated from within the body. Some nanoparticles might just qualify as transient foreign invaders, driven rapidly out of the body and diminishing efficacy. Furthermore, the non-biodegradable nanoparticles can be sequestered in tissues causing worry over long-term toxicity and chronic inflammation. An optimal balance between stability and biodegradability of nanoparticle, is essential to make them safe enzymes in vaccines and immunotherapies.
4. **Manufacturing and Scalability**: It is challenging to manufacture nanoparticle-based vaccines/immunotherapies. Synthesis of nanoparticles must be performed with great accuracy in order to achieve similar size, shapes, and surface properties that are required for biological functions. Yet that consistency in practice

proves hard, especially scaling up from the research and development lab-scale production to the industrial manufacturing. Changes in nanoparticle properties will likely result in the use of heterogeneous therapeutic results and impact the regulatory drug approval.

5. **Stability and Storage**: Nanoparticles are often sensitive to environmental conditions, which can affect their stability during storage or once administered into the body. Factors like temperature, pH, and biological fluids can cause nanoparticles to aggregate, degrade, or lose their functional properties. This instability can compromise the delivery of antigens in vaccines or therapeutic agents in immunotherapies, reducing their efficacy. Developing nanoparticles that remain stable during storage and under physiological conditions is essential for their practical application.
6. **Regulatory & Clinical Approval**: The approval and regulatory pathways for approval have different dynamics when it comes to novel NP-based vaccines, as well immunotherapies (compared to conventional drugs or vaccines). This includes the need for extensive safety and efficacy data, including human toxicity, immunogenicity studies and biodistribution. The standardised protocols for assessing nanoparticles are not available at all to complicate the approval process even more. NP-based formulations have challenges in clinical evaluation too because the limited knowledge about interactions of nanoparticles with human biology requires extensive testing and hence many more numerous expensive studies as with traditional formulations.
7. **Price and Accessibility**: The advanced technology and complex manufacturing of nanoparticles which go into NP-based vaccines, require more sophisticated methods than traditional treatments. That is a cause for alarm in terms of access, particularly with the cost of novel therapies in low and middle income.

## 10.7 Conclusion and Future Direction

Nanocarriers have been a revolutionary platform for the enhancement of vaccine and immunotherapy efficacy through enhanced antigen preservation, exposure-controlled release and focused delivery to immune cells. Because they can improve the stability of therapeutic, such as antigens, modifiers of immune (aminogens) that have been promising candidate molecules for both infectious diseases and cancer immunotherapy. Particularly effective at this are lipid nanoparticles, polymeric nanoparticles and viral-like particles for improving intracellular antigen uptake by antigen-presenting cells (APCs), hence eliciting strong and sustained immune responses. They can even contain adjuvants as well, to intensify immune activation and efficacy. In the future research will be centered on developing personalized nanovaccines to optimize delivery to specific immune cells, like dendritic cells and T cells, as well as Nanoparticles as multifunctional carriers are termed to deliver antigens along with adjuvants and/or checkpoints inhibitors. While the progress, there still exist problems with translating into clinical practice and regulatory approval as well as scalable manufacturing. Such hurdles will be necessary to pave the way in enabling nanocarrier based vaccines and immunotherapies, for disease applications such as cancer as well emerging infectious threats.

## References

Albertsen CH, Kulkarni JA, Witzigmann D, Lind M, Petersson K, Simonsen JB (2022) The role of lipid components in lipid nanoparticles for vaccines and gene therapy. Adv Drug Deliv Rev 188:114416

Aljabali AA, Obeid MA, Bashatwah RM, Serrano-Aroca Á, Mishra V, Mishra Y, El-Tanani M, Hromić-Jahjefendić A, Kapoor DN, Goyal R, Naikoo GA (2023) Nanomaterials and their impact on the immune system. Int J Mol Sci 24(3):2008

Bezbaruah R, Chavda VP, Nongrang L, Alom S, Deka K, Kalita T, Ali F, Bhattacharjee B, Vora L (2022) Nanoparticle-based delivery systems for vaccines. Vaccine 10(11):1946

Bhagchandani S, Johnson JA, Irvine DJ (2021) Evolution of Toll-like receptor 7/8 agonist therapeutics and their delivery approaches: from antiviral formulations to vaccine adjuvants. Adv Drug Deliv Rev 175:113803

Chen D, Liu X, Lu X, Tian J (2023) Nanoparticle drug delivery systems for synergistic delivery of tumor therapy. Front Pharmacol 14:1111991

“Clinicaltrials.gov.” [Online]. Available: https://clinicaltrials.gov/. Accessed 14 Sept 2024

Desai N (2012) Challenges in development of nanoparticle-based therapeutics. AAPS J 14(2):282–295

Donaldson B, Lateef Z, Walker GF, Young SL, Ward VK (2018) Virus-like particle vaccines: immunology and formulation for clinical translation. Expert Rev Vaccines 17(9):833–849

Edis Z, Wang J, Waqas MK, Ijaz M, Ijaz M (2021) Nanocarriers-mediated drug delivery systems for anticancer agents: an overview and perspectives. Int J Nanomedicine:1313–1330

Elumalai K, Srinivasan S, Shanmugam A (2024) Review of the efficacy of nanoparticle-based drug delivery systems for cancer treatment. Biomed Technol 5:109–122

Facciolà A, Visalli G, Laganà A, Di Pietro A (2022) An overview of vaccine adjuvants: current evidence and future perspectives. Vaccine 10(5):819

Farooq MU, Lawrie CH, Deng NN (2024) Engineering nanoparticles for cancer immunotherapy: current achievements, key considerations and future perspectives. Chem Eng J:150356

Grego EA, Siddoway AC, Uz M, Liu L, Christiansen JC, Ross KA, Kelly SM, Mallapragada SK, Wannemuehler MJ, Narasimhan B (2020) Polymeric nanoparticle-based vaccine adjuvants and delivery vehicles. Nanopart Ration Vaccine Des:29–76

Gupta R, Arora K, Roy SS, Joseph A, Rastogi R, Arora NM, Kundu PK (2023) Platforms, advances, and technical challenges in virus-like particles-based vaccines. Front Immunol 14:1123805

Haleem A, Javaid M, Singh RP, Rab S, Suman R (2023) Applications of nanotechnology in medical field: a brief review. Global Health J 7(2):70–77

Han Y, Liu D, Li L (2020) PD-1/PD-L1 pathway: current researches in cancer. Am J Cancer Res 10(3):727

Hardenberg G, Brouwer C, van Gemerden R, Jones NJ, Marriott AC, Rip J (2024) Polymeric nanoparticle-based mRNA vaccine is protective against influenza virus infection in ferrets. Mol Ther Nucleic Acids 35(1)

Hess KL, Medintz IL, Jewell CM (2019) Designing inorganic nanomaterials for vaccines and immunotherapies. Nano Today 27:73–98

Hou X, Zaks T, Langer R, Dong Y (2021) Lipid nanoparticles for mRNA delivery. Nat Rev Mater 6(12):1078–1094

Huang H, Liu R, Yang J, Dai J, Fan S, Pi J, Wei Y, Guo X (2023) Gold nanoparticles: construction for drug delivery and application in cancer immunotherapy. Pharmaceutics 15(7):1868

Huang X, He T, Liang X, Xiang Z, Liu C, Zhou S, Luo R, Bai L, Kou X, Li X, Wu R (2024) Advances and applications of nanoparticles in cancer therapy. MedComm–Oncology 3(1):e67
Irvine DJ, Dane EL (2020) Enhancing cancer immunotherapy with nanomedicine. Nat Rev Immunol 20(5):321–334
Joudeh N, Linke D (2022) Nanoparticle classification, physicochemical properties, characterization, and applications: a comprehensive review for biologists. J Nanobiotechnol 20(1):262
Kutikuppala LVS, Kourampi I, Kanagala RS, Bhattacharjee P, Boppana SH (2024) Prospects and challenges in developing mRNA vaccines for infectious diseases and oncogenic viruses. Med Sci 12(2):28
Lee CK, Atibalentja DF, Yao LE, Park J, Kuruvilla S, Felsher DW (2022) Anti-PD-L1 F (ab) conjugated PEG-PLGA nanoparticle enhances immune checkpoint therapy. Nano 6(3):243
Lôbo GC, Paiva KL, Silva ALG, Simões MM, Radicchi MA, Báo SN (2021) Nanocarriers used in drug delivery to enhance immune system in cancer therapy. Pharmaceutics 13(8):1167
Lu H, Zhang S, Wang J, Chen Q (2021) A review on polymer and lipid-based nanocarriers and its application to nano-pharmaceutical and food-based systems. Front Nutr 2021(8):783831
Malik S, Muhammad K, Waheed Y (2023) Emerging applications of nanotechnology in healthcare and medicine. Molecules 28(18):6624
Mitchell MJ, Billingsley MM, Haley RM, Wechsler ME, Peppas NA, Langer R (2021) Engineering precision nanoparticles for drug delivery. Nat Rev Drug Discov 20(2):101–124
Moni SS, Abdelwahab SI, Jabeen A, Elmobark ME, Aqaili D, Gohal G, Oraibi B, Farasani AM, Jerah AA, Alnajai MMA, Mohammad Alowayni AMH (2023) Advancements in vaccine adjuvants: the journey from alum to nano formulations. Vaccine 11(11):1704
Munn DH, Bronte V (2016) Immune suppressive mechanisms in the tumor microenvironment. Curr Opin Immunol 39:1–6
Nandedkar TD (2012) Nanotechnology a path to nanovaccine. Int J Pharma Bio Sci 3(2):290–292
Niżnik Ł, Noga M, Kobylarz D, Frydrych A, Krośniak A, Kapka-Skrzypczak L, Jurowski K (2024) Gold nanoparticles (AuNPs) – toxicity, safety and green synthesis: a critical review. Int J Mol Sci 25(7):4057
Pati R, Shevtsov M, Sonawane A (2018) Nanoparticle vaccines against infectious diseases. Front Immunol 9:2224
Perotti M (2020) Design and evaluation of a self-assembling nanoparticle vaccine against human cytomegalovirus. Doctoral dissertation, ETH Zurich
Pusic K, Aguilar Z, McLoughlin J, Kobuch S, Xu H, Tsang M, Wang A, Hui G (2013) Iron oxide nanoparticles as a clinically acceptable delivery platform for a recombinant blood-stage human malaria vaccine. FASEB J 27(3):1153
Rana I, Oh J, Baig J, Moon JH, Son S, Nam J (2023) Nanocarriers for cancer nano-immunotherapy. Drug Deliv Transl Res 13(7):1936–1954
Ray P, Haideri N, Haque I, Mohammed O, Chakraborty S, Banerjee S, Quadir M, Brinker AE, Banerjee SK (2021) The impact of nanoparticles on the immune system: a gray zone of nanomedicine. J Immunol Sci 5(1)
Sanmamed MF, Chen L (2018) A paradigm shift in cancer immunotherapy: from enhancement to normalization. Cell 175(2):313–326
Shah A, Dobrovolskaia MA (2018) Immunological effects of iron oxide nanoparticles and iron-based complex drug formulations: therapeutic benefits, toxicity, mechanistic insights, and translational considerations. Nanomedicine 14(3):977–990
Sharma S, Parveen R, Chatterji BP (2021) Toxicology of nanoparticles in drug delivery. Curr Pathobiol Rep:1–12
Shiravand Y, Khodadadi F, Kashani SMA, Hosseini-Fard SR, Hosseini S, Sadeghirad H, Ladwa R, O'Byrne K, Kulasinghe A (2022) Immune checkpoint inhibitors in cancer therapy. Curr Oncol 29(5):3044–3060
Silva JM, Zupancic E, Vandermeulen G, Oliveira VG, Salgado A, Videira M, Gaspar M, Graca L, Préat V, Florindo HF (2015) In vivo delivery of peptides and Toll-like receptor ligands by

mannose-functionalized polymeric nanoparticles induces prophylactic and therapeutic anti-tumor immune responses in a melanoma model. J Control Release 198:91–103
Smith AA, Gale EC, Roth GA, Maikawa CL, Correa S, Yu AC, Appel EA (2020) Nanoparticles presenting potent TLR7/8 agonists enhance anti-PD-L1 immunotherapy in cancer treatment. Biomacromolecules 21(9):3704–3712
Swetha K, Kotla NG, Tunki L, Jayaraj A, Bhargava SK, Hu H, Bonam SR, Kurapati R (2023) Recent advances in the lipid nanoparticle-mediated delivery of mRNA vaccines. Vaccine 11(3):658
Tao W, Hurst BL, Shakya AK, Uddin MJ, Ingrole RS, Hernandez-Sanabria M, Arya RP, Bimler L, Paust S, Tarbet EB, Gill HS (2017) Consensus M2e peptide conjugated to gold nanoparticles confers protection against H1N1, H3N2 and H5N1 influenza A viruses. Antivir Res 141:62–72
Tariq H, Batool S, Asif S, Ali M, Abbasi BH (2022) Virus-like particles: revolutionary platforms for developing vaccines against emerging infectious diseases. Front Microbiol 12:790121
Tenchov R, Bird R, Curtze AE, Zhou Q (2021) Lipid nanoparticles – from liposomes to mRNA vaccine delivery, a landscape of research diversity and advancement. ACS Nano 15(11):16982–17015
Thi TTH, Suys EJ, Lee JS, Nguyen DH, Park KD, Truong NP (2021) Lipid-based nanoparticles in the clinic and clinical trials: from cancer nanomedicine to COVID-19 vaccines. Vaccine 9(4):359
Tie Y, Tang F, Wei YQ, Wei XW (2022) Immunosuppressive cells in cancer: mechanisms and potential therapeutic targets. J Hematol Oncol 15(1):61
Tornesello AL, Tagliamonte M, Tornesello ML, Buonaguro FM, Buonaguro L (2020) Nanoparticles to improve the efficacy of peptide-based cancer vaccines. Cancer 12(4):1049
Tornesello AL, Tagliamonte M, Buonaguro FM, Tornesello ML, Buonaguro L (2022) Virus-like particles as preventive and therapeutic cancer vaccines. Vaccine 10(2):227
Venditti I, Cartoni A, Cerra S, Fioravanti R, Salamone TA, Sciubba F, Tabocchini MA, Dini V, Battocchio C, Iucci G, Carlini L (2022) Hydrophilic gold nanoparticles as anti-PD-L1 antibody carriers: synthesis and interface properties. Part Part Syst Charact 39(4):2100282
Wang N, Zhou Y, Xu Y, Ren X, Zhou S, Shang Q, Jiang Y, Luan Y (2020) Molecular engineering of anti-PD-L1 peptide and photosensitizer for immune checkpoint blockade photodynamic-immunotherapy. Chem Eng J 400:125995
Wilson B, Geetha KM (2022) Lipid nanoparticles in the development of mRNA vaccines for COVID-19. J Drug Deliv Sci Technol 74:103553
Yao Y, Zhou Y, Liu L, Xu Y, Chen Q, Wang Y, Wu S, Deng Y, Zhang J, Shao A (2020) Nanoparticle-based drug delivery in cancer therapy and its role in overcoming drug resistance. Front Mol Biosci 7:193
Yetisgin AA, Cetinel S, Zuvin M, Kosar A, Kutlu O (2020) Therapeutic nanoparticles and their targeted delivery applications. Molecules 25(9):2193
Zhang G, Tang T, Chen Y, Huang X, Liang T (2023) mRNA vaccines in disease prevention and treatment. Signal Transduct Target Ther 8(1):365
Zhu Y, Yu X, Thamphiwatana SD, Zheng Y, Pang Z (2020) Nanomedicines modulating tumor immunosuppressive cells to enhance cancer immunotherapy. Acta Pharm Sin B 10(11):2054–2074
Zielińska A, Costa B, Ferreira MV, Miguéis D, Louros JM, Durazzo A, Lucarini M, Eder PV, Chaud M, Morsink M, Willemen N (2020) Nanotoxicology and nanosafety: safety-by-design and testing at a glance. Int J Environ Res Public Health 17(13):4657

# Chapter 11
# Nanovaccines for Infectious Diseases

**Marriam Zargham and Muhammad Farhan Khan**

**Abstract** The development of nanoparticle-based vaccines and immunotherapies represents a groundbreaking amelioration in biomedicine, offering novel disease prevention and treatment strategies. This chapter sheds light on a comprehensive overview of the design, mechanisms, and applications of nanofabricates in enhancing the efficacy of vaccines and immunotherapeutic agents. Key topics include the unique properties of nanoparticles that make them suitable for delivering antigens and adjuvants, such as their ability to improve stability, enhance cellular uptake, and facilitate targeted delivery to immune cells. The chapter explores various types of nanoparticulates, including polymeric nanoforms, liposomes, metal nanoparticulates, and virus-like particles, highlighting their role in inducing robust and long-lasting immune responses. The chapter also aims to provide readers an in-depth comprehension of the present situation and potential of vaccines and immunotherapies based on nanoparticles, emphasizing their importance in promoting global health and preventing various diseases.

**Keywords** Adaptive · Adjuvants · Antibodies. Antiviral · Antigen presenting cell · Autoimmune · Biofilms · Checkpoint inhibitor · COVID-19 · Humoral · Immunosuppressors · Immunotherapies · Innate · Malignancies · Nanofabricates · Nanovaccines · Polymeric nanoforms · T cells · Virus-like particles

## 11.1 Introduction

The rapid increase in the world's population has a significant impact on both global health and economic values. The likelihood of infectious illness transmission climbs sharply with population density. Living in deteriorating environmental conditions

M. Zargham · M. F. Khan (✉)
University of Health Sciences, Lahore, Pakistan
e-mail: mfarhankhan1987@uhs.edu.pk

 
V. Khanna et al. (eds.), *Nanotechnologies for Drug Delivery and Biopharmaceutical Development*, Progress in Drug Research 77,
https://doi.org/10.1007/978-3-032-00796-4_11

and traveling globally all contribute to the spread of infections, which raises the possibility of pandemics and epidemics. New infections and the resurgence of other venerable diseases are being facilitated by urbanization, environmental changes, and human behavior. This highlights the need for creative healthcare solutions to satisfy the needs of the world's expanding population.

One of the biggest advances in public health in the past century is vaccines. Vaccinations are thought to save two to three million lives annually. They have successfully exterminated smallpox and significantly decreased the prevalence of several serious illnesses, including measles and polio. Currently, licensed vaccinations are available to prevent more than 30 distinct infectious diseases (Micoli et al. 2021). To impede infectious diseases and save lives, vaccinations are one of the most potent weapons that exist in modern medicine. Vaccines give lifelong protection against diseases that were historically common and frequently fatal by boosting the immune system's competence to identify and fight detrimental microorganisms (Gary and Weiner 2020).

Viral vaccines such as those for measles, mumps, rubella, chickenpox, and hepatitis A that were made from suitable cell culture substrates prevented 10 million deaths between the middle of the 1960s and 2015. Nonetheless, vaccine-preventable diseases still affirm the lives of almost 1.5 million children under the age of five every year. This is mostly because these young children cannot obtain the requisite childhood vaccinations. On the other hand, because legitimate and effective preventive vaccinations are now widely available, deadly diseases like rabies, smallpox, and polio, which have killed hundreds of millions of people in the last few centuries, have largely been contained. According to WHO estimates, the current generation of immunizations saves 2–3 million deaths around the world each year (Ward and Van Damme 2018).

The vaccination's physiological systems are well known. Upon vaccination, the immune system gets triggered and both adaptive and innate immune responses are elicited. As a consequence, the humoral response develops antibodies or, in the case of repeated exposure, memory cells that will be competent to identify the same antigen. The durability and potency of immunizations can be increased by administering repeated injections regularly. Depending on the biology of the infection and the disease that has to be averted, a vaccination may obligate to trigger certain humoral (i.e., antibodies) or cell-mediated (i.e., T cells) adaptive immunological responses effectively. Therefore, it's critical to comprehend how vaccines function to forecast their effectiveness, toxicity profile, and anticipated benefits for both the vaccinated and the broader public (Soares-Weiser et al. 2019).

Here are concise descriptions of different categories of vaccines (Table 11.1) (Kashte et al. 2021; Cheng et al. 2020).

**Table 11.1** Types of vaccines

| Vaccines | Mechanism | Application (Fries et al.) |
|---|---|---|
| Live attenuated vaccines | Incapacitated variant of the infection that does not infect healthy individuals but is still liable to replicating. | Work for measles, mumps, chicken pox, etc. |
| DNA vaccines | Plasmid-based vaccines involve the introduction of the pathogen genetic material to initiate the immune system | COVID 19 |
| Inactivated vaccines | The pathogen is killed but able to trigger immunity in the body | Hep A, Rabies |
| mRNA vaccines | Involved in the application of mRNA for protein synthesis from pathogen triggering an immune response | COVID 19 |
| Vector vaccines | Nonpathogenic viral vectors are used to deliver pathogenic genetic material to initiate immunity | COVID 19 |
| Toxoid vaccines | Utilized inactivated toxins from pathogens for immune response triggering | Diphtheria, Tetanus |
| Polysaccharide | Use pathogenic units like protein or sugar for immune response generation | Hep B, Pneumonia |

### *11.1.1 History of Vaccine Development*

The Chinese are acknowledged for creating the first smallpox vaccination and prevention techniques, which go back to around 1000 A.D. A Chinese statesman's son reputedly received a smallpox vaccination by blowing powdered wounds into his nose. Another approach for inoculation involved taking fluid out of an infected person's pustules and applying it to a healthy person's skin lesion. This practice made its way to Europe, later adopted by Turkey in 1672. It was not until 1721, after six centuries, that the discrepancy was brought to Great Britain (Santos and Almeida 2021). In 1796, Edward Jenner claimed that milkmaids who had previously been infected with cowpox naturally were immune to smallpox in the United Kingdom. He found that susceptible hosts could not get smallpox when vaccinated with modest amounts of pus from cowpox lesions, which likely embraced a vaccinia-related virus. In 1798, a vaccine against smallpox emerged. The tetanus and diphtheria toxoid vaccines were invented in 1923 as a result of techniques discovered by Alexander Glenny and Barbara Hopkins to use formaldehyde to inactivate bacterial toxins (Yang et al. 2022).

In the 1960s, the Walter Reed Army Institute of Research manufactured vaccines exploiting capsular polysaccharides of encapsulated organisms, including influenza type b, Hemophilus, meningococci, and later pneumococci. To defend several serotype variations of polysaccharide capsules, polyvalent vaccines were developed. To ameliorate its efficacy, particularly for infants, these vaccines were subsequently coupled to carrier proteins, which activate T-cells that aid in the induction of memory B-cells. (Kayser and Ramzan 2021). By the end of the twentieth century, genome sequencing and molecular genetics progression had enabled vaccine manufacturing against RNA viruses with several epitope variations. The live and disabled influenza and live rotavirus vaccines are two examples of these vaccinations. DNA excision and modification enabled the hepatitis B viral vectors to recruit the surface antigen. The human papillomavirus (Jamal) vaccine exhibits improved immunogenicity because each virus's L1 antigen creates virus-like particles (Yigit et al. 2021).

### *11.1.2 Immunotherapies Development*

Different immunotherapies have been employed to harness the body's immune system to fight against several diseases, like cancer, and other autoimmune diseases, infections, etc. The table below (Table 11.2) enlists some of the important therapies used widely for combating diseases (Varadé et al. 2021) (Lv et al. 2022; Billan et al. 2020).

The notion that febrile infections in cancer patients were occasionally linked to cancer remission emerged in the eighteenth century. Leading New York surgeon William Coley (1862–1936), considered as father of cancer immunotherapy became interested in the field after witnessing a cancer patient have two bouts of erysipelas brought on by an acute Streptococcus pyogenes infection, which resulted in an entire remission of the ailments. Coley then gave cancer patients injections of streptococcal cultures that Robert Koch had donated, and in certain instances, he saw a regression of the tumor. His work was published in 1863 (Waldman et al. 2020). By the mid of 1950–1960s research in the field of immunotherapies through the discovery of signaling proteins and cytokines paved the way for producing interleukins and interferons for immunization against cancer and other infections by 1975 monoclonal antibody discovery led to the opening of modern technology of

**Table 11.2** Different Immunotherapies

| Types of immunotherapies | Mechanism | Application |
|---|---|---|
| Cytokine therapy | Synthetic-based cytokines like interferons and interleukins, tumor necrosis factors, are used to leverage the body's immune system against diseases | Melanoma<br>Hepatitis C |
| Monoclinal antibodies | Synthetic antibodies target antigens and cancerous cells, mark them, and block their signals | Non-Hodgkin's lymphoma, breast cancer |
| Cancer vaccines | Implant-specific cancer antigens trigger the immune system to recognize and combat cancer cells. | Prostate cancer |
| Checkpoint inhibitors | Disable proteins that prohibit immune cells from targeting cancer cells, such as PD-1/PD-L1 or CTLA-4 | Kidney cancer, lung cancer |
| CAR T cell therapy | Adoptive cell transfer, in which the T lymphocytes of a patient are manipulated to detect and combat cancer cells more efficiently. | Lymphoma, leukemia |

immunotherapies. By 1992 first IL-2 interferon approved for treating kidney metastasis (Waldman et al. 2020).

By 2010, a cancer vaccine, sipuleucel-T (Provenge), was approved for prostate cancer immunization. In 2011, the first checkpoint ipilimumab discovery for melanoma treatment led to new pathways for research in neoplasia treatment. The first CAR T-cell treatments, tisagenlecleucel (Kymriah) for leukemia and axicabtagene ciloleucel (Yescarta) for lymphoma, were authorized by the FDA in 2017 (Salas-Benito et al. 2021). A patient's T cells are programmed to seek out cancer cells in these therapies through genetic engineering. In 2020, the COVID-19 pandemic led to the discovery of mRNA-based vaccines for immunizing the body against the SARS-CoV-2 virus (Wallis et al. 2023).

Many oncological-based treatments using engineered viruses and a combination of different therapies are also in the treatment pipeline for immunizing human beings for various cancers, infections, and other autoimmune disorders. This immunotherapy history illustrates the understanding and utilization of the immune system to cure illness. Immunotherapy has changed the face of medicine and continues to remain a viable treatment option for several diseases, from the early discoveries of vaccines to state-of-the-art therapies, to the same degree as checkpoint inhibitors and CAR T-cells (Butterfield and Najjar 2024).

## 11.2 Nanotechnology-Based Vaccines and Other Immunotherapies

The advancement of nanotechnology also reconstituted the industry of pharmaceutics and therapeutics. It amplifies vaccine and therapeutics development with more effectiveness and high-precision nanoparticles engineered according to the requirement for safe delivery and protection of therapeutic agents to boost the body's immune system.

Nanomanipulators can also be engineered for contained release, extending immune exposure and enhancing responses, or tailored to target particular immune cells for a more intended effect. Certain nanofabricates function as adjuvants, boosting immune responses without additional substances, and their multifunctionality enables them to transmit multiple components, such as adjuvants, immune stimulators, or antigens, which provide more potent and complex vaccines. By administering pharmaceuticals or antigens directly to targeted cells, nanoforms in immunotherapies improve the efficacy of treatments for cancer and autoimmune ailments by boosting the immune system's capacity to combat tumors or control inflammation. Nanofabrication has tremendous promise for future more effective, customized vaccinations and immunotherapies, despite concerns with safety and scalability (Gowd et al. 2022; Jaha et al. 2024).

**Table 11.3** Different types of nanomaterial for therapeutics

| Types of nanoparticles | Employment in vaccines | Employment in other therapeutics |
|---|---|---|
| Gold nanomaterial (gold-based particles) | Cancer vaccine under research, malarial vaccine | Cancer therapy targeting tumor cells |
| Carbon nanotube (cylindrical structure of carbon nanomaterial) | The tuberculosis vaccine is under research | Drug delivery system for neurogenerative disorder and cancer treatment |
| Liposomes (encapsulated lipid bilayer spherical vesicles | In the research stages for the delivery of antigen or adjuvant in experimental vaccines | Anti-inflammatory drugs for autoimmune disorder |
| Magnetic nanoforms (iron oxide-based) | N/A | Cancer immunotherapy also tracks the immune response |
| Polymeric nanoparticles (like poly(lactic-co-glycolic) acid0) | HIV vaccine, cancer vaccine | (cytokine) IL-2 for chemoimmunotherapy |
| Virosome (virus-like without genetic material) | HVB vaccine, influenza vaccine | N/A |
| Lipid nanoparticles (fat-based for genetic material transformation) | mRNA vaccines of COVID-19 | CAR-T cell delivery-based therapy, gene delivery for cancer therapy |

Typical preparations for nanomedicine include nanocrystals, polymers, liposomes, dendrimers, 2D materials, micelles, nanotubes, and other nanofabricates. Different kinds of nanoforms were employed amid these formulations based on nanoparticulates, principally because they showed deviating capacities for tumor accumulation, cancer cells uptake, and cellular reactions. Furthermore, nanoforms facilitated combination treatments such as radio-immunotherapy, photo-immunotherapy, chemo-immunotherapy, etc. Nano-based cancer immunotherapy is now perceived as a crucial component of cancer treatment (Zargham et al. 2024a).

Here is a table (Table 11.3) detailing some forms of nanomaterials employed in vaccines and therapeutic development (Gowd et al. 2022; Cremolini et al. 2021; Song et al. 2022):

## 11.3 Nanomaterial-Based Vaccines

The administration of nanoformulates as a vaccine delivery technology has drawn particular attention in recent decades to optimize vaccination potency. These nanoparticles may consist of nonmetal inorganics, metal, and various virus-like particles, polymers, and lipids that have undergone testing, some of which have been approved for application in humans and animals. In light of the properties of nanomaterials, it is now possible to target specific antigen-presenting cells and enhance vaccination procedures to produce protection (Zargham et al. 2024b). Conserving the antigens from early proteolytic destruction, modulating antigen release, boosting antigen impregnation and processing by antigen-presenting cells, and ensuring

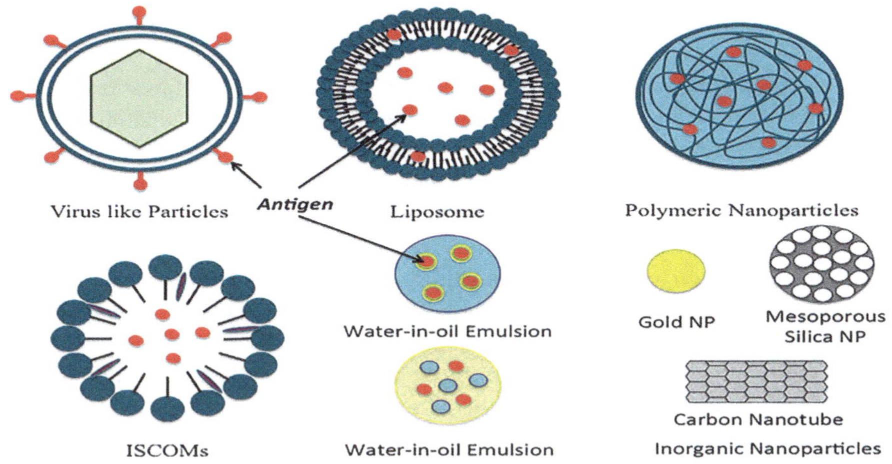


**Fig. 11.1** Different nanoforms for vaccinations. (Chowdhury and Kundu 2023) MDPI CC BY 4.0

their safety for usage in humans and animals are the primary functions of the nanoparticles. The physicochemical traits of the nanoforms could also be altered (Poon and Patel 2020) to target distinct targets. Figure 11.1 given below shows different nanoforms of vaccinations.

The therapeutic effectiveness of nanovaccines is significantly influenced by the aspects of nanofabricates, including surface, shape, charge size, and hydrophobicity. Immune responses are strengthened by the agglomeration of smaller particles (20–50 nm) in lymph nodes. Spherical nanoforms often elicit more immunological responses than other forms. Hydrophilic particles may have a lesser percussion than hydrophobic ones, which are more readily absorbed by immune cells and evoke a more vigorous immunological response. The surface charge of nanomedicine is also important; negatively charged nanoparticles have been established to elicit stronger immune responses, whereas positively charged nanoparticles are rapidly absorbed by dendritic cells but may trigger inflammatory reactions (Manju et al. 2023).

In the era of nanotechnology, an exceptionally potent nanovaccine has been developed for COVID-19, a virus produced by SARS-CoV-2. Researchers designed mRNA vaccines, such as Moderna (mRNA-1273) and Pfizer-BioNTech (BNT162b2), by enclosing and protecting the mRNA encoding the viral spike (S) protein with lipid nanoparticles (LNPs). These nanoparticles proved easier to penetrate host cells, translating the mRNA and eliciting a strong immunological response. Strong neutralizing antibody production and effective antigen presentation were rendered possible by the LNPs' improved cellular absorption and stability. With over 90% efficacy in clinical studies, this approach constitutes a major accomplishment in vaccination science.

### 11.3.1 Nanovaccine Delivery and Cellular Uptake

Nanovaccines function by precisely and carefully administering adjuvants and antigens to immune cells, enhancing immune responses while eliminating adverse effects. Their primary mechanisms of action include immunological activation, cellular absorption, antigen processing, and antigen transport, which culminate in the development of protective adaptive immunity.

Antigens are adequately transmitted by nano-vaccines to antigen-presenting cells (APCs), which are pivotal for inducing immunological responses and include dendritic cells, macrophages, and B cells. In addition to active targeting through surface modifications exploiting ligands that bind to APC receptors, these nanocarriers accelerate antigen absorption through passive targeting modalities such as endocytosis and enhanced permeability. In these reactions, DCs are significant. Particularly in cases of intracellular bacterial or viral infections, activated CD8+ T cells, also known as cytotoxic T lymphocytes, or CTLs, are pivotal for the direct destruction of infected cells (Khan 2024).

The fundamental barrier connecting the innate and adaptive immune systems comprises antigen-presenting cells (APCs), especially dendritic cells (DCs). After DCs relocate from peripheral tissues to lymph nodes, they activate naïve T cells, controlled by CCR7 and its ligands, CCL19 and CCL21. Lymph nodes, which include various APCs such as CD8+ DCs and CD169+ macrophages, are pivotal for immunological responses because they cross-present antigens to T cells while stimulating the production of antibodies in B cells. Direct transit of antigens to lymph nodes optimizes immune responses and prevents early antigen presentation that may result in tolerance (Heine et al. 2024).

### 11.3.2 Strategies Involve for Targeting APC's

Targeting specific molecules at APC receptors optimizes antigen recuperation and presentation, which boosts immune responses. As evidenced by CD11c-targeted porous silicon nanoparticles (pSiNPs) loaded with rapamycin, which increased the production of regulatory T cells to promote immunological tolerance in organ transplants, CD11c is an important receptor that is mostly found on DCs and amplifies strong T cell responses. While Langerin and the mannose receptor aid in DC circulation and antigen uptake, receptors including CD44, DEC205, and DC-SIGN facilitate antigen capture and cross-presentation. TLRs play a crucial role in bridging the disparity between innate and adaptive immunity by promoting the maturation of DCs (Shaw et al. 2024).

#### 11.3.2.1 Antigen Cross-Presentation and Action

The processing and expression of exogenous antigens on MHC class I molecules, known as cross-presentation, discovered in 1976, it is a crucial step that activates CD8+ T cells and elicits strong immune responses against intracellular infections and malignancies. Certain nanovaccines may relocate antigens to the cytosol, permitting MHC class I loading and boosting cytotoxic T lymphocyte (CTL) responses, even though exogenous antigens are typically concentrated on MHC class II presentation, which stimulates CD4+ T cells. The two main mechanisms by which this antigen evasion occurs are membrane fusion and endosomal swelling and rupture. Cationic polyethyleneimine (PEI)-based nanovaccines take advantage of the proton-sponge effect, which occurs when protons released by vacuolar ATPase (v-ATPase) are absorbed by extraneous amino groups, resulting in an influx of water and chloride ions (Dewangan and Tomar 2022). This causes endosome rupture and osmotic swelling, which permits the antigen to penetrate the cytosol, avoid lysosomal degradation, and ensure transmission via the MHC class I route. The membrane fusion mechanism, which is applied by pH-sensitive liposome-based nanovaccines, happens once the acidic maturing endosome environment destabilizes the liposomal membrane, inducing it to merge with the endosomal membrane and release the antigen into the cytosol. After the antigen enters the cytosol, it is battered by proteases into peptide fragments, which are then channeled into the endoplasmic reticulum by proteins called Transporter Associated with Antigen processing (Ambika & Singh). After being loaded onto MHC class I molecules, the antigenic peptides circulated to the dendritic cell (Bruce et al.) surface, where CD8+ T lymphocytes can detect them and initiate a strong immunological response (Achmad et al. 2022).

By stimulating T and B cell activation, nanovaccines assist the body in acquiring an enduring immunological memory that enables it to react promptly and effectively to future infections. In short, nanovaccines are immensely effective for infectious diseases, cancer immunotherapy, and customized vaccines because they augment immune responses by targeting APCs, enhancing antigen presentation, promoting strong T and B cell activation, and eliciting memory immunity. Figure 11.2 given below describes the scheme of liposome nanovaccines delivery in humans.

### *11.3.3 Nanovaccine vs Conventional Vaccines: Advantages and Disadvantages*

Nanovaccines prove more potent than conventional vaccines, particularly in optimizing antigen delivery and modifying immune responses. Nanovaccines augment the efficacy of immune activation by enhancing antigen uptake by antigen-presenting cells (APCs), such as dendritic cells (DCs), due to their nanoscale size. Additionally, they facilitate cross-presentation, which is vital for strong CD8+ T cell responses to

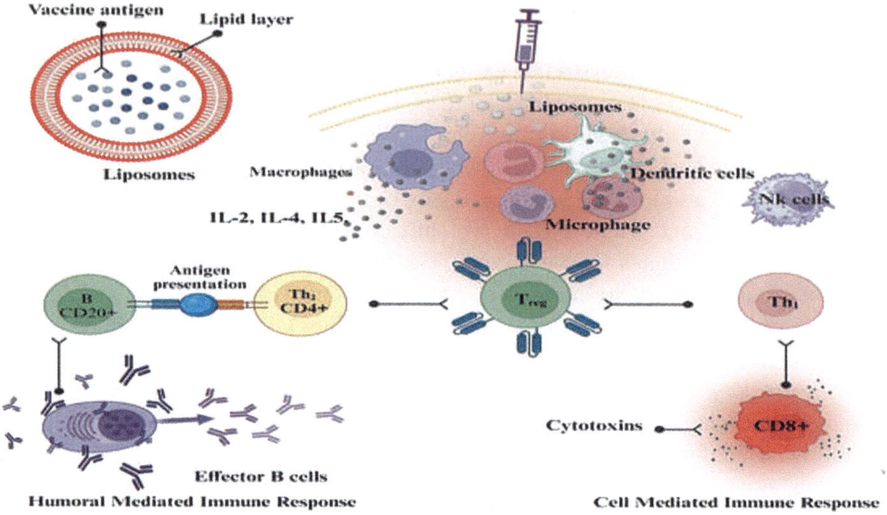


**Fig. 11.2** Scheme of liposome nanovaccines delivery in humans. (Moni et al. 2023) MDPI CC BY 4.0

intracellular infections and malignancies by enabling foreign antigens to be processed via the MHC class I pathway. Additionally, tailored and extended antigen release delivered through nanovaccines can reduce the need for multiple booster doses. Their improved resilience minimizes the need for cold-chain logistics, a major disadvantage of conventional vaccinations, and often facilitates storage at room temperature. To enable customized immune response modification, nanovaccines can also be tailored to include many antigens and immune-boosting ingredients into a single formulation. To improve patient compliance, some nanovaccines can also be administered mucosal or without a needle (Yadav et al. 2018).

Considering these benefits, when compared to conventional vaccinations, nanovaccines face several obstacles. Large-scale manufacturing is made harder and costlier by its complex production process, a prerequisite for advanced nanotechnology. Another difficulty is getting regulatory authorization in light of the special properties of nanoparticles, which necessitate extensive safety and toxicological evaluations. Some nanomaterials may cause long-term biocompatibility challenges, which could result in organ accumulation or inadvertent immune system overactivation. Furthermore, nanovaccines are more expensive than conventional vaccinations that use well-established manufacturing technologies due to the significant financial resources required for their research and development (Ahmad et al. 2024). Nanovaccines are still in the early stages of widespread clinical validation, whereas traditional vaccinations have numerous years of clinical trial evidence supporting their safety and effectiveness. Additionally, even though nanovaccines are intended to enhance antigen targeting to lymph nodes, precise delivery remains a hurdle. Generally speaking, classic vaccinations persist as a more affordable and accessible option for mass vaccination campaigns, even though nanovaccines exhibit significant promise for future immunization measures (Dey and Singh 2023).

## 11.4 Nanovaccines for Viral Infections

For two reasons, certain viral infections are not treatable with traditional vaccinations. First, the in vitro culture of the virus is the first stage in manufacturing attenuated or inactivated virus-based vaccines. Another is the fragile stability and in vivo, abasement of vaccines incorporating peptide antigens or mRNAs encoding antigens confine their applicability. The therapeutic value of nanovaccines can be increased by using nanomaterials as carriers to shield antigens from deterioration and boost immunological responses. Positive preclinical results have been reported for HIV, SARS-CoV-2, or respiratory viruses since the first nanovaccine against the hepatitis B virus was licensed in 1986. Nano innovations have also been used to harvest vaccines against human papillomaviruses and the hepatitis E virus (Gareev et al. 2022; Li et al. 2021). The ability of nano-based vaccines to deliver viral antigens to target immune cells, activate DCs (dendritic cells), effectively activate B cells and elicit humoral immune responses, and mimic viruses in size and structure (pathogen-mimicking VLPs) are some of the distinguishing characteristics that contribute to their immunological advantages. Studies suggest that nanomaterials inhibit viral reproduction and cell invasion. Moreover, many surface binding sites allow them to engage with target molecules, which trap and deactivate viruses (Zeng et al. 2021).

Here, below (Table 11.4) enlist the names of vaccines produced for viral infections (Chakravarty and Vora 2021; Kim et al. 2021b; Li et al. 2021).

Different nanofabricates have a significant impact on antiviral activity for various diseases. To prevent viruses like HIV-1, HBV, and Zika from penetrating healthy cells, viruses are trapped using nanodecoys, which imitate cell membrane receptors. Viral surface proteins interact with nanomaterials like gold and silver nanoparticles (AgNms) to restrict viral entry, preventing viruses like herpes, influenza, and HIV from attaching and penetrating (Jackman et al. 2021). Many nanomaterials, including as silver nanoparticles (AgNPs) and gold nanoparticles (AuNPs), obstruct viral surface proteins, such as HIV's gp120, which the virus needs to attach to CD4+ T cells (Kim et al. 2021a). By adhering to viral glycoproteins, AgNPs additionally inhibit the entry of viruses like the flu, and herpes simplex (HSV). This impending capacity is enhanced and viruses like hepatitis A, norovirus, HSV-1, and HSV-2 are repelled by modified nanomaterials such as poly quaternary phosphonium oligochitosan (PQPOC)-decorated AgNPs and glycosaminoglycan-coated silicon nanoparticles. In addition, carbon-based nanotechnologies, such as carbon dots and fullerenes, constrain the passage of viruses by either engaging with receptors like histo blood group antigens (HBGAs) to avert norovirus invasion or by blocking host cell receptors like DC-SIGN, which is necessary for the internalization of the Ebola virus (Bezbaruah et al. 2022).

**Table 11.4** Vaccines for viral infections

| Virus/antigen | Nm based vaccine | Trial result |
|---|---|---|
| HBV/HbsAg | VLP (virus-like protein) | For clinical use |
| HEV p239 (aa 368–606) | VLP (virus-like protein) | For clinical use |
| MERS-CoV | Spike protein nanofabricate | Employed for substantial titers of neutralizing antibodies and triggering Th2 immune response. |
| | Ferritin base VLP | CD4+<br>T-cell stimulation and IFN-/TNF- responses |
| SARS-CoV-2 (mRNA encoded length spikes) | LNP (lipid nanomaterial) | Phase I trial that produced a high titer of neutralizing antibodies (NCT04283461); phase III trial is presently underway (NCT04470427). |
| HIV (Env antigen) | Polystyrene nanofabricates | Cause an increase in antibody secretion in mice |
| Influenza virus (killed swine influenza virus antigens) | Cationic alpha-D-glucan nanomanipulator with TLR3 agonist poly(I:C) as adjuvant | Actuate elevated concentrations of secretory IgA antibodies specific to cross-reactive viruses in the nasal passage and lungs, and viral-neutralizing antibodies in bronchoalveolar lavage fluid. |
| Dengue virus (DENV) DENV-2 E protein | PLGA Nano Forms | Compared to the free E protein, enhanced the anti-E IgG titer and the antibodies' ability to neutralize mice |

Traditional vaccines for establishing immunity against HIV are not much effective in eliciting a strong immune response to tackle this challenge, (Xu et al. 2012) investigated the possible adjuvants for DNA vaccines, investigators modified the surfaces of gold nanorods (NRs) with three distinct molecules: poly (diallyldimethylammonium chloride) (PDDAC), cetyltrimethylammonium bromide (CTAB), and polyethyleneimine (PEI) to boost immune activation. Cellular and humoral immune responses were markedly enhanced by PDDAC- and PEI-modified gold nanorods in contrast to a naked HIV-1 Env plasmid DNA vaccination. The potential of these altered NRs as a nano-adjuvant for designing an HIV vaccine was demonstrated by the stimulation of antigen-presenting cell activation and increased T-cell proliferation. The vital role of surface chemistry when developing low-toxicity, highly effective nanomaterials for immunotherapy and vaccination delivery is emphasized in this work.

Viral replication is hampered by nanomaterials by direct interactions with viral genomes or proteins, or by provoking an intracellular milieu that is inhospitable to viral replication. While fullerene derivatives limit HIV-1 maturation by restricting Gag polyprotein processing, gold nanoparticles (AuNPs) bind to viral RNA or replicative proteins to hinder FMDV replication (Curley and Putnam 2022). Furthermore, to forestall viral propagation and cell death, organic nanomaterials such as PASomes control the amounts of reactive oxygen species (ROS). Better research on managing virus-infected intracellular pathways, like the MEK/ERK cascade, is required to augment the design of antiviral nanomaterials. Furthermore, viruses like the norovirus and influenza can be neutralized by using nanomaterials like graphene oxide and titanium dioxide (TiO2) to physically destroy or photocatalytically oxidize the virus's envelope or capsid (Fries et al. 2021). Hence nanomaterial-based vaccines proved to be more promising, reliable, target specific, and potent against viral diseases.

## 11.5 Nano-Based Vaccine for Bacterial Diseases

Antibacterial resistance is the key obstacle offered by bacteria toward mankind. Moreover, many bacteria create intricate communities in their surroundings known as biofilms. Nanomedicine provides promising accouterments to overcome these obstacles, providing a better cure against bacterial infections (Hamad and Salman 2023). Nanoparticles (Zhang et al.) are efficacious as vaccination vectors for transmitting antigenic proteins that trigger both humoral and cellular immune responses. They have potential to treat refractory bacterial infections by delivering outer membrane vesicles (OMVs) that stimulate immunity, such as carbapenem-resistant Klebsiella pneumoniae (CRKP). After being triggered by antigen-presenting cells via the PPAR-$\gamma$ pathway, NFs carrying polysorbitol transporter (PST) and pneumococcal surface protein A (PspA) produce long-term immunity against Streptococcus pneumoniae via T helper (Th)2 and T follicular helper (Tfh) cells. NPs can function as mucosal adjuvants in subunit vaccinations likewise. Nanoforms sheathed with PEGylated lipid bilayers (NP@Lipo) or extracellular vesicle membranes (NP@EV) show prospective ability to boost drug delivery to afflicted organs and target bacteria in Staphylococcus aureus infections (Shahmoradi et al. 2023).

Melioidosis and glanders are caused by facultative intracellular pathogens Burkholderia mallei (Bm) and B. pseudomallei (Bpm). There are currently no effective vaccinations to prevent glanders or human melioidosis. However, due to the germs' seeming perfect traits for malevolent use as a biothreat weapon, there has been a renewed focus on creating vaccinations against Burkholderia. Furthermore, a vaccine will be advantageous for immunizing vulnerable groups in regions where melioidosis and glanders are endemic. The study was conducted by Tapia et al. (2021). The primary objective of this research is to formulate a multicomponent vaccine that targets these infectious agents, which pose a risk to public health as well as a biothreat. In mice infection models, the investigation assessed glycoconjugate antigens correlated with AuNPs and revealed that intranasal administration of these formulations provided substantial immunity against lethal infections. Immunity was augmented as a result of strong humoral responses specific to the antigen. Additionally, proteins and AuNPs successfully reduced EHEC colonization and elicited robust mucosal and systemic immune responses (Tapia et al. 2021).

Summarising some nanoparticle-based vaccines for bacterial diseases in the table given below (Table 11.5) (Pati et al. 2018; Tafaghodi et al. 2020).

**Table 11.5** Vaccine for bacterial infection

| Bacteria/antigen | Nanocomposite of vaccine | Clinical trials result |
|---|---|---|
| Staphylococcus aureus | PLGA loaded with bacterial proteins | Proved to be the enhancer of antibodies and T-cell level in animal models |
| Salmonella Typhi | Chitosan-based nanomaterial along with outer membrane vesicles | Enhanced IgA and IgG level in a mouse model against typhoid |
| Bacillus anthrax | Protein antigen with alum-based nanoform coat | Rousing immune response against anthrax toxins |
| Mycobacterium tuberculosis | Liposomal nanoform-based encapsulated MTB antigen | Involved in increasing the Th1 cytokine level and antigen presentation enhanced against tuberculosis |
| Klebsiella pneumoniae | Outer membrane vesicles NFs | Presents a possible vaccination and protective immunity for incurable CRKP (Carbapenem-resistant K. pneumoniae) infections |
| Streptococcus pneumoniae | Polysorbitol transporter nanoform | Long-term defense immunity by utilizing T follicular helper (Tfh) and T helper (Th2) cells |

Systems based on nanomaterials are also persuasive adjuvants that augment cellular and humoral (antibody-mediated) immune responses. Fighting bacterial infections is complex, particularly those brought on by drug-resistant strains. For instance, methicillin-resistant S. aureus (MRSA) can be effectively treated with nanocapsules with extracellular vesicles from Staphylococcus aureus (S. aureus) because they proliferate the targeted delivery of antibiotics to infected tissues. Furthermore, by functioning as transporters and antimicrobial agents, nanofabricates like silver nanoforms AgNP improve the vaccine's effectiveness even more by directly inhibiting germs while also boosting the immune system (Lakshminarayanan et al. 2018).

Bacterial biofilms are a key source of implant-related infections and treatment failures; nanotechnology offers potential strategies to tackle them. Antimicrobial drugs and human defenses are thwarted by biofilms generated by bacteria such as S. aureus and E. coli. On biomaterial implants, such as orthopedic implants that release antibiotics and catheters coated in silver or copper, nanoparticles (NPs) have been used to stop biofilm growth. NPs can also adorn biofilm penetration, enabling a more proficient delivery of medications like doxycycline. Iron oxide, zinc oxide nanoforms, sulfur-functionalized fullerenes, and other nanoform-based treatments have endorsed effectiveness in preventing the formation of biofilms and interfering with the virulence factors of bacteria. To aid early disease detection and diagnosis nanoforms biosensors have been designed to detect those biofilms with further addition nanotechnology also introduces a tissue support system for tissue regeneration in biofilm-affected areas, Materials employed for this purpose are calcium phosphate and graphene oxide provide the adorning tool as an antimicrobial agent and tissue healing proving to nanofabricate base vaccine more proficient than

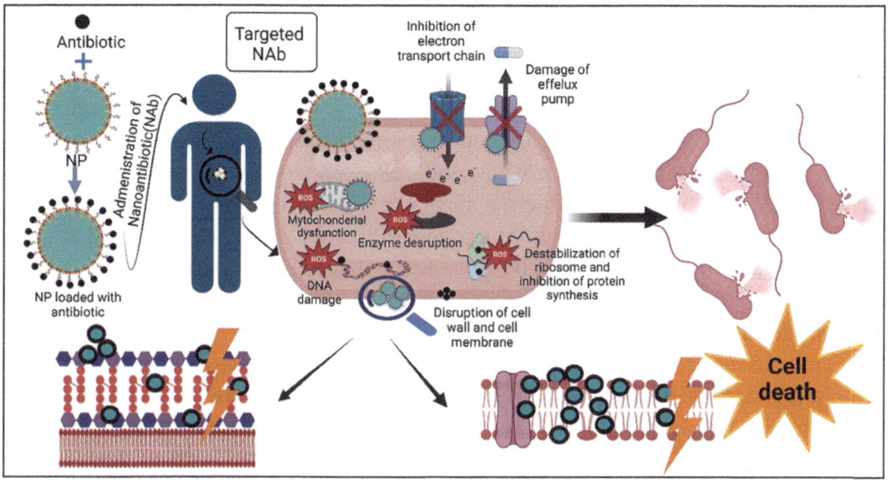


**Fig. 11.3** Illustration of np carrying antibody and its action against bacterial cell. (Hetta et al. 2023) MDPI, CCBY 4.0

conventional technologies (Ye and Chen 2022). Figure 11.3 given below gives the illustration of np carrying antibody and its action against bacterial cell.

## 11.6 Nanovaccines for Cancer Therapy

Despite these tailored and precision-based anti-cancer strategies, nanovaccine holds promising potential, but their are some impediments such as immunosuppression, inadequate immunological stimulation, and T-cell fatigue that persist in impeding their effectiveness. Identifying effective antigens, lymphoid organs combating immune evasion, successfully targeting immune cells and the tumor microenvironment, improvising immunogenicity, and evading lysosomal degradation are frequently prominent prehensions for the limited clinical effectiveness of cancer vaccines.

Tumor antigens and adjuvants can be administered synchronously via nanovaccines. In particular, stimulus-responsive nanometer-based nanovaccines have a knack for preserving the structural integrity of serum and selectively triggering the release of precise payloads into antigen-presenting cells in lymph nodes upon exposure to particular stimuli inside the tumor immunological milieu. Furthermore, in addition to resolving issues with antigen and adjuvant delivery and release, minimize off-target side effects, enhancing safety, and strengthening immune control. Adjuvants and antigens can agglomerate in lymph nodes more effectively when subjugated as nanovaccines of the right size. However, by varying the receptors that can link to the highly expressed receptors, the nanovaccine can be made to target lymph nodes (Xie et al. 2022).

Cancer vaccines like TheraCys®, Provenge®, and Imlygic® have acquired FDA approval, although many newer vaccinations are inadequately immunogenic. Using melanoma antigens and Toll-like receptor ligands, studies have corroborated that encasing antigens in NPs can activate cytotoxic T cells to combat malignancies. Also, NFs can carry nucleic acids and cytokines, inhibiting their breakdown and proliferating their distribution to immune cells. Enhancing immune responses to malignancies enables more effective and prolonged activity. To revamp the effectiveness of cancer vaccines, cytokines, nucleic acids such as DNA or RNA, and tumor-associated antigens have all been encapsulated in nanoparticulates (Yan et al. 2019). Figure 11.4 shows the encasing antigens in NPs can activate cytotoxic T cells to combat malignancies.

Tumor-associated antigens have been delivered using virus-like particles (VLPs) which have also been employed in conjunction with immunotherapy, chemotherapy, and radiation therapy. When combined with checkpoint blockade therapy, scientists invented nanodiscs that imitate high-density lipoproteins and deliver adjuvants and antigens to lymphoid organs, boosting immune responses and drastically boosting CTL production (Yan et al. 2019). One of the biggest challenges in oncology is breast cancer, particularly HER2-positive cases. A recent study was done by (Hu and Steinmetz 2021) aimed at developing a vaccination against the human epidermal growth factor receptor-2 (HER2) employing virus-like particles (VLPs). Researchers established the CH401 peptide epitope, which is extracted from HER2, using the Physalis mottle virus (PhMV) as a VLP platform.

To enhance immune responses, Toll-like receptor 9 (TLR9) agonists were embedded in the VLPs of some formulations of the vaccine, which was created using copper-free click chemistry and infusion encapsulation approach. The vaccine's potential for clinical usage was demonstrated by preclinical investigations employing murine models, which revealed that vaccination with these VLP constructs led to a substantial delay in tumor growth and prolonged survival. The

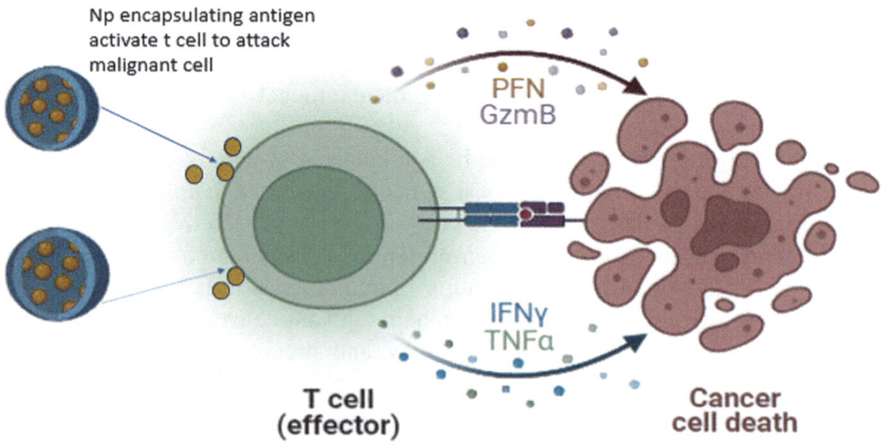


**Fig. 11.4** Encasing antigens in NPs can activate cytotoxic T cells to combat malignancies

findings emphasize the potency of VLP-based nanovaccines to induce potent antitumor immunity, offering a promising platform for breast cancer treatment.

Moreover, exposing tumor antigens in situ is another promising strategy to prevent the spread and recurrence of cancer. Damage-associated molecular patterns can be rendered visible by immunogenic cell death (ICD), which recruits antigen-presenting cells (APCs) and stimulates T lymphocytes specific to tumors. This approach of deploying biomimetic nanoparticles to boost the immune response has been used in numerous investigations. However, given the extreme heterogeneity of cancers, vaccinations against a single antigen might not be enough (Jasrotia et al. 2022). Nanofabricates containing whole tumor cell lysates can be employed to augment antigen presentation. In order to activate tumor-specific T lymphocytes, nanomaterials containing peptide epitopes and costimulatory molecules have also been used to create artificial antigen-presenting cells (aAPCs). These aAPCs, which are derived from substances like membrane coatings or iron oxide, improve immunological responses and have the promise to develop exceptionally effective bespoke tumor vaccines (Guo et al. 2023).

Subsequent research has examined hybrid membrane delivery systems, for the accumulation of antigens and adjuvants, the encouragement of dendritic cell augmentation, T cell stimulation, and the suppression of tumor enlargement. These systems use the membranes of bacteria and tumor cells. These strategies highlight the immense promise of nanomaterials in the manufacturing of cancer vaccines (Kroll et al. 2019).

## 11.7 Nanovaccines for Parasitic Diseases

Multiple types of nanoformulates have been used to treat and anticipate several serious parasitic disorders, including schistosomiasis, coccidiosis, leishmaniasis, malaria, toxoplasmosis, and anaplasmosis. There is currently no nanovaccines counter towards any parasite that is sold competitively (Feng et al. 2022). Highly desirable Th1-mediated immune protection against leishmaniasis has been demonstrated by the prevalence of vaccines constructed from nanoformualtes. Using the Self-Assembling Protein Nanoparticulates approach (SAPN), an epitope-based nanovaccine against Toxoplasma sp. has recently been produced with promise.

A study conducted by Papagiannaros et al. (2005) demonstrated that Liposomal miltefosine was more effective than the unencapsulated drug against miltefosine-resistant Leishmania promastigotes. Experimental visceral leishmaniasis (VL) demonstrated that liposomal amphotericin B (L-AmB) was significantly more potent than other amphotericin B formulations, with about 350 times the activity of AME and 2–5 times that of non-liposomal AmB. 99% of the amastigotes in the liver of squirrel monkeys infected with L. donovani were eliminated by L-AmB, which additionally boosted the therapeutic index. Although the efficacy for multiple lipid-based AmB formulations, such as Amphocil® and Amphotec®, was lesser than that of L-AmB, they were, however, more successful and less harmful than the conventional AmB-deoxycholate (D-AmB) formulation. No liver or renal damage was reported in the liposomal AmB formulation high in ergosterol (Kalsome™10), and the parasite load was considerably reduced. Furthermore, sugar-adorned liposomal systems were developed for accurate delivery to macrophages, especially those that contain galactose, mannose, and fucose (Hazra et al. 2024).

Similarly, implied vaccine antigens have displayed insulative immunity in laboratory experiments, and malaria (Anaplasma sp.) nanovaccines have seen significant proficiency. The parasite vectors of human, animal, and zoonotic illnesses (flies, ticks, and mosquitoes) could be targeted using this strategy. To create antiparasitic vaccines, it is imperative that we better coordinate and make use of the potential of mucosal immunity created by nanovaccines (Zaheer et al. 2021).

## 11.8 mRNA-Based Nanovaccines

Based on nanotechnology, mRNA vaccines encapsulate and insulate mRNA from deterioration while promoting its transport to target cells through tailored nanoparticles made of lipids or polymers. Using these nanoencapsulations, mRNA can enter immune cells and indoctrinate into protein antigens that trigger immunological reactions. mRNA vaccines have the advantage and quick emergence as a major benefit. As affirmed by the development of COVID-19 vaccines, mRNA vaccines are manufactured in a laboratory by manipulating the genetic sequences of the pathogen, in contrast to conventional vaccinations that call for the cultivation of pathogens.

mRNA vaccines are administered by injecting mRNA-containing nanoforms, which are then absorbed by immune cells like macrophages and dendritic cells. Translated into protein antigens inside the cells, the mRNA triggers humoral (Yang et al. 2022) and cellular (Abdollahiyan et al. 2021) immune responses. Long-lasting immunity and infection neutralization depend on these reactions. Lipid nanofabricates (LNFs) are notably used in Pfizer and Moderna's COVID-19 vaccines to transport mRNA encoding the SARS-CoV-2 spike protein, validating the efficacy of this technology (Kumaravel 2024).

In addition to COVID-19, research is being conducted concerning mRNA vaccines for cancer immunotherapy and other infectious diseases like HIV, Zika, and influenza. Tumor-specific antigens, sometimes called tumor-associated antigens or neoantigens, are found and mRNA sequences encoding these antigens are created for cancer vaccines. Adjuvants and cytokines, chemicals that stimulate the immune system, may also be supplemented with vaccines. In antigen-presenting cells (APCs), where the mRNA is converted into tumor antigens, LNPs conserve the mRNA and improve its apportionment. The cytotoxic T lymphocytes triggered by these antigens identify and eject cancerous cells. mRNA vaccine also involves generating immune memory to avoid disease recurrence. Further studies and investigations are progressing to improve mRNA-based nanovaccines as safer therapeutics for combating diseases (H. Cheng and Li 2022).

One of the foremost oncology challenges today is colon cancer, which calls for novel treatment approaches. To augment cancer immunotherapy, researchers (Shi et al. 2025) designed DMP-Lac (DOTAP-mPEG–PCL-Lactobacillus reuteri lysate), a novel bifunctional mRNA delivery method. The beneficial gut bacterium L. reuteri exhibits immunomodulatory characteristics making its lysate an ideal component for developing a cutting-edge mRNA delivery system. L. reuteri lysate and IL-23A mRNA were successfully co-delivered by the DMP-Lac system, yielding an impressive mRNA delivery efficiency of 75.56% ± 0.85%. With a tumor inhibition rate of 97.92% in CT26 abdominal and lung metastasis models, this formulation considerably enhanced immune system activation and maturation, leading to remarkable tumor inhibition. Further analysis using protein chip technology demonstrated that DMP functioned as an immunological adjuvant, modulating immune cells, whereas IL-23A mRNA modified downstream components, contributing to the therapeutic effects. These findings highlight the DMP-Lac/IL-23A complex's potent anticancer properties as a tumor-targeted nanomedicine. In conclusion, this study offers an innovative approach to immunogene therapy for colon cancer, demonstrating that bifunctional mRNA nanoformulations can significantly boost immune responses and suppress tumor growth, paving the way for further advancements in cancer treatment.

The abeyant of nano-vaccines in the lab and on the clinical scale can exert the envelope toward a more immunogenic, endured, resilient, and environmentally friendly approach to treating infectious and non-infectious disorders. Integrity is the basic factor that ought to be taken in scrutiny for the commercial manufacture of nanovaccines in terms of intended surface qualities during manufacturing and storage under field settings. Nanovaccines have opened up an unbounded window of opportunity for successfully preventing pathogenic, malignant, and non-infectious diseases in immune-tolerant individuals. Expanded focus on research in conjunction with business sectors may result in the swift commercialization of nanovaccines. Below is the Fig. 11.5 to show mRNA-based vaccines.

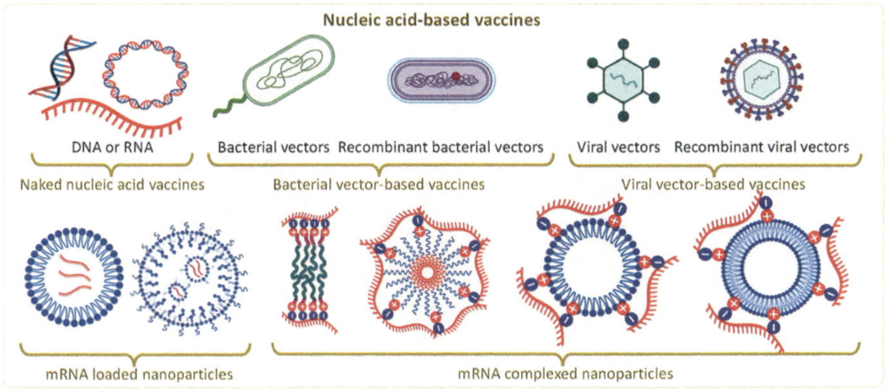


**Fig. 11.5** mRNA-based vaccines. (Pozharov and Minko 2023) MDPI, CC BY 4.0

## 11.9 Nano-based Different Immunotherapies beyond Vaccines

Since nanoformualtes have distinct physicochemical depictions that allow them to transport medicinal products to targeted cells, elude immune surveillance, and alter immune responses, they are currently employed for distinctive immunotherapeutic interventions outside of vaccine applications (Imran and Alsayeqh 2022). Here, discuss some marvels of nanoforms-based immunotherapies in combating different disease types. Figure 11.6 given below describes Nanoparticle for different immunotherapies.

The significant clinical success of immunotherapy can be attributed to delivery tactics that are competent to target tumors and/or immune cells of interest more precisely while reducing off-target effects. The apparent solution to this problem is in the delivery systems based on nanofabricates. Nanomatters are well-suited to surmount the obstacle that stands in the way of the efficacious delivery of immunotherapeutic drugs because of their intrinsic qualities and capacity to serve numerous functions. A great deal of accomplishment in preclinical and clinical studies has been demonstrated by extensive research on nanoparticulates for immunotherapeutic administration, particularly in the last ten years (Zhao et al. 2021).

Through active targeting which implicates functionalizing nanocarriers with tumor-affinity ligands or passive targeting which molds improved permeability and retention effects, nanocarriers can effectively ferry immune drugs into tumors in

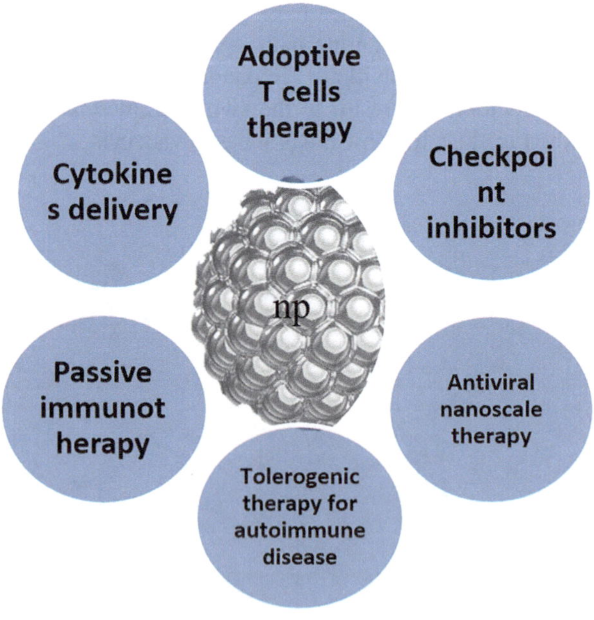


**Fig. 11.6** Nanoparticle for different immunotherapies

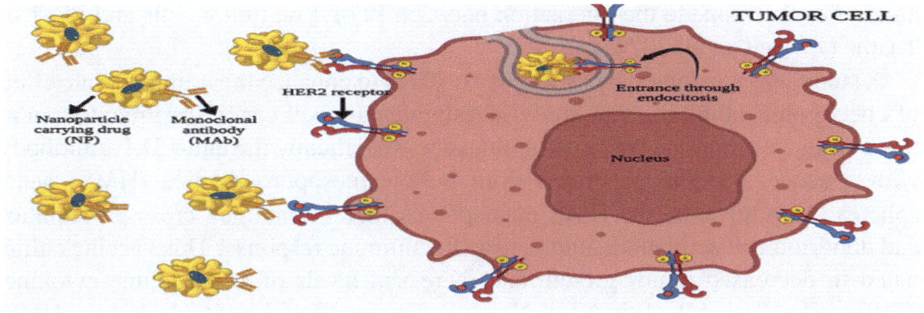


**Fig. 11.7** Representation of tumor cell targeted by nanoparticles carrying Mab. (Pavelić et al. 2023) MDPI, CC BY 4.0

cancer immunotherapy. Moreover, because of their size, they may cluster in lymphoid organs and immunological locations and can be engineered more specifically to target immune cells. By augmenting immunoadjuvants' local dispensation into lymphoid organs and abbreviating systemic toxicity, nano-based formulations have boosted immunological function. They are additionally utilized to deliver antibodies that activate T cells and antigen-presenting cells (APCs), showing clinical benefits in tumors such as melanoma, lung cancer, kidney cancer, and Hodgkin's lymphoma (Nasirmoghadas et al. 2021). Nano formulation-based immunotherapy activates systemic antitumor immunity and circumvents obstructions such as the blood-brain barrier and tumor vasculature, in contrast to standard nanomedicine, which has limitations with tumor infiltration. The treatment targets immune cells in lymphoid organs and other places. With immunological modulation, this innovative strategy increases anticancer activity and provides a multimodal therapeutic benefit (Nadukkandy et al. 2022). Figure 11.7 below represents the tumour cell targeted by nanoparticles carrying Mab.

### *11.9.1 Immunological Checkpoint Inhibitors*

Immunological checkpoint inhibitors prevent cancer cells from exploiting immunological checkpoints to avert destruction by the immune system. Checkpoint inhibitors such as anti-CTLA-4, anti-PD-1, and anti-PD-L1 antibodies can be encapsulated and delivered directly to tumors via nanoparticles, which minimizes off-target adverse effects and boosts local immune stimulation. Furthermore, nanoscale particles can target immune checkpoints at the epigenetic level. For instance, in animal models for cancer, mesoporous silica nanoparticles loaded with a GSK3 inhibitor lessened systemic toxicity while reducing PD-1 expression and demonstrating efficacy comparable to PD-1 antibodies. In other instances, reactivating cytotoxic T cells to target the tumor, lipid nanoforms (LNFs) have been created to carry PD-L1 antibodies that impede the interaction between PD-L1 on tumor cells and PD-1 on T cells (Z. Yang et al. 2020).

A study was conducted by (Wang et al. 2022) to enhance the immunogenic effect of checkpoint inhibitors, this study investigates a novel cancer treatment strategy that conjugates immune checkpoint blockage, specifically the anti-PD-L1 antibody, with a cancer vaccine generated from hollow mesoporous silica (HMS) nanospheres. As adjuvants, the HMS nanospheres increase antigen cross-presentation and dendritic cell activation, augmenting the immune response. This vaccine culminated in decreased tumor growth and increased levels of key immune cytokines (TNF-α, IL-1β, and IL-2) in mice. Notably, the anti-PD-L1 antibody and the HMS-based cancer vaccination blended to enable \a far lower dose of the immune checkpoint inhibitor without sacrificing potency. When combined with the HMS vaccine, a 1/8th dose (25 μg/mouse) of anti-PD-L1 produced anti-tumor effects comparable to the full 200 μg/mouse dose administered independently. Furthermore, a significant rise in CD4+ and CD8+ T cell populations was seen in the treated mice, indicating a robust immunological response. Hence the findings highlight the promising and remarkable approach to augmenting cancer therapy with nano-conjugated checkpoint inhibitors.

### *11.9.2 Adoptive T Cell Therapy*

In adoptive cell transfer therapy, nanocarriers have been exploited to improve T cell activity and endurance. One method, known as "backpacking" T cells, is adding cytokines to nanoformualtes to adorn T cell function and proliferation. This results in prolonged, targeted stimulation with low systemic toxicity. For instance, it has been endorsed that multilamellar lipid nanoparticles and protein nanogels containing IL-15 superagonist and IL-21 cytokines help to facilitate T-cell activation. Immunoliposomes that carry small-molecule TGF-β inhibitors to transplanted T cells have also fostered T cell expansion by obstructing signals that suppress the immune system. A recent breakthrough involves the in-situ formulation of CAR T cells using polymeric nanomanipulators functionalized to transit leukemia-targeting CAR genes directly to T cells. These nanoparticles' versatility in T cell-based immunotherapies is illustrated by the fact that they have also been linked for genome editing to boost the function of CAR T cells (Rana et al. 2023). The investigation was done for the optimization of artificial-based antigen-presenting cells designed to augment adoptive T-cell therapy, conducted by (Hou et al. 2024). they applied glucose-crosslinked bovine serum albumin (BSA) nanoparticulates as artificial antigen-presenting cells (aAPCs) to improve adoptive T cell treatment. By presenting antigens and co-stimulatory signals, these nanoformulates are intended to mimic natural APCs and effectively activate T cells. According to the study, these resilient, particle-based aAPCs are far better than earlier designs at promoting T cell activity and proliferation, which renders them a desirable tool for immune-based therapies like cancer immunotherapy. This study introduces a more structured and efficient strategy to activate T cells by boosting the resilience and efficacy of aAPCs, opening the door to novel immunotherapies. Combining different nanoformualtes and other cancer therapeutics approaches can enable tackling the proliferating stigma of various invasive and non-invasive cancers.

#### 11.9.2.1 Nanoforms Based Therapies for Autoimmune Disorders

Autoimmune disorders arise when the immune system mistakenly attacks its own body cells and tissues. Nanoform designing is also utilized in this area for selective suppression of immune response towards body cells while saving the rest of the immune system, providing more precise and less toxicity than traditional therapies.

#### 11.9.2.2 (a) Nanomaterials for Tolerogenic Therapy for Autoimmune Disease

Since tolerogenic dendritic cells (Tol DCs) are implicated in the induction and maintenance of immunological tolerance, they are being investigated as a potential treatment for autoimmune disorders. Targeting antigen-presenting cells or suppressing self-reactive lymphocytes are two ways that can promote immunological tolerance. By applying the mechanisms like T cell anergy, apoptosis, and the proliferation of regulatory T or B cells, tolerance is predestined to be impelled by co-delivering pathogenic antigens and immunomodulatory chemicals to endorse tolerance. Recent research indicates that nanoparticles can reprogram immune cells such as macrophages, dendritic cells, B cells, monocytes, and neutrophils in situ. One such research conducted on an animal model is the application of Myelin peptide-loaded poly(lactic-co-glycolic acid) (PLGA) nanocarriers to stimulate the growth of regulatory T cells (Tregs) to elicit immunological tolerance in multiple sclerosis patients (Benne et al. 2022). Researchers (Rhodes et al. 2023) created rapamycin-loaded biodegradable microparticles infused with major histocompatibility complex class II molecules that display myelin peptides and a biased interleukin-2 fusion protein. These modified particles are purposed to target regulatory T cells (Tregs) tailored to myelin. Tested in a multiple sclerosis mouse model, the medication encouraged Treg growth, resulting in full recovery in 38% of cases and sustained disease reversion in all treated mice. This strategy explains how bioengineered particles can boost antigen-specific Tregs in autoimmune disorders, thereby inducing immunological tolerance. For type 1 diabetes medication, lipid-based nanoparticles expressing insulin peptides are being explored to reprogram the immune system to tolerate insulin-producing beta cells and prohibit them from being destroyed. This study was conducted to find antigen-specific tolerance for autoimmune diseases (LaMothe et al. 2018).

#### 11.9.2.3 (b) Cytokine Delivery

Cytokines that reconcile the immune response in autoimmune disorders can be transmitted via nanoparticles like IL-10 or TGF-β, which can be delivered via nanomaterials to mitigate inflammation and repress autoreactive immune cells. A distinct method was utilized by employing nanostructures with varying functions; some unleashed their cargo inside antigen-presenting cells (APCs), while others released extracellular anti-inflammatory cytokines such as GM-CSF and TGF-β. In animal trials, this dual-release method prevented type 1 diabetes (T1D) and EAE, exhibiting immunoregulatory benefits.

The study by (Venkatesan et al. 2022) established a nanoplatform that includes gold nanocages (AuNCs) tailored with interleukin-4 (IL-4) and PEGylated folic acid (FA). Methylprednisolone (MP) is loaded into the nanopores to target M1 macrophages, vital players in the progression of rheumatoid arthritis (RA). M1 macrophages' overexpressed folate receptors were the target of the IL-4@AuNCs conjugates (PEGylated FA, IL-4, and MP), which effectively altered their phenotype from pro-inflammatory (M1) to anti-inflammatory (M2). In vitro and in vivo, this modulation drastically decreased the levels of inflammatory cytokines (IL-6, TNF-α, and IL-1β), hence reducing joint inflammation and damage. Additionally, the AuNCs conjugates served as contrast agents for CT and photoacoustic (PA) imaging, facilitating real-time treatment process monitoring. The most effective treatment in a rat model of adjuvant-induced arthritis (Araújo et al.) was IL-4@AuNCs, which was followed by MP-AuNCs, based upon ELISA, histological, and Western blot evaluations. The study concludes that the prospective use of IL-4@AuNCs as a theranostic (therapeutic + diagnostic) nanoplatform for RA, offering a focused, immune-modulating therapy that reduces inflammation and makes disease imaging monitoring simple. However, further challenges should be mitigated for clinical manufacturing of these therapeutic nanoparticle carriers.

## 11.10 Challenges and Future Perspective

Although nanotechnology has great promise for medicine and immunotherapies, there are certain obstacles to overcome when integrating various nanoformulations into vaccines and immunotherapies. Enroll in a few of the challenges listed below:

### *11.10.1 Challenges*

Biocompatibility and toxicity are genuine concerns while employing nanoforms for vaccination and immunization, they may cause inflammation or long-term toxicity, especially if accumulated in the organs, need to be degraded and eliminated from the body.

NPs' special qualities and prospective employs in various industries, such as food packaging, cosmetics, and medicine, can make regulatory approval processes for them extremely complex. Under the current statutory authorities, the U.S. Food and Drug Administration (FDA) regulates NPs according to the particular product category (e.g., medicines, biologics, devices). The FDA mandates exhaustive safety, quality, and efficacy assessments for nanomedicine, which frequently entails laboratory and clinical research procedures (Ahata and Akçapınar 2023).

Certain nanocarriers may cause an immunological response that is not desired, which may end up in allergic reactions or severe inflammation. Certain nanocomplexes cause oxidative stress, and damage DNA, which may lead to genotoxicity and mutations. It is difficult to balance the immune-stimulating effects of nanoparticulate without inducing adverse consequences (Mba et al. 2023).

Another difficulty is global harmonization, a process that ensures that societal considerations and norms are the same in all nations. The U.S. FDA established specific regulations, such as preliminary research and post-marketing surveillance, to ensure the effectiveness, safety, and quality of vaccines based on nanocomplexes.

It nevertheless remains complicated to establish that nanoparticles exclusively interact with the appropriate cells or tissues, even though they can be functionalized to target specific cells of interest. Interactions that are not explicit can have off-target effects or compromise efficacy.

Another hurdle is the intricacy of nanoparticle design, which necessitates meticulous production to produce the ideal item with the right pharmacological, biological, and physicochemical properties. Another difficulty is getting past biological barriers to distribute treatments to the intended locations. An additional challenge for mRNA-based nanovaccines consists of selecting the right target antigen, producing and purifying mRNA, and ensuring that mRNA is effectively encapsulated in lipid nanoparticles.

The use of nanotechnology in immune modulation is also a costly and complex process, scaling up of procedures needs to be established to meet the commercial demand of their application.

To preserve stability, many formulations including nanoscale entities may need to be stored under extreme conditions, such as at low temperatures, making distributive tasks more challenging, especially in areas with inadequate cold-chain infrastructure.

For proper regulations and standardization, the manufacturing process must be evaluated, the procedure and products must be validated, and a proper clinical trial setting must be established before the product is publicly available.

### 11.10.2 Future Perspective

Nanotechnology promises more future insights into the fields of immunological regulation and medicine once the obstacles are mastered. Personalized vaccinations and immunotherapies could be developed via innovations in nanotechnology.

Nanoparticle-based formulations could be tailored for the best possible outcomes in therapy by employing patient-specific biomarkers and immune profiles, ensuring optimal targeting and efficacy.

In the future, nanoformulation platforms might integrate several therapeutic objectives, such as delivering immunomodulatory drugs and vaccinations at a precise instant to boost immune responses. This "all-in-one" strategy may increase therapy performance while requiring fewer interventions.

In the future, multivalent nanovaccines, such as universal flu or COVID-19 vaccines, will offer broad protection against various viral strains by transporting many antigens. The efficiency of vaccines may be enhanced via self-amplifying mRNA nanovaccines, which could culminate in prolonged antigen expression.

Artificial intelligence (AI) has evolved into a disruptive force in the development of vaccines based on nanocomplexes, changing multiple phases of the fabrication process from initial conception to clinical use. The invention of vaccines has been greatly hastened by AI technology, especially machine learning and deep learning methodologies, which have boosted productivity and predicted accuracy. Researchers have been able to anticipate immune responses, optimize antigen design, and improve the overall effectiveness of NP-based vaccine platforms.

In the future, the fabrication of DNA and mRNA vaccines is expected to rely on nano entities. The success of COVID-19 mRNA vaccines suggests that nanoscale platforms could be enhanced to enhance the reliability, stability, and efficacy of nucleic acid-based vaccinations against cancer and other infectious illnesses.

Therapeutic components can be administered via nanoscale delivery systems designed to abstain from resistance mechanisms in viral infections or cancer and retain their curative effects long after traditional remedies have run their course.

With further research and progression, cost-effective, biodegradable, more reliable immune vaccines and other modulators can be manufactured to tackle growing diseases.

## 11.11 Conclusions

Nanotechnology revolutionized the fields of medicine and immunity, making the treatment of cancer and other non-curable diseases possible. Many viral and bacterial disease treatments shifted to nanovaccines and other nano immunocomponents. Exceptional advancements have been made in treating the area of cancer, metastasis, HIV AIDS, etc., with the evolution of nanotechnology. Compared to conventional approaches, nano-based vaccine platforms offer improved immunogenicity, targeted delivery, and prolonged antigen release, among other benefits. Further streamlining vaccine administration and enhancing efficacy is the capacity to integrate multiple antigens and adjuvants into a single nanoparticle formulation. Moreover, introduction of AI in nanovaccines enhances predictive accuracy and optimizes manufacturing systems. The potential to combine several adjuvants and antigens into a single nanoparticle formulation enhances vaccine effectiveness and streamlines administration. But some challenges like durability, nanotoxicity, affordability, complex manufacturing processes, etc. still brought obstacles in the application of nanofabricates for immunization, which need to be addressed and sorted properly. Maintaining innovation in nanomaterial engineering, for medicine and pharmaceutics manufacturing and clinical trials will be essential to achieving the potential of nanoparticle-based immunotherapies, which will be less hazardous, more efficient, and more accessible treatments for various diseases.

## References


Abdollahiyan P, Oroojalian F, Mokhtarzadeh A (2021) The triad of nanotechnology, cell signalling, and scaffold implantation for the successful repair of damaged organs: an overview on soft-tissue engineering. J Control Release 332:460–492

Achmad H, Ibrahim YS, Al-Taee MM, Gabr GA, Riaz MW, Alshahrani SH et al (2022) Nanovaccines in cancer immunotherapy: focusing on dendritic cell targeting. Int Immunopharmacol 113:109434

Ahata B, Akçapınar GB (2023) CCHFV vaccine development, current challenges, limitations, and future directions. Front Immunol 14:1238882

Ahmad S, Ahmad MA, Umar F, Munir I, Iqbal N, Ahmad N et al (2024) Green Nano-synthesis: Salix alba bark-derived zinc oxide nanoparticle and their nematicidal efficacy against root knot nematode Meloidogyne incognita. Adv Life Sci 10(4):675–681

Benne N, Ter Braake D, Stoppelenburg AJ, Broere F (2022) Nanoparticles for inducing antigen-specific T cell tolerance in autoimmune diseases. Front Immunol 13:864403

Bezbaruah R, Chavda VP, Nongrang L, Alom S, Deka K, Kalita T et al (2022) Nanoparticle-based delivery systems for vaccines. Vaccines 10(11):1946

Billan S, Kaidar-Person O, Gil Z (2020) Treatment after progression in the era of immunotherapy. Lancet Oncol 21(10):e463–e476

Butterfield LH, Najjar YG (2024) Immunotherapy combination approaches: mechanisms, biomarkers and clinical observations. Nat Rev Immunol 24(6):399–416

Chakravarty M, Vora A (2021) Nanotechnology-based antiviral therapeutics. Drug Deliv Transl Res 11:748–787

Cheng H, Li Z (2022) Advances and perspectives of pharmaceutical nanotechnology in mRNA therapy. Pharm Nanotechnol 10(5):328–333

Cheng L, Wang Y, Du J (2020) Human papillomavirus vaccines: an updated review. Vaccine 8(3):391

Chowdhury N, Kundu A (2023) Nanotechnology platform for advancing vaccine development against the COVID-19 virus. Diseases 11(4):177

Cremolini C, Vitale E, Rastaldo R, Giachino C (2021) Advanced nanotechnology for enhancing immune checkpoint blockade therapy. Nano 11(3):661

Curley SM, Putnam D (2022) Biological nanoparticles in vaccine development. Front Bioeng Biotechnol 10:867119

Dewangan HK, Tomar S (2022) Nanovaccine for transdermal delivery system. J Drug Deliv Sci Technol 67:102988

Dey S, Singh GB (2023) Green nanotechnology approaches in vaccinology: advantages and disadvantages in biomedical sciences. In: Nanovaccinology: clinical application of nanostructured materials research to translational medicine. Springer, pp 281–299

Feng C, Li Y, Ferdows BE, Patel DN, Ouyang J, Tang Z et al (2022) Emerging vaccine nanotechnology: from defense against infection to sniping cancer. Acta Pharm Sin B 12(5):2206–2223

Fries CN, Curvino EJ, Chen J-L, Permar SR, Fouda GG, Collier JH (2021) Advances in nanomaterial vaccine strategies to address infectious diseases impacting global health. Nat Nanotechnol 16(4):1–14

Gareev KG, Grouzdev DS, Koziaeva VV, Sitkov NO, Gao H, Zimina TM, Shevtsov M (2022) Biomimetic nanomaterials: diversity, technology, and biomedical applications. Nano 12(14):2485

Gary EN, Weiner DB (2020) DNA vaccines: prime time is now. Curr Opin Immunol 65:21–27

Gowd V, Ahmad A, Tarique M, Suhail M, Zughaibi TA, Tabrez S, Khan R. (2022) Advancement of cancer immunotherapy using nanoparticles-based nanomedicine. Paper presented at the Seminars in cancer biology

Guo J, Tang L, Li K, Ma Q, Luo S, Cheng R, Liu C (2023) Application of nanotechnology in therapeutic cancer vaccines. Adv NanoBiomed Res 3(7):2200122

Hamad TG, Salman TA (2023) Investigation of the impact of environmentally friendly prepared alumina nanoparticles on bacterial activity. Adv Life Sci 10:95–100

Hazra R, Kar A, Mahapatra SK (2024) Liposomal dual delivery systems in visceral leishmaniasis enhance the synergistic effects of combination therapy: a promise for the future. Asian Pac J Trop Med 17(2):47–60

Heine R, Thielen F, Mathijssen R, van Leeuwen R, Franken M, Uyl-de Groot C (2024) Applying a cost-based pricing model for innovative cancer treatments subject to indication expansion: a case study for pembrolizumab and daratumumab. PLoS One 19(2):e0293264

Hetta HF, Ramadan YN, Al-Harbi AI, Ahmed EA, Battah B, Abd Ellah NH et al (2023) Nanotechnology as a promising approach to combat multidrug resistant bacteria: a comprehensive review and future perspectives. Biomedicine 11(2):413

Hou F, Guo Z, Ho MT, Hui Y, Zhao C-X (2024) Particle-based artificial antigen-presenting cell systems for T cell activation in adoptive T cell therapy. ACS Nano 18(12):8571–8599

Hu H, Steinmetz N (2021) Development of a virus-like particle-based anti-HER2 breast cancer vaccine. Cancer 13:2909

Imran M, Alsayeqh AF (2022) Antiparasitic activity of chemically synthesized magnesium oxide nanoparticles against small ruminant haemonchosis. Adv Life Sci 9(3):356–362

Jackman JA, Yoon BK, Ouyang L, Wang N, Ferhan AR, Kim J et al (2021) Biomimetic nanomaterial strategies for virus targeting: antiviral therapies and vaccines. Adv Funct Mater 31(12):2008352

Jaha HF, Anwar Y, Almaaqar S, Khan T, Khan SB (2024) Iron-based nanoparticles synthesis, characterization, and antimicrobial effectiveness. Adv Life Sci 11(2):525–532

Jasrotia R, Dhanjal DS, Bhardwaj S, Sharma P, Chopra C, Singh R et al (2022) Nanotechnology based vaccines: cervical cancer management and perspectives. J Drug Deliv Sci Technol 71:103351

Kashte S, Gulbake A, El-Amin SF III, Gupta A (2021) COVID-19 vaccines: rapid development, implications, challenges and future prospects. Hum Cell 34(3):711–733

Kayser V, Ramzan I (2021) Vaccines and vaccination: history and emerging issues. Hum Vaccin Immunother 17(12):5255–5268

Khan PA (2024) Advancing cancer care: a deep dive into immunotherapy strategies: immunotherapy unveiled exploring revolutionary cancer treatments. Pristyn Research Solutions

Kim D, Wu Y, Kim YB, Oh Y-K (2021a) Advances in vaccine delivery systems against viral infectious diseases. Drug Deliv Transl Res 11:1401–1419

Kim E, Lim EK, Park G, Park C, Lim JW, Lee H et al (2021b) Advanced nanomaterials for preparedness against (re-) emerging viral diseases. Adv Mater 33(47):2005927

Kroll AV, Jiang Y, Zhou J, Holay M, Fang RH, Zhang L (2019) Biomimetic nanoparticle vaccines for cancer therapy. Adv Biosyst 3(1):1800219

Kumaravel JJ (2024) Advancements in nanotechnology: mRNA vaccines for infectious diseases and cancer immunotherapy driving the future of global health. Oral Oncol Rep 10:100454

Lakshminarayanan R, Ye E, Young DJ, Li Z, Loh XJ (2018) Recent advances in the development of antimicrobial nanoparticles for combating resistant pathogens. Adv Healthc Mater 7(13):1701400
LaMothe RA, Kolte PN, Vo T, Ferrari JD, Gelsinger TC, Wong J et al (2018) Tolerogenic nanoparticles induce antigen-specific regulatory T cells and provide therapeutic efficacy and transferrable tolerance against experimental autoimmune encephalomyelitis. Front Immunol 9:281
Li Y, Xiao Y, Chen Y, Huang K (2021) Nano-based approaches in the development of antiviral agents and vaccines. Life Sci 265:118761
Lv B, Wang Y, Ma D, Cheng W, Liu J, Yong T et al (2022) Immunotherapy: reshape the tumor immune microenvironment. Front Immunol 13:844142
Manju K, Raj SN, Ranjini H, Nayaka SC, Ashwini P, Satish S et al (2023) Nanovaccines to combat drug resistance: the next-generation immunisation. Future J Pharm Sci 9(1):64
Mba IE, Sharndama HC, Anyaegbunam ZKG, Anekpo CC, Amadi BC, Morumda D et al (2023) Vaccine development for bacterial pathogens: advances, challenges and prospects. Trop Med Int Health 28(4):275–299
Micoli F, Bagnoli F, Rappuoli R, Serruto D (2021) The role of vaccines in combatting antimicrobial resistance. Nat Rev Microbiol 19(5):287–302
Moni SS, Abdelwahab SI, Jabeen A, Elmobark ME, Aqaili D, Gohal G et al (2023) Advancements in vaccine adjuvants: the journey from alum to nano formulations. Vaccine 11(11):1704
Nadukkandy AS, Ganjoo E, Singh A, Dinesh Kumar L (2022) Tracing new landscapes in the arena of nanoparticle-based cancer immunotherapy. Front Nanotechnol 4:911063
Nasirmoghadas P, Mousakhani A, Behzad F, Beheshtkhoo N, Hassanzadeh A, Nikoo M et al (2021) Nanoparticles in cancer immunotherapies: an innovative strategy. Biotechnol Prog 37(2):e3070
Papagiannaros A, Dimas K, Papaioannou GT, Demetzos C (2005) Doxorubicin–PAMAM dendrimer complex attached to liposomes: cytotoxic studies against human cancer cell lines. Int J Pharm 302(1–2):29–38
Pati R, Shevtsov M, Sonawane A (2018) Nanoparticle vaccines against infectious diseases. Front Immunol 9:2224
Pavelić K, Pavelić SK, Bulog A, Agaj A, Rojnić B, Čolić M, Trivanović D (2023) Nanoparticles in medicine: current status in cancer treatment. Int J Mol Sci 24(16):12827
Poon C, Patel AA (2020) Organic and inorganic nanoparticle vaccines for prevention of infectious diseases. Nano Express 1(1):012001
Pozharov VP, Minko T (2023) Nanotechnology-based RNA vaccines: fundamentals, advantages and challenges. Pharmaceutics 15(1):194
Rana I, Oh J, Baig J, Moon JH, Son S, Nam J (2023) Nanocarriers for cancer nano-immunotherapy. Drug Deliv Transl Res 13(7):1936–1954
Rhodes KR, Tzeng SY, Iglesias M, Lee D, Storm K, Neshat SY et al (2023) Bioengineered particles expand myelin-specific regulatory T cells and reverse autoreactivity in a mouse model of multiple sclerosis. Sci Adv 9(22):eadd8693
Salas-Benito D, Pérez-Gracia JL, Ponz-Sarvisé M, Rodriguez-Ruiz ME, Martínez-Forero I, Castañón E et al (2021) Paradigms on immunotherapy combinations with chemotherapy. Cancer Discov 11(6):1353–1367
Santos P, Almeida F (2021) Exosome-based vaccines: history, current state, and clinical trials. Front Immunol 12:711565
Shahmoradi MK, Amini Nogorani M, Mansouri F, Zarei L (2023) Combination of zinc nanoparticles with chitosan scaffolds increased cytokine genes on wound healing of infected rats with methicillin-resistant Staphylococcus aureus (MRSA). Adv Life Sci 10:93–98
Shaw TI, Wagner J, Tian L, Wickman E, Poudel S, Wang J et al (2024) Discovery of immunotherapy targets for pediatric solid and brain tumors by exon-level expression. Nat Commun 15(1):3732
Shi G, Xu Y, Qiu H, Cao F, Xiao Z-X, Zhang C, Zha G-F (2025) Personalized membrane protein vaccine based on a lipid nanoparticle delivery system prevents postoperative recurrence in colorectal cancer models. Acta Biomater 192:315–327

Soares-Weiser K, Bergman H, Henschke N, Pitan F, Cunliffe N (2019) Vaccines for preventing rotavirus diarrhoea: vaccines in use. Cochrane Database Syst Rev 10

Song Y, Huang Y, Zhou F, Ding J, Zhou W (2022) Macrophage-targeted nanomedicine for chronic diseases immunotherapy. Chin Chem Lett 33(2):597–612

Tafaghodi M, Khademi F, Firouzi Z (2020) Polymer-based nanoparticles as delivery systems for treatment and vaccination of tuberculosis. In: Nanotechnology based approaches for tuberculosis treatment. Elsevier, pp 123–142

Tapia D, Sanchez-Villamil JI, Stevenson HL, Torres AG (2021) Multicomponent gold-linked glycoconjugate vaccine elicits antigen-specific humoral and mixed TH1-TH17 immunity, correlated with increased protection against Burkholderia pseudomallei. MBio 12(3). https://doi.org/10.1128/mbio.01227-01221

Varadé J, Magadán S, González-Fernández Á (2021) Human immunology and immunotherapy: main achievements and challenges. Cell Mol Immunol 18(4):805–828

Venkatesan R, Xiong H, Yao Y, Nakkala JR, Zhou T, Li S et al (2022) Immuno-modulating theranostic gold nanocages for the treatment of rheumatoid arthritis in vivo. Chem Eng J 446:136868

Waldman AD, Fritz JM, Lenardo MJ (2020) A guide to cancer immunotherapy: from T cell basic science to clinical practice. Nat Rev Immunol 20(11):651–668

Wallis RS, O'Garra A, Sher A, Wack A (2023) Host-directed immunotherapy of viral and bacterial infections: past, present and future. Nat Rev Immunol 23(2):121–133

Wang X, Li X, Ito A, Sogo Y, Ohno T (2022) Synergistic anti-tumor efficacy of a hollow mesoporous silica-based cancer vaccine and an immune checkpoint inhibitor at the local site. Acta Biomater 145:235–245

Ward JW, Van Damme P (2018) Hepatitis B vaccines. In: Hepatitis B virus and liver disease. Springer Singapore, pp 91–117

Xie X, Song T, Feng Y, Zhang H, Yang G, Wu C et al (2022) Nanotechnology-based multifunctional vaccines for cancer immunotherapy. Chem Eng J 437:135505

Xu L, Liu Y, Chen Z, Li W, Liu Y, Wang L et al (2012) Surface-engineered gold nanorods: promising DNA vaccine adjuvant for HIV-1 treatment. Nano Lett 12(4):2003–2012

Yadav HK, Dibi M, Mohammad A, Srouji AE (2018) Nanovaccines formulation and applications-a review. J Drug Deliv Sci Technol 44:380–387

Yan S, Zhao P, Yu T, Gu N (2019) Current applications and future prospects of nanotechnology in cancer immunotherapy. Cancer Biol Med 16(3):486–497

Yang Z, Ma Y, Zhao H, Yuan Y, Kim BY (2020) Nanotechnology platforms for cancer immunotherapy. Wiley Interdiscip Rev Nanomed Nanobiotechnol 12(2):e1590

Yang YJ, Murphy EA, Singh S, Sukhu AC, Wolfe I, Adurty S et al (2022) Association of gestational age at coronavirus disease 2019 (COVID-19) vaccination, history of severe acute respiratory syndrome coronavirus 2 (SARS-CoV-2) infection, and a vaccine booster dose with maternal and umbilical cord antibody levels at delivery. Obstet Gynecol 139(3):373–380

Ye J, Chen X (2022) Current promising strategies against antibiotic-resistant bacterial infections. Antibiotics 12(1):67

Yigit M, Ozkaya-Parlakay A, Senel E (2021) Evaluation of COVID-19 vaccine refusal in parents. Pediatr Infect Dis J 40(4):e134–e136

Zaheer T, Pal K, Zaheer I (2021) Topical review on nano-vaccinology: biochemical promises and key challenges. Process Biochem 100:237–244

Zargham M, Khan MF, Munawar A (2024a) Hybrid nanomaterial employment in clinical and therapeutic applications. In: Technological applications of nano-hybrid composites. IGI Global Scientific Publishing, pp 247–269

Zargham M, Khan MF, Rehman RA, Rakha A, Nazir R, Salman M et al (2024b) Role of doped ZnO variants for the development of latent fingerprint. Inorg Chem Commun 162:112269

Zeng C, Hou X, Bohmer M, Dong Y (2021) Advances of nanomaterials-based strategies for fighting against COVID-19. Viewpoints 2(4):20200180

Zhao Y, Bilal M, Qindeel M, Khan MI, Dhama K, Iqbal HM (2021) Nanotechnology-based immunotherapies to combat cancer metastasis. Mol Biol Rep 48(9):6563–6580

# Chapter 12
# Comparative Pathways of Traditional Vaccines and Nanoparticle Innovations in Immunization Science

**Love Singla**

**Abstract** Nanoparticle-based vaccines and immunotherapies take the lead in therapeutic innovations as they achieve higher efficiency and lower toxicity than their traditional counterparts. Nanoparticles as antigen/drug carriers are capable to be delivered to target cells and improve immune responses and selective therapeutic action. Such small molecules with large surface areas have the potential to controlled stability, bioavailability, and sustained, hence are adaptable for the treatment of infectious diseases, cancer, and autoimmune diseases.

Such nanoparticles are capable of mimicking pathogens through presenting antigens, inducing an egalitarian humoral and cellular immune response. Targeted delivery with ligands, minimizing off-target toxicity. New generation nanoparticles have been engineered to deliver multiple antigens, adjuvants or therapeutic molecules simultaneously, potentially providing a fit for combinatorial therapies in complex diseases.

Given their high clinical potential, however, regulatory difficulties, nanotoxicity, and scaling up manufacturing challenges, implementation remains insufficient. Researchers continue to study and conduct clinical trials to address these hindrances, bringing nanoparticle-based approaches closer to real-world applications. These innovations have the potential to revolutionize global health in the coming years, allowing to prevent and treat a wide range of diseases with more precision and effectiveness.

**Keywords** Immune response · Nanomaterial · Humoral immune response · Multifunctionality · Surface modifications

L. Singla (✉)
Department of Microbiology, Maharaja Agrasen University, Baddi, Himachal Pradesh, India

 
V. Khanna et al. (eds.), *Nanotechnologies for Drug Delivery and Biopharmaceutical Development*, Progress in Drug Research 77,
https://doi.org/10.1007/978-3-032-00796-4_12

## 12.1 Introduction

For decades, vaccines and immunotherapies have been the cornerstone of medicine in saving countless lives by controlling diseases from polio to cancer. Vaccines can only do so much, while they have completely revolutionized the control of pandemics and global health broadly speaking, more complex diseases like some cancers or autoimmune disorders remain beyond their reach. On the other hand, nanoparticles present special features that they owe to their structure (small size and high surface area), which allow them to be used as vaccines and immunotherapies with a much lower energy cost. Thus, we often engineer particles (typically made of lipids, polymers or metals) that function as carriers for antigens or therapeutic agents and so they can be directed to very specific cellular populations within the body. A precision such as this in the field of medicine: will enhance an immune response and at the same time limit detrimental off-target effects that has historically been burdened on many therapy measures.

## 12.2 Traditional Vaccines

Old-fashioned vaccines, refined over centuries, continue to be a potent weapon in the fight against disease. Vaccines have played a fundamental role in the eradication or control of diseases such as smallpox, polio, measles and diphtheria. The new mRNA and vector-based platforms are promising, but conventional vaccines are and will remain the mainstay of the world's immunization agenda and will do so because of their long history of proven use, clear mechanism of action and strong manufacturing process (Plotkin 2014; Vaccines and Immunizations | Vaccines and Immunizations | CDC n.d.).

## 12.3 Traditional Vaccines; Types and Mechanisms

Conventional vaccines are categorized mainly into: live attenuated vaccines, inactivated vaccines, toxoid vaccines, subunit vaccines and conjugate vaccines. Each uses a different approach to safely expose the immune system to antigens, priming it to recognize and attack the actual pathogen.

1. **Live Attenuated Vaccines**: Live attenuated vaccines are made using pathogens that have been weakened in a laboratory. These altered organisms still replicate in the host, but have been stripped of much of their virulence. The vaccine strains present in the host resemble the natural infection and promote a strong and sustained immune response, largely driven by antibody-dependent and cellular immunity. Examples include measles, mumps, rubella (MMR) vaccine, varicella (chickenpox) vaccine, oral polio vaccine (OPV) Their capacity to stimulate

strong protection often means that one or two doses can offer protection well into life. However, as they contain live organisms they are associated with a small risk of disease causing in immune-compromised patients as well as a theoretical risk of reversion to a virulent state (Plotkin 2014).

2. **Inactivated vaccines**: A solution is the inactivated vaccine, which is constructed by killing the pathogen, either with the aid of heat or formaldehyde treatment. And because the pathogen can no longer replicate, these are safer, particularly for immunocompromised people. But because inactivated vaccines cannot replicate, they generally elicit a lower quality immune response than live attenuated vaccines. This typically needs booster shots in order to maintain immunity. This works for the IPV (the inactivated polio vaccine) and the hepatitis A vaccine. Inactivated vaccines generally require an adjuvant (i.e., a substance that enhances the immune response), although the positive safety profile of inactivated vaccines makes inactivated vaccines a central figure in immunization programs (Vaccines and Immunizations | Vaccines and Immunizations | CDC n.d.).
3. **Toxoid Vaccines**: A few select something specific, e.g., some bacterial infections are mediated by bacteria toxins, not the bacteria. This neutralization process leads to the development of toxoid vaccines, which involve inactivation of the toxin while preserving its antigenic properties. By generating antibodies that ward off the toxin, the vaccine serves the host by preventing the consequences of the actual bacterial toxin. Examples of toxoid vaccines are diphtheria toxoid and tetanus toxoid. These toxins are "inactivated", usually with formaldehyde, in a manner that they can no longer invoke disease, but still elicit a protective immune response (Vaccines and Immunizations | Vaccines and Immunizations | CDC n.d.).
4. **Subunit Vaccines**: Subunit vaccines include only (subunits) of the pathogen, such as proteins or polysaccharides, rather than the complete microorganism. The vaccines eliminate (isolate) certain antigens to decrease the number of adverse reactions and to shape the immune response toward the most important part of the pathogen. Subunit vaccines are extremely safe, as they do not contain live pathogens, and well tolerated, but the immune responses elicited are often weaker than those generated by whole-pathogen vaccines. This limitation is typically compensated by adding adjuvants and administering booster doses. The hepatitis B vaccine is an example of a subunit vaccine (Orensteina and Ahmedb 2017).
5. **Conjugate Vaccines**: Many bacteria express polysaccharide antigens on their surface that lack significant immunogenicity when alone, particularly in young children. Conjugate vaccines overcome that limitation; by chemically attaching these polysaccharides to a protein carrier. This conjugation greatly enhances the immunogenicity of the antigen due to recruitment of a more potent and durable immune response. The greatest effect has been seen with the use of conjugated vaccines with polymers against Haemophilus influenzae type b (Hib) and Streptococcus pneumoniae, which account for most cases of bacterial meningitis and pneumonia in children (Kobayashi et al. 2024).

## 12.4 Nanoparticle-Based Vaccines: Key Features, Mechanisms, and Advantages over Traditional Vaccines

Nanoparticles based vaccines is defined as the mimicry of the natural infection process that makes nanoparticle strategies so potent. The mode of antigen presentation might allow two special features in global immunity (Bezbaruah et al. 2022). On one hand, nanoparticles could mount a stronger immune response and at the same time, it will be more durable when it being induced like host expectation. This triggers both the humoral and cellular immunity (Rincon-Restrepo 2016). Together, these two signals are critical for combatting not just infectious diseases but also complex pathologies (e.g., cancer) because to recognize and eliminate abhorrent cells (i.e. malignant cells), whole repertoires of immune cells have to be educated (E. Li et al. 2023; Lin et al. 2024). One additional advantage to using nanoparticles for treatment is that such nanoparticle-based therapies can be personalized. The use of nanoparticles as a base is amenable to formulation for different needs in medicine. This makes it possible to attach ligands (molecules that bind specially with those on the surface of target cells). The mechanism will help the nanoparticle to reach its target like a tumor in cancer therapy or immune cells for vaccine response. The scientists hope this level of personalization would increase the treatment's effectiveness and lessen its potential side effects, a major drawback from other immunotherapies or vaccine attempts. Table 12.1 shows the contrast between traditional vaccines and nanoparticles-based vaccines.

Immunotherapy using nanoparticles can support a variety of action. They are being evaluated for use with everything from anti-viral agents to cancer therapeutics. For instance, nanoparticles can be made to provide numerous antigens or therapeutic agents together, potentially opening up the possibility for combination therapies that address multiple aspects of ailments like cancer and immune diseases (Rios-Ibarra et al. 2024; Yasamineh et al. 2022). They are adaptable nanoparticles that function in combating pathogens both as a preventive vaccine and also therapeutics. However much like the pioneering scientific technologies that came before them, nanoparticle vaccines and immunotherapies certainly are not without their challenges. Menstrual bioengineering confronts many of the same challenges, such as clinical translation and scale-up manufacturing, not to mention regulatory clearance. Challenges are being overcome by research and ongoing clinical trials as the frontier of what is possible continues to expand, with new breakthroughs bringing us all one step closer towards unlocking this technology's potential that will indeed change everything. Though headwinds in the space remain, this generation of cell and gene therapies offers powerful new hope for radically improved precision medicine interventions against some of today's intractable diseases. By making their way through the research pipeline, these therapies have myriad potentials; not only do they enhance our ability to prevent and treat diseases but lead to better patient outcomes as well allowing us live healthier lives.

**Table 12.1** Comparison between traditional and nanoparticle-based vaccines (Bezbaruah et al. 2022)

| Feature | Traditional vaccines | Nanoparticle-based vaccines |
|---|---|---|
| Size | Large molecules or whole pathogens | 1–100 nm particles |
| Composition | Derived from whole pathogens or their components (live attenuated, inactivated, subunit, toxoid, conjugate) | Utilizes nanoscale carriers (lipid, polymeric, inorganic) that encapsulate antigens/adjuvants |
| Delivery mechanism | Conventional injection, often slow absorption | Direct targeting to specific cells or tissues |
| Safety profile | Proven efficacy but risks include potential reversion (live vaccines) and limitations for immunocompromised patients. | No live pathogens involved; reduced risk of vaccine-induced disease and lower systemic toxicity |
| Antigen presentation | Random or less targeted | Precisely controlled antigen display |
| Stability | Prone to degradation, requires cold storage | Enhanced stability, less temperature-sensitive |
| Immune response | Primarily humoral (antibody-mediated) | Humoral and cellular (T-cell) immunity |
| Multifunctionality | Limited | Can carry multiple antigens and adjuvants |
| Adjuvant use | Often required separately | Adjuvants can be built into the nanoparticle design |
| Design flexibility | Limited to natural antigen properties; slower updates for emerging strains | Highly modular and rapidly re-engineered for multiple antigens or new targets, ideal for emerging threats |
| Targeted delivery | Relies on natural uptake by APCs with less precise targeting | Surface modifications (ligands/antibodies) enable precise targeting to dendritic cells and lymph nodes |
| Manufacturing complexity | Well-established but involves complex processes like culturing and inactivating pathogens; time-consuming and costly | Rapid design via nanofabrication; scalable production potential though scale-up challenges remain |
| Dosing frequency | Often requires multiple doses or boosters due to quick antigen clearance | Controlled release kinetics allow prolonged antigen presentation, reducing booster frequency |
| Applicability | Widely used for many infectious diseases; alternatives needed for immunocompromised or rapidly mutating pathogens. | Versatile platform for infectious diseases, personalized medicine, and cancer immunotherapy via tailored antigen delivery |

## 12.5 Unique Features and Mechanism of Nanoparticle-Based Vaccines

Nanoparticle size for vaccine delivery is usually between 10 and 200 nm. They have a small size with high surface area-to-volume ratio which makes the perfect candidates for the delivery of immunogenic molecules. These characteristics enable high precision engineering of nanoparticles, and hence they provide a versatile platform that can encapsulate or adsorb antigens, adjuvants or even genetic material.

1. **Versatile Carrier**: Nanoparticle-based vaccines can be one of the most important because they can function as a multifunctional carrier. Different types of nanoparticles are used for the vaccine delivery, which can be:

    - Lipid Nanoparticles (LNPs)—They are the one of key components of many modern mRNA vaccines. LNPs protect the fragile mRNA from breakdown and promote uptake by host cells to drive efficient translation into antigenic proteins (Pardi et al. 2018).
    - Polymeric Nanoparticles: These nanoparticles have their components deriving from biocompatible polymers such as PLGA (poly(lactic-co-glycolic acid)) that allows for sustained and controlled release of antigens and adjuvants for a prolonged time period. This slow-release mechanism can not only maintain immune activation but also enhance memory response (Dong et al. 2024).
    - Inorganic Nanoparticles: Gold or silica nanoparticles provide a structural backbone and can be functionalized with different ligands that target specific immune cells. Its auroral optical and magnetic features equally lend opportunity in tracing and imaging of vaccine deliverance (Hess et al. 2019).

2. **Targeted Delivery and Cellular Uptake**: Antigen formulations based on nanoparticles can be optimized for the targeted delivery of the antigens to specific cell types, worth mentioning are primarily APCs, e.g. dendritic cells. Modification of the surface (e.g., by attaching certain ligands or antibodies) improves the receptor-mediated endocytosis. A nanoparticle cargo internalized by dendritic cells is processed and presented on major histocompatibility complex (MHC) molecules, leading to humoral and cellular immune responses. Such spare-dose delivery minimizes off-target effects of the vaccine and contributes to improved overall immunogenicity (Pati et al. 2018).
3. **Increased Accumulation and Cross-Presentation in Lymph Nodes**: Reduced dimensions of nanomaterials support their rapid and effective drainage into lymphatic vessels resulting in the robust accumulation in lymph nodes where immune dairy products are generated. Dendritic cells within these nodes process the delivered antigens, triggering a cascade of immune events, including T-cell activation and B-cell maturation. Furthermore, it has been reported that nanoparticles also support cross-presentation of exogenous antigens on MHC class I molecules, which is critical in evoking cytotoxic T lymphocyte responses towards intracellular pathogens and malignant neoplastic cells (T. Zhao et al. 2023).
4. **Controlled Release Kinetics**: One significant benefit of using nanoparticle-based systems is their capacity to regulate the release kinetics of both the antigen and any accompanying adjuvants. Meticulously tuning the nanoparticle's composition and topology enables controlled and prolonged delivery of vaccine constituents at the cellular membrane and cytoplasm. Such a controlled release

not only prolongs immune stimulation but also minimizes the need for booster doses. Moreover, nanoparticles keep antigens stable, which is particularly relevant for temperature-sensitive elements (Pati et al. 2018).

5. **Adjuvant Co-delivery**: Traditional vaccines often necessitate the incorporation of adjuvants that enhance the immune response. Nanoparticles are able to co-deliver antigens and adjuvants with a single delivery system for concurrent uptake by immune cells. This approach of co-delivery can lead to a synergistic effect, greatly increasing both the quantity and quality of the immune response. Additionally, presentation of antigen and adjuvant in close proximity within the nanoparticle may mimic pathogen-associated molecular patterns (PAMPs) and activate innate immune receptors, providing a convenient way to promote a stronger adaptive response (Pardi et al. 2018).
6. **Surface Chemistry Modification**: The chemistry of nanoparticles can be easily tailored to improve their biocompatibility, bioavailability, circulation half-life, and lessening of the reticuloendothelial clearance. Polyethylene glycol (PEG) or other biocompatible polymers could coat the surface to mitigate degradation and the concern of immunogenicity related to the carrier. Along with this, the surface of the nanoparticle can be functionalized with targeting moieties to enhance the uptake in target cells, further, improving the efficacy and safety profile of the vaccine (Hess et al. 2019).

## 12.6 Fundamental Principles of Nanoparticle-Based Vaccines and Immunotherapies

A personalized vaccine or immunotherapy based on nanoparticles represents one of the high impact transformative applications that will revolutionize modern medicine through leveraging properties and capabilities delivered by these nanoparticle systems. This is necessary as one needs to have the knowledge of how exactly these nanoparticles work and what benefits do they offer compared to traditional methods. Primarily, they are nanoparticles (nano = miniscule) of very small size typical within 1–100 nm. Their high surface-per-volume ratio gives them their special physical and chemical features not seen in any other material. These unique properties of nanoparticles which is gained while making of nanomaterials from various materials that are ideal for use in vaccines and immunotherapies. Table 12.2 illustrates the details of various kinds nanoparticles in terms their characteristics, merits and demerits.

Their small size and large surface-to-volume ratio assists nanoparticles to interact with biological systems differently than normal materials. These characteristics are expected for the improved immunotherapy and vaccine applications which can provide a more accurate, sustained therapeutic pathway that cannot be executed by solute molecular preparations. Figure 12.1 represents different properties of nanoparticles making them unique for its importance in medicine.

**Table 12.2** Different types of nanoparticles, including their advantages, common applications, and disadvantages

| Nanoparticle type | Material | Key properties | Advantages | Disadvantages | Common applications | References |
|---|---|---|---|---|---|---|
| Lipid-based nanoparticles (LNPs) | Lipids (similar to cell membranes) | Biocompatible, capable of encapsulating mRNA, efficient for intracellular delivery | Mimic natural cell membranes, ideal for RNA and gene delivery, high biocompatibility | Sensitive to environmental conditions (e.g., temperature, pH), possible immune system recognition | mRNA vaccines (e.g., COVID-19 vaccines), gene therapy | Aldosari et al. (2021); Chatzikleanthous et al. (2021) |
| Polymeric nanoparticles | Biodegradable polymers (e.g., PLGA) | Controlled release, biodegradable, customizable surface properties | Controlled and sustained release, can be tailored for different drugs, biodegradable | Complex manufacturing processes, potential for unpredictable drug release rates | Sustained drug delivery, cancer therapy, immunotherapy | Chan et al. (2021); Grego et al. (2020) |
| Metallic nanoparticles | Gold, silver, or other metals | High stability, potential adjuvant properties, can enhance immune response | Highly stable, can provide enhanced imaging, can act as vaccine adjuvants to boost immune response | Potential toxicity, bioaccumulation risks, expensive to produce | Vaccine adjuvants, diagnostic imaging, antimicrobial agents | Diaz-Arévalo and Zeng (2020); Filipić et al. (2023) |
| Inorganic nanoparticles | Silicon, carbon nanotubes, or other inorganics | High surface area, stable, tunable size and shape, suitable for various delivery systems | High structural stability, large surface area for drug loading, tunable size and properties | Limited biodegradability, potential toxicity, challenging to remove from the body | Drug and vaccine delivery, biosensors, imaging | Mao and Yoo (2024); Tsikourkitoudi et al. (2022) |
| Protein-based nanoparticles | Natural or engineered proteins | Biocompatible, can self-assemble, can be engineered to target specific cells | Biocompatibility can target specific tissues, can trigger an immune response when needed | Stability concerns, expensive to produce, can trigger unintended immune responses | Vaccine development, targeted drug delivery, cancer therapy | Guerrini et al. (2022); Nguyen and Tolia (2021) |

| | | | | | | |
|---|---|---|---|---|---|---|
| Polysaccharide nanoparticles | Natural polysaccharides (e.g., chitosan) | Biodegradable, low toxicity, ability to encapsulate various drugs | Biodegradable, low toxicity, naturally occurring, potential for targeted delivery | Limited mechanical strength, challenges with large-scale production | Drug delivery, wound healing, vaccine delivery | Zeng et al. (2021); Zhang et al. (2021) |
| Dendrimers | Highly branched synthetic polymers | Uniform structure, highly tunable, multiple functional groups for drug attachment | Highly tunable, multiple attachment sites for drugs or targeting molecules, uniform size and structure | Complex synthesis, potential toxicity, limited biodegradability | Drug delivery, gene therapy, imaging | Aurelia Chis et al. (2020); Mittal et al. (2021) |
| Hydrogel nanoparticles | Cross-linked polymer networks | Hydrophilic, high-water content, can swell and respond to stimuli | High biocompatibility, responsive to environmental changes (e.g., pH, temperature), controlled drug release | Mechanical weakness, and slow drug release can limit efficacy, and potential for limited stability in the body | Tissue engineering, drug delivery, wound care | Basu et al. (2021); Yang et al. (2021) |

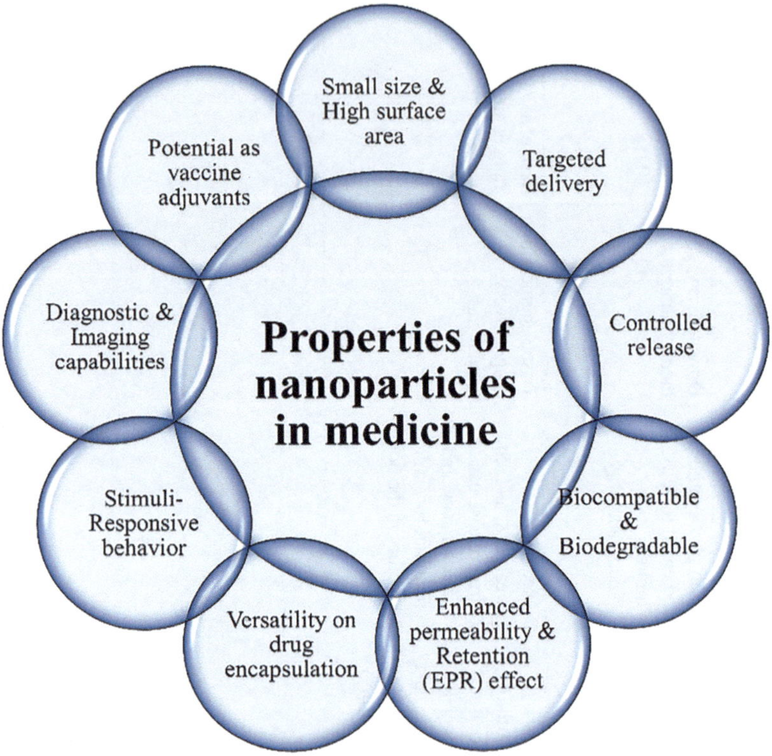


**Fig. 12.1** Properties of nanoparticles useful in medicines

## 12.7 Mechanisms of Action in Immune Response Enhancement Using Nanoparticles

Nanoparticles are able to boost immune responses through a variety of mechanisms that include greater role in antigen presentation; improved cellular uptake which leads to more effective T-cell targeted immunity. These features have made nanoparticles an ideal platform for vaccines and immunotherapies, creating potent protective immune responses against infections or tumors. The primary mechanisms through which they do this are:

1. **Immuno-stimulant by Antigen Presentation:**
   (a) **Mechanism**: Facilitation of efficient antigen presentation to immune cells (i.e., dendritic cells, and macrophages). E.g.: exosomes can fuse/adsorb antigens thus protecting them from degradation and deliver intact antigens for the targeting cell (Milicic et al. 2022)

   (b) **Implication**: Thereby, the T cell and adaptive immune response is further stimulated which results in a robust called memory formation/autoimmunity program (Díaz-Dinamarca et al. 2022).

2. **Improved Cellular Uptake:**
   (a) **Method of antigen presentation**: Nanoparticles are internalized tightly and to a great extent by appropriate immune response cells (antigen-presenting cell; APC) such as macrophages and dendritic cells due mainly to their small size alongside surface characteristics (F. Wang et al. 2021).
   (b) **Effect**: Upon entry, the nanoparticles release their payload like antigens or immune potentiating agents That uptake, in turn, enhances immune system activation by improving the processing and presentation of antigen to T cells.

3. **Immune Response Stimulation:**
   (a) **Mechanism**—different nanoparticles can act as PRR agonists to stimulate the innate immune system through Toll-like receptors (TLRs) in various immune cells
   (b) **Effect**: This in turn releasing cytokines and chemokines sets the stage for more immune cells to come join onboarding & launch primary immunity of body fighting against it (Baruah et al. 2023).

4. **Depot Effect (Slow and Steady Release of Antigen):**
   (a) **Mechanism**—Nanoparticles can function as a depot either at the site of injection or lymph by releasing antigens in a controlled manner, over an extended duration.
   (b) **Effect**—This slow release keeps the immune system around that antigen which causes a prolonged and stronger response of this immune response to prepare long-term immunity (Y. Wang et al. 2018).

5. **Co-administration of Antigens with Adjuvants:**
   (a) **Mechanism**: Nanoparticles are able to co-deliver antigens and adjuvants in a single formulation to immune cells, engaging these two stimulants simultaneously.
   (b) **Impact**: The immune system receives the trigger for response (adjuvant) and something to attack at once-this optimizes its action of activation (Z. Zhao et al. 2019).

6. **Enhanced Lymphatic Targeting:**
   (a) **Mechanism**: Nanoparticles can be designed to homing selectivity to lymphoid tissues where nearly all immune responses are initiated
   (b) **Effect**: Nanoparticles home to the lymphatic and deliver antigens directly within immune cell-rich sites, thereby amplifying immune elicitation globally (Peng et al. 2021).

7. **Induction of Cytotoxic T-Cell Response**
   (a) **Mechanism**: Nanoparticles were observed to act as adjuvants and mediators for the process of cross-presentation that required T cell activation with dendritic cells likely due to their ability in enhancing MHC class I presentation, which was necessary for CTL effector functions.
   (b) **Function**: This is a critical step in clearing infected or cancerous cells as, once activated, CTLs are capable of directly killing these cells leading to an effective immune response (E. Li et al. 2023).

## 12.8 Nanoengineering Design for Vaccine and Immunotherapy Applications

Designing nanoparticles for combinatorial vaccines and immunotherapies is a highly complex task that demands in-depth knowledge of material science, biology-even at an atomic structural level-chemistry. Use of suitable of materials and to improve its properties can help scientists make better particles, release at therapeutic agents using these nanoparticles in an inflammation stronger body immune response while suffering from lower transforms. Customizing nanoparticle delivery systems for specific medicinal purposes also increases their potential to treat multifaceted diseases. With further expanding fields, nanoparticle-based therapies are anticipated to be critical for the future of personalized medicine and global healthcare. The selection of materials is the first step in nanoparticle engineering. This requires biocompatible materials to be used in the oils, which will not cause any adverse reactions with biological systems. Some of the most popular are:

1. **Lipids**: Lipid nanoparticles (LNPs) are commonly applied to mRNA vaccines (e.g., COVID-19 vaccine). They emulate cell membranes, which is how they integrate with target cells and deliver their contents (Aldosari et al. 2021; Chatzikleanthous et al. 2021).
2. **Polymers**: Biodegradable polymers, for example, polylactic acid (PLA) or Poly (lactic-co-glycolic acid) (PLGA), have chosen to degrade in the body and provide sustained release of therapeutic agents with time (Chan et al. 2021).
3. **Metals**: Metal nanoparticles, like gold and silver, are also utilized in immunotherapy due to their stability as well as having unique properties that can be used for imaging or photothermal, etc. (Akbari et al. 2018).
4. **Inorganic materials**: Silicon nanoparticles may be used for their strength and high demonstration, which designates a large surface area available to carry multiple beings or as sustained release opponents (Mao and Yoo 2024; Tsikourkitoudi et al. 2022).

After the choice of material for nanoparticles, it is further developed by designing stability, enabling bioavailability or controlled release. Of the utmost importance, these are properties to guarantee that a therapeutic agent is able to reach targeted

**Table 12.3** Factors affecting the efficacy of NP-based vaccines and immunotherapies

| Design factor | Impact on efficacy |
|---|---|
| Material choice | Determines biocompatibility, degradation rate, and immune response. |
| Size | Affects the penetration of biological barriers and interaction with immune cells. |
| Surface coating | Enhances circulation time and reduces immune clearance. |
| Encapsulation | Protects the therapeutic agent from degradation and allows for controlled release. |
| Charge | Modulates cellular uptake and interactions with biological membranes. |

cells exactly when and where it should without getting degraded or released to soon. Some of the major techniques affecting the efficiency of nanoparticle-based vaccines and immunotherapies are summarized in Table 12.3.

Their surface properties, size, and payload can be modified to design nanoparticle delivery systems based on specific diseases or treatments. The personalization is sir one of the things that make nanoparticle-based therapies quite unique (Bugybayeva et al. 2024; Yu et al. 2023).

1. Nanoparticles can be targeted to specific cell types using ligands on their surface. For instance, nanoparticles engineered for cancer therapy could be coated with ligands that attach to receptors highly expressed on the surface of tumor cells and deliver therapeutic agents directly at the site where they are most needed (the tumor), reducing off-target effects.
2. In most treatments, it is important that the therapeutic agent be released gradually over time rather than all at once. Special types of nanoparticles can be engineered to break down slowly, so the drug or vaccine is released over an extended period. More specifically, this can be incredibly advantageous in immunotherapies where there is a great need to maintain an immune response over time for treatment efficacy.
3. Nanoparticles can simultaneously deliver multiple agents (e.g., antigens, adjuvants, and therapeutic molecules) to the target site. This versatility is particularly useful for combination therapies, which must simultaneously target two or more aspects of disease in a single treatment. For example, a nanoparticle capable of targeting an antigen to the immune system for cancer immunotherapy and at the same time delivering a drug to eliminate the cancer cells.

**Surface Modifications and Targeting Strategies for Nanoparticles** Nanoparticles need to traverse the labyrinth of the human body where the immune system, biological barriers and off-target effects can interfere with their intended purpose. Surface modification approaches are crucial to guarantee that nanoparticles arrive at their site of action with full potency. Nanoparticles can be designed to avoid the immune system, bind target cells, and enhance treatment response simply by changing surface properties (Ren et al. 2024; Takechi-Haraya et al. 2022).

Surface modification, one of the most important parts of nanoparticles-based vaccines and immunotherapies, provide specific direction, lower side effects and applicability of the treatment. To increase targeting, strategies such as ligand attachment, PEGylation or charge modification can minimize nanoparticle off-target effects. The future of nanoparticle technology that will provide the optimum surface modifications to be incorporated into new, safe and effective therapies for complicated disease will likely be equally as important in continuing to evolve. Among the most important advantages of surface modifications is the reduction of off-target effects. In pharmaceuticals, off-target effects are when a drug or therapeutic agent affects cells or tissues other than the intended target tissue. For vaccines and immunotherapies this could lead to inadvertently destroying normal cells, or systemic immune activation. These risks can be decreased by surface modification of nanoparticles to direct the therapeutic agent directly to the cells of interest and avoid unwanted exposure (out of interest).

1. **Ligand Attachment for Targeted Delivery**: Ligand attachment is the primary method of surface modification. Ligands are molecules that bind to specific receptors on target cells. In this way, by connecting ligands to the nanoparticles surface scientists restrict the nanoparticles binding to only those cells that expresses the respective receptors, thus enhancing the specificity of vaccine/drug delivery.

   (a) **Antibody-Conjugated Nanoparticles**: A typical method involves the conjugation of antibodies on the nanoparticle surfaces. These antibodies are intended to attach infiltrating cells to the antigens on the target cells surface, e.g., cancer cells. In cancer immunotherapy, for instance, nanoparticles might be loaded with antibodies specifically targeting tumor cells, so that their therapeutic payload is delivered right to the cancerous tissue. This approach reduces side effects on healthy cells (Kumari et al. 2023).
   (b) **Peptide and Protein Ligands**: In addition to antibodies, targeting ligands can include peptides and proteins. Those ligands target specific protein/receptors which are up-regulated on a disease such as tumor or auto immune disease inflamed tissue. Tissue targeting can also be achieved by engineering peptide ligands enabling recognition of specific types of tissues directing nanoparticles to designated sites. (Herrera-Barrera et al. 2023).
   (c) **Folic Acid Conjugation**: Folic acid is one of the most used as ligands as many cancer cells upregulate folate receptors. Folic acid coated nanoparticles will specifically attach to those receptors, targeting the release of anticancer drugs or immunotherapeutic agents to the tumor site more efficiently than other nanoparticles (Jamshidi et al. 2024).

2. **Reducing Off-Target Effects**: One larger hurdle for immunotherapies and vaccines to overcome is not triggering the immune system where it is not wanted. Systemic immune activation, for example, can trigger global inflammation, leading typically to adverse events such as fever, swelling, or even organ injury.

Surface modification of nanoparticles can alleviate these off-target effects, allowing for more precise delivery of therapeutics at the cellular or tissue level.

(a) **PEGylation (Polyethylene Glycol Coating)**: PEGylation is a well-characterized strategy to decrease off-target effects. For instance, when nanoparticles are coated in polyethylene glycol (PEG), scientists can "cloak" them from the immune system, meaning they spend longer in the bloodstream without being spotted and attacked by immune cells. PEGylation is crucial for therapeutic applications that depend on the ability of the nanoparticle to survive circulation and reach targeted tissues (e.g., brain or tumor) (Chen et al. 2023; Cheng et al. 2022).

(b) **Charge Alteration**: Surface charge of nanoparticles may sometimes even influence their interactions with cells. Positive-charge nanoparticles are better taken up by cells, whereas negatively charged particles might be taken up slower. Finally, the surface charge of nanoparticles can also be modified, resulting again in a further reduction of off-target effects through reduced interactions with non-target cells (Karunakaran et al. 2023).

3. **Precision in Treatment Delivery**: Surface-modified nanoparticles merge both the spatial resolution of delivered treatment and the multiplex resolution of delivered treatment, into a single modality, a desirable feature of delivery schemes for complex therapies targeting diseases like cancer or autoimmune disorders. Therapeutic agents are dispensed directly to the specific tissue in need of care rather than to the entire cytotoxic region, minimizing the expansive side effects and increasing the local effects of treatment; all aimed at precision targeting.

(a) **Multifunctional Nanoparticles**: The nanoparticles can be multi-functional carriers contain two or more therapeutic agents or target two or more pathways are also multi-functional due to surface modifications. Such a particle could, for example, include an antigen to trigger the adaptive immune system and a cell-killing drug, for targeting a tumor. The versatility of these multifunctional devices stems from the ability of surface modifications to bind different molecules to the same nanoparticle in a highly selective manner, which ultimately increases the efficacy and versatility of the treatment (Z. Cheng et al. 2012).

(b) **Localized Immune Activation**: In the context of vaccines, surface-modified nanoparticles can encourage the elicitation of antigens in a way that only induces immune response in specific tissues or cell types. This localized immune activation steers the antibody response towards the tumor site, which reduces the likelihood of system-wide side effects, such as inflammation (Kokate et al. 2016).

## 12.9 Multifunctional Nanoparticles in Vaccines and Immunotherapies

Multifunctional nanoparticles are considered as a smart new class of drug carriers for cancer therapy, autoimmune diseases and infectious diseases. Conventional treatments usually entail administration of a single agent toward a specific aspect of the ailment, whereas multifunctional nanoparticles are engineered to multitask. By packaging a number of therapeutic agents into a single delivery system for co-delivery, they use a more holistic approach to therapy that targets multiple aspects of a disease. The ability to deliver combination therapies along with the reduction of side effects as well as the ability to develop both complementary and enhanced humoral and cellular immune responses represent a new frontier in the field of medicine. Multifunctional nanoparticles are likely to become an important modality of future vaccines and immunotherapies.

1. **Development and Design of Multifunctional Nanoparticles**: Multifunctional nanoparticles can be formulated with multiple therapeutics, such as antigens, adjuvants and drugs, within a single delivery platform. This function allows combination therapies to simultaneously target multiple features of a disease. For example, one nanoparticle can carry one immune -stimulating antigen, one immune-potentiating adjuvant, and one therapeutic agent that preferentially kills the tumor cell or pathogens directly. Various materials and surface modifications of these nanoparticles could be described to provide an innovative architecture to assist varieties of biopharmaceutical functions (Min et al. 2023) These materials include:

   (a) **Core Materials**: Different materials may be used to construct the nanoparticle core, such as lipids, polymers, or metals. Each core material has its advantages. Thus, for example, lipids are thermodynamically biocompatible and they are therefore well-suited for drug delivery (vaccines), metals, e. g. gold, suitable for therapeutic bands and imaging.

   (b) **Surface Functionalization**: The nanoparticle surface can be modified with ligands or molecules which bind based on cell/tissue. Nanoparticle functionalization enables cargo delivery to the target. For example, a ligand that specifically binds to cancer may be conjugated to the nanoparticle to guarantee localization to cancerous cells.

   (c) **Encapsulation & Layering**: The nanoparticle can be designed with several layers that perform different functions. For instance, the anisotropic outer layer could secrete an antigen to activate the immune response, and the inner layer could deliver a therapeutic agent to infectious or malignant cells.

2. **Advantages of Multifunctional Nanoparticles**: The application of multifunctional nanoparticles addresses many limitations associated with conventional therapies that impact their ability to combat complex diseases (Yin et al. 2023; Yuan et al. 2023). Here is a list of some of the most important benefits:

(a) **Combination Therapies**: Classical therapies only target a single feature of a disease; however, many diseases will likely need a multi-faceted therapy. These multifunctional nanoparticles can deliver two or more therapeutic agents, such as an immune-activating antigen as well as a cancer-destructive drug, allowing for simultaneous targeting of different mechanisms of disease.
(b) **Targeted Delivery**: The modifications on the NP surface create the possibility of carrying target specific delivery avoiding the binding process with healthy cells yet binding only at zones of disease completion. This is highly beneficial since it lowers the probability of off-target effects as well as systemic effects, which is particularly vital in non-specific toxicity related to chemotherapies.
(c) **Controlled Release**: A multifunctional nanoparticle can be developed in a way to release its therapeutic agents in a controlled manner over time, It provides controlled release, which also means that it could have a prolonged effect, potentially minimizing the number of doses needed.
(d) **Minimized adverse events**: These nanoparticles can directly target the cell or even the tissue, which helps decrease the duration of unwanted side effects of conventional therapeutics where the drug acts on healthy and affected cell types. In cancer immunotherapy for instance, nanoparticles can synergise with cytotoxic agents and be loaded to target tumour cells while causing the least effect on neighbouring healthy tissue.
(e) **Immunomodulation**: In vaccines, multifunctional nanoparticles can deliver antigens to promote humoral (antibody-mediated) and/or cellular immunity (T-cell-mediated). These multifunctional nanoparticles combine these two forms of immune activation to elicit both a stronger and longer-lasting immune response which is critical for the prevention or treatment of diseases such as cancer or viral infections.

3. **Applications in Combination Therapies**: Multifunctional nanosystems have gained immense interest as the ideal candidates for combination therapy. These treatment approaches include two or more modalities that address various disease mechanisms. Multifunctional nanoparticles are especially appealing as vehicles for the delivery of combination therapy, as more than one agent can be delivered at a time (W. H. Chen et al. 2019; Y. Liu et al. 2023; Shao et al. 2018).

(a) **Cancer Therapy**: Combination therapies are especially advantageous in the case of cancer, which is one of the most classical examples of a disease. In cancer immunotherapy, multifunctional nanoparticles facilitate co-delivery of an immune-activating antigen and a chemotherapy drug to the tumor site. There, the antigen ignites the immune system's response to the cancer, just as the chemotherapy drug kills the cancer cells. And together the two increase the chances of killing off the tumor while sparing as much normal tissue as possible.

(b) **Infectious diseases**: Multifunctional nanoparticles are also currently being investigated as vaccines against different infectious diseases. For instance, they can present multiple antigens from a pathogen such as a virus or bacterium, promoting a diverse immune response that targets different parts of the invading organism. Adjuvants can also be included in the nanoparticles to increase immune response and confer more durable and stronger immunity.

(c) **Autoimmune Disorders**: Multifunctional nanoparticles can be used to target therapeutic agents to the tissues damaged by an autoimmune disease, whilst concomitantly suppressing the aberrant immune response that caused tissue injury; Having the targeted approach of balancing immune modulation and tissue repair with a treatment you can tailor to the patient goes a long way to providing synergy in therapy making for a complete treatment strategy.

## 12.10 Clinical Translation and Challenges of Nanoparticle-Based Vaccines and Immunotherapies

Nanoparticle-based vaccines and immunotherapies have tremendous clinical translation potential, but numerous barriers must first be surmounted. Before these groundbreaking therapies can be adopted widely, they face a number of major challenges, including regulatory approval, safety issues, and the difficulties of large-scale production. Yet so far successes in trials and real-world examples, given via COVID-19 mRNA vaccines, prove that with further research and funding, therapies based on nanoparticles are a promising ingredient to the medicine of the future. As the field matures, it will be necessary to meet these challenges to realize the full potential of nanoparticles to revolutionize disease prevention and treatment.

Nanoparticles have shown remarkable promise in preclinical studies and have been demonstrated to deliver antigens or therapeutic agents with enhanced precision and efficacy. However, transitioning these findings to clinical trials is a meticulous process. Nanoparticle-based vaccines and immunotherapies are under clinical investigation for a number of conditions like cancer, infectious diseases (including COVID-19), and autoimmune diseases. Example technologies, such as mRNA vaccines that utilize lipid nanoparticles (LNPs) to deliver genetic material, reached the clinic very quickly, with the Pfizer-BioNtech and Moderna vaccines leading the way during the COVID-19 pandemic. These vaccines demonstrate there is a real-world application for nanoparticles beyond discovery, as they translate fundamental research insights into a clinical practice ultimately saving lives and providing a molecular-based solution to a viral infection. But these successes also reveal unique regulatory, safety, and manufacturing issues associated with nanoparticle-based therapies.

1. **Regulatory Considerations and Hurdles**: Regulatory approval of nanoparticle-based therapies is consequently far more complicated than that of traditional vaccines and drugs. The small size, surface characteristics, and routing in the body monomers have demand for the safety and efficiency. Policy bodies like the U.S. Food and Drug Administration (FDA) and the European Medicines Agency (EMA) are still responding to these emerging technologies, developing regulatory structures for these new products that account for the specific dangers perceived to be posed by nanoparticles (Costa et al. 2024; Guerrini et al. 2022; Liu et al. 2022).
2. Getting regulators to approve nanoparticle-based treatments is much more complicated than it is for traditional vaccines and drugs. Combatting potential safety and efficacy issues posed by the small size, distinctive surface properties, and biological nature of nanoparticles. Regulatory bodies such as U.S. Foods and Drug Administration (FDA) and European Medicines Agency (EMA) have not been suitable for these novel technologies and developing regulations that are more suitable for the nanomaterials having their own pros and cons.
   (a) **Safety and Efficacy Assessments**: Nanoparticles should be evaluated for biocompatibility, stability, and efficacy. A key challenge is the appearance and behaviour of the nanoparticle in the body: how long is it in circulation for, where does it accumulate, how does it become eliminated and how does it behave in the body overall.
   (b) **Long-Term Effects**: A long term effects of human body for nanoparticle is still a most vital concern. In particular, the safety of nanoparticles may not be fully investigated when they accumulate in a tissue or an organ over the long term, despite preliminary findings from short-term studies suggesting that the nanoparticles are safe. Regulatory approval processes will probably change when we find out more about the long-term safety of these materials.
   (c) **Toxicity Concerns**: Owing to their small size, nanoparticles can behave and interact with biological systems as whole organisms rather than particulate matter—which is how conventional therapies do. Metallic nanoparticles are beneficial in targeting the cancer cells but toxic if they are retained in normal tissues. Lipid nanoparticles have been well tolerated when used in vaccines; however, there are patients who may be at risk of inflammatory and/or hypersensitivity responses. Explaining away any concern for potential toxicity is a key to regulatory acceptance, however.
3. **Manufacturing Complexities**: Scaling Up: Getting nanoparticle-based vaccines and therapies into mass production is also a significant challenge. The process of producing nanoparticles is very complicated and needs to have a high degree of control in terms of size, surface modifications, and composition. Instead, nanoparticles must be manufactured to exact specifications, unlike traditional pharmaceuticals, in order to ensure consistency and safety. It is a major challenge to manufacture nanoparticles at scale while ensuring the same quality

and stability required. This uniformity is vital for safety and efficacy since even minor size differences can result in a very different response in the body, For example.

Furthermore, not only nanoparticles being potential molecular-based therapeutics, they need specialized instruments and facilities for the production that are very high and expensive. Plus, in order for these therapies to be sufficient to treat the global population, it poses another logistical hurdle. The fast-tracking of mRNA vaccines during the COVID19 pandemic has proven that it is indeed possible to manufacture not just mRNA in the nanogram level premises that are common for all the aforementioned vaccine modalities. However, this paradigm only holds true if the infrastructure investment is massive and the supply chains are functional for vaccines based on nanoparticles.

Note that all of this remains in the research and development phase, and issues related to scalability, safety, and regulatory acceptance are still being ironed out, but the ongoing development of nanoparticle-based vaccines and therapies remains an important topic for consideration. A number of strategies are being followed to tackle toxicity issues, with better manufacturing processes, and clearer regulatory pathways. For example, scientists are investigating biodegradable nanoparticle materials that the body can more easily process and excrete, lowering long-run safety concerns.

## 12.11 Emerging Trends and Innovations in Nanoparticle-Based Vaccines and Immunotherapies

Vaccines and immunotherapies based on nanoparticles are currently leading-edge, generating novel candidates for the treatment and prevention of diseases. Since more and more research are being made on this field, few emerging trends and innovations are shaping the future direction nanoparticle applications in healthcare. Such advances will offer the potential to make treatments more effective, safe, and accurate for a broad spectrum of diseases from cancer and autoimmune disease to viral infection. With emerging trends and innovations in nanoparticle-based vaccines and immunotherapies, a promising future is emerging through enhancing their efficacy, safety, and personalization. These include personalized medicine, design biomimetic, multi-functional platform, RNA-based therapy and biodegradable materials which drive the future directions of innovation in this domain. These trends, as research further deepens our understanding, will likely result in new and advanced therapies for chronic disease, and with it, the promise of better health around the globe.

1. **Personalized Nanoparticle Vaccines and Immunotherapies**: One of the more exciting trends in nanoparticle technology is the development of vaccines and immunotherapies that could be tailored to an individual patient. For example,

personalized medicines for a particular patient are meant to target specific characteristics of that patient (genetic, immune, or disease-related). This versatility allows nanoparticles to be used for personalized approaches based on various types of therapeutic agents. For example, they are loaded with cancer antigens of patients for personalized cancer immunotherapy. This generates an immune response that is more customized to the patient, improving the odds of an effective outcome. Nanoparticles can also be personalized for delivery of several therapeutic agents at the same time, customizing treatment for each patient based on their specific disease trajectory (Q. Li et al. 2021).

2. **Bioinspired Nanoparticles**: The other trend which is on the edge is the bioinspired preparation of nanoparticles mimicking the natural biological systems. The nanoparticles serve as a virulent mimicry of naturally occurring virus, bacteria, or cells, thus resulting in immune evasion, improved cellular delivery, and a more robust immune response. For instance, virus-like nanoparticles (VLPs) are being investigated as potential vaccines, which can mimic the structure of the virus without containing any of the infectious genetic material. These VLPs afford an ideal platform for the display of viral antigens to the immune system and can elicit robust humoral and cellular immunity. Another exciting avenue of bioinspired development are nanoparticles designed to mimic immune cells themselves which increase localization and activation of the immune response, and may improve targeting of cancers and infections (Geraldi et al. 2018).
3. **Multifunctional and Combination Therapies**: Nanoparticles are increasingly being engineered as multi-functional platforms capable of delivering multiple therapeutics in a single nanoparticle. It is increasingly apparent within combination therapies that use multiple drugs or therapeutics to treat various features of disease. Nanoparticles can encapsulate immune-stimulating antigens, chemotherapy agents, and adjuvants within one platform that targets both the disease and the immune system at the same time. As an example, multifunctional nanoparticles that simultaneously deliver a chemotherapy drug and an immune checkpoint inhibitor for cancer therapy. That combination enhances the immune system assault on cancer cells while providing the cytotoxic agents to kill the tumor cells. This makes for a more holistic therapy that prevents the cancer cells from bypassing the immune response (evasion tools) and killing them directly (to avoid destroying the body) (Yin et al. 2023).
4. **Nanoparticles for RNA Delivery**: While mRNA vaccines has proven crucial in combating the COVID-19 pandemic, their exciting success have piqued interest in other medical uses of nanoparticles for RNA delivery. Lipid nanoparticles (LNP) have been successfully employed to deliver mRNA for COVID-19 vaccines, and further, LNPs are being studied in various RNA-based therapies. These consist of siRNA/microRNA (small interfering RNA and microRNA)-based therapeutics that modulate the expression of causative or pathogenic genes/nanoparticle delivery systems for gene silencing and transfer for applications in cancer and genetic disorders. By shielding the RNA against degradation, which also facilitates its specific target cell delivery, the nanoparticles pave the way for genetic-level clinical therapies. These delivery systems will be further

developed focusing on different diseases including rare genetic diseases, neurodegenerative diseases and metabolic diseases (Aldosari et al. 2021).

5. **Biodegradable Nanoparticles**: Nanoparticle design is on track to include biodegradability as a major innovation. They are biodegradable nanoparticles that are proven to degrade naturally within the body, minimizing the risk of long-term bioaccumulation and toxicity. Polymeric materials such as polylactic-co-glycolic acid (PLGA) are often used to make nanoparticles that fully degrade into nontoxic products that can exit the body. This is a crucial step for therapies, such as immunotherapies, that are used multiple times, as we do not want the nanoparticles to build up over time, and have unintended consequences," Trivedi said. Nanoparticles that are biodegradable can be applied pharmaceutical for drug delivery, gene therapy, sustained-release vaccines. This ability for long, controllable release of therapeutics while minimizing chronic safety risks potentially makes them useful materials for developing future therapeutics (Haque et al. 2023).

## 12.12 Conclusion

In conclusion, the promising implications for vaccine and immunotherapy based on nanoparticles represent a significant step forward in the field of therapeutics. By delivering antigens and drugs to target cells, these nanoscale carriers are translating into increased efficiency and reduced toxicities over traditional treatments. These platforms offer a high surface area, enabling precise manipulation of stability, bioavailability, and controlled release, allowing for a high degree of versatility, enabling them to address complicated diseases including infections, cancer, and autoimmune illnesses. Also, it enables balanced and vigorous humoral and cellular immunity due to the antigen presentation capability of nanoparticles mimicking pathogens. With targeting ligands used in a strategic manner to minimize off-target effects and the possibility of multiple antigens, adjuvants and therapeutic molecules to be incorporated into a single platform, this approach allows for ground-breaking combinatorial therapy strategies.

However, substantial obstacles remain, such as regulatory challenges, potential nanotoxicity, and manufacturing scalability challenges that have hindered broader adoption. Further studies and well-controlled clinical trials are needed to overcome these hurdles for safe and effective applications. With methodical solutions to some of these global health challenges, nanoparticle-based strategies will progressively revolutionise health paradigms worldwide, enabling precise treatments and preventative measures that span a broad range of diseases. Their continued development could potentially revolutionize patient outcomes and the future of medicine around the globe.

# References


Akbari A, Amini M, Tarassoli A, Eftekhari-Sis B, Ghasemian N, Jabbari E (2018) Transition metal oxide nanoparticles as efficient catalysts in oxidation reactions. Nano Struct Nano Objects 14:19–48. https://doi.org/10.1016/J.NANOSO.2018.01.006

Aldosari BN, Alfagih IM, Almurshedi AS (2021) Lipid nanoparticles as delivery systems for RNA-based vaccines. Pharmaceutics 13(2):206. https://doi.org/10.3390/PHARMACEUTICS13020206

Aurelia Chis A, Dobrea C, Morgovan C, Arseniu AM, Rus LL, Butuca A et al (2020) Applications and limitations of dendrimers in biomedicine. Molecules 25(17):3982. https://doi.org/10.3390/MOLECULES25173982

Baruah C, Das P, Devi P, Saikia PM, Deka B (2023) The emergence of nanovaccines as a new paradigm in virological vaccinology: a review. Open Explor 3(4):361–383. https://doi.org/10.37349/EI.2023.00107

Basu P, Saha N, Saha T, Saha P (2021) Polymeric hydrogel based systems for vaccine delivery: a review. Polymer 230:124088. https://doi.org/10.1016/J.POLYMER.2021.124088

Bezbaruah R, Chavda VP, Nongrang L, Alom S, Deka K, Kalita T et al (2022) Nanoparticle-based delivery systems for vaccines. Vaccines 10(11):1946. https://doi.org/10.3390/VACCINES10111946

Bugybayeva D, Dumkliang E, Patil V, Yadagiri G, Suresh R, Singh M et al (2024) Evaluation of efficacy of surface coated versus encapsulated influenza antigens in mannose–chitosan nanoparticle-based intranasal vaccine in swine. Vaccine 12(6):647. https://doi.org/10.3390/VACCINES12060647/S1

Chan Y, Ng SW, Singh SK, Gulati M, Gupta G, Chaudhary SK et al (2021) Revolutionizing polymer-based nanoparticle-linked vaccines for targeting respiratory viruses: a perspective. Life Sci 280:119744. https://doi.org/10.1016/J.LFS.2021.119744

Chatzikleanthous D, O'Hagan DT, Adamo R (2021) Lipid-based nanoparticles for delivery of vaccine adjuvants and antigens: toward multicomponent vaccines. Mol Pharm 18(8):2867–2888. https://doi.org/10.1021/ACS.MOLPHARMACEUT.1C00447/ASSET/IMAGES/LARGE/MP1C00447_0004.JPEG

Chen WH, Luo GF, Zhang XZ (2019) Recent advances in subcellular targeted cancer therapy based on functional materials. Adv Mater 31(3):1802725. https://doi.org/10.1002/ADMA.201802725

Chen J, Xu Y, Zhou M, Xu S, Varley AJ, Golubovic A et al (2023) Combinatorial design of ionizable lipid nanoparticles for muscle-selective mRNA delivery with minimized off-target effects. Proc Natl Acad Sci USA 120(50):e2309472120. https://doi.org/10.1073/PNAS.2309472120

Cheng Z, Al Zaki A, Hui JZ, Muzykantov VR, Tsourkas A (2012) Multifunctional nanoparticles: cost versus benefit of adding targeting and imaging capabilities. Science 338(6109):903–910. https://doi.org/10.1126/SCIENCE.1226338

Cheng B, Ahn HH, Nam H, Jiang Z, Gao FJ, Minn I, Pomper MG (2022) A unique core–shell structured, glycol chitosan-based nanoparticle achieves cancer-selective gene delivery with reduced off-target effects. Pharmaceutics 14(2):373. https://doi.org/10.3390/PHARMACEUTICS14020373/S1

Costa B, Gouveia MJ, Vale N (2024) Safety and efficacy of antiviral drugs and vaccines in pregnant women: insights from physiologically based pharmacokinetic modeling and integration of viral infection dynamics. Vaccine 12(7):782. https://doi.org/10.3390/VACCINES12070782

Diaz-Arévalo D, Zeng M (2020) Nanoparticle-based vaccines: opportunities and limitations. In: Nanopharmaceuticals: volume 1: expectations and realities of multifunctional drug delivery systems. Elsevier, Amsterdam, pp 135–150. https://doi.org/10.1016/B978-0-12-817778-5.00007-5

Díaz-Dinamarca DA, Salazar ML, Castillo BN, Manubens A, Vasquez AE, Salazar F, Becker MI (2022) Protein-based adjuvants for vaccines as immunomodulators of the innate and adaptive immune response: current knowledge, challenges, and future opportunities. Pharmaceutics 14(8):1671. https://doi.org/10.3390/PHARMACEUTICS14081671

Dong C, Zhu W, Wei L, Kim JK, Ma Y, Kang SM, Wang BZ (2024) Enhancing cross-protection against influenza by heterologous sequential immunization with mRNA LNP and protein nanoparticle vaccines. Nat Commun 15(1):5800. https://doi.org/10.1038/S41467-024-50087-5

Filipić B, Pantelić I, Nikolić I, Majhen D, Stojić-Vukanić Z, Savić S, Krajišnik D (2023) Nanoparticle-based adjuvants and delivery systems for modern vaccines. Vaccines 11(7):1172. https://doi.org/10.3390/VACCINES11071172

Geraldi NR, Dodd LE, Xu BB, Wood D, Wells GG, McHale G, Newton MI (2018) Bioinspired nanoparticle spray-coating for superhydrophobic flexible materials with oil/water separation capabilities. Bioinspir Biomim 13(2):024001. https://doi.org/10.1088/1748-3190/AAA1C1

Grego EA, Siddoway AC, Uz M, Liu L, Christiansen JC, Ross KA et al (2020) Polymeric nanoparticle-based vaccine adjuvants and delivery vehicles. Curr Top Microbiol Immunol 433:29–76. https://doi.org/10.1007/82_2020_226

Guerrini G, Magrì D, Gioria S, Medaglini D, Calzolai L (2022) Characterization of nanoparticles-based vaccines for COVID-19. Nat Nanotechnol 17(6):570–576. https://doi.org/10.1038/s41565-022-01129-w

Haque K, Rana M, Khan MSS, Tufael (2023) Biodegradable nanoparticles for sustainable drug delivery. Biosens Nanotheranostics 2(1):1–18. https://doi.org/10.25163/BIOSENSORS.217334

Herrera-Barrera M, Ryals RC, Gautam M, Jozic A, Landry M, Korzun T et al (2023) Peptide-guided lipid nanoparticles deliver mRNA to the neural retina of rodents and nonhuman primates. Sci Adv 9(2):eadd4623. https://doi.org/10.1126/SCIADV.ADD4623/SUPPL_FILE/SCIADV.ADD4623_TABLE_S1.ZIP

Hess KL, Medintz IL, Jewell CM (2019) Designing inorganic nanomaterials for vaccines and immunotherapies. Nano Today 27:73–98. https://doi.org/10.1016/J.NANTOD.2019.04.005

Jamshidi N, Tarighatnia A, Ghaziyani MF, Sajadian F, Olad-Ghaffari M, Nader ND (2024) Folic acid-conjugated Fe-Au-based nanoparticles for dual detection of breast cancer cells by magnetic resonance imaging and computed tomography. Front Biomed Technol 11(1):122–129. https://doi.org/10.18502/FBT.V11I1.14519

Karunakaran B, Gupta R, Patel P, Salave S, Sharma A, Desai D et al (2023) Emerging trends in lipid-based vaccine delivery: a special focus on developmental strategies, fabrication methods, and applications. Vaccine 11(3):661. https://doi.org/10.3390/VACCINES11030661

Kobayashi M, Leidner AJ, Gierke R, Farrar JL, Morgan RL, Campos-Outcalt D et al (2024) Use of 21-valent pneumococcal conjugate vaccine among U.S. adults: recommendations of the Advisory Committee on Immunization Practices—United States, 2024. MMWR Morb Mortal Wkly Rep 73(36):793–798. https://doi.org/10.15585/MMWR.MM7336A3

Kokate RA, Chaudhary P, Sun X, Thamake SI, Maji S, Chib R et al (2016) Rationalizing the use of functionalized poly-lactic-co-glycolic acid nanoparticles for dendritic cell-based targeted anticancer therapy. Nanomedicine 11(5):479–494. https://doi.org/10.2217/NNM.15.213

Kumari M, Acharya A, Krishnamurthy PT (2023) Antibody-conjugated nanoparticles for target-specific drug delivery of chemotherapeutics. Beilstein J Nanotechnol 14(1):912–926. https://doi.org/10.3762/BJNANO.14.75

Li Q, Li J, Song S, Chen W, Shen X, Li S, Xing D (2021) Nanoparticle-mediated tumor vaccines for personalized therapy: preparing tumor antigens in vivo or ex vivo? J Mater Chem B 9(10):2352–2366. https://doi.org/10.1039/D0TB02915G

Li E, Butkovich N, Tucker JA, Nelson EL, Wang SW (2023) Evaluating anti-tumor immune responses of protein nanoparticle-based cancer vaccines. Methods Mol Biol 2671:321–333. https://doi.org/10.1007/978-1-0716-3222-2_18

Lin Y, Chen X, Wang K, Liang L, Zhang H (2024) An overview of nanoparticle-based delivery platforms for mRNA vaccines for treating cancer. Vaccine 12(7):727. https://doi.org/10.3390/VACCINES12070727

Liu MA, Zhou T, Sheets RL, Meyer H, Knezevic I (2022) WHO informal consultation on regulatory considerations for evaluation of the quality, safety and efficacy of RNA-based prophylactic vaccines for infectious diseases, 20–22 April 2021. Emerg Microbes Infect 11(1):384–391. https://doi.org/10.1080/22221751.2022.2026742

Liu Y, Liang Y, Lei P, Zhang Z, Chen Y, Liu Y et al (2023) Multifunctional superparticles for magnetically targeted NIR-II imaging and photodynamic therapy. Adv Sci 10(2):2203669. https://doi.org/10.1002/ADVS.202203669

Mao W, Yoo HS (2024) Inorganic nanoparticle functionalization strategies in immunotherapeutic applications. Biomater Res 28:0086. https://doi.org/10.34133/BMR.0086/ASSET/EF0D16C8-B01D-4427-9E28-E2F6E9D8129B/ASSETS/GRAPHIC/BMR.0086.FIG.004.JPG

Milicic A, Reinke S, Fergusson J, Lindblad EB, Hu K, Thakur A et al (2022) Adjuvants, immunomodulators, and adaptogens. In: Vaccinology and methods in vaccine research. Academic Press, London, pp 223–280. https://doi.org/10.1016/B978-0-323-91146-7.00009-3

Min SH, Lei W, Jun CJ, Yan ZS, Guang YX, Tong Z et al (2023) Design strategy and research progress of multifunctional nanoparticles in lung cancer therapy. Expert Opin Investig Drugs 32(8):723–739. https://doi.org/10.1080/13543784.2023.2254683

Mittal P, Saharan A, Verma R, Altalbawy FMA, Alfaidi MA, Batiha GES et al (2021) Dendrimers: a new race of pharmaceutical nanocarriers. Biomed Res Int 2021(1):8844030. https://doi.org/10.1155/2021/8844030

Nguyen B, Tolia NH (2021) Protein-based antigen presentation platforms for nanoparticle vaccines. NPJ Vaccines 6(1):1–11. https://doi.org/10.1038/s41541-021-00330-7

Orensteina WA, Ahmedb R (2017) Simply put: vaccination saves lives. Proc Natl Acad Sci USA 114(16):4031–4033. https://doi.org/10.1073/PNAS.1704507114/ASSET/B07A5589-95D2-4DB7-97E6-18AF1702798F/ASSETS/GRAPHIC/PNAS.1704507114FIG01.JPEG

Pardi N, Hogan MJ, Porter FW, Weissman D (2018) mRNA vaccines—a new era in vaccinology. Nat Rev Drug Discov 17(4):261–279. https://doi.org/10.1038/nrd.2017.243

Pati R, Shevtsov M, Sonawane A (2018) Nanoparticle vaccines against infectious diseases. Front Immunol 9:385476. https://doi.org/10.3389/FIMMU.2018.02224/PDF

Peng X, Wang J, Zhou F, Liu Q, Zhang Z (2021) Nanoparticle-based approaches to target the lymphatic system for antitumor treatment. Cell Mol Life Sci 78(12):5139–5161. https://doi.org/10.1007/S00018-021-03842-6

Plotkin S (2014) History of vaccination. Proc Natl Acad Sci 111(34):12283–12287. https://doi.org/10.1073/PNAS.1400472111

Ren M, Abdullah SW, Pei C, Guo H, Sun S (2024) Use of virus-like particles and nanoparticle-based vaccines for combating picornavirus infections. Vet Res 55(1):1–22. https://doi.org/10.1186/S13567-024-01383-X

Rincon-Restrepo M (2016) Engineering nanoparticle-based vaccines: implications for the quality of humoral and cellular immunity. Doctoral thesis, Swiss Federal Institute of Technology, Lausanne. https://doi.org/10.5075/EPFL-THESIS-6848

Rios-Ibarra CP, Salinas-Santander M, Orozco-Nunnelly DA, Bravo-Madrigal J (2024) Nanoparticle-based antiviral strategies to combat the influenza virus (Review). Biomed Rep 20(4):65. https://doi.org/10.3892/BR.2024.1753

Shao H, Ma T, Hong D, Xu W, Li S, Arellano RS, Xu K (2018) Combination of multifunctional nanoparticles and interventional techniques makes each other going further in the field of cancer treatment. Curr Drug Metab 19(2):94–99. https://doi.org/10.2174/1389200218666170710185906

Takechi-Haraya Y, Ohgita T, Demizu Y, Saito H, Izutsu K, Sakai-Kato K (2022) Current status and challenges of analytical methods for evaluation of size and surface modification of nanoparticle-based drug formulations. AAPS PharmSciTech 23(5):1–20. https://doi.org/10.1208/S12249-022-02303-Y

Tsikourkitoudi V, Henriques-Normark B, Sotiriou GA (2022) Inorganic nanoparticle engineering against bacterial infections. Curr Opin Chem Eng 38:100872. https://doi.org/10.1016/J.COCHE.2022.100872

Vaccines & Immunizations | Vaccines & Immunizations | CDC (n.d.) Retrieved March 1, 2025, from https://www.cdc.gov/vaccines/

Wang Y, Deng L, Kang SM, Wang BZ (2018) Universal influenza vaccines: from viruses to nanoparticles. Expert Rev Vaccines 17(11):967–976. https://doi.org/10.1080/14760584.2018.1541408
Wang F, Ullah A, Fan X, Xu Z, Zong R, Wang X, Chen G (2021) Delivery of nanoparticle antigens to antigen-presenting cells: from extracellular specific targeting to intracellular responsive presentation. J Control Release 333:107–128. https://doi.org/10.1016/J.JCONREL.2021.03.027
Yang A, Bai Y, Dong X, Ma T, Zhu D, Mei L, Lv F (2021) Hydrogel/nanoadjuvant-mediated combined cell vaccines for cancer immunotherapy. Acta Biomater 133:257–267. https://doi.org/10.1016/J.ACTBIO.2021.08.014
Yasamineh S, Kalajahi HG, Yasamineh P, Yazdani Y, Gholizadeh O, Tabatabaie R et al (2022) An overview on nanoparticle-based strategies to fight viral infections with a focus on COVID-19. J Nanobiotechnol 20(1):1–26. https://doi.org/10.1186/S12951-022-01625-0
Yin J, Su X, Yan S, Shen J (2023) Multifunctional nanoparticles and nanopesticides in agricultural application. Nanomaterials 13(7):1255. https://doi.org/10.3390/NANO13071255
Yu Y-B, Liu Y, Liang H, Dong X, Yang X-Y, Li S, Guo Z (2023) A nanoparticle-based anticaries vaccine enhances the persistent immune response to inhibit Streptococcus mutans and prevent caries. Microbiol Spectr 11(2):e0432822. https://doi.org/10.1128/SPECTRUM.04328-22
Yuan T, Wang T, Zhang J, Liu P, Xu J, Gu Z et al (2023) Robust and multifunctional nanoparticles assembled from natural polyphenols and metformin for efficient spinal cord regeneration. ACS Nano 17(18):18562–18575. https://doi.org/10.1021/ACSNANO.3C06991/SUPPL_FILE/NN3C06991_SI_001.PDF
Zeng Y, Xiang Y, Sheng R, Tomás H, Rodrigues J, Gu Z et al (2021) Polysaccharide-based nanomedicines for cancer immunotherapy: a review. Bioact Mater 6(10):3358–3382. https://doi.org/10.1016/J.BIOACTMAT.2021.03.008
Zhang X, Zhang Z, Xia N, Zhao Q (2021) Carbohydrate-containing nanoparticles as vaccine adjuvants. Expert Rev Vaccines 20(7):797–810. https://doi.org/10.1080/14760584.2021.1939688
Zhao Z, Hu Y, Harmon T, Pentel P, Ehrich M, Zhang C (2019) Effect of adjuvant release rate on the immunogenicity of nanoparticle-based vaccines: a case study with a nanoparticle-based nicotine vaccine. Mol Pharm 16(6):2766–2775. https://doi.org/10.1021/ACS.MOLPHARMACEUT.9B00279
Zhao T, Cai Y, Jiang Y, He X, Wei Y, Yu Y, Tian X (2023) Vaccine adjuvants: mechanisms and platforms. Signal Transduct Target Ther 8(1):1–24. https://doi.org/10.1038/s41392-023-01557-7

# Chapter 13
# Nanoencapsulation of Bioactive Compounds

**Manish Kumar, Tarun Virmani, and Kamla Pathak**

**Abstract** Bioactive compounds of natural, microbial and animal origin exhibit multifunctional bioactivity against various health disorders. However, most of bioactive compounds have limited solubility, poor permeation, low stability towards GIT enzymes or follow high first pass effect. Besides this, it may exhibit non selective distribution to all organs and toxic effects on healthy cells. To subside these challenges and improve the therapeutic efficacy of bioactive compound, nanoencapsulation technique can be employed. Formulation of nanocarriers for encapsulation of bioactive molecules results in size dependent increase in aqueous solubility and permeability. Nanoencapsulation also provided stealth to harsh conditions or environment of body thereby improving the stability of bioactive compound. Nanocarriers also facilitate precise delivery of bioactive compound to target site minimizing non selective distribution to undesirable site and thus reducing undesirable side effects. Present chapter enlists the fabrication technique of bioactive compounds in nanocarriers and related benefits. The applications in bioactive compound encapsulated nanocarriers in pharmaceutical and biomedical fields was also discussed.

**Keywords** Nanoencapsulation · Drug delivery · Natural compound · Nanomedicine

---

M. Kumar · T. Virmani
Amity Institute of Pharmacy, Amity University, Greater Noida Campus, Uttar Pradesh, India

K. Pathak (✉)
Faculty of Pharmacy, Uttar Pradesh University of Medical Sciences, Saifai, Etawah, Uttar Pradesh, India



V. Khanna et al. (eds.), *Nanotechnologies for Drug Delivery and Biopharmaceutical Development*, Progress in Drug Research 77,
https://doi.org/10.1007/978-3-032-00796-4_13

## 13.1 Introduction

Nanoencapsulation is the technique of loading bioactive compounds in the core or shell of nanocarriers ranging 1–1000 nm (Pugazhendhi et al. 2024). Nanocarriers are usually consists of natural polymer or synthetic polymers or blend of both (M. Kumar et al. 2022b), or phospholipids (Borges et al. 2020), or proteins (R. Zhang et al. 2022). The physicochemical properties of bioactive compounds can be improved by their encapsulation in nanocarriers. It not only overcome the solubility and permeability issues majorly observed for bioactive compounds but also provide it protection towards harsh microenvironment or unfavourable events that may result in degradation, deterioration or loss (Saadi et al. 2023). Nanocarriers can also modulate the release kinetic of bioactive compounds to enhance its translation. This may facilitate the distribution and accumulation of bioactive compound to the desired target site while minimizing undesirable side effects. The nanoencapsulation can also modify the organoleptic properties bioactive compound improving the palatability and patient compliance (Pachuau et al. 2021; Saadi et al. 2023). A number of nanocarriers have been employed for encapsulating bioactive compounds among which few are dendrimers, polymeric nanoparticles (Wong et al. 2020), polymeric micelles, lipid-based nanoparticles (Chaudhari et al. 2020), nanoemulsions (Aguzzi et al. 2023), nanocomposites (Castro et al. 2019), nanocapsules (Wardana et al. 2023), nanocrystals (M. Kumar et al. 2024a), mesoporous silica nanoparticles (Fang et al. 2023), etc.

## 13.2 Nanotechnology for Bioactive Compounds Encapsulation

The encapsulation of bioactive compounds in nanocarriers compound can be achieved by various methods used for preparation of respective nanocarriers. Few methods are discussed below with examples as shown in Fig. 13.1 and Table 13.1.

### *13.2.1 Extrusion Technology*

The extrusion methods like hot-melt extrusion, co-extrusion, and electro-static extrusion can be used for encapsulation of bioactive compounds in nanoparticles. The extrusion technology results in nanoparticles of size as per equipment specifications that determine the encapsulation efficiency of bioactive compounds (Bamidele and Emmambux 2021).

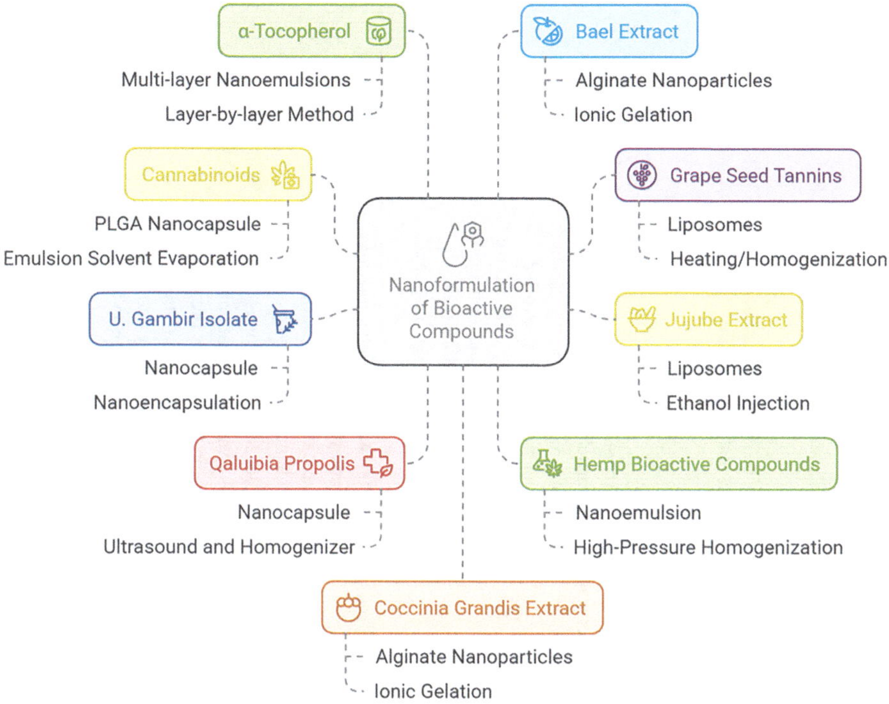


**Fig. 13.1** Nanocarriers for encapsulation of bioactive compounds and their preparation methods

### *13.2.2 Nano-emulsions*

Nano-emulsions can be employed for encapsulating bioactive compounds like essential oils (Jemaa et al. 2019). It can be prepared by two methods; either high-energy/top down methods or low-energy/down-up methods (Banasaz et al. 2020). The top-down methods usually are high-pressure homogenization, ultrasonication, or milling while down-up methods include emulsification, solvent diffusion, and phase inversion techniques.

### *13.2.3 Nano Spray Drying*

This method is exploited for large-scale production of bioactive compounds encapsulated nanocarriers. The product is as dry powder with advantages of economic, high encapsulation, more stability, high reproducibility and drug homogeneity (Jafari et al. 2023). This technology can employ spray mesh technology for the design of nanocarriers of varying particle size (Jayaprakash et al. 2023).

**Table 13.1** Bioactive compound encapsulated nanoformulations

| Bioactive compound | Nanoformulation | Components | Method of preparation | Advantages | References |
|---|---|---|---|---|---|
| *Phytoconstituent(s)* | | | | | |
| Cannabinoids (tetrahydrocannabinol and cannabidiol) | PLGA nanocapsule | Poly(lactic-co-glycolic acid) (PLGA), functionalized with pectin, alginate and chitosan | Emulsion solvent evaporation technique | 150–700 nm, maintained cannabinoid profile | Villate et al. (2024) |
| Grape seed tannins | Liposomes | Soy lecithin, tannins | Heating/ homogenization method | About 570 nm homogeneous particles with a surface charge of −37 mV, improved strength and wettability, controlled release, stability of bioactive compounds, protection against external conditions (O2, pH, temperature, etc.), long shelf life. | Vidal et al. (2024) |
| Jujube (*Ziziphus jujuba Mill.*) extract | Liposomes | Nanoliposomes, pea protein isolates as coating material | Ethanol injection method | 100–400 nm, high antioxidant activity, high stability, increased bioavailability, enhanced bioactivity. | Parhizkary et al. (2024) |
| U. gambir isolate (catechin) | Nanocapsule | Alginate, *Uncaria gambir* | Nanoencapsulation technique | Bioactive compounds had low bioavailability; nanoparticles of about 80 nm were prepared. Nanoencapsulation increased the bioavailability and stability, avoided deterioration against pH, temperature, and salinity, provided controlled release, good anticancer activity, and improved bioactivity. | Wardana et al. (2023) |

| | | | | | |
|---|---|---|---|---|---|
| *Qaluibia Propolis* | Nanocapsule | Sodium alginate, Propolis extracts | Ultrasound and high-speed homogenizer methods | 20–40 nm, high-temperature stability and better cytotoxic effects against human cancer cells (i.e. PC3, MCF7, and HePG2). | Amin et al. (2023) |
| Hemp (*Cannabis sativa L.*) bioactive compounds (terpenes and cannabinoids)/essential oil | Nanoemulsion | Polysorbate 80, ethyl oleate, essential oil | High-pressure homogenization method | 100–150 nm, high stability and shelf life, higher anticancer activity, and safe. | Aguzzi et al. (2023) |
| α-Tocopherol | Multi-layer nanoemulsions | Lecithin as primary-NEs, chitosan as secondary-NEs, and dextran sulfate as tertiary-Nes. | Layer-by-layer method | 90–200 nm, provided stability, improved bioavailability, good adhesion, higher cellular internalization, and increased antioxidant activity. | Hong et al. (2024) |
| Bael extract (Aeglemarmelos L. Correa) | Alginate nanoparticles | Alginate | Ionic gelation method | Provided controlled release, enhanced bioactivity (antioxidant, anti-inflammatory, and antidiabetic) | De Silva et al. (2023a) |
| Cocciniagrandis L. extract | Alginate nanoparticles | Alginate, aqueous extract, $CaCl_2$ | Ionic gelation | Higher encapsulation, particle size of about 70 nm, enhanced antioxidant activity. | DE Silva et al. (2023b) |
| Pomegranate peel extract rich in polyphenols | Nano-phytosomes | Phosphatidylcholine, gamma-oryzanol | Thin film hydration/ sonication method | Formed stable nanosystem of about 60 nm and surface charge of −32 mV, high encapsulation of 95%, preserved the antioxidant activity, sustained the release, | Andishmand et al. (2024) |
| Betalains from pitaya (Stenocereus thurberi) | Gelatin nanoparticles | Gelatin | Co-axial electrospray technique | About 800 nm, increased antioxidant activity and maintained bioactivity. | Figueroa-Enriquez et al. (2023) |

(continued)

**Table 13.1** (continued)

| Bioactive compound | Nanoformulation | Components | Method of preparation | Advantages | References |
|---|---|---|---|---|---|
| Curcumin | Nanoparticles | Lecithin, chitosan and olive oil | Ultrasonication and probe sonication | About 30 nm and negative zeta potentials of −28 to −34 mV, higher encapsulation of 91%, biocompatible, biodegradable | Atwan and Al-Ogaidi (2024) |
| β-Carotene | Nanoencapsulate | Zein protein | Electrospraying technique | Higher encapsulation of about 81%, improved dissolution, increased permeability | Mahalakshmi et al. (2020) |
| Cinnamaldehyde | Nanoparticles | Chitosan hydrochloride, sulfobutyl ether-β-cyclodextrin | Cross-linking method | High encapsulation (86%), enhanced thermal stability, amorphous in nature, released slowly, suitable for enhancing solubility, avoiding degradation, and prolonging residence in nanoparticles. | Zhu et al. (2020) |
| Borage seed oil, peppermint oil and lycopene | Water/oil/water nanoemulsions | Gum ultra, soy protein isolate, pectin, whey protein isolate | Emulsification | High encapsulation, shielded from oxidation, high stability, controlled release. | Rehman et al. (2021) |
| Eggplant (Solanum melongena L.) Peel extract | Nanofibers | Gelatin | Electrospinning | 600 nm diameter, high encapsulation (90%). | Estrella-Osuna et al. (2022) |

| | | | | | |
|---|---|---|---|---|---|
| Bioactive peptides | Polymeric, lipid and other nanocarriers | Lipid, carbohydrate and protein | Nanoencapsulation methods. | Methods like HPH, electrospraying, electrospinning, supercritical fluid, emulsification, ionic gelation, ultrasonication, coacervation, and nanoprecipitation can be used. Overcame limitations of undesired organoleptic property, low solubility, chemical-instability, and low-bioavailability. Better bioactivity. | Mazloomi et al. (2024) |
| Organoselenium compound 5′-Seleno-(phenyl)-3′-(ferulic-amido)-thymidine and paclitaxel | pH-responsive nanoparticles | PCL, lysine-based surfactant (77KS) as a pH-responsive adjuvant | Nanoprecipitation methods | Higher antiproliferative activity in multi drug resistance tumor cells | Mathes et al. (2024) |
| Bioactive compound/ siRNA | Nanovehicles | siRNA, natural products, Nanosized encapsulants | – | Encapsulation improved solubility and bioavailability of bioactive, and therapeutic action. Co-delivery provided a synergistic effect. | Ashrafizadeh et al. (2020) |

### 13.2.4 Electrospraying

Electrospraying allows the encapsulation of bioactive compounds preserving their therapeutic activity while protecting from harsh process factors. The formulation composition and process constraints may govern bioactive compound loading and their release from nanoparticles as well as size, shape and surface characteristics that further decide physicochemical properties of payload (Drosou et al. 2017). Electrospraying is an advanced drying technique for heat-sensitive and labile bioactive compounds with solubility in different solvent systems (organic or non-organic solvents) to encapsulate bioactive compounds. These electric-assisted techniques economical and takes short drying time to gives stable and monodispersed particles with well-defined shape and high encapsulation efficiency (Jayaprakash et al. 2023; Tanhaei et al. 2021).

### 13.2.5 High-Pressure Homogenization (HPH)

The process is high energy top down method for the formulation of bioactive compounds encapsulated nanocarriers (Aouf et al. 2020). It involves high shear forces, cavitation and turbulence that reduces the particle size to give nanocarrier of desired size (Levy et al. 2021). The stabilizer solution and bioactive compound is usually forced through a narrow orifice under a high pressure to form the bioactive compound encapsulated nanocarriers.

### 13.2.6 Ultrasonication

Ultrasonication is another widely used technique for the production of bioactive compound nanoparticles as an alternative to high-pressure homogenization. It utilizes high-frequency sound waves to produce cavitation effects and intense shear forces that break down the bioactive compound's larger particles into nano-size particles (Yang et al. 2021). The technique is followed at a lower temperature and in the presence of a stabilizer to avoid particle growth and formulation of stable nanoparticles (M. Kumar et al. 2024a).

### 13.2.7 Microfluidization

Microfluidization is an advanced technique for the production of bioactive compound encapsulated nanoparticles (Ganesan et al. 2018). The method involves the use of high-pressure and controlled fluid dynamics (microchannels) for the production of nanosized particles due to intense shear, impact, and cavitation forces (Ganesan et al. 2018; Maa and Hsu 1999). The method is particularly used for sensitive bioactive to obtain uniform and stable nanoparticles.

## 13.3 Mechanism of Encapsulation

Various types of interactions like hydrophobic, electrostatic, van der Wall forces, hydrogen bonding as well as covalent bonding between bioactive compound and polymer are used while formulation of nanoparticles can result in encapsulation of bioactive compound in nanoparticles. For example; hydrophobic interaction resulted in the encapsulation of ferulic acid in the porous nanosponge cavity of cyclodextrin (Rezaei et al. 2019b). The essential oil of *L. origanoides*is loaded in voids of Chitosan/ferulic acid nanogel via the electrostatic interaction of essential oil components with protonated amine groups of hydrogel (Almeida et al. 2018). The hydrogen bond interaction was responsible for the encapsulation of eggplant peel extract in gelatin nanofibers (Estrella-Osuna et al. 2022). The interaction type may also guide the solubility, permeability, release, and stability of bioactive compounds. The other broad mechanism involved in design of bioactive compounds encapsulated nanocarriers involve mainly physical and chemical interactions (Fig. 13.2).

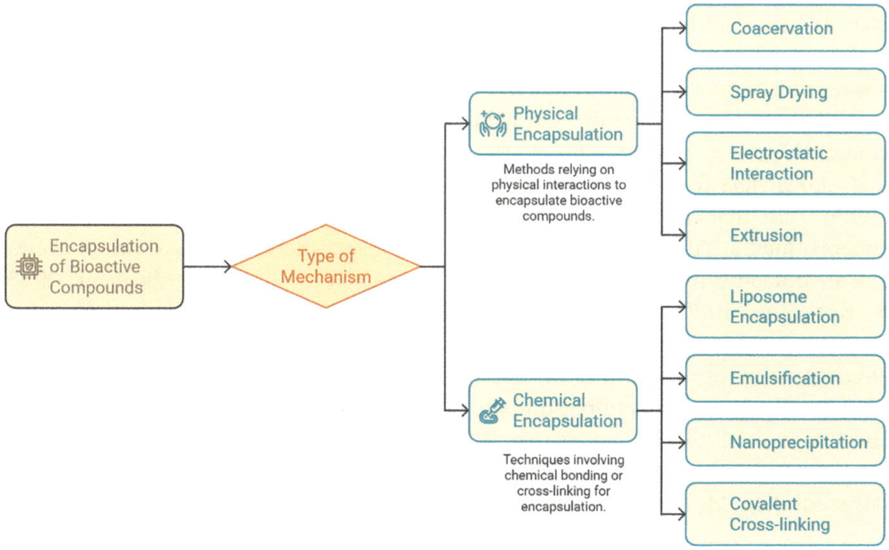


**Fig. 13.2** Mechanism relying on encapsulation of bioactive compounds

## 13.4 Advantage of Nanoencapsulation of Bioactive Molecule and Mechanism

Phytoconstituents like curcumin, quercetin, naringenin, genistein, apigenin, kaempferol, thymoquinone, epigallocatechin, etc., demonstrated potent therapeutic effects against various health disorders (Yıldırım et al. 2023). Despite this, inherent bioactivity may not translate into therapeutic benefits due to poor solubility, low stability, high volatility, or sensitivity to environmental factors such as light, temperature, or oxygen (Almeida et al. 2018; M. Huang et al. 2019; Mahalakshmi et al. 2020; Rezaei et al. 2019b). Encapsulation of bioactive compounds in nanoparticles can help to surmount these challenges (Yıldırım et al. 2023). Table 13.1 enlists a few examples of nano encapsulated bioactive molecules and their associated advantages. The nanoencapsulation role in improving the properties of bioactive molecules is as discussed below.

### *13.4.1 Improving Solubility*

Nanoparticles play a crucial role in solubility improvement of bioactive molecules of BCS Class II and IV, a common challenge in pharmaceutical product development (Liu et al. 2024). Ferulic acid encapsulation in cyclodextrinn nonosponge observed to increase the aqueous solubility by approximately 15 folds (Rezaei et al. 2019b). This could be due to size dependent increase in surface area and hydrophilic surface of prepared nanocarriers. According to the Noyes-Whitney equation, higher dissolution of poorly soluble bioactive molecules with nanosized carriers can results in higher drug bioavailability and thus improving the therapeutic efficacy (Loh et al. 2015).

Nanocarriers for encapsulation of particle size may increase the surface area to volume ratio that allows higher interaction of bioactive compound with surrounding aqueous media (Alshora et al. 2016). This in turn accelerate the dissolution of bioactive molecule improving its apparent aqueous solubility (Csicsák et al. 2023; R. Kumar et al. 2022c). Nanocarriers can also maintain the crystalline bioactive compound in molecularly dispersed or amorphous state thereby improving the aqueous solubility (Manjit et al. 2024). The nanocarriers can maintain the supersaturation state and thus improving the oral bioavailability of bioactive compounds (Brouwers et al. 2009). Nanocarriers surface modified with hydrophilic materials can enhance the wettability of bioactive molecules responsible for its higher aqueous solubility (M. Kumar et al. 2024b).

### 13.4.2 Improving Permeability

Nanocarriers can improve the permeation of bioactive compound (especially BCS class III and IV) across biological barriers attributed to their size and surface characteristics (Babadi et al. 2020; Ghasemiyeh and Mohammadi-Samani 2020; Yu et al. 2021). Nanoparticles allow passage of bioactive molecules through pores as well as via endocytosis to cross the biological barriers. The nanoencapsulated bioactive compounds can easily cross biological barriers like blood-brain barriers (Mishra et al. 2023). Polymeric and lipid-based nanosystems have shown their role in improving the permeation of bioactive molecules (Arpicco et al. 2016; Becker Peres et al. 2016; Xu et al. 2019). Nanoencapsulation can also avoid P-glycoprotein (P-gp) efflux and occurrence of multi-drug resistances (Dewanjee et al. 2017).

### 13.4.3 Improving Stability

Nanoencapsulation can provide stability to bioactive molecules towards external microenvironments for avoiding their degradation and improving efficacy (Taouzinet et al. 2023). The protection can be from internal or external factors like pH, temperature, enzymes, photons or humidity that have negative impact on the quality, therapeutic efficacy and safety of bioactive compounds (Hcini et al. 2021; Rehman et al. 2021). For example; baicalein nanoencapsulation in liposomes can increase the stability by approximately 53% (Aliakbari et al. 2018). Encapsulating bioactive compounds in nanoparticles creates a stealth or protective barrier that prevents negative interactions with the external microenvironment (L. Huang et al. 2024). Nanoencapsulation prevents the breakdown of bioactive compounds by limiting direct interaction with harmful environmental conditions such as oxygen or moisture (oxidation or hydrolysis), pH shift, and enzymatic destruction (Stojiljković et al. 2019). Ellagic acid encapsulation in liposomes proved enhanced stability at variable pH, avoiding degradation and enhancing antioxidant activity for superior hepatoprotective effects than the free form (Stojiljković et al. 2019). Nanoliposomal encapsulation of quercetin resulted in increased structural stability towards gastrointestinal conditions, resulting in higher bioavailability and better bioactivity (Rodriguez et al. 2019). Neohesperidin's physicochemical stability under storage, heat, pH, ionic, UV, oxidative, and serum conditions was also enhanced by its encapsulation in liposomes (Karim et al. 2020).

During storage, the effectiveness of certain bioactive chemicals can be diminished due to evaporation or volatilization. Enclosing volatile compounds in nanoparticles prevents evaporation and maintains their concentration over time, resulting in the stabilization of volatile components. Cinnamaldehyde is an antibacterial component prone to oxidation and volatilization during storage. Encapsulation of cinnamaldehyde in nanoparticles showed to increase the solubility, as well as avoided oxidative degradation, reduced its volatilization and thus prolonged the shelf-life

(Zhu et al. 2020). Towards similar application, nanoemulsions have shown their potential in improving the stability of volatile oils (Alsakhawy et al. 2024; Aguzzi et al. 2023).

### 13.4.4 Controlling Drug Release

Nanoencapsulation can modify the release kinetic of bioactive compound in systemic circulation or/and extracellular or intracellular microenvironment (Jha et al. 2024b, c; Noore et al. 2024). The nanoencapsulation modified release of bioactive compounds in body fluid or target site is advantageous in higher accumulation at target site (Rehman et al. 2021; Rezaei et al. 2019b). The release can vary with type of polymer or excipients used and also nanocarriers-type (Martínez-Ballesta et al. 2018; Siddiqui et al. 2023). The stimuli responsive nanocarriers release the active molecules as response to particular stimulus like pH, temperature, redox, enzymes or photons, providing precise and selective delivery (Rezaei et al. 2022).

## 13.5 Increasing Systemic Circulation

A primary challenge faced by bioactive compounds is their low systemic circulation owing to rapid degradation, fast clearance by the liver and kidneys, and immunogenicity. Biological functions like gastrointestinal digestion or cellular metabolism can clear the bioactive compounds prior to their reach at target site (Martínez-Ballesta et al. 2018). Encapsulating them in nanoparticles provide stealth from extreme pH (such as the acidic environment of the stomach), enzymes or temperature changes, to avoid premature degradation of bioactive compounds for improving their systemic circulation (Vlasova et al. 2016).

The major organs like liver, spleen, and kidneys can also clear the bioactive compounds form the bloodstream by treating them as foreign molecules. Nanoparticles can be engineered to avoid detection of such molecules by the mononuclear phagocyte system (MPS) and resist opsonization (the process where blood proteins mark particles for clearance). Surface modification like PEGylation, surface charge or ligand functionalization reduces recognition by immune cells and avoid nonspecific uptake, delay opsonization, and improve tissue specific accumulation, thus prolonging the systemic circulation (Alexis et al. 2008; Vanti 2021). Nanoparticles are typically designed to be within the optimal size of range (10–150 nm) for avoiding rapid clearance by the kidneys (which clear smaller particles) and uptake by the liver (which clears larger particles) (Y. Huang et al. 2021; Mok 2024).

### 13.5.1 Minimizing Toxicity or Undesirable Events

The unique properties of nanoparticles like small size, surface functionality, and ability to carry both hydrophilic and hydrophobic drugs, may help to mitigate the toxicity and adverse effects associated with coarse bioactive compound (Rezaei et al. 2019a; Vrignaud et al. 2011). The inherent role of nanoparticles in increasing solubility and permeability improves the bioavailability of bioactive compounds which in turn reduce the dose requirement and thus toxicity (Fan et al. 2018; M. Kumar et al. 2020). By preventing rapid clearance and degradation, nanoparticles prolong the systemic circulation that further decrease the need for higher doses that could cause toxicity (H. Zhang et al. 2015).

## 13.6 Providing Site-Specific Delivery

Nanoparticles can be designed to deliver the bioactive compound at the target site, such as within cancerous tissue or at an inflamed site. This can be attained by designing surface modified nanocarriers with high selectivity towards target site. To attain site specific delivery, the surface functionalized of nanocarriers can be done with targeting ligands like antibodies, peptides, or small molecules (Jha et al. 2024a; M. Kumar et al. 2020; M. Kumar and Rajnikanth 2020; Manzari et al. 2021). This results in selective accumulation of bioactive compounds in specific cells via receptors present on its surface towards which the targeted ligand have high affinity (Jha et al. 2024b; Yadav et al. 2021). This results in precise delivery of bioactive molecule at target site, thereby minimizing off target distribution, minimize undesirable effects, avoid wastage and improving therapeutic efficacy.

Nanoparticles can be engineered using stimuli responsive moiety that allows release the payload as a response to a particular stimulus. (e.g., pH, temperature, or enzyme-triggered release) allowing for more precise control over drug delivery (Jha et al. 2024b). Braga et al. (2020) prepared Goniothalamin (*Goniothalamus* species) encapsulated nanoparticles for pH responsive delivery. Goniothalamin (GTN) exhibit anticancer activity against several cancers but has low solubility and toxic effects on healthy cells. Encapsulation of GTN in nanoparticles can overcome these challenges, improving the pharmacokinetic performance. Moreover, pH-responsive acetylated dextran (Ac-Dex) was used for design of nanoparticles (about 100 nm) that exhibited high encapsulation efficiency of the bioactive compound. The nanoparticles were stable at physiological pH of 7.4 improving the systemic circulation and providing cancer-site specific drug delivery. The particle disassembled at acidic pH 5.5 releasing payload in a sustained manner. The nanoparticles exhibited higher internalization in cancer cells and thus demonstrated higher cytotoxicity in cancer cells compared to free GNT. The histological assessment also revealed toxicity-free result (Braga et al. 2020).

## 13.7 Applications of Nano-encapsulated Bioactive Compounds

Natural compounds have been utilized widely in management of various health disorders like microbial infection, inflammation, wounds, cancer, Alzheimer's, diabetes, cardiovascular diseases, etc., (Enrico 2019; Stoleru et al. 2024; Watkins et al. 2015). Natural compound-encapsulated nanoparticles led in more effective treatments, serving as an encouragement towards bio-nanomedicine.

### *13.7.1 Skin Disorder*

Phyto-antioxidants like polyphenols (catechin, quercetin or gallic acid) can be efficiently encapsulated in nanoemulsions for topical delivery and skin protection/skin rejuvenation applications (Romes et al. 2021). Nanocarriers encapsulated with natural compounds improve the dermal delivery of payload by facilitating their intercellular and intracellular delivery into or across skin (M. Kumar et al. 2022b).

### *13.7.2 Microbial Infections*

Natural compounds like; essential oils, plant extracts, peptides of animal origin, bacteriocins etc. with antimicrobial activity can be used for treatment of microbial infections (Stan et al. 2021). Natural compound encapsulated nanoparticles can be prepared for improving the stability, bioavailability, cellular internalization, pharmacokinetics and safety of natural compounds. Lemongrass essential oil (LEO) with antibacterial activity is volatile, instable and low bioavailable. Encapsulation of LEO in zein-sodium caseinate nanoparticles of about 243 nm was stable with high payload potential (85%). Nanoparticle enhanced the thermal stability, increased cell wall permeability, provided sustained release and improved bioavailability of LMO. Resultant LMO encapsulated nanoparticle exhibited potent antimicrobial activity against Staphylococcus epidermidis, *Enterococcus faecalis*, *Escherichia coli*, and *Klebsiella pneumoniae* compared to free LMO (Alsakhawy et al. 2024). The essential oils of *Origanum glandulosum* Desf. can be used as an alternative to antibiotics but with limited application due to hydrophobicity and negative interaction with the environmental conditions. The encapsulation for the oil into nanocapsules by was performed high-speed homogenization or into nanoemulsion by high-pressure homogenization was done to conserve major volatile components of oil. The nanoemulsion of about 50 nm and nanocapsules of about 120 nm exhibited potent antibacterial activity against *E. coli*, *S. aureus*, and *A. baumannii* (Bouaouina et al. 2022). Nanoparticles can effectively eradicate the microbes by facilitating the accumulation of bioactive at infection site and cellular internalization, responsible for potent antimicrobial action. Zou et al. (2023) prepared glabridin encapsulated nanoparticles responsive to pH of infected site. Prepared nanoparticle received targeted delivery to infected macrophages for eradicating intracellular MRSA effectively (Zou et al. 2023).

### 13.7.3 Neurodegenerative and Inflammatory-Disorder

Neurodegenerative disorders like Huntington's, Alzheimer's, and Parkinson's experiences mitochondrial dysfunction following inflammation, oxidative stress, and apoptosis that results in neuronal cytotoxicity and degeneration. Natural compounds with antioxidant, anti-inflammatory and neuroprotective action can be encapsulated in nanocarriers to favor their delivery at target site (Tavan et al. 2024). Nanoencapsulation of magnoflorine, an aporphine quaternary alkaloid with antioxidant action in chitosan-collagen nanocapsules resulted in a system that effectively managed the neurodegeneration. Synthesized nanocapsules were about10 nm of spherical shape, provided sustained release, exhibited excellent anti-oxidant action with no major sign of cytotoxicity on SH-SY5Y cells (Bashir et al. 2022). Bioactive compounds like cordyceps and turmeric with anti-oxidant effects can be used for neurological disorders. Encapsulation of cordyceps and turmeric extracts in nanoparticles improve cellular internalization and delivery to target cells (neuroblastoma cells: SH-SY5Y), responsible for better therapeutic effect (Kaokaen et al. 2023).

Natural compound with anti-inflammatory action can be encapsulated in nanoparticles for improving their stability and bioavailability (Rakotondrabe et al. 2023). Natural compound loaded nanocarrier for neuroinflammation have been employed to overcome limitations (like, poor pharmacokinetic and stability) and aid selective delivery to target site (Khadka et al. 2020). Plant isolates (like thymoquinone, resveratrol, curcumin, quercetin, embelin, silymarin), peptides (tuftsin, vasoactive intestinal peptide), insect-derived bioactives, nucleic acids, etc., encapsulated in nanocarriers have been employed for treatment of Inflammatory Bowel Disease (Khare et al. 2020). Moreover, targeted and stimuli responsive nanoparticles have further open the avenues for effective management of such disorders (M. Kumar et al. 2022a; Taghipour et al. 2018).

### 13.7.4 Hepatic Disorder

Hydroxytyrosol (HT) is a natural polyphenol that can provide protection against hepatic steatosis or NAFLD owing to its lipid-lowering, anti-oxidant and anti-inflammatory actions. However, HT delivery to hepatocytes has seldom achieved the required attention. The encapsulation of HT in nanoformulation is a key method for improving the intracellular uptake in hepatocytes. Prepared nanogels based

HT-loaded nanoscaffolds were non-toxic to cells, did not induce oxidative stress and significantly decreased triglyceride levels. Formulated nanosystem exhibited superior performance for better management of NAFLD (Mauri et al. 2021). In another hepatic disorder like hepatocytes injury, pelargonidin-*3-O*-glucoside can provide protection but are unstable in physiological environment. Their encapsulation in nanoliposomes with chitosan and/or pectin functionalization resulted in a system that aided hepatocellular delivery of hydrophilic compounds, providing protection against palmitic acid (PA)-induced hepatocytes injury in L02 cells with greater biological activity (Karim et al. 2022).

### *13.7.5 Wounds*

Natural compound encapsulated nanoformulation have shown tremendous potential in sacr-free wound healing (Berry et al. 2024). Plant extracts or phytochemicals encapsulated in nanoparticles aid drug delivery at wound area by releasing the payload in sustained manner for fastening the healing process (angiogenesis, neovascularization, and re-epithelialization) and providing prolonged wound healing activity (A. K. Das et al. 2024; Hajialyani et al. 2018). Natural compound like naringenin encapsulated in nanofibrous scaffolds found to shorten the wound healing time where nanofibrous scaffolds not only improve the solubility of drug to maintain concentration gradient but also act as carrier supporting retention at site and providing controlled drug release for effective management of chronic wounds (Madhukiran et al. 2021; Manjit et al. 2024).

### *13.7.6 Cancer Therapy*

Natural compounds of plants, bacteria, fungi, and marine origin have shown role in cancer treatment (Andreani et al. 2024). Natural compounds like curcumin, honokiol, formononetin, hesperidin, gossypol, sulforaphane, resveratrol, emodin, allicin, morusin, ursolic acid, arnicolide, epigallocatechin gallate, genistein, anthocyanins, deoxyshikonin, apigenin, quercetin, lycopene, piperine, taxanes, berberine, betanin, camptothecin, brucine, crytolepine, lycorine, papaverine, vincristine, flavonoids, lignans, saponins, tannins etc., are few reported examples of bioactives with promising anticancer activity targeting various cancer signalling pathways that participates in cancer progression, angiogenesis, apoptosis, and metastasis (Auti et al. 2024; Ghosh et al. 2024; Hashem et al. 2022; Kubczak et al. 2021). However, despite their importance in cancer therapy, they have limited clinical applications due to poor aqueous solubility, low permeability, and low stability. Nanoencapsulation can therefore be used to increase solubility, permeability, and consequently bioavailability of bioactive compound. Nanoencapsulation has also been shown to improve the shelf life, pharmacokinetics, and biodistribution of bioactive compounds, consequently increasing anticancer activity (Chavda et al. 2022; R. Das et al. 2022; Jha et al. 2022).

## 13.8 Conclusion

Bioactive compounds are multifunctional with their role in management of diverse health disorders. The low clinical applicability of bioactive compounds is attributed to their poor aqueous solubility, low permeability, non-selective distribution, high dose requirement, systemic side effects and instability. To overcome these, nanoencapsulation technique can be used as a promising and effective strategy. Besides this, nanoencapsulation can overcome biological barriers to reach target site for more effective bioactivity. Nanoencapsulation technology equipped with targeting approach via use of targeting ligand and stimulus-responsive materials, a more precise delivery can be achieved. This not only improve the translational and therapeutic potency of bioactive compounds but also subside toxicity and undesirable effects. Nanoencapsulation can also increase the organoleptic properties of bioactive compounds and thus increasing patient compatibility. Nanoencapsulation is an economic and highly effective methods of designing a safe and effective nanomedicine with superior organoleptic, physicochemical, pharmacokinetics, and pharmacodynamics properties. In conclusion, nanoencapsulation can be used as a promising approach for broadening the application of bioactive compounds in pharmaceuticals, biomedical fields and cosmetics.

## References

Aguzzi C, Perinelli DR, Cespi M, Zeppa L, Mazzara E, Maggi F, Petrelli R, Bonacucina G, Nabissi M (2023) Encapsulation of hemp (Cannabis sativa L.) essential oils into nanoemulsions for potential therapeutic applications: assessment of cytotoxicological profiles. Molecules 28(18):6479. https://doi.org/10.3390/molecules28186479

Alexis F, Pridgen E, Molnar LK, Farokhzad OC (2008) Factors affecting the clearance and biodistribution of polymeric nanoparticles. Mol Pharm 5(4):505–515. https://doi.org/10.1021/mp800051m

Aliakbari F, Shabani AA, Bardania H, Mohammad-Beigi H, Tayaranian Marvian A, Dehghani Esmatabad F, Vafaei AA, Shojaosadati SA, Saboury AA, Christiansen G, Otzen DE, Morshedi D (2018) Formulation and anti-neurotoxic activity of baicalein-incorporating neutral nanoliposome. Colloids Surf B: Biointerfaces 161:578–587. https://doi.org/10.1016/j.colsurfb.2017.11.023

Almeida RR, Silva Damasceno ET, de Carvalho SYB, de Carvalho GSG, Gontijo LAP, de Lima Guimarães LG (2018) Chitosan nanogels condensed to ferulic acid for the essential oil of *Lippia origanoides* Kunth encapsulation. Carbohydr Polym 188:268–275. https://doi.org/10.1016/j.carbpol.2018.01.103

Alsakhawy SA, Baghdadi HH, El-Shenawy MA, El-Hosseiny LS (2024) Enhancement of lemongrass essential oil physicochemical properties and antibacterial activity by encapsulation in zein-caseinate nanocomposite. Sci Rep 14(1):17278. https://doi.org/10.1038/s41598-024-67273-6

Alshora DH, Ibrahim MA, Alanazi FK (2016) Chapter 6—Nanotechnology from particle size reduction to enhancing aqueous solubility. In: Grumezescu AM (ed) Surface chemistry of nanobiomaterials. William Andrew Publishing, pp 163–191. https://doi.org/10.1016/B978-0-323-42861-3.00006-6

Amin AA, Mahmoud KF, Salama MF, Longo V, Pozzo L, Seliem EI, Ibrahim MA (2023) Characterization and stability evaluation of Egyptian propolis extract nano-capsules and their application. Sci Rep 13(1):16065. https://www.nature.com/articles/s41598-023-42025-0

Andishmand H, Yousefi M, Jafari N, Azadmard-Damirchi S, Homayouni-Rad A, Torbati M, Hamishehkar H (2024) Designing and fabrication of colloidal nano-phytosomes with gamma-oryzanol and phosphatidylcholine for encapsulation and delivery of polyphenol-rich extract from pomegranate peel. Int J Biol Macromol 256(Pt 2):128501. https://doi.org/10.1016/j.ijbiomac.2023.128501

Andreani T, Cheng R, Elbadri K, Ferro C, Menezes T, dos Santos MR, Pereira CM, Santos HA (2024) Natural compounds-based nanomedicines for cancer treatment: future directions and challenges. Drug Deliv Transl Res 14(10):2845–2916. https://doi.org/10.1007/s13346-024-01649-z

Aouf A, Ali H, Al-Khalifa AR, Mahmoud KF, Farouk A (2020) Influence of nanoencapsulation using high-pressure homogenization on the volatile constituents and anticancer and antioxidant activities of Algerian Saccocalyx satureioides Coss. et Durieu. Molecules 25(20):4756. https://doi.org/10.3390/molecules25204756

Arpicco S, Battaglia L, Brusa P, Cavalli R, Chirio D, Dosio F, Gallarate M, Milla P, Peira E, Rocco F, Sapino S, Stella B, Ugazio E, Ceruti M (2016) Recent studies on the delivery of hydrophilic drugs in nanoparticulate systems. J Drug Deliv Sci Technol 32:298–312. https://doi.org/10.1016/j.jddst.2015.09.004

Ashrafizadeh M, Zarrabi A, Hushmandi K, Hashemi F, Moghadam ER, Raei M, Kalantari M, Tavakol S, Mohammadinejad R, Najafi M, Tay FR, Makvandi P (2020) Progress in natural compounds/siRNA co-delivery employing nanovehicles for cancer therapy. ACS Comb Sci 22(12):669. https://doi.org/10.1021/acscombsci.0c00099

Atwan QS, Al-Ogaidi I (2024) Improving the targeted delivery of curcumin to esophageal cancer cells via a novel formulation of biodegradable lecithin/chitosan nanoparticles with downregulated miR-20a and miR-21 expression. Nanotechnology 35(13):135103. https://doi.org/10.1088/1361-6528/ad15b9

Auti A, Tathode M, Marino MM, Vitiello A, Ballini A, Miele F, Mazzone V, Ambrosino A, Boccellino M, Auti A, Tathode M, Marino MM, Vitiello A, Ballini A, Miele F, Mazzone V, Ambrosino A, Boccellino M (2024) Nature's weapons: bioactive compounds as anti-cancer agents. AIMS Public Health 11(3):747–772. https://doi.org/10.3934/publichealth.2024038

Babadi D, Dadashzadeh S, Osouli M, Daryabari MS, Haeri A (2020) Nanoformulation strategies for improving intestinal permeability of drugs: a more precise look at permeability assessment methods and pharmacokinetic properties changes. J Control Release 321:669–709. https://doi.org/10.1016/j.jconrel.2020.02.041

Bamidele OP, Emmambux MN (2021) Encapsulation of bioactive compounds by "extrusion" technologies: a review. Crit Rev Food Sci Nutr 61(18):3100–3118. https://doi.org/10.1080/10408398.2020.1793724

Banasaz S, Morozova K, Ferrentino G, Scampicchio M (2020) Encapsulation of lipid-soluble bioactives by nanoemulsions. Molecules 25(17):3966. https://doi.org/10.3390/molecules25173966

Bashir DJ, Manzoor S, Khan IA, Bashir M, Agarwal NB, Rastogi S, Arora I, Samim M (2022) Nanonization of magnoflorine-encapsulated novel chitosan-collagen nanocapsules for neurodegenerative diseases: in vitro evaluation. ACS Omega 7(8):6472–6480. https://doi.org/10.1021/acsomega.1c04459

Becker Peres L, Becker Peres L, de Araújo PHH, Sayer C (2016) Solid lipid nanoparticles for encapsulation of hydrophilic drugs by an organic solvent free double emulsion technique. Colloids Surf B Biointerfaces 140:317–323. https://doi.org/10.1016/j.colsurfb.2015.12.033

Berry CE, Brenac C, Gonzalez CE, Kendig CB, Le T, An N, Griffin MF (2024) Natural compounds and biomimetic engineering to influence fibroblast behavior in wound healing. Int J Mol Sci 25(6):Article 6. https://doi.org/10.3390/ijms25063274

Borges A, de Freitas V, Mateus N, Fernandes I, Oliveira J (2020) Solid lipid nanoparticles as carriers of natural phenolic compounds. Antioxidants 9(10):Article 10. https://doi.org/10.3390/antiox9100998

Bouaouina S, Aouf A, Touati A, Ali H, Elkhadragy M, Yehia H, Farouk A (2022) Effect of nanoencapsulation on the antimicrobial and antibiofilm activities of Algerian Origanum glandulosum Desf. against multidrug-resistant clinical isolates. Nanomaterials 12(15):2630. https://doi.org/10.3390/nano12152630

Braga CB, Kido LA, Lima EN, Lamas CA, Cagnon VHA, Ornelas C, Pilli RA (2020) Enhancing the anticancer activity and selectivity of goniothalamin using pH-sensitive acetalated dextran (Ac-Dex) nanoparticles: a promising platform for delivery of natural compounds. ACS Biomater Sci Eng 6(5):2929–2942. https://doi.org/10.1021/acsbiomaterials.0c00057

Brouwers J, Brewster ME, Augustijns P (2009) Supersaturating drug delivery systems: the answer to solubility-limited oral bioavailability? J Pharm Sci 98(8):2549–2572. https://doi.org/10.1002/jps.21650

Castro PM, Sarmento B, Madureira AR, Pintado ME (2019) Chapter 20—Organic nanocomposites for the delivery of bioactive molecules. In: Hasnain MS, Nayak AK (eds) Natural polysaccharides in drug delivery and biomedical applications. Academic Press, pp 471–493. https://doi.org/10.1016/B978-0-12-817055-7.00020-0

Chaudhari VS, Murty US, Banerjee S (2020) Lipidic nanomaterials to deliver natural compounds against cancer: a review. Environ Chem Lett 18(6):1803–1812. https://idp.springer.com/authorize/casa?redirect_uri=https://link.springer.com/article/10.1007/s10311-020-01042-5&casa_token=PUxbkGxV9PoAAAAA:WiPa-xgKmIyjcIitMtDncHBItuwxjrHCKMDVRawNPMIMFCJ5nWTAd9UzRokEoFyZ817KRphiNc4UYgMX-7w

Chavda VP, Patel AB, Mistry KJ, Suthar SF, Wu Z-X, Chen Z-S, Hou K (2022) Nano-drug delivery systems entrapping natural bioactive compounds for cancer: recent progress and future challenges. Front Oncol 12:867655. https://doi.org/10.3389/fonc.2022.867655

Csicsák D, Szolláth R, Kádár S, Ambrus R, Bartos C, Balogh E, Antal I, Köteles I, Tőzsér P, Bárdos V, Horváth P, Borbás E, Takács-Novák K, Sinkó B, Völgyi G (2023) The effect of the particle size reduction on the biorelevant solubility and dissolution of poorly soluble drugs with different acid-base character. Pharmaceutics 15(1):278. https://doi.org/10.3390/pharmaceutics15010278

Das R, Kumar H, Choithramani A, Bothra G, Shard A (2022) Polymeric nanoparticles to entrap natural drugs for cancer therapy. In: Padhi S, Behera A, Lichtfouse E (eds) Polymeric nanoparticles for the treatment of solid tumors. Springer International Publishing, pp 167–211. https://doi.org/10.1007/978-3-031-14848-4_6

Das AK, Gandhi H, Panwar V, Awasthi A, Singh SK, Kumar D (2024) Chapter 14—Natural product-loaded nanoparticles for wound healing. In: Solanki PR, Kumar A, Pratap Singh R, Singh J, Singh KRB (eds) Nanotechnological aspects for next-generation wound management. Academic Press, pp 319–331. https://doi.org/10.1016/B978-0-323-99165-0.00004-6

De Silva ND, Attanayake AP, Karunaratne DN, Arawwawala LDAM, Pamunuwa GK (2023a) Bael (Aegle marmelos L. Correa) fruit extracts encapsulated alginate nanoparticles as a potential dietary supplement with improved bioactivities. J Food Sci 88(12):4942–4961. https://doi.org/10.1111/1750-3841.16827

DE Silva WND, Attanayake AP, Arawwawala LDAM, Karunaratne DN, Pamunuwa GK (2023b) In vitro antioxidant activity of alginate nanoparticles encapsulating the aqueous extract of Coccinia grandis L. Turk J Chem 47(4):715–725. https://doi.org/10.55730/1300-0527.3573

Dewanjee S, Dua TK, Bhattacharjee N, Das A, Gangopadhyay M, Khanra R, Joardar S, Riaz M, Feo VD, Zia-Ul-Haq M (2017) Natural products as alternative choices for P-glycoprotein (P-gp) inhibition. Molecules 22(6):Article 6. https://doi.org/10.3390/molecules22060871

Drosou CG, Krokida MK, Biliaderis CG (2017) Encapsulation of bioactive compounds through electrospinning/electrospraying and spray drying: a comparative assessment of food-related applications. Dry Technol 35(2):139–162. https://doi.org/10.1080/07373937.2016.1162797

Enrico C (2019) Chapter 3—Nanotechnology-based drug delivery of natural compounds and phytochemicals for the treatment of cancer and other diseases. In: Atta-ur-Rahman (ed) Studies in natural products chemistry, vol 62. Elsevier, pp 91–123. https://doi.org/10.1016/B978-0-444-64185-4.00003-4

Estrella-Osuna DE, Tapia-Hernández JA, Ruíz-Cruz S, Márquez-Ríos E, Ornelas-Paz JDJ, Del-Toro-Sánchez CL, Ocaño-Higuera VM, Rodríguez-Félix F, Estrada-Alvarado MI, Cira-Chávez LA (2022) Nanoencapsulation of eggplant (Solanum melongena L.) peel extract in electrospun gelatin nanofiber: preparation, characterization, and in vitro release. Nanomaterials 12(13):2303. https://doi.org/10.3390/nano12132303

Fan M, Geng S, Liu Y, Wang J, Wang Y, Zhong J, Yan Z, Yu L (2018) Nanocrystal technology as a strategy to improve drug bioavailability and antitumor efficacy for the cancer treatment. Curr Pharm Des 24(21):2416–2424. https://doi.org/10.2174/1381612824666180515154109

Fang L, Zhou H, Cheng L, Wang Y, Liu F, Wang S (2023) The application of mesoporous silica nanoparticles as a drug delivery vehicle in oral disease treatment. Front Cell Infect Microbiol 13:1124411. https://doi.org/10.3389/fcimb.2023.1124411

Figueroa-Enriquez CE, Rodríguez-Félix F, Plascencia-Jatomea M, Sánchez-Escalante A, Vargas-López JM, Tapia-Hernández JA, Canizales-Rodríguez DF, Castro-Enriquez DD, Ruiz-Cruz S, Santos-Sauceda I, Burruel-Ibarra SE, Pompa-Ramos JL (2023) Nanoparticles of betalain-gelatin with antioxidant properties by coaxial electrospraying: preparation and characterization. ACS Omega 8(44):41156–41168. https://doi.org/10.1021/acsomega.3c04021

Ganesan P, Karthivashan G, Park SY, Kim J, Choi D-K (2018) Microfluidization trends in the development of nanodelivery systems and applications in chronic disease treatments. Int J Nanomedicine 13:6109–6121. https://doi.org/10.2147/IJN.S178077

Ghasemiyeh P, Mohammadi-Samani S (2020) Potential of nanoparticles as permeation enhancers and targeted delivery options for skin: advantages and disadvantages. Drug Des Devel Ther 14:3271–3289. https://doi.org/10.2147/DDDT.S264648

Ghosh S, Das SK, Sinha K, Ghosh B, Sen K, Ghosh N, Sil PC (2024) The emerging role of natural products in cancer treatment. Arch Toxicol 98(8):2353–2391. https://doi.org/10.1007/s00204-024-03786-3

Hajialyani M, Tewari D, Sobarzo-Sánchez E, Nabavi SM, Farzaei MH, Abdollahi M (2018) Natural product-based nanomedicines for wound healing purposes: therapeutic targets and drug delivery systems. Int J Nanomedicine 13:5023–5043. https://doi.org/10.2147/IJN.S174072

Hashem S, Ali TA, Akhtar S, Nisar S, Sageena G, Ali S, Al-Mannai S, Therachiyil L, Mir R, Elfaki I, Mir MM, Jamal F, Masoodi T, Uddin S, Singh M, Haris M, Macha M, Bhat AA (2022) Targeting cancer signaling pathways by natural products: exploring promising anti-cancer agents. Biomed Pharmacother 150:113054. https://doi.org/10.1016/j.biopha.2022.113054

Hcini K, Lozano-Pérez AA, Luis Cenis J, Quílez M, José Jordán M (2021) Extraction and encapsulation of phenolic compounds of Tunisian Rosemary (Rosmarinus officinalis L.) extracts in silk fibroin nanoparticles. Plants 10(11):2312. https://doi.org/10.3390/plants10112312

Hong SE, Lee J-S, Lee HG (2024) α-Tocopherol-loaded multi-layer nanoemulsion using chitosan, and dextran sulfate: cellular uptake, antioxidant activity, and *in vitro* bioaccessibility. Int J Biol Macromol 254:127819. https://doi.org/10.1016/j.ijbiomac.2023.127819

Huang M, Liang C, Tan C, Huang S, Ying R, Wang Y, Wang Z, Zhang Y (2019) Liposome co-encapsulation as a strategy for the delivery of curcumin and resveratrol. Food Funct 10(10):6447–6458. https://doi.org/10.1039/c9fo01338e

Huang Y, Wang J, Jiang K, Chung EJ (2021) Improving kidney targeting: the influence of nanoparticle physicochemical properties on kidney interactions. J Control Release 334:127–137. https://doi.org/10.1016/j.jconrel.2021.04.016

Huang L, Huang X-H, Yang X, Hu J-Q, Zhu Y-Z, Yan P-Y, Xie Y (2024) Novel nano-drug delivery system for natural products and their application. Pharmacol Res 201:107100. https://doi.org/10.1016/j.phrs.2024.107100

Jafari S, Jafari SM, Ebrahimi M, Kijpatanasilp I, Assatarakul K (2023) A decade overview and prospect of spray drying encapsulation of bioactives from fruit products: characterization, food application and *in vitro* gastrointestinal digestion. Food Hydrocoll 134:108068. https://doi.org/10.1016/j.foodhyd.2022.108068

Jayaprakash P, Maudhuit A, Gaiani C, Desobry S (2023) Encapsulation of bioactive compounds using competitive emerging techniques: electrospraying, nano spray drying, and electrostatic spray drying. J Food Eng 339:111260. https://doi.org/10.1016/j.jfoodeng.2022.111260

Jemaa MB, Falleh H, Ksouri R, Jemaa MB, Falleh H, Ksouri R (2019) Encapsulation of natural bioactive compounds: nanoemulsion formulation to enhance essential oils activities. In: Microencapsulation—processes, technologies and industrial applications. IntechOpen. https://doi.org/10.5772/intechopen.84183

Jha A, Kumar M, Mishra B (2022) Marine biopolymer-based anticancer drug delivery systems. In: Marine biomaterials: drug delivery and therapeutic applications. Springer, Singapore, pp 351–401

Jha A, Goswami P, Kumar M, Bharti K, Manjit M, Satpute AP, Gupta A, Moorkoth S, Koch B, Mishra B (2024a) Cetuximab functionalized chitosan/hyaluronic acid-based nanoparticles loaded with cabazitaxel enhances anti-tumor efficacy in DMBA-induced breast cancer model in rats through spatial targeting. Chem Phys Impact 9:100750. https://www.sciencedirect.com/science/article/pii/S2667022424002949

Jha A, Kumar M, Bharti K, Manjit M, Mishra B (2024b) Biopolymer-based tumor microenvironment-responsive nanomedicine for targeted cancer therapy. Nanomedicine 19(7):633–651. https://doi.org/10.2217/nnm-2023-0302

Jha A, Kumar M, Goswami P, Manjit M, Bharti K, Koch B, Mishra B (2024c) Hyaluronic acid-oleylamine and chitosan-oleic acid conjugate-based hybrid nanoparticle delivery via. Dissolving microneedles for enhanced treatment efficacy in localized breast cancer. Biomater Adv 160:213865. https://doi.org/10.1016/j.bioadv.2024.213865

Kaokaen P, Sorraksa N, Phonchai R, Chaicharoenaudomrung N, Kunhorm P, Noisa P (2023) Enhancing neurological competence of nanoencapsulated cordyceps/turmeric extracts in human neuroblastoma SH-SY5Y cells. Cell Mol Bioeng 16(1):81–93. https://doi.org/10.1007/s12195-022-00752-w

Karim N, Shishir MRI, Chen W (2020) Surface decoration of neohesperidin-loaded nanoliposome using chitosan and pectin for improving stability and controlled release. Int J Biol Macromol 164:2903–2914. https://doi.org/10.1016/j.ijbiomac.2020.08.174

Karim N, Shishir MRI, Li Y, Zineb OY, Mo J, Tangpong J, Chen W (2022) Pelargonidin-3-O-glucoside encapsulated pectin-chitosan-nanoliposomes recovers palmitic acid-induced hepatocytes injury. Antioxidants 11(4):623. https://doi.org/10.3390/antiox11040623

Khadka B, Lee J-Y, Park DH, Kim K-T, Bae J-S (2020) The role of natural compounds and their nanocarriers in the treatment of CNS inflammation. Biomolecules 10(10):Article 10. https://doi.org/10.3390/biom10101401

Khare T, Palakurthi SS, Shah BM, Palakurthi S, Khare S (2020) Natural product-based nanomedicine in treatment of inflammatory bowel disease. Int J Mol Sci 21(11):Article 11. https://doi.org/10.3390/ijms21113956

Kubczak M, Szustka A, Rogalińska M (2021) Molecular targets of natural compounds with anticancer properties. Int J Mol Sci 22(24):Article 24. https://doi.org/10.3390/ijms222413659

Kumar M, Rajnikanth PS (2020) A mini-review on HER2 positive breast cancer and its metastasis: resistance and treatment strategies. Curr Nanomed 10(1):36–47. https://doi.org/10.2174/2468187310666191223141038

Kumar M, Shanthi N, Rajnikanth PS, Mahato AK (2020) Authors review on drug nanocrystals: a progress to targeted delivery. Curr Nanomed 10(3):248–270

Kumar M, Jha A, Bharti K, Parmar G, Mishra B (2022a) Advances in lipid-based pulmonary nanomedicine for the management of inflammatory lung disorders. Nanomedicine 17(12):913–934. https://doi.org/10.2217/nnm-2021-0389

Kumar M, Jha A, Mishra B (2022b) Marine biopolymers for transdermal drug delivery. In: Marine biomaterials: drug delivery and therapeutic applications. Springer, Singapore, pp 157–207

Kumar R, Thakur AK, Chaudhari P, Banerjee N (2022c) Particle size reduction techniques of pharmaceutical compounds for the enhancement of their dissolution rate and bioavailability. J Pharm Innov 17(2):333–352. https://doi.org/10.1007/s12247-020-09530-5

Kumar M, Jha A, Bharti K, Manjit M, Kumbhar P, Dhapte-Pawar V, Mishra B (2024a) Lipid-coated nanocrystals of paclitaxel as dry powder for inhalation: characterization, in-vitro performance, and pharmacokinetic assessment. Colloids Surf B: Biointerfaces 237:113865. https://doi.org/10.1016/j.colsurfb.2024.113865

Kumar M, Jha A, Bharti K, Manjit, Mishra B (2024b) Fucoidan-based bosutinib nanocrystals for pulmonary drug delivery: solid state characterization and *in-vitro* assessment. Chem Phys Impact 8:100644. https://doi.org/10.1016/j.chphi.2024.100644

Levy R, Okun Z, Shpigelman A (2021) High-pressure homogenization: principles and applications beyond microbial inactivation. Food Eng Rev 13(3):490–508. https://doi.org/10.1007/s12393-020-09239-8

Liu Y, Liang Y, Yuhong J, Xin P, Han JL, Du Y, Yu X, Zhu R, Zhang M, Chen W, Ma Y (2024) Advances in nanotechnology for enhancing the solubility and bioavailability of poorly soluble drugs. Drug Des Devel Ther 18:1469–1495. https://doi.org/10.2147/DDDT.S447496

Loh ZH, Samanta AK, Sia Heng PW (2015) Overview of milling techniques for improving the solubility of poorly water-soluble drugs. Asian J Pharm Sci 10(4):255–274. https://doi.org/10.1016/j.ajps.2014.12.006

Maa Y-F, Hsu CC (1999) Performance of sonication and microfluidization for liquid–liquid emulsification. Pharm Dev Technol 4(2):233–240. https://doi.org/10.1081/PDT-100101357

Madhukiran DR, Jha A, Kumar M, Ajmal G, Bonde GV, Mishra B (2021) Electrospun nanofiber-based drug delivery platform: advances in diabetic foot ulcer management. Expert Opin Drug Deliv 18(1):25–42. https://doi.org/10.1080/17425247.2021.1823966

Mahalakshmi L, Maria Leena M, Moses JA, Anandharamakrishnan C (2020) Micro- and nano-encapsulation of β-carotene in zein protein: size-dependent release and absorption behavior. Food Funct 11(2):1647–1660. https://doi.org/10.1039/c9fo02088h

Manjit M, Kumar K, Kumar M, Jha A, Bharti K, Tiwari P, Tilak R, Singh V, Koch B, Mishra B (2024) Fabrication of gelatin coated polycaprolactone nanofiber scaffolds co-loaded with luliconazole and naringenin for treatment of Candida infected diabetic wounds. Int J Biol Macromol 261:129621. https://www.sciencedirect.com/science/article/pii/S0141813024004240

Manzari MT, Shamay Y, Kiguchi H, Rosen N, Scaltriti M, Heller DA (2021) Targeted drug delivery strategies for precision medicines. Nat Rev Mater 6(4):Article 4. https://doi.org/10.1038/s41578-020-00269-6

Martínez-Ballesta MC, Gil-Izquierdo Á, García-Viguera C, Domínguez-Perles R (2018) Nanoparticles and controlled delivery for bioactive compounds: outlining challenges for new "smart-foods" for health. Foods 7(5):Article 5. https://doi.org/10.3390/foods7050072

Mathes D, Macedo LB, Pieta TB, Maia BC, Rodrigues OED, Leal JG, Wendt M, Rolim CMB, Mitjans M, Nogueira-Librelotto DR (2024) Co-delivery of an innovative organoselenium compound and paclitaxel by pH-responsive PCL nanoparticles to synergistically overcome multidrug resistance in cancer. Pharmaceutics 16(5):590. https://doi.org/10.3390/pharmaceutics16050590

Mauri E, Gori M, Giannitelli SM, Zancla A, Mozetic P, Abbruzzese F, Merendino N, Gigli G, Rossi F, Trombetta M, Rainer A (2021) Nano-encapsulation of hydroxytyrosol into formulated nanogels improves therapeutic effects against hepatic steatosis: an in vitro study. Mater Sci Eng C Mater Biol Appl 124:112080. https://doi.org/10.1016/j.msec.2021.112080

Mazloomi N, Safari B, Can Karaca A, Karimzadeh L, Moghadasi S, Ghanbari M, Assadpour E, Sarabandi K, Jafari SM (2024) Loading bioactive peptides within different nanocarriers to enhance their functionality and bioavailability; in vitro and in vivo studies. Adv Colloid Interf Sci 334:103318. https://doi.org/10.1016/j.cis.2024.103318

Mishra A, Kumar R, Mishra J, Dutta K, Ahlawat P, Kumar A, Dhanasekaran S, Gupta AK, Sinha S, Bishi DK, Gupta PK, Nayak S (2023) Strategies facilitating the permeation of nanoparticles through blood-brain barrier: an insight towards the development of brain-targeted drug delivery system. J Drug Deliv Sci Technol 86:104694. https://doi.org/10.1016/j.jddst.2023.104694

Mok ZH (2024) The effect of particle size on drug bioavailability in various parts of the body. Pharm Sci Adv 2:100031. https://doi.org/10.1016/j.pscia.2023.100031

Noore S, Pathania S, Fuciños P, O'Donnell CP, Tiwari BK (2024) Controlled release and target delivery of nanoencapsulated compounds. In: Noore S, Pathania S, Fuciños P, O'Donnell CP, Tiwari BK (eds) Nanocarriers for controlled release and target delivery of bioactive compounds. Springer Nature Switzerland, pp 63–65. https://doi.org/10.1007/978-3-031-57488-7_8

Pachuau L, Laldinchhana, Roy PK, Zothantluanga JH, Ray S, Das S (2021) Encapsulation of bioactive compound and its therapeutic potential. In: Pal D, Nayak AK (eds) Bioactive natural products for pharmaceutical applications. Springer International Publishing, pp 687–714. https://doi.org/10.1007/978-3-030-54027-2_20

Parhizkary M, Jafari SM, Assadpour E, Enayati A, Kashiri M (2024) Pea protein-coated nanoliposomal encapsulation of jujube phenolic extract with different stabilizers; characterization and in vitro release. Food Chem X 23:101771. https://doi.org/10.1016/j.fochx.2024.101771

Pugazhendhi A, Alshehri MA, Kandasamy S, Sarangi PK, Sharma A (2024) Deciphering the importance of nanoencapsulation to improve the availability of bioactive molecules in food sources to the human body. Food Chem 464:141762. https://doi.org/10.1016/j.foodchem.2024.141762

Rakotondrabe TF, Fan M-X, Muema FW, Guo M-Q (2023) Modulating inflammation-mediated diseases via natural phenolic compounds loaded in nanocarrier systems. Pharmaceutics 15(2):Article 2. https://doi.org/10.3390/pharmaceutics15020699

Rehman A, Tong Q, Korma SA, Han W, Karim A, Sharif HR, Ali A, Yaqoob S, Khalifa SA, Cacciotti I (2021) Influence of diverse natural biopolymers on the physicochemical characteristics of borage seed oil-peppermint oil loaded W/O/W nanoemulsions entrapped with lycopene. Nanotechnology 32(50):505302. https://doi.org/10.1088/1361-6528/ac22de

Rezaei A, Fathi M, Jafari SM (2019a) Nanoencapsulation of hydrophobic and low-soluble food bioactive compounds within different nanocarriers. Food Hydrocoll 88:146–162. https://doi.org/10.1016/j.foodhyd.2018.10.003

Rezaei A, Varshosaz J, Fesharaki M, Farhang A, Jafari SM (2019b) Improving the solubility and in vitro cytotoxicity (anticancer activity) of ferulic acid by loading it into cyclodextrin nanosponges. Int J Nanomedicine 14:4589–4599. https://doi.org/10.2147/IJN.S206350

Rezaei A, Rafieian F, Akbari-Alavijeh S, Kharazmi MS, Jafari SM (2022) Release of bioactive compounds from delivery systems by stimuli-responsive approaches; triggering factors, mechanisms, and applications. Adv Colloid Interf Sci 307:102728. https://doi.org/10.1016/j.cis.2022.102728

Rodriguez EB, Almeda RA, Vidallon MLP, Reyes CT (2019) Enhanced bioactivity and efficient delivery of quercetin through nanoliposomal encapsulation using rice bran phospholipids. J Sci Food Agric 99(4):1980–1989. https://doi.org/10.1002/jsfa.9396

Romes NB, Abdul Wahab R, Abdul Hamid M (2021) The role of bioactive phytoconstituents-loaded nanoemulsions for skin improvement: a review. Biotechnol Biotechnol Equip 35(1):711–730. https://doi.org/10.1080/13102818.2021.1915869

Saadi S, Nacer NE, Chenaker H, Ariffin AA, Ghazali HM, Saari N, Mohammed AS, Anwar F, Hamid AA (2023) A review on trends in microencapsulation of bioactive compounds: coating materials, design, and applications. Eur Food Res Technol 249(12):3123–3139. https://doi.org/10.1007/s00217-023-04354-2

Siddiqui SA, Singh S, Bahmid NA, Mehany T, Shyu DJH, Assadpour E, Malekjani N, Castro-Muñoz R, Jafari SM (2023) Release of encapsulated bioactive compounds from active packaging/coating materials and its modeling: a systematic review. Colloid Interface 7(2):Article 2. https://doi.org/10.3390/colloids7020025

Stan D, Enciu A-M, Mateescu AL, Ion AC, Brezeanu AC, Stan D, Tanase C (2021) Natural compounds with antimicrobial and antiviral effect and nanocarriers used for their transportation. Front Pharmacol 12:723233. https://doi.org/10.3389/fphar.2021.723233

Stojiljković N, Ilić S, Stojanović N, Janković-Veličković L, Stojnev S, Kocić G, Radenković G, Arsić I, Stojanović M, Petković M (2019) Nanoliposome-encapsulated ellagic acid prevents cyclophosphamide-induced rat liver damage. Mol Cell Biochem 458(1):185–195. https://doi.org/10.1007/s11010-019-03541-8

Stoleru OA, Burlec AF, Mircea C, Felea MG, Macovei I, Hăncianu M, Corciovă A (2024) Multiple nanotechnological approaches using natural compounds for diabetes management. J Diabetes Metab Disord 23(1):267–287. https://doi.org/10.1007/s40200-023-01376-1

Taghipour YD, Bahramsoltani R, Marques AM, Naseri R, Rahimi R, Haratipour P, Panah AI, Farzaei MH, Abdollahi M (2018) A systematic review of nano formulation of natural products for the treatment of inflammatory bowel disease: drug delivery and pharmacological targets. DARU J Pharm Sci 26(2):229–239. https://doi.org/10.1007/s40199-018-0222-4

Tanhaei A, Mohammadi M, Hamishehkar H, Hamblin MR (2021) Electrospraying as a novel method of particle engineering for drug delivery vehicles. J Control Release 330:851–865. https://doi.org/10.1016/j.jconrel.2020.10.059

Taouzinet L, Djaoudene O, Fatmi S, Bouiche C, Amrane-Abider M, Bougherra H, Rezgui F, Madani K (2023) Trends of nanoencapsulation strategy for natural compounds in the food industry. Processes 11(5):Article 5. https://doi.org/10.3390/pr11051459

Tavan M, Hanachi P, de la Luz Cádiz-Gurrea M, Segura Carretero A, Mirjalili MH (2024) Natural phenolic compounds with neuroprotective effects. Neurochem Res 49(2):306–326. https://idp.springer.com/authorize/casa?redirect_uri=https://link.springer.com/article/10.1007/s11064-023-04046-z&casa_token=l3sbUZwn4cQAAAAA:ard09RlDkoQTzpo3BYU_C2O_UgtsYdxtmLH8VHQ5qHI1G07yIn_blQNv-eFPwsrbFO-a4nMC6xQjbhIEVJc

Vanti G (2021) Recent strategies in nanodelivery systems for natural products: a review. Environ Chem Lett 19(6):4311–4326. https://doi.org/10.1007/s10311-021-01276-x

Vidal C, Lopez-Polo J, Osorio FA (2024) Physical properties of cellulose derivative-based edible films elaborated with liposomes encapsulating grape seed tannins. Antioxidants 13(8):989. https://doi.org/10.3390/antiox13080989

Villate A, Barreto GP, Nicolás MS, Aizpurua-Olaizola O, Olivares M, Usobiaga A (2024) Development, characterization and in vitro gastrointestinal release of PLGA nanoparticles loaded with full-spectrum cannabis extracts. AAPS PharmSciTech 25(5):120. https://doi.org/10.1208/s12249-024-02836-4

Vlasova II, Kapralov AA, Michael ZP, Burkert SC, Shurin MR, Star A, Shvedova AA, Kagan VE (2016) Enzymatic oxidative biodegradation of nanoparticles: mechanisms, significance and applications. Toxicol Appl Pharmacol 299:58–69. https://doi.org/10.1016/j.taap.2016.01.002

Vrignaud S, Benoit J-P, Saulnier P (2011) Strategies for the nanoencapsulation of hydrophilic molecules in polymer-based nanoparticles. Biomaterials 32(33):8593–8604. https://doi.org/10.1016/j.biomaterials.2011.07.057

Wardana AP, Aminah NS, Kristanti AN, Fahmi MZ, Zahrah HI, Widiyastuti W, Ajiz HA, Zubaidah U, Wiratama PA, Takaya Y (2023) Nano Uncaria gambir as chemopreventive agent against breast cancer. Int J Nanomedicine 18:4471–4484. https://doi.org/10.2147/IJN.S403385

Watkins R, Wu L, Zhang C, Davis RM, Xu B (2015) Natural product-based nanomedicine: recent advances and issues. Int J Nanomedicine 10:6055–6074. https://doi.org/10.2147/IJN.S92162

Wong KH, Lu A, Chen X, Yang Z (2020) Natural ingredient-based polymeric nanoparticles for cancer treatment. Molecules 25(16):Article 16. https://doi.org/10.3390/molecules25163620

Xu J, Chen Y, Jiang X, Gui Z, Zhang L (2019) Development of hydrophilic drug encapsulation and controlled release using a modified nanoprecipitation method. Processes 7(6):Article 6. https://doi.org/10.3390/pr7060331

Yadav UP, Rhuthuparna M, Vasudev K, Suman P, Munshi A, Kumar S, Singh S (2021) Natural compound-based nanoparticles to target free radicals in cancer. In: Chakraborti S (ed) Handbook of oxidative stress in cancer: therapeutic aspects. Springer Nature, pp 1–14. https://doi.org/10.1007/978-981-16-1247-3_276-1

Yang G, Lin W, Lai H, Tong J, Lei J, Yuan M, Zhang Y, Cui C (2021) Understanding the relationship between particle size and ultrasonic treatment during the synthesis of metal nanoparticles. Ultrason Sonochem 73:105497. https://doi.org/10.1016/j.ultsonch.2021.105497

Yıldırım M, Sessevmez M, Poyraz S, Düzgüneş N (2023) Recent strategies for cancer therapy: polymer nanoparticles carrying medicinally important phytochemicals and their cellular targets. Pharmaceutics 15(11):Article 11. https://doi.org/10.3390/pharmaceutics15112566

Yu Y-Q, Yang X, Wu X-F, Fan Y-B (2021) Enhancing permeation of drug molecules across the skin via delivery in nanocarriers: novel strategies for effective transdermal applications. Front Bioeng Biotechnol 9:646554. https://doi.org/10.3389/fbioe.2021.646554

Zhang H, Hu H, Zhang H, Dai W, Wang X, Wang X, Zhang Q (2015) Effects of PEGylated paclitaxel nanocrystals on breast cancer and its lung metastasis. Nanoscale 7(24):10790–10800. https://doi.org/10.1039/c4nr07450e

Zhang R, Han Y, Xie W, Liu F, Chen S (2022) Advances in protein-based nanocarriers of bioactive compounds: from microscopic molecular principles to macroscopical structural and functional attributes. J Agric Food Chem 70(21):6354–6367. https://doi.org/10.1021/acs.jafc.2c01936

Zhu W, Wu J, Guo X, Sun X, Li Q, Wang J, Chen L (2020) Development and physicochemical characterization of chitosan hydrochloride/sulfobutyl ether-β-cyclodextrin nanoparticles for cinnamaldehyde entrapment. J Food Biochem 44(6):e13197. https://doi.org/10.1111/jfbc.13197

Zou X, Cai S, Wang T, Zheng S, Cui X, Hao J, Chen X, Liu Y, Zhang Z, Li Y (2023) Natural antibacterial agent-based nanoparticles for effective treatment of intracellular MRSA infection. Acta Biomater 169:410–421. https://doi.org/10.1016/j.actbio.2023.08.004

# Part IV
# Nanotechnology in Diagnostics and Imaging

# Chapter 14
# Recent Advancements in Prodrug Strategies for Targeted Cancer Therapy

**Chae Bin Lee, Lukáš Tenora, Arina Ranjit, Kateřina Tenorová, and Rana Rais**

**Abstract** Cancer therapies have improved significantly over the last few decades. However, most anticancer agents, while effective, often cause severe side effects to healthy tissues. To address these challenges, prodrug strategies have gained considerable attention. Prodrugs with enhanced physicochemical and pharmacological properties have facilitated the FDA approval for many drug candidates. In the context of cancer therapy, prodrugs are typically designed to be stable and inert in systemic circulation, becoming activated following distribution to cancer cells. These strategies improve the targeted delivery of cancer therapies, enhance effectiveness, reduce off-target effects, thereby offering improved tolerability and safety. This review provides an update on the recent advances in prodrug design for anticancer agents, with a focus on small molecules. It highlights prodrug approaches based on receptor-targeted, transporter-targeted, enzyme-activated, and those that take advantage of the unique properties of the tumor microenvironment.

**Keywords** Prodrugs · Cancer therapy · Receptor · Transporter · Enzyme · Cancer microenvironment

C. B. Lee · L. Tenora · A. Ranjit · K. Tenorová
Department of Neurology, Baltimore, MD, USA

Johns Hopkins Drug Discovery, Baltimore, MD, USA

R. Rais (✉)
Department of Neurology, Baltimore, MD, USA

Johns Hopkins Drug Discovery, Baltimore, MD, USA

Department of Pharmacology and Molecular Sciences, Baltimore, MD, USA

Department of Oncology, Johns Hopkins School of Medicine, Baltimore, MD, USA
e-mail: rrais2@jhmi.edu

 
V. Khanna et al. (eds.), *Nanotechnologies for Drug Delivery and Biopharmaceutical Development*, Progress in Drug Research 77,
https://doi.org/10.1007/978-3-032-00796-4_14

## Abbreviations

| | |
|---|---|
| ADC | Antibody drug conjugate |
| ADM | Adriamycin |
| ASGPR | Asialoglycoprotein receptor |
| CA-4 | Combretastatin A4 |
| CTSL | Cathepsin L |
| DNA | Deoxyribonucleic acid |
| DOX | Doxorubicin |
| DON | 6-Diazo-5-oxo-L-norleucine |
| ECM | Extracellular matrix |
| FA | Folic acid |
| FDA | Food and Drug Administration |
| FDG | 2-Fluoro-2-deoxyglucose |
| FR | Folic receptors |
| FRET | Förster resonance energy transfer |
| Gal | Galactose |
| GalNAc | N-Acetyl-galactosamine |
| GLUT1 | Glucose transporter type 1 |
| GSH | Glutathione |
| HDAC | Histone deacetylase |
| HER2 | Human epidermal growth factor receptor 2 |
| HIF | Hypoxia-inducible factor |
| HPMA | N-(2-Hydroxypropyl) methacrylamide |
| ICLs | Interstrand cross-links |
| KSPi | Kinesin spindle inhibitor |
| LAT1 | L- type amino acid transporter 1 |
| MMAE | Monomethyl auristatin E |
| MMP | Matrix metalloproteinases |
| PAA | Poly(acrylic acid) |
| PABC | *p*-Aminobenzylcarbamate |
| PDEAEM | Poly(N,N′-dimethylaminoethyl methacrylate) |
| PEG | Polyethylene glycol |
| Phis | Poly(L-histidine) |
| PMAA | Poly(methacrylic acid) |
| PMMA | Poly(methyl methacrylate) |
| PSMA | Prostate-specific membrane antigen |
| PTX | Paclitaxel |
| PVP | Poly(4-vinylpyridine) |
| QM | Quinone methide |
| ROS | Reactive oxygen species |
| SMDC | Small molecule drug conjugate |
| SMVT | Sodium-dependent multivitamin transporter |
| uPA | Urokinase plasminogen activator |
| VDA | Vascular-disrupting agent |

## 14.1 Introduction

Over the last few decades, cancer treatments have significantly advanced with the discovery of several successful therapies, from small molecules to antibodies. However, small molecule chemotherapeutic agents continue to be the first-line therapy for cancer treatment. Conventional chemotherapy drugs, including alkylating agents, antimetabolites, anthracyclines, topoisomerase inhibitors, and plant alkaloids, primarily work by disrupting cell division or DNA synthesis and preventing their replication (Connors and Knox 1995). While this works effectively for cancer cells, the lack of selectivity of conventional chemotherapeutics for cancer versus host tissue often leads to significant side effects, thus hampering their therapeutic potential (Baudino 2015).

These limitations have been addressed by the development of targeted therapies that exploit specific features of cancer cells. For example, tamoxifen, discovered several years ago, binds to the estrogen receptor and inhibits the growth of estrogen receptor-positive breast cancer. This discovery led to significant advancements in the field of targeted therapies (Jordan 2008; Santhosh et al. 2015). However, its high specificity for estrogen receptors limited its use, underscoring the need for new strategies. Novel approaches have thus been explored to develop drugs with broader selectivity to enable targeting various cancer types.

Prodrugs are chemically modified derivatives, that release active compounds through enzymatic or chemical transformation (Rautio et al. 2008). Prodrugs are widely used to improve the physicochemical, biopharmaceutic, or pharmacokinetic properties of active compounds. In the context of cancer therapy, prodrug strategies have been applied to target cancer cells by exploiting overexpressed receptors or enzymes, transporters, and unique cancer features such as low extracellular pH and discrete microenvironment properties. This chapter is organized into four sections based on various targets as illustrated in Fig. 14.1 and outlined below:

- **Receptor-targeted prodrugs**, which bind to receptors upregulated in cancer cells including prostate-specific membrane antigens (PSMA), folate receptors (FR), and asialoglycoprotein receptors (ASGPR).
- **Transporter-targeted prodrugs**, which are designed to target overexpressed transporters such as glucose transporter 1 (GLUT1), L-type amino acid transporter 1 (LAT1), and sodium-dependent multivitamin transporter (SMVT).
- **Enzyme-activated prodrugs**, which are designed to be activated via the overexpressed enzymes in cancer such as cathepsins, legumain, matrix metalloproteinases (MMPs), serine proteases, and other proteases.
- **Microenvironment-activated prodrugs**, designed to respond to tumor microenvironment and conditions like low pH, elevated reactive oxygen species (ROS), or high glutathione (GSH) levels.

The general scheme for targeted prodrugs is illustrated in Fig. 14.2. The prodrug design features specific functionalities including a targeting moiety, a cleavable linker, and a cytotoxic agent, for preferential tumor distribution and/or tumor specific activation for drug release. The targeting moieties, such as PSMA ligands, folic

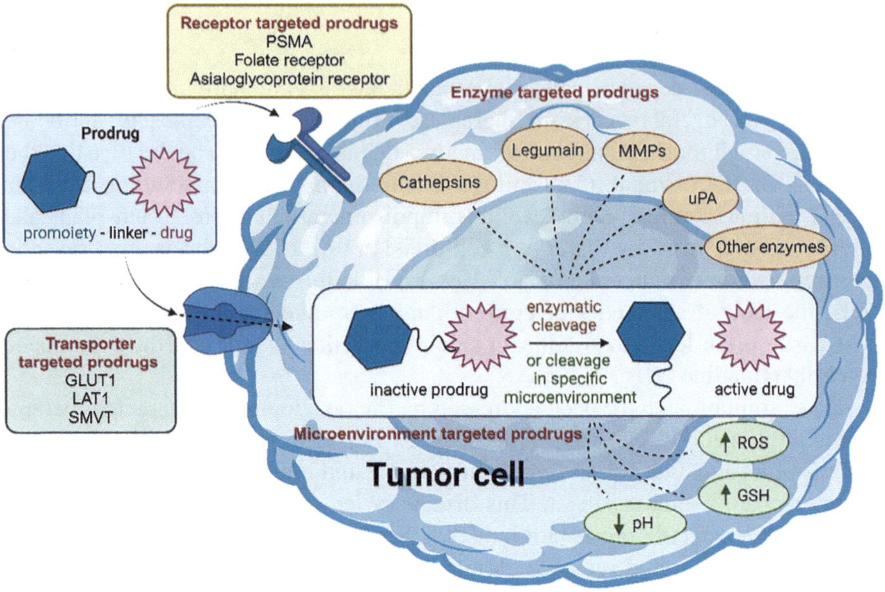


**Fig. 14.1** General outline illustrating the targeting of prodrugs to tumor cells via receptors, transporters, and enzymes as well as in tumor micrenvironment. Receptor-targeted prodrugs utilize upregulated recptors auch as PSMA, folate, and asialoglycoprotein receptors. Transporter-targeted prodrugs use overexpressed transport systems such as GLUT1, LAT1, and SMVT. Enzyme-targeted prodrugs employ proteases such cathepsins, legumain, MMPs, uPA, and other enzymes. The process involves a prodrug (with a promoiety-linker-drug) entering the tumor cell, where enzymatic cleavage or specific microenvironmental conditions convert it into an active drug. Microenvironment factors include ROS, GSH, and acidic pH

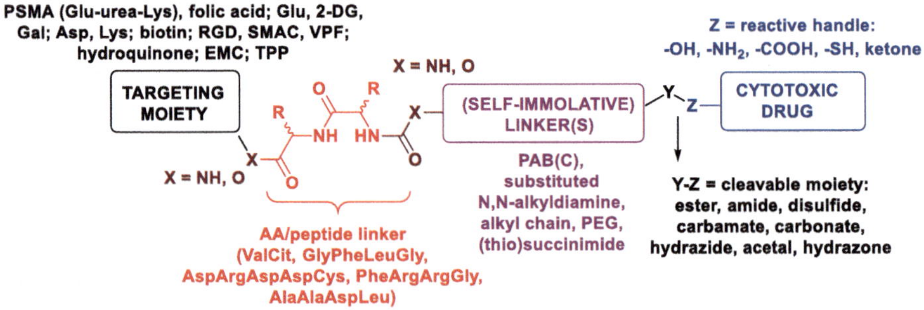


**Fig. 14.2** Chemical structure illustrating the concept of cancer-targeted prodrug systems. The structure includes a targeting moiety, a self-immolative linker, and a cytotoxic drug. The targeting moiety options include PSMA substrate, folic acid, and others. The linker used includes PAB(C) or substituted N,N-alkyldiamine, amongst others. The cytotoxic drug is connected via a reactive handle, Z, such as OH, -NH2, -COOH, -SH, or ketonevia and ester, amide, disulfide, and more. The peptide linker options include ValCit, GlyPheLeuGly, and other peptide sequence

acid (FA), peptides (e.g., RGD, SMAC), glucose, or its analogs and amino acids enable uptake of prodrugs in tumors by binding specifically to overexpressed receptors or transport systems. A linker, generally a peptide sequence, like Val-Cit or Gly-Phe-Leu-Gly, serves as a spacer and is cleaved by cancer enriched enzymes (e.g., cathepsins), in the tumor environment, causing drug release. Lastly, the cleavable linker, designed as a self-immolative system, rapidly disintegrates following activation, ensuring effective release of the cytotoxic drug. This approach helps to minimize systemic toxicity while maximizing preferential drug delivery to cancer cells, as discussed in the examples below.

## 14.2 Receptor Targeted Prodrugs

Over the past decade, several receptors have been identified that are specifically overexpressed in certain types of cancers, such as PSMA for prostate cancer, and human epidermal growth factor receptor 2 (HER2) for breast cancer (Bacus et al. 1994), as well as those that are generally expressed such as folate (Parker et al. 2005) and transferrin receptors (Inoue et al. 1993). These receptors are considered promising targets for delivering cancer therapeutics. Strategies designed to target these receptors ensure that the prodrugs remain metabolically stable within the systemic circulation, minimizing their premature release. However, once bound to the targeted receptor, undergo internalization via endocytosis, and subsequently release the cytotoxic payload upon activation within the cancer cell to achieve specificity (Fig. 14.3).

### *14.2.1 Prostate-Specific Membrane Antigen*

PSMA is a membrane-bound glycoprotein expressed primarily in prostate epithelium and is also significantly overexpressed in prostate cancer cells (Perner et al. 2007; Silver et al. 1997). PSMA expression is also observed in the endothelial cells of tumor vasculature in various non-prostate solid tumors (Chang et al. 1999; Perner et al. 2007; Silver et al. 1997), as well as in kidney, salivary glands (Nedelcovych et al. 2025), brain (Sacha et al. 2007) and inflamed intestinal epithelium (Peters 2024). PSMA serves as a marker for prostate cancer as its expression levels correlate with the prostate cancer progression and disease severity (Perner et al. 2007). The high expression of PSMA, as well as its strong catalytic activity for its substrates, makes it an ideal target for anticancer prodrug strategies (Fig. 14.3).

PSMA undergoes receptor-mediated endocytosis upon ligand binding, enabling the release of therapeutic agents into cells (Liu et al. 2008b). The marked upregulation and internalization have been leveraged to deliver various types of PSMA-targeting therapeutics, including radiopharmaceutical therapies, antibody-conjugates (ADCs), aptamers, and small molecule drug conjugates (SMDCs). In particular, radiopharmaceutical agents such as $^{68}$Ga-PSMA-11, $^{18}$F-DFCPyL, and

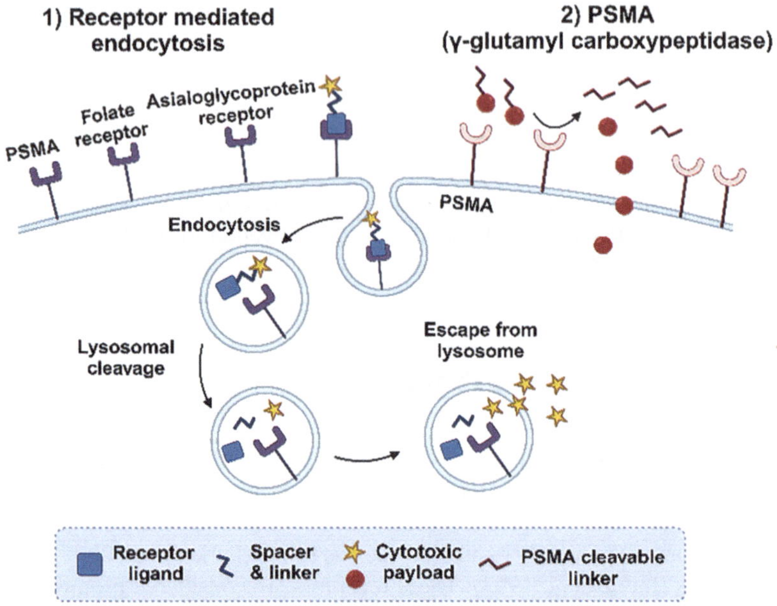


**Fig. 14.3** Illustration of receptor-mediated endocytosis and PSMA (γ-glutamyl carboxypeptidase) processes. On the left, receptors such as PSMA, folate, and asialoglycoprotein bind to their respective ligands, initiating endocytosis. The internalized complex undergoes lysosomal cleavage, releasing cytotoxic payload. On the right,,prodrugs can undergo PSMA specific γ-glutamyl carboxypeptidase mediated activation by utilizing PSMA cleavable linkers. Key elements include receptor ligands, spacers, linkers, and cytotoxic payloads, depicted with specific symbols

$^{177}$Lu-PSMA-618 have successfully utilized this strategy with promising clinical results as both imaging agents and therapeutics for the treatment of prostate cancer. While successful, these radiopharmaceutical agents have clinical limitations, including radionuclide-induced toxicity to non-prostatic tissues such as kidney and salivary glands (Kratochwil et al. 2018; Rahbar et al. 2017). Similarly, ADCs have shown promising therapeutic effects in preclinical studies but have faced challenges such as off-target toxicity and immunogenicity (Mirzaei et al. 2024). For example, the PSMA-targeted ADC MLN2704 caused peripheral neuropathy and neurotoxicity with limited efficacy in phase I/II clinical trials, due to rapid catabolism and premature release of the cytotoxic agent (Nakajima 2024).

Several SMDCs are also being designed and evaluated based on this strategy comprising a specific ligand, a linker, and a cytotoxic drug, as illustrated in Fig. 14.2, to help minimize off-target effects and enhance therapeutic efficacy (Perner et al. 2007). Various cytotoxic agents including tubulysin B hydrazide (Leamon et al. 2019), paclitaxel (PTX) (Lv et al. 2018), docetaxel (Machulkin et al. 2022b), and monomethyl auristatin E (MMAE) (Boinapally et al. 2021; Lahnif et al. 2023; Machulkin et al. 2021; Olatunji et al. 2022; Pal et al. 2019a) have been conjugated to PSMA ligands such as DUPA (2-[3-(1,3-dicarboxypropyl)ureido]pentanedioic acid), urea-based moieties, and phosphoramide-based ligands (Table 14.1, Sect.

**Table 14.1** Receptor targeted cancer prodrug examples

| Active moiety | Prodrug strategy (X = active moiety) | Preclinical Outcome | Phase | References |
|---|---|---|---|---|
| 14.2.1 Prostate-specific membrane antigen receptor | | | | |
| 12-ADT | Glu-Glu(γ)-Glu(γ)-Glu(γ)-Asp(β)-X | Regression of tumors in LNCaP, SN12C, MCF-7 xenografts | II (*NCT01777594*, *NCT02067156*) | Denmeade et al. (2012) |
| PTX | DUPA-ss-X | Inhibition of tumor growth with PTX, without toxicity in 22RV1 xenograft | Preclinical | Lv et al. (2018) |
| Tubulysin B hydrazide | Glu-urea-Lys-peptide spacer-disulfide-X | Complete remission in LNCaP xenograft | I (*NCT02202447*) | Leamon et al. (2019) |
| DOX | Glu-urea-Lys-alkylamide linkers-Phe-Phe-hydrazone-X | Inhibition of tumor growth in PC3-PIP xenograft | Preclinical | Ivanenkov et al. (2019) |
| MMAE | Glu-urea-Lys and transthyretin ligand connected via spacers to Val-Cit-PABC-X | Reduction in tumor volume in LNCaP xenograft | Preclinical | Pal et al. (2019a) |
| MMAE | Glu-urea-Lys-alkyl linker-Val-Cit-PABC-X | Extended survival rate in PSMA-PC3 xenograft model without toxicity | Preclinical | Boinapally et al. (2021) |
| Docetaxel | Glu-urea-Lys-alkyl linkers-Phe-Tyr-triazole linker-X | Inhibition of tumor growth in 22RV1 xenograft | Preclinical | Machulkin et al. (2022b) |
| MMAE | Glu-urea-Lys-alkyl linkers-Phe-Tyr-triazole linker-Val-Cit-PABC-X | Inhibition tumor growth in 22RV1 xenograft | Preclinical | Machulkin et al. (2021) |
| MMAE | CTT1298-alkyl and PEG linkers with albumin binding moiety-DBCO-Val-Cit-PABC or phosphoramidate-X | Effective PC3-PIP xenograft | Preclinical | Olatunji et al. (2022) |
| Abiraterone | Glu-urea-Lys-3-chlorobenzyl-alkyl linkers-Phe-Tyr-triazole linker-Val-Cit-PABC-X | Inhibition of tumor growth in 22RV1 xenograft | Preclinical | Machulkin et al. (2022a) |

(continued)

**Table 14.1** (continued)

| Active moiety | Prodrug strategy (X = active moiety) | Preclinical Outcome | Phase | References |
|---|---|---|---|---|
| MMAE | Glu-urea-Lys-naphtyl-cyclohexyl linker (KuE-617)-squaric acid-Val-Cit-PABC-X | Dose dependent tumor growth inhibition in LNCaP xenograft | Preclinical | Lahnif et al. (2023) |
| 4-Methyl quinazoline PI3K inhibitor | Glu-urea-Lys-naphtyl-cyclohexyl linker (KuE-617)-X | Dose dependent tumor growth inhibition in 22RV1 xenograft | Preclinical | Peng et al. (2024) |
| 14.2.2 Folate receptor | | | | |
| Vintafolide | FA-peptide linker (Asp-Arg-Asp-Asp-Cys)-ss-X | Therapeutic effects in lung and ovarian cancer | III (*NCT01170650*) | Vergote and Leamon (2015) |
| PTX | FA and albumin binding moiety (Evans blue attached via thiosuccinimide linker)-X | Prolonged blood circulation, and inhibition of tumor growth in MDA-MB-231 xenograft | Preclinical | Shan et al. (2018) |
| Rotenol | FA-ester-X | Inhibition of tumor growth in 4 T1 xenograft | Preclinical | Hong et al. (2024) |
| DOX | FA-AMA-hydrazide-X | Inhibition of tumor growth in MDA-MB-231 xenograft | Preclinical | Liao et al. (2021) |
| EC2491 Pyrrolobenzodiazepine dimer | FA-peptide linker (Asp-Arg-Asp-Asp-Cys)-succinimide-PEG spacer-penicillamine ss linker-X (precursor) | Complete responses in KB tumor xenograft | Preclinical | Reddy et al. (2020) |
| 14.2.3 Asialoglycoprotein receptor | | | | |
| W-105 (β-elemene derivative) | (Gal-triazole)-mono/di/trihydroxylalkyl linkage-disulfide-carbamate linker-X | Inhibition of tumor growth and accumulation in liver in H22 xenograft | Preclinical | Wang et al. (2021) |
| Betulin | GalNAc-alkyl-triazole-X (mono/bivalent) | Selective liver accumulation in balb/c mouse | Preclinical | Yamansarov et al. (2021) |
| Cisplatin | Pullulan (polysaccharide)-succinate-O-X (polymeric prodrug) | Inhibition of tumor growth in MHCC-97H xenograft | Preclinical | Wang et al. (2015) |

Abbreviations: *12-ADT* 8-*O*-(12-aminododecanoyl)-8-*O*-debutanoyl thapsigargin, *DUPA-ss* Glu-urea-Glu-1,2-diaminoethyl-disulfide, *MMAE* monomethyl auristatin E, *PABC p*-aminobenzyl carbamate, *CTT1298* PSMA targeting phosphoramidate inhibitor, *DBCO* dibenzocyclooctyne, *DAVLBH* desacetylvinblastine hydrazide, *FA-AMA-hydrazide* folate-aminocaproic acid-hydrazide, *GalNAc* N-acetyl-D-galactosamine

14.2.1). EC1169 represents one such example that consists of a urea-based Glu-urea-Lys targeting ligand, a disulfide-bond-based linkage, and tubulysin B hydrazide as a cytotoxic agent. It achieved a 100% response in PSMA-positive LNCaP xenograft models (Leamon et al. 2019). Additionally, EC1169 was inactive in PSMA-negative KB cell xenograft models at the same dose. Based on these promising preclinical results, EC1169 has advanced to a multi-center, open-label, non-randomized phase I clinical trial in patients with recurrent metastatic castration-resistant prostate cancer (*NCT02202447*).

The second strategy for PSMA targeting is based on its highly efficient protease activity as glutamate carboxypeptidase II (GCPII) and N-acetyl-L-aspartyl-L--glutamate peptidase I (NAALADase I) (Fig. 14.3). PSMA catalyzes the hydrolytic cleavage of α- or γ-linked glutamates from peptides or small molecules (Ghosh and Heston 2004). Employing this hydrolysis function, numerous studies (Table 14.1) have utilized γ-glutamated or α-glutamated peptide linkers for prodrug design. One notable example is G-202, with a PSMA cleavable linker (Asp-γGlu-γGlu-γGlu-Glu) and thapsigargin derivative [8-*O*-(12-aminododecanoyl)-8-*O*-debutanoyl thapsigargin] that demonstrated over 50% tumor regression in a human prostate cancer xenograft model (Denmeade et al. 2012). G-202 also progressed to a phase I dose-escalation study in patients with advanced solid tumors. Following the phase I study, phase II trials were conducted to investigate the safety and efficacy of G-202 in patients with recurrent or progressive glioblastoma (*NCT01777594*) and hepatocellular carcinoma receiving sorafenib therapy (*NCT02067156*) (Aloysius and Hu 2015). G-202 demonstrated clinical activity in hepatocellular carcinoma by reducing tumor blood flow, however no patients experienced a complete or partial response. Additionally, there is no further information available regarding phase III clinical trials (Mahalingam et al. 2019).

PSMA-targeted therapies have demonstrated strong potential in the treatment of prostate cancer, with ongoing clinical investigations exploring radiopharmaceuticals, ADCs, and immunotherapies (Afshar-Oromieh et al. 2013; Autio et al. 2018; Petrylak et al. 2019). While no PSMA-targeted SMDCs are in the clinic yet, PSMA remains a promising target, with its expression in prostate cancer and neovasculature of non-prostatic cancers (Wang and Kiess 2023). However, PSMAs high expression in several non-cancer tissues including salivary glands, kidneys, and small bowel (Afshar-Oromieh et al. 2013), leads to significant challenges due to off-target toxicities specifically in kidneys and salivary glands (Ling et al. 2022; Zechmann et al. 2014). Overcoming these off-target effects will be critical in future design of prodrugs for the successful development of PSMA-targeting small molecule therapeutics.

### *14.2.2 Folate Receptor*

Folate receptors (FR) regulate the uptake of folic acid (FA, folate) into cells, and play an important role in DNA synthesis and repair, purine and thymidine biosynthesis, as well as methylation (Stanger 2002), thereby supporting rapid proliferation of cancer cells (Fernandez et al. 2018). Folates are highly potent and bind with strong affinity to FR ($K_d$ = 0.1–1 nM). FR exist in four isoforms, namely FRα, FRβ, FRγ, and FRδ (Leamon et al. 2014; Lynn et al. 2015), among which FRα exhibits about 100–300 fold higher expression in various cancer types, including breast (Parker et al. 2005), lung (Franklin et al. 1994), kidney (Weitman et al. 1992), and ovarian cancers (Holm et al. 1995). These properties together with the straightforward chemistry of folate, make FRα an interesting target for cancer therapeutics. (Fernandez et al. 2018).

Drugs conjugated to FA (FA drug conjugates) target FR via receptor-mediated endocytosis. The preferred conjugation site for high-affinity binding is the γ-carboxylic acid site of FA (McMullon et al. 2024). Effective targeting also requires the incorporation of bioreleasable linkers, to facilitate cytotoxic drug release while maintaining optimal solubility. Furthermore, ensuring spatial separation between the ligand and payload, and keeping the drug size below 2000 Da, are critical for optimal tumor penetration (Vlahov and Leamon 2012).

The structural features of the FA-drug conjugates, similar to those of PSMA-targeted prodrug designs, consist of a targeting ligand (FA), a hydrophilic spacer, a bioreleasable linker, and a cytotoxic drug. Common spacers include polysaccharides, peptides, and polyethylene glycol (PEG) chains which afford a wide variety of linkages to accommodate various drug types. Linkers designed for cancer-specific activation include: pH-sensitive linker, enzyme-sensitive linker, self-immolative or redox-sensitive disulfide linker, and acyl hydrazone linker. These linkers are employed for enhanced tumor selectivity and tolerability of these therapeutic agents (Hong et al. 2024; Liao et al. 2021; Vergote and Leamon 2015). Many of these conjugates have demonstrated *in vitro* cytotoxicity across a wide range of cancer cell lines and have progressed to preclinical animal models (Table 14.1, Sect. 14.2.2). Notably, some FA drug conjugates have also advanced to clinical trials (Cheung et al. 2016; Morris et al. 2014). One of the most promising FA drug conjugates is exemplified by vintafolide, a water-soluble conjugate of desacetyl vinblastine monohydrazine (DAVLBH). Vintafolide exhibited excellent results both as a single agent and in combination with doxorubicin (DOX), advancing to phase II clinical trials for ovarian (*NCT00507741*), non-small cell lung cancers (*NCT01577654*), and a randomized open-label phase II study for platinum-resistant ovarian cancer (*NCT00722592*). In 2010, vintafolide entered phase III clinical trials for an advanced form of platinum resistant ovarian cancer overexpressing FRα, but it failed to show efficacy in progression-free survival (*NCT01170650*). Similarly, EC-0489, was designed as a next-generation version of vintafolide, by incorporating carbohydrate segments into the spacer. This modification aimed to reduce the hepatic clearance of vintafolide, resulting in a 70% decrease in bile clearance. EC-0489 is being currently evaluated in phase I clinical trials (*NCT00852189*).

Currently, none of the FA drug conjugates have completed all stages of clinical development or received regulatory approval. However, vintafolide has demonstrated significant advancements, suggesting that FA conjugated therapies hold great potential. Despite vintafolide encountering some challenges through phase II and III clinical trials, it has expanded the possibility of further research and development in this area. With advancements in targeted drug delivery and improvements in conjugation strategies, it is likely that future FA-drug conjugates may succeed and offer new effective cancer therapies.

### *14.2.3 Asialoglycoprotein Receptor*

Besides PSMA and FA receptors, other receptors that have gained attention for cancer-targeted prodrug strategies include the asialoglycoprotein receptor (ASGPR) (D'Souza and Devarajan 2015), mannose receptor (Dalle Vedove et al. 2018), integrin receptor ($\alpha v\beta 3$) (Liu et al. 2008a), and carbonic anhydrase IX receptor (Kunjiappan et al. 2021; Thiry et al. 2006). ASGPR, also known as the Ashwell-Morell receptor, is a C-type lectin primarily expressed on the sinusoidal surface of hepatocytes, with an estimated $1.35 \times 10^8$ receptors per gram of liver tissue (Kedderis and Held 1996). In addition to the liver, colorectal carcinoma cells originating from liver metastases also exhibit high ASGPR expression (D'Souza and Devarajan 2015; Witzigmann et al. 2016). Interestingly, its expression is limited in non-hepatic tissues such as the thyroid, intestinal epithelial cells and renal cells (D'Souza and Devarajan 2015; Park et al. 1998). This profile validates ASGPR as an excellent target for liver-specific drug delivery, ideally for hepatocellular carcinomas. Natural ASGPR ligands such as lactose, galactosamine, sterylglucoside, hyaluronic acid, and pullulan have been investigated as potential ASGPR-targeting moieties for liver-specific delivery (D'Souza and Devarajan 2015; Maitani et al. 2001). Additionally, several synthetic ASGPR ligands have been designed, including gal-chitosan-conjugated polymers, gal-poly-glutamic acid, gal-poly-L-lysine, and gal-cholesterol (D'Souza and Devarajan 2015). In addition to small molecule prodrugs, polymers and nanoparticle systems exploiting ASGPR ligands have been evaluated for targeted delivery of therapeutics (Chen et al. 2014; Oh et al. 2016; Xue et al. 2016). The success of this strategy is exemplified by ASGPR-targeted galactosamine-conjugated poly N-(2-hydroxypropyl) methacrylamide (HPMA)-DOX, referred to as PK2. This conjugate is linked through a tetrapeptide spacer (Gly-Phe-Leu-Gly) which is cleaved by lysosomal cathepsins, enabling liver-targeted delivery of DOX. This approach has shown promising efficacy in patients with solid hepatic neoplasms (Seymour et al. 2002).

Although a promising target, none of the ASGPR-targeted conjugates have gained approval to date. As listed in Table 14.1, Sect. 14.2.3, numerous efforts are ongoing to develop novel ASGPR ligands, such as bicyclic bridged ketal small molecules (Sanhueza et al. 2017). Importantly, ASGPR-targeting is not limited to only cancer therapy; it holds potential for other liver-related diseases, making it ideal for developing innovative ASGPR-targeted systems.

## 14.3 Transporter Targeted Prodrugs

Transporters are membrane-bound proteins that mediate the uptake of vital substrates such as amino acids and peptides, nucleosides, and sugars for proper cell functioning (Guan 2022; Lee 2000). Compared to normal cells, cancer cells demonstrate a high metabolic demand for nutrients and amino acids. To overcome these significant demands, cancer cells overexpress various transport systems, such as those for glucose, amino acids, peptides, and vitamins, to support rapid growth and proliferation (Ganapathy et al. 2009). This overexpression profile (Fig. 14.4) enables exploitation of these transport systems as promising targets for cancer therapy (Almasi and El Hiani 2020). This section highlights advancements in transporter-mediated prodrugs for cancer-targeted therapy (Table 14.2).

### *14.3.1 Glucose Transporter*

Cancer cells exhibit high demand for glucose uptake and glycolytic oxidation to support rapid cell proliferation, cell transformation, and tumor formation (Hanahan and Weinberg 2011). This increased demand is met by overexpression of glucose transporters primarily GLUT1, that is upregulated in several cancer types including breast (Brown and Wahl 1993), lung (Brown et al. 1999), renal (Suganuma et al.

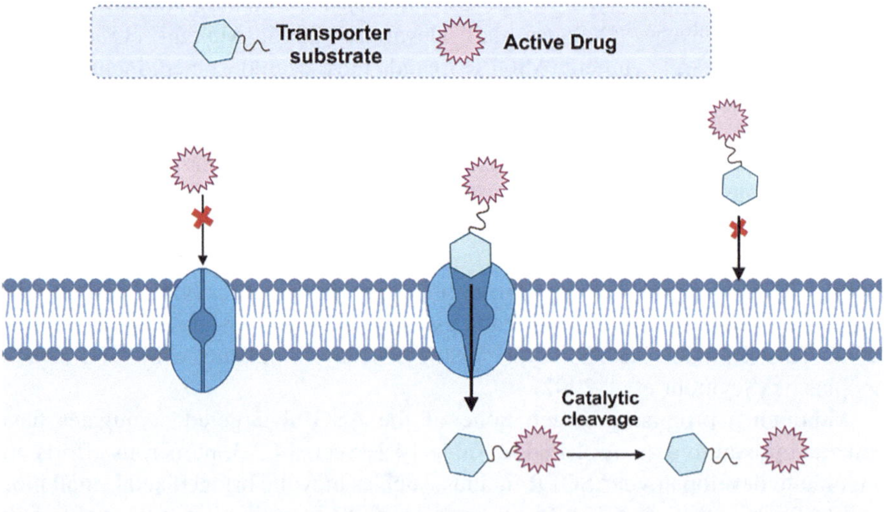


**Fig. 14.4** Schematic illustration of transporter-targeted prodrugs for cancer therapy. Prodrugs targeting transporters such as glucose, L-type amino acid, and sodium-dependent multivitamin transporters are overexpressed in cancer. Active drugs are released via catalytic cleavage once a prodrug is internalized in the cells

**Table 14.2** Transporter targeted cancer prodrugs

| Active moiety | Prodrug strategy (X = active moiety) | Preclinical Outcome | Phase | References |
|---|---|---|---|---|
| 14.3.1 Glucose transporter | | | | |
| Ifosfamide | Glufosfamide (Glu-ifosfamide) | Therapeutic effects in pancreatic cancer, non-small cell lung cancer, and ovarian cancer | III (*NCT00099294*) | Briasoulis et al. (2000), Briasoulis et al. (2003), Pohl et al. (1995), Stuben et al. (1996), and Ciuleanu et al. (2009) |
| Adriamycin | 2-DG–SUC–X | Antitumor activity in SKOV3 xenograft without toxicity | Preclinical | Cao et al. (2013) |
| Chlorambucil | 3,4,6-tris-*O*-Ac-FDG-4-aminophenol-Gly-X | Increase of MTD, tumor growth inhibition in B16FO and CT-26 xenograft | Preclinical | Miot-Noirault et al. (2011) |
| Methotrexate | Glu-X | Tumor growth inhibition in 4T1 xenograft and reduced liver and lung toxicity | Preclinical | Wozniak et al. (2021) |
| Oxaliplatin | Glu-X-HDN | Tumor growth inhibition in MCF-7 xenograft and reduced toxicity | Preclinical | Ma et al. (2017) |
| 14.3.2 L-type amino acid transporter | | | | |
| Melphalan | Melphalan flufenamide | Approved in 2021 by FDA for multiple myelomas but later withdrawn | Clinical | Chauhan et al. (2013), Huledal et al. (2024), and Olivier and Prasad (2022) |

(continued)

**Table 14.2** (continued)

| Active moiety | Prodrug strategy (X = active moiety) | Preclinical Outcome | Phase | References |
|---|---|---|---|---|
| DOX | L-Asp-X | Increased half-life, stronger inhibition of tumor growth in HepG2 and HCT 116 xenograft | Preclinical | Wu et al. (2015) |
| Methotrexate | L-Lys-X | Increased half-life, increased brain penetration in mice and lesser toxic effects | Preclinical | Singh and Subudhi (2016) |
| 14.3.3 Sodium-dependent multivitamin transporter | | | | |
| Camptothecin | Biotin-ss-X | Tumor inhibition in MGC-803 xenograft xenografted with gastric cancer and reduced side effects | Preclinical | Liang et al. (2018) |
| Ruthenium | $[Ru(\eta5\text{-}CpH)(P(C_6H_5)_3)(2,2'\text{-bipy-}4,4'\text{-PLA-biotin})]^+[CF_3SO_3]^-$ | Anti-cancer effects in MDA-MB-231 xenograft | Preclinical | Corte-Real et al. (2022) |

Abbreviations: *2-DG-SUC* 2-amino-2-deoxy-D-glucose-succinate, *Glu-benzyloxycarbonyl* N-(β-D-glucopyranosylbenzyloxycarbonyl), *3,4,6-tris-O-Ac-FDG-4-aminophenol-Gly* N-{2-[4-(3,4,6-tri-*O*-acetyl-2-deoxy-2-fluoro-β- D glucopyranosyloxy)phenyl]amino-2-oxoethyl, *Glu-X-HDN* glyocosylated-oxalipaltin-hexadecanoic, *melphalan flufenamide* 4-fluoro-L-phenylalaninemelphalan ethyl ester, *ss* disulfide, $[Ru(\eta^5\text{-}CpH)(P(C_6H_5)_3)(2,2'\text{-bipy-}4,4'\text{-PLA-biotin})][CF_3SO_3]$ Ru(II) organometallic complexes using biotin and polylactide, *MTD* Maximum tolerated dose

2007), colorectal (Younes et al. 1996), and pancreatic cancer (Lu et al. 2016; Szablewski 2013). The dependence of cancer cells on GLUT1 makes it attractive for targeting anticancer therapies. One approach involves chemically conjugating desired drugs to glucose and using it as a shuttling system for uptake by cancer cells via GLUT1 (Calvaresi and Hergenrother 2013).

Designing effective glycoconjugates requires careful structural considerations. For glucose, it is essential that the hydroxyl groups at positions 1 and 3, along with the pyran oxygen at position 5, must remain in a closed conformation to stabilize hydrogen bonding within the transporter (Barnett et al. 1973; Mueckler and Makepeace 2009). Besides glucose, other sugars like D-xylose, 2-deoxy-D-galactose, L-arabinose, D-ribose, D-fucose, and D-lyxose are also substrates for GLUT1 and used as shuttling systems for targeting (Lefevre 1961).

Based on these principles, glufosfamide, a glycoconjugate prodrug that combines glucose linked via hydroxyl group at position 1 to phosphorodiamidate moiety of ifosfamide mustard was reported. The prodrug was evaluated in several aggressive tumor models (Stuben et al. 1996) and supported by pharmacokinetic studies showing tumor localization of the active drug for at least 24 h (Stuben et al. 1996). Clinical trials in pancreatic cancer, glioblastoma, and non-small cell lung carcinoma exhibited hints of efficacy, but glufosfamide was not further advanced due to limited improvements in outcomes (Chiorean et al. 2010). Even decades later, glufosfamide remains one of the most successful glycoconjugated prodrugs identified to date.

Classical anticancer agents utilizing this prodrug approach include taxoids, anthracyclines, and alkylating agents. For example, 2-amino-2-deoxy-D-glucose conjugated adriamycin (ADM) with a succinic acid linker (2-DG–SUC–ADM) showed a significant decrease in organ toxicity while maintaining antitumor efficacy in ovarian adenocarcinoma-bearing mice (Cao et al. 2013). Similarly, 2-fluoro-2-deoxyglucose attached to chlorambucil through a glycine or 4-aminophenol linker resulted in improved antitumor activity of oxaliplatin and good tolerance in mice bearing melanoma and colon adenocarcinoma tumors (Miot-Noirault et al. 2011). Wozniak et al. designed a D-glucose conjugate of methotrexate via a glycosidic linker that demonstrated inhibition of breast cancer growth and attenuated liver, and lung toxicity compared to underivatized methotrexate (Wozniak et al. 2021). Further, Ma et al. conjugated glucoside with oxaliplatin using hexadecenoic acid as a linker, known to improve stability and lipophilicity of the drug. In addition, these conjugates were substrates for both GLUT1 and organic cation transporters (OCT), resulting in improved targeting in a pre-clinical breast cancer model as evidenced by tumor growth inhibition, reduced toxicity, and improved dose tolerance (Ma et al. 2017).

Although GLUT1 mediated prodrugs have achieved success in preclinical settings, very few have advanced to clinical trials. This can be largely attributed to the lack of preclinical models that can replicate the heterogeneity of human GLUT1 transporters (Montel-Hagen et al. 2009; Seyhan 2019). Furthermore, the nonspecific binding of glycoconjugate promoeities to other proteins, has also impeded the progress due to increased risk of off-target effects (Ernst and Magnani 2009).

Therefore, GLUT1-mediated prodrugs need a strategic approach in the design and evaluation of prodrugs, and suitable preclinical models mimicking human GLUT1 expression. These considerations can significantly improve the therapeutic outcomes in both preclinical and clinical settings.

### 14.3.2 L-type Amino Acid Transporter

Another interesting feature of rapidly proliferating cancer cells is their increased uptake of amino acids, to satiate their increased metabolic demands. This is achieved via the upregulation of amino acid transporters that belong to the solute carrier (SLC) super family (Hafliger and Charles 2019). Among these, LAT1 is found to be predominantly overexpressed in various solid tumors (Liang et al. 2011), including prostate (Zhao et al. 2023), lung (Kaira et al. 2008), bladder (Maimaiti et al. 2020), brain (Kobayashi et al. 2008), uterine (Sato et al. 2020), and gastrointestinal cancers (Ichinoe et al. 2011), making it a potential therapeutic target for delivering anticancer agents (Puris et al. 2020).

Several reports have outlined the key features for targeting the LAT1 transport system. A structure activity relationship study by Ylikangas et al. identified that both the amino and carboxylic acid functionalities are important for binding to LAT1 (Ylikangas et al. 2014). Moreover, it was reported that large, neutral, and aromatic amino acids, bind more strongly to LAT1 than aliphatic amino acids, and LAT1 is preferentially stereoselective for L-amino acids versus D-amino acids (Chien et al. 2018). A successful application of this strategy is exemplified through melphalan flufenamide, a dipeptide with 4-fluorophenylalanine ethyl ester linked to the active drug melphalan via amide bond (Chauhan et al. 2013). In preclinical studies, melphalan flufenamide significantly inhibited tumor growth and improved survival in human plasmacytoma xenograft mouse models compared to mice administered equimolar doses of melphalan. This LAT1-mediated prodrug demonstrated superior therapeutic effects, due to improved lipophilicity and facilitated cellular uptake, with rapid hydrolysis by aminopeptidase in the cytoplasm. This led to preferential drug accumulation in the cancer cells (Huledal et al. 2024). This prodrug, based on the success of its preclinical data, progressed into clinical trials, however, it was later withdrawn from the market due to insufficient benefits in the confirmatory trials (Olivier and Prasad 2022).

Similarly, a LAT1-targeted DOX prodrug was synthesized by attaching aspartate to the $NH_2$-terminal group of DOX via an amide bond, retaining free carboxyl and amino groups on the α-carbon of the aspartate residue. This prodrug exhibited a significant increase in half-life and tumor accumulation in hepatic and colon tumor-bearing mice compared to the active drug. Moreover, the prodrug demonstrated stronger tumor growth inhibition in a hepatic cancer mouse model (Wu et al. 2015). Similarly, a methotrexate prodrug conjugated to L-lysine via an amide linkage was developed to target brain cancer, utilizing LAT1-mediated delivery to cross the blood-brain barrier. The amide linkage improved stability, doubling the half-life, increasing brain exposure by eight-fold, and enhancing relative brain distribution by over five-fold compared to methotrexate in mice. Additionally, toxicity studies in rats showed reduced hemolytic toxicity for the lysine conjugate; however, no *in vivo* efficacy studies were conducted. Moreover LAT1-mediated delivery was not confirmed (Singh and Subudhi 2016).

Despite its promise, only a few LAT1-mediated prodrugs have advanced to clinical trials. LAT1 is generally expressed across other tissues and its broad substrate specificity overlaps with other nutrient transporters, complicating absolute targeting (Puris et al. 2020). This shows that a rational design approach is essential to fully leverage the benefits of amino acid transporter-mediated prodrugs.

### *14.3.3 Sodium-Dependent Multivitamin Transporter*

Sodium-dependent multivitamin transporter (SMVT) is an important transmembrane protein responsible for the transport of vitamins like vitamins B7 and B5 and other cofactors (Quick and Shi 2015). Cancer cells are reported to overexpress SMVT in several aggressive cancers like ovarian (Yellepeddi et al. 2011), lung (Azari et al. 2023), prostate (Patel et al. 2012), and breast cancer (Shi et al. 2014), making it an attractive target for cancer therapy (Maiti and Paira 2018).

A noteworthy example of a SMVT-mediated prodrug, is a biotinylated polymer ruthenium conjugate, which showed suppression of tumor growth and reduced liver toxicity in breast cancer xenograft nude mice (Corte-Real et al. 2022). Similarly, a biotin conjugated camptothecin prodrug demonstrated tumor growth inhibition, and decreased toxicity in gastric cancer xenografts through its tumor-specific localization (Liang et al. 2018). Additionally a biotin-conjugated rhodamine polymer, showed accumulation in tumors compared to normal tissues, as well as allowed easier imaging of both primary and metastatic tumor, highlighting its potential in both targeted therapy and as an imaging agent (Russell-Jones et al. 2011). The active targeting of SMVT has been achieved using nanostructured lipid carriers functionalized with biotin via PEG or polymers, encapsulating anticancer drugs such as DOX (Singh et al. 2017; Soininen et al. 2012), PTX (Rompicharla et al. 2019), daunomycin (Schnyder et al. 2005) and inorganic ruthenium (Corte-Real et al. 2022). However, this review focuses primarily on small molecule prodrug approaches and thus nanostructured carriers are outside the scope and not discussed in detail.

## 14.4 Enzyme Activatable Prodrugs

Proteases represent approximately 2% of the total human genome (Rawlings et al. 2018) and play a critical role in essential biological processes (Puente et al. 2003). Dysregulation of protease pathways is linked to various pathological conditions,

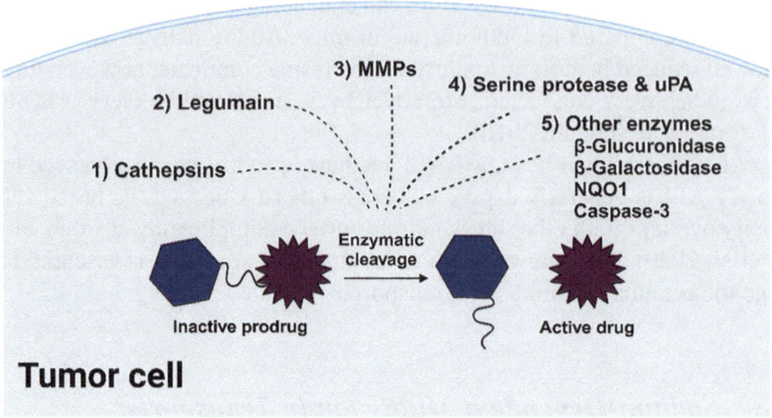


**Fig. 14.5** Schematic illustration of enzyme-activable prodrugs for cancer therapy. Various enzymes, including cathepsins, legumain, matrix metalloproteases, serine proteases, urokinase plasminogen activator, β-glucuronidase, β-galactosidase, NAD(P)H:quinone oxidoreductase 1 (NQO1), and caspase-3, are overexpressed in cancer cells, and utilized for the activation of rationally designed prodrugs

including neurodegenerative disorders (Broder and Becker-Pauly 2013), blood disorders (Mahajan and Prabhakar 2016), inflammation (de Almeida and Dufour 2022), and cancer (Amini et al. 2022). In tumor microenvironment, several proteases are highly upregulated, driving cancer progression (Hanahan and Weinberg 2011) (Fig. 14.5). This property makes tumor enriched proteases a promising strategy for tumor-targeted prodrugs. Here, we discuss the recent development of novel prodrugs targeting cancer-specific proteases (Table 14.3).

## *14.4.1 Cathepsins*

Cathepsins are lysosomal proteases with diverse structures and functions and are classified based on the amino acids in their active sites, including serine (cathepsins A and G), aspartic (cathepsin D and E) and cysteine proteases (cathepsins B, C, F, H, K, L, O, S, V, X and W) (Brix et al. 2008; Patel et al. 2018; Turk et al. 2012). Among these, cathepsin B is the most widely studied therapeutic target, overexpressed in several types of tumors, such as prostate, ovarian, hepatocellular, and breast carcinoma (Berquin and Sloane 1996; Gabrijelcic et al. 1992; Heidtmann et al. 1997; Herszényi et al. 2008; Koblinski et al. 2002; Kos et al. 1998; Leto et al. 1997; Maciewicz et al. 1989; Miyake et al. 2004).

In recent decades numerous prodrugs with cathepsin cleavable linkers have been synthesized and tested in preclinical and clinical studies. Most linkers include peptide sequences like Val-Cit (Caculitan et al. 2017; Huang et al. 2023; Ye et al. 2021), Val-Ala (Fu et al. 2019; Kahl et al. 2019; Lee 2021; Pal et al. 2019b), Phe-Lys

**Table 14.3** Development of cancer specific enzymes targeted prodrugs

| Active moiety | Prodrug strategy (X = active moiety) | Preclinical Outcome | Phase | References |
|---|---|---|---|---|
| 14.4.1 Cysteine cathepsins | | | | |
| DOX | FRRG-X | Tumor growth inhibition in HT-29 xenograft without toxicity | Preclinical | Shim et al. (2019) |
| DOX | SMAC-FRRG-X | Tumor growth inhibition in MCF-7 xenograft, increased survival rate without toxicity | Preclinical | Shim et al. (2020) |
| DOX | VPF-FRRG-X | Tumor growth inhibition in CT-26 xenograft | Preclinical | Choi et al. (2021) |
| DOX | FRRL-X | Tumor growth inhibition in HT-29 xenograft with increased survival rate without toxicity | Preclinical | Shim et al. (2022) |
| Gemcitabine | RGD-GFLG-X | Tumor growth inhibition in BxPC-3 xenograft with reduced systemic toxicity | Preclinical | Wang et al. (2024) |
| DON | Ad-Lys(Ac)-X-*i*Pr | Increased tumor accumulation in EL4 xenograft with reduced distribution to GI | Preclinical | Tenora et al. (2019) |
| 14.4.2 Legumain | | | | |
| DOX | Succinate-Ala-Ala-Asp-Leu-X | Tumor growth inhibition in multidrug-resistant tumor, extended survival | Preclinical | Wu et al. (2006) |
| MMAE | αvβ3 integrin inhibitor-Ala-Ala-Asn-PABC-X | Tumor growth inhibition in 4T1 murine breast cancer, D121 Lewis lung carcinoma, and MDA-MB-435 human breast cancer | Preclinical | Liu et al. (2012) |

**Table 14.3** (continued)

| Active moiety | Prodrug strategy (X = active moiety) | Preclinical Outcome | Phase | References |
|---|---|---|---|---|
| KSPi | ADC-X | Tumor growth inhibition in NCI-H292 and Ku-19-19 xenografts | Preclinical | Lerchen et al. (2020) |
| 14.4.3 Matrix metalloproteases | | | | |
| Azademethylcolchicine | Fluoresceinamine-thiourea-nonapeptide-X | Tumor growth inhibition in A549, MCF7 and DLD1 immunodeficient xenografts without toxicity | Preclinical | Gill et al. (2014) |
| DOX | Ac-γE-ProCitGlyHofTyrLeu-X | Tumor growth inhibition in HT2080 xenografts with reduced toxicity | Preclinical | Albright et al. (2005) |
| 14.4.4 Serine protease/Urokinase plasminogen activator | | | | |
| Si113 | D-Ala-Leu-Lys-alkyl linker-X | Tumor growth inhibition in HepG2 and A2780TC xenografts with extended half-life | Preclinical | Rango et al. (2021) |
| DON | DRP-104 (Ac-Trp-X-*i*Pr) | Preferential activation in tumor while deactivated in GI tissue, complete tumor regression | II (*NCT04471415*) | Rais et al. (2022) |
| 14.4.5 Other enzymes | | | | |
| 14.4.5.1 β-Glucuronidase | | | | |
| CA-4 | β-Glu-X analogs | Tumor growth inhibition in MDA-MB-231 xenografts without cardiotoxicity | Preclinical | Peng et al. (2022) |

| | | | | |
|---|---|---|---|---|
| SN-38 | Mal-Glu-X | Tumor growth inhibition in HCT116 xenografts with prolonged half-life and increased exposure | Preclinical | Huang et al. (2021) |
| 14.4.5.2 β-Galactosidase | | | | |
| MMAE | Gal-PABC-X-folic acid | Tumor growth inhibition and prolonged survival rate in KB xenograft without any detectable toxicity | Preclinical | Legigan et al. (2012) |
| DOX | Gal-X | Tumor growth inhibition in HT-29 xenografts | Preclinical | Sharma et al. (2018) |
| 14.4.5.3 NQO1 | | | | |
| CA-4 | Hydroquinone-SI linker-X | Tumor growth inhibition in HepG2 xenografts without toxicity | Preclinical | Zhang et al. (2020) |
| 5-Fluorouracil | Hydroquinone-N-methyl-PABC-X | Tumor growth inhibition in A549 xenografts with increased survival rate and safety | Preclinical | Zhang et al. (2018) |
| SN-38 | Hydroquinone-PABC (biotin)-X | Tumor growth inhibition in A549 xenografts with reduced drug resistance induction response | Preclinical | Shin et al. (2016) |
| 14.4.5.4 Caspase-3 | | | | |

**Table 14.3** (continued)

| Active moiety | Prodrug strategy (X = active moiety) | Preclinical Outcome | Phase | References |
|---|---|---|---|---|
| DOX | EMC-KGDEVD-PABC-X | Prolonged half-life with enhanced tumor accumulation; without systemic toxicity *in vivo* | Preclinical | Chung et al. (2016) |
| MMAE | EMC-KGDEVD-PABC-X | Extended circulation time; tumor growth inhibition in SCC7 xenograft with good safety profile | Preclinical | Chung et al. (2019) |

Abbreviations: *FRRG* tetrapeptide Phe-Arg-Arg-Gly, *SMAC* pro-apoptotic peptide drug Ala-Val-Pro-Ile-Ala-Gln, *VPF* photosensitizer verteporfin, *FRRL* tetrapeptide Phe-Arg-Arg-Leu, *RGD-GFLG* tripeptide Arg-Gly-Asp and tetrapeptide Gly-Phe-Leu-Gly, *DON* diazo-6-oxo-L-norleucine, *DMG* dimethylglycine, *KSPi* kinesin spindle inhibitor, *ADC* antibody drug conjugate, *MMAE* monomethyl auristatin E, *CA-4* combrestatin A4, *SN-38* 7-ethyl-10-hydroxycamptothecin, *Mal-Glu* maltose-glucose, *Gal* galactose, *PABC* *p*-aminobenzyl carbamate, *SI linker* self-immolate linker, *EMC* ε-maleimidocaproic acid spacer, *KGDEVD* hexapeptide Ac-Lys-Gly-Asp-Glu-Val-Asp

(Doronina et al. 2003; Yang et al. 2011), Gly-Phe-Gly-Gly (Ogitani et al. 2016a; Ogitani et al. 2016b) or Gly-Phe-Leu-Gly (Shi et al. 2023; Wu et al. 2022; Yang et al. 2017). These linkers have been successfully attached to multiple active drugs including DOX (Cheng et al. 2015; Zhi et al. 2019), MMAE (Deeks 2019; Markham 2021; Shi et al. 2022), PTX (Liang et al. 2012; Pan et al. 2019; Satsangi et al. 2014), or gemcitabine (Maksimenko et al. 2013; Wu et al. 2020b). Furthermore, presence of various self-immolative linkers between peptide sequence and cytotoxic drug, especially *p*-aminobenzylcarbamate (PABC) or its variations, has shown improved cathepsin cleavage efficiency (Dal Corso et al. 2019).

Despite the widespread use of linkers discussed above, several drawbacks have been recently reported, particularly with the most widely used Val-Cit linker. These include premature payload release, hydrophobicity-induced aggregation and a limited drug to antibody ratio (Dorywalska et al. 2016; Hamblett et al. 2004). To address these limitations, several strategies incorporating hydrophilic polymer scaffolds were reported, such as polysarcosine (Viricel et al. 2019), polyacetals (Toader et al. 2023), 3′-amino-α-cyclodextrin components (Evans et al. 2022), PEG (Giese et al. 2021), unstructured hydrophilic polypeptide XTEN (Haeckel et al. 2016) and glutamic acid (Anami et al. 2018; Ha et al. 2022). Furthermore, novel linkers with modified PABC moiety, for example PEGylated *m*-amide *p*-aminobenzyl carbamate (Poudel et al. 2020), or tandem-cleavage linkers bearing additional glucuronide moiety (Chuprakov et al. 2021) have received attention in recently published reports. In the latest study, Watanabe et al. (Watanabe et al. 2024) combined hydrophilic glutamic acid and PABC moiety with exo-cleavable peptide linker to the ADC-cytotoxic payload conjugates. Prodrugs containing MMAE or exatecan payload were evaluated *in vivo* and demonstrated reduced premature payload release, higher drug-to-antibody ratios, and minimized aggregation.

Overall, this review focuses on recently published examples with emphasis on carrier-free conjugates, supported by promising *in vitro* and *in vivo* results. Shim and collaborators (Shim et al. 2019) discovered carrier-free nanoparticles of cathepsin-B-cleavable peptide-conjugated DOX. Among the peptide linkers tested, the most promising results were exhibited by the tetrapeptide, Phe-Arg-Arg-Gly (FRRG), linkage. Prodrug FRRG-DOX confirmed successful tumor-targeting in *in vivo* studies using human-colon adenocarcinoma (HT-29) tumor-bearing mice and enhanced tolerability. Moreover, the synthetic strategy was scalable enabling mass production for advancing these conjugates into clinical trials. In subsequent studies, a modified version of these prodrugs was designed by the addition of a pro-apoptotic peptide (SMAC) to FRRG-DOX. Prodrug SMAC-FRRG-DOX nanoparticles efficiently inhibited pulmonary metastatic tumor growth by inhibiting drug resistance-inducing proteins (Shim et al. 2020). In addition, photosensitizer verteporfin (VPF) containing prodrug nanoparticles VPF-FRRG-DOX, were used for the heterogeneous treatment using chemo/photodynamic therapy under visible light irradiation (Choi et al. 2021). In the most recent study, researchers focused on further optimization of carrier-free DOX prodrug nanoparticles (Shim et al. 2022). This was achieved by screening tetra- or pentapeptide linkers of which tetrapeptide Phe-Arg-Arg-Leu-DOX (FRRL-DOX) showed improved anticancer efficacy (2.1-fold), tumor-specific accumulation (2.3–16.3-fold) and no visible signs of toxicity at doses up to 10 mg/kg (equiv. of DOX).

Our lab recently developed a prodrug strategy for 6-diazo-5-oxo-L-norleucine (DON), based on a similar but distinct prodrug strategy. DON is a glutamine antagonist with well-established antitumor activity in preclinical and clinical studies (Lemberg et al. 2018). Despite its promise, DON's clinical development was discontinued due to gastrointestinal (GI) toxicity issues, as the GI system is also highly dependent on glutamine. To address this, we designed prodrugs with ε-acetyl-Lys moiety (Ueki et al. 2013; Ueki et al. 2016), which allows selective cleavage by histone deacetylases (HDACs) and cathepsin L (CTSL). The lead prodrug JHU607, contained a bulky, hydrophobic adamantyl moiety, which along with the targeting functionality, enhanced penetration and preferential activation in tumor, exhibiting an 11-fold higher DON exposure in tumor versus GI tissue in an EL4 tumor mouse model (Tenora et al. 2019). Interestingly, unlike other similar published prodrugs (Gondi and Rao 2013; Sudhan and Siemann 2015), JHU607 was exclusively activated by cathepsins B and L via hydrolysis of the adamantane-1-carbonyl lysine moiety, while remaining resistant to HDAC activation.

Wang et al. recently published a cathepsin-B sensitive gemcitabine prodrug containing arginyl-glycyl-aspartic acid (RGD) targeting ligand (Wang et al. 2024) as a potential treatment for pancreatic cancer. Gemcitabine conjugate, connected to the drug via Gly-Phe-Leu-Gly (GFLG) peptide linker, achieved significantly improved stability, enhanced uptake by pancreatic cells due to the active targeting RGD group, enhanced antitumor activity, and reduced systemic toxicity *versus* the free drug.

Overall, cathepsin-targeted prodrugs represent a promising approach to enhancing the selective delivery and efficacy of anticancer agents. Although a thorough understanding of prodrug design is essential for success, ongoing advancements in linker design and carrier-free formulations are creating new opportunities for the successful clinical translation of these prodrugs.

### *14.4.2 Legumain*

Legumain is an asparagine endopeptidase that is abundant in the lysosomes (Vidmar et al. 2017) and secreted into tumor microenvironment by cancer cells and by tumor associated macrophages. There is evidence that legumain is involved in remodeling tumor extracellular matrix (ECM), angiogenesis, and metastasis, as well as showing high substrate specificity (Dall and Brandstetter 2016; Khan et al. 2023). In comparison to other lysosomal proteins, legumain exclusively cleaves peptides that contain amino acid asparagine at P1 position (Chen et al. 1997). Legumain is overexpressed in various solid tumors, typically in gastric (Wang et al. 2020), breast (Lewen et al. 2008) and colorectal carcinoma (Haugen et al. 2015). Additionally, legumain is a potential biomarker for several tumors. These properties make legumain an ideal target for prodrug activation in cancer therapy (Haugen et al. 2015).

Several older studies have been published in support of this strategy for small molecule prodrugs, activation by legumain. For example, Wu et al. (Wu et al. 2006) synthesized cell-impermeable LEG-3-DOX prodrug, for activation exclusively in the tumor microenvironment. LEG-3 contains tetrapeptide linker Leu-Asp-Ala-Ala with terminal succinate moiety to be hydrolyzed to Leu-DOX by extracellular legumain. LEG-3-DOX completely arrested growth of multidrug-resistant tumor *in vivo* and extended survival without evidence of myelosuppression or cardiac toxicity. Similarly, Stern et al. (Stern et al. 2009) strategized a legumain-cleavable prodrug conjugate of etoposide. The promising candidate carried carbonylbenzyloxy-Ala-Ala-Asp-ethylenediamine moiety. *In vitro* studies using human embryonic kidney 293 (HEK 293) cells confirmed these prodrugs were shown to be effectively cleaved by legumain. Liu and collaborators (Liu et al. 2012) designed a prodrug using a legumain cleavable Ala-Ala-Asn-PABC linker, MMAE as the active drug, and αvβ3 integrin inhibitor, to bind cell surface glycoprotein integrin αvβ3. Prodrug efficacy was confirmed in various *in vivo* cancer models, including 4 T1 murine breast cancer, D121 Lewis lung carcinoma, and MDA-MB-435 human breast cancer. Notably, compared to free MMAE, its prodrugs exhibited lower toxicity.

Despite these promising results, recent focus has largely shifted towards using this approach for ADCs. Lerchen et al. (Lerchen et al. 2020) developed a new generation of protease-cleavable ADCs utilizing legumain-cleavable linkers for specific and highly efficient payload release. They designed and synthesized TWEAK receptor (TWEAKR)-targeting ADCs conjugates with several peptide linkers (variations of D-/L-Ala-Ala-Asn; D-/L-Asn) and small molecule kinesin spindle inhibitor (KSPi) as the payload (Karpov et al. 2019; Lerchen et al. 2018). Prodrugs showed high potency *in vivo* in TWEAKR expressing NCI-H292 NSCLC and Ku-19-19 urothelial cancer models, resulting in long lasting tumor regression. Furthermore, preferential activation in tumors compared to healthy tissue was confirmed (Lerchen et al. 2020).

Legumain exhibits a broad specificity to different peptide linkers, such as those with just a single asparagine residue. This has enabled design of a wide variety of linkers for the generation of ADCs (Lerchen et al. 2020). Miller et al. (Miller et al. 2021) used a FRET-based (Förster resonance energy transfer) assay in high throughput enzyme screening, and evaluated cleavage efficiency of 75 peptide pairs in lysosomal extracts and human and mouse plasma. This study helped identify a series of asparagine containing linkers that were efficiently hydrolyzed by legumain while stable towards cathepsin B. The most promising Asn-Asn linker used as a part of MMAE containing ADC proved to be stable in mouse and human plasma and exhibited potent and selective cytotoxicity, comparable to analogs with Val-Cit linker. Furthermore, the Asn-containing linkers were shown to be completely stable to human neutrophil elastase, an enzyme thought to be responsible for the neutropenia and thrombocytopenia associated with Val-Cit-PABC-MMAE ADCs (Miller et al. 2021).

Although still in the early stages of development, these findings highlight the significant potential of legumain-cleavable linkers for safer and more effective

ADCs and perhaps in future can help revamp small molecule agents as improved cancer therapeutics.

### 14.4.3 Matrix Metalloproteinases

Matrix metalloproteinases (MMPs) are zinc-dependent endopeptidases which under physiological conditions, degrade ECM and regulate inflammatory processes, as well as play a key role in tumor progression and metastasis (Hey and Linder 2024; Maurya et al. 2024; Sakamoto and Seiki 2017; Zhu et al. 2024). With over 20 distinct MMPs identified in human metabolism, these enzymes regulate several pathological processes, making them a promising target for cancer therapeutics. Prodrug strategies targeting this class of protease may help with tumor specific delivery.

The prodrug ICT2588 is based on vascular-disrupting agent (VDA) azademethylcolchicine (Gill et al. 2014) and conjugated with fluoresceinamine-thiourea-nonapeptide linker, designed for selective cleavage by tumor-specific MMP-14. This strategy helps eliminate cardiovascular toxicity commonly associated with other VDAs (Mita et al. 2013; van Heeckeren et al. 2006). Preclinical evaluations in human colorectal, lung, breast, and prostate cancer models showed significant therapeutic potential of ICT2588, which was superior to azademethylcolchicine and comparable to other clinically approved compounds (Gill et al. 2014). Recently, Barnieh et al. (Barnieh et al. 2022) confirmed the prodrug's activation in cells and its impact on the cell cycle, supporting its therapeutic promise. However, these have not advanced into clinic.

Similarly, a prodrug of DOX with unique peptide linker, Ac-γE-PCitG-HofYL-DOX, was designed to be specifically cleaved by MMPs in the middle of the hexapeptide (Albright et al. 2005). The resulting tripeptide-DOX intermediate was sequentially cleaved by extracellular proteases to smaller Leu-DOX, which was finally activated to DOX (Albright et al. 2005). This prodrug was preclinically evaluated in mice with HT1080 xenografts, where exhibited a ten-fold increased tumor/heart ratio of DOX and significantly reduced toxicity (Albright et al. 2005), eliminating the off-target toxicity associated with DOX.

Despite promising results in the preclinical studies with the prodrugs discussed above, none of these derivatives have advanced to clinical trials. However, further follow-up research and carefully designed conjugates could lead to compounds with reduced toxicity and improved therapeutic potential.

### 14.4.4 Serine Protease/Urokinase Plasminogen Activator

A serine endopeptidase urokinase-type plasminogen activator (uPA) is localized in extracellular space and involves the plasminogen activator system. Active plasmin is released by cleavage of plasminogen by uPA, which causes degradation of fibrin,

ECM, and activation of growth factors. Blasi et al. (Blasi and Carmeliet 2002) discovered that plasmin can also degrade the extracellular matrix by activation of various matrix metalloproteinase zymogens. uPA acts as a regulator for tumor progression and metastasis in cancer (Mahmood et al. 2018). There is strong evidence that uPA reaches high concentrations in cancer types like breast or colorectal cancer (Look et al. 2002; Skelly et al. 1997).

Rango et al. (Rango et al. 2021) synthesized a prodrug, ProSI113-TP, containing pyrazolo[3,4-*d*]pyrimidine-based anticancer drug Si113 (D'Antona et al. 2015), that has been identified as a potent inhibitor of Sgk1 (serum and glucocorticoid-regulated kinase 1)(Ortuso et al. 2014). Si113 showed promising *in vivo* results in hepatocellular (Talarico et al. 2015) and ovarian carcinoma (D'Antona et al. 2019) models, but extremely poor water solubility which prevents its further development. Recently developed prodrug ProSI113-TP overcame this limitation by introducing plasmin-cleavable tripeptide D-Ala-Leu-Lys. *In vivo* pharmacokinetic properties and preliminary tissue distribution confirmed a better profile compared to Si113. In advance, antitumor efficacy using HCC xenograft murine model showed significant reduction in tumor growth (approx. 90%). Furthermore, prodrug ProSi113-TP was evaluated for treatment in combination with PTX in a paclitaxel-resistant ovarian carcinoma xenografted murine model. The results achieved were encouraging, showing a reduction of tumor volume by greater than 95%.

Another example of a serine protease-activated prodrug, DRP-104, was synthesized by our laboratory and is currently in clinical trials (Rais et al. 2022). This dipeptide consists of DON-isopropyl ester and N-acetyl-L-tryptophan attached via amide bond. DRP-104 showed excellent stability in plasma and gut homogenate, as well as strong antitumor efficacy (Rais et al. 2022). Moreover, by enhancing T-cell viability and proliferation, DRP-104 was found to robustly enhance antitumor immune response. Interestingly, DRP-104 showed differential metabolism in tumor (target) versus gut (toxicity site). In tumor, DRP-104 was activated by "unknown" serine proteases to the active DON, while in the GI tissue it was inactivated to an inert metabolite "M1" by action of carboxylesterase enzymes (Rais et al. 2022). This distinct active/inactive metabolism profile gave DRP-104 an exceptional *in vivo* pharmacokinetic profile in tumor-bearing mice with six-fold greater exposure of DON in tumor versus plasma and 11-fold higher exposure versus GI tissue (Rais et al. 2022). Based on these promising results, DRP-104 (also named sirpiglenastat) was selected as a lead candidate for clinical trials in solid tumors (*NCT04471415*) and fibrolamellar hepatocellular carcinoma (*NCT06027086,* in combination with durvalumab*)*.

In conclusion, rationally designed prodrugs activated by serine proteases hold significant potential for the treatment of various cancers. Recently developed and tested prodrugs, such as ProSI113-TP and DRP-104, exemplify this promising approach. However, detailed knowledge about specific serine proteases and their cleavage mechanisms is crucial for achieving fully rational drug design.

### 14.4.5 Other Enzymes

In addition to the enzymes for prodrug targets discussed above, other important enzymes play a significant role in cancer-related pathological processes, including angiogenesis, invasion, and metastasis (Baig et al. 2019). Amongst these, some enzymes that are significantly upregulated in tumors include β-glucuronidase (Cheng et al. 2021), β-galactosidase (Asanuma et al. 2015), carboxylesterases (Li et al. 2017), NAD(P)H:quinone oxidoreductase 1 (NQO1) (Wu et al. 2018) or caspase-3 (Chung et al. 2019). These enzymes are not only involved in tumor progression but also exploited for selectively targeting anticancer agents to cancer cells. By leveraging their tumor-specific overexpression, these enzymes can act as triggers for the activation of prodrugs at the tumor site. Recent studies that highlight growing emphasis on the prodrugs targeting these overexpressed enzymes in cancer are summarized in Table 14.3 (*although not discussed in detail*).

## 14.5 Cancer Specific Microenvironment

Tumor cells have characteristics that differ from normal cells, such as distinct organizational structure, metabolism, and microenvironment (Fig. 14.6). Cancer cells rely on passive diffusion for gas exchange and nutrient transport. As they reach a certain size, the lack of oxygen and the accumulation of metabolic waste creates a hostile microenvironment which is hypoxic and acidic (Anderson and Simon 2020). Additionally, it is known that cancer cells exhibit high levels of reactive oxygen species (ROS) and glutathione (GSH). In this section, we will describe the development of novel prodrugs which target cancer-specific microenvironment (Table 14.4).

### 14.5.1 pH Sensitive Prodrugs

Cancer cells generally exhibit a lower pH in the microenvironment compared to normal cells due to their elevated metabolic rate. Glycolysis in cancer cells occurs at an accelerated rate, independent of oxygen supply, often resulting in a state of hypoxia. Hypoxia, in turn, stimulates the production of lactic acid and carbonic acid, contributing to the acidic microenvironment pH (Lu et al. 2002). Particularly, hypoxia-inducible factor (HIF) activates carbonic anhydrases IX and XII, which catalyze the reversible conversion of carbon dioxide and water into carbonic acid. This acid diffuses out of the cell membrane, causing an accumulation of hydrogen ions ($H^+$) in the tumor microenvironment. Additionally, the impaired lymphatic drainage in tumor tissues prevents the effective removal of these acids, further exacerbating the acidic pH range of 4.0–6.5. (Patel and Sant 2016).

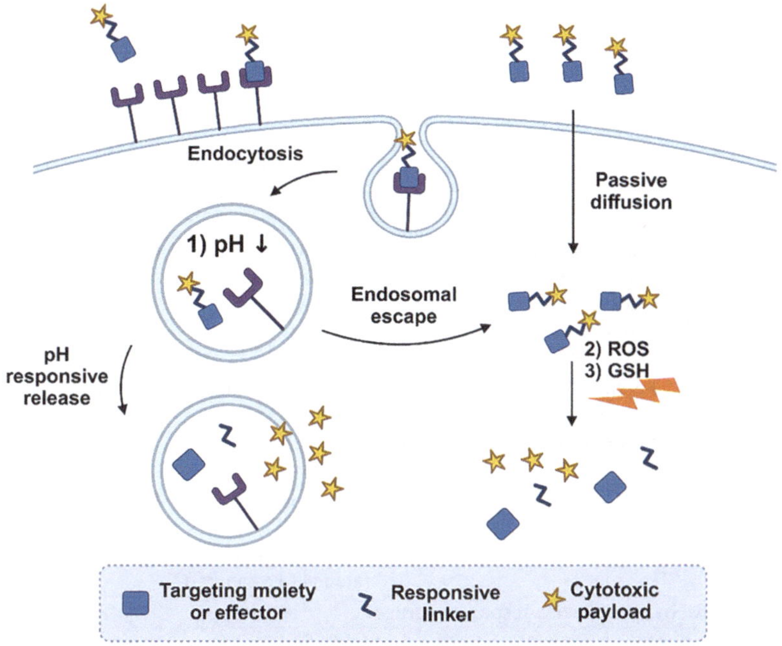


**Fig. 14.6** Illustration of prodrug concept for cancer-specific microenvironment targeting. Strategies to target the cancer microenvironment include: (1) pH-responsive drug release; (2) reactive oxygen species (ROS)-responsive drug release; and (3) glutathione (GSH)-responsive drug release

There are two common strategies for targeting the acidic tumor microenvironment. The first involves using acid-sensitive chemical bonds conjugated with anticancer therapeutics. Several pH-responsive chemical bonds, including hydrazone, imine, amide, ester, acetal, cis-aconityl, and phenyl vinyl ether linkers, have been employed (Nazli et al. 2024). INNO-206 is one such prodrug conjugated with an acid-labile hydrazone bond (Kratz et al. 2013), employing albumin as a carrier to target tumors. Albumin accumulates in solid tumors due to the hypervasculature, impaired drainage, and defective vascular architectural characteristic of tumor tissues (Kratz and Beyer 1998). With this design, INNO-206 demonstrated potent anticancer effects in xenograft models across various tumor cell lines, compared to free DOX (Graeser et al. 2010). INNO-206 subsequently progressed to clinical trials and resulted in improved progression-free survival and response rates compared to free DOX; however, it did not show a survival benefit in treatment-naïve patients with advanced sarcoma. Unfortunately, in the phase III clinical trial, INNO-206 further failed to demonstrate benefits in progression-free or overall survival in this population (Chawla et al. 2017; Cranmer 2019). However, due to a lower cardiotoxicity profile of INNO-206 compared to DOX, additional clinical trials are ongoing to investigate its efficacy in combination with other medications (*NCT04390399*).

**Table 14.4** Cancer specific microenvironment targeted prodrugs

| Active drugs | Prodrug strategy (X = active moiety) | Preclinical Outcome | Phase | References |
|---|---|---|---|---|
| 14.5.1. pH-responsive prodrugs | | | | |
| DOX | Albumin-thiol-(6-maleimidocaproyl)-hydrazone-X | Superior antitumor effects in MDA-MB 435, MCF-7 and M3366 xenograft | Phase III | Gong et al. (2018) and Kratz et al. (2002) |
| Cabazitaxel | Polymeric micelles consist of two copolymer blocks: first PEG, enzyme-responsive peptide, cholesterol; second DUPA, PEG, cholesterol | Inhibition of tumor growth in MDA-MB-231 xenograft | Preclinical | Barve et al. (2020) |
| DOX | Peptides T10 (HAIYPRHGGC) or T15 (THRPPMWSPVWPGGC)-EMCH-O-X | Inhibition of tumor growth in MCF-7/ADR xenograft, longer half-life | Preclinical | Sheng et al. (2015) |
| PTX | PAP-ss-alkyl chain-carbamate-X | Inhibition of tumor growth and prolonged survival rate in4T1 xenograft | Preclinical | Tang et al. (2024) |
| 14.5.2. Reactive oxygen species-responsive drugs | | | | |
| Nitrogen mustards | Trisubstituted benzyl ring with boronic acid, *N*,*N*-(2-chloroethyl)amino and methyl (λ-sulfaneyl)alaninate moieties | Inhibition of tumor growth inhibition in MDA-MB-468 xenograft without toxicity | Preclinical | Chen et al. (2018) |
| Quinone methide | Amino ferrocene-basedprodrug with dialkylamino-residue and 4-phenylboronic acid pinacol ester | Inhibition of tumor growth inhibition in Nemeth-Kellner lymphoma model | Preclinical | Daum et al. (2017) |
| Quinone methide and Cinnamaldehyde | QCA—Dual stimuli responsive hybrid drug with boronate & acetal linkers | Inhibition of tumor growth inhibition in SW620 and DU145 xenograft without toxicity | Preclinical | Noh et al. (2015) |
| Belinostat | 2-(4-(λ-Methyl) phenyl)-4,4,5,5-tetramethyl-1,3,2-dioxaborolane-X | Inhibition of tumor growth in MCF xenograft | Preclinical | Zheng et al. (2018) |
| PTX | Albumin-conjugated maleimide group-sulfide or disulfide-X | Inhibition of tumor growth in 4 T1 xenograft | Preclinical | Yang et al. (2018) |
| Camptothecin | GR-BCP-PEG-X (thiol and sulfone activable groups) | Inhibition of tumor growth in H22 xenograft | Preclinical | Yin et al. (2020) |
| Camptothecin | Boronic acid (position 10)-X | Inhibition of tumor growth in U87MG xenograft | Preclinical | Wang et al. (2016) |

(continued)

**Table 14.4** (continued)

| Active drugs | Prodrug strategy (X = active moiety) | Preclinical Outcome | Phase | References |
|---|---|---|---|---|
| 14.5.3. Glutathione-responsive prodrugs | | | | |
| Camptothecin | Biotin-ss-X | Inhibition of tumor growth inhibition without toxicity in MGC803 xenograft | Preclinical | Liang et al. (2018) |
| Floxuridine (X1) and dimethyl cantharidin (X2) | TPP moiety-triazole spacer-ss-X2-ss-X1 | Inhibition of tumor growth inhibition without toxicity in HCT116 xenograft | Preclinical | Jangili et al. (2022) |
| DOX (X1) or cisplatin (X2) | TPP moiety-organoarsenic moiety-ss-X1/X2 | Inhibition of tumor growth inhibition in HepG2 xenograft | Preclinical | Luo et al. (2021) |
| Coibamide A | Cyclic-RGD-ss-p-aminobenzyl-X | Inhibition of tumor growth in A549 xenograft | Preclinical | Wu et al. (2021) |
| Camptothecin | TC6-carbonate-X | Inhibition of tumor growth in H22 xenograft | Preclinical | Liu et al. (2022) |

Abbreviations: *DUPA* L-Glu-urea-L-Glu-1,2-diaminoethyl, *EMCH* 6-maleimidocaproyl hydrazone, *PAP-ss* all-D enantiomer $_{D}$(KLAKLAK)$_2$ peptide (PAP) connected to active drug via disulfide bond, *ss* disulfide bond, *QCA* 4-(1,3,2-dioxaborinan-2-yl)benzyl ((5-methyl-2-styryl-1,3-dioxan-5-yl)methyl) carbonate, *GR-BCP-PEG* GSH and ROS dual-responsive amphiphilic diblock copolymer prodrug with poly (ethylene glycol) linker, *TPP* triphenylphosphonium, *cyclic-RGD* cyclic pentapeptide with the cRGDyK sequence, *TC6* tricyclic structure containing α,β-unsaturated ketone responsive groups, activable and releasable by glutathione (thiol-Michael addition-based conjugate)

The second approach involves utilizing pH-responsive nanoparticles with acid-labile linker and pH-sensitive polymers such as poly(acrylic acid) (PAA), poly(methacrylic acid) (PMAA), poly(N,N′-dimethylaminoethyl methacrylate) (PDEAEM), poly(4-vinylpyridine) (PVP), or poly(L-histidine) (Phis). However, since the focus herein is on small molecule prodrug strategies, an in-depth discussion of nanostructured carriers is not included.

Although pH-responsive prodrugs have been actively explored, several challenges remain for successful translation. First, pH variation within tumors leads to uneven release of active compounds and inconsistent therapeutic effects. Additionally, one-third of tumor pH values overlap with those of normal tissues, making cancer-specific targeted release more challenging (Gerweck 1998; Gerweck and Seetharaman 1996). However, since INNO-206 successfully progressed to clinical trials, other pH-responsive prodrugs incorporating nanoparticles (*NCT03168061*), antibodies (*NCT01101594*), and polymers (*NCT00415298*) have also been explored in clinical trials. These advancements underscore the importance of this strategy, which, with rational design, could be effectively exploited in combination with other approaches described in this review.

### 14.5.2 Reactive Oxygen Species-Activated Prodrugs

ROS generated as by-product of aerobic metabolism, are molecules with unpaired electrons that confer reactivity towards various biological targets. Examples include the superoxide anion ($O^{2-}$), hydrogen peroxide ($H_2O_2$), and hydroxyl radicals (HO•). Increasing evidence supports that cancer cells produce higher levels of ROS than normal cells, leading to oxidative damage and apoptosis (Hecht et al. 2016). Thus, ROS-activated prodrugs offer a promising strategy for targeting cancer. To be cleaved or activated by ROS, its structural framework must contain two functional domains, i.e. a ROS-accepting moiety ('trigger') and an active drug 'effector' (Peng and Gandhi 2012). The trigger unit should be incorporated with a linker system, resulting in an increase in the cytotoxic potency of the effector. Initially, the trigger unit suppresses the cytotoxicity of the effector; however, once metabolized by ROS, the effector is released and subsequently exhibits cytotoxic effects (Peng and Gandhi 2012). Boronic acids/esters are one of the most common trigger units activated by $H_2O_2$. Several studies have employed boronic acid/esters as triggers in conjugation with cytotoxic compounds such as camptothecin, belinostat, and vorinostat (Table 14.4, Sect 5.2) (Wang et al. 2016; Zheng et al. 2018).

Another approach to targeting ROS involves using alkylating agents that are activated by ROS. One of the most common ROS-activated alkylating agents is nitrogen mustards. Nitrogen mustards are typically linked to electron-withdrawing groups, such as a carboxamide linker or boronic acid, which mask their toxicity by reducing electron density in normal tissues. Once a target site reaches high levels of ROS, $H_2O_2$ triggers the release of nitrogen mustard, and facilitates the formation of DNA interstrand cross-link (ICLs)(Peiro Cadahia et al. 2019; Zhang et al. 2017).

Chen et al. developed a ROS-activated DNA cross-linking agent employing nitrogen mustards, which exhibited significant anti tumor effects in a triple-negative breast cancer xenograft model. This demonstrated greater inhibition of tumor growth compared to control mice (Chen et al. 2018). Several studies have additionally developed prodrugs using quinone methide (QM) as an alkylating agent. The potency of QM prodrugs was shown to increase by conjugation with electron-donating groups. For example, Noh et al. developed prodrugs that are responsive to $H_2O_2$ and acidic pH. This group employed QM conjugated with boronate as a trigger and showed significant anticancer effects, that were comparable to camptothecin (Noh et al. 2015).

Aminoferrocene-based prodrugs (Daum et al. 2018; Daum et al. 2017; Marzenell et al. 2013; Schikora et al. 2015) additionally can act as ROS amplifiers and generate QM. Along with aminoferrocene-based prodrugs, the thiazolidinone moiety (Previtali et al. 2020) has also been investigated as a ROS-responsive agent. Under high ROS conditions, this moiety undergoes hydrolysis to release a bioactive compound with a free carboxylic acid.

Although many ROS-activated prodrugs have been developed, none have achieved success in the clinic. The main challenge limiting their progression is the adaptation of cancer cells to high ROS levels through enhanced antioxidative defense or autophagy induction (Gonzalez et al. 2014; Sznarkowska et al. 2017). Therefore, suppressing intracellular antioxidant mechanisms and increasing ROS levels have emerged as promising strategies for ROS-mediated cancer therapy (Hagen et al. 2012). In addition, ongoing research supports advancing ROS-responsive prodrugs, as outlined in Table 14.4, Sect. 14.5.2.

### *14.5.3 Glutathione-Responsive Prodrugs*

GSH tripeptide containing endogenous thiol bond, has an essential role in regulating the redox balance and defense against oxidative stress (Hwang et al. 1992; Traverso et al. 2013). The thiol group exhibits strong nucleophilicity, reductive capacity, and metal-chelating ability, enabling GSH to play an essential role in various cellular functions, such as maintenance of redox balance by serving as an electron donor for multiple antioxidant systems (Anderson 1985). Thus, the cellular antioxidant capacity heavily relies on the levels of GSH. GSH is present in almost all mammalian tissues, especially in the liver, and it showed higher concentrations in the cancer cells (2–8 mM) than other tissues (plasma: 1–2 μM). Thus, cellular antioxidant capacity depends on GSH levels. (Wu et al. 2004). Consequently, these differences between normal tissue and cancer cells make GSH a potential therapeutic target.

There are several strategies to target GSH. First, disulfide-based prodrugs are widely used to target cancer cells conjugated with cytotoxic compounds. The disulfide linker offers the advantage of stability and inactivity in plasma and extracellular environment with low GSH levels, but can be readily cleaved by GSH, enabling targeted release in GSH-rich tumor environments (Fernandes and Ramos 2004; Nagy 2013). For example, Jangili et al. (Jangili et al. 2022) developed prodrugs which target DNA damaged cancer cells employing floxuridine and protein phosphatase 2A (PP2A) inhibitor conjugated with GSH-sensitive disulfide linker. This prodrug successfully demonstrated significant inhibition of tumor growth while free floxuridine and PP2A inhibitor showed minor inhibition of tumor growth (Jangili et al. 2022). Beyond disulfide-based prodrugs, several other strategies target GSH, including diselenide-based prodrugs (Li et al. 2020), Michael acceptor-based prodrugs (Liu et al. 2022), and sulfonamide or sulfonate-based prodrugs (Liu and Cao 2020) (Zhao et al. 2024).

In addition to primarily targeting GSH, dual-responsive prodrugs that respond to both factors like ROS and GSH have also been developed. One example of prodrug, targeting GSH and ROS simultaneously consists of camptothecin, PEG, and PMAA, linked via thioether bond (Yin et al. 2020). This dual-targeting compound demonstrated significant inhibition of tumor growth (2.0-fold decrease), while single-targeting compounds for ROS (8.3-fold increase) and GSH (5.1-fold decrease) showed less pronounced effects on tumor growth inhibition.

Overall, only a limited number of GSH-responsive prodrugs (e.g., TLK286 reached phase III in 2008) have progressed to clinical trials. However, as suggested above, advancements in dual-targeting strategies (Yin et al. 2020), GSH-responsive polymers (Xia et al. 2022), and nanotechnology (Wu et al. 2020a) hold promise to enhance clinical development by overcoming the limitations of single-targeting prodrugs.

## 14.6 Conclusions

Overall, this review aimed to comprehensively discuss prodrug strategies for various cancer targets, over the past few decades. While most of the prodrug strategies discussed have shown promising results in preclinical settings, their clinical translation has been limited. Advancements in understanding cancer biology, including tumor heterogeneity amongst species, and tumor microenvironment, can be leveraged to improve clinical translation through improved targeting strategies.

## References

Afshar-Oromieh A, Malcher A, Eder M, Eisenhut M, Linhart HG, Hadaschik BA, Holland-Letz T, Giesel FL, Kratochwil C, Haufe S, Haberkorn U, Zechmann CM (2013) Reply to Reske et al.: PET imaging with a [68Ga]gallium-labelled PSMA ligand for the diagnosis of prostate cancer: biodistribution in humans and first evaluation of tumour lesions. Eur J Nucl Med Mol Imaging 40(6):971–972. https://doi.org/10.1007/s00259-013-2386-y

Albright CF, Graciani N, Han W, Yue E, Stein R, Lai Z, Diamond M, Dowling R, Grimminger L, Zhang S-Y, Behrens D, Musselman A, Bruckner R, Zhang M, Jiang X, Hu D, Higley A, DiMeo S, Rafalski M et al (2005) Matrix metalloproteinase–activated doxorubicin prodrugs inhibit HT1080 xenograft growth better than doxorubicin with less toxicity. Mol Cancer Ther 4(5):751–760. https://doi.org/10.1158/1535-7163.Mct-05-0006

Almasi S, El Hiani Y (2020) Exploring the therapeutic potential of membrane transport proteins: focus on cancer and Chemoresistance. Cancers (Basel) 12(6):1624. https://doi.org/10.3390/cancers12061624

Aloysius H, Hu L (2015) Targeted prodrug approaches for hormone refractory prostate cancer. Med Res Rev 35(3):554–585. https://doi.org/10.1002/med.21333

Amini AP, Kirkpatrick JD, Wang CS, Jaeger AM, Su S, Naranjo S, Zhong Q, Cabana CM, Jacks T, Bhatia SN (2022) Multiscale profiling of protease activity in cancer. Nat Commun 13(1):5745. https://doi.org/10.1038/s41467-022-32988-5

Anami Y, Yamazaki CM, Xiong W, Gui X, Zhang N, An Z, Tsuchikama K (2018) Glutamic acid-valine-citrulline linkers ensure stability and efficacy of antibody-drug conjugates in mice. Nat Commun 9(1):2512. https://doi.org/10.1038/s41467-018-04982-3

Anderson ME (1985) Determination of glutathione and glutathione disulfide in biological samples. Methods Enzymol 113:548–555. https://doi.org/10.1016/s0076-6879(85)13073-9

Anderson NM, Simon MC (2020) The tumor microenvironment. Curr Biol 30(16):R921–R925. https://doi.org/10.1016/j.cub.2020.06.081

Asanuma D, Sakabe M, Kamiya M, Yamamoto K, Hiratake J, Ogawa M, Kosaka N, Choyke PL, Nagano T, Kobayashi H, Urano Y (2015) Sensitive β-galactosidase-targeting fluorescence probe for visualizing small peritoneal metastatic tumours in vivo. Nat Commun 6:6463. https://doi.org/10.1038/ncomms7463

Autio KA, Dreicer R, Anderson J, Garcia JA, Alva A, Hart LL, Milowsky MI, Posadas EM, Ryan CJ, Graf RP, Dittamore R, Schreiber NA, Summa JM, Youssoufian H, Morris MJ, Scher HI (2018) Safety and efficacy of BIND-014, a docetaxel nanoparticle targeting prostate-specific membrane antigen for patients with metastatic castration-resistant prostate cancer: a phase 2 clinical trial. JAMA Oncol 4(10):1344–1351. https://doi.org/10.1001/jamaoncol.2018.2168

Azari F, Kennedy GT, Chang A, Nadeem B, Bou-Samra P, Chang A, Segil A, Bernstein E, Sullivan NT, Eruslanov E, Delikatny J, Singhal S (2023) Sodium multivitamin transporter-targeted Fluorochrome facilitates enhanced metabolic evaluation of tumors through coenzyme-R dependent intracellular signaling pathways. Mol Imaging Biol 25(3):569–585. https://doi.org/10.1007/s11307-022-01792-4

Bacus SS, Zelnick CR, Plowman G, Yarden Y (1994) Expression of the erbB-2 family of growth factor receptors and their ligands in breast cancers. Implication for tumor biology and clinical behavior. Am J Clin Pathol 102(4 Suppl 1):S13–S24. https://www.ncbi.nlm.nih.gov/pubmed/7942609

Baig MH, Adil M, Khan R, Dhadi S, Ahmad K, Rabbani G, Bashir T, Imran MA, Husain FM, Lee EJ, Kamal MA, Choi I (2019) Enzyme targeting strategies for prevention and treatment of cancer: implications for cancer therapy. Semin Cancer Biol 56:1–11. https://doi.org/10.1016/j.semcancer.2017.12.003

Barnett J, Holman G, Munday K (1973) Structural requirements for binding to the sugar-transport system of the human erythrocyte. Biochem J 131(2):211–221

Barnieh FM, Ribeiro Morais G, Garland H, Loadman PM, Falconer RA (2022) Targeted delivery of a colchicine analogue provides synergy with ATR inhibition in cancer cells. Biochem Pharmacol 201:115095. https://doi.org/10.1016/j.bcp.2022.115095

Barve A, Jain A, Liu H, Zhao Z, Cheng K (2020) Enzyme-responsive polymeric micelles of cabazitaxel for prostate cancer targeted therapy. Acta Biomater 113:501–511. https://doi.org/10.1016/j.actbio.2020.06.019

Baudino TA (2015) Targeted cancer therapy: the next generation of cancer treatment. Curr Drug Discov Technol 12(1):3–20. https://doi.org/10.2174/1570163812666150602144310

Berquin IM, Sloane BF (1996) Cathepsin B expression in human tumors. Adv Exp Med Biol 389:281–294. https://doi.org/10.1007/978-1-4613-0335-0_35

Blasi F, Carmeliet P (2002) uPAR: a versatile signalling orchestrator. Nat Rev Mol Cell Biol 3(12):932–943. https://doi.org/10.1038/nrm977

Boinapally S, Ahn HH, Cheng B, Brummet M, Nam H, Gabrielson KL, Banerjee SR, Minn I, Pomper MG (2021) A prostate-specific membrane antigen (PSMA)-targeted prodrug with a favorable in vivo toxicity profile. Sci Rep 11(1):7114. https://doi.org/10.1038/s41598-021-86551-1

Briasoulis E, Judson I, Pavlidis N, Beale P, Wanders J, Groot Y, Veerman G, Schuessler M, Niebch G, Siamopoulos K, Tzamakou E, Rammou D, Wolf L, Walker R, Hanauske A (2000) Phase I trial of 6-hour infusion of glufosfamide, a new alkylating agent with potentially enhanced selectivity for tumors that overexpress transmembrane glucose transporters: a study of the European Organization for Research and Treatment of cancer early clinical studies group. J Clin Oncol 18(20):3535–3544. https://doi.org/10.1200/JCO.2000.18.20.3535

Briasoulis E, Pavlidis N, Terret C, Bauer J, Fiedler W, Schoffski P, Raoul JL, Hess D, Selvais R, Lacombe D, Bachmann P, Fumoleau P (2003) Glufosfamide administered using a 1-hour infusion given as first-line treatment for advanced pancreatic cancer. A phase II trial of the EORTC-new drug development group. Eur J Cancer 39(16):2334–2340. https://doi.org/10.1016/s0959-8049(03)00629-4

Brix K, Dunkhorst A, Mayer K, Jordans S (2008) Cysteine cathepsins: cellular roadmap to different functions. Biochimie 90(2):194–207. https://doi.org/10.1016/j.biochi.2007.07.024

Broder C, Becker-Pauly C (2013) The metalloproteases meprin α and meprin β: unique enzymes in inflammation, neurodegeneration, cancer and fibrosis. Biochem J 450(2):253–264. https://doi.org/10.1042/bj20121751

Brown RS, Wahl RL (1993) Overexpression of Glut-1 glucose transporter in human breast cancer. An immunohistochemical study. *Cancer* 72(10):2979–2985. https://doi.org/10.1002/1097-0142(19931115)72:10<2979::aid-cncr2820721020>3.0.co;2-x
Brown RS, Leung JY, Kison PV, Zasadny KR, Flint A, Wahl RL (1999) Glucose transporters and FDG uptake in untreated primary human non-small cell lung cancer. J Nucl Med 40(4):556–565. https://www.ncbi.nlm.nih.gov/pubmed/10210213
Caculitan NG, Dela Cruz Chuh J, Ma Y, Zhang D, Kozak KR, Liu Y, Pillow TH, Sadowsky J, Cheung TK, Phung Q, Haley B, Lee BC, Akita RW, Sliwkowski MX, Polson AG (2017) Cathepsin B is dispensable for cellular processing of Cathepsin B-cleavable antibody-drug conjugates. Cancer Res 77(24):7027–7037. https://doi.org/10.1158/0008-5472.Can-17-2391
Calvaresi EC, Hergenrother PJ (2013) Glucose conjugation for the specific targeting and treatment of cancer. Chem Sci 4(6):2319–2333. https://doi.org/10.1039/C3SC22205E
Cao J, Cui S, Li S, Du C, Tian J, Wan S, Qian Z, Gu Y, Chen WR, Wang G (2013) Targeted cancer therapy with a 2-deoxyglucose-based adriamycin complex. Cancer Res 73(4):1362–1373. https://doi.org/10.1158/0008-5472.CAN-12-2072
Chang SS, O'Keefe DS, Bacich DJ, Reuter VE, Heston WD, Gaudin PB (1999) Prostate-specific membrane antigen is produced in tumor-associated neovasculature. Clin Cancer Res 5(10):2674–2681. https://www.ncbi.nlm.nih.gov/pubmed/10537328
Chauhan D, Ray A, Viktorsson K, Spira J, Paba-Prada C, Munshi N, Richardson P, Lewensohn R, Anderson KC (2013) In vitro and in vivo antitumor activity of a novel alkylating agent, melphalan-flufenamide, against multiple myeloma cells. Clin Cancer Res 19(11):3019–3031. https://doi.org/10.1158/1078-0432.CCR-12-3752
Chawla SP, Ganjoo KN, Schuetze S, Papai Z, Tine BAV, Choy E, Liebner DA, Agulnik M, Chawla S, Wieland S, Levitt DJ (2017) Phase III study of aldoxorubicin vs investigators' choice as treatment for relapsed/refractory soft tissue sarcomas. J Clin Oncol 35(15_suppl):11000–11000. https://doi.org/10.1200/JCO.2017.35.15_suppl.11000
Chen JM, Dando PM, Rawlings ND, Brown MA, Young NE, Stevens RA, Hewitt E, Watts C, Barrett AJ (1997) Cloning, isolation, and characterization of mammalian legumain, an asparaginyl endopeptidase. J Biol Chem 272(12):8090–8098. https://doi.org/10.1074/jbc.272.12.8090
Chen S, Tam YY, Lin PJ, Leung AK, Tam YK, Cullis PR (2014) Development of lipid nanoparticle formulations of siRNA for hepatocyte gene silencing following subcutaneous administration. J Control Release 196:106–112. https://doi.org/10.1016/j.jconrel.2014.09.025
Chen W, Fan H, Balakrishnan K, Wang Y, Sun H, Fan Y, Gandhi V, Arnold LA, Peng X (2018) Discovery and optimization of novel hydrogen peroxide activated aromatic nitrogen mustard derivatives as highly potent anticancer agents. J Med Chem 61(20):9132–9145. https://doi.org/10.1021/acs.jmedchem.8b00559
Cheng YJ, Luo GF, Zhu JY, Xu XD, Zeng X, Cheng DB, Li YM, Wu Y, Zhang XZ, Zhuo RX, He F (2015) Enzyme-induced and tumor-targeted drug delivery system based on multifunctional mesoporous silica nanoparticles. ACS Appl Mater Interfaces 7(17):9078–9087. https://doi.org/10.1021/acsami.5b00752
Cheng Z, Huang Y, Shen Q, Zhao Y, Wang L, Yu J, Lu W (2021) A camptothecin-based, albumin-binding prodrug enhances efficacy and safety in vivo. Eur J Med Chem 226:113851. https://doi.org/10.1016/j.ejmech.2021.113851
Cheung A, Bax HJ, Josephs DH, Ilieva KM, Pellizzari G, Opzoomer J, Bloomfield J, Fittall M, Grigoriadis A, Figini M, Canevari S, Spicer JF, Tutt AN, Karagiannis SN (2016) Targeting folate receptor alpha for cancer treatment. Oncotarget 7(32):52553–52574. https://doi.org/10.18632/oncotarget.9651
Chien HC, Colas C, Finke K, Springer S, Stoner L, Zur AA, Venteicher B, Campbell J, Hall C, Flint A, Augustyn E, Hernandez C, Heeren N, Hansen L, Anthony A, Bauer J, Fotiadis D, Schlessinger A, Giacomini KM, Thomas AA (2018) Reevaluating the substrate specificity of the L-type amino acid transporter (LAT1). J Med Chem 61(16):7358–7373. https://doi.org/10.1021/acs.jmedchem.8b01007

Chiorean EG, Dragovich T, Hamm J, Barrios CH, Gorini CF, Langmuir VK, Kroll S, Jung DT, Tidmarsh GT, Loehrer PJ (2010) A phase 2 trial of glufosfamide in combination with gemcitabine in chemotherapy-naive pancreatic adenocarcinoma. Am J Clin Oncol 33(2):111–116. https://doi.org/10.1097/COC.0b013e3181979204

Choi J, Shim MK, Yang S, Hwang HS, Cho H, Kim J, Yun WS, Moon Y, Kim J, Yoon HY, Kim K (2021) Visible-light-triggered prodrug nanoparticles combine chemotherapy and photodynamic therapy to potentiate checkpoint blockade cancer immunotherapy. ACS Nano 15(7):12086–12098. https://doi.org/10.1021/acsnano.1c03416

Chung SW, Choi JU, Lee BS, Byun J, Jeon OC, Kim SW, Kim IS, Kim SY, Byun Y (2016) Albumin-binding caspase-cleavable prodrug that is selectively activated in radiation exposed local tumor. Biomaterials 94:1–8. https://doi.org/10.1016/j.biomaterials.2016.03.043

Chung SW, Cho YS, Choi JU, Kim HR, Won TH, Kim SY, Byun Y (2019) Highly potent monomethyl auristatin E prodrug activated by caspase-3 for the chemoradiotherapy of triple-negative breast cancer. Biomaterials 192:109–117. https://doi.org/10.1016/j.biomaterials.2018.11.001

Chuprakov S, Ogunkoya AO, Barfield RM, Bauzon M, Hickle C, Kim YC, Yeo D, Zhang F, Rabuka D, Drake PM (2021) Tandem-cleavage linkers improve the in vivo stability and tolerability of antibody-drug conjugates. Bioconjug Chem 32(4):746–754. https://doi.org/10.1021/acs.bioconjchem.1c00029

Ciuleanu TE, Pavlovsky AV, Bodoky G, Garin AM, Langmuir VK, Kroll S, Tidmarsh GT (2009) A randomised Phase III trial of glufosfamide compared with best supportive care in metastatic pancreatic adenocarcinoma previously treated with gemcitabine. Eur J Cancer 45(9):1589–1596. https://doi.org/10.1016/j.ejca.2008.12.022

Connors TA, Knox RJ (1995) Prodrugs in cancer chemotherapy. Stem Cells 13(5):501–511. https://doi.org/10.1002/stem.5530130507

Corte-Real L, Bras AR, Pilon A, Mendes N, Ribeiro AS, Martins TD, Farinha JPS, Oliveira MC, Gartner F, Garcia MH, Preto A, Valente A (2022) Biotinylated polymer-ruthenium conjugates: in Vitro and in vivo studies in a triple-negative breast cancer model. Pharmaceutics 14(7):1388. https://doi.org/10.3390/pharmaceutics14071388

Cranmer LD (2019) Spotlight on aldoxorubicin (INNO-206) and its potential in the treatment of soft tissue sarcomas: evidence to date. Onco Targets Ther 12:2047–2062. https://doi.org/10.2147/OTT.S145539

Dal Corso A, Pignataro L, Belvisi L, Gennari C (2019) Innovative linker strategies for tumor-targeted drug conjugates. Chemistry 25(65):14740–14757. https://doi.org/10.1002/chem.201903127

Dall E, Brandstetter H (2016) Structure and function of legumain in health and disease. Biochimie 122:126–150. https://doi.org/10.1016/j.biochi.2015.09.022

Dalle Vedove E, Costabile G, Merkel OM (2018) Mannose and Mannose-6-phosphate receptor-targeted drug delivery systems and their application in cancer therapy. Adv Healthc Mater 7(14):e1701398. https://doi.org/10.1002/adhm.201701398

D'Antona L, Amato R, Talarico C, Ortuso F, Menniti M, Dattilo V, Iuliano R, Gigliotti F, Artese A, Costa G, Schenone S, Musumeci F, Abbruzzese C, Botta L, Trapasso F, Alcaro S, Paggi MG, Perrotti N (2015) SI113, a specific inhibitor of the Sgk1 kinase activity that counteracts cancer cell proliferation. Cell Physiol Biochem 35(5):2006–2018. https://doi.org/10.1159/000374008

D'Antona L, Dattilo V, Catalogna G, Scumaci D, Fiumara CV, Musumeci F, Perrotti G, Schenone S, Tallerico R, Spoleti CB, Costa N, Iuliano R, Cuda G, Amato R, Perrotti N (2019) In preclinical model of ovarian cancer, the SGK1 inhibitor SI113 counteracts the development of paclitaxel resistance and restores drug sensitivity. Transl Oncol 12(8):1045–1055. https://doi.org/10.1016/j.tranon.2019.05.008

Daum S, Reshetnikov MSV, Sisa M, Dumych T, Lootsik MD, Bilyy R, Bila E, Janko C, Alexiou C, Herrmann M, Sellner L, Mokhir A (2017) Lysosome-targeting amplifiers of reactive oxygen species as anticancer prodrugs. Angew Chem Int Ed Engl 56(49):15545–15549. https://doi.org/10.1002/anie.201706585

Daum S, Babiy S, Konovalova H, Hofer W, Shtemenko A, Shtemenko N, Janko C, Alexiou C, Mokhir A (2018) Tuning the structure of aminoferrocene-based anticancer prodrugs to prevent their aggregation in aqueous solution. J Inorg Biochem 178:9–17. https://doi.org/10.1016/j.jinorgbio.2017.08.038

de Almeida LGN, Dufour A (2022) Chapter 6 – Proteases are cut out to regulate acute and chronic inflammation. In: Zelanis A (ed) Proteolytic signaling in health and disease. Academic, pp 93–112. https://doi.org/10.1016/B978-0-323-85696-6.00003-2

Deeks ED (2019) Polatuzumab Vedotin: First Global Approval. Drugs 79(13):1467–1475. https://doi.org/10.1007/s40265-019-01175-0

Denmeade SR, Mhaka AM, Rosen DM, Brennen WN, Dalrymple S, Dach I, Olesen C, Gurel B, Demarzo AM, Wilding G, Carducci MA, Dionne CA, Moller JV, Nissen P, Christensen SB, Isaacs JT (2012) Engineering a prostate-specific membrane antigen-activated tumor endothelial cell prodrug for cancer therapy. Sci Transl Med 4(140):140ra186. https://doi.org/10.1126/scitranslmed.3003886

Doronina SO, Toki BE, Torgov MY, Mendelsohn BA, Cerveny CG, Chace DF, DeBlanc RL, Gearing RP, Bovee TD, Siegall CB, Francisco JA, Wahl AF, Meyer DL, Senter PD (2003) Development of potent monoclonal antibody auristatin conjugates for cancer therapy. Nat Biotechnol 21(7):778–784. https://doi.org/10.1038/nbt832

Dorywalska M, Dushin R, Moine L, Farias SE, Zhou D, Navaratnam T, Lui V, Hasa-Moreno A, Casas MG, Tran TT, Delaria K, Liu SH, Foletti D, O'Donnell CJ, Pons J, Shelton DL, Rajpal A, Strop P (2016) Molecular basis of valine-Citrulline-PABC linker instability in site-specific ADCs and its mitigation by linker design. Mol Cancer Ther 15(5):958–970. https://doi.org/10.1158/1535-7163.Mct-15-1004

D'Souza AA, Devarajan PV (2015) Asialoglycoprotein receptor mediated hepatocyte targeting – strategies and applications. J Control Release 203:126–139. https://doi.org/10.1016/j.jconrel.2015.02.022

Ernst B, Magnani JL (2009) From carbohydrate leads to glycomimetic drugs. Nat Rev Drug Discov 8(8):661–677. https://doi.org/10.1038/nrd2852

Evans N, Grygorash R, Williams P, Kyle A, Kantner T, Pathak R, Sheng X, Simoes F, Makwana H, Resende R, de Juan E, Jenkins A, Morris D, Michelet A, Jewitt F, Rudge F, Camper N, Manin A, McDowell W et al (2022) Incorporation of hydrophilic macrocycles into drug-linker reagents produces antibody-drug conjugates with enhanced in vivo performance. Front Pharmacol 13:764540. https://doi.org/10.3389/fphar.2022.764540

Fernandes PA, Ramos MJ (2004) Theoretical insights into the mechanism for thiol/disulfide exchange. Chemistry 10(1):257–266. https://doi.org/10.1002/chem.200305343

Fernandez M, Javaid F, Chudasama V (2018) Advances in targeting the folate receptor in the treatment/imaging of cancers. Chem Sci 9(4):790–810. https://doi.org/10.1039/c7sc04004k

Franklin WA, Waintrub M, Edwards D, Christensen K, Prendegrast P, Woods J, Bunn PA, Kolhouse JF (1994) New anti-lung-cancer antibody cluster 12 reacts with human folate receptors present on adenocarcinoma. Int J Cancer Suppl 8:89–95. https://doi.org/10.1002/ijc.2910570719

Fu Y, Urban DJ, Nani RR, Zhang YF, Li N, Fu H, Shah H, Gorka AP, Guha R, Chen L, Hall MD, Schnermann MJ, Ho M (2019) Glypican-3-specific antibody drug conjugates targeting hepatocellular carcinoma. Hepatology 70(2):563–576. https://doi.org/10.1002/hep.30326

Gabrijelcic D, Svetic B, Spaić D, Skrk J, Budihna M, Dolenc I, Popovic T, Cotic V, Turk V (1992) Cathepsins B, H and L in human breast carcinoma. Eur J Clin Chem Clin Biochem 30(2):69–74. http://europepmc.org/abstract/MED/1316176

Ganapathy V, Thangaraju M, Prasad PD (2009) Nutrient transporters in cancer: relevance to Warburg hypothesis and beyond. Pharmacol Ther 121(1):29–40. https://doi.org/10.1016/j.pharmthera.2008.09.005

Gerweck LE (1998) Tumor pH: implications for treatment and novel drug design. Semin Radiat Oncol 8(3):176–182. https://doi.org/10.1016/s1053-4296(98)80043-x

Gerweck LE, Seetharaman K (1996) Cellular pH gradient in tumor versus normal tissue: potential exploitation for the treatment of cancer. Cancer Res 56(6):1194–1198. https://www.ncbi.nlm.nih.gov/pubmed/8640796

Ghosh A, Heston WD (2004) Tumor target prostate specific membrane antigen (PSMA) and its regulation in prostate cancer. J Cell Biochem 91(3):528–539. https://doi.org/10.1002/jcb.10661

Giese M, Davis PD, Woodman RH, Hermanson G, Pokora A, Vermillion M (2021) Linker architectures as steric auxiliaries for altering enzyme-mediated payload release from bioconjugates. Bioconjug Chem 32(10):2257–2267. https://doi.org/10.1021/acs.bioconjchem.1c00429

Gill JH, Loadman PM, Shnyder SD, Cooper P, Atkinson JM, Ribeiro Morais G, Patterson LH, Falconer RA (2014) Tumor-targeted prodrug ICT2588 demonstrates therapeutic activity against solid tumors and reduced potential for cardiovascular toxicity. Mol Pharm 11(4):1294–1300. https://doi.org/10.1021/mp400760b

Gondi CS, Rao JS (2013) Cathepsin B as a cancer target. Expert Opin Ther Targets 17(3):281–291. https://doi.org/10.1517/14728222.2013.740461

Gong J, Yan J, Forscher C, Hendifar A (2018) Aldoxorubicin: a tumor-targeted doxorubicin conjugate for relapsed or refractory soft tissue sarcomas. Drug Des Devel Ther 12:777–786. https://doi.org/10.2147/DDDT.S140638

Gonzalez Y, Aryal B, Chehab L, Rao VA (2014) Atg7- and Keap1-dependent autophagy protects breast cancer cell lines against mitoquinone-induced oxidative stress. Oncotarget 5(6):1526–1537. https://doi.org/10.18632/oncotarget.1715

Graeser R, Esser N, Unger H, Fichtner I, Zhu A, Unger C, Kratz F (2010) INNO-206, the (6-maleimidocaproyl hydrazone derivative of doxorubicin), shows superior antitumor efficacy compared to doxorubicin in different tumor xenograft models and in an orthotopic pancreas carcinoma model. Investig New Drugs 28(1):14–19. https://doi.org/10.1007/s10637-008-9208-2

Guan L (2022) Structure and mechanism of membrane transporters. Sci Rep 12(1):13248. https://doi.org/10.1038/s41598-022-17524-1

Ha SYY, Anami Y, Yamazaki CM, Xiong W, Haase CM, Olson SD, Lee J, Ueno NT, Zhang N, An Z, Tsuchikama K (2022) An enzymatically cleavable tripeptide linker for maximizing the therapeutic index of antibody-drug conjugates. Mol Cancer Ther 21(9):1449–1461. https://doi.org/10.1158/1535-7163.Mct-22-0362

Haeckel A, Appler F, Ariza de Schellenberger A, Schellenberger E (2016) XTEN as biological alternative to PEGylation allows complete expression of a protease-Activatable Killin-based cytostatic. PLoS One 11(6):e0157193. https://doi.org/10.1371/journal.pone.0157193

Hafliger P, Charles RP (2019) The L-type amino acid transporter LAT1-An emerging target in cancer. Int J Mol Sci 20(10):2428. https://doi.org/10.3390/ijms20102428

Hagen H, Marzenell P, Jentzsch E, Wenz F, Veldwijk MR, Mokhir A (2012) Aminoferrocene-based prodrugs activated by reactive oxygen species. J Med Chem 55(2):924–934. https://doi.org/10.1021/jm2014937

Hamblett KJ, Senter PD, Chace DF, Sun MM, Lenox J, Cerveny CG, Kissler KM, Bernhardt SX, Kopcha AK, Zabinski RF, Meyer DL, Francisco JA (2004) Effects of drug loading on the antitumor activity of a monoclonal antibody drug conjugate. Clin Cancer Res 10(20):7063–7070. https://doi.org/10.1158/1078-0432.Ccr-04-0789

Hanahan D, Weinberg RA (2011) Hallmarks of cancer: the next generation. Cell 144(5):646–674. https://doi.org/10.1016/j.cell.2011.02.013

Haugen MH, Boye K, Nesland JM, Pettersen SJ, Egeland EV, Tamhane T, Brix K, Maelandsmo GM, Flatmark K (2015) High expression of the cysteine proteinase legumain in colorectal cancer – implications for therapeutic targeting. Eur J Cancer 51(1):9–17. https://doi.org/10.1016/j.ejca.2014.10.020

Hecht F, Cazarin JM, Lima CE, Faria CC, Leitao AA, Ferreira AC, Carvalho DP, Fortunato RS (2016) Redox homeostasis of breast cancer lineages contributes to differential cell death response to exogenous hydrogen peroxide. Life Sci 158:7–13. https://doi.org/10.1016/j.lfs.2016.06.016

Heidtmann H-H, Salge U, Abrahamson M, Bencina M, Kastelic L, Kopitar-Jerala N, Turk V, Lah TT (1997) Cathepsin B and cysteine proteinase inhibitors in human lung cancer cell lines. Clin Exp Metastasis 15(4):368–381. https://doi.org/10.1023/A:1018494020001

Herszényi L, Farinati F, Cardin R, István G, Molnár LD, Hritz I, De Paoli M, Plebani M, Tulassay Z (2008) Tumor marker utility and prognostic relevance of cathepsin B, cathepsin L, urokinase-type plasminogen activator, plasminogen activator inhibitor type-1, CEA and CA 19-9 in colorectal cancer. BMC Cancer 8:194. https://doi.org/10.1186/1471-2407-8-194

Hey S, Linder S (2024) Matrix metalloproteinases at a glance. J Cell Sci 137(2):jcs261898. https://doi.org/10.1242/jcs.261898

Holm J, Hansen SI, Hoier-Madsen M, Helkjaer PE, Bzorek M (1995) Folate receptor in malignant effusions of ovarian carcinoma. APMIS 103(9):663–670. https://doi.org/10.1111/j.1699-0463.1995.tb01420.x

Hong M, Wang J, Chen H, Qi J, Ji Q, Liu X, Yue Q, Li L, Cheng S (2024) Synthesis and biological evaluation of folic acid-rotenol conjugate as a potent targeted anticancer prodrug. Eur J Pharmacol 970:176482. https://doi.org/10.1016/j.ejphar.2024.176482

Huang Y, Wang L, Cheng Z, Yang B, Yu J, Chen Y, Lu W (2021) SN38-based albumin-binding prodrug for efficient targeted cancer chemotherapy. J Control Release 339:297–306. https://doi.org/10.1016/j.jconrel.2021.09.040

Huang CH, Chang E, Zheng L, Raj JGJ, Wu W, Pisani LJ, Daldrup-Link HE (2023) Tumor protease-activated theranostic nanoparticles for MRI-guided glioblastoma therapy. Theranostics 13(6):1745–1758. https://doi.org/10.7150/thno.79342

Huledal G, Ruiz-Garcia A, Kawakatsu S, Wang X, Sjoberg P, Gullbo J, Pekar D, Norin S, Jerling M (2024) Pharmacokinetics and metabolism of Melflufen, an alkylating peptide-drug conjugate, in patients with relapsed refractory multiple myeloma. J Clin Pharmacol 64(2):240–252. https://doi.org/10.1002/jcph.2355

Hwang C, Sinskey AJ, Lodish HF (1992) Oxidized redox state of glutathione in the endoplasmic reticulum. Science 257(5076):1496–1502. https://doi.org/10.1126/science.1523409

Ichinoe M, Mikami T, Yoshida T, Igawa I, Tsuruta T, Nakada N, Anzai N, Suzuki Y, Endou H, Okayasu I (2011) High expression of L-type amino-acid transporter 1 (LAT1) in gastric carcinomas: comparison with non-cancerous lesions. Pathol Int 61(5):281–289. https://doi.org/10.1111/j.1440-1827.2011.02650.x

Inoue T, Cavanaugh PG, Steck PA, Brunner N, Nicolson GL (1993) Differences in transferrin response and numbers of transferrin receptors in rat and human mammary carcinoma lines of different metastatic potentials. J Cell Physiol 156(1):212–217. https://doi.org/10.1002/jcp.1041560128

Ivanenkov YA, Machulkin AE, Garanina AS, Skvortsov DA, Uspenskaya AA, Deyneka EV, Trofimenko AV, Beloglazkina EK, Zyk NV, Koteliansky VE, Bezrukov DS, Aladinskaya AV, Vorobyeva NS, Puchinina MM, Riabykh GK, Sofronova AA, Malyshev AS, Majouga AG (2019) Synthesis and biological evaluation of doxorubicin-containing conjugate targeting PSMA. Bioorg Med Chem Lett 29(10):1246–1255. https://doi.org/10.1016/j.bmcl.2019.01.040

Jangili P, Kong N, Kim JH, Zhou J, Liu H, Zhang X, Tao W, Kim JS (2022) DNA-damage-response-targeting mitochondria-activated multifunctional prodrug strategy for self-defensive tumor therapy. Angew Chem Int Ed Engl 61(16):e202117075. https://doi.org/10.1002/anie.202117075

Jordan VC (2008) Tamoxifen: catalyst for the change to targeted therapy. Eur J Cancer 44(1):30–38. https://doi.org/10.1016/j.ejca.2007.11.002

Kahl BS, Hamadani M, Radford J, Carlo-Stella C, Caimi P, Reid E, Feingold JM, Ardeshna KM, Solh M, Heffner LT, Ungar D, He S, Boni J, Havenith K, O'Connor OA (2019) A phase I study of ADCT-402 (Loncastuximab Tesirine), a novel Pyrrolobenzodiazepine-based antibody-drug conjugate, in relapsed/refractory B-cell non-Hodgkin lymphoma. Clin Cancer Res 25(23):6986–6994. https://doi.org/10.1158/1078-0432.Ccr-19-0711

Kaira K, Oriuchi N, Imai H, Shimizu K, Yanagitani N, Sunaga N, Hisada T, Kawashima O, Iijima H, Ishizuka T, Kanai Y, Endou H, Nakajima T, Mori M (2008) Expression of L-type amino acid transporter 1 (LAT1) in neuroendocrine tumors of the lung. Pathol Res Pract 204(8):553–561. https://doi.org/10.1016/j.prp.2008.02.003

Karpov AS, Nieto-Oberhuber CM, Abrams T, Beng-Louka E, Blanco E, Chamoin S, Chene P, Dacquignies I, Daniel D, Dillon MP, Doumampouom-Metoul L, Drosos N, Fedoseev P, Furegati M, Granda B, Grotzfeld RM, Hess Clark S, Joly E, Jones D et al (2019) Discovery of potent and selective antibody-drug conjugates with Eg5 inhibitors through linker and payload optimization. ACS Med Chem Lett 10(12):1674–1679. https://doi.org/10.1021/acsmedchemlett.9b00468

Kedderis GL, Held SD (1996) Prediction of furan pharmacokinetics from hepatocyte studies: comparison of bioactivation and hepatic dosimetry in rats, mice, and humans. Toxicol Appl Pharmacol 140(1):124–130. https://doi.org/10.1006/taap.1996.0205

Khan SU, Khan IM, Khan MU, Ud Din MA, Khan MZ, Khan NM, Liu Y (2023) Role of LGMN in tumor development and its progression and connection with the tumor microenvironment. Front Mol Biosci 10:1121964. https://doi.org/10.3389/fmolb.2023.1121964

Kobayashi K, Ohnishi A, Promsuk J, Shimizu S, Kanai Y, Shiokawa Y, Nagane M (2008) Enhanced tumor growth elicited by L-type amino acid transporter 1 in human malignant glioma cells. Neurosurgery 62(2):493–503.; discussion 503-494. https://doi.org/10.1227/01.neu.0000316018.51292.19

Koblinski JE, Dosescu J, Sameni M, Moin K, Clark K, Sloane BF (2002) Interaction of human breast fibroblasts with collagen I increases secretion of Procathepsin B *. J Biol Chem 277(35):32220–32227. https://doi.org/10.1074/jbc.M204708200

Kos J, Nielsen HJ, Krasovec M, Christensen IJ, Cimerman N, Stephens RW, Brünner N (1998) Prognostic values of cathepsin B and carcinoembryonic antigen in sera of patients with colorectal cancer. Clin Cancer Res 4(6):1511–1516

Kratochwil C, Bruchertseifer F, Rathke H, Hohenfellner M, Giesel FL, Haberkorn U, Morgenstern A (2018) Targeted alpha-therapy of metastatic castration-resistant prostate cancer with (225) ac-PSMA-617: swimmer-plot analysis suggests efficacy regarding duration of tumor control. J Nucl Med 59(5):795–802. https://doi.org/10.2967/jnumed.117.203539

Kratz F, Beyer U (1998) Serum proteins as drug carriers of anticancer agents: a review. Drug Deliv 5(4):281–299. https://doi.org/10.3109/10717549809065759

Kratz F, Warnecke A, Scheuermann K, Stockmar C, Schwab J, Lazar P, Druckes P, Esser N, Drevs J, Rognan D, Bissantz C, Hinderling C, Folkers G, Fichtner I, Unger C (2002) Probing the cysteine-34 position of endogenous serum albumin with thiol-binding doxorubicin derivatives. Improved efficacy of an acid-sensitive doxorubicin derivative with specific albumin-binding properties compared to that of the parent compound. J Med Chem 45(25):5523–5533. https://doi.org/10.1021/jm020276c

Kratz F, Azab S, Zeisig R, Fichtner I, Warnecke A (2013) Evaluation of combination therapy schedules of doxorubicin and an acid-sensitive albumin-binding prodrug of doxorubicin in the MIA PaCa-2 pancreatic xenograft model. Int J Pharm 441(1–2):499–506. https://doi.org/10.1016/j.ijpharm.2012.11.003

Kunjiappan S, Pavadai P, Vellaichamy S, Ram Kumar Pandian S, Ravishankar V, Palanisamy P, Govindaraj S, Srinivasan G, Premanand A, Sankaranarayanan M, Theivendren P (2021) Surface receptor-mediated targeted drug delivery systems for enhanced cancer treatment: A state-of-the-art review. Drug Dev Res 82(3):309–340. https://doi.org/10.1002/ddr.21758

Lahnif H, Grus T, Salvanou EA, Deligianni E, Stellas D, Bouziotis P, Rosch F (2023) Old drug, new delivery strategy: MMAE repackaged. Int J Mol Sci 24(10):8543. https://doi.org/10.3390/ijms24108543

Leamon CP, Vlahov IR, Reddy JA, Vetzel M, Santhapuram HK, You F, Bloomfield A, Dorton R, Nelson M, Kleindl P, Vaughn JF, Westrick E (2014) Folate-vinca alkaloid conjugates for cancer therapy: a structure-activity relationship. Bioconjug Chem 25(3):560–568. https://doi.org/10.1021/bc400441s

Leamon CP, Reddy JA, Bloomfield A, Dorton R, Nelson M, Vetzel M, Kleindl P, Hahn S, Wang K, Vlahov IR (2019) Prostate-specific membrane antigen-specific antitumor activity of a self-Immolative Tubulysin conjugate. Bioconjug Chem 30(6):1805–1813. https://doi.org/10.1021/acs.bioconjchem.9b00335

Lee VH (2000) Membrane transporters. Eur J Pharm Sci 11(Suppl 2):S41–S50. https://doi.org/10.1016/s0928-0987(00)00163-9

Lee A (2021) Loncastuximab Tesirine: first approval. Drugs 81(10):1229–1233. https://doi.org/10.1007/s40265-021-01550-w

Lefevre PG (1961) Sugar transport in the red blood cell: structure-activity relationships in substrates and antagonists. Pharmacol Rev 13:39–70. https://www.ncbi.nlm.nih.gov/pubmed/13760340

Legigan T, Clarhaut J, Tranoy-Opalinski I, Monvoisin A, Renoux B, Thomas M, Le Pape A, Lerondel S, Papot S (2012) The first generation of β-galactosidase-responsive prodrugs designed for the selective treatment of solid tumors in prodrug monotherapy. Angew Chem Int Ed Engl 51(46):11606–11610. https://doi.org/10.1002/anie.201204935

Lemberg KM, Vornov JJ, Rais R, Slusher BS (2018) We're not "DON" yet: optimal dosing and prodrug delivery of 6-Diazo-5-oxo-L-norleucine. Mol Cancer Ther 17(9):1824–1832. https://doi.org/10.1158/1535-7163.Mct-17-1148

Lerchen HG, Wittrock S, Stelte-Ludwig B, Sommer A, Berndt S, Griebenow N, Rebstock AS, Johannes S, Cancho-Grande Y, Mahlert C, Greven S, Terjung C (2018) Antibody-drug conjugates with pyrrole-based KSP inhibitors as the payload class. Angew Chem Int Ed Engl 57(46):15243–15247. https://doi.org/10.1002/anie.201807619

Lerchen HG, Stelte-Ludwig B, Sommer A, Berndt S, Rebstock AS, Johannes S, Mahlert C, Greven S, Dietz L, Jörißen H (2020) Tailored linker chemistries for the efficient and selective activation of ADCs with KSPi payloads. Bioconjug Chem 31(8):1893–1898. https://doi.org/10.1021/acs.bioconjchem.0c00357

Leto G, Tumminello FM, Pizzolanti G, Montalto G, Soresi M, Gebbia N (1997) Lysosomal cathepsins B and L and Stefin A blood levels in patients with hepatocellular carcinoma and/or liver cirrhosis: potential clinical implications. Oncology 54(1):79–83. https://doi.org/10.1159/000227666

Lewen S, Zhou H, Hu HD, Cheng T, Markowitz D, Reisfeld RA, Xiang R, Luo Y (2008) A Legumain-based minigene vaccine targets the tumor stroma and suppresses breast cancer growth and angiogenesis. Cancer Immunol Immunother 57(4):507–515. https://doi.org/10.1007/s00262-007-0389-x

Li J, Shi J, Medina JE, Zhou J, Du X, Wang H, Yang C, Liu J, Yang Z, Dinulescu DM, Xu B (2017) Selectively inducing cancer cell death by intracellular enzyme-instructed self-assembly (EISA) of dipeptide derivatives. Adv Healthc Mater 6(15). https://doi.org/10.1002/adhm.201601400

Li X, Hou Y, Zhao J, Li J, Wang S, Fang J (2020) Combination of chemotherapy and oxidative stress to enhance cancer cell apoptosis. Chem Sci 11(12):3215–3222. https://doi.org/10.1039/c9sc05997k

Liang Z, Cho HT, Williams L, Zhu A, Liang K, Huang K, Wu H, Jiang C, Hong S, Crowe R, Goodman MM, Shim H (2011) Potential biomarker of L-type amino acid transporter 1 in breast cancer progression. Nucl Med Mol Imaging 45(2):93–102. https://doi.org/10.1007/s13139-010-0068-2

Liang L, Lin SW, Dai W, Lu JK, Yang TY, Xiang Y, Zhang Y, Li RT, Zhang Q (2012) Novel cathepsin B-sensitive paclitaxel conjugate: higher water solubility, better efficacy and lower toxicity. J Control Release 160(3):618–629. https://doi.org/10.1016/j.jconrel.2012.02.020

Liang Y, Huang W, Zeng D, Huang X, Chan L, Mei C, Feng P, Tan CH, Chen T (2018) Cancer-targeted design of bioresponsive prodrug with enhanced cellular uptake to achieve precise cancer therapy. Drug Deliv 25(1):1350–1361. https://doi.org/10.1080/10717544.2018.1477862

Liao TT, Han JF, Zhang FY, Na R, Ye WL (2021) Enhanced anti-tumor effect of folate-targeted FA-AMA-hyd-DOX conjugate in a xenograft model of human breast cancer. Molecules 26(23):7110. https://doi.org/10.3390/molecules26237110

Ling SW, de Blois E, Hooijman E, van der Veldt A, Brabander T (2022) Advances in (177) Lu-PSMA and (225)Ac-PSMA radionuclide therapy for metastatic castration-resistant prostate cancer. Pharmaceutics 14(10):2166. https://doi.org/10.3390/pharmaceutics14102166

Liu J, Cao C (2020) Evaluation of a GSH-targeting prodrug via a sulfonamide-induced "integrative" platform for selective cancer therapy. Analyst 145(14):4901–4905. https://doi.org/10.1039/d0an00627k

Liu Z, Wang F, Chen X (2008a) Integrin alpha(v)beta(3)-targeted cancer therapy. Drug Dev Res 69(6):329–339. https://doi.org/10.1002/ddr.20265

Liu T, Wu LY, Kazak M, Berkman CE (2008b) Cell-surface labeling and internalization by a fluorescent inhibitor of prostate-specific membrane antigen. Prostate 68(9):955–964. https://doi.org/10.1002/pros.20753

Liu Y, Bajjuri KM, Liu C, Sinha SC (2012) Targeting cell surface alpha(v)beta(3) integrin increases therapeutic efficacies of a legumain protease-activated auristatin prodrug. Mol Pharm 9(1):168–175. https://doi.org/10.1021/mp200434n

Liu Z, Chen S, Wang H, Zhao Y, Dong S (2022) Thiol-Michael addition based conjugate for glutathione activation and release. Bioorg Chem 129:106221. https://doi.org/10.1016/j.bioorg.2022.106221

Look MP, van Putten WL, Duffy MJ, Harbeck N, Christensen IJ, Thomssen C, Kates R, Spyratos F, Ferno M, Eppenberger-Castori S, Sweep CG, Ulm K, Peyrat JP, Martin PM, Magdelenat H, Brunner N, Duggan C, Lisboa BW, Bendahl PO et al (2002) Pooled analysis of prognostic impact of urokinase-type plasminogen activator and its inhibitor PAI-1 in 8377 breast cancer patients. J Natl Cancer Inst 94(2):116–128. https://doi.org/10.1093/jnci/94.2.116

Lu H, Forbes RA, Verma A (2002) Hypoxia-inducible factor 1 activation by aerobic glycolysis implicates the Warburg effect in carcinogenesis. J Biol Chem 277(26):23111–23115. https://doi.org/10.1074/jbc.M202487200

Lu K, Yang J, Li DC, He SB, Zhu DM, Zhang LF, Zhang XU, Chen XC, Zhang B, Zhou J (2016) Expression and clinical significance of glucose transporter-1 in pancreatic cancer. Oncol Lett 12(1):243–249. https://doi.org/10.3892/ol.2016.4586

Luo X, Gong X, Su L, Lin H, Yang Z, Yan X, Gao J (2021) Activatable mitochondria-targeting Organoarsenic prodrugs for bioenergetic cancer therapy. Angew Chem Int Ed Engl 60(3):1403–1410. https://doi.org/10.1002/anie.202012237

Lv Q, Yang J, Zhang R, Yang Z, Yang Z, Wang Y, Xu Y, He Z (2018) Prostate-specific membrane antigen targeted therapy of prostate cancer using a DUPA-paclitaxel conjugate. Mol Pharm 15(5):1842–1852. https://doi.org/10.1021/acs.molpharmaceut.8b00026

Lynn RC, Poussin M, Kalota A, Feng Y, Low PS, Dimitrov DS, Powell DJ Jr (2015) Targeting of folate receptor beta on acute myeloid leukemia blasts with chimeric antigen receptor-expressing T cells. Blood 125(22):3466–3476. https://doi.org/10.1182/blood-2014-11-612721

Ma J, Wang Q, Huang Z, Yang X, Nie Q, Hao W, Wang PG, Wang X (2017) Glycosylated platinum(IV) complexes as substrates for glucose transporters (GLUTs) and organic cation transporters (OCTs) exhibited cancer targeting and human serum albumin binding properties for drug delivery. J Med Chem 60(13):5736–5748. https://doi.org/10.1021/acs.jmedchem.7b00433

Machulkin AE, Uspenskaya AA, Zyk NU, Nimenko EA, Ber AP, Petrov SA, Polshakov VI, Shafikov RR, Skvortsov DA, Plotnikova EA, Pankratov AA, Smirnova GB, Borisova YA, Pokrovsky VS, Kolmogorov VS, Vaneev AN, Khudyakov AD, Chepikova OE, Kovalev S et al (2021) Synthesis, characterization, and preclinical evaluation of a small-molecule prostate-specific membrane antigen-targeted monomethyl Auristatin E conjugate. J Med Chem 64(23):17123–17145. https://doi.org/10.1021/acs.jmedchem.1c01157

Machulkin AE, Nimenko EA, Zyk NU, Uspenskaia AA, Smirnova GB, Khan II, Pokrovsky VS, Vaneev AN, Timoshenko RV, Mamed-Nabizade VV, Zavertkina MV, Erofeev A, Gorelkin P, Majouga AG, Zyk NV, Khazanova ES, Beloglazkina EK (2022a) Synthesis and preclinical evaluation of small-molecule prostate-specific membrane antigen-targeted Abiraterone conjugate. Molecules 27(24):8795. https://doi.org/10.3390/molecules27248795

Machulkin AE, Uspenskaya AA, Zyk NY, Nimenko EA, Ber AP, Petrov SA, Shafikov RR, Skvortsov DA, Smirnova GB, Borisova YA, Pokrovsky VS, Kolmogorov VS, Vaneev AN, Ivanenkov YA, Khudyakov AD, Kovalev SV, Erofeev AS, Gorelkin PV, Beloglazkina EK et al (2022b) PSMA-targeted small-molecule docetaxel conjugate: synthesis and preclinical evaluation. Eur J Med Chem 227:113936. https://doi.org/10.1016/j.ejmech.2021.113936

Maciewicz RA, Wardale RJ, Etherington DJ, Paraskeva C (1989) Immunodetection of cathepsins B and L present in and secreted from human pre-malignant and malignant colorectal tumour cell lines. Int J Cancer 43(3):478–486. https://doi.org/10.1002/ijc.2910430323

Mahajan C, Prabhakar H (2016) Chapter 17 – Coagulopathy. In: Prabhakar H (ed) Complications in Neuroanesthesia. Academic, pp 133–140. https://doi.org/10.1016/B978-0-12-804075-1.00017-1

Mahalingam D, Peguero J, Cen P, Arora SP, Sarantopoulos J, Rowe J, Allgood V, Tubb B, Campos L (2019) A phase II, multicenter, single-arm study of Mipsagargin (G-202) as a second-line therapy following Sorafenib for adult patients with progressive advanced hepatocellular carcinoma. Cancers (Basel) 11(6):833. https://doi.org/10.3390/cancers11060833

Mahmood N, Mihalcioiu C, Rabbani SA (2018) Multifaceted role of the Urokinase-type plasminogen activator (uPA) and its receptor (uPAR): diagnostic, prognostic, and therapeutic applications. Front Oncol 8:24. https://doi.org/10.3389/fonc.2018.00024

Maimaiti M, Sakamoto S, Yamada Y, Sugiura M, Rii J, Takeuchi N, Imamura Y, Furihata T, Ando K, Higuchi K, Xu M, Sazuka T, Nakamura K, Kaneda A, Kanai Y, Kyprianou N, Ikehara Y, Anzai N, Ichikawa T (2020) Expression of L-type amino acid transporter 1 as a molecular target for prognostic and therapeutic indicators in bladder carcinoma. Sci Rep 10(1):1292. https://doi.org/10.1038/s41598-020-58136-x

Maitani Y, Kawano K, Yamada K, Nagai T, Takayama K (2001) Efficiency of liposomes surface-modified with soybean-derived sterylglucoside as a liver targeting carrier in HepG2 cells. J Control Release 75(3):381–389. https://doi.org/10.1016/s0168-3659(01)00411-4

Maiti S, Paira P (2018) Biotin conjugated organic molecules and proteins for cancer therapy: A review. Eur J Med Chem 145:206–223. https://doi.org/10.1016/j.ejmech.2018.01.001

Maksimenko A, Mougin J, Mura S, Sliwinski E, Lepeltier E, Bourgaux C, Lepêtre S, Zouhiri F, Desmaële D, Couvreur P (2013) Polyisoprenoyl gemcitabine conjugates self assemble as nanoparticles, useful for cancer therapy. Cancer Lett 334(2):346–353. https://doi.org/10.1016/j.canlet.2012.08.023

Markham A (2021) Tisotumab Vedotin: first approval. Drugs 81(18):2141–2147. https://doi.org/10.1007/s40265-021-01633-8

Marzenell P, Hagen H, Sellner L, Zenz T, Grinyte R, Pavlov V, Daum S, Mokhir A (2013) Aminoferrocene-based prodrugs and their effects on human normal and cancer cells as well as bacterial cells. J Med Chem 56(17):6935–6944. https://doi.org/10.1021/jm400754c

Maurya S, Prasad D, Mukherjee S (2024) Matrix Metalloproteinases in Oral cancer pathogenesis and their use in therapy. Anti Cancer Agents Med Chem 24(1):3–17. https://doi.org/10.2174/0118715206270002231108071917

McMullon GT, Ezdoglian A, Booth AC, Jimenez-Royo P, Murphy PS, Jansen G, van der Laken CJ, Faulkner S (2024) Synthesis and characterization of folic acid-conjugated terbium complexes as luminescent probes for targeting folate receptor-expressing cells. J Med Chem 67(16):14062–14076. https://doi.org/10.1021/acs.jmedchem.4c00919

Miller JT, Vitro CN, Fang S, Benjamin SR, Tumey LN (2021) Enzyme-agnostic lysosomal screen identifies new Legumain-cleavable ADC linkers. Bioconjug Chem 32(4):842–858. https://doi.org/10.1021/acs.bioconjchem.1c00124

Miot-Noirault E, Reux B, Debiton E, Madelmont JC, Chezal JM, Coudert P, Weber V (2011) Preclinical investigation of tolerance and antitumour activity of new fluorodeoxyglucose-coupled chlorambucil alkylating agents. Investig New Drugs 29(3):424–433. https://doi.org/10.1007/s10637-009-9371-0

Mirzaei Y, Hussein Mer A, Fattah Maran B, Omidvar L, Misamogooe F, Amirkhani Z, Javaheri Haghighi N, Bagheri N, Keshtkaran Z, Rezaei B, Bargrizaneh F, Jahandideh S, Barpour N,

Shahsavarani H, Bazyari A, Abdollahpour-Alitappeh M (2024) Clinical and preclinical advances in PSMA-directed antibody-drug conjugates (ADCs): current status and hope for the future. Bioorg Chem 153:107803. https://doi.org/10.1016/j.bioorg.2024.107803

Mita MM, Sargsyan L, Mita AC, Spear M (2013) Vascular-disrupting agents in oncology. Expert Opin Investig Drugs 22(3):317–328

Miyake H, Hara I, Eto H (2004) Serum level of cathepsin B and its density in men with prostate cancer as novel markers of disease progression. Anticancer Res 24(4):2573–2578

Montel-Hagen A, Kinet S, Manel N, Mongellaz C, Prohaska R, Battini J-L, Delaunay J, Sitbon M, Taylor N (2009) Response: species diversity in GLUT expression and function. Cell 137(2):201–202

Morris RT, Joyrich RN, Naumann RW, Shah NP, Maurer AH, Strauss HW, Uszler JM, Symanowski JT, Ellis PR, Harb WA (2014) Phase II study of treatment of advanced ovarian cancer with folate-receptor-targeted therapeutic (vintafolide) and companion SPECT-based imaging agent (99mTc-etarfolatide). Ann Oncol 25(4):852–858. https://doi.org/10.1093/annonc/mdu024

Mueckler M, Makepeace C (2009) Model of the exofacial substrate-binding site and helical folding of the human Glut1 glucose transporter based on scanning mutagenesis. Biochemistry 48(25):5934–5942. https://doi.org/10.1021/bi900521n

Nagy P (2013) Kinetics and mechanisms of thiol-disulfide exchange covering direct substitution and thiol oxidation-mediated pathways. Antioxid Redox Signal 18(13):1623–1641. https://doi.org/10.1089/ars.2012.4973

Nakajima R (2024) Targeted therapy for prostate cancer by prostate-specific membrane antigen-targeted small-molecule drug conjugates. Chem Pharm Bull (Tokyo) 72(2):136–142. https://doi.org/10.1248/cpb.c23-00535

Nazli A, Irshad Khan MZ, Racz A, Beni S (2024) Acid-sensitive prodrugs; a promising approach for site-specific and targeted drug release. Eur J Med Chem 276:116699. https://doi.org/10.1016/j.ejmech.2024.116699

Nedelcovych MT, Dash RP, Wu Y, Choi EY, Lapidus RS, Majer P, Jancarik A, Abou D, Penet MF, Nikolopoulou A, Amor-Coarasa A, Babich J, Thorek DL, Rais R, Kratochwil C, Slusher BS (2025) JHU-2545 preferentially shields salivary glands and kidneys during PSMA-targeted imaging. Eur J Nucl Med Mol Imaging 52:1631–1641. https://doi.org/10.1007/s00259-024-07044-7

Noh J, Kwon B, Han E, Park M, Yang W, Cho W, Yoo W, Khang G, Lee D (2015) Amplification of oxidative stress by a dual stimuli-responsive hybrid drug enhances cancer cell death. Nat Commun 6:6907. https://doi.org/10.1038/ncomms7907

Ogitani Y, Aida T, Hagihara K, Yamaguchi J, Ishii C, Harada N, Soma M, Okamoto H, Oitate M, Arakawa S, Hirai T, Atsumi R, Nakada T, Hayakawa I, Abe Y, Agatsuma T (2016a) DS-8201a, A novel HER2-targeting ADC with a novel DNA topoisomerase I inhibitor, demonstrates a promising antitumor efficacy with differentiation from T-DM1. Clin Cancer Res 22(20):5097–5108. https://doi.org/10.1158/1078-0432.Ccr-15-2822

Ogitani Y, Hagihara K, Oitate M, Naito H, Agatsuma T (2016b) Bystander killing effect of DS-8201a, a novel anti-human epidermal growth factor receptor 2 antibody-drug conjugate, in tumors with human epidermal growth factor receptor 2 heterogeneity. Cancer Sci 107(7):1039–1046. https://doi.org/10.1111/cas.12966

Oh HR, Jo HY, Park JS, Kim DE, Cho JY, Kim PH, Kim KS (2016) Galactosylated liposomes for targeted co-delivery of doxorubicin/Vimentin siRNA to hepatocellular carcinoma. Nanomaterials (Basel) 6(8):141. https://doi.org/10.3390/nano6080141

Olatunji FP, Pun M, Herman JW, Romero O, Maniatopoulos M, Latoche JD, Parise RA, Guo J, Beumer JH, Anderson CJ, Berkman CE (2022) Modular smart molecules for PSMA-targeted chemotherapy. Mol Cancer Ther 21(11):1701–1709. https://doi.org/10.1158/1535-7163.MCT-22-0160

Olivier T, Prasad V (2022) The approval and withdrawal of melphalan flufenamide (melflufen): implications for the state of the FDA. Transl Oncol 18:101374. https://doi.org/10.1016/j.tranon.2022.101374

Ortuso F, Amato R, Artese A, D'Antona L, Costa G, Talarico C, Gigliotti F, Bianco C, Trapasso F, Schenone S, Musumeci F, Botta L, Perrotti N, Alcaro S (2014) In silico identification and biological evaluation of novel selective serum/glucocorticoid-inducible kinase 1 inhibitors based on the pyrazolo-pyrimidine scaffold. J Chem Inf Model 54(7):1828–1832. https://doi.org/10.1021/ci500235f

Pal A, Albusairi W, Liu F, Tuhin MTH, Miller M, Liang D, Joo H, Amin TU, Wilson EA, Faridi JS, Park M, Alhamadsheh MM (2019a) Hydrophilic small molecules that harness transthyretin to enhance the safety and efficacy of targeted chemotherapeutic agents. Mol Pharm 16(7):3237–3252. https://doi.org/10.1021/acs.molpharmaceut.9b00432

Pal SK, Forero-Torres A, Thompson JA, Morris JC, Chhabra S, Hoimes CJ, Vogelzang NJ, Boyd T, Bergerot PG, Adashek JJ, Li H, Yu X, Gartner EM, Carret AS, Smith DC (2019b) A phase 1 trial of SGN-CD70A in patients with CD70-positive, metastatic renal cell carcinoma. Cancer 125(7):1124–1132. https://doi.org/10.1002/cncr.31912

Pan X, Chen J, Yang M, Wu J, He G, Yin Y, He M, Xu W, Xu P, Cai W, Zhang F (2019) Enzyme/pH dual-responsive polymer prodrug nanoparticles based on 10-hydroxycamptothecin-carboxymethylchitosan for enhanced drug stability and anticancer efficacy. Eur Polym J 117:372–381. https://doi.org/10.1016/j.eurpolymj.2019.04.050

Park JH, Cho EW, Shin SY, Lee YJ, Kim KL (1998) Detection of the asialoglycoprotein receptor on cell lines of extrahepatic origin. Biochem Biophys Res Commun 244(1):304–311. https://doi.org/10.1006/bbrc.1998.8256

Parker N, Turk MJ, Westrick E, Lewis JD, Low PS, Leamon CP (2005) Folate receptor expression in carcinomas and normal tissues determined by a quantitative radioligand binding assay. Anal Biochem 338(2):284–293. https://doi.org/10.1016/j.ab.2004.12.026

Patel A, Sant S (2016) Hypoxic tumor microenvironment: opportunities to develop targeted therapies. Biotechnol Adv 34(5):803–812. https://doi.org/10.1016/j.biotechadv.2016.04.005

Patel M, Vadlapatla RK, Shah S, Mitra AK (2012) Molecular expression and functional activity of sodium dependent multivitamin transporter in human prostate cancer cells. Int J Pharm 436(1–2):324–331. https://doi.org/10.1016/j.ijpharm.2012.06.011

Patel S, Homaei A, El-Seedi HR, Akhtar N (2018) Cathepsins: proteases that are vital for survival but can also be fatal. Biomed Pharmacother 105:526–532. https://doi.org/10.1016/j.biopha.2018.05.148

Peiro Cadahia J, Previtali V, Troelsen NS, Clausen MH (2019) Prodrug strategies for targeted therapy triggered by reactive oxygen species. Medchemcomm 10(9):1531–1549. https://doi.org/10.1039/c9md00169g

Peng X, Gandhi V (2012) ROS-activated anticancer prodrugs: a new strategy for tumor-specific damage. Ther Deliv 3(7):823–833. https://doi.org/10.4155/tde.12.61

Peng Y, Shi Z, Liang Y, Ding K, Wang Y (2022) Targeting the tumor microenvironment by an enzyme-responsive prodrug of tubulin destabilizer for triple-negative breast cancer therapy with high safety. Eur J Med Chem 236:114344. https://doi.org/10.1016/j.ejmech.2022.114344

Peng S, Li H, Cui W, Xiong T, Hu J, Qi H, Lin S, Wu D, Ji M, Xu H (2024) Design, synthesis and biological evaluation of a novel PSMA-PI3K small molecule drug conjugate. RSC Med Chem 15(10):3485–3494. https://doi.org/10.1039/d4md00246f

Perner S, Hofer MD, Kim R, Shah RB, Li H, Moller P, Hautmann RE, Gschwend JE, Kuefer R, Rubin MA (2007) Prostate-specific membrane antigen expression as a predictor of prostate cancer progression. Hum Pathol 38(5):696–701. https://doi.org/10.1016/j.humpath.2006.11.012

Peters DE (2024) Targeting glutamate carboxypeptidase II in IBD. Adv Pharmacol 101:265–285. https://doi.org/10.1016/bs.apha.2024.10.001

Petrylak DP, Kantoff P, Vogelzang NJ, Mega A, Fleming MT, Stephenson JJ Jr, Frank R, Shore ND, Dreicer R, McClay EF, Berry WR, Agarwal M, DiPippo VA, Rotshteyn Y, Stambler N, Olson WC, Morris SA, Israel RJ (2019) Phase 1 study of PSMA ADC, an antibody-drug conjugate targeting prostate-specific membrane antigen, in chemotherapy-refractory prostate cancer. Prostate 79(6):604–613. https://doi.org/10.1002/pros.23765

Pohl J, Bertram B, Hilgard P, Nowrousian MR, Stuben J, Wiessler M (1995) D-19575–a sugar-linked isophosphoramide mustard derivative exploiting transmembrane glucose transport. Cancer Chemother Pharmacol 35(5):364–370. https://doi.org/10.1007/s002800050248

Poudel YB, Chowdari NS, Cheng H, Iwuagwu CI, King HD, Kotapati S, Passmore D, Rampulla R, Mathur A, Vite G, Gangwar S (2020) Chemical modification of linkers provides stable linker-payloads for the generation of antibody-drug conjugates. ACS Med Chem Lett 11(11):2190–2194. https://doi.org/10.1021/acsmedchemlett.0c00325

Previtali V, Petrovic K, Peiro Cadahia J, Troelsen NS, Clausen MH (2020) Auxiliary in vitro and in vivo biological evaluation of hydrogen peroxide sensitive prodrugs of methotrexate and aminopterin for the treatment of rheumatoid arthritis. Bioorg Med Chem 28(2):115247. https://doi.org/10.1016/j.bmc.2019.115247

Puente XS, Sánchez LM, Overall CM, López-Otín C (2003) Human and mouse proteases: a comparative genomic approach. Nat Rev Genet 4(7):544–558. https://doi.org/10.1038/nrg1111

Puris E, Gynther M, Auriola S, Huttunen KM (2020) L-type amino acid transporter 1 as a target for drug delivery. Pharm Res 37(5):88. https://doi.org/10.1007/s11095-020-02826-8

Quick M, Shi L (2015) The sodium/multivitamin transporter: a multipotent system with therapeutic implications. Vitam Horm 98:63–100. https://doi.org/10.1016/bs.vh.2014.12.003

Rahbar K, Ahmadzadehfar H, Kratochwil C, Haberkorn U, Schafers M, Essler M, Baum RP, Kulkarni HR, Schmidt M, Drzezga A, Bartenstein P, Pfestroff A, Luster M, Lutzen U, Marx M, Prasad V, Brenner W, Heinzel A, Mottaghy FM et al (2017) German multicenter study investigating 177Lu-PSMA-617 Radioligand therapy in advanced prostate cancer patients. J Nucl Med 58(1):85–90. https://doi.org/10.2967/jnumed.116.183194

Rais R, Lemberg KM, Tenora L, Arwood ML, Pal A, Alt J, Wu Y, Lam J, Aguilar JMH, Zhao L, Peters DE, Tallon C, Pandey R, Thomas AG, Dash RP, Seiwert T, Majer P, Leone RD, Powell JD, Slusher BS (2022) Discovery of DRP-104, a tumor-targeted metabolic inhibitor prodrug. Sci Adv 8(46):eabq5925. https://doi.org/10.1126/sciadv.abq5925

Rango E, D'Antona L, Iovenitti G, Brai A, Mancini A, Zamperini C, Trivisani CI, Marianelli S, Fallacara AL, Molinari A, Cianciusi A, Schenone S, Perrotti N, Dreassi E, Botta M (2021) Si113-prodrugs selectively activated by plasmin against hepatocellular and ovarian carcinoma. Eur J Med Chem 223:113653. https://doi.org/10.1016/j.ejmech.2021.113653

Rautio J, Kumpulainen H, Heimbach T, Oliyai R, Oh D, Jarvinen T, Savolainen J (2008) Prodrugs: design and clinical applications. Nat Rev Drug Discov 7(3):255–270. https://doi.org/10.1038/nrd2468

Rawlings ND, Barrett AJ, Thomas PD, Huang X, Bateman A, Finn RD (2018) The MEROPS database of proteolytic enzymes, their substrates and inhibitors in 2017 and a comparison with peptidases in the PANTHER database. Nucleic Acids Res 46(D1):D624–d632. https://doi.org/10.1093/nar/gkx1134

Reddy JA, Nelson M, Dircksen C, Vetzel M, Johnson T, Cross V, Westrick E, Qi L, Hahn S, Santhapuram HK, Parham G, Wang K, Vaughn JF, Felten A, Pugh M, Lu J, Klein P, Vlahov IR, Leamon CP (2020) Pre-clinical studies of EC2629, a highly potent folate- receptor-targeted DNA crosslinking agent. Sci Rep 10(1):12772. https://doi.org/10.1038/s41598-020-69682-9

Rompicharla SVK, Kumari P, Bhatt H, Ghosh B, Biswas S (2019) Biotin functionalized PEGylated poly(amidoamine) dendrimer conjugate for active targeting of paclitaxel in cancer. Int J Pharm 557:329–341. https://doi.org/10.1016/j.ijpharm.2018.12.069

Russell-Jones G, McTavish K, McEwan J (2011) Preliminary studies on the selective accumulation of vitamin-targeted polymers within tumors. J Drug Target 19(2):133–139. https://doi.org/10.3109/10611861003734027

Sacha P, Zamecnik J, Barinka C, Hlouchova K, Vicha A, Mlcochova P, Hilgert I, Eckschlager T, Konvalinka J (2007) Expression of glutamate carboxypeptidase II in human brain. Neuroscience 144(4):1361–1372. https://doi.org/10.1016/j.neuroscience.2006.10.022

Sakamoto T, Seiki M (2017) Integrated functions of membrane-type 1 matrix metalloproteinase in regulating cancer malignancy: beyond a proteinase. Cancer Sci 108(6):1095–1100. https://doi.org/10.1111/cas.13231

Sanhueza CA, Baksh MM, Thuma B, Roy MD, Dutta S, Preville C, Chrunyk BA, Beaumont K, Dullea R, Ammirati M, Liu S, Gebhard D, Finley JE, Salatto CT, King-Ahmad A, Stock I, Atkinson K, Reidich B, Lin W et al (2017) Efficient liver targeting by polyvalent display of a compact ligand for the Asialoglycoprotein receptor. J Am Chem Soc 139(9):3528–3536. https://doi.org/10.1021/jacs.6b12964

Santhosh S, Kumar P, Ramprasad V, Chaudhuri A (2015) Evolution of targeted therapies in cancer: opportunities and challenges in the clinic. Future Oncol 11(2):279–293. https://doi.org/10.2217/fon.14.198

Sato K, Miyamoto M, Takano M, Tsuda H (2020) Correlation of high LAT1 expression with the prognosis of endometrioid carcinoma of the uterine corpus. Virchows Arch 477(3):421–427. https://doi.org/10.1007/s00428-020-02781-0

Satsangi A, Roy SS, Satsangi RK, Vadlamudi RK, Ong JL (2014) Design of a paclitaxel prodrug conjugate for active targeting of an enzyme upregulated in breast cancer cells. Mol Pharm 11(6):1906–1918. https://doi.org/10.1021/mp500128k

Schikora M, Reznikov A, Chaykovskaya L, Sachinska O, Polyakova L, Mokhir A (2015) Activity of aminoferrocene-based prodrugs against prostate cancer. Bioorg Med Chem Lett 25(17):3447–3450. https://doi.org/10.1016/j.bmcl.2015.07.013

Schnyder A, Krahenbuhl S, Drewe J, Huwyler J (2005) Targeting of daunomycin using biotinylated immunoliposomes: pharmacokinetics, tissue distribution and in vitro pharmacological effects. J Drug Target 13(5):325–335. https://doi.org/10.1080/10611860500206674

Seyhan AA (2019) Lost in translation: the valley of death across preclinical and clinical divide–identification of problems and overcoming obstacles. Transl Med Commun 4(1):1–19

Seymour LW, Ferry DR, Anderson D, Hesslewood S, Julyan PJ, Poyner R, Doran J, Young AM, Burtles S, Kerr DJ, Cancer Research Campaign Phase, I. I. I. C. T. c (2002) Hepatic drug targeting: phase I evaluation of polymer-bound doxorubicin. J Clin Oncol 20(6):1668–1676. https://doi.org/10.1200/JCO.2002.20.6.1668

Shan L, Zhuo X, Zhang F, Dai Y, Zhu G, Yung BC, Fan W, Zhai K, Jacobson O, Kiesewetter DO, Ma Y, Gao G, Chen X (2018) A paclitaxel prodrug with bifunctional folate and albumin binding moieties for both passive and active targeted cancer therapy. Theranostics 8(7):2018–2030. https://doi.org/10.7150/thno.24382

Sharma A, Kim EJ, Shi H, Lee JY, Chung BG, Kim JS (2018) Development of a theranostic prodrug for colon cancer therapy by combining ligand-targeted delivery and enzyme-stimulated activation. Biomaterials 155:145–151. https://doi.org/10.1016/j.biomaterials.2017.11.019

Sheng Y, Xu J, You Y, Xu F, Chen Y (2015) Acid-sensitive peptide-conjugated doxorubicin mediates the lysosomal pathway of apoptosis and reverses drug resistance in breast cancer. Mol Pharm 12(7):2217–2228. https://doi.org/10.1021/mp500386y

Shi JF, Wu P, Jiang ZH, Wei XY (2014) Synthesis and tumor cell growth inhibitory activity of biotinylated annonaceous acetogenins. Eur J Med Chem 71:219–228. https://doi.org/10.1016/j.ejmech.2013.11.012

Shi F, Liu Y, Zhou X, Shen P, Xue R, Zhang M (2022) Disitamab vedotin: a novel antibody-drug conjugates for cancer therapy. Drug Deliv 29(1):1335–1344. https://doi.org/10.1080/10717544.2022.2069883

Shi Z, Yang Y, Guo Z, Feng S, Wan Y (2023) A cathepsin B/GSH dual-responsive fluorinated peptide for effective siRNA delivery to cancer cells. Bioorg Chem 135:106485. https://doi.org/10.1016/j.bioorg.2023.106485

Shim MK, Park J, Yoon HY, Lee S, Um W, Kim JH, Kang SW, Seo JW, Hyun SW, Park JH, Byun Y, Kwon IC, Kim K (2019) Carrier-free nanoparticles of cathepsin B-cleavable peptide-conjugated doxorubicin prodrug for cancer targeting therapy. J Control Release 294:376–389. https://doi.org/10.1016/j.jconrel.2018.11.032

Shim MK, Moon Y, Yang S, Kim J, Cho H, Lim S, Yoon HY, Seong JK, Kim K (2020) Cancer-specific drug-drug nanoparticles of pro-apoptotic and cathepsin B-cleavable peptide-conjugated doxorubicin for drug-resistant cancer therapy. Biomaterials 261:120347. https://doi.org/10.1016/j.biomaterials.2020.120347

Shim N, Jeon SI, Yang S, Park JY, Jo M, Kim J, Choi J, Yun WS, Kim J, Lee Y, Shim MK, Kim Y, Kim K (2022) Comparative study of cathepsin B-cleavable linkers for the optimal design of cathepsin B-specific doxorubicin prodrug nanoparticles for targeted cancer therapy. Biomaterials 289:121806. https://doi.org/10.1016/j.biomaterials.2022.121806

Shin WS, Han J, Verwilst P, Kumar R, Kim JH, Kim JS (2016) Cancer targeted enzymatic Theranostic prodrug: precise diagnosis and chemotherapy. Bioconjug Chem 27(5):1419–1426. https://doi.org/10.1021/acs.bioconjchem.6b00184

Silver DA, Pellicer I, Fair WR, Heston WD, Cordon-Cardo C (1997) Prostate-specific membrane antigen expression in normal and malignant human tissues. Clin Cancer Res 3(1):81–85. https://www.ncbi.nlm.nih.gov/pubmed/9815541

Singh VK, Subudhi BB (2016) Development and characterization of lysine-methotrexate conjugate for enhanced brain delivery. Drug Deliv 23(7):2327–2337. https://doi.org/10.3109/10717544.2014.984369

Singh Y, Durga Rao Viswanadham KK, Kumar Jajoriya A, Meher JG, Raval K, Jaiswal S, Dewangan J, Bora HK, Rath SK, Lal J, Mishra DP, Chourasia MK (2017) Click Biotinylation of PLGA template for biotin receptor oriented delivery of doxorubicin hydrochloride in 4T1 cell-induced breast cancer. Mol Pharm 14(8):2749–2765. https://doi.org/10.1021/acs.molpharmaceut.7b00310

Skelly MM, Troy A, Duffy MJ, Mulcahy HE, Duggan C, Connell TG, O'Donoghue DP, Sheahan K (1997) Urokinase-type plasminogen activator in colorectal cancer: relationship with clinicopathological features and patient outcome. Clin Cancer Res 3(10):1837–1840. https://www.ncbi.nlm.nih.gov/pubmed/9815571

Soininen SK, Lehtolainen-Dalkilic P, Karppinen T, Puustinen T, Dragneva G, Kaikkonen MU, Jauhiainen M, Allart B, Selwood DL, Wirth T, Lesch HP, Maatta AM, Monkkonen J, Yla-Herttuala S, Ruponen M (2012) Targeted delivery via avidin fusion protein: intracellular fate of biotinylated doxorubicin derivative and cellular uptake kinetics and biodistribution of biotinylated liposomes. Eur J Pharm Sci 47(5):848–856. https://doi.org/10.1016/j.ejps.2012.09.001

Stanger O (2002) Physiology of folic acid in health and disease. Curr Drug Metab 3(2):211–223. https://doi.org/10.2174/1389200024605163

Stern L, Perry R, Ofek P, Many A, Shabat D, Satchi-Fainaro R (2009) A novel antitumor prodrug platform designed to be cleaved by the endoprotease legumain. Bioconjug Chem 20(3):500–510. https://doi.org/10.1021/bc800448u

Stuben J, Port R, Bertram B, Bollow U, Hull WE, Schaper M, Pohl J, Wiessler M (1996) Pharmacokinetics and whole-body distribution of the new chemotherapeutic agent beta-D-glucosylisophosphoramide mustard and its effects on the incorporation of [methyl-3H]-thymidine in various tissues of the rat. Cancer Chemother Pharmacol 38(4):355–365. https://doi.org/10.1007/s002800050495

Sudhan DR, Siemann DW (2015) Cathepsin L targeting in cancer treatment. Pharmacol Ther 155:105–116. https://doi.org/10.1016/j.pharmthera.2015.08.007

Suganuma N, Segade F, Matsuzu K, Bowden DW (2007) Differential expression of facilitative glucose transporters in normal and tumour kidney tissues. BJU Int 99(5):1143–1149. https://doi.org/10.1111/j.1464-410X.2007.06765.x

Szablewski L (2013) Expression of glucose transporters in cancers. Biochim Biophys Acta 1835(2):164–169. https://doi.org/10.1016/j.bbcan.2012.12.004

Sznarkowska A, Kostecka A, Meller K, Bielawski KP (2017) Inhibition of cancer antioxidant defense by natural compounds. Oncotarget 8(9):15996–16016. https://doi.org/10.18632/oncotarget.13723

Talarico C, D'Antona L, Scumaci D, Barone A, Gigliotti F, Fiumara CV, Dattilo V, Gallo E, Visca P, Ortuso F, Abbruzzese C, Botta L, Schenone S, Cuda G, Alcaro S, Bianco C, Lavia P, Paggi MG, Perrotti N, Amato R (2015) Preclinical model in HCC: the SGK1 kinase inhibitor SI113 blocks tumor progression in vitro and in vivo and synergizes with radiotherapy. Oncotarget 6(35):37511–37525. https://doi.org/10.18632/oncotarget.5527

Tang T, Liu N, Wang L, Zuo K, Zhu X (2024) A dual bispecific hydrolysis peptide-drug conjugate responsive to micro-acidic and reduction circumstance promotes antitumor efficacy in triple-negative breast cancer. Chembiochem 25(18):e202400426. https://doi.org/10.1002/cbic.202400426

Tenora L, Alt J, Dash RP, Gadiano AJ, Novotná K, Veeravalli V, Lam J, Kirkpatrick QR, Lemberg KM, Majer P, Rais R, Slusher BS (2019) Tumor-targeted delivery of 6-Diazo-5-oxo-l-norleucine (DON) using substituted acetylated lysine prodrugs. J Med Chem 62(7):3524–3538. https://doi.org/10.1021/acs.jmedchem.8b02009

Thiry A, Dogne JM, Masereel B, Supuran CT (2006) Targeting tumor-associated carbonic anhydrase IX in cancer therapy. Trends Pharmacol Sci 27(11):566–573. https://doi.org/10.1016/j.tips.2006.09.002

Toader D, Fessler SP, Collins SD, Conlon PR, Bollu R, Catcott KC, Chin CN, Dirksen A, Du B, Duvall JR, Higgins S, Kozytska MV, Bellovoda K, Faircloth C, Lee D, Li F, Qin L, Routhier C, Shaw P et al (2023) Discovery and preclinical characterization of XMT-1660, an optimized B7-H4-targeted antibody-drug conjugate for the treatment of cancer. Mol Cancer Ther 22(9):999–1012. https://doi.org/10.1158/1535-7163.Mct-22-0786

Traverso N, Ricciarelli R, Nitti M, Marengo B, Furfaro AL, Pronzato MA, Marinari UM, Domenicotti C (2013) Role of glutathione in cancer progression and chemoresistance. Oxidative Med Cell Longev 2013:972913. https://doi.org/10.1155/2013/972913

Turk V, Stoka V, Vasiljeva O, Renko M, Sun T, Turk B, Turk D (2012) Cysteine cathepsins: from structure, function and regulation to new frontiers. Biochim Biophys Acta 1824(1):68–88. https://doi.org/10.1016/j.bbapap.2011.10.002

Ueki N, Lee S, Sampson NS, Hayman MJ (2013) Selective cancer targeting with prodrugs activated by histone deacetylases and a tumour-associated protease. Nat Commun 4:2735. https://doi.org/10.1038/ncomms3735

Ueki N, Wang W, Swenson C, McNaughton C, Sampson NS, Hayman MJ (2016) Synthesis and preclinical evaluation of a highly improved anticancer prodrug activated by histone deacetylases and Cathepsin L. Theranostics 6(6):808–816. https://doi.org/10.7150/thno.13826

van Heeckeren WJ, Bhakta S, Ortiz J, Duerk J, Cooney MM, Dowlati A, McCrae K, Remick SC (2006) Promise of new vascular-disrupting agents balanced with cardiac toxicity: is it time for oncologists to get to know their cardiologists? vol 24. American Society of Clinical Oncology, pp 1485–1488

Vergote I, Leamon CP (2015) Vintafolide: a novel targeted therapy for the treatment of folate receptor expressing tumors. Ther Adv Med Oncol 7(4):206–218. https://doi.org/10.1177/1758834015584763

Vidmar R, Vizovišek M, Turk D, Turk B, Fonović M (2017) Protease cleavage site fingerprinting by label-free in-gel degradomics reveals pH-dependent specificity switch of legumain. EMBO J 36(16):2455–2465. https://doi.org/10.15252/embj.201796750

Viricel W, Fournet G, Beaumel S, Perrial E, Papot S, Dumontet C, Joseph B (2019) Monodisperse polysarcosine-based highly-loaded antibody-drug conjugates. Chem Sci 10(14):4048–4053. https://doi.org/10.1039/c9sc00285e

Vlahov IR, Leamon CP (2012) Engineering folate-drug conjugates to target cancer: from chemistry to clinic. Bioconjug Chem 23(7):1357–1369. https://doi.org/10.1021/bc2005522

Wang JH, Kiess AP (2023) PSMA-targeted therapy for non-prostate cancers. Front Oncol 13:1220586. https://doi.org/10.3389/fonc.2023.1220586

Wang Y, Liu Y, Liu Y, Zhou W, Wang H, Wan G, Sun D, Zhang N, Wang Y (2015) A polymeric prodrug of cisplatin based on pullulan for the targeted therapy against hepatocellular carcinoma. Int J Pharm 483(1–2):89–100. https://doi.org/10.1016/j.ijpharm.2015.02.027

Wang L, Xie S, Ma L, Chen Y, Lu W (2016) 10-Boronic acid substituted camptothecin as prodrug of SN-38. Eur J Med Chem 116:84–89. https://doi.org/10.1016/j.ejmech.2016.03.063

Wang H, Chen B, Lin Y, Zhou Y, Li X (2020) Legumain promotes gastric cancer progression through tumor-associated macrophages in vitro and in vivo. Int J Biol Sci 16(1):172–180. https://doi.org/10.7150/ijbs.36467

Wang M, Li Z, Liu F, Yi Q, Pu C, Li Y, Luo T, Liang J, Wang J (2021) Development of Asialoglycoprotein-mediated hepatocyte-targeting antitumor prodrugs triggered by glutathione. J Med Chem 64(19):14793–14808. https://doi.org/10.1021/acs.jmedchem.1c01365

Wang S, Cen D, Zhang C (2024) A Cathepsin B-sensitive gemcitabine prodrug for enhanced pancreatic cancer therapy. J Pharm Sci 113(7):1927–1933. https://doi.org/10.1016/j.xphs.2024.03.018

Watanabe T, Arashida N, Fujii T, Shikida N, Ito K, Shimbo K, Seki T, Iwai Y, Hirama R, Hatada N, Nakayama A, Okuzumi T, Matsuda Y (2024) Exo-cleavable linkers: enhanced stability and therapeutic efficacy in antibody-drug conjugates. J Med Chem 67(20):18124–18138. https://doi.org/10.1021/acs.jmedchem.4c01251

Weitman SD, Lark RH, Coney LR, Fort DW, Frasca V, Zurawski VR Jr, Kamen BA (1992) Distribution of the folate receptor GP38 in normal and malignant cell lines and tissues. Cancer Res 52(12):3396–3401. https://www.ncbi.nlm.nih.gov/pubmed/1596899

Witzigmann D, Quagliata L, Schenk SH, Quintavalle C, Terracciano LM, Huwyler J (2016) Variable asialoglycoprotein receptor 1 expression in liver disease: implications for therapeutic intervention. Hepatol Res 46(7):686–696. https://doi.org/10.1111/hepr.12599

Wozniak M, Pastuch-Gawolek G, Makuch S, Wisniewski J, Krenacs T, Hamar P, Gamian A, Szeja W, Szkudlarek D, Krawczyk M, Agrawal S (2021) In Vitro and in vivo efficacy of a novel glucose-methotrexate conjugate in targeted cancer treatment. Int J Mol Sci 22(4):1748. https://doi.org/10.3390/ijms22041748

Wu G, Fang YZ, Yang S, Lupton JR, Turner ND (2004) Glutathione metabolism and its implications for health. J Nutr 134(3):489–492. https://doi.org/10.1093/jn/134.3.489

Wu W, Luo Y, Sun C, Liu Y, Kuo P, Varga J, Xiang R, Reisfeld R, Janda KD, Edgington TS, Liu C (2006) Targeting cell-impermeable prodrug activation to tumor microenvironment eradicates multiple drug-resistant neoplasms. Cancer Res 66(2):970–980. https://doi.org/10.1158/0008-5472.CAN-05-2591

Wu W, Dong Y, Gao J, Gong M, Zhang X, Kong W, Li Y, Zeng Y, Si D, Wei Z, Ci X, Jiang L, Li W, Li Q, Yi X, Liu C (2015) Aspartate-modified doxorubicin on its N-terminal increases drug accumulation in LAT1-overexpressing tumors. Cancer Sci 106(6):747–756. https://doi.org/10.1111/cas.12672

Wu X, Li X, Li Z, Yu Y, You Q, Zhang X (2018) Discovery of Nonquinone substrates for NAD(P)H: Quinone oxidoreductase 1 (NQO1) as effective intracellular ROS generators for the treatment of drug-resistant non-small-cell lung cancer. J Med Chem 61(24):11280–11297. https://doi.org/10.1021/acs.jmedchem.8b01424

Wu W, Chen M, Luo T, Fan Y, Zhang J, Zhang Y, Zhang Q, Sapin-Minet A, Gaucher C, Xia X (2020a) ROS and GSH-responsive S-nitrosoglutathione functionalized polymeric nanoparticles to overcome multidrug resistance in cancer. Acta Biomater 103:259–271. https://doi.org/10.1016/j.actbio.2019.12.016

Wu L, Zhang F, Chen X, Wan J, Wang Y, Li T, Wang H (2020b) Self-assembled gemcitabine prodrug nanoparticles show enhanced efficacy against patient-derived pancreatic ductal adenocarcinoma. ACS Appl Mater Interfaces 12(3):3327–3340. https://doi.org/10.1021/acsami.9b16209

Wu C, Cheng Z, Lu D, Liu K, Cheng Y, Wang P, Zhou Y, Li M, Shao X, Li H, Su W, Fang L (2021) Novel N-methylated Cyclodepsipeptide prodrugs for targeted cancer therapy. J Med Chem 64(2):991–1000. https://doi.org/10.1021/acs.jmedchem.0c01387

Wu ZY, Shen JM, Lang H, Yue T, Sun C (2022) pH/enzyme dual sensitive and nucleus-targeting dendrimer nanoparticles to enhance the antitumour activity of doxorubicin. Pharm Dev Technol 27(3):357–371. https://doi.org/10.1080/10837450.2022.2055569

Xia M, Guo Z, Liu X, Wang Y, Xiao C (2022) A glutathione-responsive sulfur dioxide polymer prodrug selectively induces ferroptosis in gastric cancer therapy. Biomater Sci 10(15):4184–4192. https://doi.org/10.1039/d2bm00678b

Xue WJ, Feng Y, Wang F, Guo YB, Li P, Wang L, Liu YF, Wang ZW, Yang YM, Mao QS (2016) Asialoglycoprotein receptor-magnetic dual targeting nanoparticles for delivery of RASSF1A to hepatocellular carcinoma. Sci Rep 6:22149. https://doi.org/10.1038/srep22149

Yamansarov EY, Lopatukhina EV, Evteev SA, Skvortsov DA, Lopukhov AV, Kovalev SV, Vaneev AN, Shkil DO, Akasov RA, Lobov AN, Naumenko VA, Pavlova EN, Ryabaya OO, Burenina OY, Ivanenkov YA, Klyachko NL, Erofeev AS, Gorelkin PV, Beloglazkina EK, Majouga AG (2021) Discovery of bivalent GalNAc-conjugated Betulin as a potent ASGPR-directed agent against hepatocellular carcinoma. Bioconjug Chem 32(4):763–781. https://doi.org/10.1021/acs.bioconjchem.1c00042

Yang Y, Aw J, Chen K, Liu F, Padmanabhan P, Hou Y, Cheng Z, Xing B (2011) Enzyme-responsive multifunctional magnetic nanoparticles for tumor intracellular drug delivery and imaging. Chem Asian J 6(6):1381–1389. https://doi.org/10.1002/asia.201000905

Yang J, Zhang R, Pan H, Li Y, Fang Y, Zhang L, Kopeček J (2017) Backbone degradable N-(2-Hydroxypropyl)methacrylamide copolymer conjugates with gemcitabine and paclitaxel: impact of molecular weight on activity toward human ovarian carcinoma xenografts. Mol Pharm 14(5):1384–1394. https://doi.org/10.1021/acs.molpharmaceut.6b01005

Yang J, Lv Q, Wei W, Yang Z, Dong J, Zhang R, Kan Q, He Z, Xu Y (2018) Bioresponsive albumin-conjugated paclitaxel prodrugs for cancer therapy. Drug Deliv 25(1):807–814. https://doi.org/10.1080/10717544.2018.1451935

Ye Z, Zhang Y, Liu Y, Liu Y, Tu J, Shen Y (2021) EGFR targeted Cetuximab-Valine-Citrulline (vc)-Doxorubicin Immunoconjugates- Loaded Bovine Serum Albumin (BSA) nanoparticles for colorectal tumor therapy. Int J Nanomedicine 16:2443–2459. https://doi.org/10.2147/ijn.S289228

Yellepeddi VK, Kumar A, Maher DM, Chauhan SC, Vangara KK, Palakurthi S (2011) Biotinylated PAMAM dendrimers for intracellular delivery of cisplatin to ovarian cancer: role of SMVT. Anticancer Res 31(3):897–906. https://www.ncbi.nlm.nih.gov/pubmed/21498711

Yin W, Ke W, Lu N, Wang Y, Japir A, Mohammed F, Wang Y, Pan Y, Ge Z (2020) Glutathione and reactive oxygen species dual-responsive block copolymer prodrugs for boosting tumor site-specific drug release and enhanced antitumor efficacy. Biomacromolecules 21(2):921–929. https://doi.org/10.1021/acs.biomac.9b01578

Ylikangas H, Malmioja K, Peura L, Gynther M, Nwachukwu EO, Leppanen J, Laine K, Rautio J, Lahtela-Kakkonen M, Huttunen KM, Poso A (2014) Quantitative insight into the design of compounds recognized by the L-type amino acid transporter 1 (LAT1). ChemMedChem 9(12):2699–2707. https://doi.org/10.1002/cmdc.201402281

Younes M, Lechago LV, Lechago J (1996) Overexpression of the human erythrocyte glucose transporter occurs as a late event in human colorectal carcinogenesis and is associated with an increased incidence of lymph node metastases. Clin Cancer Res 2(7):1151–1154. https://www.ncbi.nlm.nih.gov/pubmed/9816281

Zechmann CM, Afshar-Oromieh A, Armor T, Stubbs JB, Mier W, Hadaschik B, Joyal J, Kopka K, Debus J, Babich JW, Haberkorn U (2014) Radiation dosimetry and first therapy results with a (124)I/ (131)I-labeled small molecule (MIP-1095) targeting PSMA for prostate cancer therapy. Eur J Nucl Med Mol Imaging 41(7):1280–1292. https://doi.org/10.1007/s00259-014-2713-y

Zhang X, Li X, You Q, Zhang X (2017) Prodrug strategy for cancer cell-specific targeting: A recent overview. Eur J Med Chem 139:542–563. https://doi.org/10.1016/j.ejmech.2017.08.010

Zhang X, Li X, Li Z, Wu X, Wu Y, You Q, Zhang X (2018) An NAD(P)H:Quinone Oxidoreductase 1 responsive and self-immolative prodrug of 5-Fluorouracil for safe and effective cancer therapy. Org Lett 20(12):3635–3638. https://doi.org/10.1021/acs.orglett.8b01409

Zhang C, Qu Y, Ma X, Li M, Li S, Li Y, Wu L (2020) NQO1-selective activated prodrugs of combretastatin A-4: synthesis and biological evaluation. Bioorg Chem 103:104200. https://doi.org/10.1016/j.bioorg.2020.104200

Zhao X, Sakamoto S, Wei J, Pae S, Saito S, Sazuka T, Imamura Y, Anzai N, Ichikawa T (2023) Contribution of the L-type amino acid transporter family in the diagnosis and treatment of prostate cancer. Int J Mol Sci 24(7):6178. https://doi.org/10.3390/ijms24076178

Zhao J, Li X, Ma T, Chang B, Zhang B, Fang J (2024) Glutathione-triggered prodrugs: design strategies, potential applications, and perspectives. Med Res Rev 44(3):1013–1054. https://doi.org/10.1002/med.22007

Zheng S, Guo S, Zhong Q, Zhang C, Liu J, Yang L, Zhang Q, Wang G (2018) Biocompatible boron-containing prodrugs of Belinostat for the potential treatment of solid tumors. ACS Med Chem Lett 9(2):149–154. https://doi.org/10.1021/acsmedchemlett.7b00504

Zhi X, Jiang Y, Xie L, Li Y, Fang CJ (2019) Gold Nanorods functionalized with Cathepsin B targeting peptide and doxorubicin for combinatorial therapy against multidrug resistance. ACS Appl Bio Mater 2(12):5697–5706. https://doi.org/10.1021/acsabm.9b00755

Zhu S, He J, Yin L, Zhou J, Lian J, Ren Y, Zhang X, Yuan J, Wang G, Li X (2024) Matrix metalloproteinases targeting in prostate cancer. Urol Oncol 42(9):275–287. https://doi.org/10.1016/j.urolonc.2024.05.002

# Chapter 15
# Nanosensors for Biomedical Applications

**Parvindar M. Sah, Harshala S. Naik, and Rajesh W. Raut**

**Abstract** Precise diagnosis of a disease is paramount for effective early treatment and the prevention of further complications. Antigens, enzymes, and proteins present in human bodily fluids as a result of infection can serve as potential hallmarks, or biomarkers, for specific diseases, aiding in diagnosis. Traditional biosensors, while widely used, have limitations, including high sample volume requirements, low sensitivity, and a purely indicative nature, often requiring further confirmation. This process is time-consuming and not cost-effective. Nanotechnology-based solutions address these limitations. The continuous development of novel nanomaterials offers faster detection, requires smaller sample sizes, and provides high precision, making them suitable for detecting a wide range of diseases, including bacterial, viral, and fungal infections. Beyond infectious diseases, these technologies can also detect conditions such as diabetes, cancer, and Alzheimer's disease. While a comprehensive study of all biomedical applications of nanobiosensors is beyond the scope of this chapter, we will focus on their use in the diagnosis of HIV, Alzheimer's disease.

**Keywords** Nano biosensors · Biomedical applications · Alzheimer · HIV

## 15.1 Introduction

Biosensors detect specific biological substances. Nanobiosensors are a type of biosensor that uses nanomaterials, giving them increased sensitivity, faster response times, and the ability to detect multiple targets. Essentially, nano biosensors are biosensors that use nanotechnology (Principles et al. 2023).

P. M. Sah · H. S. Naik · R. W. Raut (✉)
Department of Botany, The Institute of Science, Dr Homi Bhabha State University, Mumbai, Maharashtra, India
e-mail: rajesh@iscm.ac.in



V. Khanna et al. (eds.), *Nanotechnologies for Drug Delivery and Biopharmaceutical Development*, Progress in Drug Research 77,
https://doi.org/10.1007/978-3-032-00796-4_15

Biosensors function based on two key components: the specific interaction between a bioreceptor and its target analyte, and the detection or visualization of this interaction via a signal-generating system, the transducer. A complete biosensor comprises both a bioreceptor and a transducer, which interprets the bioreceptor-analyte reaction. Biosensors utilize the highly specific interaction between a bioreceptor and its target analyte (a disease-specific biomarker) for disease detection. This "lock-and-key" fit, exemplified by the binding of a SARS-CoV-2 Spike protein (analyte) to its corresponding antibody (bioreceptor), ensures accurate detection even in complex biological samples. This principle applies to a wide range of diseases, from cancer (e.g., PSA) and heart disease (e.g., troponin) to infectious diseases (e.g., viral antigens) and autoimmune disorders (e.g., autoantibodies). The resulting bioreceptor-analyte binding event is then detected through various methods, such as optical, electrochemical, or mass-based changes, enabling highly sensitive, specific, and often rapid point-of-care diagnostics (Altug et al. 2022).

Transducers, which convert biological interactions into measurable signals, are central to biosensor classification. Five major categories of biosensors exist, each defined by its transducer type: electrochemical, optical, piezoelectric, thermal, and magnetic. Electrochemical biosensors detect changes in electrical properties like current (amperometric), voltage (potentiometric), or impedance (impedimetric), as exemplified by glucose meters. Optical biosensors measure alterations in light properties such as absorbance, fluorescence, or refractive index, and are used in some pregnancy tests, often employing techniques like Surface Plasmon Resonance (SPR) or fluorescence. Piezoelectric biosensors sense changes in mass by measuring shifts in crystal vibrational frequency, finding applications in some gas sensors. Thermal biosensors, though less common, detect heat changes associated with reactions. Finally, magnetic biosensors, an emerging high-sensitivity technology, measure changes in magnetic fields through the use of magnetic particles (Ahmed et al. 2014).

Beyond classification by transducer type, biosensors can be categorized in several other ways. Based on the biorecognition element, biosensors can be enzyme-based (like glucose sensors), antibody-based (like ELISA), nucleic acid-based (like PCR diagnostics), cell-based (like toxicity sensors), Aptamer-based, or microorganism-based. Classified by application field, they can be medical/diagnostic (like glucose meters), used for environmental monitoring, food safety, industrial monitoring, or security and defense. Regarding signal generation, biosensors can be direct or indirect (requiring a secondary label or amplification, like ELISA). Based on format, they can be label-free (like SPR), label-based (like fluorescence biosensors), disposable (like many point-of-care tests), or reusable. A single biosensor can often fall into multiple categories. A single biosensor can belong to multiple categories. For example, a glucose meter could be categorized as electrochemical, enzyme-based, medical/diagnostic, and disposable.

Nanotechnology is a fascinating field of science that utilizes tiny metal nanoparticles to develop new technologies. Nanoparticles, by definition, range in size from 1 to 100 nanometers. When metals are reduced to this nanoscale, they exhibit different properties compared to their ionic or bulk metallic forms. These altered properties make nanomaterials suitable for a wide range of applications (Nehra and Pal Singh 2015).

Nanotechnology finds major applications in diverse fields, including agriculture, medicine, electronics, waste and hazardous material management, environmental remediation, and food science and packaging. Each of these fields utilizes nanotechnology in different ways, depending on the specific application. For example, nanosensors are employed in agriculture, food packaging, medicine, and environmental monitoring for the detection of pathogens or contaminants. These nanosensors facilitate easier and more efficient detection compared to traditional methods (Joudeh and Linke 2022). Nanosensors use the unique properties of nanomaterials (like a high surface-to-volume ratio) to detect tiny amounts of specific substances. When the target substance interacts with the nanomaterial, it causes a change in the material's properties (electrical, optical, mechanical), which is then converted into a detectable signal. This allows for highly sensitive and specific detection (Principles et al. 2023).

Nanosensors utilize nanomaterials, including various types like carbon dots, quantum dots, and metal ions. Quantum dots are particularly common for detection and imaging. These nanomaterials can be structured in one, two, or three dimensions (Darwish et al. 2024). Nanomaterials play a vital role in improving how well biosensors can detect specific molecules in complex biological samples. This is important because it allows for the identification of things like glucose, proteins, nucleic acids, and pathogens within substances like blood, saliva, and urine. The use of nanomaterials makes these biosensors more sensitive and selective, which is essential for accurate detection (Paramasivam et al. 2021; Zhu et al. 2015). The Fig. 15.1 shows fluorescent quantum dot for detection of $Fe^{2*}$ in living cell.

Enhanced healthcare delivery hinges on optimized health management, which can be realized through prompt decision-making informed by rapid diagnostics, intelligent data analysis, and informatics. This, in turn, requires the creation of sophisticated therapeutics, analytical instruments, and diagnostic platforms, notably point-of-care devices, for individualized patient care. Point-of-care testing facilitates rapid analyte detection at the patient's location, thus improving disease diagnosis, monitoring, and management, leading to more timely medical interventions and, consequently, enhanced patient outcomes. The continuous innovation in point-of-care device development is establishing the foundation for future generations of diagnostic technologies (Misra et al. 2022; Wasilewski et al. 2024).

Nano biosensors represent a significant advancement in point-of-care diagnostics, offering the potential for transformative improvements in disease detection and management. These devices leverage the unique properties of nanomaterials to achieve highly sensitive and specific detection of biomarkers associated with diverse health conditions. This capability allows for the early identification of indicators for diseases ranging from malignancies and metabolic disorders to infectious and autoimmune conditions. For instance, nano biosensors can detect tumor-associated markers in biological fluids, enabling earlier cancer diagnoses and personalized treatment plans. They can also facilitate continuous glucose monitoring for individuals with diabetes, empowering them to better manage their condition and

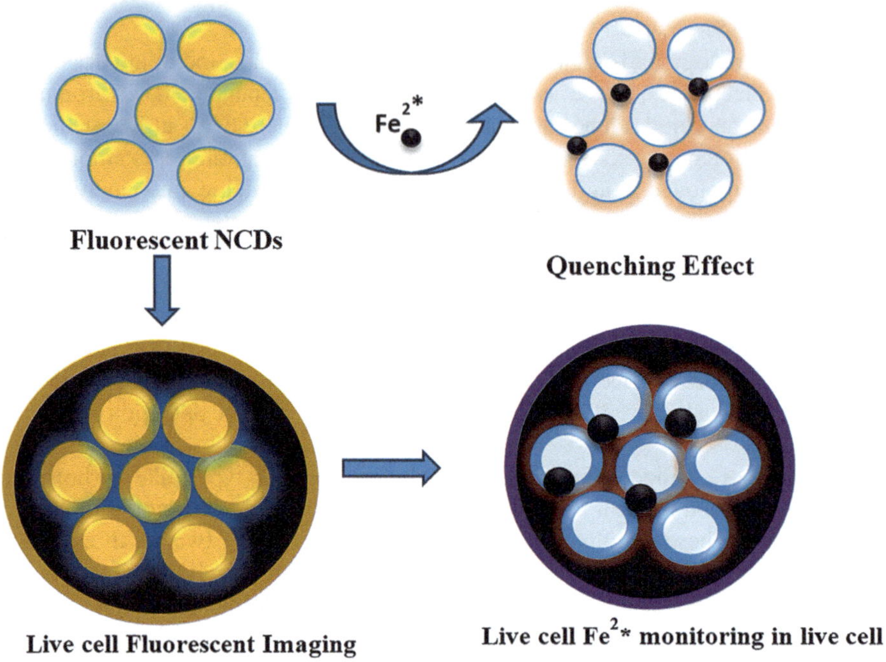


**Fig. 15.1** Fluorescent quantum dot for detection of $Fe^{2*}$ in living cell

minimize the risk of complications. Furthermore, nano biosensors offer rapid pathogen identification in infectious diseases, enabling timely interventions and outbreak prevention. In the context of autoimmune disorders, these devices can detect autoantibodies or other relevant markers, supporting early diagnoses and tailored treatment approaches. Even in conditions like HIV, nano biosensors can accurately quantify viral load, a critical factor in monitoring disease progression and the effectiveness of therapies.

The potential applications of nano biosensors extend beyond these specific examples to encompass a broad spectrum of health concerns, including cardiovascular diseases, neurological disorders, and even environmental monitoring. Their compact size, portability, and potential for low-cost production make them particularly well-suited for point-of-care applications, facilitating rapid and decentralized diagnostics, especially in areas with limited resources. While obstacles related to standardization, scalability, and regulatory pathways remain, nano biosensors hold tremendous promise for revolutionizing healthcare delivery and enhancing patient outcomes across a wide range of diseases.

Imagine a world where diagnostic testing is no longer confined to centralized labs but readily available wherever patients need it most. This vision is being realized through the combined advancements of point-of-care (POC) and lab-on-a-chip (LOC) technologies, though these two approaches have distinct characteristics.

POC devices encompass a wide spectrum of diagnostic tools, from simple, single-use tests like home pregnancy kits to sophisticated, handheld devices capable of complex analyses. They share a common goal: bringing rapid and convenient testing directly to the patient, regardless of location. LOC devices, a specialized subset within the POC category, take this concept a step further. These miniature laboratories on a chip utilize microfluidic technology to integrate and automate multiple laboratory processes—sample preparation, analysis, and detection—onto a single, tiny platform (Tavakoli et al. 2022).

In essence, LOC devices represent a more advanced and highly automated form of POC diagnostics. While simpler POC tests might involve some manual steps and are often designed for self-administration, LOC devices, due to their intricate nature, typically require operation by trained personnel in clinical or point-of-care settings. Despite these differences, both POC and LOC technologies share the overarching objective of democratizing access to diagnostic information, enabling faster, more convenient, and ultimately more effective healthcare delivery (Kumari et al. 2024).

Nanoparticle synthesis is fundamental to the development of advanced nanobiosensors, but it presents a complex interplay of challenges and opportunities. Achieving precise control over nanoparticle size, shape, and composition is crucial, as these parameters directly influence their interactions with target biomolecules and the efficiency of signal transduction. Furthermore, ensuring long-term stability and biocompatibility are essential for both in vitro and in vivo applications. Various synthetic methodologies exist, including chemical reduction, green synthesis, hydrothermal/solvothermal synthesis, and template-assisted synthesis, each with its own advantages and limitations in terms of yield, purity, and control over nanoparticle properties. Careful consideration must be given to factors such as precursor selection, reaction conditions, and the use of stabilizing agents to optimize the synthesis process (Rahman et al. 2024; Zhang et al. 2024).

Despite these challenges, the unique properties of nanoparticles make them invaluable for biosensing. Their high surface-to-volume ratio enhances interactions with target molecules, while their unique optical, electrical, and magnetic properties enable diverse detection strategies. Quantum dots, with their size-tunable fluorescence, offer high sensitivity and multiplexing capabilities. Magnetic nanoparticles can be used for targeted drug delivery and diagnostics, while plasmonic nanoparticles enhance biosensing through surface plasmon resonance. Ongoing research focuses on overcoming the limitations of current synthetic methods and developing new approaches to create nanoparticles with tailored properties for specific biomedical applications, ultimately driving progress in disease diagnostics, personalized medicine, and healthcare (Srinivasan and Rana 2024; Kulkarni et al. 2022a).

This chapter examines Nano biosensors, a powerful class of biosensors that utilize nanomaterials to significantly enhance detection capabilities. It begins by outlining the fundamental principles of biosensing, emphasizing the crucial role of bio receptor-analyte interactions and the transduction of these interactions into measurable signals. A core focus is the diverse range of detection techniques employed in biosensors, including electrochemical, optical, piezoelectric, thermal, and magnetic methods, each offering unique advantages for specific applications. The chapter then explores how nanotechnology, specifically through the incorporation of nanomaterials, revolutionizes biosensor performance, leading to increased sensitivity, faster response times, and multiplexing capabilities. Nanoparticle synthesis, a critical aspect of Nano biosensor development, is discussed in detail, covering various synthetic strategies and the challenges associated with controlling nanoparticle properties. Finally, the chapter connects these advancements to the broader context of point-of-care diagnostics, highlighting the transformative potential of Nano biosensors for rapid, decentralized, and personalized healthcare.

## 15.2 Biosensors

Biosensors are small devices that detect and measure specific biological substances. They combine a biological recognition element (like antibodies) with a transducer (which converts the interaction into a measurable signal) (Pohanka 2018; Radi 2011; Haniff et al. 2021). This allows for highly specific and sensitive detection of targets like DNA or proteins in various samples, including bodily fluids. They are categorized by their transducer type (e.g., electrochemical, optical) and offer potential for rapid diagnostics and continuous monitoring, particularly with non-invasive liquid biopsies (Labuda et al. 2010; Thévenot et al. 2001). Figure 15.2 below shows Schematic diagram of a typical biosensor.

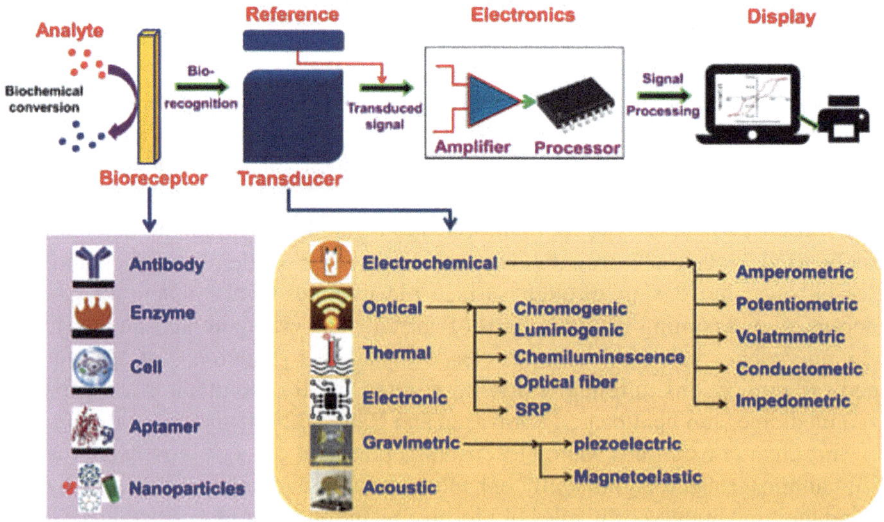


**Fig. 15.2** Schematic diagram of a typical biosensor, including bioreceptor, transducer, and electronic system (amplifier and processor), and display (PC or printer). Various types of bioreceptors and transducers used in biosensors are also shown. (Adapted from Das et al. (2019) licensed under CC BY 4.0 (2021, Sensor))

### 15.2.1 Basic Principles of Biosensor

Biosensors represent a sophisticated class of analytical devices designed to precisely detect and quantify specific analytes by harnessing the interaction between a highly selective biological recognition element and a transducer. This interaction initiates a transduction cascade, transforming the biochemical event into a measurable physical signal (Roosan et al. 2021; Mishra et al. 2024; Asci Erkocyigit et al. 2023).

1. **Biological Recognition Element**: The core of a biosensor's specificity lies in its biological recognition element, which is meticulously chosen for its strong affinity toward the analyte. This element could be an enzyme, antibody, nucleic acid, aptamer, or other biomolecules that exhibit remarkable specificity toward the analyte's molecular structure (Abuzeid et al. 2024; Lucia-Tamudo et al. 2023). These recognition molecules function as a "lock," to which only a specific analyte "key" can bind, thus initiating the detection process.
2. **Immobilization and Interface Engineering**: The immobilization of the recognition element onto the transducer's surface is pivotal, as it ensures proximity and stable orientation between the biomolecule and the detection interface. Various surface chemistries, cross-linkers, or encapsulation techniques are employed to anchor the recognition element securely. This structural integration enhances the biosensor's robustness, prolonging its shelf-life and activity by maintaining the biomolecule's conformation and binding functionality (Song et al. 2021).
3. **Transducer Mechanism**: The transducer is the conversion platform, translating the specific biochemical interaction into an interpretable signal, typically electrical, optical, or mechanical in nature (Bhalla et al. 2016). Transducers must be designed with high sensitivity and selectivity, as well as signal optimization measures to amplify the detection output while minimizing interference:

   (a) **Optical Transducers**: These involve mechanisms like surface plasmon resonance, fluorescence, or luminescence, sensitive to minute changes in optical properties.
   (b) **Electrochemical Transducers**: They capture electrical signals generated by redox reactions or conductivity changes, essential in enzymatic biosensors.
   (c) **Mechanical Transducers**: These use mass or elasticity shifts, such as in piezoelectric biosensors, which are sensitive to small mass changes induced by analyte binding.

4. **Signal Processing and Analysis**: The signal derived from the transducer undergoes complex processing to extract quantitative or qualitative information on the analyte concentration. Advanced digital signal processing, statistical filtering, and spectral analysis methods are applied based on the type and nature of the signal (e.g., amplitude, wavelength, phase shift) (Andreescu and Marty 2006). This phase is crucial, as accurate data interpretation underpins the sensor's reliability in diverse environments.

5. **Applications and Advancements**: With unparalleled sensitivity and specificity, biosensors are increasingly implemented in medical diagnostics, environmental monitoring, food safety, and biotechnology. Notably, innovations in nanotechnology and biomolecular engineering continue to refine biosensor performance, enabling detection at ultra-trace levels and in real-time applications across challenging sample matrices.

## 15.3 Classification of Biosensors

The classification of biosensors is indeed extensive, reflecting the field's complexity and adaptability to diverse applications. Table 15.1 represents the classification of biosensor.

Additionally, the biosensors can be classified based on the following criteria:

1. **Bioreceptor-Analyte Pairs-Based classifications**: Some biosensors are tailored to specific analyte-biological recognition pairs (e.g., glucose-oxidase for glucose detection, horseradish peroxidase for hydrogen peroxide detection).
2. **Detection System-Based Classifications**:
    1. **Optical Detection Systems**: Provide high sensitivity, widely used in immunoassays and gene expression studies.
    2. **Electrical and Electronic Systems**: Common in wearable biosensors due to their compact design and low power requirement.
    3. **Thermal Detection Systems**: Employed where reactions have a significant thermal output; they are selective but less commonly used than optical or electrochemical systems.
    4. **Mechanical Detection Systems**: Include microcantilever sensors that detect molecular binding events through mechanical shifts; applicable in both biomedical and environmental studies.
    5. **Magnetic Detection Systems**: Utilized for applications that require minimized noise and high sensitivity, often paired with magnetic nanoparticles for better specificity.
3. **Technological Innovations-Based**:
    1. **Nanotechnology-Enhanced Biosensors**: Use nanomaterials like carbon nanotubes, quantum dots, and nanoparticles to improve sensitivity and allow multiplexed detection in applications like cancer diagnostics.
    2. **Surface Plasmon Resonance (SPR)**: Allows for real-time, label-free monitoring of binding events, widely used in drug discovery and kinetic studies.
    3. **Lab-on-a-Chip**: Miniaturized systems that integrate multiple functions (sampling, detection, data processing) on a single chip, crucial for point-of-care diagnostics.
    4. **Deployable and Wearable Biosensors**: Designed for continuous, real-time monitoring, often incorporated into wearable devices for health monitoring.

**Table 15.1** Classification of biosensor

| Classification of biosensors | | |
|---|---|---|
| **Bio receptor-based** | **Enzymatic biosensors** | Utilizing enzyme-substrate specificity, enzymatic biosensors are often applied in glucose, lactate, and cholesterol monitoring. |
| | **Immunosensors** | Leveraging antigen-antibody interactions, immunosensors provide high precision, making them suitable for pathogen, toxin, and cancer biomarker detection. |
| | **Aptameric-based and nucleic acid biosensors** | Aptamers are crucial tools in disease detection, particularly in medicine. These small molecules, composed of DNA, RNA, or peptides, exhibit high binding affinity and exceptional sensitivity for their target molecules. |
| | **Microbial/whole-cell biosensors** | Whole-cell biosensors use living cells (bacteria, yeast, or mammalian) to detect specific molecules or environmental changes by measuring cellular responses like metabolic activity, proliferation, or gene expression. |
| **Transducer-based** | **Electrochemical biosensors** | (a) Potentiometric: Measure potential difference between electrodes, proportional to ion concentration. Simple, real-time, but prone to interference. Used in pH meters and ion-selective measurements.<br>(b) Amperometric: Measure current from redox reactions, proportional to analyte concentration. Ideal for biological redox reactions (e.g., glucose, lactate).<br>(c) Conductometric: Measure conductivity changes due to biochemical reactions. Simple, label-free, but less sensitive and susceptible to interference. Used in various applications.<br>(d) Impedance-based: Measure changes in electrical impedance, highly sensitive to small amounts of analytes (e.g., pathogens, cells). Complex and potentially expensive. |
| | **Electronic biosensors** | Electronic biosensors, like Filed Effect Transistor (FETs), detect biological interactions using electronic signals, offering high sensitivity and fast response for targets like DNA and proteins. |
| | **Thermal biosensors** | Another category, thermal biosensors, measures temperature changes resulting from reactions. These biosensors are particularly effective when the reaction involves a significant heat release or absorption |
| | **Optical biosensors** | (a) SPR: Measures refractive index changes for real-time, label-free biomolecular interaction analysis.<br>(b) Fluorescence: Uses fluorescent tags for analyte detection, common in DNA/RNA analysis.<br>(c) Colorimetric/luminescent: Provides visible signals for rapid diagnostics.<br>(d) Mass-based: Detects mass variations (e.g., QCM) for molecular binding and pollutant detection. |

## 15.4 Biosensors Based on Bioreceptors

Biosensors can be broadly classified based on the biorecognition mechanism into two main types: **catalytic biosensors** and **affinity (non-catalytic) biosensors**.

(a) **Catalytic Biosensors**: In these biosensors, the interaction between the analyte and the bioreceptor initiates a biochemical reaction that produces a new compound. The reaction product serves as the measurable output of the biosensor. Catalytic biosensors commonly utilize biorecognition elements like enzymes, microorganisms, tissues, or whole cells, as these components can actively catalyze reactions with the analyte (Nguyen et al. 2019).

(b) **Affinity (Non-Catalytic) Biosensors**: Unlike catalytic biosensors, affinity biosensors rely on a binding event between the analyte and the bioreceptor that does not generate a new product. Instead, this binding is typically irreversible, forming a stable complex. This class of biosensors includes components like antibodies, cellular receptors, and nucleic acids that specifically bind to target analytes. They are especially useful for detecting specific molecules, such as proteins or DNA sequences, with high specificity and selectivity (Spychalska et al. 2020).

### *15.4.1 Enzyme-Based Biosensors*

Enzyme-based biosensors detect specific substances by leveraging the interaction between an enzyme and its target molecule (substrate). For example, urea biosensors use the enzyme urease to break down urea into ammonia and carbon dioxide. The resulting ammonia (detected electrochemically) or pH change (measured with a pH electrode) is proportional to the original urea concentration, enabling accurate measurement in samples like blood or urine (Micha 2017).

Glucose biosensors typically use glucose oxidase or dehydrogenase to oxidize glucose. This reaction consumes oxygen or produces hydrogen peroxide, which are then detected electrochemically (hydrogen peroxide) or with an oxygen electrode. The resulting signal correlates directly with glucose concentration for accurate measurement in samples like blood or urine (Azam et al. 2023; Cordeiro et al. 2018).

For instance, Cordeiro et al. developed a W-Au-based amperometric glucose biosensor for real-time brain glucose monitoring, while Uygun et al. designed a nanoparticle-enhanced potentiometric urea biosensor with a detection limit of 0.77 μM ((Kevin) Lai et al. 2017). The integration of nanomaterials with enzymes has further improved biosensor performance by enhancing stability, electron transfer, and overall efficiency, making enzyme-based biosensors vital tools in diagnostics and environmental monitoring (Kladnik 2012).

### 15.4.2 Immunosensors

Antibodies are highly specific biorecognition elements widely used in biosensors due to their strong antigen-antibody interactions and broad applicability. Structurally, antibodies are immunoglobulins (Ig) with a characteristic "Y" shape, comprising two heavy and two light polypeptide chains linked by disulfide bonds. They are classified into five types—IgG, IgM, IgA, IgD, and IgE—based on variations in heavy chain (Su et al. 2023). Biosensors utilizing antibodies as ligands or based on antigen-antibody interactions are termed immunosensors. These are categorized into label-free and labelled types. Label-free immunosensors detect physical changes caused by antigen-antibody complex formation, while labelled immunosensors incorporate detectable tags to enhance sensitivity (Police Patil et al. 2023). For instance, Madurro et al. developed a label-free immunosensors for ovarian cancer detection, achieving a detection limit of 1.45 U/mL with a linear range of 5–80 U/mL (Janik-Karpinska et al. 2022). Similarly, Marquette et al. created electrical and optical biosensors for label-free detection of Aflatoxin B1 (AFB1) using gold nano bipyramids (Zhang et al. 2022).

### 15.4.3 Aptamer-Based and Nucleic Acid Biosensors

Aptamers are synthetic single-stranded nucleic acids (DNA or RNA) known for their high selectivity and ability to bind target molecules. Aptamer-based biosensor detection leverages the unique properties of aptamers. Aptamers and antibodies both play roles in molecular recognition, but their characteristics and production methods differ significantly. Antibodies, Y-shaped proteins produced by the immune system, target specific antigens. However, their production relies on animal subjects and is limited to substances capable of triggering an immune response, restricting the range of target molecules (Das et al. 2019).

Aptamers, developed in vitro, offer a compelling alternative to traditional antibodies, eliminating the need for animals and targeting a wider range of molecules, including non-immunogenic substances. The controlled SELEX process ensures consistency and reliability. Furthermore, aptamers bind diverse targets, from ions and small molecules to complex proteins and cells (Tombelli et al. 2005; Stoltenburg et al. 2007). This versatility makes aptamers applicable to a wide range of diseases, including inflammatory conditions, metabolic disorders, neurodegenerative diseases, cardiovascular events, cancers, and infectious diseases. They can detect biomarkers such as IL-6 (inflammation), HbA1c (diabetes), dopamine (Parkinson's), and cancer-related molecules, promising early and accurate diagnosis (Stoltenburg et al. 2007).

Aptamer-based diagnostics are rapidly expanding, targeting diverse diseases. In metabolic disorders, they detect biomarkers for diabetes (HbA1c, GHSA) and obesity (vaspin). They also show promise in neurodegenerative diseases (Alzheimer's, Parkinson's, Huntington's) by targeting disease-specific proteins and neurotransmitters like dopamine. Aptamers are also being developed for detecting heart attack biomarkers, various cancer markers (Nucleolin, PSA, HER2), and infectious diseases (bacterial like E. coli, Salmonella, and viral like Influenza, RSV, SARS-CoV-2, HIV), highlighting their potential to revolutionize disease diagnostics (Tuerk and Gold 1990).

Aptamer-based biosensors use diverse detection methods, often enhanced by nanomaterials like quantum dots (QDs) and gold nanoparticles (GNPs). QDs, ideal for real-time tracking, are linked with aptamers to detect targets like cancer cells, proteins, and bacterial spores. For example, aptamer-capped QDs can rapidly detect thrombin with high sensitivity. GNPs, biocompatible and with adjustable optical properties, are commonly used in electrochemical impedance spectroscopy (EIS) biosensors, offering high sensitivity for detecting various molecules (Chu et al. 2006).

Aptamer-based platforms offer impressive detection capabilities. For example, an aptamer-quantum dot system for thrombin detection achieved a 1 nM limit, demonstrating high sensitivity for protein detection. Similarly, an electrochemical impedance spectroscopy (EIS) biosensor using gold nanoparticles and aptamers for L-arginine reached a broad linear range (0.1 pM to 0.1 mM) and an extremely low detection limit (0.01 pM). These examples highlight the versatility of combining aptamers with nanomaterials, enabling quantification of target molecules across many orders of magnitude and detecting even trace amounts (Ayed et al. 2021).

### *15.4.4 Whole-Cell-Based Biosensors*

Whole-cell biosensors utilize living cells (microorganisms or mammalian) to detect specific target substances. When the target interacts with the cell, it triggers a biological response, like a change in metabolism or gene expression, which is then converted into a measurable signal by a transducer. These biosensors offer high sensitivity, specificity, real-time monitoring, and cost-effectiveness, making them valuable for environmental monitoring, food safety, biomedical diagnostics, and drug discovery. However, challenges remain regarding cell stability, reproducibility, and the complexity of engineering cells with desired sensing capabilities (Kylilis et al. 2019).

Reporter genes are essential components of whole-cell biosensors, acting as signal generators. They encode proteins that produce a detectable signal, such as light, fluorescence, or a colored product. These genes are introduced into cells under the control of a target-sensitive promoter. When the target substance is present, it activates the promoter, turning on the reporter gene and leading to the production of the reporter protein. The amount of this protein is proportional to the target's concentration. Examples include the *lux* operon for bioluminescence, GFP for fluorescence, and beta-galactosidase for enzymatic activity. Reporter genes offer signal amplification, high specificity due to promoter selection, and versatility for detecting diverse targets (Lei 2015).

One example of their utility is the development of a label-free optical whole-cell biosensor using *E. coli* for the detection of parathyroid insecticides. This particular system, as demonstrated by Polizzi et al., achieved a detection limit of 3 ng/mL for 3-PBA within a linear range of 0.01–2 ng/mL. This example illustrates the potential of whole-cell biosensors as valuable tools for identifying various analytes, including contaminants, across multiple disciplines (Berepiki et al. 2020; Lim et al. 2022).

## 15.5 Transducers Based Biosensors

### *15.5.1 Electrochemical Biosensors*

A significant category of biosensors, electrochemical biosensors, operates by exploiting the electrochemical behavior of both the target analyte and the transducer. These biosensors are favored due to their high sensitivity, selectivity, and versatility in detecting a wide range of substances. The fundamental mechanism involves the binding of the analyte to a bioreceptor located on the transducer. This binding event initiates an electrochemical reaction, resulting in a measurable signal. This signal, which can manifest as changes in voltage, current, impedance, or capacitance, is then used to determine the analyte's concentration. These biosensors can be further classified according to the specific electrochemical transduction method employed (Wei and Ivaska 2006).

Electrochemical biosensors can be categorized based on the specific mechanisms by which the bio recognition event is transduced into a measurable signal. These categories include:

#### 15.5.1.1 Potentiometric Biosensors

- Measure the potential difference between a working electrode and a reference electrode at zero current.
- The signal is generated by the interaction of the analyte with the bioreceptor, causing charge accumulation.
- Devices like ion-selective electrodes and ion-sensitive field-effect transistors are commonly employed to convert biochemical reactions into potential signals (Grieshaber et al. 2008).

#### 15.5.1.2 Amperometric Biosensors

- Operate in two or three-electrode configurations, detecting the current produced from electrochemical oxidation or reduction of electroactive species at the working electrode.
- A constant potential is applied, and the resulting current is directly proportional to the analyte concentration in the solution.
- These biosensors offer advantages such as high sensitivity, rapid response, precision, and linear output, making them suitable for large-scale production.
- However, they can suffer from poor selectivity and interference by other electroactive substances (Chaubey and Malhotra 2002).

#### 15.5.1.3 Conductometric Biosensors

- Measure changes in electrical conductance between electrode pairs caused by electrochemical reactions, reflecting alterations in the analyte's conductivity.
- Commonly used for monitoring metabolic processes in biological systems (Pisoschi 2016).

#### 15.5.1.4 Impedimetric Biosensors

- Measure electrical impedance at the electrode/electrolyte interface using a small sinusoidal excitation voltage (Herzog 2014).
- Electrochemical Impedance Spectroscopy (EIS) measures these changes, which are then correlated to analyte concentration. These label-free biosensors offer high sensitivity, real-time monitoring, and versatility, making them attractive for biomedical applications like disease diagnostics (cancer, infectious diseases, cardiovascular), drug discovery, and personalized medicine. While challenges remain regarding specificity, sensitivity in complex media, and miniaturization, ongoing research is addressing these through improved bio recognition elements, nanomaterials, and microfluidics, positioning impedimetric biosensors as a powerful tool in future biomedical diagnostics (Herzog 2014).

#### 15.5.1.5 Voltammetric Biosensors

- Detect analytes by monitoring current variations during controlled changes in the applied potential.
- Advantages include high sensitivity and the ability to simultaneously detect multiple analytes (Naresh and Lee 2021).

### 15.5.2 Optical Biosensors

Optical biosensors are sophisticated analytical devices that integrate a biorecognition element with an optical transducer system to detect analytes with high sensitivity and specificity. They operate by generating signals proportional to the analyte concentration, enabling real-time and often label-free detection. These biosensors utilize various biorecognition elements, including enzymes, antibodies, aptamers, whole cells, and tissues (Abbasi et al. 2024). The transduction mechanism triggers changes in light properties such as absorption, reflection, transmission, refraction, phase, amplitude, frequency, or polarization, which are influenced by the interaction between the biorecognition element and the analyte. Based on their working principle, optical biosensors are categorized into label-free and label-based systems. Label-free biosensors generate signals directly from the interaction of the analyte with the transducer, while label-based biosensors rely on optical markers like colorimetric, fluorescence, or luminescent signals (Sun et al. 2023). These biosensors are designed using various optical principles, such as surface plasmon resonance (SPR), evanescent wave fluorescence, optical waveguide interferometry, chemiluminescence, fluorescence, refractive index modulation, and surface-enhanced Raman scattering (SERS). Among these, the most commonly used types include fluorescence-based, chemiluminescence-based, SPR-based, and optical fiber-based biosensors (Touhami 2017).

#### 15.5.2.1 Fluorescence-Based Optical Biosensors

Fluorescence-based optical biosensors employ the principle of fluorescence, where target substances are tagged with fluorescent labels to enable their identification. These biosensors are highly regarded for their remarkable sensitivity, high specificity, and fast response times, making them well-suited for various applications such as medical diagnostics, environmental monitoring, and food safety analysis. A range of fluorescent markers, including quantum dots (QDs), organic dyes, and fluorescent proteins, are essential components in these sensors (Kozitsina et al. 2018).

Fluorescence-based biosensors typically function through three main approaches: fluorescence quenching (signal reduction), fluorescence enhancement (signal increase), and fluorescence resonance energy transfer (FRET). FRET-based biosensors have proven particularly valuable for investigating cellular processes due to their high sensitivity. In FRET, energy is transferred non-radiatively from an excited donor molecule to an acceptor molecule in its ground state via long-range multipole interactions. This mechanism enables FRET-based biosensors to detect even minute molecular changes with angstrom-to-nanometer precision (Zhang et al. 2018a).

FRET-based optical biosensors, with their unique properties, find widespread use in clinical settings, including cancer diagnosis and aptamer analysis. For instance, a research team led by Liu developed a FRET-based sensor that utilizes a combination of carbon dots and gold nanorods to detect lead ions. This sensor demonstrated a linear detection range from 0 to 155 μM and an impressive detection limit of 0.05 μM (Wei et al. 2021).

#### 15.5.2.2 Chemiluminescence-Based Optical Biosensors

The generation of light through chemical reactions, known as chemiluminescence, has become a focus of research in biosensing. Its appeal stems from its ease of use, high sensitivity, ability to detect low concentrations of target substances, and relatively inexpensive equipment requirements. Recent progress has significantly expanded the potential applications of this technique (McCaughey et al. 2024).

Chemiluminescence-based biosensors have seen significant advancements, enabling the detection of target molecules with exceptional precision. For example, He and colleagues developed a biosensor that utilizes graphene oxide for DNA detection. This sensor demonstrated a linear detection range spanning from 0.1 to 3 nM, with an impressively low detection limit of 34 pM (Chiticaru et al. 2019).

The integration of advanced materials, such as MXenes, is also gaining momentum in chemiluminescence-based sensing. These two-dimensional nanomaterials possess exceptional optical and electrochemical characteristics, making them ideal for this application. MXenes, used in forms like nanosheets and quantum dots, contribute to improved sensitivity and stability in biosensors. Furthermore, electrochemiluminescence (ECL), a specialized form of chemiluminescence, has become a crucial technique in health diagnostics. Modern ECL biosensors, incorporating innovative reagents and bio-interfaces, are enabling more effective detection of biomarkers, including cancer indicators and viral antigens (Kailasa et al. 2021).

#### 15.5.2.3 SPR-Based Biosensors

Surface plasmon resonance (SPR) biosensors are a label-free technology used to detect molecular interactions. They function by measuring shifts in the refractive index near a metal surface. This is achieved by exploiting the surface plasmon resonance phenomenon. When polarized light strikes a metal surface at the interface between two media with different refractive indices, it generates collective oscillations of electrons, known as surface plasmons. At a specific angle of incidence, called the resonance angle, the reflected light intensity dramatically decreases. This angle is highly sensitive to changes in the refractive index at the metal surface, which are directly related to the mass of molecules interacting with the sensor. Therefore, by precisely measuring variations in this resonance angle, SPR biosensors can detect and quantify target analytes based on how they affect the refractive index at the biorecognition element attached to the transducer (Castillo-Henríquez et al. 2020).

Surface plasmon resonance (SPR) biosensors offer a label-free approach to detecting molecular interactions. They operate by measuring alterations in the refractive index in close proximity to a metal surface. This measurement is based on the principle of surface plasmon resonance. Specifically, when polarized light interacts with a metal surface separating two media with distinct refractive indices, it generates collective electron oscillations known as surface plasmons. A characteristic angle of incidence, termed the resonance angle, results in a significant reduction in the intensity of reflected light. Critically, this angle is highly sensitive to even minute changes in the refractive index at the metal surface, which are directly proportional to the mass of interacting molecules. Consequently, by accurately tracking shifts in this resonance angle, SPR biosensors can identify and quantify target analytes by observing their impact on the refractive index at the biorecognition element, which is coupled to the transducer (Damborský et al. 2016).

SPR technology continues to evolve, leading to more sophisticated and sensitive biosensors. Recent studies have demonstrated the potential of SPR in various applications. For instance, researchers have used SPR with gold nanosatellites to create a DNA biosensor capable of detecting specific markers in donkey meat with high sensitivity and precision. Another group developed a silver-based LSPR biosensor for detecting *E. coli* endotoxin, achieving an exceptionally low detection limit (Mariani and Minunni 2014).

Furthermore, SPR technology is being enhanced by integrating it with other advanced techniques. Combining SPR with artificial intelligence and machine learning allows for real-time data analysis, which significantly improves detection accuracy and speeds up decision-making. Hybrid systems that merge SPR with electrochemical sensors have also shown remarkable results in detecting cancer biomarkers and viruses. These advancements highlight the significant impact of SPR and LSPR technologies on the future of biosensing (Mariani and Minunni 2014).

#### 15.5.2.4 Optical Fiber-Based Optical Biosensors

Optical fiber biosensors utilize the properties of light within optical fibers to detect and quantify biological targets, including cells, proteins, and aptamers. These biosensors are increasingly recognized as a compelling alternative to traditional biomolecule detection methods. They provide enhanced sensitivity, a compact form factor, and the capability for real-time monitoring of biological processes. A key sensing mechanism employed in many of these devices is evanescent field sensing, a technique commonly implemented using tapered optical fibers (Valenzuela-Amaro et al. 2023).

The propagation of light within an optical fiber, under conditions of total internal reflection, results in the generation of an evanescent wave at the interface. This evanescent field, which diminishes exponentially with increasing distance from the fiber surface, can be utilized to induce fluorescence in molecules positioned in close proximity to the fiber's sensing area (Lyu et al. 2022).

Tapered optical fibers are frequently combined with various optical transduction methods, such as refractive index changes, absorption, fluorescence, and surface plasmon resonance (SPR), making them adaptable for use in biosensing. As an

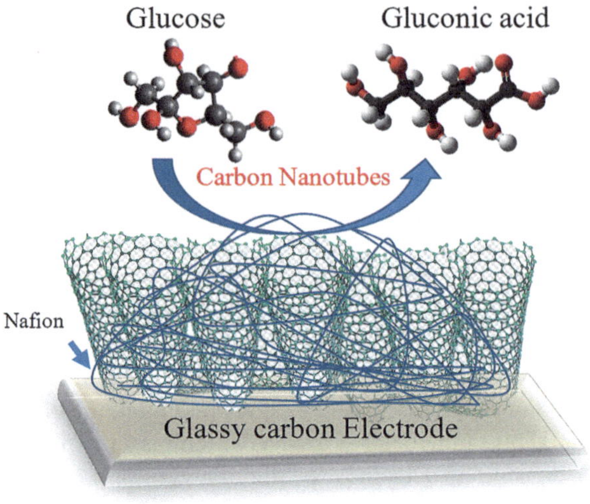


**Fig. 15.3** Carbon nanotube based biosensor for detection of Glucose oxidase

example, researchers have developed a fiber optic sensor using a complex of poly (N-isopropyl acrylamide)-co-acrylamide P (NIPAAm-co-AAm) magnetic immobilized glucose oxidase (GOD) to detect both cholesterol and glucose. This system proved highly effective, detecting cholesterol at 38 °C within a concentration range of 25–250 mg/dL, and glucose at 25 °C within a range of 50–700 mg/dL (Lin et al. 2019; Norris and Hosticka 2002). The carbon nanotube based biosensor for detection of Glucose oxidase show in Fig. 15.3.

Optical fiber biosensors are constantly being improved with the incorporation of new and innovative materials. The use of materials like graphene and carbon quantum dots has led to significant gains in both sensitivity and stability for these biosensors. Furthermore, the combination of optical fibers with plasmonic nanomaterials has enabled the detection of extremely low concentrations of biomolecules, including important targets like cancer markers and viral proteins. Another exciting development is the integration of machine learning algorithms into "smart" optical fiber systems. This allows for real-time data analysis and opens up new possibilities in areas such as personalized healthcare, environmental monitoring, and food safety. These ongoing advancements demonstrate the dynamic nature of optical fiber biosensing and its potential for future applications (Baruah et al. 2024).

### *15.5.3 Gravimetric Biosensors*

Gravimetric biosensors are mass-sensitive devices that generate measurable signals in response to changes in the mass of bound biomolecules, such as proteins or antibodies, on their surface. These sensors employ thin piezoelectric quartz crystals, which vibrate at a specific frequency influenced by the mass of the detected material and the applied current. Common gravimetric transducers include piezoelectric biosensors, quartz crystal microbalance (QCM) sensors, and magnetoelastic sensors (MES) (Cali et al. 2020).

Piezoelectric biosensors utilize crystals capable of elastic deformation when subjected to an alternating electric field. This deformation produces waves within the crystal. When analytes bind or detach from the crystal's biorecognition-coated surface, the resulting change in resonant frequency indicates the occurrence of molecular interactions. Yang et al. developed a piezoelectric biosensor with a lead titanate zirconate (PZT) ceramic resonator for label-free cancer biomarker detection, achieving a sensitivity of 0.25 ng/mL within a 30-minute detection window for small sample volumes (Inacio et al. 2008).

QCM sensors, another gravimetric technology, function on the piezoelectric principle using a thin quartz crystal sandwiched between electrodes. Any mass change on the crystal's surface alters its resonance frequency. Lee et al. constructed a QCM biosensor for miR-21 detection, demonstrating exceptional sensitivity with a detection limit of 0.87 pM and a broad linear range from 0.1 pM to 10 μM. This biosensor also proved effective in analyzing serum samples, making it a promising tool for clinical diagnostics (Skládal 2024).

Magnetoelastic sensors (MES) are gaining attention due to their wireless, passive nature and their ability to measure mechanical parameters such as force, pressure, and strain. MES operates on the principle of magnetostriction, where mechanical deformation arises from an applied magnetic field. When time-varying magnetic fields generate magnetoelastic vibrations, they produce elastic waves that generate detectable magnetic flux. MES are particularly valuable for biomedical applications owing to their cost-effectiveness, longevity, and compact design. Atalay et al. used an MES to detect $Fe_3O_4$ magnetic nanoparticles, achieving a detection sensitivity as low as $1.1 \times 10^9$ particles, equivalent to 0.025 mg or 1 μL of magnetic nanoparticles (Li et al. 2016).

Recent advancements in gravimetric biosensors include integration with nanomaterials like graphene and gold nanoparticles to improve sensitivity and specificity. Additionally, hybrid systems combining QCM or piezoelectric sensors with electrochemical detection mechanisms are being explored for applications in cancer diagnostics and infectious disease monitoring, paving the way for next-generation biosensing platforms (Yoo et al. 2021).

### *15.5.4 Thermal Biosensors*

Thermal biosensors operate by exploiting the exothermic or endothermic characteristics of biochemical reactions, measuring the heat energy absorbed or released during such processes. The heat change (ΔT) recorded by these sensors is directly related to the enthalpy change (ΔH) and the quantity of product molecules (np) formed, and inversely related to the system's heat capacity (Cp), expressed as

$\Delta T = -(np\ \Delta H)/Cp$ (Lakshmipriya and Gopinath 2018). Calorimetric-based biosensors detect these heat changes to monitor reaction dynamics or biomolecular structural changes in solution. However, traditional calorimetric techniques face challenges like long experimental durations and low specificity in temperature measurements, making it difficult to distinguish between specific and non-specific heat changes (Sivarajasekar et al. 2019). The advent of enzyme thermistors, which integrate flow injection analysis with immobilized biocatalysts and heat-sensing components, has largely addressed these limitations. Thermal biosensors typically use a flow injection analysis system with immobilized enzymes in a reactor. A differential temperature measurement across the enzyme reactor is achieved using thermistors or thermopiles. This setup minimizes ambient temperature interference, enhancing specificity. The thermal signals generated during enzymatic catalysis are proportional to the substrate concentration, making it possible to quantify metabolites at clinically relevant levels with enthalpy changes between −10 and −200 kJ/mol (Kulkarni et al. 2022b).

**Commonly Used Transducers**

1. **Thermistors:** These measure absolute temperature through changes in electrical resistance with temperature but have limited sensitivity.
2. **Thermopiles:** Comprising a series of thermocouple junctions, these detect temperature differences between regions with improved sensitivity.

Emerging technologies, such as **microelectromechanical systems (MEMS)**, are now integrated into thermal biosensors. MEMS-based sensors offer advantages like miniaturization, high sensitivity, low thermal mass, and low power consumption, enabling rapid and parallel measurements of small sample volumes. MEMS technology is particularly beneficial in metabolic monitoring, providing cost-effective and scalable solutions with enhanced thermal isolation and linear response ranges (Wang 2022).

## 15.6 Recent Innovations

- **Bimetallic strips**, liquid gas expansion systems, and **pyroelectric systems** that convert temperature changes into electrical signals are being increasingly explored.
- High-sensitivity thermocouples and MEMS thermal sensors have been applied in clinical diagnostics, allowing precise detection of biomolecular interactions and enzymatic reactions.
- Advances in MEMS fabrication enable the development of compact biosensors capable of monitoring multiple samples simultaneously, paving the way for applications in personalized medicine and environmental analysis.

Optical fiber biosensors are constantly being improved with the incorporation of new and innovative materials. The use of materials like graphene and carbon quantum dots has led to significant gains in both sensitivity and stability for these biosensors. Furthermore, the combination of optical fibers with plasmonic nanomaterials has enabled the detection of extremely low concentrations of biomolecules, including important targets like cancer markers and viral proteins. Another exciting development is the integration of machine learning algorithms into "smart" optical fiber systems. This allows for real-time data analysis and opens up new possibilities in areas such as personalized healthcare, environmental monitoring, and food safety. These ongoing advancements demonstrate the dynamic nature of optical fiber biosensing and its potential for future applications (Lim and Lee 2020).

### 15.6.1 Electronic Biosensors

Silicon nanowire (SiNW) field-effect transistor (FET) biosensors utilize the fundamental principles of FET operation for biomolecular detection. FETs, three-terminal devices comprising a source, drain, and gate, control current flow through a semiconductor channel by applying an electric field at the gate. This field modulates the channel's impedance, thus regulating the current between the source and drain. The growing interest in FET-based biosensors stems from their ability to directly transduce biomolecular interactions at the sensor surface into detectable electrical signals. Specifically, when a target biomolecule (analyte) binds to its corresponding bioreceptor, which is immobilized on the SiNW surface, it perturbs the local surface potential. This potential shift affects the conductivity of the SiNW channel, resulting in a measurable change in the source-drain current. Critically, the high input impedance characteristic of these semiconductor devices enables the detection of even subtle electrical changes induced by analyte-bioreceptor binding, thereby contributing to the high sensitivity of these biosensors (Nath 2024; Molaie 2016).

FET-based biosensors offer advantages over other sensing methods, including high sensitivity, spatial resolution, and the ability to provide real-time analysis. However, they face challenges, especially in in vitro applications, such as stability and selectivity issues. Among the common FET sensing platforms, ion-sensitive field-effect transistors (ISFETs) and metal-oxide-semiconductor field-effect transistors (MOSFETs) differ based on design considerations, such as gate voltage application methods, material choice, and the channel region's configuration (Molaie 2016). For example, Zhu et al. developed a ZnO Nanorod-based FET glucose sensor that employs AC frequency mixing for detection, as opposed to the traditional DC-based three-electrode measurement method (Estrela et al. 2005). This sensor achieved a high sensitivity of 1.6 mA/ (μm·cm$^2$) and demonstrated a glucose detection limit of 1.0 μM, alongside impressive long-term stability (38 hours), showcasing its potential for practical use in biosensing applications (Amen et al. 2022).

### *15.6.2 Acoustic Biosensors*

Acoustic biosensors function by detecting changes in the physical properties of an acoustic wave, which are directly related to the quantity of analyte absorbed onto the sensor surface (Durmu et al. 2013). Piezoelectric materials are often employed in these sensors due to their ability to generate and transmit acoustic waves in a frequency-dependent manner. The resonant frequency for acoustic wave propagation is influenced by the physical characteristics and dimensions of the piezoelectric crystal. Any variation in the mass of the material deposited on the crystal's surface causes a measurable shift in the crystal's natural resonant frequency (Castillo-Henríquez et al. 2020). Acoustic transducers are categorized into two types based on their mass balance: Bulk-Acoustic Wave (BAW) devices and Surface-Acoustic Wave (SAW) devices. BAW devices transmit acoustic waves across the crystal, from one surface to another, while SAW devices propagate waves along a single crystal surface, moving from one point to another. While these devices are well understood in gas-phase applications, their behavior in liquid environments is less thoroughly explored. Over the years, the piezoelectric mechanism has been well-established and presents a promising alternative for transducing signals in affinity biosensors (Gruhl et al. 2009). However, challenges such as non-specific binding and suboptimal sensitivity need to be addressed before they can be fully utilized (Gruhl et al. 2009).

## 15.7 Nanomaterials Fabrication for Enhanced Biosensing

To address the growing demands across various sectors, innovative approaches to sensor technology have emerged, significantly driven by advancements in nanotechnology and nanoscience. Nanotechnology has found applications across a wide range of industries, including physics, chemistry, biotechnology, bioscience, medical science, healthcare, food engineering, aerospace, electronics, energy, and environmental studies. The ability to manipulate and control materials at the atomic and molecular scale (nanometer range), along with a deeper understanding of the processes occurring at the nanoscale, has opened new pathways for the development of advanced biosensors.

A key factor influencing the properties of nanomaterials is their dimensionality, which dictates their physical, chemical, biological, electrical, and optical characteristics. Nanomaterials are categorized based on their dimensional size: zero-dimensional (0D), one-dimensional (1D), two-dimensional (2D), and three-dimensional (3D) materials. Materials that exhibit nanoscale dimensions in all three spatial directions are classified as 0D nanomaterials (such as nanoparticles (NPs) and quantum dots (QDs)). Materials with two nanoscale dimensions and one much larger dimension are considered 1D nanomaterials (such as nanowires (NWs), nanorods (NRs), nanotubes (NTs), and nanobelts). In contrast, 2D nanomaterials are those with just one nanoscale dimension, such as nanoprisms, nanoplates, nanosheets, and carbon nanotubes (CNTs). Bulk materials that do not conform to the nanoscale in any dimension (>100 nm) are classified as 3D nanomaterials, which include nanoballs, dendritic structures, nanocoils, nanopillars, and nanoflowers (Gómez-Gómez et al. 2023).

Nanomaterials exhibit unique properties due to their nanoscale dimensions, leading to enhanced physical, chemical, and biological behaviors. This characteristic is fundamental to the advancements in nanotechnology. Two primary strategies exist for nanomaterial synthesis: top-down and bottom-up. The top-down approach involves the reduction of bulk materials into nanoscale structures using techniques like lithography, laser ablation, ion milling, and chemical etching. Conversely, the bottom-up approach builds nanomaterials by assembling individual atoms or molecules into desired nanostructures. Methods employed in this bottom-up fabrication include molecular beam epitaxy, vapor deposition, and chemical synthesis routes for creating supramolecular assemblies, self-assembled monolayers, and protein-polymer nanocomposites (Nadtoka et al. 2023).

### *15.7.1 Development and Types of Nanobiosensors*

Nanotechnology has revolutionized biosensors by enabling the use of diverse nanomaterials (NMs) like nanoparticles, nanowires, and carbon nanotubes, leading to enhanced sensitivity. Nano biosensors, utilizing nanoscale components for signal processing, offer advantages over traditional biosensors, including adaptability to various applications like monitoring confined environments, detecting subcellular markers, and identifying trace hazardous substances (Abdel-Karim et al. 2020).

Nanomaterials significantly improve biosensing. Nanoparticles amplify signals, nanotubes enhance specificity, nanowires enable efficient charge transport, and quantum dots boost optical responses, all contributing to more capable biosensing technology (Abdel-Karim et al. 2020). Table 15.2 represent various type of nanomaterial used in biomedical application.

### *15.7.2 Nanoparticle-Based Biosensors*

Nanoparticles (NPs) are integral to a variety of biomedical applications, particularly in the development of biosensors for health diagnostics, imaging, drug delivery, and therapy, owing to their unique properties. Their small size and shape allow them to exhibit distinctive physical and chemical characteristics that are highly sensitive to the binding of target biomolecules (Parvin et al. 2024). These characteristics make NPs ideal for bioanalytical applications, where they are often utilized to enhance the sensitivity and specificity of electrochemical catalysis in biosensors. Catalytically

**Table 15.2** Various type of nanomaterial used in biomedical application

| Types of nanomaterial | Types | Application |
|---|---|---|
| 1. Nanoparticle-based biosensor | Gold | Used for the detection of various diseases (Baranwal et al. 2022). |
| | Silver | Multifunctional colorimetric probe using biosynthesized Ag NPs for the detection of glucose (Arregui et al. 2014). |
| | Platinum | A non-enzymatic glucose sensor combined with Pt NPs and zinc oxide nanostructures, is used for the detection of glucose in whole blood and blood serum (Imran et al. 2022). |
| | Palladium | Pd-decorated graphene oxide (rGO) ball nanocomposite for detection (Mohammadi et al. 2020). |
| | Copper | Cu/rGO-decorated material used for glucose detection (Contributor B). |
| | Ohers | Noble metal NPs, Rhodium, Ruthenium, Iridium, Osmium used for detection (Baranwal et al. 2022) |
| Metal oxide nanoparticle | Nickel oxide (NiO) | Used in FET-based glucose biosensors for high sensitivity and wide linear range (Zhang et al. 2023), and in ECL biosensors for glucose detection (Du et al. 2022). |
| | Cobalt oxide ($Co_3O_4$) | Used in screen-printed glutamate biosensors for high sensitivity (Bhargava et al. 2018), and in potentiometric urea biosensors for good sensitivity over a wide range (Kulkarni et al. 2022b). |
| | Iron oxide ($Fe_2O_3$) and Manganese oxide ($MnO_2$) | $Fe_2O_3$ used in magnetic biosensors for anticancer drug detection (Koh and Josephson 2009), $MnO_2$ used with graphene nanoribbons in a disposable biosensor for high sensitivity and low detection limit (Poltavets et al. 2023). |
| | Zinc oxide (ZnO) | Used in carbon monoxide sensors showing good response (Hjiri et al. 2020). |
| | Molybdenum oxide ($MoO_3$) | Used for detecting $H_2O_2$ in chemical reactions (Rashmi et al. 2020). |
| | Titanium oxide ($TiO_2$) | Used in optical immunosensors for detecting *Salmonella typhimurium* (Zhang et al. 2018b). |

(continued)

**Table 15.2** (continued)

| Types of nanomaterial | Types | Application |
|---|---|---|
| Other | Carbon nanotubes (CNTs) | CNT-based biosensors are used for detecting a range of analytes, including glucose, fructose, galactose, neurotransmitters, amino acids, cancer biomarkers, immunoglobulins, insulin, DNA, and various cells and microorganisms (Singh 2011; Luo et al. 2018; Janssen et al. 2019; Tang et al. 2006; Cho et al. 2020; Dudina et al. 2019). |
| | Quantum dots (QDs) | Quantum dots are used in optical biosensors for detecting organic compounds, pharmaceutical analytes, and biomolecules like nucleic acids, proteins, enzymes, amino acids, carbohydrates, and neurotransmitters, and also in electrochemical sensors for detecting copper ions (Banigo et al. 2020; Ma et al. 2018; Shtepliuk et al. 2020; Kalkal et al. 2020; Amesh and Govindaraju 2022; Cui et al. 2019). |
| | Nanowires (NWs) | Nanowires are used for highly sensitive, selective, and reliable electrical sensing of biological binding events, and also for plasmonic biosensing and direct genome detection (Arora et al. 2013; Kim et al. 2019). |
| | Nanorods | Nanorods are used for detecting nucleic acids and basic biochemical markers like glucose and hydrogen peroxide, and also for lead ion detection and continuous glucose monitoring (Chavali and Nikolova 2019; Ahmad et al. 2022) |

active NMs like transition metal oxides are developed as Nano enzymes, facilitating the catalysis of biochemical reactions on biosensors.

Metal and noble metal NPs, including gold (Au), silver (Ag), platinum (Pt), palladium (Pd), cobalt (Co), iron (Fe), and copper (Cu), as well as metal oxide NPs (ZnO, $TiO_2$, $SnO_2$, $MnO_2$), exhibit remarkable optical, electronic, magnetic, chemical, mechanical, and catalytic properties. To improve the performance of NPs in biosensing applications, they are often coated with various materials, such as metal oxides, silica, polymers, graphene, fibers, or dendrimers (Baranwal et al. 2022).

Gold NPs, in particular, are widely studied due to their unique physicochemical properties, including ease of synthesis, chemical stability, biocompatibility, and broad electrochemical potential range. These advantages make them highly sought after in biosensor design. Wu et al. demonstrated the use of gold NP-based electrochemical sensors for sensitive detection of uranyl ions in natural water, achieving a detection range of 2.4–480 μg $L^{-1}$ and a detection limit of 0.3 μg $L^{-1}$ using anodic stripping voltammetry (Nouman et al. 2023). In another study, Luo et al. developed a "turn-on" fluorescent sensor using gold NPs and graphene quantum dots (GQDs) for detecting $Pb^{2+}$ ions, with a detection range of 50 nM to 4 μM and a detection limit of 16.7 nM (Kadhim et al. 2023).

Silver NPs (Ag NPs) have attracted attention for their strong surface-enhanced Raman scattering (SERS) effect, biocompatibility, high conductivity, and catalytic activity. Rivero et al. fabricated an optical fiber sensor using both localized surface plasmon resonance (LSPR) and lossy-mode resonance (LMR) based on Ag NPs. The sensor demonstrated high sensitivity (0.943 nm per RH%) and a large dynamic range (42.4 nm for relative humidity changes between 25% and 70%), making it suitable for monitoring human breathing (Arregui et al. 2014). Mehdinia et al. developed a multifunctional colorimetric probe using biosynthesized Ag NPs for detecting $Fe^{2+}$, $H_2O_2$, and glucose. The sensor achieved detection limits of 0.54 μM for $Fe^{2+}$, 0.032 μM for $H_2O_2$, and 0.29 μM for glucose (Arregui et al. 2014).

Platinum NPs (Pt NPs) have gained interest in electrochemical biosensors due to their superior electrocatalytic activity. Liu et al. created a Pt NP-based ammonia sensor with a sensing response of 1467 at 1000 ppm $NH_3$/air at 250 °C, showing rapid response times. Dharuman et al. developed a non-enzymatic glucose sensor using Pt NPs and zinc oxide nanostructures, achieving a wide linear detection range (250 μM to 110 mM) and demonstrated its reusability in whole blood and blood serum (Imran et al. 2022).

Palladium NPs (Pd NPs) are particularly appealing for biomedical applications because of their high catalytic and sensing activities. Pd NPs are more abundant and cost-effective than gold and platinum, making them ideal for sensor applications. Ye et al. demonstrated Pd/Co-NCNTs for hydrazine oxidation with a high sensitivity of 343.909 μA $mM^{-1}$ and a low detection limit of 0.007 μM (Lee 2023). Swihart et al. developed a 3D Pd-decorated reduced graphene oxide (rGO) ball nanocomposite for hydrogen detection at room temperature, achieving a sensitivity of 14.8% and fast response and recovery times (73 s and 126 s, respectively) for $H_2$ concentrations ranging from 0.0025% to 2% (Mohammadi et al. 2020).

Copper NPs (Cu NPs) are gaining prominence in electrochemical sensors due to their excellent conductivity, stability, electrocatalytic properties, and cost-effectiveness. Huang et al. fabricated a Cu NP-based electrochemical glucose sensor on a flexible graphite sheet, exhibiting high sensitivity (7254.1 μA $mM^{-1}$ $cm^{-2}$) and a low detection limit of 1.05 μM for glucose, with excellent anti-interference properties against common substances like sodium chloride and dopamine (Contributor B). Roushani et al. developed a sensor using Cu NPs incorporated into a multi-walled carbon nanotube (MWCNT)/ionic liquid (IL)/chitosan (Chit)/rutin nanocomposite for detecting $H_2O_2$, with a linear detection range from 0.35 to 2500 μM and a detection limit of 0.11 μM. Zhao et al. designed a Cu/rGO-decorated buckypaper electrode for glucose detection, achieving a linear range of $0.1^{-2}$ mM and a detection limit of 11 μM.

#### 15.7.2.1 Metal Oxide-Based Biosensors

Over the past decade, metal oxide-based nanomaterials have garnered significant attention for their versatile applications across electrochemistry, magnetism, catalysis, and sensing, primarily due to their exceptional electrical, chemical, and physical properties. These metal oxide nanoparticles (NPs) have proven to be efficient electrocatalysts for biosensing a wide array of biological analytes, owing to their low cost, strong electrocatalytic activity, and impressive organic capture capabilities. Commonly used metal oxide NPs include copper oxide (CuO), nickel oxide (NiO), iron oxide ($Fe_2O_3$), cobalt oxide ($Co_3O_4$), manganese oxide ($MnO_2$), zinc oxide (ZnO), titanium oxide ($TiO_2$), tin oxide ($SnO_2$), cadmium oxide (CdO), molybdenum oxide ($MoO_3$), and cerium oxide ($CeO_2$) (Shi et al. 2014).

Nickel oxide nanoparticles (NiO NPs) stand out for their excellent electrical, magnetic, optical, thermal, catalytic, and mechanical properties. These NPs have found diverse uses as catalysts, thermistors, sensors, and additives, particularly in gas sensing and ceramics. As a p-type semiconductor with a direct bandgap of 3.56 eV, NiO exhibits both superparamagnetic and super antiferromagnetic properties depending on particle size and oxidation state (Hashem et al. 2016). For improved glucose monitoring, Duan et al. (Zhang et al. 2023) developed a novel FET-based biosensor employing bimetallic Ni/Cu metal-organic frameworks. Their device offers high sensitivity, with a 0.51 μM detection limit, and a wide linear range (1–20 mM). Kamyabi et al. also contributed an electroluminescence (ECL) biosensor, immobilizing glucose oxidase within a NiO NP-decorated nickel foam structure (Du et al. 2022).

Cobalt oxide nanoparticles ($Co_3O_4$ NPs) are also of great interest due to their remarkable physical, magnetic, optical, and chemical properties. These NPs have applications in catalysis, solar energy absorption, gas sensing, lithium-ion batteries, supercapacitors, and various other fields. $Co_3O_4$ is a p-type semiconductor with an optical bandgap between 1.66 and 2.19 eV (Bhargava et al. 2018). Chu et al. fabricated a screen-printed glutamate biosensor using porous $Co_3O_4$ nanocubic crystals, demonstrating a high sensitivity of 20.12 μA $mM^{-1}$ $cm^{-2}$ and a linear range from 10 to 600 μM. Wazir et al. developed a potentiometric urea biosensor on a glass filter paper, immobilizing urease enzyme on a cobalt oxide-chitosan nanocomposite. The sensor exhibited good sensitivity over a wide concentration range from $1 \times 10^{-4}$ to $8 \times 10^{-2}$ M (Kulkarni et al. 2022b). Iron oxide ($Fe_2O_3$) and manganese oxide ($MnO_2$)-based NPs are well-known for their magnetic properties and bioanalytical applications, offering higher electron transfer rates ideal for electrochemical biosensing (Koh and Josephson 2009). Phan et al. demonstrated the use of soft ferromagnetic amorphous ribbons with nano-hole-patterned surfaces to develop highly sensitive magnetic biosensors for anticancer drug detection, utilizing superparamagnetic NPs. Zhang et al. developed a label-free NMR biosensor for direct detection of Salmonella with a detection limit of $2.6 \times 10^4$ CFU $mL^{-1}$ (Li et al. 2020). Stankovic et al. created a disposable biosensor based on graphene nanoribbons supported by $MnO_2$ NPs, achieving a sensitivity of 56.32 μA $mmol^{-1}$ $cm^{-2}$ and a detection limit of 0.05 mmol $L^{-1}$ (Poltavets et al. 2023). Zinc oxide (ZnO), titanium oxide ($TiO_2$), and tin oxide ($SnO_2$) NPs have also seen significant research attention. ZnO NPs are valued for their high conductivity, biocompatibility, and resistance to oxidation, with ZnO being an n-type semiconductor possessing a wide bandgap of 3.37 eV (Liang et al. 2016). Hjiri et al. developed a carbon monoxide sensor using ZnO NPs synthesized via the sol-gel method, which showed a response of 74% to

80 ppm of CO at 250 °C (Hjiri et al. 2020). $TiO_2$ NPs, known for their broad range of applications in medical, biomedical, and pharmaceutical fields, were utilized by Tereshchenko et al. in an optical immunosensors for detecting Salmonella typhimurium with sensitivity ranging from $10^3$ to $10^5$ cL $mL^{-1}$ (Zhang et al. 2018b). Molybdenum oxide ($MoO_3$) NPs, noted for their rapid response capabilities, were recently employed by Ravi Kumar et al. for detecting $H_2O_2$ in chemical reactions (Rashmi et al. 2020).

#### 15.7.2.2 Carbon Nanotube-Based Biosensors

Carbon nanotubes (CNTs) are an exciting class of one-dimensional nanomaterials (1D NMs) that have garnered extensive attention in various fields, including biosensing, diagnostics, tissue engineering, drug delivery, cell tracking, and biomolecule detection. These structures are hollow cylindrical tubes made of one, two, or several concentric layers of graphite, capped by fullerene hemispheres, classified as single-walled, double-walled, or multi-walled CNTs (SWCNTs, DWCNTs, and MWCNTs). CNTs are prized for their unique structural properties, such as high electrical and mechanical strength, excellent thermal conductivity, chemical stability, superior electrocatalytic activity, minimal surface fouling, low over-voltage, and a high aspect ratio (surface-to-volume ratio) (Singh 2011).

The high surface-to-volume ratio of CNTs, coupled with their exceptional electron transport properties, means their electrical conductance can be significantly altered by even minor surface changes, such as the binding of macromolecules. These characteristics make CNT-based biosensors highly sensitive for detecting a range of analytes across healthcare, industrial, environmental, and food quality applications. They have been widely used for electrochemical sensing, particularly in glucose monitoring, as well as for detecting other substances such as fructose, galactose, neurotransmitters, amino acids, cancer biomarkers, immunoglobulins, insulin, DNA, and various cells and microorganisms.

Among the different CNT types, multi-walled carbon nanotubes (MWCNTs) are the most commonly used in biosensing applications. These 1D nanomaterials enable real-time, highly sensitive, label-free bioelectronics detection and offer extensive redundancy in nanosensor arrays (Singh 2011). For instance, Cui et al. developed a wearable amperometric biosensor incorporated into gloves, creating an innovative platform for lactate detection (Luo et al. 2018). A carbon nanotube (CNT)-based biosensor for bovine serum albumin (BSA) detection was developed by Janssen et al. The sensor exhibited a detection limit of 2.89 ng/mL, highlighting its high sensitivity for protein detection (Janssen et al. 2019).

Several research efforts have focused on developing advanced DNA and neurotransmitter biosensors using carbon nanotubes (CNTs) and other innovative approaches (Tang et al. 2006). For example, (Cho et al. 2020) constructed a DNA sensor using single-walled carbon nanotubes (SWNTs) and elucidated the sensing mechanism. Their work provided valuable insights into the interaction between SWNTs and DNA, which is crucial for advancements in biotechnology and

DNA-based manipulation of nanotubes. In another study, (Cho et al. 2020) developed a DNA sensor employing metallic floating electrodes with adjustable responses, demonstrating that increasing the number of electrodes leads to improved sensor performance. Furthermore, (Dudina et al. 2019) created a CNT-based biosensor system-on-a-chip for neurotransmitter detection. This system integrated CNT sensors with complementary metal-oxide-semiconductor (CMOS) chips, showcasing its potential for sophisticated biomedical applications, including lab-on-a-chip (LoC) systems designed for neuronal culture.

#### 15.7.2.3 Quantum Dot-Based Biosensors

Quantum dots (QDs) are zero-dimensional (0D) nanomaterials composed of inorganic nanocrystals (NCs). They possess unique optical properties, such as broad excitation spectra, narrow and size-tunable emission spectra, high photochemical stability, and resistance to photo bleaching (Banigo et al. 2020). These unique properties make them an excellent alternative to traditional fluorophores, particularly in the development of optical biosensors for detecting a wide variety of organic compounds, pharmaceutical analytes, and biomolecules such as nucleic acids, proteins, enzymes, amino acids, carbohydrates, and neurotransmitters. Additionally, QDs have gained significant attention in in vivo cancer detection due to their ultra-high sensitivity, specificity, miniaturized size, cost-effectiveness, and ability to detect multiple analytes simultaneously (Ma et al. 2018).

Cui et al. developed a straightforward electrochemical sensor for detecting copper ions ($Cu^{2+}$) using graphene quantum dots (GQDs) combined with graphene. This hybrid material significantly improved sensor performance, achieving a low detection limit of 1.34 nM across a broad linear range of 0.015–8.775 μM (Shtepliuk et al. 2020). Packirisamy et al. created fluorescent turn-on biosensors for ultrasensitive detection of small cell lung cancer biomarkers. The biosensors used biofunctionalized GQDs as the energy donor and gold nanoparticles (AuNPs) as the energy acceptor, resulting in a rapid response time of 16 minutes and a broad linear detection range of 0.1 pg $mL^{-1}$ to 1000 ng $mL^{-1}$, with an impressive detection limit of 0.09 pg $mL^{-1}$ (Kalkal et al. 2020). Xiao et al. demonstrated the use of CdTe/CdS/ZnS core/shell/shell quantum dots for the fluorescent detection of L-ascorbic acid. Their biosensor exhibited a linear detection range of $8.0 \times 10^{-9}$–$1.0 \times 10^{-7}$ M and a detection limit of $1.8 \times 10^{-9}$ M (Amesh and Govindaraju 2022). Sun et al. introduced a "turn-on" magnetic fluorescent biosensor combining GQDs, $Fe_3O_4$ nanoparticles, and molybdenum disulfide ($MoS_2$) nanosheets to detect EpCAM, with a detection range from 2 to 64 nM and a detection limit of 1.19 nM. This sensor is particularly effective for the rapid, efficient, and sensitive detection of circulating tumor cells (CTCs) (Cui et al. 2019).

#### 15.7.2.4 Gold Nanoparticle-Based Biosensors

Gold nanoparticle-based biosensors (AuNPs) are characterized by their inertness, biocompatibility, non-toxicity, ease of synthesis, and the ability to modify their surface functionality and adjust their size and shape (Momeni et al. 2022). These features have sparked significant interest in utilizing AuNPs to interface biorecognition systems with signal transduction mechanisms, making them ideal for the development of functional biosensing devices. AuNPs are widely applied in biosensors that produce signals proportional to the analyte concentration, offering a range of applications. Notable uses include viral diagnostics (such as during the COVID-19 pandemic), clinical diagnostics, monitoring of clinical processes, bioreactors, and the production and release of drugs. Key benefits of biosensors include their rapid response time and ease of use. AuNPs-based biosensors are categorized into optical, electrochemical, and piezoelectric types (Purohit et al. 2020). The following section provides an overview of these different GNP-based biosensors, with an emphasis on the impact of AuNPs on the biosensing process. The Fig. 15.4 shows role of gold nanoparticle in detection of COVID.

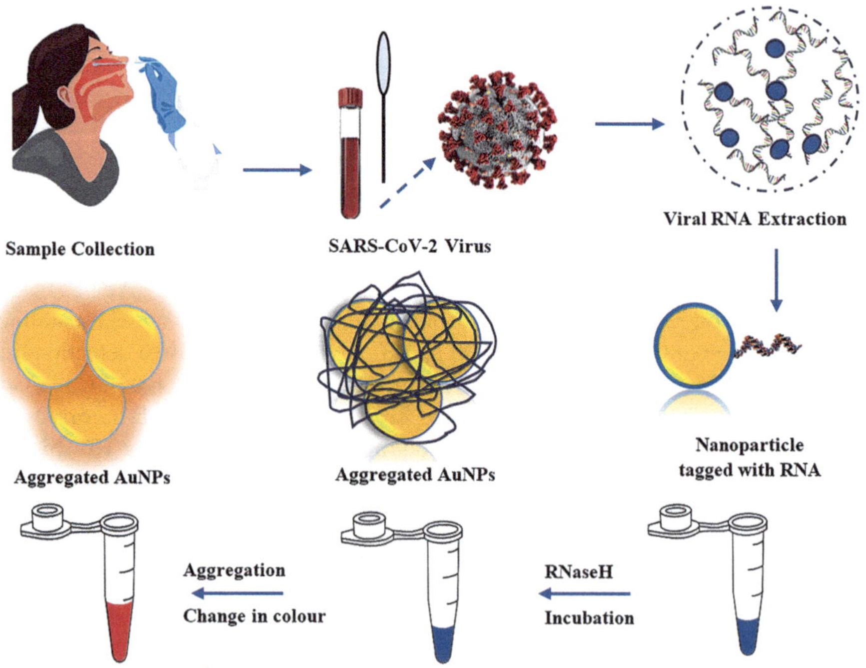


**Fig. 15.4** Selective naked-eye detection of positive COVID-19 cases

#### 15.7.2.5 Nanorod–Based Biosensor

Nanorods are frequently employed as electrochemical modifiers due to their ability to facilitate highly specific processes. These Nanorods are typically composed of materials such as gold, graphene, manganese, zinc, iron oxide, or combinations of these substances. Their primary applications include detecting nucleic acids and basic biochemical markers like glucose and hydrogen peroxide. For example, Liu et al. developed a novel CDs/Au Nanorod (NR) assembly-based Förster Resonance Energy Transfer (FRET) sensor for lead ion detection, achieving a linear detection range from 0 to 155 μM with a limit of detection of 0.05 μM. Zhu et al. demonstrated a ZnO Nanorod-based Field-Effect Transistor (FET) biosensor for continuous glucose monitoring using alternating current (AC) frequency mixing (Chavali and Nikolova 2019). Several recent biosensor developments highlight advancements in sensitivity, stability, and target analyte range. One sensor demonstrated high sensitivity (1.6 mA), a 1 μM detection limit, and remarkable stability during 38 hours of continuous monitoring. In another study, Sun et al. employed gold nanorods (GNRs) and graphene oxide (GO) to improve the sensitivity of a wavelength modulation surface plasmon resonance (SPR) biosensor for bovine IgG detection. Their GNR/GO-based sensor detected bovine IgG concentrations between 0.075 and 40.0 μg/mL. Furthermore, Hahn et al. developed a vertically aligned ZnO nanorod field-effect transistor (FET) biosensor for phosphate detection, achieving a sensitivity of 80.57 $\mu A/mM/cm^2$ and a broad linear range from 0.1 to 7.0 mM (Ahmad et al. 2022).

#### 15.7.2.6 Nanowire-Based Biosensors

Nanowires (NWs) offer ideal conditions for developing highly sensitive, selective, and reliable electrical sensors for detecting biological binding events. Their optical and electrical properties are exceptionally reproducible, making them valuable for precise measurements. In one-dimensional systems like NWs and nanotubes (NTs), the flow of current is highly sensitive to minor disturbances, as the flow occurs very close to the surface. The diameters of NWs are well-suited for sensing biological and chemical species, offering excellent transduction properties that convert small signals into readable outputs. The tunable conductivity of semiconducting NWs, combined with their ability to capture analytes on their surface, enables direct, label-free electrical measurements. These NW-based sensors typically operate on the principle of ion-selective Field-Effect Transistors (FETs), where the interaction of external charges with semiconductor carriers enhances sensitivity, particularly at low ionic strengths (Arora et al. 2013). For example, Park et al. developed a fiber optic sensor using ZnO NWs and AuNPs, achieving highly sensitive plasmonic biosensing. Priolo et al. demonstrated a PCR-free, label-free optical biosensor based on silicon NWs for direct genome detection, with a detection limit of 2 copies per reaction for synthetic genomes and 20 copies per reaction for genomes extracted from human blood. Additionally, Nuzaihan et al. created a silicon NW-based biosensor with molecular gate control to electrically detect Dengue virus (DENV) DNA, achieving a remarkable detection limit of 2.0 fM with a sensitivity of 45.0 μA/M (Kim et al. 2019).

## 15.8 Types of Biosensors Used for HIV and Alzheimer's Diagnosis

### *15.8.1 Biosensors for HIV Detection*

Human immunodeficiency virus (HIV), the causative agent of acquired immunodeficiency syndrome (AIDS), poses a significant global health challenge. Conventional HIV testing kits, while effective in detecting antibodies against HIV, fail to identify infections during the early stages. To address this limitation, nanomaterial-based biosensors have emerged as transformative tools for early HIV detection (Farzin et al. 2020). These biosensors utilize advanced nanomaterials, including metal nanoparticles (e.g., gold nanoparticles), quantum dots, graphene, and metal nanoclusters, to enhance signal transduction and enable ultra-sensitive detection of key biomarkers such as $CD^{4+}$ T cells, HIV genes, p24 core protein, p17 peptide, and HIV-1 virus-like particles (VLPs). Leveraging mechanisms like electrochemistry, fluorescence, surface plasmon resonance (SPR), surface-enhanced Raman spectroscopy (SERS), and colorimetry, these biosensors provide rapid, accurate, and specific results. However, translating these advancements into point-of-care (POC) technologies remains a priority. The focus is on developing affordable, robust, and portable devices capable of delivering accurate diagnostics in diverse settings, thereby advancing personalized medicine and improving HIV management (León-Reyes et al. 2017). The Fig. 15.5 shows surface plasmon based assay for detection of HIV antigen.

The application of nanotechnology in biosensing has significantly advanced HIV diagnostics, offering solutions for both direct and indirect detection methods. Direct detection involves identifying intact HIV particles or virus-like particles (VLPs), which mimic the structure of authentic viruses without carrying infectious genetic material. Virus-like particles (VLPs) are attractive for biosensor development due to their structural resemblance to viruses but without infectivity. Lee et al. developed an electrochemical sensor using gold nanoparticles on indium tin oxide (ITO) glass, achieving a 600 fg/mL detection limit for HIV-1 VLPs. An LSPR-based biosensor with antibody-modified gold dots achieved a 200 fg/mL detection limit, enabling quantitative analysis via UV-Vis spectral shifts. Indirect detection methods target HIV-related proteins, nucleic acids, or host antibodies (Zhang et al. 2020).

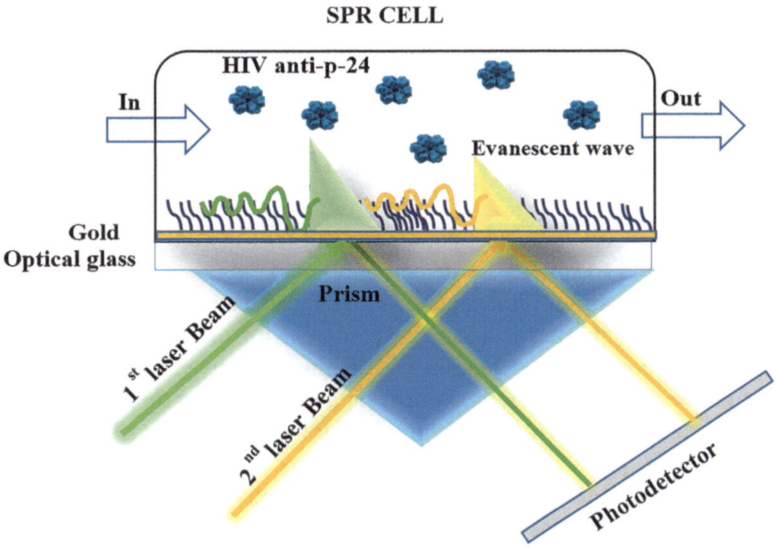


**Fig. 15.5** Surface plasmon Resonance Assay for Label free and selective detection of HIV-1p24 protein

## 15.8.2 Electrochemical Biosensors

Electrochemical biosensors have transformed HIV diagnostics, capitalizing on progress in microelectronics and nanotechnology to achieve exceptional sensitivity and specificity. These sensors, initially developed for simpler analytes like glucose and urea, have been adapted to detect complex biomolecules relevant to HIV, such as proteins, nucleic acids, and antibodies. Their appeal stems from their low manufacturing costs, compatibility with electronic systems, and ability to analyze diverse biological fluids, even those where optical methods might be challenging (Lyons et al. 1964).

Electrochemical biosensors offer several key advantages for HIV detection, including robustness, easy miniaturization and portability, and the capacity for real-time analysis with small sample volumes. These characteristics make them well-suited for point-of-care diagnostics. Recent progress includes the incorporation of nanomaterials like gold nanoparticles, graphene, and carbon nanotubes onto electrodes to improve electron transfer and signal amplification. Furthermore, the use of aptamers, antibodies, and enzymes for immobilization has led to increased specificity and reusability (Grieshaber et al. 2008).

Label-free detection and the use of novel redox probes have simplified and made electrochemical biosensor assays faster and less expensive by eliminating the need for extra reagents. These advancements have enabled HIV-specific biosensors to detect viral proteins or antibodies at femtogram levels, allowing for early and precise diagnosis after infection. The robust performance of these systems in complex clinical samples suggests their potential for broad implementation, particularly in resource-constrained environments (Krishnan et al. 2023).

### 15.8.3 Fluorescent Biosensors for HIV Detection

Fluorescent biosensors have become a powerful tool in HIV diagnostics, offering high sensitivity and specificity for detecting biomarkers. Their rapid response times and adaptability to various bioreceptor systems have made fluorescence-based sensing a central technique in biosensor development for the last three decades. The integration of fluorescence detection with highly selective bioreceptors has positioned fluorescent biosensors as a superior alternative to other optical methods, offering unmatched performance in complex biological samples (Frenea-Robin and Marchalot 2022).

Modern fluorescent biosensors increasingly utilize nanomaterials as fluorophores and quenchers. Materials like quantum dots, graphene oxide, and gold nanoparticles enhance signal strength and quenching efficiency, leading to highly sensitive detection of HIV biomarkers. These biosensors can precisely identify key HIV components, including viral proteins, nucleic acids, and antibodies. Quantum dot-based fluorescent systems, for instance, offer excellent photostability and adjustable emission spectra, making them well-suited for simultaneously detecting multiple HIV markers (Brokowski 2019).

These nanomaterial-enhanced biosensors offer the significant advantages of real-time operation and requiring only small sample volumes, thereby shortening diagnostic turnaround times. Their suitability for point-of-care use also makes them accessible in settings with limited resources (Ugo and Moretto 2017).

### 15.8.4 Electromagnetic Piezoelectric Biosensors for HIV Detection

Acoustic sensing, including electromagnetic piezoelectric biosensors, shows promise for HIV diagnostics, but limited pressure detection remains a challenge. Researchers are addressing this by developing improved sensing platforms and using advanced materials to increase sensitivity and reliability. Self-assembled monolayer (SAM) chemistry is crucial in piezoelectric biosensor fabrication, enabling precise biomolecule immobilization for highly specific and reproducible HIV biomarker detection (Souza and Agrawal 2024).

Sheikh et al. made a significant advancement in HIV diagnostics by developing a label-free piezoelectric acoustic immunosensor capable of distinguishing HIV-2 from HIV-1 in serum. Their innovative sensor design employed a mixed self-assembled monolayer (SAM) on a quartz wafer, functionalized with HIV-2 epitopes, enabling specific detection of HIV-2 (Sheikh et al. 2011).

### 15.8.5 Surface-Enhanced Raman Scattering (SERS) for HIV Detection

Surface-enhanced Raman scattering (SERS) has emerged as a powerful tool for ultrasensitive HIV detection, leveraging molecules with large Raman cross-sections and the high SERS activity of silver (Ag) and gold (Au) nanoparticles (NPs). Traditionally, Raman labels and capture probes are adsorbed on NP surfaces or conjugated before being adsorbed, facilitating signal enhancement (Huh et al. 2009). Using the SERS properties of AuNPs, Hu et al. (2010) devised a Nano junction-based biosensing method for detecting HIV gene-related DNA. This innovative subattomolar assay relied on multilayer metal-molecule-metal nano junctions formed by complementary DNA probes acting as structural bricks. These junctions created SERS "hot spots" by clustering AuNPs, reducing the distance between nanoparticles and Raman labels, and significantly amplifying the Raman signal due to distance-dependent electromagnetic enhancement. The platform exhibited remarkable sensitivity, detecting HIV-1 DNA sequences at concentrations as low as $10^{-19}$ M, well below femtomolar levels, while also discriminating single-base mismatches. The SERS signal showed near-linear growth from 0 to $10^{-13}$ M before saturating at higher concentrations.

In another approach, Fu et al. (2016) developed a SERS-based lateral flow assay utilizing Raman reporter-labeled AuNPs as nanotags. This assay enabled quantitative detection of HIV-1 DNA markers by monitoring characteristic Raman peaks of DNA-conjugated AuNPs. With optimized conditions, the assay demonstrated a linear relationship between Raman intensity and DNA concentrations in the range of 8 pg/mL to 64 ng/mL, achieving an impressive detection limit of 0.24 pg/mL.

These advancements underscore the potential of SERS technology in achieving unprecedented sensitivity and specificity in HIV diagnostics, offering a promising pathway for early and precise detection. Silver-containing metal nanostructures exhibiting strong and stable SERS signals have emerged as powerful tools for ultrasensitive bio detection. Su et al. (2017) developed multicolour DNA-mediated Au–Ag nano-mushrooms, acting as ready-to-use SERS Nano probes for multiplexed DNA/RNA detection. These Nano probes leveraged the hybridization properties of DNA/RNA with exposed oligonucleotides on AuNPs, enabling them to simultaneously recognize multiple DNA targets. This study was the first to establish an efficient protocol where thiol-containing Raman reporters were co-assembled with probe DNA, overcoming the challenge of labeling Raman reporters directly onto oligonucleotides. The approach allowed for the successful simultaneous detection of three virus DNA targets within a single sample.

Raman spectroscopy also shows potential for rapid screening in clinical applications, such as blood transfusions or blood banks, where label-free detection of HIV-1 p24 antigen is critical. Otange et al. (2017) demonstrated the use of conductive silver paste-coated glass slides as Raman spectroscopy substrates, facilitating the label-free detection of HIV-1 p24 antigen in blood plasma. The characteristic Raman spectrum of HIV-1 p24 exhibited distinct bands corresponding to RNA and protein components, confirming the antigen's presence with high sensitivity and specificity. These advances underscore the potential of SERS-based platforms for rapid, accurate, and non-invasive detection of HIV biomarkers.

### *15.8.6 POC Diagnostics and Mobile Phone Integration for HIV Detection*

The fast and simple monitoring of infections remains a critical challenge in the diagnosis and treatment of infectious diseases. Recent debates have centered on the value and feasibility of laboratory monitoring, particularly in resource-limited settings. However, the emergence of simple, accurate, and affordable monitoring tools, such as rapid point-of-care (POC) assays, has the potential to revolutionize diagnostic strategies.

POC tests offer significant advantages, including shortened time-to-result and the ability to be performed at the bedside or in remote care centres, without requiring extensive training or advanced infrastructure (Moyo et al. 2016). Commercial POC tests are already widely available for diagnosing bacterial, viral, and parasitic infections, including malaria (Bissonnette and Bergeron 2012). The evolution of POC technologies may further expand, incorporating mobile phone-based detection methods, which have recently emerged as a promising trend.

Mobile phones, as nearly ubiquitous and low-cost communication tools, possess numerous advantages such as portability, intelligent functionality, and internet connectivity, making them ideal for real-time POC diagnostics. The integration of mobile phones into POC devices eliminates the need for professional instruments, facilitating low-cost, long-range data transmission and enhancing accessibility in resource-limited regions. Efforts have focused on developing mobile phone-based POC devices (van Aalen et al. 2024).

In this context, (Liu et al. 2021) introduced a low-cost, portable immuno-testing platform utilizing a mobile phone and a micropit array chip (μPAC). These μPACs, fabricated on transparent cyclic olefin copolymer sheets, enabled optical detection and solvent compatibility. The device featured covalently immobilized antibodies targeting the p24 antigen, allowing for sandwich immunoassay detection. Images captured by a mobile phone were sent to a server for quantitative analysis, achieving a colorimetric immunoassay with detection limits of 190 pg $mL^{-1}$ in buffer and 650 pg $mL^{-1}$ in human serum. The μPAC's low cost, ease of use, and global mobile network coverage position this device as a promising diagnostic tool, especially in resource-limited settings, where convenience and portability are essential.

## 15.9 Nanomaterial-Based Biosensors for Detection of Alzheimer's Disease

Alzheimer's disease (AD) is a debilitating neurodegenerative disorder and the leading cause of dementia worldwide. In 2019, more than 50 million people were affected by AD, a number projected to rise to 152 million by 2050, according to Alzheimer's disease International (ADI) and the World Health Organization (WHO) (Lynch 2020). While AD predominantly impacts older adults, approximately 4% of cases occur in individuals under 65 years, making it a multifactorial disease (A. D. International and M. University 2021). The condition often begins with mild cognitive impairment (MCI), characterized by subtle memory and non-memory-related impairments, eventually progressing to dementia. This advanced stage results in significant cognitive decline, memory loss, disorientation, and the inability to perform daily activities, deeply affecting both patients and their caregivers emotionally and economically (Joe and Ringman 2019). In 2015, the global economic burden of AD was estimated at $818 billion USD, accounting for 1.1% of the global GDP, with predictions indicating a rise to $2 trillion USD by 2030 (Prince et al. 2016). Recognizing the urgency, the G8 political forum in 2013 identified the development of a therapy for AD by 2025 as a global priority (Burns and Iliffe 2015). Despite advances in research on disease-modifying drugs, no effective cure currently exists, highlighting the critical importance of early detection for improving therapeutic outcomes. Existing diagnostic methods include cognitive tests and imaging techniques such as magnetic resonance imaging (MRI), positron emission tomography (PET) (Kaushik et al. 2016), and near-infrared imaging (NIR). Cerebrospinal fluid (CSF) and blood plasma biomarkers are often assessed via techniques like immunohistochemistry and ELISA. However, these methods present challenges due to their lengthy processing times, high costs, invasive nature, and limitations for broad early screening (Shui et al. 2023).

To overcome the limitations of traditional biomarker analysis, electrochemical biosensors offer a compelling alternative for diagnostics. These devices provide high sensitivity, cost-effectiveness, and require only small sample volumes, minimizing patient discomfort. The integration of nanomaterials into these biosensors has led to further performance enhancements, specifically in sensitivity, specificity, and reliability for Alzheimer's disease (AD) biomarker detection. This review examines recent progress in electrochemical biosensors for AD diagnosis, highlighting the significant impact of nanomaterials. The creation of affordable, user-friendly, and non-invasive diagnostic tools is crucial for addressing the increasing global prevalence of AD.

### 15.9.1 Immunosensors Targeting Biomarkers (e.g., β-amyloid, Tau Proteins)

Alzheimer's disease (AD) is a complex neurodegenerative disorder characterized by a gradual progression, making early diagnosis difficult. The initial symptoms, such as mild cognitive impairment and dementia, can manifest differently in each individual, which complicates diagnosis based solely on cognitive assessments and can result in misdiagnosis and delayed treatment (Shui et al. 2023). This variability underscores the need for dependable biomarkers for early detection. The most commonly recognized pathological features of AD are amyloid-beta (Aβ) plaques and intracellular neurofibrillary tangles (NFTs), which are composed of hyperphosphorylated tau protein. Aβ plaques consist of aggregated peptides derived from the amyloid precursor protein (APP) through either the amyloidogenic or non-amyloidogenic proteolytic pathways (Shui et al. 2023).

In the amyloid genic pathway, APP is cleaved by β-secretase (BACE1) and γ-secretase, generating Aβ peptides like Aβ (1-40) and Aβ (1-42) (O'Brien and Wong 2011). The accumulation of these peptides due to imbalances in production or clearance leads to the formation of toxic oligomers (Aβo), protofibrils, and fibrils, which aggregate into extracellular senile plaques (Chen et al. 2017). These plaques were initially thought to be the primary cause of AD, as per the "amyloid cascade hypothesis," but subsequent studies revealed that soluble Aβo in cerebrospinal fluid (CSF) also contributes to neurotoxicity and cognitive decline, even in the absence of plaques.

Another critical hallmark is the formation of NFTs, resulting from hyper phosphorylation of tau protein, a microtubule-associated protein essential for neuronal function. In AD, hyper phosphorylated tau aggregates into paired helical filaments (PHFs) and NFTs, disrupting axonal transport and contributing to neuronal degeneration (Iqbal et al. 2010). These tangles are often found near Aβ plaques, further implicating their role in AD pathogenesis. Beyond Aβ and tau, apolipoprotein E4 (ApoE4) has been identified as a major genetic risk factor for AD, particularly for early-onset cases. ApoE4, which binds to Aβ, promotes pathological β-sheet formation, exacerbating amyloid aggregation (Iqbal et al. 1998). Additionally, oxidative stress markers, neurotransmitters like acetylcholine and dopamine, and proteins such as p53 and α-1-antitrypsin have been explored as potential biomarkers for AD detection (Mahley et al. 2009).

Emerging evidence has also linked novel biomarkers like microRNAs (miRNAs), BACE1, and salivary markers such as lactoferrin to AD, broadening the scope of diagnostic research (Eriksson et al. 1995). Saliva is gaining attention as a non-invasive sample source, though further studies are needed to validate its application in biosensing. Biomarkers have primarily been detected in CSF and blood, but the increasing focus on saliva represents a promising shift toward less invasive diagnostic methods (Eriksson et al. 1995). Electrochemical biosensors have shown immense potential in detecting these biomarkers due to their sensitivity, cost-effectiveness, and ability to analyze small sample volumes (Serafín et al. 2020). These advancements pave the way for innovative, accessible, and early diagnostic solutions for Alzheimer's disease.

### 15.9.2 Electrochemical Biosensor for Alzheimer's Disease Diagnostics

The increasing incidence of AD has turned it into the most studied neurodegenerative disease. The presence of AD biomarkers in blood and CSF has promoted the development of electrochemical biosensors for their detection, as they allow the abstention of quantitative results in 5 minutes to 2 hours (Valkova and Pohanka 2021). However, reaching the low cut-off values of the main AD biomarkers in human samples and improving selectivity and time-stability of biosensors are the heading challenges here (Forlenza et al. 2015). For that purpose, nanomaterials have been introduced in biosensors either as labels or electrode modifiers. As it can be deduced from the works revised in this review, peptides are, by far, the most widely detected biomarkers followed by proteins, as Aβ peptide is, till now, the mainly accepted biomarker. With regard to biological recognition elements, antibodies are predominant even though the use of aptamers is remarkable as well as the combination of these two elements in sandwich type biosensors. From nanomaterials, AuNPs are, without no doubt, the most frequently used materials, both as electrode modifiers and as labels, followed by carbon-based nanomaterials. It is worth remarking that the use of combinations of different nanomaterials, known as nanocomposites, is increasing as in the majority of cases they present a synergistic effect. Below are described the most representative electrochemical biosensors for the detection of AD biomarkers in which nanomaterials are involved, classified according to the analyte measured: DNA or RNA, peptides, proteins, neurotransmitters and oxidative stress biomarkers (Azimzadeh et al. 2017), and attending to the nanomaterials used and the contribution they made to biosensors performance.

### 15.9.3 Nanobiosensors for Early Detection and Improved Accuracy

Gold nanoparticles (AuNPs) are highly effective as electrode modifiers, enhancing biosensor performance for detecting Alzheimer's disease (AD) biomarkers. Wu et al. developed an electrochemical impedance spectroscopy (EIS) immunosensors to detect Aβ (1-42) peptide (Vais et al. 2021). By utilizing an anodic aluminium oxide template to electrodeposit a gold film, followed by AuNP immobilization, they increased the biosensor's surface area, enabling improved analyte capture. Monoclonal antibodies specific to Aβ (1-42) were attached using 11-mercaptoundecanoic acid and EDC/NHS chemistry (Carneiro et al. 2017).

Tested with Fe (CN) $6^{3-/4-}$ as an electroactive indicator, this sensor achieved a limit of detection (LOD) of 22.2 fM, with SEM and AFM analysis confirming the formation of Aβ aggregates. Similarly, Carneiro et al. developed a label-free immunosensors using a gold electrode modified with a mercaptopropionic acid monolayer and electrodeposited AuNPs. This sensor employed EIS and square wave voltammetry (SWV), achieving a LOD of 1.15 pM. While its sensitivity was lower than Wu et al.'s immunosensors, it demonstrated the potential of AuNPs for surface functionalization (Rajasree et al. 2024). However, performance in real sample applications was not evaluated. Amor-Gutiérrez et al. extended the application of AuNPs in biosensing by designing an immunosensors for detecting unfolded p53 in blood samples. Using screen-printed carbon electrodes (SPCEs) modified with AuNPs, they immobilized monoclonal antibodies specific to p53, achieving a LOD of 0.05 nM through a competitive assay and linear sweep voltammetry (LSV) (Rama et al. 2017). This sensor proved effective in detecting unfolded p53 in blood samples from patients with mild cognitive impairment and AD, with results comparable to ELISA. AuNPs have also been employed in sandwich-type immunosensors. Zhou et al. developed an aptamer-based sandwich biosensor to detect Aβ oligomers (Aβo), utilizing nanocomposites of AuNPs and copper metal-organic frameworks (AuNPs/Cu-MOFs) as labels and gold nanoflowers (AuNFs) as electrode modifiers. Although the sensor achieved a LOD of 0.45 nM (Leite et al. 2023), its sensitivity did not significantly surpass previous approaches. Another notable contribution came from Lien et al., who developed an impedimetric immunosensors for detecting Aβ peptides using carbon-printed electrodes modified with protein G. Their approach included successive electrode modifications with AuNPs and a self-assembled monolayer (SAM) of 16-mercaptohexadecanoic acid, significantly reducing the LOD to 0.57 nM (Sethi et al. 2020) compared to unmodified electrodes. Qin et al. demonstrated the potential of cellular prion protein (PrPc) peptides as bioreceptors for Aβ oligomers. Using AuNPs embedded in poly (3,4-ethylene dioxythiophene) (PEDOT) layers, they developed an impedance biosensor capable of detecting Aβo with a remarkable LOD of $10^{-2}$ fM, the lowest reported for Aβ detection (Qin et al. 2019). This underscores the advantages of specific proteins like PrPc in improving sensitivity and selectivity in biosensing.

### *15.9.4 Carbon Nanotubes and Nano Membranes*

Carbon nanotubes (CNTs), with their exceptional electrochemical and physical properties, have been utilized for biosensing applications. Moreira et al. developed a biosensor for Aβ oligomers using CNTs and molecularly imprinted polymers (MIPs) (Paramsothy 2024). By employing copper oxide nanoparticle-modified CNTs as conductive substrates, they achieved an impressive LOD of 88.6 fM 102, demonstrating the potential of CNTs in enhancing electrocatalytic activity and electron transferability. Nano membranes, thin structures with high selectivity and durability, have also been used in AD biosensing. (Koklu et al. 2021) introduced Nano membranes of poly (styrene-b-4-vinylpyridine) in conjunction with organic electrochemical transistors (OECTs) to detect Aβ aggregates (Rivnay et al. 2018). Functionalized with Congo red, the Nano membrane selectively captured Aβ aggregates, blocking ion transport and enabling detection with a LOD of 2.21 pM. While this sensitivity was not the lowest, the use of Nano membranes offers potential for multianalyte detection.

### *15.9.5 Nickel Ferrite Nanoparticles and Advanced Nanocomposites*

Nickel ferrite nanoparticles ($NiFe_2O_4$ NPs), known for their magnetic and electrochemical properties, have been used in biosensing. Devi et al. (2020) developed a label-free immunosensors combining Au/$NiFe_2O_4$ NPs, graphene oxide (GO), and chitosan on a glassy carbon electrode. This biosensor achieved a LOD of 0.66 pM in cerebrospinal fluid samples, showcasing the synergistic benefits of combining multiple nanomaterials for enhanced performance (Toyos-Rodríguez et al. 2020). Gold nanoparticles have also been extensively used as labels in biosensing. Yoo et al. designed a sandwich-type immunosensors with antibody-functionalized AuNPs, achieving a LOD of 22.15 fM for Aβ peptides. Magnetic bead platforms combined with AuNPs have further expanded the scope of AD biomarker detection. Iglesias-Mayor et al. used Au@Pt/Au NPs as labels in an immunosensors for unfolded p53, enabling chronoamperometric detection in neutral pH environments with a LOD of 66 nM (Hu et al. 2017). This innovative design simplified the sensing process by eliminating the need for additional reagents. Zhou et al. also explored a biosensor approach, using carboxyl graphene and AuNP-thionine labels for Aβ oligomer detection, reporting a limit of detection (LOD) of 100 pM. Similarly, Xia et al. leveraged the affinity of gold nanoparticles (AuNPs) for PrPc peptides, achieving selective Aβ oligomer detection in complex biological matrices. Their biosensor reached an LOD of 45 pM, highlighting the impact of both the detection strategy and the nanomaterial properties on biosensor performance (Hu et al. 2017).

## 15.10 Conclusion

Optical biosensors, particularly those incorporating nanotechnology, hold tremendous potential for revolutionizing disease diagnostics. The ability to detect minute concentrations of biomarkers in complex biological samples offers a pathway to earlier and more personalized interventions, particularly for debilitating conditions like neurodegenerative diseases. While challenges in scalability, consistency, biocompatibility, and user integration remain, the ongoing research and development in this field are rapidly paving the way for practical, point-of-care diagnostic tools. The integration of artificial intelligence and machine learning with these biosensors promises further enhancements in diagnostic accuracy and data interpretation. As we move forward, addressing the regulatory and ethical considerations surrounding these technologies will be crucial for their successful translation into widespread clinical use. The future of disease diagnostics is bright, with optical biosensors poised to play a central role in transforming healthcare and improving patient outcomes globally.

## References

A. D. International and M. University (2021) World Alzheimer report 2021. Alzheimer's Dis Int:2–314. [Online]. Available: https://www.alzint.org/resource/world-alzheimer-report-2021/

Abbasi R, Hu X, Zhang A, Dummer I, Wachsmann-Hogiu S (2024) Optical image sensors for smart analytical chemiluminescence biosensors. Bioengineering 11(9):912. [Online]. Available: https://www.mdpi.com/2306-5354/11/9/912

Abdel-Karim R, Reda Y, Abdel-Fattah A (2020) Review—nanostructured materials-based nanosensors. J Electrochem Soc 167(3):037554. https://doi.org/10.1149/1945-7111/ab67aa

Abuzeid HR, Abdelaal AF, Elsharkawy S, Ali GAM (2024) Handbook of nanosensors. Springer Nature Switzerland

Ahmad F et al (2022) Unique properties of surface-functionalized nanoparticles for bio-application: functionalization mechanisms and importance in application. Nanomaterials 12(8). https://doi.org/10.3390/nano12081333

Ahmed A, Rushworth JV, Hirst NA, Millner PA (2014) Biosensors for whole-cell bacterial detection. Clin Microbiol Rev 27(3):631–646. https://doi.org/10.1128/CMR.00120-13

Altug H, Oh SH, Maier SA, Homola J (2022) Advances and applications of nanophotonic biosensors. Nat Nanotechnol 17(1):5–16. https://doi.org/10.1038/s41565-021-01045-5

Amen MT, Pham TTT, Cheah E, Tran DP, Thierry B (2022) Metal-oxide FET biosensor for point-of-care testing: overview and perspective. Molecules 27(22). https://doi.org/10.3390/molecules27227952

Amesh M, Govindaraju T (2022) Multipronged diagnostic and therapeutic strategies for Alzheimer's disease. Chem Sci 13:13657–13689. https://doi.org/10.1039/d2sc03932j

Andreescu S, Marty JL (2006) Twenty years research in cholinesterase biosensors: from basic research to practical applications. Biomol Eng 23(1):1–15. https://doi.org/10.1016/j.bioeng.2006.01.001

Arora P, Sindhu A, Dilbaghi N, Chaudhury A (2013) Engineered multifunctional nanowires as novel biosensing tools for highly sensitive detection. Appl Nanosci 3(5):363–372. https://doi.org/10.1007/s13204-012-0142-4

Arregui FJ et al (2014) Fiber-optic lossy mode resonance sensors. Procedia Eng 87:3–8. https://doi.org/10.1016/j.proeng.2014.11.253

Asci Erkocyigit B, Ozufuklar O, Yardim A, Guler Celik E, Timur S (2023) Biomarker detection in early diagnosis of cancer: recent achievements in point-of-care devices based on paper microfluidics. Biosensors 13(3). https://doi.org/10.3390/bios13030387

Ayed Z, Malhotra S, Dobhal G, Goreham RV (2021) Aptamer conjugated indium phosphide quantum dots with a zinc sulphide shell as photoluminescent labels for acinetobacter baumannii. Nanomaterials 11(12). https://doi.org/10.3390/nano11123317

Azam T, Bukhari SH, Liaqat U, Miran W (2023) Emerging methods in biosensing of immunoglobin G—a review. Sensors 23(2). https://doi.org/10.3390/s23020676

Azimzadeh M, Rahaie M, Nasirizadeh N, Daneshpour M, Naderi-Manesh H (2017) Electrochemical miRNA biosensors: the benefits of nanotechnology. Nanomedicine Res J 2(1):36–48. https://doi.org/10.22034/NMRJ.2017.23336

Banigo AT, Azeez TO, Ejeta KO, Lateef A, Ajuogu E (2020) Nanobiosensors: applications in biomedical technology. IOP Conf Ser Mater Sci Eng 805(1). https://doi.org/10.1088/1757-899X/805/1/012028

Baranwal J, Barse B, Gatto G, Broncova G (2022) Electrochemical sensors and their applications: a review. Chemosensors 10(9):363

Baruah A et al (2024) Biomedical applications of graphene-based nanomaterials: recent progress, challenges, and prospects in highly sensitive biosensors. Discov Nano 19(1). https://doi.org/10.1186/s11671-024-04032-6

Berepiki A, Kent R, MacHado LFM, Dixon N (2020) Development of high-performance whole cell biosensors aided by statistical modeling. ACS Synth Biol 9(3):576–589. https://doi.org/10.1021/acssynbio.9b00448

Bhalla N, Jolly P, Formisano N, Estrela P (2016) Introduction to biosensors. Essays Biochem 60(1):1–8. https://doi.org/10.1042/EBC20150001

Bhargava R, Khan S, Ahmad N, Ansari MMN (2018) Investigation of structural, optical and electrical properties of Co3O4 nanoparticles. AIP Conf Proc 1953:1–5. https://doi.org/10.1063/1.5032369

Bissonnette L, Bergeron MG (2012) Infectious disease management through point-of-care personalized medicine molecular diagnostic technologies. J Pers Med 2(2):50–70. https://doi.org/10.3390/jpm2020050

Brokowski CAM (2019) 乳鼠心肌提取 HHS Public Access. Physiol Behav 176(5):139–148. https://doi.org/10.1007/s00216-018-1224-2.Graphene

Burns A, Iliffe S (2015) Alzheimer's disease. BMJ 338(7692):467–471

Cali K, Tuccori E, Persaud KC (2020) Gravimetric biosensors, vol 642, 1st edn. Elsevier Inc.

Carneiro P, Loureiro J, Delerue-Matos C, Morais S, do Carmo Pereira M (2017) Alzheimer's disease: development of a sensitive label-free electrochemical immunosensor for detection of amyloid beta peptide. Sensors Actuators B Chem 239:157–165. https://doi.org/10.1016/j.snb.2016.07.181

Castillo-Henríquez L, Brenes-Acuña M, Castro-Rojas A, Cordero-Salmerón R, Lopretti-Correa M, Vega-Baudrit JR (2020) Biosensors for the detection of bacterial and viral clinical pathogens. Sensors (Switzerland) 20(23):1–26. https://doi.org/10.3390/s20236926

Chaubey A, Malhotra BD (2002) Review mediated biosensors. Biosens Bioelectron 7:441–456

Chavali MS, Nikolova MP (2019) Metal oxide nanoparticles and their applications in nanotechnology. SN Appl Sci 1(6):1–30. https://doi.org/10.1007/s42452-019-0592-3

Chen GF et al (2017) Amyloid beta: structure, biology and structure-based therapeutic development. Acta Pharmacol Sin 38(9):1205–1235. https://doi.org/10.1038/aps.2017.28

Chiticaru EA, Pilan L, Damian CM, Vasile E, Burns JS, Ionita M (2019) Influence of graphene oxide concentration when fabricating an electrochemical biosensor for DNA detection. Biosensors 9(4):1–18. https://doi.org/10.3390/bios9040113

Cho Y, Ba VAP, Jeong JY, Choi Y, Hong S (2020) Ion-selective carbon nanotube field-effect transistors for monitoring drug effects on nicotinic acetylcholine receptor activation in live cells. Sensors (Switzerland) 20(13):1–13. https://doi.org/10.3390/s20133680

Chu TC et al (2006) Labeling tumor cells with fluorescent nanocrystal-aptamer bioconjugates. Biosens Bioelectron 21(10):1859–1866. https://doi.org/10.1016/j.bios.2005.12.015

Contributor B (2021) Nanostructured Materials-Enabled Biosensors. Nanostructured Mater Biosens 1:1–27

Cordeiro CA, Sias A, Koster T, Westerink BHC, Cremers TIFH (2018) In vivo 'real-time' monitoring of glucose in the brain with an amperometric enzyme-based biosensor based on gold coated tungsten (W-Au) microelectrodes. Sensors Actuators B Chem 263:605–613. https://doi.org/10.1016/j.snb.2018.02.116

Cui F et al (2019) A novel magnetic fluorescent biosensor based on graphene quantum dots for rapid, efficient, and sensitive separation and detection of circulating tumor cells. Anal Bioanal Chem 411(5):985–995. https://doi.org/10.1007/s00216-018-1501-0

Damborský P, Švitel J, Katrlík J (2016) Optical biosensors. Essays Biochem 60(1):91–100. https://doi.org/10.1042/EBC20150010

Darwish MA, Abd-Elaziem W, Elsheikh A, Zayed AA (2024) Advancements in nanomaterials for nanosensors: a comprehensive review. Nanoscale Adv 6(16):4015–4046. https://doi.org/10.1039/d4na00214h

Das P, Fatehbasharzad P, Colombo M, Fiandra L, Prosperi D (2019) Multifunctional magnetic gold nanomaterials for cancer. Trends Biotechnol 37(9):995–1010. https://doi.org/10.1016/j.tibtech.2019.02.005

Devi R, Gogoi S, Dutta HS, Bordoloi M, Sanghi SK, Khan R (2020) Au/NiFe2O4 nanoparticle-decorated graphene oxide nanosheets for electrochemical immunosensing of amyloid beta peptide. Nanoscale Adv 2(1):239–248. https://doi.org/10.1039/c9na00578a

Du Y, Zhang X, Liu P, Yu DG, Ge R (2022) Electrospun nanofiber-based glucose sensors for glucose detection. Front Chem 10(August):1–31. https://doi.org/10.3389/fchem.2022.944428

Dudina A, Seichepine F, Chen Y, Stettler A, Hierlemann A, Frey U (2019) Monolithic CMOS sensor platform featuring an array of 9’216 carbon-nanotube-sensor elements and low-noise, wide-bandwidth and wide-dynamic-range readout circuitry. Sensors Actuators B Chem 279:255–266. https://doi.org/10.1016/j.snb.2018.10.004

Durmu NG, Lin RL, Kozberg M, Dermici D, Khademhosseini A, Demirci U (2013) Encyclopedia of microfluidics and nanofluidics. Enc Microfluid Nanofluid:1–15. https://doi.org/10.1007/978-3-642-27758-0

Eriksson S, Janciauskiene S, Lannfelt L (1995) α1-Antichymotrypsin regulates Alzheimer β-amyloid peptide fibril formation. Proc Natl Acad Sci USA 92(6):2313–2317. https://doi.org/10.1073/pnas.92.6.2313

Estrela P, Migliorato P, Takiguchi H, Fukushima H, Nebashi S (2005) Electrical detection of biomolecular interactions with metal-insulator- semiconductor diodes. Biosens Bioelectron 20(8 SPEC. ISS):1580–1586. https://doi.org/10.1016/j.bios.2004.08.010

Farzin L, Shamsipur M, Samandari L, Sheibani S (2020) HIV biosensors for early diagnosis of infection: the intertwine of nanotechnology with sensing strategies. Talanta 206(August):120201. https://doi.org/10.1016/j.talanta.2019.120201

Forlenza OV et al (2015) Cerebrospinal fluid biomarkers in Alzheimer’s disease: diagnostic accuracy and prediction of dementia. Alzheimer’s Dement Diagn Assess Dis Monit 1(4):455–463. https://doi.org/10.1016/j.dadm.2015.09.003

Frenea-Robin M, Marchalot J (2022) Basic principles and recent advances in magnetic cell separation. Magnetochemistry 8(1). https://doi.org/10.3390/magnetochemistry8010011

Fu X, Cheng Z, Yu J, Choo P, Chen L, Choo J (2016) A SERS-based lateral flow assay biosensor for highly sensitive detection of HIV-1 DNA. Biosens Bioelectron 78:530–537. https://doi.org/10.1016/j.bios.2015.11.099

Gómez-Gómez AL, Martínez-Ayala AL, del Rosario Moguel Concha D, Borges-Martínez JE, de Jesús Perea-Flores M, Dávila-Ortiz G (2023) Relationship of nanomaterials’ structure based on their application in the food industry: physicochemical and techno-functional characteristics. Appl Sci 13(12). https://doi.org/10.3390/app13127167

Grieshaber D, MacKenzie R, Vörös J, Reimhult E (2008) Electrochemical biosensors – sensor principles and architectures. Sensors 8(3):1400–1458. https://doi.org/10.3390/s8031400

Gruhl FJ, Rapp BE, Rapp M, Länge K (2009) Surface acoustic wave (SAW) biosensor chip system – a promising alternative for biomedical applications. IFMBE Proc 25(8):73–76. https://doi.org/10.1007/978-3-642-03887-7_20

Haniff M, Vajravelu A, Muhammad T, Mohamed AG (2021) Fundamentals of biosensor – an introduction with its applications in the engineering perspective. Int J Adv Eng Manag (IJAEM) 3(5):249–257. https://doi.org/10.35629/5252-0305249257

Hashem M et al (2016) Fabrication and characterization of semiconductor nickel oxide (NiO) nanoparticles manufactured using a facile thermal treatment. Results Phys 6(November):1024–1030. https://doi.org/10.1016/j.rinp.2016.11.031
Herzog G (2014) Richard C. Alkire, Dieter M. Kolb and Jacek Lipkowski (Eds): Bioelectrochemistry. Fundamentals, applications and recent developments. Chromatographia 77(1–2):201–201. https://doi.org/10.1007/s10337-013-2547-y
Hjiri M, Bahanan F, Aida MS, El Mir L, Neri G (2020) High performance CO gas sensor based on ZnO nanoparticles. J Inorg Organomet Polym Mater 30(10):4063–4071. https://doi.org/10.1007/s10904-020-01553-2
Hu J, Zheng PC, Jiang JH, Shen GL, Yu RQ, Liu GK (2010) Sub-attomolar HIV-1 DNA detection using surface-enhanced Raman spectroscopy. Analyst 135(5):1084–1089. https://doi.org/10.1039/b920358c
Hu T, Lu S, Chen C, Sun J, Yang X (2017) Colorimetric sandwich immunosensor for Aβ(1-42) based on dual antibody-modified gold nanoparticles. Sensors Actuators B Chem 243:792–799. https://doi.org/10.1016/j.snb.2016.12.052
Huh YS, Chung AJ, Erickson D (2009) Surface enhanced Raman spectroscopy and its application to molecular and cellular analysis. Microfluid Nanofluid 6(3):285–297. https://doi.org/10.1007/s10404-008-0392-3
Imran H, Alam A, Dharuman V, Lim S (2022) Fabrication of enzyme-free and rapid electrochemical detection of glucose sensor based on ZnO rod and Ru doped carbon nitride modified gold transducer. Nanomaterials 12(10). https://doi.org/10.3390/nano12101778
Inacio P, Marat-Mendes JN, Neagu E, Dias CJ (2008) Optimization of a biosensor based on a piezoelectric polymer film. Proc Int Symp Electrets (February). https://doi.org/10.1109/ISE.2008.4814063
Iqbal K et al (1998) Mechanisms of neurofibrillary degeneration and the formation of neurofibrillary tangles. J Neural Transm Suppl 53:169–180. https://doi.org/10.1007/978-3-7091-6467-9_15
Iqbal K, Liu F, Gong C-X, Grundke-Iqbal I (2010) Tau in Alzheimer disease and related tauopathies. Curr Alzheimer Res 7(8):656–664. https://doi.org/10.2174/156720510793611592
Janik-Karpinska E et al (2022) Immunosensors—the future of pathogen real-time detection. Sensors 22(24):1–20. https://doi.org/10.3390/s22249757
Janssen J, Lambeta M, White P, Byagowi A (2019) Carbon nanotube-based electrochemical biosensor for label-free protein detection. Biosensors 9(4). https://doi.org/10.3390/bios9040144
Joe E, Ringman JM (2019) Cognitive symptoms of Alzheimer's disease: clinical management and prevention. BMJ 367:1–14. https://doi.org/10.1136/bmj.l6217
Joudeh N, Linke D (2022) Nanoparticle classification, physicochemical properties, characterization, and applications: a comprehensive review for biologists. J Nanobiotechnol 20(1):1–29. https://doi.org/10.1186/s12951-022-01477-8
Kadhim MM et al (2023) Evaluation of a biosensor-based graphene oxide-DNA nanohybrid for lung cancer. RSC Adv 13(4):2487–2500. https://doi.org/10.1039/d2ra05808a
Kailasa SK, Joshi DJ, Koduru JR, Malek NI (2021) Review on MXenes-based nanomaterials for sustainable opportunities in energy storage, sensing and electrocatalytic reactions. J Mol Liq 342:117524. https://doi.org/10.1016/j.molliq.2021.117524
Kalkal A, Pradhan R, Kadian S, Manik G, Packirisamy G (2020) Biofunctionalized graphene quantum dots based fluorescent biosensor toward efficient detection of small cell lung cancer. ACS Appl Bio Mater 3(8):4922–4932. https://doi.org/10.1021/acsabm.0c00427
Kaushik A, Jayant RD, Tiwari S, Vashist A, Nair M (2016) Nano-biosensors to detect beta-amyloid for Alzheimer's disease management. Biosens Bioelectron 80:273–287. https://doi.org/10.1016/j.bios.2016.01.065
(Kevin) Lai CY, Foot PJS, Brown JW, Spearman P (2017) A urea potentiometric biosensor based on a thiophene copolymer. Biosensors 7(1):1–13. https://doi.org/10.3390/bios7010013
Kim HM, Park JH, Lee SK (2019) Fiber optic sensor based on ZnO nanowires decorated by Au nanoparticles for improved plasmonic biosensor. Sci Rep 9(1):1–9. https://doi.org/10.1038/s41598-019-52056-1

Kladnik D (2012) Slovenian geography and geographical names. Geogr Vestn 84(1):237–249. https://doi.org/10.1016/j.jaci.2009.09.046.Structure

Koh I, Josephson L (2009) Magnetic Nanoparticle Sensors. Sensors 9(10):8130–8145. https://doi.org/10.3390/s91008130

Koklu A et al (2021) Microfluidic integrated organic electrochemical transistor with a nanoporous membrane for amyloid-β detection. ACS Nano 15(5):8130–8141. https://doi.org/10.1021/acsnano.0c09893

Kozitsina AN, Svalova TS, Malysheva NN, Okhokhonin AV, Vidrevich MB, Brainina KZ (2018) Sensors based on bio and biomimetic receptors in medical diagnostic, environment, and food analysis. Biosensors 8(2):1–34. https://doi.org/10.3390/bios8020035

Krishnan SK, Nataraj N, Meyyappan M, Pal U (2023) Graphene-based field-effect transistors in biosensing and neural interfacing applications: recent advances and prospects. Anal Chem 95(5):2590–2622. https://doi.org/10.1021/acs.analchem.2c03399

Kulkarni MB, Ayachit NH, Aminabhavi TM (2022a) Recent advancements in nanobiosensors: current trends, challenges, applications, and future scope. Biosensors 12(10). https://doi.org/10.3390/bios12100892

Kulkarni MB, Ayachit NH, Aminabhavi TM (2022b) Biosensors and microfluidic biosensors: from fabrication to application. Biosensors 12(7). https://doi.org/10.3390/bios12070543

Kumari M, Gupta V, Kumar N, Arun RK (2024) Microfluidics-based nanobiosensors for healthcare monitoring. Mol Biotechnol 66(3):378–401. https://doi.org/10.1007/s12033-023-00760-9

Kylilis N et al (2019) Whole-cell biosensor with tunable limit of detection enables low-cost agglutination assays for medical diagnostic applications. ACS Sensors 4(2):370–378. https://doi.org/10.1021/acssensors.8b01163

Labuda J et al (2010) Electrochemical nucleic acid-based biosensors: concepts, terms, and methodology (IUPAC Technical Report). Pure Appl Chem 82(5):1161–1187. https://doi.org/10.1351/PAC-REP-09-08-16

Lakshmipriya T, Gopinath SCB (2018) An introduction to biosensors and biomolecules. Elsevier Inc.

León-Reyes G, Maida-Claros RF, Urrutia-Medina AX, et al (2017) Oxidative profiles of LDL and HDL isolated from women with preeclampsia. Lipids Health Dis 16:1–9. https://doi.org/10.1186/s12944-017-0480-z

Lee J (2023) Carbon nanotube-based biosensors using fusion technologies with biologicals & chemicals for food assessment. Biosensors 13(2). https://doi.org/10.3390/bios13020183

Lei Y (2015) Synthetic strategies in molecular imprinting. Adv Biochem Eng Biotechnol 123(July 2015):127–141. https://doi.org/10.1007/10_2015_313

Leite JP, Figueira F, Mendes RF, Almeida Paz FA, Gales L (2023) Metal-organic frameworks as sensors for human amyloid diseases. ACS Sensors 8(3):1033–1053. https://doi.org/10.1021/acssensors.2c02741

Li X, Wu X, Shi P, Ye ZG (2016) Lead-free piezoelectric diaphragm biosensors based on micromachining technology and chemical solution deposition. Sensors (Switzerland) 16(1):1–10. https://doi.org/10.3390/s16010069

Li T et al (2020) A novel low-field NMR biosensor based on dendritic superparamagnetic iron oxide nanoparticles for the rapid detection of Salmonella in milk. Lwt 133(August):110149. https://doi.org/10.1016/j.lwt.2020.110149

Liang H, Liu B, Yuan Q, Liu J (2016) Magnetic iron oxide nanoparticle seeded growth of nucleotide coordinated polymers. ACS Appl Mater Interfaces 8(24):15615–15622. https://doi.org/10.1021/acsami.6b04038

Lim JY, Lee SS (2020) Sensitive detection of microRNA using QCM biosensors: sandwich hybridization and signal amplification by TiO2 nanoparticles. Anal Methods 12(42):5103–5109. https://doi.org/10.1039/d0ay01481h

Lim B, Yin Y, Ye H, Cui Z, Papachristodoulou A, Huang WE (2022) Reprogramming synthetic cells for targeted cancer therapy. ACS Synth Biol 11(3):1349–1360. https://doi.org/10.1021/acssynbio.1c00631

Lin H, Li M, Ding L, Huang J (2019) A fiber optic biosensor based on hydrogel-immobilized enzyme complex for continuous determination of cholesterol and glucose. Appl Biochem Biotechnol 187(4):1569–1580. https://doi.org/10.1007/s12010-018-2897-x

Liu Y, Zhan L, Qin Z, Sackrison J, Bischof JC (2021) Ultrasensitive and highly specific lateral flow assays for point-of-care diagnosis. ACS Nano 15(3):3593–3611. https://doi.org/10.1021/acsnano.0c10035

Lucia-Tamudo J, Nogueira JJ, Díaz-Tendero S (2023) An efficient multilayer approach to model DNA-based nanobiosensors. J Phys Chem B 127(7):1513–1525. https://doi.org/10.1021/acs.jpcb.2c07225

Luo X, Shi W, Yu H, Xie Z, Li K, Cui Y (2018) Wearable carbon nanotube-based biosensors on gloves for lactate. Sensors (Switzerland) 18(10). https://doi.org/10.3390/s18103398

Lynch C (2020) World Alzheimer report 2019: attitudes to dementia, a global survey. Alzheimers Dement 16(S10). https://doi.org/10.1002/alz.038255

Lyons C, Clark LC, McDowell H, Mcarthur K (1964) Cerebral venous oxygen content during carotid thrombintimectomy. Ann Surg 160(4):561–567. https://doi.org/10.1097/00000658-196410000-00001

Lyu S, Wu Z, Shi X, Wu Q (2022) Optical fiber biosensors for protein detection: a review. Photonics 9:987

Ma F, Li CC, Zhang CY (2018) Development of quantum dot-based biosensors: principles and applications. J Mater Chem B 6(39):6173–6190. https://doi.org/10.1039/c8tb01869c

Mahley RW, Weisgraber KH, Huang Y (2009) Apolipoprotein E: structure determines function, from atherosclerosis to Alzheimer's disease to AIDS. J Lipid Res 50(SUPPL):S183–S188. https://doi.org/10.1194/jlr.R800069-JLR200

Mariani S, Minunni M (2014) Surface plasmon resonance applications in clinical analysis. Anal Bioanal Chem 406(9–10):2303–2323. https://doi.org/10.1007/s00216-014-7647-5

McCaughey CS et al (2024) A label-free approach for relative spatial quantitation of c-di-GMP in microbial biofilms. Anal Chem 96(21):8308–8316. https://doi.org/10.1021/acs.analchem.3c04687

Micha R (2017) 乳鼠心肌提取 HHS Public Access. Physiol Behav 176(1):100–106. https://doi.org/10.1177/0022146515594631.Marriage

Mishra M, Ahmed R, Das DK, Das Pramanik D, Dash SK, Pramanik A (2024) Recent advancements in the application of circulating tumor DNA as biomarkers for early detection of cancers. ACS Biomater Sci Eng 10(8):4740–4756. https://doi.org/10.1021/acsbiomaterials.4c00606

Misra R, Acharya S, Sushmitha N (2022) Nanobiosensor-based diagnostic tools in viral infections: special emphasis on Covid-19. Rev Med Virol 32(2):1–13. https://doi.org/10.1002/rmv.2267

Mohammadi MM, Kumar A, Liu J, Liu Y, Thundat T, Swihart MT (2020) Hydrogen sensing at room temperature using flame-synthesized palladium-decorated crumpled reduced graphene oxide nanocomposites. ACS Sensors 5(8):2344–2350. https://doi.org/10.1021/acssensors.0c01040

Molaie M (2016) Field effect transistor nanobiosensors: state-of-the-art and key challenges as point of care testing devices. Nanomed J 3(32):69–82. https://doi.org/10.7508/nmj.2016.02.001

Momeni N, Javadifar K, Patrick MA, Hasan MH, Chowdhury F (2022) Review on gold nanoparticles-based biosensors in clinical and non-clinical applications. Int J Eng Mater Manuf 7(1):1–12. https://doi.org/10.26776/ijemm.07.01.2022.01

Moyo S, Mohammed T, Wirth KE, Prague M, Bennett K, Holme P (2016) Point-of-care Cepheid Xpert HIV-1 viral load test in rural African. J Clin Microbiol 54(12):3050–3055. https://doi.org/10.1128/JCM.01594-16.Editor

Nadtoka O, Virych P, Krysa V, Chumachenko V, Kutsevol N (2023) Hydrogel materials for biomedical application: a review. Springer Proc Phys 280:291–318. https://doi.org/10.1007/978-3-031-18104-7_21

Naresh V, Lee N (2021) A review on biosensors and recent development of nanostructured materials-enabled biosensors. Sensors 21:1109

Nath S (2024) Advancements in food quality monitoring: integrating biosensors for precision detection. Sustain Food Technol 2(4):976–992. https://doi.org/10.1039/d4fb00094c

Nehra A, Pal Singh K (2015) Current trends in nanomaterial embedded field effect transistor-based biosensor. Biosens Bioelectron 74:731–743. https://doi.org/10.1016/j.bios.2015.07.030

Nguyen HH, Lee SH, Lee UJ, Fermin CD, Kim M (2019) Immobilized enzymes in biosensor applications. Materials (Basel) 12(1):1–34. https://doi.org/10.3390/ma12010121

Norris PM, Hosticka B (2002) Temperature measurements. Mechatronics Handb:19-73–19-88. https://doi.org/10.1542/peds.90.4.650

Nouman M et al (2023) 311 Journal of Chemical Reviews article info. Chem Rev 5(3):311–340. [Online]. Available: https://doi.org/10.22034/JCR.2023.300453.1223

O'Brien RJ, Wong PC (2011) Amyloid precursor protein processing and Alzheimer's disease. Annu Rev Neurosci 34:185–204. https://doi.org/10.1146/annurev-neuro-061010-113613

Otange BO, Birech Z, Okonda J, Rop R (2017) Conductive silver paste smeared glass substrates for label-free Raman spectroscopic detection of HIV-1 and HIV-1 p24 antigen in blood plasma. Anal Bioanal Chem 409(12):3253–3259. https://doi.org/10.1007/s00216-017-0267-0

Paramasivam G, Palem VV, Sundaram T, Sundaram V, Kishore SC, Bellucci S (2021) Nanomaterials: synthesis and applications in theranostics. Nanomaterials 11(12). https://doi.org/10.3390/nano11123228

Paramsothy M (2024) Carbon nanotubes and nanosheets for sustainable solutions. Nanomaterials 14(20):1–6. https://doi.org/10.3390/nano14201649

Parvin N, Kumar V, Mandal TK, Joo SW (2024) Advancements in nanoporous materials for biomedical imaging and diagnostics. J Funct Biomater 15(8). https://doi.org/10.3390/jfb15080226

Pisoschi AM (2016) Potentiometric biosensors: concept and analytical applications-an editorial. Biochem Anal Biochem 5(3):19–20. https://doi.org/10.4172/2161-1009.1000e164

Pohanka M (2018) Overview of piezoelectric biosensors, immunosensors and DNA sensors and their applications. Materials (Basel) 11(3). https://doi.org/10.3390/ma11030448

Police Patil AV, Chuang YS, Li C, Wu CC (2023) Recent advances in electrochemical immunosensors with nanomaterial assistance for signal amplification. Biosensors 13(1). https://doi.org/10.3390/bios13010125

Poltavets V et al (2023) Formation of MnO2-based electrodes with high catalytic activity on the surface of ITO for glucose detection, pp 1–23

Prince M, Ali GC, Guerchet M, Prina AM, Albanese E, Wu YT (2016) Recent global trends in the prevalence and incidence of dementia, and survival with dementia. Alzheimers Res Ther 8(1). https://doi.org/10.1186/s13195-016-0188-8

Principles D, Ramesh M, Janani R, Deepa C, Rajeshkumar L (2023) Nanotechnology-enabled biosensors: a review of fundamentals. Biosensors 13(40):1–32

Purohit B, Vernekar PR, Shetti NP, Chandra P (2020) Biosensor nanoengineering: design, operation, and implementation for biomolecular analysis. Sensors Int 1(September):100040. https://doi.org/10.1016/j.sintl.2020.100040

Qin J, Cho M, Lee Y (2019) Ultrasensitive detection of amyloid-β using cellular prion protein on the highly conductive Au nanoparticles-poly(3,4-ethylene dioxythiophene)-poly(thiophene-3-acetic acid) composite electrode. Anal Chem 91(17):11259–11265. https://doi.org/10.1021/acs.analchem.9b02266

Radi A-E (2011) Electrochemical aptamer-based biosensors: recent advances and perspectives. Int J Electrochem 2011:1–17. https://doi.org/10.4061/2011/863196

Rahman M et al (2024) A review on nanomaterial-based additive manufacturing: dynamics in properties, prospects, and challenges. Prog Addit Manuf 9(4):1197–1224. https://doi.org/10.1007/s40964-023-00514-8

Rajasree RS, Pede SV, Kharat R, Pooja Sharma S, Gopika GS, Bansode S (2024) Alzheimer's disease (AD) classification using MRI: a deep ensemble model with modified local pattern feature set. Intell Decis Technol 18:1–21. https://doi.org/10.3233/idt-230524

Rama EC, Costa-García A, Fernández-Abedul MT (2017) Pin-based electrochemical glucose sensor with multiplexing possibilities. Biosens Bioelectron 88:34–40. https://doi.org/10.1016/j.bios.2016.06.068

Rashmi BN, Harlapur SF, Ravikumar CR et al (2020) MoO3nanoparticles based electrodes as novel electrochemical sensors for the detection of H2O2. Mater Today Proc 46:5931–5935. https://doi.org/10.1016/j.matpr.2020.09.161

Rivnay J, Inal S, Salleo A, Owens RM, Berggren M, Malliaras GG (2018) Organic electrochemical transistors. Nat Rev Mater 3. https://doi.org/10.1038/natrevmats.2017.86

Roosan MR et al (2021) Usefulness of circulating tumor DNA in identifying somatic mutations and tracking tumor evolution in patients with non-small cell lung cancer. Chest 160(3):1095–1107. https://doi.org/10.1016/j.chest.2021.04.016

Serafín V et al (2020) Enlightening the advancements in electrochemical bioanalysis for the diagnosis of Alzheimer's disease and other neurodegenerative disorders. J Pharm Biomed Anal 189. https://doi.org/10.1016/j.jpba.2020.113437

Sethi J, Van Bulck M, Suhail A, Safarzadeh M, Perez-Castillo A, Pan G (2020) A label-free biosensor based on graphene and reduced graphene oxide dual-layer for electrochemical determination of beta-amyloid biomarkers. Microchim Acta 187(5):288. https://doi.org/10.1007/s00604-020-04267-x

Sheikh S, Blaszykowski C, Thompson M (2011) Label-free detection of HIV-2 antibodies in serum with an ultra-high frequency acoustic wave sensor. Talanta 85(1):816–819. https://doi.org/10.1016/j.talanta.2011.04.008

Shi X, Gu W, Li B, Chen N, Zhao K, Xian Y (2014) Enzymatic biosensors based on the use of metal oxide nanoparticles. Microchim Acta 181(1–2):1–22. https://doi.org/10.1007/s00604-013-1069-5

Shtepliuk I, Vagin M, Yakimova R (2020) Electrochemical deposition of copper on epitaxial graphene. Appl Sci 10(4). https://doi.org/10.3390/app10041405

Shui S, Buckley S, Scheller L, Correia BE (2023) Rational design of small-molecule responsive protein switches. Protein Sci 32:1–12. https://doi.org/10.1002/pro.4774

Singh RP (2011) Prospects of nanobiomaterials for biosensing. Int J Electrochem 2011:1–30. https://doi.org/10.4061/2011/125487

Sivarajasekar N, Dharshni R, Dharshini R (2019) Thermal biosensors and their applications. Am Int J Res Sci (December):50. [Online]. Available: http://www.iasir.net

Skládal P (2024) Piezoelectric biosensors: shedding light on principles and applications. Microchim Acta 191(4):184. https://doi.org/10.1007/s00604-024-06257-9

Song M et al (2021) Materials and methods of biosensor interfaces with stability. Front Mater 7(January):1–11. https://doi.org/10.3389/fmats.2020.583739

Souza KJ, Agrawal DK (2024) Employing toehold-mediated DNA strand displacement reactions for biomedical applications. Med-X 2(1):1–19. https://doi.org/10.1007/s44258-024-00015-5

Spychalska K, Zajac D, Baluta S, Halicka K, Cabaj J (2020) Functional polymers structures for (bio)sensing application-a review. Polymers (Basel) 12(5). https://doi.org/10.3390/POLYM12051154

Srinivasan LV, Rana SS (2024) A critical review of various synthesis methods of nanoparticles and their applications in biomedical, regenerative medicine, food packaging, and environment. Discov Appl Sci 6(7). https://doi.org/10.1007/s42452-024-06040-8

Stoltenburg R, Reinemann C, Strehlitz B (2007) SELEX-A (r)evolutionary method to generate high-affinity nucleic acid ligands. Biomol Eng 24(4):381–403. https://doi.org/10.1016/j.bioeng.2007.06.001

Su J et al (2017) Multicolor gold-silver nano-mushrooms as ready-to-use SERS probes for ultrasensitive and multiplex DNA/miRNA detection. Anal Chem 89(4):2531–2538. https://doi.org/10.1021/acs.analchem.6b04729

Su Q, Shi W, Huang X, Yin S, Yang X, Lu X (2023) Recent advances of nanobody applications in diagnosis and detection. MedComm Biomater Appl 2(3). https://doi.org/10.1002/mba2.54

Sun G, Wei X, Zhang D, Huang L, Liu H, Fang H (2023) Immobilization of enzyme electrochemical biosensors and their application to food bioprocess monitoring. Biosensors 13:886

Tang X, Bansaruntip S, Nakayama N, Yenilmez E, Chang YI, Wang Q (2006) Carbon nanotube DNA sensor and sensing mechanism. Nano Lett 6(8):1632–1636. https://doi.org/10.1021/nl060613v

Tavakoli H, Mohammadi S, Li X, Fu G, Li XJ (2022) Microfluidic platforms integrated with nano-sensors for point-of-care bioanalysis. TrAC – Trends Anal Chem 157:1–49. https://doi.org/10.1016/j.trac.2022.116806

Thévenot DR, Toth K, Durst RA, Wilson GS (2001) Electrochemical biosensors: recommended definitions and classification. Anal Lett 34(5):635–659. https://doi.org/10.1081/AL-100103209
Tombelli S, Minunni M, Mascini M (2005) Analytical applications of aptamers. Biosens Bioelectron 20(12):2424–2434. https://doi.org/10.1016/j.bios.2004.11.006
Touhami A (2017) Biosensors and nanobiosensors: design and applications. Biosens Nanobiosensors 1:391–462. https://doi.org/10.1142/9789813147263_0011
Toyos-Rodríguez C, García-Alonso FJ, de la Escosura-Muñiz A (2020) Electrochemical biosensors based on nanomaterials for early detection of Alzheimer's disease. Sensors (Switzerland) 20(17):1–43. https://doi.org/10.3390/s20174748
Tuerk C, Gold L (1990) Systematic evolution of ligands by exponential enrichment:RNA ligands to bacteriophage T4 DNA ploymerase. Science 249(4968):505–510
Ugo P, Moretto LM (2017) Electrochemical immunosensors and aptasensors. Chemosensors 5(2):13
Vais RD, Yadegari H, Heli H, Sattarahmady N (2021) A β-amyloid(1-42) biosensor based on molecularly imprinted poly-pyrrole for early diagnosis of Alzheimer's disease. J Biomed Phys Eng 11(2):215–228. https://doi.org/10.31661/jbpe.v0i0.1070
Valenzuela-Amaro HM et al (2023) Emerging applications of nanobiosensors in pathogen detection in water and food. Biosensors 13(10). https://doi.org/10.3390/bios13100922
Valkova P, Pohanka M (2021) Novel trends in electrochemical biosensors for early diagnosis of Alzheimer's disease. Int J Anal Chem 2021. https://doi.org/10.1155/2021/9984876
van Aalen EA et al (2024) Turning antibodies into ratiometric bioluminescent sensors for competition-based homogeneous immunoassays. ACS Sensors 9(3):1401–1409. https://doi.org/10.1021/acssensors.3c02478
Wang Z (2022) Manufacturing and characterization of... ERA, pp 1–15. [Online]. Available: https://era.library.ualberta.ca/items/fcde84b0-f795-4d78-8bac-22862c79a519
Wasilewski T, Kamysz W, Gębicki J (2024) AI-assisted detection of biomarkers by sensors and biosensors for early diagnosis and monitoring. Biosensors 14(7). https://doi.org/10.3390/bios14070356
Wei D, Ivaska A (2006) Electrochemical biosensors based on polyaniline. Chem Anal 51(6):839–852
Wei M, Deng T, Zhang Q, Cheng Z, Li S (2021) Seed-mediated synthesis of gold nanorods at low concentrations of CTAB, vol 6, pp 9188–9195. https://doi.org/10.1021/acsomega.1c00510
Yoo J, Jeong H, Park SK, Park S, Lee JS (2021) Interdigitated electrode biosensor based on plasma-deposited TiO2 nanoparticles for detecting DNA. Biosensors 11(7). https://doi.org/10.3390/bios11070212
Zhang B, Banal JL, Jones DJ, Tang BZ, Ghiggino KP, Wong WWH (2018a) Aggregation-induced emission-mediated spectral downconversion in luminescent solar concentrators. Mater Chem Front 2(3):615–619. https://doi.org/10.1039/c7qm00598a
Zhang D, Jiang C, Zhou X (2018b) Fabrication of Pd-decorated TiO2/MoS2 ternary nanocomposite for enhanced benzene gas sensing performance at room temperature. Talanta 182:324–332
Zhang N, Li C, Jiang S, Du L (2020) Recent advances in the development of virus-like particle-based flavivirus vaccines. Vaccines 8(3):1–20. https://doi.org/10.3390/vaccines8030481
Zhang Y, Lin T, Shen Y, Li H (2022) A high-performance self-supporting electrochemical biosensor to detect aflatoxin B1. Biosensors 12(10):1–12. https://doi.org/10.3390/bios12100897
Zhang Y et al (2023) MOF-based materials for glucose detection. Chemosensors 11(8). https://doi.org/10.3390/chemosensors11080429
Zhang Z, Liu X, Zhou H, Xu S, Lee C (2024) Advances in machine-learning enhanced nanosensors: from cloud artificial intelligence toward future edge computing at chip level. Small Struct 5(4). https://doi.org/10.1002/sstr.202300325
Zhu C, Yang G, Li H, Du D, Lin Y (2015) Electrochemical sensors and biosensors based on nanomaterials and nanostructures. Anal Chem 87(1):230–249. https://doi.org/10.1021/ac5039863

# Chapter 16
# Recent Advances of Nanomaterials in Cosmetics

**Saranya Thekkila-Veedu, Dhilin Pathayappurakkal Mohanan, Joseph Pushpa Sweety, P. Navya Sree, Dhanalekshmi Unnikrishnan Meenakshi, and Sri Durga Devi Nagarajan**

**Abstract** The cosmetics industry is a dynamic, ever-changing one, particularly in light of the introduction of new technology that have produced innovative formulas with both medicinal and cosmetic advantages. Significant improvements and developments in cosmetic formulations and delivery systems could be facilitated by nanotechnology. The application of nanotechnology in cosmetic formulations is still a relatively new, highly promising, and thoroughly studied field. Nanocosmetics demonstrates its functions through a variety of delivery mechanisms, such as lipid nanocarrier systems, metallic or polymeric nanoparticles, nanocapsules, dendrimers, and nanosponges. The limitations of traditional cosmetics are addressed and new beneficial characteristics are introduced through the use of nanotechnology in cosmetics. Nanomaterials improve product performance and customer needs. Applications involving skin, hair, nails, lips, and teeth have been extensively studied in the fields of nanocosmetics. This has led to the replacement of many traditional cosmeceuticals with nanocosmeceuticals. However, nanotoxicological studies have raised concerns about the potential health hazards of nanocosmeceuticals, particularly regarding skin penetration and subsequent toxic effects. This review summarizes the various nanotechnology-based approaches currently used in the delivery of cosmetics and nanomaterials used in cosmetics. It discusses the advantages as well as

S. Thekkila-Veedu · D. P. Mohanan
Department of Pharmaceutics, Dr. Moopen's College of Pharmacy, Wayanad, Kerala, India

J. P. Sweety
Department of Chemical Engineering, Indian Institute of Technology, Madras, Chennai, Tamil Nadu, India

P. Navya Sree
Indira Gandhi Centre for Atomic Research, Kalpakkam, Tamil Nadu, India

D. U. Meenakshi
College of Pharmacy, National University of Science and Technology, Muscat, Oman

S. D. D. Nagarajan (✉)
Sydney pharmacy school, Faculty of Medicine and Health, The University of Sydney, Sydney, Australia

 
V. Khanna et al. (eds.), *Nanotechnologies for Drug Delivery and Biopharmaceutical Development*, Progress in Drug Research 77,
https://doi.org/10.1007/978-3-032-00796-4_16

the potential health and environmental risks associated with these products. The research also looks at the regulatory status of cosmeceuticals, comparing and contrasting national and international norms.

**Keywords** Nanocosmetics · Nanomaterials · Nanocarrier · Nanocapsules · Nanotoxicology · Regulatory status

## 16.1 Introduction

Cosmetics is world filled with beauty kits and grooming junks. The term "cosmetic" is any particular substance which is used to either clean, enhance, or beautify the god's natural creation. On the other hand, any cosmetic product should be more effective, more agreeable to use, more attractive and longer lasting.

In the past, people have included plant-based extracts in their daily diets, which strengthens cells' innate ability to defend against radiation and oxidative stress. In addition to food, they applied a variety of natural skincare treatments. A few of these include lignin, ascorbic acid, and anthocyanins as sunscreen whitening agents; sunflower seed oil, walnut oil, and grape seed oil were utilised for barrier repair functions; curcumin from turmeric, resveratrol, and flavonoids from various plants as antioxidants; and many more. Remnants of popular natural moisturisers like coconut oil and raspberry oil are still in use (Xie et al. 2024; Thibane et al. 2019). Excellent anti-fungal neem oil has long been used as a hair care product to eradicate dandruff and fungal diseases. Coconut oil and raspberry oil are well-known natural moisturizers still in practice. Neem oil is an excellent anti-fungal agent has long been used as a hair care product to eradicate dandruff and fungal diseases. Brushing with neem sticks is still a common practice. Later on, people began to value their appearance more because of their jobs, lifestyles, and surroundings (Hughes et al. 2019). Tooth decay, ageing, fungal infections, problems with dandruff and hair loss, and hyperpigmentation are the key concerns that everyone can afford to ignore. The main challenging factor in the field of cosmetics is fighting against aging mechanism, which is either caused by endogenous as well as exogenous factors. Healthy skin barrier is a key to protect from photoaging i.e. mainly occurred by the sunlight exposure, dehydration, radiation and reactive oxygen species. To overcome these issues, researchers started working on the preparation of various natural and chemical-based products, called cosmetics. Cosmetics broadly includes personal hygiene products for cleansing our body such as using shampoos for hair growth, body wash, face wash, and soap for skin care, toothpaste and mouth wash, and many more. This is how the cosmetics industry has developed into a hugely successful one (Gupta et al. 2022). The first cosmetic products to use nanotechnology were placed on the EU market in 1986, and their use has since increased. India is ranked fourth globally in terms of revenue from the beauty and personal care sector in 2023.

Cosmetic formulations are created by carefully combining various ingredients to produce a meaningful result. Novel drug delivery systems in cosmetics deals with either nanoparticles or nanopolymers or nanoemulsions as carriers for the active ingredients. This newer kind of technological approach reach its maximum attention due to its ease of delivering the natural cosmetic ingredient to the desired skin layer at the right time and right dose level. Also, the significant point to notice is that nanotechnology will aid in improving the usage of cosmetic products through increasing its overall performance, efficiency and targeted delivery.

The development of advanced drug delivery techniques is made possible in large part by nanocosmeceuticals, particularly when they take the form of nanoemulsions or nanocapsules. The most crucial factor when it comes to skincare products is pH adjustment; if not, the user will be harmed. In order to produce goods that are both ethically and environmentally sound, it is highly desired to use sustainable substances. UV-protecting substances in sunscreen and anti-aging products include para-aminobenzoic acid and its derivatives, cinnamates, salicylates, camphor derivatives, avobenzone, triazones, and benzophenones.

Some of the important nano-based formulations have been highlighted which are playing crucial role in enhancing cosmetic applications. First comes the nanoemulsion, which is widely being used in the fields of food, cosmetics, as well as pharmaceutical industry. The important property of its small droplet size in nanoscale range helps in delivering the lipophilic compounds through penetrating the skin layer. This kind of formulation is preferred commonly for making body lotions, skin creams, and sunscreen lotions. The main drawback of this nanoemulsion is that it is thermodynamically unstable due to its very small radius. The other significant type is the nanoliposome-based cosmetic. Nanoliposomes are widely reported for its skin penetration property, due to their interaction with the cell membrane (Kim et al. 2019). Recently, enhanced anti-aging effect of nanoliposomes has been found when using nanoliposomes for delivering the bioactive peptide molecules. This study addressed the natural biological phenomenon of skin aging using nanoliposomes loaded with carnosine and palmitoyl tripeptide-5 and acetyl hexapeptide-3. These nanoscale system could protect applicant's face in terms of reduced volume and area of wrinkles, with improved skin elasticity (Han et al. 2020). Cao et al. (2022) have studied the benefits of fibrous clays in dermopharmaceutical and cosmetic application. These fibrous clays are excellent nanofillers or nanocarriers for sensitive drugs and cosmetic ingredient (Cao et al. 2022).

Skin pigmentation is a main concern nowadays among the working women who are travelling in the daytime. Recently, multifunctional melanin-based nanoparticles were studied for its nanomedical as well as nanocosmetic applications. In this study, melanin shows excellent property of protective action against UV and oxidants, when it is in the form of nanopigment. In addition to that, multifunctionality of melanin and its analogous PDA were elucidated which includes, its prominent drug loading capacity, metal chelation property, and easy functionalization. While scientists are trying to mimic the behavior of natural melanin pigment, this research group has used PDA nanoparticles, which can mimic the behavior of natural melanosomes and replicating the actions of endocytosis, accumulation, and the creation of so-called micro parasols that shield cells from UV harm (Mavridi-Printezi et al. 2020). Nanohydrogels, nanocapsules, and dendrimers are other important nanoformulations being used in the cosmetic industry.

### 16.1.1 Advantages of Nano DDS Over Traditional Cosmetics

Nano-based cosmetic products have shown improved physicochemical along with the biological properties from the classical products. For example, nanoemulsions are optically transparent with defined surface tension suitable for creaming application. High retention time, more durability, deeper skin penetration and targeted performance of nano-based cosmetics are highlighted by many studies (Dubey et al. 2022). Nanofibers, used in making face masks and cleansers, are usually one-dimensional material with increased surface area and high aspect ratio compared to the traditional one. Metal-based inorganic nanocosmetics are playing crucial role in commercial market; however, more care not be taken in case of safer formulation, preventing metal-based cytotoxic effects. Earlier, $ZnO_2$ and $TiO_2$ are commonly used in sunscreens to protect skin against UV; nowadays $ZnO_2$ and $TiO_2$ nanoparticles are often used as they shown higher protection compared to the $TiO_2$ pigment. Nowadays, nanopigments are used in the manufacturing of a lot of lipsticks. In contrast to earlier lipsticks that eventually chip or flake, these are a more homogenised formulation. Though lots of advantages of nanocosmetics have been highlighted, they are not environmental friendly in case of improper disposal and management. However, natural and plant-based extracts found in traditional cosmetic products are friendly to the environment and way of life (Bocca et al. 2014).

### 16.1.2 Their Pharmacokinetics and Efficacy

Once chemically derived cosmetics enter our bodies, they must interact with our bodies' enzymes in order to be consumed. The active centre of an enzyme is the region where the interaction takes place which should be optically active and three-dimensional. *In vitro or in vivo* characterization would also reveal the potential action of the cosmetic product. Accordingly, a research study has been carried out for evaluating the *in vitro* and *in vivo* safety of a new cosmetic material derived from the by-product of *Agave sisalana*. Here, polysaccharide-enriched fraction was formulated as nanoemulsion to improve skin moisturizing. Transepidermal water loss is the important parameter to study about the skin hydration (Maria et al. 2017).

## 16.2 Overview of Nanomaterials in Cosmetics

The first industry to use nanotechnology was the cosmetics sector, which created nanoparticles of size ranges 1 to 100 nm. After the first liposome-based anti-aging cream, CaptureTM, hit the market in 1986, several well-known companies began using nanomaterials in skincare products. Increased surface area for enhanced skin penetration, introduce new lipstick and nail polish colors; long-lasting makeup effects; and powerful sun blockers with a translucent, crystal-clear effect on all skin types are the main goals of nanomaterials. Furthermore, a large variety of nanocosmetic products are on the market, illustrating the diverse applications of nanotechnology in the cosmetics sector. Eye liner, bronzer, body firming lotion, exfoliating scrub, and styling gel are a few examples of these products.

Cosmetic nanomaterials fall into two categories: inorganic and organic. Because of their hydrophilicity, stability, biocompatibility, and non-toxicity, inorganic nanoparticles in particular are used more frequently in cosmetics than organic ones. Another significant benefit of inorganic nanoparticles is their ease of surface modification, which enables a variety of formulations to produce cosmetic effects while lowering the possibility of adverse effects. The issues with organic nanoparticles can be resolved by inorganic ones, which have greater chemical stability and cause less skin irritation, itching, and allergy. In this regard, inorganic nanoparticles are becoming increasingly important in dermopharmacy and the cosmetics sector. Furthermore, these benefits are enhanced by their lower production costs.

### *16.2.1 Inorganic Nanoparticles*

Metals or metal oxides are the building blocks of inorganic nanoparticles. These particles have higher hydrophilicity, biocompatibility, stability, and safety compared to natural nanoparticles. These nanoparticles are distinct from natural nanoparticles, which are composed of polymers, due to their inorganic source, which is composed of elements such as Ag, Au, Ti, and others. Although they can be used as a rheology modifier or carrier to enhance cosmetic performance, their primary function is as active ingredients for UV and antimicrobial protection (Salvioni et al. 2021). The following figures depicts the different types of inorganic particles and distribution in cosmetic formulations.

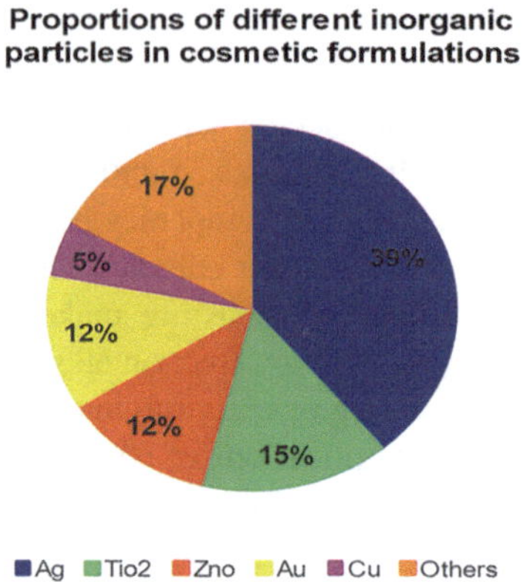

## 16.2.2 *Organic Nanoparticles*

Organic nanoparticles are biodegradable and nontoxic which are composed of various formulations at nanoscale made of natural or synthetic agents. Lipid, surfactant, and polymer-based nanostructures are the most commonly utilized organic nanocarriers in the cosmetic products. The primary goal of nanocrystal formulation is to enhance the solubility of substances. The following figure depicts the different type of organic nanoparticles.

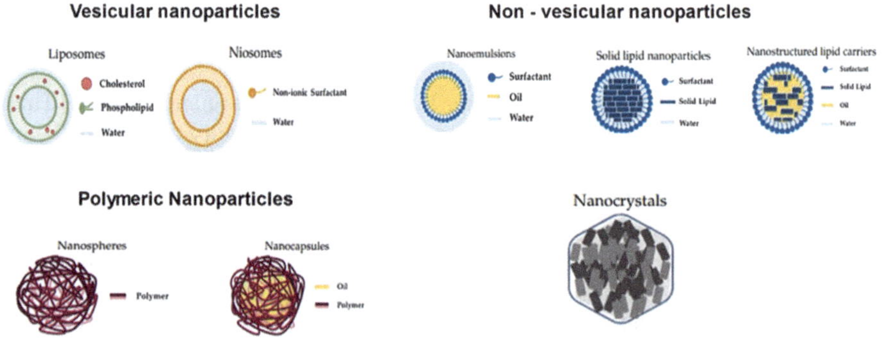


# 16.3 Nanodrug Delivery Systems in Cosmetics

Delivery systems are engineered technologies which are utilized to transport an active ingredient in a controlled and targeted manner. The human skin serves as a barrier against foreign molecules. Delivery systems can enhance drug permeability, regulating its concentration in both the product and the skin. Nanotechnology provides innovative pathways to various challenges in the medical and pharmaceutical fields for the past few decades. This same principle has been employed in the realm of cosmetics, leading to the development of new formulations referred to as nanocosmeceuticals, which offer tailored solutions for cosmeceutical issues. The unique advantages of nanocosmeceuticals are due to their smaller size, which provides them with novel characteristics such as enhanced solubility, transparency, chemical reactivity, and stability. Nanomaterials like liposomes, ethosomes, solid lipid nanoparticles, nanocapsules, dendrimers, nanocrystals, cubosomes, and nanoemulsions are commonly employed. Presently, nanocosmeceuticals are widely available in the market (Gupta et al. 2022).

## *16.3.1 Types of Nanodrug Delivery Systems*

### 16.3.1.1 Lipid- and Surfactant-Derived Nanoparticles

#### Vesicular Systems

Multilamellar vesicles (MLVs), small unilamellar vesicles (SUVs), and large unilamellar vesicles (LUVs) are the three general classifications for vesicular nanoparticles (Van Hoogevest and Fahr 2019). While SUVs (10–100 nm) and LUVs (100–500 nm) are made from MLVs using techniques like extrusion, sonication, and high-shear homogenization, MLVs (0.5–10 μm) are easily made by dispersing raw materials in an aqueous phase (Marianecci et al. 2014). In cosmetics, submicron nanocarriers are usually preferred (Van Tran et al. 2019). These systems similarities to naturally occurring cellular membranes and vesicles prompted the cosmetics industry to employ them in 1986. These nanoparticles improve the penetration of active ingredients and effectively deliver them into the deeper skin layer because vesicles are crucial for molecular transport between cells. These engineered vesicles can hold both lipid soluble and waters soluble active ingredients because of their amphiphilic nature. Furthermore, these vesicles lipidic and surfactant constituents are biocompatible and largely biodegradable (Carita et al. 2018).

The most studied nanomaterial in cosmetics is liposomes, which are composed of phospholipids (primarily phosphatidylcholine derivatives) (Khezri et al. 2018). The composition of bilayer influences how well they deliver their contents. Unsaturated phospholipids increase the permeability of nanocarriers, while saturated phospholipids containing cholesterol stiffen the bilayer structure (Akbarzadeh et al. 2013). This particular kind of nanocarrier has low toxicity, is extremely biocompatible and biodegradable, and is highly resistant to the enzymatic breakdown of medications. Its use in pharmaceutical and cosmetic products has become more common due to these qualities. Additional benefits include the ability to encapsulate both hydrophilic and hydrophobic particles, simplicity in preparation, and good topical product applicability. Nevertheless, there are certain drawbacks to this system, including inadequate penetration, challenging large-scale production, and poor physical and chemical stability.

Ethosomes and transferosomes, which are highly deformable vesicles, are prepared by incorporating ethanol (ranging from 20–45% by weight) or softening surfactants (known as edge activators) into liposome formulations. This modification has resulted in vesicles with improved flexibility, leading to superior skin penetration, a fact that has been widely demonstrated (Das et al. 2018). In contrast, niosomes are composed of non-ionic surfactants, such as alkyl esters (Spans and Tweens) and alkyl ethers (Brijs) (Chen et al. 2019a, b). Niosomes were produced as an economical replacement to liposomes due to the availability and affordability of their raw materials. They also show better long-term stability when compared to vesicles based on phospholipids (Salvioni et al. 2021). A research study conducted by Lu et al. demonstrated that the quercetin-loaded niosomes, showcased enhanced skin penetration, sustained release over time, and significantly increased transdermal penetration with a retention rate of 2.95 times higher than that of quercetin solution. Furthermore, a study by Radmard et al. reported that a formula containing arbutin-loaded niosomes is not toxic and did not provoke skin irritation, while simultaneously improving arbutin delivery to the skin. As a result, arbutin loaded niosomes could be a novel method for treating hyperpigmentation conditions.

**Non-vesicular Systems**

This category includes several distinct heterogeneous systems made up of immiscible phases (i.e. hydrophilic and lipophilic) and surfactants. Microemulsion and high-pressure homogenization are commonly used to prepare solid lipid NPs (SLNs) and nanostructured lipid carriers (NLCs), which have an inner lipid core that is solid at body temperature. In SLNs, solid lipids alone make up the lipid mixture (e.g., long-chain glycerides, fatty acids, waxes) (Geszke-Moritz and Moritz 2016), whereas in NLCs the lipophilic matrix is composed of solid and liquid lipids (e.g., short-chain glycerides) in a ratio ranging between 70:30 and 99.9:0.1 (Khosa et al. 2018). In addition to its occlusive properties, which facilitate skin hydration and drug penetration, this type of nanoparticle is being studied for its potential as a carrier of chemically unstable compounds (Guimarães and Ré 2011).

**Solid Lipid Nanoparticles (SLNs)**

These lipids are biodegradable, low toxicity, and capable of encasing a variety of active substances with a range of characteristics, including hydrophilic, lipophilic, and poor water solubility (Arora et al. 2012). When these nanoparticles are encapsulated, the enzymatic breakdown of cosmetic agents is inhibited, the controlled release of active agents is prolonged, and the entry of these materials into the stratum corneum is improved. Addition of resveratrol to SLNs made from *Theobroma grandiflorum* natural seed butter enhanced antioxidant activity by 20% while also improving the active ingredient's penetration and retention in human skin. In addition, the resveratrol-containing SLNs showed a more than two-fold increase in resveratrol concentration in the stratum corneum when compared to an ethanolic solution of resveratrol. SLNs' inherent qualities, including their targeted drug action, intensified skin penetration, decreased toxicity, enhanced bioavailability, and nanoscale geometry, make them suitable for use in cosmeceuticals (Abbasi et al. 2020). Additionally, these nanoparticles' occlusive qualities improve the skin hydration. The techniques used for preparation and the physico-chemical properties of the active ingredients can affect the efficacy of SLNs (Jensen et al. 2010). The transdermal delivery of SLNs loaded tazarotene demonstrated to have superior tolerability to the commercial formulation to treat psoriasis.

**Nanostructured Lipid Carriers**

The limitations of SLNs were addressed by the development of nanostructured lipid carriers (NLCs), which are superior in terms of stability, safety, drug encapsulation, skin permeability, skin hydration and UV protection. Due to their enhanced adhesion with the stratum corneum and biphasic release pattern, these nanoparticles increase the drug penetration through the dermal layer. NLCs also offer improved UV protection with few adverse effects and maintain their stability over time (Raju et al. 2020). Enhanced bioavailability, film formation, and controlled occlusion are the valuable qualities of NLCs which make them a valuable tool for cosmetic applications. Furthermore, toxicity concerns are minimized because NLCs are made with physiological and biocompatible lipids. In order to increase the effectiveness of the cosmetic agent, a transdermal formulation of NLC based isoliquiritigenin with ceramide was developed. The results of in vivo testing of quercetin nanoencapsulation revealed improved skin barrier function and hydration while having an additive effect on the sun protection factor. Rifampicin and clindamycin phosphate were encapsulated into NLCs for the treatment of hidradenitis suppurativa, a chronic inflammatory disease linked to an irreversible blockage of the pilosebaceous units. The prepared formulation is non-irritant and exhibited controlled release profiles. In summary, a number of studies have demonstrated that NLCs can deliver various cosmetic ingredients to the appropriate body parts while overcoming their limitations, resulting in long-term stability.

**Nanoemulsion**

A nanoemulsion possesses a number of desirable characteristics, including high solubilization capacity, low viscosity, high interfacial area, high kinetic stability, merging textures, enhanced drug penetration, improved hydration and controlled release. The sunflower oil nanoemulsion with a higher sunprotection factor than an emulsion was demonstrated. Furthermore, the ability of a lavender essential oil and licorice extract nanoemulsion cream to heal deep skin wounds using a murine model were also investigated. Increased collagen deposition, quicker re-epithelialization, and elevated antioxidant activity were all observed in the wound area after the application of the nanoemulsion.

**Polymeric Nanoparticles**

Increased drug stability, co-encapsulation capability, and possible side effect reduction are the primary advantages of polymeric nanoparticles. Among the synthetic polymers, biodegradable aliphatic polyesters (e.g., polylactic acid (PLA), poly-lactic-co-glycolic acid (PLGA) and poly-caprolactone (PCL)), biocompatible polyacrylates and cellulose derivatives are commonly employed, whereas extensively studied natural polymers include chitosan, alginate and gelatin (Souto et al. 2020). The application of polymers derived from natural sources is gaining popularity because it is thought to be safer and more environmentally friendly than the synthetic polymers, despite their drawbacks, which include high manufacturing costs, high variability, and low product purity.

Nanocapsules are reservoirs made up of a polymeric shell enclosing a liquid core, usually filled with oil and stabilized by a surfactant coating. The bioactive agent is typically housed in the inner core, and its release is regulated by the polymeric counterpart (Salvioni et al. 2021). The ability of nanocapsules to control the drug penetration into deeper layers of the skin, leaving only a thin surface layer, makes them popular in sunscreen formulations. In order to lessen the compatibility problems of active agents in the formulation, they are also frequently used to cover up some of the distinctive smells of active ingredients that could negatively impact customer satisfaction. Nevertheless, employing this kind of nanocarrier may result in certain adverse effects. The polymeric matrix that makes up nanospheres allows for the entrapment or adsorption of a variety of active ingredients (Santos et al. 2019a, b). Although these NPs can hold a wide range of hydrophilic and hydrophobic molecules, their primary use is to enhance the functionality of chemically labile (retinoids and antioxidants), poorly water soluble (organic filters) and volatile agents (fragrances) (Guterres et al. 2019).

Dendrimers are a novel class of nanoparticles for nanocosmeceuticals because of their superior skin penetration. These polymeric nanoparticles are distinguished by their extremely ordered structure, which is made up of repeated branches that branch out from a central point and end in various neutral, cationic, or anionic functional groups on the surface (Chis et al. 2020).

**Nanocrystals**

The cosmetic industry is well aware of the problem of poorly water-soluble agents. This limitation affects many natural compounds and organic filters. The production of nanocrystals is suggested as a remedy for this issue since the drug solubility is greatly increased by the large surface area and low crystallinity exhibited. The stabilization of nanocrystals through surfactant/polymeric coating can be achieved through top-down or bottom-up methods, though the latter is preferred (e.g., wet milling and high-pressure homogenization) are generally preferred (Souto et al. 2020). The significant adhesiveness of nanocrystals, which results from their high encapsulating capacity and surface area, is one of their intriguing characteristics. This leads to prolonged retention at the target site. 18β-glycyrrhetinic acid (GA) nanocrystals were prepared to address the substance's low water solubility. The use of nanocrystals as promising nanocarriers for vitamin C and quercetin in cosmetics have also been documented in the literature.

**Gold Nanoparticles (Au NPs) Nanoparticles**

They have numerous applications (e.g., colorant, anti-aging, preservative, carrier) due to their remarkable stability, high loading capacity and easy manipulability of morphological properties (Cao et al. 2016). Toothpaste could benefit from the addition of nanogold, as it helps eradicate oral bacteria (Raj et al. 2012). They are naturally inert, highly stable, non-cytotoxic, and biocompatible. Because of their shape, small size, crystallinity, and large surface area, these nanoparticles can be delivered to the target site with ease. Furthermore, the gold nanoparticles' antibacterial and antioxidant qualities allow lotion production that improve skin texture and postpone aging. Because of their potent antibacterial and antifungal qualities, as well as their exceptionally high drug-loading capacity, nanogold particles are valuable materials in the cosmeceutical sector (Yeh et al. 2012). Gold nanoparticles are used in beauty care to boost blood circulation, enhance skin elasticity and firmness, give antiseptic and anti-inflammatory qualities, and vitalize skin metabolism (Thakor et al. 2011).

**Silver Nanoparticles (Ag NPs)**
The use of silver nanoparticles in cosmeceutical products is particularly pertinent to their inherent antimicrobial, antifungal, anti-inflammatory, and preservative qualities. The use of silver nanoparticles in cosmetics—like underarm deodorant—offers all-day protection. In terms of silver nanoparticles, their antibacterial action results from both the bonding of silver ions to the respiratory chains and the alteration in the penetration of the bacterial cell wall. Reactive oxygen species (ROS), which are produced in greater quantities as a result of this association, hinder the growth and survival of bacteria (Abbasi et al. 2020).

**Titanium Oxide and Zinc Oxide Nanoparticles ($TiO_2$ and ZnO NPs)**
Nanoparticles, which are produced by nanoengineering zinc oxide and titanium dioxide, have various benefits in cosmetic applications. Improved skin transparency is one obvious advantage, which solves the long-standing problem of the white cast frequently connected to conventional formulations with larger mineral particles (Nath et al. 2021). Customers looking for sun protection without the unfavorable aesthetic effects will find nanoparticles appealing since they allow for a more even and sophisticated cosmetic application. Improved UV light absorption and scattering characteristics are made possible by the use of nanomaterials. This suggests that zinc oxide and titanium dioxide nanoparticles can provide enhanced protection against UVA and UVB radiation, reducing the risk of skin damage, early aging, and skin cancer.

**Silica Nanoparticles**
Their principal ingredient is amorphous silica, which is typically utilized in rinse-off and leave-on cosmetics to enhance texture and/or impart a matte (opaque) finish in the area that the product is used. Apart from their oil-absorbing qualities, silica nanoparticles are highly appreciated for their ability to work wonders in cosmetic products for people with combination or oily skin (Bamford et al. 2021). For a longer wearing time and a matte finish, these nanoparticles help regulate excessive sebum production. An important development in formulation technology is the incorporation of silica nanomaterials into cosmetics, which provide better sun protection, oil control, and efficient delivery of active ingredients.

**Copper Nanoparticles**

An example of copper nanoparticles with anti-aging qualities is an NP-impregnated pillow that reduces wrinkles in face (Baek et al. 2012). Nevertheless, there are very few skincare products on the market that contain these NPs. Since copper is essential for the synthesis of skin extracellular matrix and has biocidal properties, copper-based nanomaterials have been studied for both biocidal and anti-aging applications (Borkow 2015). They are not commonly used in skincare, and their role is still debatable.

Although aluminium oxide nanoparticles are also on the EU-approved list of nanomaterials, they are utilized in cosmetics (concealer, foundation, and mineral foundations, for example) rather than skincare products. Numerous carbon-based nanomaterials, including nanotubes, fullerenes, and nanodiamonds, were investigated for cosmetic applications (Fytianos et al. 2020). While the first two have been used in skincare, the latter is mostly used as a pigment carrier in haircare along with other non-toxic and low-cost nanotubes (e.g. naturally occurring halloysite) (Santos et al. 2019a, b).

### *16.3.2 Formulation and Characterisation of Cosmetic Products*

Once the nanoparticles (NPs) are synthesized, they must be included in a solvent that suits cosmetic applications. Since NPs are rarely applied directly to the skin in aqueous dispersions, selection of suitable vehicle to formulate them is necessary. The stability of nanoparticles (NPs) in skincare products is dependent on both the formulation and the medium composition in which they are dispersed. Moreover, the selection of vehicle depends on the type of cosmetic formulation i.e., day or night cream, body lotion, concentrated serum, etc.

The type and strength of the interactions between the particles, dispersant, emulsifiers, and thickeners determine the physical stability. Buffers, antioxidants and chelating agents are included to maintain chemical stability. Preservatives added for microbiological stability. Customers are more satisfied with the product's organoleptic qualities when dyes and fragrances are used. Liposome-containing cosmetics are difficult to wash off because of their high substantivity. In particular, liposomes can raise the SC's lipid content, which will improve skin hydration and reduce dryness (Betz et al. 2005).

Liposomes can be added to gels or O/W emulsions with a low surfactant concentration because they are primarily made in aqueous systems. Damage to the structure of liposomes can result from high surfactant concentrations. One of the most widely used colloids was methylcellulose, followed by hydroxyethylcellulose, hydroxypropylmethylcellulose, and carboxymethyl cellulose. In terms of consistency and stability, acrylic acid polymers (Carbopol®) produced better results. Later, stable emulsions without surfactants were made possible by the use of polymeric emulsifiers like Pemulen®, which are copolymers of acrylic acid and C10–C30 alkyl acrylate crosslinked with allyl pentaerythritol (Kocic et al. 2020). A combination of surfactants sourced from vegetables was used to introduce innovations in liposome formulation in more recent times (Filipović et al. 2016).

Similar to liposome preparations, niosome-containing products are typically low-viscosity aqueous dispersions that are unsafe to apply topically due to their propensity to leak from the site of administration. This disadvantage is typically addressed by adding gelling polymers in amounts that ensure suitable spreading property; additionally, the gel creates an occlusion effect on the administration site that stops water evaporation. This promotes penetration of formulation and improve skin hydration (Chen et al. 2019a, b).

When a suitable stabilizing agent is not used, SLNs may aggregate. Hydrogels are promising delivery systems for the cutaneous administration of these NPs, according to an analysis of SLN and NLC formulations. According to the authors, the lipids found in both systems may exhibit a synergistic hydration in O/W creams (Garcês et al. 2018). The addition of SLNs to a cream improved its occlusive qualities without altering product characteristics, which is beneficial from a marketing perspective. Another significant characteristic of SLNs is that, while low concentration dispersions frequently aggregate, highly concentrated particle dispersions show good physical stability during storage. Desired consistency can be exhibited by concentrated NLC or SLN dispersions in any case. This leads to strategic cost savings because the final product consistency can now be achieved without the addition of thickening agents to the dispersion (Müller et al. 2002).

The majority of nanoemulsion based cosmetic formulations are of the O/W type, and their consistency can be enhanced by adding hydrocolloids like xanthan gum, Carbopol, cellulose derivatives, etc., or by increasing oil content (in certain situations, up to 25%) (Sonneville-Aubrun et al. 2004). The finished product in this instance is referred to as a "nanoemulgel" (Ahmad et al. 2018). In contrast, W/O nanoemulsions can be combined using the emulsion inversion point method to form an O/W cream or to formulate them as oleogels (Al-Edresi and Baie 2009). Utilizing suitable solutions to disperse the nanosized active ingredient is necessary for the production of nanocrystals for cosmetic applications. Because small nanocrystals are more effective than larger particles, only small amounts of nanosuspension are typically needed. To formulate the finished product, nanosuspensions can be mixed directly into a gel or cream. In order to prevent the particle growth and aggregation during storage (Ostwald ripening), it is necessary to avoid fluctuation in temperatures while adding the nanosuspension to a cream. Ionized additives like electrolytes must also be avoided to stop nanocrystal agglomeration (Pelikh et al. 2019).

Consumer satisfaction with sunscreen formulations has increased with the use of inorganic NPs such as zinc oxide and titanium dioxide. Their physical stability has proven to be quite difficult, though. Before adding inorganic nanoparticles (NPs) to the final product—be it a lotion, cream, or gel—they should be evenly distributed into an appropriate carrier. The main purpose of polymeric surfactants is colloidal stabilization. The hydrophobic chain of the polymer should be highly miscible in the medium whereas the hydrophilic chain is specifically made to favor anchoring onto the NP surface. The majority of sunblock formulas are O/W emulsions. The particles in this instance may be in the internal phase, the continuous phase, or both. The non-aqueous dispersion in W/O emulsions stays in the oily phase (Tadros 2018).

Prior to being included in a cosmetic formulation, the NPs are typically evaluated for morphology, size distribution, zeta potential, encapsulation, and/or loading efficiency. In contrast, a thorough evaluation of the entire cosmetic product is required to verify that it satisfies the necessary quality standards. The selection of suitable ingredients and the right manufacturing process should be the first steps in this evaluation, which should be followed by stability studies (Edwards 2018). The final assessment of a product involves microbiological, chemical, and physical analysis. Physical analysis considers viscosity, pH, and organoleptic characteristics such as color, odor, and phase separation. If there are active ingredients in the product, it is helpful to assess their concentration and rate of release from the formulation using specific analytical tests. Sun protection factor (SPF) and water resistance measurements are necessary for evaluating the quality of sunscreen formulations. Additionally, microbiological stability needs to be assessed to ensure that product quality standards are met both during product preparation and during the time for consumer storage and use (Salvioni et al. 2021).

## *16.3.3 Safety Assessment Testing Methods*

### 16.3.3.1 In Vitro Toxicity Assessment

This method is highly advantageous, because of its reduced cost, quicker result prediction, and less ethical concerns. Apoptosis, oxidative stress, and proliferation assays are carried out in this process.

### 16.3.3.2 Oxidative Stress Assay

The formation of reactive oxygen species and reactive nitrogen species is a well-known consequence of nanoparticle exposure. The technique for identifying these RNS and ROS species makes use of the reaction between 2,2,6,6-tetramethylpiperidine and oxygen-stable radicals. An X-band paramagnetic resonance is used in this technique. Because of its high cost, this method is not applied very often. Fluorescent probes are therefore employed because they are economical. However, the probe's inability to engage with a wide range of species is its flaw. This could produce deceptive outcomes. Therefore, 2′,7′ dichlorofluorescein, a non-fluorescent probe, is used to solve this issue. Lipid peroxidation is another metric used to quantify oxidative stress which is measured by malondialdehyde based TBA assay and C11 BIODIPY assay. Nitro blue tetrazolium assay by superoxide dismutase and amplex red assay by lipid hydroperoxide are additional assays that are employed.

#### 16.3.3.3 Proliferation Assay

This assay evaluates the metabolically active cells in order to quantify and estimate cellular metabolism. The most often utilized tetrazolium salt in evaluation is 2,5-diphenyltetrazolium bromide. This method has the advantage of producing rapid yields, reproducible results, and requiring little modification of model cells.

#### 16.3.3.4 Apoptosis Assay

It is commonly known that apoptosis is a marker used to evaluate the toxicity of nanoparticles. Apoptosis can be detected by Annexin-V, TdT-mediated dUTP-biotin nick end labeling (TUNEL), and comet assays. The DNA laddering technique is another method that is employed to visualize the endonuclease cleavage site of apoptosis. The induction of apoptosis can be confirmed by any atypical shrinkage caused by DNA fragmentation. Because it can actively distinguish between necrosis and apoptosis, agarose gel electrophoresis is also utilized in the DNA laddering procedure. The structure of a number of genomic fragments that are obtained during this electrophoresis indicates the type of cells, Regular ladder-shaped patterns signify apoptosis, while irregular patterns represent necrotic cells. Propidium iodide and Annexin V are the primary markers that are employed. The comet assay is another sensitive method that is employed to find out whether the material being tested has the ability to cause mutations. Both single-stranded and double-stranded cell breaks are readily detectable by this method (Lohani et al. 2014).

## 16.4 Biotechnology Derived Cosmetics

Biotechnology in cosmetics refers to the use of biological processes such as microbial fermentation, genetic engineering, and cellular extraction to create active substances that are closely aligned with the biological demands of the skin. Unlike synthetic chemicals, which can cause skin irritation or long-term adverse effects, biotechnology enables the creation of bioidentical substances. These mirror the skin's natural structure and function, resulting in greater compatibility and fewer side effects.

### *16.4.1 Bioengineered Peptides and Proteins*

Biotechnology enables the synthesis of specific peptides and proteins used in cosmetics for wound healing, anti-aging, hair loss and pigmentation. Peptides act as messengers, stimulating skin to perform functions like repairing, hydrating, or boosting collagen production. Bioactive peptides are classified into signal peptide, carrier peptide, neurotransmitter inhibitor peptide and enzyme inhibitor peptide. These bioactive peptides can be synthesized chemically or derived naturally from plant, animal or marine sources. The main challenges of these peptides are optimized production conditions to maintain peptide bioactivity, stability, and efficacy during manufacturing and storage. Delivering peptides effectively to the skin is difficult due to their size and low permeability. This can be addressed with low-molecular-weight peptides to permeate the skin and advanced delivery methods like encapsulation. Additionally, peptides often have poor solubility, complex formulation and require low concentrations for stability. To produce pure synthetic peptide, the production costs are high and detecting peptides in complex formulations remains challenging (Ngoc et al. 2023).

### *16.4.2 Stem Cell Technology in Cosmetics*

Stem cells are used to develop products that promote rejuvenation and regeneration. Stem cell condition media is used to develop topical creams, serums, intradermal injections. The cells when grown in condition media release certain factors like cytokines, growth factors, exosomes which are purified for the formulations (Al-Sowayan and Al-Shareeda 2021). Adipose derived stem cells are majorly used to treat skin conditions like rejuvenation, burns, photoaging and immune related skin diseases. The method of extracting fat and isolation of stem cells affects the quality of the product (Semsarzadeh and Khetarpal 2022). The use of animal stem cells still has legislative, ethical challenges hence plant stem cells are used for commercial production. These are grown in microbe free environment and engineered to produce animal proteins (Trehan et al. 2017).

### *16.4.3 Probiotics for Skin Health*

Probiotics in cosmetics are live beneficial bacteria or yeast strains that help balance the skin's microbiome. Restoring the skin's natural bacterial environment, probiotics strengthen its protective barrier, reduce inflammation, and decrease sensitivity. They are especially effective in products designed for sensitive, acne-prone, or inflamed skin. By targeted alteration of skin microbiota, probiotics can also enhance hydration and soothe irritation, helping the skin stay balanced and resilient (Arora et al. 2024). The use of probiotics or extracts in creams, lotions, serums, cleaning agent gives the product anti-inflammatory, anti-aging, antioxidant and skin whitening properties. However, the method of manufacturing these probiotic derivatives has profound effect on these properties. It has been found that few extracts exhibited better efficacy when produced by fermentation (Dou et al. 2023).

### 16.4.4 Biomimetic Ingredients in Cosmetics

Biomimetic ingredients are synthetically created compounds that mimic natural skin substances. Biomimetic ingredients are designed to integrate easily with the skin with being highly effective and unlikely to cause irritation. Since they can be synthesized, they're often more sustainable than traditional sources, as they don't require harvesting from animals or plants. These ingredients include hyaluronic acid, squalene, collagen (Zhao et al. 2024).

### 16.4.5 DNA Repair Enzymes in Cosmetics for Anti-aging

DNA repair enzymes, derived from sources like marine organisms, are used in anti-aging and sun protection products to aid in cellular repair. These enzymes work to fix DNA damage caused by UV exposure, reducing signs of aging such as wrinkles and hyperpigmentation. The sunscreens contain combination of antioxidants and DNA repair enzymes providing duel action and reducing photoaging due to exposure to UV light. Photolyase and T4 endonuclease V are two most commonly used enzymes. Photolyase enzyme identifies the thymine dimers formed due to UV radiation and repairs these sites. T4 endonuclease V repairs cyclobutane pyrimidine dimers in DNA caused due to UV radiation and retains the extra cellular matrix composition, hence maintain the skin and decreasing the process of photoaging (Gupta et al. 2023a, b).

### 16.4.6 Encapsulation Technologies for Enhanced Stability

Few active ingredients in cosmetics are highly sensitive and to retain their structures these are encapsulated into biocarriers for controlled release and thus improving the stability of these active ingredients. One such study was conducted on Vitamin C which is highly sensitive encapsulated in dendrimers. Liposomal-encapsulated vitamin C serums stay stable and effective longer, reaching deeper skin layers for better brightening and anti-aging effects. Encapsulation technology is also used in fragrances to allow scents to last longer on the skin (Bacha et al. 2022).

### 16.4.7 Synthetic Biology for Collagen

Collagen was traditionally derived from animals, but through technological innovation, collagen may now be produced through microbial fermentation. The companies can create collagen that is free of animal products and ethical while still having the same chemical makeup as collagen generated from animals by using genetically modified yeast or bacteria (Báez et al. 2005).

### 16.4.8 Polar Lipids in Cosmetics

The primary purpose in these polar lipids, such as ceramides, sphingolipids, and glycolipids, stems from the fact that they have structural features similar to skin lipids, allowing for good compatibility and efficiency. Similarly, techniques for obtaining such lipids should be guided by the concepts of green chemistry and sustainability. The highlighted needs include biodegradability, biocompatibility, efficiency, quality, and profitability. Polar lipids, such as phospholipids or glycolipids, are amphiphilic and can function as co-emulsifiers in water/oil and oil/water system cosmetic formulations, as well as thickeners, dispersants, and wetting agents. Polar lipids derived from natural resources, due to their amphiphilic qualities and structures, have a wide range of cosmetic applications, both functional and physiological. These also can act as surfactants or emulsifiers (Traversier et al. 2018).

### 16.4.9 Nanomaterials for Targeted Delivery

Their small size enables effective skin permeation, delivering ingredients directly to specific areas in the skin's deeper layers. This makes ingredients like antioxidants, vitamins, or moisturizing agents more potent, as they reach where they're most needed. Additionally, nanomaterials provide controlled release, so products last longer and are less likely to irritate the skin. Nanotechnology is critical in improving the efficacy of biotechnology-derived compounds. Nanotechnology delivers active chemicals in nanoscale particles, ensuring deeper penetration and longer efficacy of components. Nanomaterials, such as nanoliposomes, nano-emulsions, and nanoparticles, are designed to penetrate certain skin layers and slowly release chemicals for long-term effects.

## 16.5 The Application of Nano Drug Delivery Systems in Cosmetic Products

In recent decades, the advancement in nanotechnology extensively influenced the transformation of food, drug, cosmetic and personal care industries. The development of novel cosmetic and cosmeceutical products in application with the nano drug delivery system (NDDS) is one of the notable innovations in the field of beauty industry. The emerging landscape of nanotechnology empowers novel cosmetic products by addressing many drawbacks of conventional cosmetics through enhanced stability, comprehensive protection, site-specific activity, wide area coverage and user compliance (Cardoza et al. 2022).

## *16.5.1 Skin Care*

### 16.5.1.1 Moisturizer

The primary skincare routine is incomplete without the application of moisturizers and face creams to hydrate and nourish the skin. They act by various mechanisms like occlusion, aquaporin formation and active hydration to retain moisture content in the epidermis. Nowadays effectiveness of moisturizers enhanced through the application of nanodrug delivery systems that allow site-specific delivery of various encapsulated active principles (Souto et al. 2020).

The nano emulsions are one of the attractive nano carriers in cosmetics due to their versatile nature to incorporate in various systems like creams, gels, sprays, foams etc. with improved stability, spread ability, skin compatibility and permeability of loaded principles into the stratum corneum. For instance, a hydroglycolic extract of *Opuntia ficus-indica* (L.) Mill was encapsulated in an o/w nano emulsion to enhance the moisturizing and antiaging effect of phytoconstituents in the extract especially the carbohydrates and phenolics. The formulation possessed an effective moisturizing effect upto 5 h of application and at least 60 days stability in accelerated stability studies (Ribeiro et al. 2015).

A randomized, double-blind study on 15 human volunteers reported a patented nanoliposome loaded with skimmed donkey milk as a moisturizer cream and anti-aging product. Nanoliposomes have the advantage of non-immunogenic nature owing to their structural similarity with cell membrane, that efficiently delivers encapsulated hydrophilic or lipophilic materials into the skin layers. The trans-epidermal water loss, electrical capacitance and pH was investigated and concluded that the developed formulation is an asset to deliver bioactive proteins in the donkey milk as a valuable moisturizer (Kocic et al. 2020). Similar way, serine encapsulated liposome was loaded into four different cosmeceutical bases such as O/W cream, W/O cream, O/W essence and hydrogel. The liposomal serine in hydrogel base exhibited exceptional water retention capacity on stratum corneum owing to the synergistic hydration potential of hyaluronic acid, Carbopol ETD 2020 and Rhesperse RM 100 in hydrogel base (Kim et al. 2015).

### 16.5.1.2 Sunscreen

The sunscreens are any preparations containing one or more active ingredients with the ability to protect the skin from harmful UV radiation through absorbing, scattering or reflecting UV rays. They play a vital role in preserving healthy skin from premature aging, tanning, sunburn and even from incidence of skin cancer by providing protection from sunlight containing high energy Ultra Violet rays especially the UVA and UVB rays (Rajasekar et al. 2024; Chavda et al. 2023). Unlike the conventional sunscreen preparations, nano technology offered tremendous benefits in novel sunscreen preparations such as significant broad spectrum photoprotection through an improved surface area to volume ratio, photostability through nano encapsulation and user compliance owing to its light weight, water resistant, transparent and no white cast nature (Cadet and Douki 2018; D'Orazio et al. 2013).

A nanostructured lipid carrier based on two natural lipids such as bacuri butter and raspberry seed oil was developed using $2^2$ Face Centered Central Composite Design to encapsulate three organic UV filters viz. Uvinul® A Plus, Uvinul® T150, and Tinosorb® S. The optimized nano formulation was loaded into a cream base at a 20% w/w level and evaluated for various parameters. The product exhibited a broad-spectrum photo protection even at a 10% reduction in total filter content when compared to conventional products. An improved encapsulation efficiency of >90% and size of 122–135 nm suggested an efficient and stable nano sunscreen formulation compared to emulsion-based controls (de Araújo et al. 2024).

In another study lipoid colloidal carrier (LCC) is fabricated to encapsulate diosmin, a flavonoid glycoside along with various essential oils such as ginger oil, rose marry oil and grape seed oil. LCCs are versatile nano carriers that improve solubility of insoluble drug, thereby skin permeability. They form a thin, non- irritant, occlusive film of nano particles that allows the penetration of encapsulated ingredients to the upper most layer of epidermis. The diosmin alone loaded LCC has lower effects compared to diosmin/essential oil combination, exhibiting more photoprotective, anti-ageing, anti-wrinkling anti-inflammatory, anti-erythemal and antioxidant activities (Kamel et al. 2017).

Some of the conventional sunscreens with UV filters containing one or two aromatic rings possessed poor thermal and photostability and potential health hazards due to their easy penetration through the epidermis, dermis and finally to the blood stream. Thus Zhou et al. 2020 constructed polydopamine grafted lignin in which the UV filter was emulsified first, followed by their cross-linking through ultrasonic cavitation to form a lignin-polydopamine nanocapsules as a bioadhessive carrier. This prevented the penetration of chemical UV filters into the deep epidermis or blood stream making them more biocompatible with an SPF factor of 195.33 (Zhou et al. 2020).

As like lipid-based carriers, potential sunscreen properties were observed in metal ion nano particles. A Carbopol 934-based phytogel loaded with Methyl anthranilate encapsulated silver nanoparticle was prepared and optimized using Box–Behnken design by Ghazwani et al. 2023. The Methyl anthranilate and silver nano particle together performed a synergistic UV blockade effect. Moreover, the formulation was excellent in spread-ability, permeability, skin retention etc. (Ghazwani et al. 2023).

### 16.5.2 Anti-aging Products

The aging of the skin is a serious concern, especially for women. The necessity for youthful skin has been directed to a lucid surge in demand for anti-aging skin care products. Several components such as peptides, retinoids, growth factors, etc. have been proven to have anti-aging properties, while their effective delivery to the skin is always challenging. Here the role of nano drug delivery systems became appreciable in the recent era (Salvioni et al. 2021).

UV radiations are the major factor underlying the photo-aging of skin. It may cause DNA and Reactive oxygen species (ROS) mediated photodamage. A series of biological events will occur in the presence of UV radiation not only to the epidermis, but to the deeper collagen and elastin tissues of dermis also. The skin wrinkles when the integrity of extracellular matrix (ECM) of the dermis lost, affecting the elasticity and rigidity of collagen. Ellagic acid is a water -soluble polyphenolic compound present in certain berries and fruits like pomegranate. It has excellent anti-aging properties owing to its antioxidant and UV shielding effects. But low bioavailability, hinders its application as anti-wrinkle product. Niosomes are vesicular systems composed of non-ionic surfactant, stabilized by cholesterol. The pomegranate fruit extract (PFE) containing Ellagic acid is encapsulated within the niosomes and coated with chitosan polymer to target the site. The drug is safely entrapped inside the poorly water-soluble niosomes, improved the bioavailability, sustainability and stability (Abd-Elghany and Mohamad 2023).

Similarly in another study, gallic acid was loaded into neutral and cationic CTAB niosomes to combat the signs of aging. Gallic acid is a hydrophilic secondary plant metabolite, called phenolic acid with a potential of free radical scavenging, anti-inflammatory and anti-melanogenesis activity. The final formulation was found to be non toxic to human skin fibroblast and melanocytes. The anti-aging activities demonstrated by melanin suppression effect, tyrosinase inhibition, and matrix metalloproteinase-2 (MMP-2) inhibition etc. were higher in gallic acid loaded CTAB noisome when compared to neutral noisome and free gallic acid (Chaikul et al. 2019).

Just like niosomes, the liposomes and ethosomes are another versatile nano delivery system for encapsulating plant-derived antioxidant molecules as a potent anti-aging material. Yücel et al. (2019) compared the efficiency of rosmarinic acid loaded ethosome and liposome for antioxidant, anti-elastase and anti-collagenase activities. The drug loaded ethosomes exhibited the greatest anti-aging activity when compared to loaded liposomes and rosmarinic acid solutions. This may be due to the improved transdermal flux of ethosomes which are composed of ethanol and phospholipids, allowing drug delivery to the deeper skin layers (Yücel et al. 2019).

### 16.5.3 Hair Care Products: Nourishing from Root to Tip

The revolutions in nano drug delivery systems also influenced hair care industries. Conventional hair care preparations like shampoos and conditioners often fail to deliver beneficial ingredients successfully due to the challenges in scalp nature and size of hair strands. This failure is addressed by nano delivery systems by permitting the active principles to penetrate in depth to the scalp and hair shaft. Various active ingredients like vitamins, amino acids, essential oils, proteins etc. have been incorporated into many nano drug delivery systems to combat different hair concerns like scalp condition, hair damage, hair loss, grey hair, hair dryness and lice infestations (Pereira-Silva et al. 2022).

Hair loss is a common problem in men and women always. One of the reasons is androgenic alopecia, which can be effectively treated by topical application of minoxidil. But the commercial products of minoxidil contains high concentrations of propylene glycol and or ethyl alcohol, the long-term usage of which produces itching, scalp dryness, allergic dermatitis, redness etc. In this sense, a new formulation free from organic solvents is essential for long term application of this medicine. Thus, Padois et al. (2011) developed a solid lipid nanoparticle-based suspension of minoxidil and compared its effectiveness with the commercial solution of minoxidil. The nano size of the SLN formulation allowed exact targeting of hair follicles, which are the site of action of minoxidil. The ex-vivo skin permeation study suggested an improved permeation compared to commercial product. But the main advantage of developed product was the excellent non corrosive nature than the commercial product due to the presence of physiologically compatible ingredients of SLN. Thus, the minoxidil loaded SLN for androgenic alopecia can be concluded as a efficient product with improved patient compliance (Padois et al. 2011).

Similarly follicular delivery of minoxidil is also achieved by encapsulating in a nanoemulsion prepared with clove oil as oil phase and Kolliphor® P188 as surfactant. The formulation was 26-fold more potent than commercial product in terms of follicular penetration (Cardoso and Barradas 2020). Dutasteride, another drug for androgenic alopecia, encapsulated in nano capsule also targeted the follicles efficiently. But chitosan coating of nano capsule was found to be some more effective. Here a simple physical stimulus of massaging procedure enhanced the targeting potential almost 5 times and 2 times better in uncoated and coated formulation respectively (Ushirobira et al. 2020).

The application of nano delivery systems also created great revolutions in the field of hair dye also. The toxic chemical dye are now being replaced by graphene-based hair dye. It provides water based strong coatings on hair strands which are long-lasting and resistant to shampoo washings. These products are advantageous since it does not contain any toxic small molecules or organic solvents common in conventional hair dyes. It also exhibited thermal dissipation and antistatic and properties (Luo et al. 2018).

### 16.5.4 Oral Care Products

Oral care is another area in cosmetic industry where nano drug delivery systems are imparting crucial roles (Abedi et al. 2024). The efficacy of conventional tooth pastes hinders efficacy due to poor adhesion and easy washing away of beneficial agents like fluorides. The nano technology surpasses all these disadvantages by improving the localization and controlled release of encapsulated ingredients and could prevent tooth decay to some level (Amaechi et al. 2019). The tooth whitening products containing nano- abrasives offer exceptional whitening effects with minimal damage to the enamel through gentle polishing (Rohmawati et al. 2021); Moreover, the nano materials in oral care products impart dental remineralization (Teixeira et al. 2018), desensitization (Gupta et al. 2023a, b) and antimicrobial actions (Junevičius et al. 2015) etc.

For instance, a green synthesized gum Arabic silver nanoparticle (GA-AgNPs) loaded tooth paste was prepared and evaluated for antimicrobial activity against oral microbes. In order to encapsulate nano particles, four commercial tooth paste were selected and evaluated against oral microbes. Finaly the tooth paste with least antimicrobial activity in micro dilution and agar disc diffusion assay was selected and incorporated the GA-AgNP. The final formulation observed improved antimicrobial activity (Ahmed et al. 2023).

A randomized, double blind clinical trial was conducted on forty-two patients underwent fixed orthodontic treatment to study the effect of Nano silver mouthwash on white spot lesions on teeth. Nano silver is well known broad spectrum antibacterial agent which interacts with the bacterial phospholipid membrane and proteins. The incorporation of nano silver in mouthwash displayed improved reduction in occurrence of white spot lesion compared to conventional fluoride and chlorhexidine mouthwash (Ali et al. 2022).

## 16.6 Risk Associated with Nanodrug Delivery Systems in Cosmetics

### 16.6.1 Health Risks

Even though the role of nano particles in the development of novel cosmetics and cosmeceuticals are inevitable, their potential health risks associated with the size, surface charge, shape, exposure rate, route of administration, amount, solubility etc. are always of prime concern. The tiny size and shape allow the nano particles to be entered easily across various cells, tissues, membranes and organs leading to toxicity and even cell death when compared to conventional delivery systems (Wani et al. 2011; Holsapple et al. 2005).

The toxicity associated with nano particle is mainly related to the fundamental changes in physicochemical properties and biological activity of nano particle compared to their larger counter parts of same material. When a material is converted to a nano size, its surface area to volume ratio is increased, hence chemical reactivity is enhanced, leading to the production of reactive oxygen species (ROS). The free radical mediated oxidative stress will eventually lead to toxicity and inflammatory and immunomodulatory responses to the proteins, DNA and cell membranes and even various human systems viz. cardiovascular, pulmonary, neurological, reticuloendothelial and endocrine systems (Brown et al. 2002).

The most unexpected exposure routes of nano materials in the human body are through transdermal, oral and inhalational routes. The nano cosmetics applied to the skin and scalp can randomly penetrate into deeper viable skin layer and enter to systemic circulations. The entry will be fastened in case of damaged skin conditions like psoriasis, wound, eczema and UV injured (Ryman-Rasmussen et al. 2006). The inhalational toxicity is mainly associated with occupational hazards, where the products are manufactured and handled by workers. Also, accidental inhalation is happened during application, disposal or recycling of the cosmetics in powder or spray forms, by the consumers (Hougaard et al. 2015). Similarly the oral ingestion is associated the transfer of cosmetics viz. creams, lotions, serum, oral care products etc. from hands to mouth (Yildirimer et al. 2011).

Another potential area of nano material hazard is due to exposure to the pregnant women. The nano particle can enter to the foetus through placenta, endometrium and yolk sac and causes serious developmental risk to the embryo due to oxidative stress and increased cytokine production. This may cause foetal anomalies, delayed neonatal development, placental damage, impaired foetal reproductive, neuro and brain developments (Teng et al. 2021; Li et al. 2014).

### 16.6.2 Environmental Risks

Nano cosmetics have the potential to cause serious risk in the environment also, imparting biochemical interference in the biological system via various mechanisms viz. formation of reactive oxygen species, creating oxidative stress, bioaccumulation, autophagy, lysosomal disfunction etc. Nanomaterials are discharged from the manufacturing units into aquatic, terrestrial and atmospheric regions will cause serious environmental hazards (Taghavi et al. 2013; Stern et al. 2012).

If nano materials with antibacterial properties enter into waste-water treatment plants, will interfere with the useful microbes employed in the reuse of water. The nano titanium dioxide presented in many cosmetics will affect the organic activity of many useful microbes in sewage treatment plants. It will also stimulate photochemical reactions in atmospheric environment and potential toxicity to the marine aquatic microbiota. In many cases, carbon, titanium dioxide, polystyrene and fullerene nanoparticles produce toxic effects not only to humans, but also to other mammals and microbes. Silver, zinc oxide and silica nano particles are reported to affect the plants and algae, interfering the seed germination and metabolic pathways, which in turn will enter to the food chain (Brunet et al. 2009; Xia et al. 2006; Gupta et al. 2022).

## 16.7 Regulatory Guidelines for NDDS in Cosmetics

Major cosmetic regulations of India include the Drug and Cosmetic Act 1940 and the Cosmetic Rules 2020. The first implies a comprehensive guide for regulating manufacturers, importers, distributors as wells as vendors. The latter primarily comprises regulatory instructions for cosmetics import and business. The two main organizational levels managing Drug Control are the state governments and central drugs standard control organization (CDSCO). Several scientific organizations, including the Royal Society, the most renowned scientific organization in Britain, and the US Food and Drug Administration (USFDA) have strictly warned against the commercialization of nanocosmetics until a full assessment of the dangers involved has been conducted (Kumud and Sanju 2018).

### *16.7.1 Central Drugs Standard Control Organization (CDSCO)*

The regulation and promulgation of cosmetics as well as the related activities are managed by CDSCO. To manage tasks including licence or registration applications, CDSCO formed the Drugs Controller General (DCGI), India, an executive sector.

Some of the CDSCO's primary responsibilities for cosmetic regulation are listed here:

1. Modifying the New Cosmetics Rules 2020, the Drugs & Cosmetics Rules, 1945, and the import registration requirements for cosmetics.
2. Examining and assessing cosmetic registration applications in accordance with the guidelines established by the Drugs and Cosmetics Act 1940, Rules 1945, and New Cosmetics Rules 2020.
3. Examining and assessing different requests for NOCs and clarifications pertaining to the import of cosmetics.
4. Writing draft answers to questions about cosmetics posed by the Parliament, RTI requests, and VIP requests.
5. Giving necessary responses to BIS requests and correspondence from the government.
6. Assisting the public with questions or concerns about the import registration of cosmetics and offering advice in this regard.
7. Address concerns regarding cosmetic standards raised by the general public, business associations, NGOs, and consumer forums.
8. The applicant's registration and import requests for cosmetics are pre-screened.
9. To reflect the current working practices, the pre-screening checklist and standard operating procedures for assessing cosmetic import and registration applications have been modified.

### 16.7.2 State Governing Bodies

Next is the Department of State Drug Control that deals with licensing of both cosmetic manufacturing as well as sales establishments. Making sure that the state's residents receive high-quality medications at a reasonable cost in accordance with GOI regulations is one of its most significant responsibilities (NPPA).

Additionally, the Bureau of Indian Standards (BIS) publishes lists of components that are permitted and prohibited for use in cosmetics in addition to providing precise requirements for cosmetics.

Some of the important regulations, acts and systems followed by countries like USA, Australia, and Canada have been mentioned in Figure.

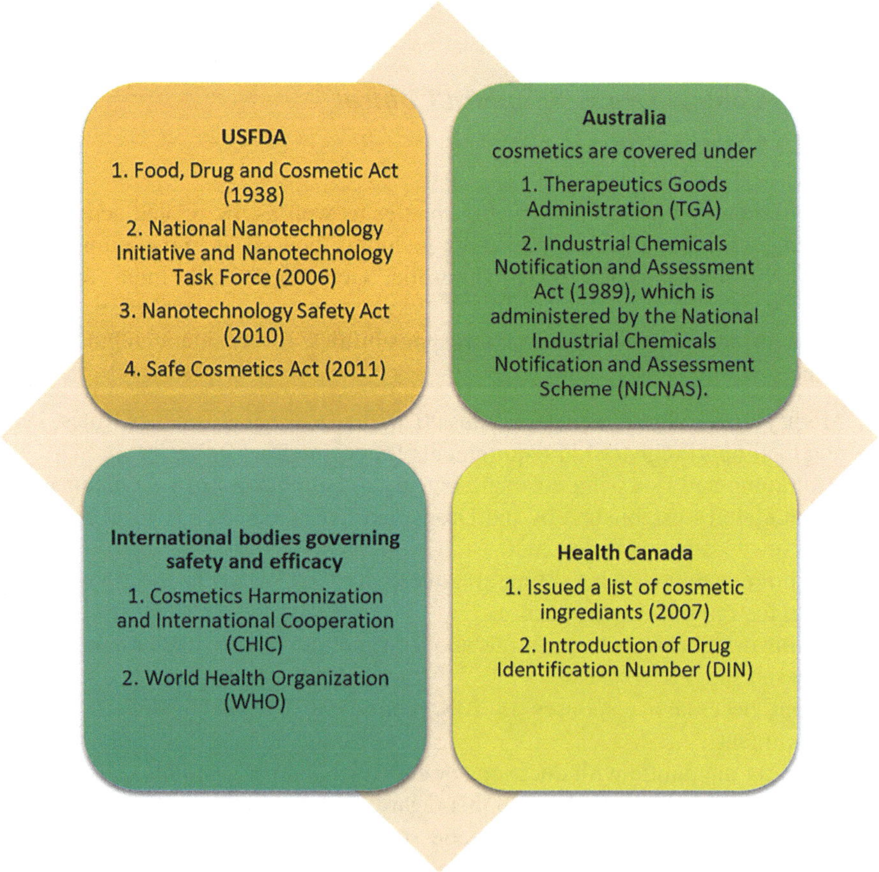


Apart from these specific organisations, commercialization of nanoscale product including nanocosmetics should be monitored by some of the important monitoring groups like Enivironment Protection Agency (EPA), Nanotechnology Interest Group (NTIG) Nanotechnology Public Engagement (NPEG), Nanotechnology Environmental and Health Implications (NEHI), Nanoscale Engineering and Technology (NSET), and Global Issues in Nanotechnology (GIN).

## 16.8 The Global Market and Patents

The global market for nano materials in cosmetics is increasing. In 2023, it was valued at over \$10.4 billion and is anticipated to keep expanding quickly and reach to \$ 26.8 billion by 2030 with a compound annual growth rate (CAGR) of 14.5%. Consumers are on the quest for things that perform properly and are safe. Brands are ready to follow the trend, resulting to a variety of choices on the shelves. From anti-aging creams to sunscreens, there's something for everyone. This demand is especially great in nations like the United States, Japan, and Europe. With more individuals becoming aware of the benefits of nano technology, sales are increasing (Annual Growth Rate of the Cosmetics Market Worldwide 2024; Nanotechnology in Cosmetics Market 2024).

Enhanced product performance attained by nanomaterials, innovations in drug delivery systems leveraging nanotechnology, consumer demand for high-quality cosmetic and personal care products, regulatory and safety developments ensuring the safe use of nanomaterials, and developments in nanotechnology research driving product development and efficacy help to explain the growth observed during the historic period. Improvements in a number of areas are responsible for the projected growth during the forecast period, including advancements in nanotechnology leading to innovative product formulations, increased consumer demand for high-performance cosmetic and personal care products, a growing awareness of skin health leading to a focus on effective skincare solutions, the rising trend of personalized beauty products tailored to individual needs, and regulatory developments ensuring safety standards in the use of nanomaterials (Mohd-Setapar et al. 2022).

The cosmetic business is trying to get patents on nano materials. Major companies like L'Oréal, Estée Lauder, and Procter & Gamble are filing patents like they're going out of style. They realise that being the first to create a breakthrough nano technology may lead to significant revenues. In the last several years, there's been an explosion of patents linked to nano materials in cosmetics. Companies are striving harder than ever to preserve their ideas. It's like a treasure hunt where everyone wants to locate the greatest and most unusual ingredients to give customers (Cosmetics Market Size, Share, Industry Trends & Analysis 2021–2027; Beauty Industry Trends & Cosmetics Marketing: Statistics and Strategies for Your Ecommerce Growth 2022). Some of the marketed nano cosmetics and patented products are listed below in Tables 16.1 and 16.2 respectively.

## 16.9 Future Perspectives and Challenges (NSS)

Use of nanomaterials in cosmetics to increase the efficacy and better delivery of the drug to treat wrinkles, spots, hair loss, pigmentation has shown to be effective. But nanomaterials in cosmetics always raise concerns for potential toxicity and biocompatibiltiy of the product. The nanomaterials are to be checked for size, shape, surface are, surface charge, skin penetrability, cellular toxicity and also the route of exposure—inhalation, ingestion, through skin using various in vitro and in vivo techniques to identify the hazard and dose-response characterization (Raj et al. 2012).

When these cosmetics are washed off into the environment, the biodegradability is a major concern. The product should not accumulate in the nature leading to biomagnification and also not harm any other living organism in the environment. In a study, the biomagnification of $nTiO_2$ was investigated from algae to daphnia finding incomplete clearance, leading to accumulation in gut and damaging epithelial cells and microvilli in daphnia (Chen et al. 2019a, b).

The cosmetics industry contribution to SDG by 2030 could be achieved by smart nanomaterials, which respond only to the stimulus. While the preference is leaning towards green cosmetics, the chemical cosmetics are to be replaced with plant derived nanomaterials to make them more environment friendly and consumer friendly. The manufacturing of nanoparticles has to be tailored according to its physico-chemical properties, to produce non-toxic materials in the best interest of public health and climate preservation (Qamar et al. 2024). Current strategies to alleviate this toxicity include introducing naturally derived cell membranes from different cell types, modifying the surface chemistry of nanoparticle by changing the hydrophobicity, charge density and most widely used is the surface coating (Zhang et al. 2022).

## 16.10 Conclusion

The use of nanotechnology in skin care and dermatology products has sparked a new wave of innovation by providing customers with cutting-edge, effective, and visually appealing solutions. To realize the full potential of nanotechnology in cosmetics while protecting the health and well-being of users will require more research and collaboration between scientists, dermatologists, and regulatory agencies as technology advances. Nanomaterials derived from plant and animal by products improve efficacy and bioavailability as nanocarriers not only enhance penetration but also regulate the release of bioactive substances. The environmental footprint of cosmetic production is reduced through the use of by-products, which also promotes environmental sustainability by reducing waste and promoting a circular economy. These materials based on naturally occurring resources, can also be cheaper than synthetic substitutes. Naturally occurring nanocarriers are also safer and more biocompatible for dermal applications, reducing the possibility of adverse reactions. However, to ensure the safety and effectiveness of cosmetics based on nanotechnology, extensive regulatory frameworks must be established. It is equally important to educate and raise awareness among consumers about the safety and benefits of cosmetics made with nanotechnology and natural by-products. Transparency and accountability can be improved through creation and implementation of sustainability metrics to assess and document the environmental impact of

**Table 16.1** Examples of marketed nano cosmetic products

| Sl. no. | Marketed product | Manufacturer | Nano material type | Active constituent used | Application | Reference |
|---|---|---|---|---|---|---|
| 1. | Eusu Niosome Makam Pom Whitening Facial Cream | EUSU | Nano-particles | Nano-particles of Indian gooseberry | Anti-aging serum | http://online.goldencosmetic.com/product/eusunanocream/ |
| 2. | Cetaphil ™ Sunscreen | Cetaphil | Liposomes | Methylene Bis-Benzotriazolyl Tetramethylbutylphenol | Sunscreen | https://www.cetaphil.com.au/sunscreens/sun-spf50%2B-kids-liposomallotion/31511.html# |
| 3. | Rehydrating Liposome Day cream | Kerstin Florian | Liposomes | Natural emollients | Moisturizer | https://www.kerstinflorian.co.uk/shop/rehydrating-liposome-day-creme/ |
| 4. | C-Vit Liposomal Serum | Sesderma | Liposomes | Ascorbyl glucoside, hyaluronic acid | Radiating serum, luminous glow and a brighter complexion | http://www.sesderma.co.in/in/categorias/familias/c-vit/c-vit-liposomal-serum.html |
| 5. | Fillderma Lips | Sesderma | Nanospheres | Hyaluronic acid, collagen | Lip Volumizer | https://www.sesderma.com/int_en/FILLDERMA-Lips-40004133.html |
| 6. | VITANICS Hair Health Shampoo | Vitacos | Nanemulsion | Silicone nanoemulsion | Anti-dandruff shampoo with hair conditioning effect | http://www.vitacos.co.kr/en/products/hair-and-body?tpf=product/view&category_code=17&code=158 |
| 7. | NanoVital VITANICS Whitening C Intensive Nano-E Serum | Vitacos | Nanemulsion | Bisabolol and stabilized Vitamin C | Skin whitening, anti-wrinkle, moisturizing, and firming effects | http://www.vitacos.co.kr/en/products/skin-care?tpf=product/view&category_code=1611&code=192 |
| 8. | Clearly It! ™ Complexion Mist | Kara Vita | Nanosphere | Vitamin C and E | Facial toner | https://product.statnano.com/product/4618/clearly-it!%C2%AE-complexion-mist |
| 9. | Pure, 24 K Gold Facial Mask Majestic Pure | Majestic | Nanogold | Nanogold | Face mask, skin cleanser, wrinkle moisturizer | https://majesticpure.com/products/24k-gold-facial-mask-8-oz?srsltid=AfmBOoqgk3h3Vop1yIyIbznQriNLg4PN9qshEMhvkaI-yg3vRF_xg9C6 |
| 10. | Shampooing Floral Repair | Identik | Niosomes | Adenosine phosphate, hydrolyzed wheat protein | Shampoo | https://www.bol.com/nl/nl/p/identik-repair-shampoo-250ml/9300000137314411/ |

(continued)

**Table 16.1** (continued)

| Sl. no. | Marketed product | Manufacturer | Nano material type | Active constituent used | Application | Reference |
|---|---|---|---|---|---|---|
| 11. | Filler Intense Cream | Cellact | DNA Nanospheres | Ceramides NG, Palmitoyl Hexapeptide-12 | Anti-aging and hydrating | https://cellactswiss.com/products/filler-intense-cream?_pos=2&_sid=bae9d5810&_ss=r |
| 12. | Nano-in Hand and Nail Moisturizing Serum | Nano-Infinity Nanotech | Nano ZnO | Zinc oxide | ZnO NPs are tiny crystals that enter skin tissue to keep hands and nails from becoming infected. | https://www.nanotechproject.tech/cpi/products/nano-in-hand-and-nail-moisturizing-serum-and-foot-moisturizing-serum/ |
| 13. | Primordiale Optimum Lip | Lancome | Nanocapsule | Vitamin E | Lip care | https://www.nanotechproject.tech/cpi/products/primordiale-optimum-lip/ |
| 14. | Fresh as a Daisy® Body Lotion | Kara Vita | Nanospheres Nanoliposome | Vitamin C and E | Moisturizer | https://product.statnano.com/product/4616/fresh-as-a-daisy%C2%AE-body-lotion |
| 15. | Capture Totale | Dior | Liposomes | Hyaluronic acid | Sunscreen | https://www.dior.com/en_int/beauty/skincare/skincare-lines/capture-totale |
| 16. | Youth Gold Lifting Serum Emulsion | Wipro Unza | Nano 24k gold | Vitamin C, 24K Gold, Niacinamide, Vitamin B3 | Provides instant hydration to smoothen wrinkles & reduces signs of aging | https://smytten.com/shop/product/serums-and-essence/youth-gold-lifting-serum-emulsion-with-vitamin-c-24k-gold/RNA0008AB1?srsltid=AfmBOorRQRspf6LlTBmlgni4TOj7pGmcRxCbAU3OcXPTJFUwG31_2G6Q |
| 17. | Hydra Flash Bronzer Daily Cream | Lancome | Nano capsules | Vitamin C, vitamin E, Aloe | Antioxidant, hydrating moisturizer | https://www.nanotechproject.tech/cpi/products/hydra-flash-r-bronzer-daily-face-moisturizer/ |
| 18. | Cosil Nano Beauty Soap | Natural Korea | Nanoparticles | Nano-silver, Arbutin, Aloe vera | Cleanser, detoxifier, exfoliator | https://www.nanotechproject.tech/cpi/products/cosil-nano-beauty-soap/ |
| 19. | GreenYarn G-Moist Soft Cloth Mask | Greenyarn LLC. | Nano particle | Nano-particles of bamboo-charcoal (Carbon) | Deodorizing, absorbs and emits far-infrared radiation, thermal regulating | https://www.nanotechproject.tech/cpi/products/greenyarn-g-moist-soft-cloth-mask/ |
| 20. | Eczemel Nano-Cream | Dermaviduals® U.S.A. | Nanoparticles | Vitamin A | Smoothing care and protection of mature skin, regeneration of the skin | https://www.nanotechproject.tech/cpi/products/eczemel-nano-cream/ |

**Table 16.2** Examples of patents in nano cosmetic products

| Title | Nano particle type | Publication number and year | Reference |
|---|---|---|---|
| Cosmetic composition containing natural fullerene and method for manufacturing the same | Nano-sized porous natural fullerenes | KR101633302B1, 2016 | https://patents.google.com/patent/KR101633302B1/en?q=(Fullerenes+in+cosmetics)&oq=Fullerenes+in+cosmetics |
| Light attenuating formulations | Diamond nanoparticles | US20180028418A1, 2019 | https://patents.google.com/patent/US20180028418A1/en?q=(nanodiamonds+in+cosmetics)&oq=nanodiamonds+in+cosmetics |
| Flexible lipidosome cosmetics comprising active peptides and preparation method thereof | Lipidosome cosmetics of active peptides | CN109700687A, 2022 | https://patents.google.com/patent/CN109700687A/en?q=(nano+peptides+in+cosmetics)&oq=nano+peptides+in+cosmetics |
| Customized compositions and uses thereof | Nanofiber | US9095524B2, 2015 | https://patents.google.com/patent/US9095524B2/en?q=(nanodiamonds+in+cosmetics)&oq=nanodiamonds+in+cosmetics |
| Manufacturing method of nano-emulsion cosmetic composition stabilized oil of high content | Nanoemulsion | KR102237462B1, 2021 | https://patents.google.com/patent/KR102237462B1/en?q=(Nanoemulsions+in+cosmetics)&oq=Nanoemulsions+in+cosmetics |
| Nanodiamond UV protectant formulations | Diamond nanoparticle | US8753614B2, 2014 | https://patents.google.com/patent/US8753614B2/en?q=(nano+sunscreen)&oq=nano+sunscreen |
| Cosmetic composition containing novel fractal particle-based gels | Nanoparticles | US9561163B2, 2017 | https://patents.google.com/patent/US9561163B2/en?q=(cosmetics)&oq=cosmetics |
| Application of the nanometer gold in face cream and the face cream that is made from it and preparation method thereof | Nanometer gold | CN107951730A, 2018 | https://patents.google.com/patent/CN107951730A/en?oq=CN107951730A |
| Nano-emulsion cosmetic composition in which high content oils are stabilized | Silicone, ester and hydrocarbon nano-emulsion | WO2017175899A1, 2017 | https://patents.google.com/patent/WO2017175899A1/en?oq=WO2017175899A1 |
| Colored nanoparticles for cosmetic and its manufacturing method | Colored noble metal nanoparticle | JP2009221140A, 2009 | https://patents.google.com/patent/JP2009221140A/en?q=(nano+gold+containing+cosmetics)&oq=nano+gold+containing+cosmetics |
| Powder type cosmetic composition | Nanoporous carrier | KR100768113B1, 2007 | https://patents.google.com/patent/KR100768113B1/en?q=(nano+gold+containing+cosmetics)&oq=nano+gold+containing+cosmetics |

cosmetic products. But there are obstacles to the advancement in this field. Technical challenges, regulatory scrutiny, and the requirement for lifecycle assessments to guarantee environmental safety all arise when nanotechnology applications are scaled up. Furthermore, there are financial obstacles that need to be overcome because of the intricacy of manufacturing procedures and the high expenses of research and development. Collaboration between scientific research, industry practices, and consumer education is essential as the sector develops. There are opportunities to address the societal ramifications of using by-products, standardize techniques for characterizing NPs, and create transparent supply chains.

There are a number of new trends and areas for research in the cosmetics industry, which is a promising field where natural byproducts and nanotechnology meet. Using nanotechnology to create customized skincare products for different skin types and conditions is one such trend and the another is the development of intelligent delivery systems that release active ingredients in response to particular environmental factors or skin conditions. Innovations in environmentally friendly and biodegradable packaging materials are also being made, which lessens the overall impact on the environment. The field of skin care could undergo a revolution as science works to fully understand the potential of nanotechnology in cosmetics, offering consumers more advanced and effective products.

## References


Abbasi BH, Fazal H, Ahmad N, Ali M, Giglioli-Guivarch N, Hano C (2020) Chapter 5 – Nanomaterials for cosmeceuticals: nanomaterials-induced advancement in cosmetics, challenges, and opportunities. In: Nanda A, Nanda S, Nguyen TA, Rajendran S, Slimani YBT-N (eds) Micro and nano technologies. Elsevier, Amsterdam, pp 79–108. ISBN: 978-0-12-822286-7

Abd-Elghany AA, Mohamad EA (2023) Chitosan-coated niosomes loaded with ellagic acid present antiaging activity in a skin cell line. ACS Omega 8(19):16620–16629

Abedi M, Ghasemi Y, Nemati MM (2024) Nanotechnology in toothpaste: fundamentals, trends, and safety. Heliyon

Ahmad J, Gautam A, Komath S, Bano M, Garg A, Jain K (2018) Topical nano-emulgel for skin disorders: formulation approach and characterization. Recent Pat Antiinfect Drug Discov 14:36–48. https://doi.org/10.2174/1574891x14666181129115213

Ahmed OA, Sibuyi NR, Fadaka AO, Maboza E, Olivier A, Madiehe AM, Meyer M, Geerts G (2023) Prospects of using gum Arabic silver nanoparticles in toothpaste to prevent dental caries. Pharmaceutics. 15(3):871

Akbarzadeh A, Rezaei-Sadabady R, Davaran S, Joo SW, Zarghami N, Hanifehpour Y et al (2013) Liposome: classification, preparation, and applications. Nanoscale Res Lett 8:102. https://doi.org/10.1186/1556-276X-8-102

Al-Edresi S, Baie S (2009) Formulation and stability of whitening VCO-in-water nanocream. Int J Pharm 373:174–178. https://doi.org/10.1016/j.ijpharm.2009.02.011

Ali A, Ismail H, Amin K (2022) Effect of nanosilver mouthwash on prevention of white spot lesions in patients undergoing fixed orthodontic treatment – a randomized double-blind clinical trial. J Dent Sci 17(1):249–255

Al-Sowayan BS, Al-Shareeda AT (2021) Stem cells and the pursuit of youth, a tale of limitless possibilities and commercial fraud. World J Biol Chem 12(4):52–56. https://doi.org/10.4331/wjbc.v12.i4.52

Amaechi BT, AbdulAzees PA, Alshareif DO, Shehata MA, Lima PP, Abdollahi A, Kalkhorani PS, Evans V (2019) Comparative efficacy of a hydroxyapatite and a fluoride toothpaste for prevention and remineralization of dental caries in children. BDJ Open 5(1):18

Annual Growth Rate of the Cosmetics Market World Wide (2024) Available online: https://www.statista.com/statistics/297070/growth-rate-of-the-global-cosmetics-market/. Accessed 20 Oct 2024

Arora N, Agarwal S, Rayasa M (2012) Latest technology advances in cosmeceuticals. Int J Pharm Sci Drug Res 4:168–182

Arora R, Kaur R, Babbar R, Dhingra S, Dhingra AK, Grewal AS (2024) Evolving advances in the cosmetic use of probiotics and postbiotics: health, regulatory and marketing aspects. Curr Pharm Biotechnol 25(11):1349–1361. https://doi.org/10.2174/1389201024666230703115732

Bacha K, Chemotti C, Monboisse JC, Robert A, Furlan AL, Smeralda W, Damblon C, Estager J, Brassart-Pasco S, Mbakidi JP, Pršić J, Bouquillon S, Deleu M (2022) Encapsulation of vitamin C by glycerol-derived dendrimers, their interaction with biomimetic models of stratum corneum and their cytotoxicity. Molecules 27(22):8022. https://doi.org/10.3390/molecules27228022

Baek JH, Yoo MA, Koh JS, Borkow G (2012) Reduction of facial wrinkles depth by sleeping on copper oxide-containing pillowcases: a double blind, placebo controlled, parallel, randomized clinical study. J Cosmet Dermatol 11:193–200. https://doi.org/10.1111/j.1473-2165.2012.00624.x

Báez J, Olsen D, Polarek JW (2005) Recombinant microbial systems for the production of human collagen and gelatin. Appl Microbiol Biotechnol 69(3):245–252. https://doi.org/10.1007/s00253-005-0180-x

Bamford E, Grahn A, Århammar C, Ajaxon I, Annerén C (2021) Mesoporous magnesium carbonate for use in powder cosmetics. Int J Cosmet Sci 43(1):57–67

Beauty Industry Trends & Cosmetics Marketing: Statistics and Strategies for Your Ecommerce Growth (2022) Available online: https://commonthreadco.com/blogs/coachs-corner/beauty-industry-cosmetics-marketing-ecommerce#trends. Accessed 20 Oct 2024

Betz G, Aeppli A, Menshutina N, Leuenberger H (2005) In vivo comparison of various liposome formulations for cosmetic application. Int J Pharm 296:44–54. https://doi.org/10.1016/j.ijpharm.2005.02.032

Bocca B, Pino A, Alimonti A, Forte G (2014) Toxic metals contained in cosmetics: a status report. Regul Toxicol Pharmacol 68(3):447–467. https://doi.org/10.1016/j.yrtph.2014.02.003

Borkow G (2015) Using copper to improve the Well-being of the skin. Curr Chem Biol 8:89–102. https://doi.org/10.2174/2212796809666150227223857

Brown JS, Zeman KL, Bennett WD (2002) Ultrafine particle deposition and clearance in the healthy and obstructed lung. Am J Respir Crit Care Med 166(9):1240–1247

Brunet L, Lyon DY, Hotze EM, Alvarez PJ, Wiesner MR (2009) Comparative photoactivity and antibacterial properties of C60 fullerenes and titanium dioxide nanoparticles. Environ Sci Technol 43(12):4355–4360

Cadet J, Douki T (2018) Formation of UV-induced DNA damage contributing to skin cancer development. Photochem Photobiol Sci 17(12):1816–1841

Cao M, Li J, Tang J, Chen C, Zhao Y (2016) Gold nanomaterials in consumer cosmetics nanoproducts: analyses, characterization, and dermal safety assessment. Small 12:5488–5496. https://doi.org/10.1002/smll.201601574

Cao L, Xie W, Cui H, Xiong Z, Tang Y, Zhang X, Feng Y (2022) Fibrous clays in dermopharmaceutical and cosmetic applications: traditional and emerging perspectives. Int J Pharm 625(June):122097. https://doi.org/10.1016/j.ijpharm.2022.122097

Cardoso SA, Barradas TN (2020) Developing formulations for drug follicular targeting: nanoemulsions loaded with minoxidil and clove oil. J Drug Deliv Sci Technol 59:101908

Cardoza C, Nagtode V, Pratap A, Mali SN (2022) Emerging applications of nanotechnology in cosmeceutical health science: latest updates. Health Sci Rev 4:100051

Carita AC, Eloy JO, Chorilli M, Lee RJ, Leonardi GR (2018) Recent advances and perspectives in liposomes for cutaneous drug delivery. Curr Med Chem 25:606–635. https://doi.org/10.2174/0929867324666171009120154

Chaikul P, Khat-Udomkiri N, Iangthanarat K, Manosroi J, Manosroi A (2019) Characteristics and in vitro anti-skin aging activity of gallic acid loaded in cationic CTAB niosome. Eur J Pharm Sci 131:39–49

Chavda VP, Acharya D, Hala V, Daware S, Vora LK (2023) Sunscreens: a comprehensive review with the application of nanotechnology. J Drug Deliv Sci Technol 86:104720

Chen S, Hanning S, Falconer J, Locke M, Wen J (2019a) Recent advances in non-ionic surfactant vesicles (niosomes): fabrication, characterization, pharmaceutical and cosmetic applications. Eur J Pharm Biopharm 144:18–39. https://doi.org/10.1016/j.ejpb.2019.08.015

Chen X, Zhu Y, Yang K, Zhu L, Lin D (2019b) Nanoparticle TiO2 size and rutile content impact bioconcentration and biomagnification from algae to daphnia. Environ Pollut 247:421–430., ISSN:0269-7491. https://doi.org/10.1016/j.envpol.2019.01.022

Chis AA, Dobrea C, Morgovan C, Arseniu AM, Rus LL, Butuca A, Juncan AM, Totan M, Vonica-Tincu AL, Cormos G et al (2020) Applications and limitations of dendrimers in biomedicine. Molecules 25:3982

Cosmetics Market Size, Share, Industry Trends & Analysis 2021–2027. Available online: https://www.alliedmarketresearch.com/cosmetics-market. Accessed 20 Oct 2024

D'Orazio J, Jarrett S, Amaro-Ortiz A, Scott T (2013) UV radiation and the skin. Int J Mol Sci 14(6):12222–12248

Das SK, Chakraborty S, Roy C, Rajabalaya R, Mohaimin AW, Khanam J (2018) Ethosomes as novel vesicular carrier: an overview of the principle, preparation and its applications. Curr Drug Deliv 15:795–817. https://doi.org/10.2174/1567201815666180116091604

de Araújo MM, Schneid AC, Oliveira MS, Mussi SV, de Freitas MN, Carvalho FC, Bernes Junior EA, Faro R, Azevedo H (2024) NLC-based sunscreen formulations with optimized proportion of encapsulated and free filters exhibit enhanced UVA and UVB photoprotection. Pharmaceutics 16(3):427

Dou J, Feng N, Guo F, Chen Z, Liang J, Wang T, Guo X, Xu Z (2023) Applications of probiotic constituents in cosmetics. Molecules 28(19):6765. https://doi.org/10.3390/molecules28196765

Dubey SK, Dey A, Singhvi G, Pandey MM, Singh V, Kesharwani P (2022) Emerging trends of nanotechnology in advanced cosmetics. Colloids Surf B: Biointerfaces 214(December 2021):112440. https://doi.org/10.1016/j.colsurfb.2022.112440

Edwards K (2018) Stability testing guidance for product safety and shelf-life insight. Cosmet Toilet. https://www.cosmeticsandtoiletries.com/testing/methoddevelopment/Stability-Testing-Guidance-for-Product-Safety-and-Shelf-life-Insight-503072831.html

Filipović M, Lukić M, Krstonošić V, Doroević S, Pantelić I, Gledović A et al (2016) Feasibility of a natural surfactant as a stabilizer for cosmetics with liposome-encapsulated plant stem cells: pre-formulation and formulation through stability studies. Tenside Surfactant Deterg 53:214–226. https://doi.org/10.3139/113.110426

Fytianos G, Rahdar A, Kyzas GZ (2020) Nanomaterials in cosmetics: recent updates. Nanomaterials 10:979. https://doi.org/10.3390/nano10050979

Garcês A, Amaral MH, Sousa Lobo JM, Silva AC (2018) Formulations based on solid lipid nanoparticles (SLN) and nanostructured lipid carriers (NLC) for cutaneous use: a review. Eur J Pharm Sci 112:159–167. https://doi.org/10.1016/j.ejps.2017.11.023

Geszke-Moritz M, Moritz M (2016) Solid lipid nanoparticles as attractive drug vehicles: composition, properties and therapeutic strategies. Mater Sci Eng C 68:982–994. https://doi.org/10.1016/j.msec.2016.05.119

Ghazwani M, Hani U, Alqarni MH, Alam A (2023) Development and characterization of methyl-anthranilate-loaded silver nanoparticles: a phytocosmetic sunscreen gel for UV protection. Pharmaceutics. 15(5):1434

Guimarães KL, Ré MI (2011) Lipid nanoparticles as carriers for cosmetic ingredients: the first (SLN) and the second generation (NLC). In: Beck R, Guterres S, Pohlmann A (eds) Nanocosmetics and nanomedicines. Springer, Berlin/Heidelberg, pp 101–122. https://doi.org/10.1007/978-3-642-19792-5_5

Gupta V, Mohapatra S, Mishra H, Farooq U, Kumar K, Ansari MJ, Aldawsari MF, Alalaiwe AS, Mirza MA, Iqbal Z (2022) Nanotechnology in cosmetics and cosmeceuticals – a review of latest advancements. Gels 8:173. [Internet]
Gupta I, Chauhan S, Amaranath BJ, Das N, Johnson L, Mehrotra V (2023a) Effect of commercially available nano-hydroxyapatite containing desensitizing toothpaste and mouthwash on dentinal tubular occlusion: a SEM analysis. J Pharm Bioallied Sci 15(Suppl 2):S1027–S1029
Gupta A, Singh AP, Singh VK, Singh PR, Jaiswal J, Kumari N, Upadhye V, Singh SC, Sinha RP (2023b) Natural sun-screening compounds and DNA-repair enzymes: photoprotection and photoaging. Catalysts 13:745. https://doi.org/10.3390/catal13040745
Guterres SS, Paese K, Pohlmann AR (2019) Polymeric nanoparticles. In: Cornier J, Keck CM, Van de Voorde M (eds) Nanocosmetics. Spinger, pp 73–94. https://doi.org/10.1007/978-3-030-16573-4_5
Han F, Luo D, Qu W, Chen D, Hong Y, Sheng J, Yang X, Liu W (2020) Nanoliposomes codelivering bioactive peptides produce enhanced anti-aging effect in human skin. J Drug Deliv Sci Technol 57(January):101693. https://doi.org/10.1016/j.jddst.2020.101693
Holsapple MP, Farland WH, Landry TD, Monteiro-Riviere NA, Carter JM, Walker NJ, Thomas KV (2005) Research strategies for safety evaluation of nanomaterials. Part II: Toxicological and safety evaluation of nanomaterials, current challenges and data needs. Toxicol Sci 88(1):12–17
Hougaard KS, Campagnolo L, Chavatte-Palmer P, Tarrade A, Rousseau-Ralliard D, Valentino S, Park MV, de Jong WH, Wolterink G, Piersma AH, Ross BL (2015) A perspective on the developmental toxicity of inhaled nanoparticles. Reprod Toxicol 56:118–140
http://online.goldencosmetic.com/product/eusunanocream/
http://www.sesderma.co.in/in/categorias/familias/c-vit/c-vit-liposomal-serum.html
https://cellactswiss.com/products/filler-intense-cream?_pos=2&_sid=bae9d5810&_ss=r
https://majesticpure.com/products/24k-gold-facial-mask-8-oz?srsltid=AfmBOoqgk3h3Vop1yIyIbznQriNLg4PN9qshEMhvkaI-yg3vRF_xg9C6
https://patents.google.com/patent/CN107951730A/en?oq=CN107951730A
https://patents.google.com/patent/CN109700687A/en?q=(nano+peptides+in+cosmetics)&oq=nano+peptides+in+cosmetics
https://patents.google.com/patent/JP2009221140A/en?q=(nano+gold+containing+cosmetics)&oq=nano+gold+containing+cosmetics
https://patents.google.com/patent/KR100768113B1/en?q=(nano+gold+containing+cosmetics)&oq=nano+gold+containing+cosmetics
https://patents.google.com/patent/KR101633302B1/en?q=(Fullerenes+in+cosmetics)&oq=Fullerenes+in+cosmetics
https://patents.google.com/patent/KR102237462B1/en?q=(Nanoemulsions+in+cosmetics)&oq=Nanoemulsions+in+cosmetics
https://patents.google.com/patent/US20180028418A1/en?q=(nanodiamonds+in+cosmetics)&oq=nanodiamonds+in+cosmetics
https://patents.google.com/patent/US8753614B2/en?q=(nano+sunscreen)&oq=nano+sunscreen
https://patents.google.com/patent/US9095524B2/en?q=(nanodiamonds+in+cosmetics)&oq=nanodiamonds+in+cosmetics
https://patents.google.com/patent/US9561163B2/en?q=(cosmetics)&oq=cosmetics
https://patents.google.com/patent/WO2017175899A1/en?oq=WO2017175899A1
https://product.statnano.com/product/4616/fresh-as-a-daisy%C2%AE-body-lotion
https://product.statnano.com/product/4618/clearly-it!%C2%AE-complexion-mist
https://www.bol.com/nl/nl/p/identik-repair-shampoo-250ml/9300000137314411/
https://www.cetaphil.com.au/sunscreens/sun-spf50%2B-kids-liposomallotion/31511.html#
https://www.dior.com/en_int/beauty/skincare/skincare-lines/capture-totale
https://www.kerstinflorian.co.uk/shop/rehydrating-liposome-day-creme/
https://www.nanotechproject.tech/cpi/products/cosil-nano-beauty-soap/
https://www.nanotechproject.tech/cpi/products/eczemel-nano-cream/
https://www.nanotechproject.tech/cpi/products/greenyarn-g-moist-soft-cloth-mask/

https://www.nanotechproject.tech/cpi/products/hydra-flash-r-bronzer-daily-face-moisturizer/
https://www.nanotechproject.tech/cpi/products/nano-in-hand-and-nail-moisturizing-serum-and-foot-moisturizing-serum
https://www.nanotechproject.tech/cpi/products/primordiale-optimum-lip/
https://www.sesderma.com/int_en/FILLDERMA-Lips-40004133.html

Hughes K, Ho R, Butaud JF, Filaire E, Ranouille E, Berthon JY, Raharivelomanana P (2019) A selection of eleven plants used as traditional Polynesian cosmetics and their development potential as anti-aging ingredients, hair growth promoters and whitening products. J Ethnopharmacol 245(April):112159. https://doi.org/10.1016/j.jep.2019.112159

Jensen LB, Magnussson E, Gunnarsson L, Vermehren C, Nielsen HM, Petersson K (2010) Corticosteroid solubility and lipid polarity control release from solid lipid nanoparticles. Int J Pharm 390:53–60

Junevičius J, Žilinskas J, Česaitis K, Česaitienė G, Gleiznys D, Maželienė Ž (2015) Antimicrobial activity of silver and gold in toothpastes: a comparative analysis. Stomatologija 17(1):9–12

Kamel R, Abbas H, Fayez A (2017) Diosmin/essential oil combination for dermal photo-protection using a lipoid colloidal carrier. J Photochem Photobiol B Biol 170:49–57

Khezri K, Saeedi M, Maleki DS (2018) Application of nanoparticles in percutaneous delivery of active ingredients in cosmetic preparations. Biomed Pharmacother 106:1499–1505. https://doi.org/10.1016/j.biopha.2018.07.084

Khosa A, Reddi S, Saha RN. Nanostructured lipid carriers for site-specific drug delivery. Biomed Pharmacother 2018; 103:598–613. doi:https://doi.org/10.1016/j.biopha.2018.04.055

Kim H, Ro J, Barua S, Hwang DS, Na SJ, Lee HS, Jeong JH, Woo S, Kim H, Hong B, Yun G (2015) Combined skin moisturization of liposomal serine incorporated in hydrogels prepared with carbopol ETD 2020, rhesperse RM 100 and hyaluronic acid. Korean J Physiol Pharmacol 19(6):543

Kim JE, Oh GH, Jang GH, Kim YM, Park YJ (2019) Transformer-ethosomes with palmitoyl pentapeptide for improved transdermal delivery. J Drug Deliv Sci Technol 52(April):460–467. https://doi.org/10.1016/j.jddst.2019.04.039

Kocic H, Stankovic M, Tirant M, Lotti T, Arsic I (2020) Favorable effect of creams with skimmed donkey milk encapsulated in nanoliposomes on skin physiology. Dermatol Ther 33(4):e13511

Kumud M, Sanju N (2018) Nanotechnology driven cosmetic products: commercial:112–121. https://doi.org/10.2174/2213476X05666180530093111

Li Y, Zhang Y, Yan B (2014) Nanotoxicity overview: nano-threat to susceptible populations. Int J Mol Sci 15(3):3671–3697

Lohani A, Verma A, Joshi H, Yadav N, Karki N (2014) Nanotechnology-based cosmeceuticals. ISRN Dermatol.:843687. https://doi.org/10.1155/2014/843687

Luo C, Zhou L, Chiou K, Huang J (2018) Multifunctional graphene hair dye. Chem 4(4):784–794

Maria S, Gomes A, Saatmãn M, Moreira A, Raphael H, Silva B, Assis D, Leite-silva VR, Alves P, Medeiros M, De Negreiros F, Alexandre H, Rocha DO, Arantes E, Santos P, Soraya V, Sales DF (2017) Industrial crops & products evaluation of in vitro and in vivo safety of the by-product of Agave sisalana as a new cosmetic raw material: development and clinical evaluation of a nanoemulsion to improve skin moisturizing. Ind Crops Prod 108(March):470–479. https://doi.org/10.1016/j.indcrop.2017.06.064

Marianecci C, Di Marzio L, Rinaldi F, Celia C, Paolino D, Alhaique F et al (2014) Niosomes from 80s to present: the state of the art. Adv Colloid Interf Sci 205:187–206. https://doi.org/10.1016/j.cis.2013.11.018

Mavridi-Printezi A, Guernelli M, Menichetti A, Montalti M (2020) Bio-applications of multifunctional melanin nanoparticles: from nanomedicine to nanocosmetics. Nano 10(11):2276. https://doi.org/10.3390/nano10112276

Mohd-Setapar SH, John CP, Mohd-Nasir H, Azim MM, Ahmad A, Alshammari MB (2022) Application of nanotechnology incorporated with natural ingredients in natural cosmetics. Cosmetics 9(6):110

Müller RH, Radtke M, Wissing SA (2002) Solid lipid nanoparticles (SLN) and nanostructured lipid carriers (NLC) in cosmetic and dermatological preparations. Adv Drug Deliv Rev 54:S131–S155. https://doi.org/10.1016/S0169-409X(02)00118-7. Elsevier

Nanotechnology in Cosmetics Market (2024) Available online https://finance.yahoo.com/news/nanotechnology-cosmetics-market-projected-reach-141500311.html. Accessed 20 Oct 2024

Nath R, Chakraborty R, Roy R, Mukherjee D, Nag S, Bhattacharya A (2021) Nanotechnology based cosmeceuticals. Int J Sci Res Sci Technol 8:94–106

Ngoc LT, Nhu J-YM, Lee Y-C (2023) Insights into bioactive peptides in cosmetics. Cosmetics 10(4):111. https://doi.org/10.3390/cosmetics10040111

Padois K, Cantiéni C, Bertholle V, Bardel C, Pirot F, Falson F (2011) Solid lipid nanoparticles suspension versus commercial solutions for dermal delivery of minoxidil. Int J Pharm 416(1):300–304

Pelikh O, Hartmann SF, Abraham AM, Keck CM (2019) Nanocrystals for dermal application. In: Cornier J, Keck CM, Van de Voorde M (eds) Nanocosmetics, pp 161–177. https://doi.org/10.1007/978-3-030-16573-48

Pereira-Silva M, Martins AM, Sousa-Oliveira I, Ribeiro HM, Veiga F, Marto J, Paiva-Santos AC (2022) Nanomaterials in hair care and treatment. Acta Biomater 142:14–35

Qamar W, Gulia S, Athar M, Ahmad R, Imam MT, Chandra P, Singh BP, Haque R, Hassan MI, Rahman S (2024) An insight into impact of nanomaterials toxicity on human health. PeerJ 12:e17807. https://doi.org/10.7717/peerj.17807

Raj S, Jose S, Sumod US, Sabitha M (2012) Nanotechnology in cosmetics: opportunities and challenges. J Pharm Bioallied Sci 4(3):186–193. https://doi.org/10.4103/0975-7406.99016

Rajasekar M, Mary J, Sivakumar M, Selvam M (2024) Recent developments in sunscreens based on chromophore compounds and nanoparticles. RSC Adv 14(4):2529–2563

Raju NS, Krishnaswami V, Vijayaraghavalu S, Kandasamy R (2020) Chapter 2 – Transdermal and bioactive nanocarriers. In: Nanda A, Nanda S, Nguyen TA, Rajendran S, Slimani YBT-N (eds) Micro and nano technologies. Elsevier, Amsterdam, pp 17–33. ISBN:978-0-12-822286-7

Ribeiro RC, Barreto SM, Ostrosky EA, Rocha-Filho PA, Veríssimo LM, Ferrari M (2015) Production and characterization of cosmetic nanoemulsions containing Opuntia ficus-indica (L.) Mill extract as moisturizing agent. Molecules 20(2):2492–2509

Rohmawati L, Setyarsih W, Hefdea A, Wulancahayani E (2021) Application titanium dioxide nanotube from natural source as tooth whitening. J Phys Conf Ser 1805(1):012012. IOP Publishing

Ryman-Rasmussen JP, Riviere JE, Monteiro-Riviere NA (2006) Penetration of intact skin by quantum dots with diverse physicochemical properties. Toxicol Sci 91(1):159–165

Salvioni L, Morelli L, Ochoa E, Labra M, Fiandra L, Paluganb L, Prosperi D, Colombo M (2021) The emerging role of nanotechnology in skincare. Adv Colloid Interf Sci 293:102437

Santos AC, Morais F, Simões A, Pereira I, Sequeira JAD, Pereira-Silva M et al (2019a) Nanotechnology for the development of new cosmetic formulations. Expert Opin Drug Deliv 16:313–330. https://doi.org/10.1080/17425247.2019.1585426

Santos AC, Panchal A, Rahman N, Pereira-Silva M, Pereira I, Veiga F et al (2019b) Evolution of hair treatment and care: prospects of nanotube-based formulations. Nanomaterials 9:903. https://doi.org/10.3390/nano9060903

Semsarzadeh N, Khetarpal S (2022) Rise of stem cell therapies in aesthetics. Clin Dermatol 40(1):49–56., ISSN:0738-081X. https://doi.org/10.1016/j.clindermatol.2021.08.012

Sonneville-Aubrun O, Simonnet JT, L'Alloret F (2004) Nanoemulsions: a new vehicle for skincare products. Adv Colloid Interf Sci 108–109:145–149. https://doi.org/10.1016/j.cis.2003.10.026

Souto EB, Fernandes AR, Martins-Gomes C, Coutinho TE, Durazzo A, Lucarini M, Souto SB, Silva AM, Santini A (2020) Nanomaterials for skin delivery of cosmeceuticals and pharmaceuticals. Appl Sci 10(5):1594

Stern ST, Adiseshaiah PP, Crist RM (2012) Autophagy and lysosomal dysfunction as emerging mechanisms of nanomaterial toxicity. Part Fibre Toxicol 9:1–7

Synthetic ARON (2016) Eur J Biomed Pharmaceut Sci 3(8):133–145

Tadros TF (2018) 15. Formulation of sunscreens for UV protection. Pharm Cosmet Pers Care Formul De Gruyter:289–304. https://doi.org/10.1515/9783110587982-017

Taghavi SM, Momenpour M, Azarian M, Ahmadian M, Souri F, Taghavi SA, Sadeghain M, Karchani M (2013) Effects of nanoparticles on the environment and outdoor workplaces. Electron Physician 5(4):706

Teixeira JA, Silva AV, Santos Júnior VE, Melo Júnior PC, Arnaud M, Lima MG, Flores MA, Stamford TC, Dias Pereira JR, Ribeiro Targino AG, Galembeck A (2018) Effects of a new nano-silver fluoride-containing dentifrice on demineralization of enamel and Streptococcus mutans adhesion and acidogenicity. Int J Dentist 2018(1):1351925

Teng C, Jiang C, Gao S, Liu X, Zhai S (2021) Fetotoxicity of nanoparticles: causes and mechanisms. Nano 11(3):791

Thakor AS, Jokerst J, Zavaleta C, Massoud TF, Gambhir SS (2011) Gold nanoparticles: a revival in precious metal administration to patients. Nano Lett 11:4029–4036

Thibane VS, Ndhlala AR, Abdelgadir HA, Finnie JF, Van Staden J (2019) The cosmetic potential of plants from the eastern Cape Province traditionally used for skincare and beauty. S Afr J Bot 122:475–483. https://doi.org/10.1016/j.sajb.2018.05.003

Traversier M, Gaslondes T, Milesi S et al (2018) Polar lipids in cosmetics: recent trends in extraction, separation, analysis and main applications. Phytochem Rev 17:1179–1210. https://doi.org/10.1007/s11101-018-9588-7

Trehan S, Michniak-Kohn B, Beri K (2017) Plant stem cells in cosmetics: current trends and future directions. Future Sci OA 3(4):FSO226. https://doi.org/10.4155/fsoa-2017-0026

Ushirobira CY, Afiune LA, Pereira MN, Cunha-Filho M, Gelfuso GM, Gratieri T (2020) Dutasteride nanocapsules for hair follicle targeting: effect of chitosan-coating and physical stimulus. Int J Biol Macromol 151:56–61

Van Hoogevest P, Fahr A (2019) Phospholipids in cosmetic carriers. In: Cornier J, Keck CM, Van de Voorde M (eds) Nanocosmetics. Springer, pp 95–140. https://doi.org/10.1007/978-3-030-16573-4_6

Van Tran V, Moon JY, Lee YC (2019) Liposomes for delivery of antioxidants in cosmeceuticals: challenges and development strategies. J Control Release 300:114–140. https://doi.org/10.1016/j.jconrel.2019.03.003

Wani MY, Hashim MA, Nabi F, Malik MA (2011) Nanotoxicity: dimensional and morphological concerns. Adv Phys Chem 2011(1):450912

Xia T, Kovochich M, Brant J, Hotze M, Sempf J, Oberley T, Sioutas C, Yeh JI, Wiesner MR, Nel AE (2006) Comparison of the abilities of ambient and manufactured nanoparticles to induce cellular toxicity according to an oxidative stress paradigm. Nano Lett 6(8):1794–1807

Xie M, Jiang Z, Lin X, Wei X (2024) Application of plant extracts cosmetics in the field of anti-aging. J Dermatol Sci Cosmet Technol 1(2):100014. https://doi.org/10.1016/j.jdsct.2024.100014

Yeh Y-C, Creran B, Rotello VM (2012) Gold nanoparticles: preparation, properties, and applications in bionanotechnology. Nanoscale 4:1871–1880

Yildirimer L, Thanh NT, Loizidou M, Seifalian AM (2011) Toxicology and clinical potential of nanoparticles. Nano Today 6(6):585–607

Yücel Ç, Şeker Karatoprak G, Değim İT (2019) Anti-aging formulation of rosmarinic acid-loaded ethosomes and liposomes. J Microencapsul 36(2):180–191

Zhang N, Xiong G, Liu Z (2022) Toxicity of metal-based nanoparticles: challenges in the nano era. Front Bioeng Biotechnol 10:1001572. https://doi.org/10.3389/fbioe.2022.1001572

Zhao M, Chi Z, Lai D, Liu K, Liu Z, Zheng Y (2024) Advances in the microbial synthesis of active ingredients in cosmetics. Sheng Wu Gong Cheng Xue Bao 40(8):2489–2512. Chinese. https://doi.org/10.13345/j.cjb.240126

Zhou Y, Qian Y, Wang J, Qiu X, Zeng H (2020) Bioinspired lignin-polydopamine nanocapsules with strong bioadhesion for long-acting and high-performance natural sunscreens. Biomacromolecules 21(8):3231–3241

# Chapter 17
# Nanotechnology in Precision Medicine

**Raj Nandini, Bhavya Sharma, and Reema Gabrani**

**Abstract** Nanotechnology has transformed precision medicine, into something truly miraculous. Now, it offers unprecedented disease diagnosis, treatment, and monitoring capabilities. Fluorescent nanoparticles, aptamer-based systems, and glucose-sensitive devices are some of the nanosensors that assist in real-time tracking of biological markers and disease advancements. Liposomes, polymeric nanoparticles, and hybrid systems increase drug delivery through improved targeting and sustained-release opportunities within nanocarriers. There is a plethora of nano-technology innovations in medical imaging and diagnosis; starting from the use of metallic nanoparticles for enhancing contrast to theranostic nanosystems that have both therapeutic and diagnostic functions. These are exciting innovations that demonstrate significant potential, but difficulties persist, such as the stability and approval of the approach. The possibilities with artificial intelligence integration and the optimization of multifunctional nanoparticles can revolutionize individualised healthcare. This review underscores the crucial role of nanotechnology in the development of precision medicine, while emphasizing the need for continued research and regulatory adaptation to ensure safe and effective clinical implementation.

**Keywords** Cancer therapy · Drug delivery · Imaging · Nanocarriers · Nanosensors

R. Nandini · B. Sharma · R. Gabrani (✉)
Department of Biotechnology, Jaypee Institute of Information Technology,
Noida, Uttar Pradesh, India
e-mail: reema.gabrani@mail.jiit.ac.in

 
V. Khanna et al. (eds.), *Nanotechnologies for Drug Delivery and Biopharmaceutical Development*, Progress in Drug Research 77,
https://doi.org/10.1007/978-3-032-00796-4_17

## 17.1 Introduction

In recent history, nanotechnology has emerged as perhaps one of the most effective tools used in precision medicine, enhancing the sensitivity of disease diagnosis, treatment, and monitoring. This interdisciplinary field combines principles from physics, chemistry, biology, and engineering to alter matter at the nanoscale, typically between 1 and 100 nm. It is known as nanomedicine when nanotechnology is applied to diagnose a disease or used in therapeutic applications (Alghamdi et al. 2022).

To get accurate therapeutic results, nanotechnology is currently being used to specialties such as immunology, neurology, cardiology, and oncology as depicted in Fig. 17.1. Accurate targeting can improve patient recovery in cardiovascular disorders, where it is being used to build improved drug delivery systems (Anjum et al. 2021). It can potentially traverse the blood-brain barrier, which makes it useful in neurology as well, particularly for targeted therapy to treat neurodegenerative diseases (Yousef Sheikh 2014). Similarly, modified nanoparticles are being used in immunotherapy to improve the administration of vaccines and immune modulators to prevent autoimmune illnesses and infectious diseases in some patients (Scheetz et al. 2019).

Nanomedicine has become especially impactful in oncology, where the introduction of nanoparticles has redefined cancer therapy. Engineered nanoparticles can supply therapeutic agents straight to tumor spots. This minimizes systemic toxicity

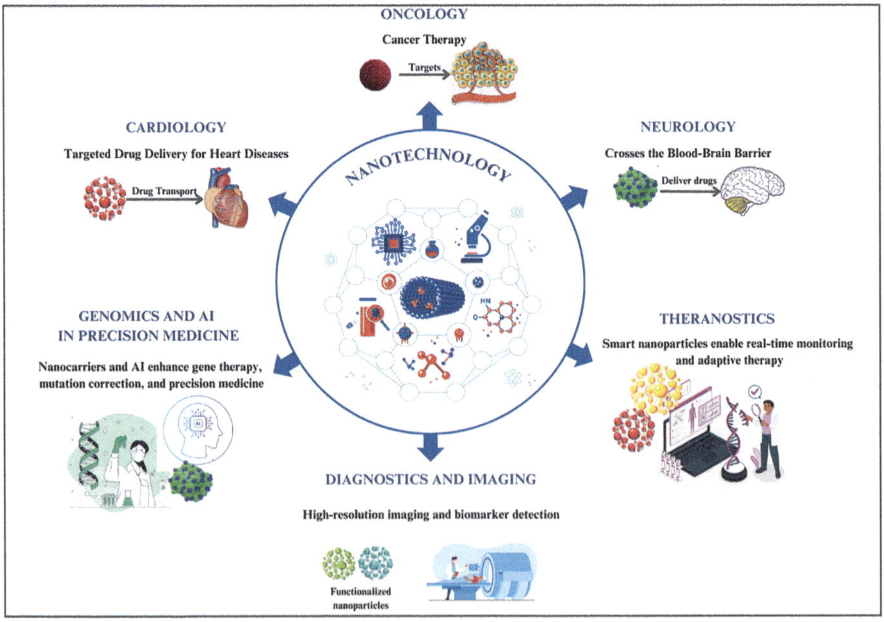


**Fig. 17.1** Applications of nanotechnology in medicine and healthcare

and improves restorative efficacy. This targeted delivery is possible by the enhanced permeability and retention (EPR) effect which means that nanoparticles can preferentially accumulate in tumor tissues because they exhibit unique vascular architecture (Choi 2011). Drug delivery nanoparticles, including lipid-based, polymeric, and inorganic formulations, are now specifically designed keeping in mind, the unique characteristics of individual tumors (Mitchell et al. 2020).

In addition to drug delivery, nanotechnology plays a key role in precision medicine's biomarker identification and diagnostic imaging. It is possible to functionalize nanoparticles with imaging agents to improve imaging methods' contrast and resolution helping the disease to be accurately assessed and identified (Pan et al. 2010). This can also be combined with techniques like molecular imaging which allows the biological processes to be monitored in real-time and help in gaining understandings into disease mechanisms (Anurogo 2023).

Another critical aspect of precision medicine that scientists believe is significant in relation to nanomedicine is genomics. It is relevant for creating nanocarriers that are specifically designed to repress or fix genes that may exhibit aberrant expression in tumors that are specific to each individual. This method lessens the possibility of negative outcomes linked to traditional treatments (Gu 2023). It can also help evaluate large datasets to find trends and forecast treatment outcomes when combined with artificial intelligence (AI) (Rawat et al. 2022).

Another feature of nanotechnology is highlighted by its ability to add to theranostics—the combination of therapeutic and diagnostic mechanisms. This paradigm modification allows for the monitoring of the therapeutic and disease states in parallel which means that clinicians are in a position to realign the direction of therapies depending on the reaction of the patient (Madamsetty et al. 2019). Research within the field is ever-evolving so, the development of multifunctional nanoparticles that can diagnose, treat, and monitor diseases will likely become a cornerstone of personalized healthcare.

## 17.2 Nanosensors

Nanosensors offer highly rapid, selective, and sensitive tools for comprehensive and real-time molecular-level diagnostics and health monitoring, essential to personalized medicine. These sensors are built with cutting-edge nanomaterials that can identify target biological markers such as certain proteins, nucleic acids, and other pathological indicators. This allows clinicians to track disease evolution and adjust therapies accordingly (Pachkawade 2023). Apart from disease detection and response monitoring, nanosensors in clinical environments are also applied in drug delivery. Here, we discuss various forms of nanosensors depending on the nanomaterials used, including fluorescent nanoparticles, aptamers, nanocomposites, and glucose-sensitive nanofibers.

### 17.2.1 Fluorescent Nanoparticles

Fluorescent nanoparticles emit fluorescence when irradiated with particular wavelengths of light and they are widely used in nanosensor constructions. One of the studies showcased the application of these nanoparticles in tracing mesenchymal stem cells' (MSC's) survival and monitoring their function. Yeo et al. bypassed the limitation of live detection and could emit fluorescent signal upon expression of biomarkers in the cell up to 30 days. The scientists could track MSC's and monitor their survival (Yeo et al. 2015). As shown in Table 17.1, these properties make them suited for target biomolecule identification as well as monitoring the activity of cells in vivo. Moreover, these kind of nanosensors can help in comprehending stem cell therapies and their effect on patients to provide clinicians with real-time data to direct treatment plans.

### 17.2.2 Protease Activity Detection Nanoparticles

Protease activity detection nanoparticles and the nanosensors can be used to monitor bacterial infections and determine the options for therapy. Buss et al. designed a real-time urine-based nanosensor with fluoro-nanomaterials to identify and differentiate protease activity in variated bacterial infections. The antibiotic outcomes could be tracked in resource-limited regions to help monitor patients at home and remotely (Buss et al. 2018).

**Table 17.1** Types of nanosensor and their applications

| Type of nanosensor | Applications | References |
|---|---|---|
| Fluorescent nanoparticles | Tracking mesenchymal stem cells (MSCs) for long-term cell survival and function; real-time post-transplant monitoring. | Yeo et al. (2015) |
| Protease activity detection nanoparticles | Identifying bacterial infections; tracking therapeutic responses using urine-based nanosensors. | Buss et al. (2018) |
| Aptamer-based nanosensors | Detecting low-abundance biomarkers; targeted imaging of cancer cells (e.g., PSMA); diagnosing diseases using dual aptamer systems. | Zhang et al. (2019); Park and Ban (2023) |
| Nanocomposite-based nanosensors | Enhanced cancer diagnostics using radiolabeling and conjugated biomarkers; measuring electron transfer resistance. | Subramanian et al. (2015) |
| Glucose-sensitive nanosensors | Continuous monitoring of blood sugar levels for diabetes management; guiding insulin therapy using nanofiber scaffolds. | Balaconis et al. (2015) |
| Photothermal nanoparticles | Cancer imaging and therapy using NIR photothermal effects; combined photodynamic and photothermal therapy; delivery of therapeutic agents via NIR irradiation. | Husni et al. (2020); Fu et al. (2020); Luan et al. (2021) |

### 17.2.3 Aptamer-Based Nanosensors

Aptamers are synthetic molecules that selectively interact with their target biomolecules, hence becoming very useful in biosensing. This high specificity implies that these biomarkers can accurately identify low concentration biomarkers making them suitable for use in early disease diagnosis. Aptamer-conjugated nanoparticles have an ability to specifically target cancer markers, such as prostate-specific membrane antigen (PSMA) (Zhang et al. 2019). Thus, one application of nanoparticles could be in targeted imaging of cancer cells. In a study, Park and Ban showcased the formation of a fluorescence nanosensor based on dual aptamers to detect periostin, a biomarker linked to a number of illnesses. This nanosensor effectively measured the high amounts of periostin in serum taken from human patients, indicating aptamers' potential for use in clinical diagnostics (Park and Ban 2023). There are many prospects for creating diagnostic instruments that are quicker, more sensitive, and more selective due to this application.

### 17.2.4 Nanocomposite-Based Nanosensors

Nanocomposites are the materials that are formed by combining several nanomaterials. They are progressively being used as a modification to nanosensors as they showcase improved effects, increased stability and sensitivity. These are particularly important in the diagnosis of cancer. Subramanian et al. depicted the applications of nanocomposites that could supplement the cancer detection mainly containing biomarkers especially as immunosensors to enhance the accuracy in cancer detection. Moreover, they also developed nanosensors to measure electron transfer resistance that could be used in diagnosing cancer in laboratories (Subramanian et al. 2015). These nanosensors are very useful in precision medicine as they combine both the imaging and therapeutic applications.

### 17.2.5 Glucose-Sensitive Nanosensors

Glucose-sensitive nanosensors are important in diabetes management and are becoming very popular. They can track the blood glucose levels continuously and are being utilised to monitor insulin therapy. A study by Balaconis et al. developed a biocompatible nanofiber scaffold and used derivatives of fluorescent labelled boronic acid to measure live glucose levels. This personalized approach is very useful to the patients as it helps in the controlling of glucose levels in real-time and modify the insulin therapy in compliance with the results (Balaconis et al. 2015).

### 17.2.6 Photothermal Nanoparticles

The utilization of photo-based nanoparticles in therapy and imaging has been evolving as an area of interest. A study by Husni et al. discusses the features of optical imaging and phototherapy and how these techniques with the help of the remarkable properties of nanomaterials contribute to increasing the efficacy of diagnosis and treatment (Husni et al. 2020). Fu et al. lists a few subcellular nano formulations like pH-responsive photothermal nanoparticles using bovine serum albumin that target the Golgi apparatus in cancer cells, activated by the acidic environment for effective therapy (Fu et al. 2020). In a review by Luan et al., the authors mention light activated nanoparticles like, polydopamine (PDA) nanoparticles which utilize NIR (near-infrared) photothermal properties to kill cancer cells (Luan et al. 2021). They enhance anti-tumor effects through combined photodynamic and photothermal therapy (PDT and PTT). Small organic nanoparticles (SPNs) have been used for photothermal therapy (PTT) under NIR irradiation, demonstrating size-independent optical properties and enabling fluorescence imaging. Gold Nanoparticles (AuNPs) too can break Au-S bonds under NIR irradiation, facilitating biomacromolecule delivery. They can be modified to deliver therapeutic agents when irradiated with NIR light (Luan et al. 2021).

## 17.3 Nanocarriers for Personalised Drug Delivery

Nanocarriers have become a revolutionary component in precision medicine, particularly for enhancing drug delivery systems. These nanomaterials, including liposomes, dendrimers, micelles, and hybrid nanoparticles (shown in Fig. 17.2), enable targeted drug distribution and ameliorate the solubility and bioavailability of hydrophobic drugs. Their aptitude to facilitate the co-transport of multiple therapeutic agents has drastically improved therapeutic outcomes while dropping systemic toxicity. Moreover, the ability to utilize nanoparticles as imaging agents allows for real-time monitoring of drug distribution and release at the target site. Lammers et al. emphasize the importance of this capability in optimizing drug delivery strategies and improving patient outcomes through personalized treatment plans (Lammers et al. 2010). Nanotechnology-based drug delivery systems (NDDSs) have garnered significant interest because of their unique physical, chemical, and biological properties. E. Güven states that these systems enable local, sustained, and targeted drug delivery, which is crucial for improving therapeutic efficiency while minimizing toxicity, especially in target orthopaedic conditions, such as bone regeneration and healing (Güven 2021). Advantages of NDDSs include controlled drug release by navigating biological barriers, continued circulation stint, and enhanced cellular uptake by delivering therapeutics directly to target cells. These can carry growth factors or anti-inflammatory drugs that promote bone healing (Güven 2021). Here, the nanocarriers used in precision medicine can be classified into organic and inorganic materials, each offering distinct benefits for targeted therapy. Nanocarriers and their applications are enumerated in Table 17.2.

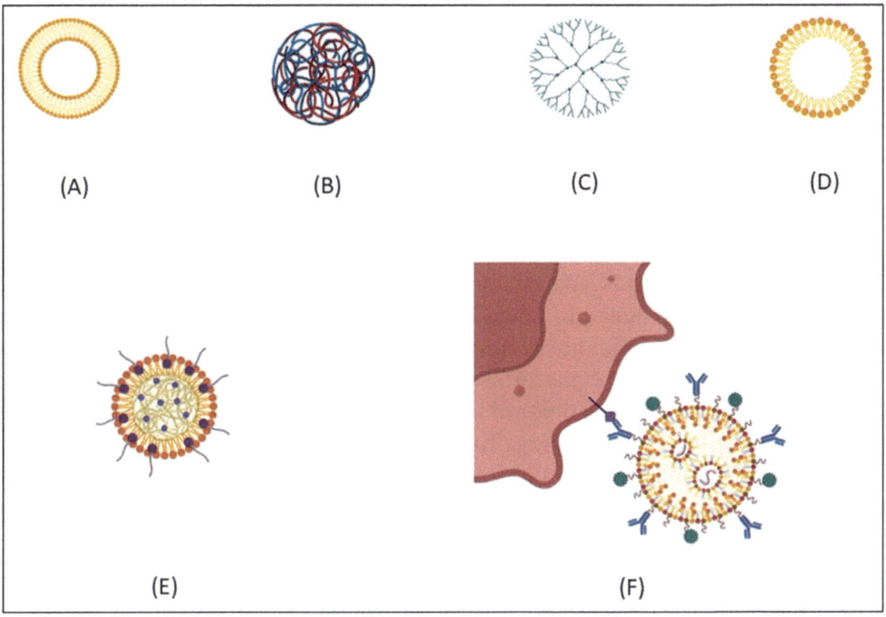


**Fig. 17.2** Different types of nanocarriers. (**a**) Liposome. (**b**) Polymeric nanoparticle made up of two different materials. (**c**) Dendrimer. (**d**) Micelle. (**e**) Hybrid nanoparticle. (**f**) Targeting nanoparticle specific to cell surface receptors

**Table 17.2** Types of nanocarrier and their applications

| Type of nanocarrier | Applications | References |
|---|---|---|
| Liposomes | Targeted drug delivery in oncology and infectious diseases; controlled release; insulin delivery | Singh et al. (2019); Šamec et al. (2020); Torchilin (2014); Caturano et al. (2024) |
| Polymeric nanoparticles | Oncology and infectious disease treatment; controlled release; oral insulin delivery | Singh et al. (2019); Caturano et al. (2024); Thangam et al. (2021) |
| Nanovesicles | Natural drug delivery vehicles; targeted delivery to inflamed tissues; shielding therapeutic payloads | Gao et al. (2016); Tran et al. (2020) |
| Dendrimers | Drug encapsulation; cancer therapies; gene delivery systems | Dai et al. (2018); Singh et al. (2019) |
| Micelles | Solubilizing hydrophobic drugs; systemic delivery; cancer treatments | Singh et al. (2019) |
| Hybrid nanoparticles | Combined organic and inorganic properties for enhanced stability and targeting | Luo et al. (2023) |
| Click-crosslinked nanocarriers | Co-delivery of multiple agents; targeted cancer treatments; overcoming chemoresistance | Dai et al. (2018) |
| Targeting nanocarriers | Dual-receptor targeting for personalized cancer treatments; delivery of siRNA and small-molecule drugs | Li et al. (2016); Hrkach et al. (2012); Li et al. (2020) |

### 17.3.1 Liposomes

Liposomes are orbicular vesicles comprised of lipid bilayers, capable of encapsulating both hydrophilic and hydrophobic drugs, as seen in Fig. 17.2a. These organic nanocarriers are favored for their biocompatibility and knack to carry a broad scope of therapeutic agents (Table 17.2). They have been widely used in oncology and infectious diseases since the targeted drug release and delivery rate is important. Their ability to deliver drugs directly to tumor cells while sparing the healthy tissues has transformed the landscape of drug administration (Singh et al. 2019). Some liposomes are also designed to release drugs such as paclitaxel specifically within tumor zones due to their acidic pH, improving targeted therapy (Šamec et al. 2020). Torchilin explains that these platforms can be used for multiple applications, including targeting cancer by using doxorubicin-containing liposomes which are designed for combined delivery and imaging (Torchilin 2014). They include nested liposomes that employ ultrasound to release the drug, providing a non-invasive process to control when and where drugs are released (Torchilin 2014). Caturano et al. noted that liposomes can encapsulate insulin, releasing it sustainably in a targeted manner without being degraded. Modifications like cell-penetrating peptides and glucose-sensitive coatings further increase their effectiveness in insulin delivery (Caturano et al. 2024).

### 17.3.2 Polymeric Nanoparticles

Polymeric nanoparticles represent another class of organic nanocarrier characterized by biocompatibility and the ability to incorporate various types of therapeutic agents as depicted in Fig. 17.2b. Such nanoparticles help to stabilize the encapsulated drugs and enable controlled release which makes them good anti-infective agents and appropriate for oncological use. Polymeric nanocarriers can also be functionalized to direct drug deliveries towards specific cells, making them more accurate (Singh et al. 2019). Suitable polymeric nanoparticles such as PLGA are biocompatible systems that encapsulate therapeutic as well as imaging agents for accurate cancer treatment and diagnosis (Thangam et al. 2021). According to the study conducted by Caturano et al., chitosan nanoparticles improved insulin absorption by adhering to mucosal surfaces and facilitating transport across intestinal barriers. Modifications to withstand acidic environments make them especially effective for oral delivery (Caturano et al. 2024). Biocompatible PLGA nanoparticles offer the potential to provide specific delivery and controlled extended-release profiles. This allows for sustained insulin release by modifying the polymer composition, enhancing glucose management (Caturano et al. 2024).

### 17.3.3 Nanovesicles

The use of extracellular vesicles and exosomes as natural drug distribution vehicles have also gained traction in recent years. Their natural directing properties and steadiness in circulation can enhance the delivery of therapeutics to diseased tissues (Gao et al. 2016). Gao et al. explored the creation of nanovesicles from activated neutrophils, which retained targeting molecules for selective binding to inflamed blood vessels, and demonstrated their potential as a drug delivery system, effectively reducing acute lung inflammation (Gao et al. 2016). The combination of exosomes with engineered ligands, such as aptamers, further enhanced their specificity and efficacy, as exosomes could shield their therapeutic payloads from degradation (Tran et al. 2020). As exosomes occur naturally, they can evade immune recognition, making them highly promising candidates for precision therapeutics (Tran et al. 2020).

### 17.3.4 Dendrimers and Micelles

Dendrimers are highly branched, tree-like structures that have multiple functional groups at their surface, similar in Fig. 17.2c. They are used in drug transport systems because of their high surface area and precise control over drug loading. Dendrimers are particularly effective in encapsulating drugs and ensuring their targeted delivery, which increases therapeutic efficacy while minimizing side effects. These nanocarriers are under consideration for application in cancer treatment and gene delivery (Dai et al. 2018).

Micelles are nanoparticles that consist of one hydrophilic head and one hydrophobic tail (amphiphilic); they can solubilize hydrophobic drugs, enhancing their solubility and their bioavailability for systemic use (shown in Fig. 17.2d). Micelles are employed for solubilization of lipophilic drugs that are poorly soluble in aqueous media. These nanocarriers are capable of enhancing the efficiency of delivering chemotherapeutic agents to tumor cells (Singh et al. 2019).

### 17.3.5 Hybrid Nanoparticles

Hybrid nanoparticles include lipids and metals to exploit the favorable features of both forms of materials, organic and inorganic as depicted in Fig. 17.2e. These have better stability and functionalization capabilities, allowing to design more effective drug delivery systems. This makes hybrid nanocarriers especially suited for cancer applications, where capabilities of both organic and inorganic components in drug delivery and targeting could be harnessed (Luo et al. 2023).

### *17.3.6 Click-Crosslinked Nanocarriers*

Click-crosslinked nanocarriers have been pushing their way into precision medicine in recent years mainly because of their ability to deliver multiplexed therapeutic agents. These systems allow for the co-delivery of microRNA therapeutics plus small-molecule anticancer drugs that may help reverse chemoresistance in tumor cells. Compared with other systems, click-crosslinked nanocarriers provide a highly programmable stimuli-driven release of drugs by incorporating different stimuli (pH, temperature, or light) for cancer treatment (Dai et al. 2018).

### *17.3.7 Targeting Nanocarriers*

Targeting nanocarriers are developed to target two specific receptors on the surface of cancer cells to enhance the selectivity of the delivery system, as seen in Fig. 17.2f. These synergistic targeting platforms use two distinct ligands to bind to overexpressed receptors on tumor cells, increasing the specificity and effectiveness of drug delivery. This approach enhances the therapeutic index of drugs and enables more personalized treatments for cancer patients (Li et al. 2016). For example, Hrkach et al. clearly showed the efficacy of a prostate-specific membrane antigen (PSMA)-targeted docetaxel nanoparticle, which enhanced drug deposition at tumor sites but reduced exposure to the rest of the body hence increasing the therapeutic window of docetaxel (Hrkach et al. 2012). This targeted approach is crucial in reducing the side effects usually related with conventional chemotherapy, which often affects healthy tissues alongside cancerous ones. Li et al. created a hypoxia-responsive nanoparticle (HRNP) to deliver siRNA targeting CDC20, an oncogene, upregulated in breast cancer tissues, correlating with tumor hypoxia, demonstrating effective silencing of the gene and potent antitumor effects (Li et al. 2020). It proved how RNAi integrated nanoparticles too, could offer a promising strategy for precision cancer therapy (Li et al. 2020).

## 17.4 Nanomaterials for Personalised Imaging and Diagnostics

The nanomaterials such as metallic nanoparticles, quantum dots, carbon-based nanomaterials, and polymeric nanoparticles, find applications in many imaging modalities that include CT, MRI, and fluorescence microscopy (Basu et al. 2023) as depicted in Fig. 17.3. Specific tumors should be targeted using quantum dots and other nanomaterials, such as gold and iron oxide nanoparticles, in vivo, to enhance

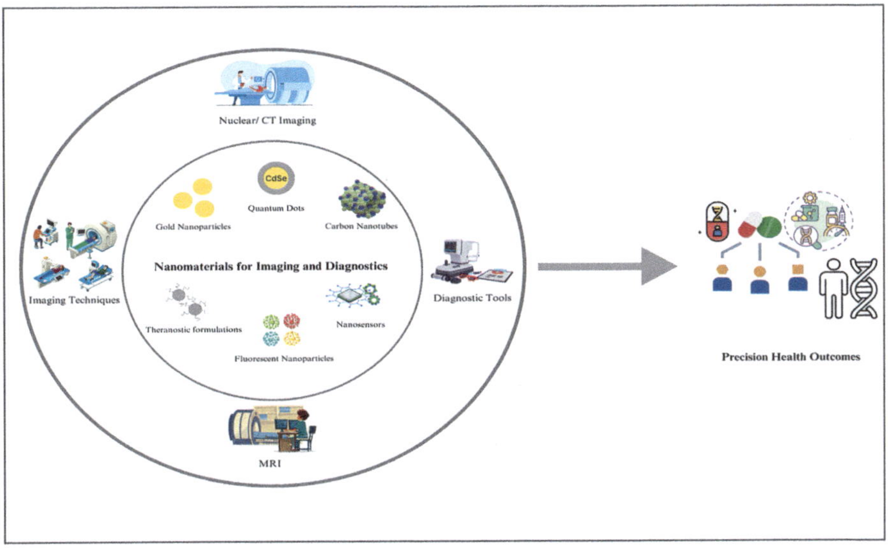


**Fig. 17.3** Role of nanomaterials in personalized imaging and diagnostics

the specificity or accuracy of cancer diagnosis and make the treatment more effective passively and actively (Jin et al. 2020a). Nanomaterials have now become imperative in precision medicine for imaging and diagnostic purposes (Table 17.3). Because of their dimensions, surface properties, and the ease with which tunability is achieved, they offer an excellent opportunity to increase sensitivity, specificity, and precision of diagnostic tools.

### *17.4.1 Metallic Nanoparticles*

Metallic nanoparticles, such as gold and silver, are widely used in imaging for their unique optical properties that enhance contrast in the major imaging techniques of CT and MRI. Gold nanoparticles represent one of the most usual nanoparticles used in diagnostic imaging because of their ability to scatter light and enhance the sensitivity of imaging systems. Similarly, silver nanoparticles were also employed because of their plasmonic characteristics, which help to improve imaging resolution (Jin et al. 2020b). With the advent of advanced imaging modalities for the early identification of genitourinary cancers employing superparamagnetic iron oxide (IO), nanotechnology has transformed urological imaging approaches, as demonstrated by Jayasimha's study (Jayasimha 2016).

**Table 17.3** Types of nanomaterial for imaging and diagnostics and their applications

| Type of nanomaterials | Applications | References |
|---|---|---|
| Metallic nanoparticles | Enhance contrast in imaging techniques like CT and MRI, especially gold and silver nanoparticles | Jin et al. (2020b); Jayasimha (2016) |
| Quantum dots | Utilized for identifying biomarker and detecting early disease via fluorescence microscopy and in vivo imaging | Abate et al. (2019) |
| Carbon-based nanomaterials | Enhance sensitivity in diagnostic tools, cross the BBB for glioblastoma treatment, and enable biosensors | Šamec et al. (2020) |
| Theranostics | Combine diagnostic and therapeutic functions, including imaging, drug delivery, and thermal therapies | Bao et al. (2013); Jin et al. (2020b); Li et al. (2019); Alafeef et al. (2020); Kunjachan et al. (2012) |
| Fluorescent nanoparticles | Guide tumor visualization during surgeries, enhance intraoperative imaging and tumor resection precision | Zhang et al. (2019); Šamec et al. (2020); Jin et al. (2020b) |
| Nanosensors | Detect low-abundance biomarkers, such as circulating tumor cells, with high sensitivity | Abate et al. (2019) |

### *17.4.2 Quantum Dots*

Quantum dots are semiconductor nanocrystals exhibiting a size-dependent fluorescence property. Because of this particular size-dependent property of emission of a variety of colors, they can serve as excellent probes in fluorescence microscopy and in vivo imaging. QDs bring together the advantages of high sensitivity, high multivalence potential, and high multiplexing capabilities; hence, these probes are becoming increasingly valuable in early disease detection and identification of biomarkers in cancer diagnosis (Abate et al. 2019).

### *17.4.3 Carbon-Based Nanomaterials*

Graphene and carbon nanotubes (CNTs) are being researched for their mechanical and electrical qualities, which make them ideal for imaging and biosensor applications. These carbon-based nanoparticles are used to increase imaging technique accuracy and the sensitivity of diagnostic equipment. Carbon-based nanoparticles are normally used owing to their good biocompatibility and ability to cross the blood-brain barrier, an important problem in glioblastoma treatment (Šamec et al. 2020). Their use into biosensors has made it possible to track drug delivery systems and disease biomarkers in real time.

### 17.4.4 Theranostics

Theranostic agents integrate therapeutic and diagnostic capabilities on a single platform, including drug delivery, molecular imaging, and heat therapy (Bao et al. 2013). The effects of treatment are significantly enhanced by these integrated agents, which allow for simultaneous medication delivery and imaging. A study by Jin et al. illustrated the usage of $^{64}$Cu-doped PdCu@Au tripods that can be used for PET imaging and utilise photothermal therapy in restricting cancer growth paving the way for its potential uses in precision medicine (Jin et al. 2020b). Li et al. investigated the development of nanomedicine agents that can detect changes in hydrogen peroxide, glutathione, and pH, which are found in tumor microenvironments and causes targeted drug release (Li et al. 2019). The nanocarriers are designed to release drugs upon elevated levels of GSH, reducing systemic exposure. Similarly, high concentrations of hydrogen peroxide in tumors mediate the release of drugs from nanomedicine systems in response to $H_2O_2$, ensuring that drug delivery is tailored to the specific conditions of each tumor cell. This precise and localized drug release, driven by stimuli-responsive mechanisms, significantly improves targeting and minimizes off-target effects.

Machine learning (ML) models are also being incorporated in the arena of theranostics in precision medicine. In order to improve the detection and accuracy of breast cancer as well as the prediction of nanoparticle cellular internalization, a study by Alafeef et al. showed the value of deep learning algorithms in processing complicated datasets (Alafeef et al. 2020). In addition to increasing diagnostic accuracy, the combination of nanotechnology and artificial intelligence makes it possible to create customized theranostic platforms based on patient profiles.

Machine learning can recognise both inter-patient and intratumoral diverseness, which are crucial for optimizing treatment regime. Kunjachan et al. highlighted the importance of advanced imaging techniques that can monitor the location of the drug in the cell and its therapeutic effectiveness (Kunjachan et al. 2012). This capability is vital for monitoring the performance of nanomedicine formulations in clinical settings, ensuring greater precision and improved treatment outcomes.

### 17.4.5 Fluorescent Nanoparticles for Intraoperative Imaging

During intraoperative imaging, fluorescent nanoparticles help surgeons in identifying malignancies and cancers. These nanoparticles produce near-infrared light and thus help surgeons to resect tumors with greater precision, minimizing damage to surrounding healthy cells. In order to enhance tumor imaging during glioblastoma procedures, fluorescent nanoparticles are frequently used in combination with intraoperative MRI (Šamec et al. 2020). This development is very helpful for improving the precision of tumor resection in cancer surgery (Jin et al. 2020b). Rare-earth upconverting nanophosphors (UCNPs) can convert near-infrared light into visible light, providing a new method for biological luminescence labelling, which enhances imaging capabilities in cancer diagnostics (Zhang et al. 2019).

### *17.4.6 Nanosensors for Biomarker Detection*

The early identification of biomarkers for a number of illnesses, including cancer, is greatly aided by nanosensors. Their high sensitivity allows them to detect even very small amounts of biomarkers in blood samples. To support early cancer detection, Abate et al. developed a portable microfluidic system that uses metal nanoparticles to identify circulating tumor cells with single-cell precision (Abate et al. 2019).

## 17.5 Ethical, Regulatory, and Safety Considerations

The integration of precision medicine and nanotechnology offers great opportunities for healthcare, but the ethical, regulatory and safety needs to be fully considered, especially regarding patient safety and patient and informed consent (Chaupoo et al. 2023).

One of the main ethical issues of nanotechnology in medicine is the unique properties of nanomaterials that can lead to unpredictable biological interactions, which can lead to unanticipated health outcomes (Halamoda-Kenzaoui et al. 2021). Due to their unique behaviour from a nanoscale perspective, inhalation or injection of nanoparticles can cause toxicity problems (Su et al. 2018). Biosafety valuations and consistent testing procedures are urgently needed to identify potential risks linked with nanomedicine (Guo et al. 2014). Regulatory organizations emphasize the importance of nanotechnology-based product safety testing (Ehmann et al. 2013).

Nanotechnology is still a nascent field where all its aspects are not fully studied paramounting to regulatory gap, hindering the development of groundbreaking treatments (Tinkle et al. 2014). Unlike traditional pharmaceutical drugs that stick to established regulatory procedures, nanomedicines possess one-of-a-kind characteristics and interactions with biological systems that can be beneficial and adverse. Nanoparticles' size, shape, and surface chemistry can significantly influence their biological behavior. The potential for cumulative exposure to nanomaterials also raises concerns about long-term health effects. Thus, all the stakeholders must possess a thorough understanding of pharmacokinetics and toxicology (Su et al. 2018). Ongoing monitoring and post-market surveillance of nanomedicines and their effects as well as tailored safety assessments for each nanomaterial is a must (Guo et al. 2014; Halamoda-Kenzaoui et al. 2018).

There is also a chance of nanotechnology being misused, particularly in areas like genetic engineering and targeted medicines (Akkam and Khatatbeh 2022). To ensure that advancements in precision medicine and not to compromise fundamental ethical values, it is essential to establish standards that govern the responsible use of nanotechnology (Akkam and Khatatbeh 2022).

Stakeholder engagement, including input from patients, healthcare providers, and regulatory agencies, is essential to ensure that the benefits of nanomedicine are equitably distributed. Ethical considerations should be integrated into the decision-making process (Akkam and Khatatbeh 2022).

## 17.6 Challenges and Future Prospects

Incorporation of nanotechnology in precision medicine still needs to overcome serious obstacles to emerge as a promising field of research. One of the greatest barriers to the clinical translation of nanomaterials themselves is the lack of standardized protocols to assess their safety (Alvarez et al. 2017). Besides, the other challenge is the complexity in tumor biology. The heterogeneity of cancer cells may mean that responses to nanoparticle-based therapies can be very different and may complicate efforts toward universally effective treatments (Shi et al. 2016). Given the complexity of nanomaterials, it is frequently challenging to produce nanomedicine in a scalable and reproducible manner (Güven 2021).

The future of nanotechnology in precision medicine is bright, with research currently in progress to overcome many of the challenges being faced. The design of new nanomaterials, like biodegradable and bioresponsive nanoparticles, might improve the safety profile of nanomedicines (Angeli and Bousquet 2023). In addition, new imaging technologies in combination with nanotechnology could allow for more precise diagnosis and monitoring of treatment responses (Sadr et al. 2023).

Successful nanomedicine approaches in the not-so-distant future are thought to be driven by amalgamated teams comprising engineers and biologists along with a clinical component (Anjum et al. 2021). Artificial intelligence and machine learning in the design and usage of nanomedicines may further allow treatment by data generated by a specific person (Dewi et al. 2022).

Collaboration between multidisciplinary teams, including engineers, biologists, and clinicians, will be essential for driving the development of effective nanomedicine strategies (Anjum et al. 2021). Furthermore, the integration of artificial intelligence and machine learning in the design and application of nanomedicines could facilitate personalized treatment plans based on individual patient data (Dewi et al. 2022).

## 17.7 Conclusion

Precision medicine is an emerging field where nanotechnology is applied to revolutionize drug delivery, diagnosis, and treatment planning. The role of various nanotechnology platforms includes nanomaterials for enhanced accuracy in imaging, nanocarriers for site-specific medication delivery with minimum side effects, and nanosensors for real-time detection of biomarkers. This approach drives medical treatment toward therapies adapted to the individual molecular and cellular characteristics of patients. It is fair to say that, so far, most nanoparticles have been able to avoid the consequences of their entry into the systemic circulation, especially in oncology applications, showing improved therapeutic efficiency and reduced systemic toxicity enabled by the EPR effect, which allows for their selective accumulation at tumor sites.

There are still a lot of issues with nanotherapeutics' long-term safety and biocompatibility, as well as several regulatory concerns. Their unique characteristics make them extremely advantageous, but they also make biological interactions with them highly unpredictable. For safe clinical translation to be realized, established protocols and thorough biosafety studies are required. The regulatory systems' restrictions are frequently a barrier for some of the novel treatments. Therefore, this shows how seriously the stringent laws governing nanotechnology are enforced in order to balance safety concerns with the rate of technological growth.

The future of nanotechnology in precision medicine thus appears to be quite fascinating, with the inclusion of AI and genomics being very useful. These can create further development of diagnostic and therapeutic multipurpose platforms, allowing for real-time monitoring and dynamic changes in interventions. The prospect that an independent set of multifunctional nanoparticles will be able to realize this diagnosis, treatment, or even simultaneous monitoring of disease probably constitutes a new frontier facing health medicine, in which continuous efforts for optimizing treatment strategy management would occur based on updated, individual patient information.

While challenges persist, nanotechnology in the improvement of precision medicine has huge potential. More interdisciplinary research, adaptation by regulatory bodies, and serious ethical considerations are required for the full realization of its potential. Only under these conditions nanomedicine can become a cornerstone for future medical practice.

## References

Abate MF et al (2019) Visual quantitative detection of circulating tumor cells with single-cell sensitivity using a portable microfluidic device. Small 15(14):e1804890. https://doi.org/10.1002/smll.201804890

Akkam Y, Khatatbeh M (2022) Future ethical implications and potential health consequences of nanotechnology research in Jordan: a scientometric analysis. Trop J Pharm Res 20(2):425–431. https://doi.org/10.4314/tjpr.v20i2.29

Alafeef M, Srivastava I, Pan D (2020) Machine learning for precision breast cancer diagnosis and prediction of the nanoparticle cellular internalization. ACS Sens 5(6):1689–1698. https://doi.org/10.1021/acssensors.0c00329

Alghamdi MA, Fallica AN, Virzì N, Kesharwani P, Pittalà V, Greish K (2022) The promise of nanotechnology in personalized medicine. J Pers Med 12(5):673. https://doi.org/10.3390/jpm12050673

Alvarez MM et al (2017) Emerging trends in micro- and nanoscale technologies in medicine: from basic discoveries to translation. ACS Nano 11(6):5195–5214. https://doi.org/10.1021/acsnano.7b01493

Angeli E, Bousquet G (2023) A boom in nanotechnologies for a high level of precision medicine. Cancer 15(9):2522–2522. https://doi.org/10.3390/cancers15092522

Anjum S et al (2021) Emerging applications of nanotechnology in healthcare systems: grand challenges and perspectives. Pharmaceuticals 14(8):707. https://doi.org/10.3390/ph14080707
Anurogo D (2023) The art of Nanoimmunoherbogenomics 5.0. Adv Tissue Eng Regen Med Open Access 9(1):19–22. https://doi.org/10.15406/atroa.2023.09.00136
Balaconis MK, Luo Y, Clark HA (2015) Glucose-sensitive nanofiber scaffolds with an improved sensing design for physiological conditions. Analyst 140(3):716–723. https://doi.org/10.1039/C4AN01775G
Bao G, Mitragotri S, Tong S (2013) Multifunctional nanoparticles for drug delivery and molecular imaging. Annu Rev Biomed Eng 15(1):253–282. https://doi.org/10.1146/annurev-bioeng-071812-152409
Basu DK, Das M, Kundu S (2023) A review of the application of nanotechnology in different spheres of life sciences. Int J Res Publ Rev 4(9):2128–2143. https://doi.org/10.55248/gengpi.4.923.92463
Buss CG, Dudani JS, Akana RTK, Fleming HE, Bhatia SN (2018) Protease activity sensors noninvasively classify bacterial infections and antibiotic responses. EBioMedicine 38:248–256. https://doi.org/10.1016/j.ebiom.2018.11.031
Caturano A et al (2024) Advances in nanomedicine for precision insulin delivery. Pharmaceuticals 17(7):945–945. https://doi.org/10.3390/ph17070945
Chaupoo AS et al (2023) Transforming horticulture: the influence of nanotechnology on crop enhancement. Int J Environ Clim Chang 13(11):3381–3400. https://doi.org/10.9734/ijecc/2023/v13i113512
Choi E-J (2011) Nanotechnology in cancer therapy: overview and applications. J Pharm Investig 41(2):59–65. https://doi.org/10.4333/kps.2011.41.2.059
Dai Y, Chen X, Zhang X (2018) Recent developments in the area of click-crosslinked nanocarriers for drug delivery. Macromol Rapid Commun 40(3):e1800541. https://doi.org/10.1002/marc.201800541
Dewi MK, Chaerunisaa AY, Muhaimin M, Joni IM (2022) Improved activity of herbal medicines through nanotechnology. Nanomaterials 12(22):4073. https://doi.org/10.3390/nano12224073
Ehmann F et al (2013) Next-generation nanomedicines and nanosimilars: EU regulators' initiatives relating to the development and evaluation of nanomedicines. Nanomedicine 8(5):849–856. https://doi.org/10.2217/nnm.13.68
Fu X et al (2020) Precise design strategies of nanomedicine for improving cancer therapeutic efficacy using subcellular targeting. Signal Transduct Target Ther 5(1):262. https://doi.org/10.1038/s41392-020-00342-0
Gao J, Chu D, Wang Z (2016) Cell membrane-formed nanovesicles for disease-targeted delivery. J Control Release 224:208–216. https://doi.org/10.1016/j.jconrel.2016.01.024
Gu S (2023) Application of advanced technologies—nanotechnology, genomics technology, and 3D printing technology—in precision anesthesia: a comprehensive narrative review. Pharmaceutics 15(9):2289. https://doi.org/10.3390/pharmaceutics15092289
Guo J-W, Lee Y-H, Huang HW, Tzou M-C, Wang Y-J, Tsai J-C (2014) Development of Taiwan's strategies for regulating nanotechnology-based pharmaceuticals harmonized with international considerations. Int J Nanomedicine 9:4773–4783. https://doi.org/10.2147/ijn.s68134
Güven E (2021) Nanotechnology-based drug delivery systems in orthopedics. Jt Dis Relat Surg 32(1):267–273. https://doi.org/10.5606/ehc.2021.80360
Halamoda-Kenzaoui B, Holzwarth U, Roebben G, Bogni A, Bremer-Hoffmann S (2018) Mapping of the available standards against the regulatory needs for nanomedicines. Wiley Interdiscip Rev Nanomed Nanobiotechnol 11(1):e1531. https://doi.org/10.1002/wnan.1531
Halamoda-Kenzaoui B et al (2021) Methodological needs in the quality and safety characterisation of nanotechnology-based health products: priorities for method development and standardisation. J Control Release 336:192–206. https://doi.org/10.1016/j.jconrel.2021.06.016
Hrkach J et al (2012) Preclinical development and clinical translation of a PSMA-targeted docetaxel nanoparticle with a differentiated pharmacological profile. Sci Transl Med 4(128):128ra39. https://doi.org/10.1126/scitranslmed.3003651

Husni P et al (2020) Photo-based nanomedicines using polymeric systems in the field of cancer imaging and therapy. Biomedicine 8(12):618. https://doi.org/10.3390/biomedicines8120618

Jayasimha S (2016) Nanotechnology in urology. Indian J Urol 33:13–18. https://doi.org/10.4103/0970-1591.194780

Jin C, Wang K, Oppong-Gyebi A, Hu J (2020a) Application of nanotechnology in cancer diagnosis and therapy – a mini-review. Int J Med Sci 17(18):2964–2973. https://doi.org/10.7150/ijms.49801

Jin G, Cheah P, Qu J, Liu L, Zhao Y (2020b) Applications of nanomaterials for theranostics of melanoma. J Nanotheranostics 1(1):39–55. https://doi.org/10.3390/jnt1010004

Kunjachan S, Jayapaul J, Mertens ME, Storm G, Kiessling F, Lammers T (2012) Theranostic systems and strategies for monitoring nanomedicine-mediated drug targeting. Curr Pharm Biotechnol 13(4):609–622. https://doi.org/10.2174/138920112799436302

Lammers T, Kiessling F, Hennink WE, Storm G (2010) Nanotheranostics and image-guided drug delivery: current concepts and future directions. Mol Pharm 7(6):1899–1912. https://doi.org/10.1021/mp100228v

Li Z, Ye E, David, Lakshminarayanan R, Loh XJ (2016) Recent advances of using hybrid nanocarriers in remotely controlled therapeutic delivery. Small 12(35):4782–4806. https://doi.org/10.1002/smll.201601129

Li Y, An L, Lin J, Tian Q, Yang S (2019) Smart nanomedicine agents for cancer, triggered by pH, glutathione, $H_2O_2$, or $H_2S$. Int J Nanomedicine 14:5729–5749. https://doi.org/10.2147/ijn.s210116

Li Y et al (2020) Dual hypoxia-targeting RNAi nanomedicine for precision cancer therapy. Nano Lett 20(7):4857–4863. https://doi.org/10.1021/acs.nanolett.0c00757

Luan X, Pan Y, Gao Y, Song Y (2021) Recent near-infrared light-activated nanomedicine toward precision cancer therapy. J Mater Chem B 9(35):7076–7099. https://doi.org/10.1039/d1tb00671a

Luo F et al (2023) Dual-ligand synergistic targeting anti-tumor nanoplatforms with cascade-responsive drug release. Pharmaceutics 15(7):2014–2014. https://doi.org/10.3390/pharmaceutics15072014

Madamsetty VS, Mukherjee A, Mukherjee S (2019) Recent trends of the bio-inspired nanoparticles in cancer theranostics. Front Pharmacol 10:1264. https://doi.org/10.3389/fphar.2019.01264

Mitchell MJ, Billingsley MM, Haley RM, Wechsler ME, Peppas NA, Langer R (2020) Engineering precision nanoparticles for drug delivery. Nat Rev Drug Discov 20(1):1–24. https://doi.org/10.1038/s41573-020-0090-8

Pachkawade V (2023) N/MEMS biosensors: an introduction. In: MEMS and microfluidics in healthcare. Lecture notes in electrical engineering. Springer, Singapore, pp 1–10. https://doi.org/10.1007/978-981-19-8714-4_1

Pan D et al (2010) Nanomedicine strategies for molecular targets with MRI and optical imaging. Future Med Chem 2(3):471–490. https://doi.org/10.4155/fmc.10.5

Park J, Ban C (2023) Development of a one-shot dual aptamer-based fluorescence nanosensor for rapid, sensitive, and label-free detection of periostin. Sci Rep 13(1):10224. https://doi.org/10.1038/s41598-023-37418-0

Rawat B, Bist AS, Supriyanti D, Elmanda V, Sari SN (2022) AI and nanotechnology for healthcare: a survey. APTISI Trans Manag 7(1):86–91. https://doi.org/10.33050/atm.v7i1.1819

Sadr S et al (2023) Challenges and prospective of enhancing hydatid cyst chemotherapy by nanotechnology and the future of nanobiosensors for diagnosis. Trop Med Infect Dis 8(11):494–494. https://doi.org/10.3390/tropicalmed8110494

Šamec N, Zottel A, Videtič Paska A, Jovčevska I (2020) Nanomedicine and immunotherapy: a step further towards precision medicine for glioblastoma. Molecules 25(3):490. https://doi.org/10.3390/molecules25030490

Scheetz L et al (2019) Engineering patient-specific cancer immunotherapies. Nat Biomed Eng 3(10):768–782. https://doi.org/10.1038/s41551-019-0436-x

Shi J, Kantoff PW, Wooster R, Farokhzad OC (2016) Cancer nanomedicine: progress, challenges and opportunities. Nat Rev Cancer 17(1):20–37. https://doi.org/10.1038/nrc.2016.108

Singh AP, Biswas A, Shukla A, Maiti P (2019) Targeted therapy in chronic diseases using nanomaterial-based drug delivery vehicles. Signal Transduct Target Ther 4(1):33. https://doi.org/10.1038/s41392-019-0068-3

Su H, Wang Y, Gu Y, Bowman L, Zhao J, Ding M (2018) Potential applications and human biosafety of nanomaterials used in nanomedicine. J Appl Toxicol 38(1):3–24. https://doi.org/10.1002/jat.3476

Subramanian AP, Jaganathan SK, Supriyanto E (2015) Overview on in vitro and in vivo investigations of nanocomposite based cancer diagnosis and therapeutics. RSC Adv 5(89):72638–72652. https://doi.org/10.1039/c5ra11912j

Thangam R, Paulmurugan R, Kang H (2021) Functionalized nanomaterials as tailored theranostic agents in brain imaging. Nanomaterials 12(1):18. https://doi.org/10.3390/nano12010018

Tinkle S et al (2014) Nanomedicines: addressing the scientific and regulatory gap. Ann N Y Acad Sci 1313(1):35–56. https://doi.org/10.1111/nyas.12403

Torchilin VP (2014) Multifunctional, stimuli-sensitive nanoparticulate systems for drug delivery. Nat Rev Drug Discov 13(11):813–827. https://doi.org/10.1038/nrd4333

Tran PHL et al (2020) Exosomes and nanoengineering: a match made for precision therapeutics. Adv Mater 32(18):e1904040. https://doi.org/10.1002/adma.201904040

Yeo D, Wiraja C, Chuah YJ, Gao Y, Xu C (2015) A nanoparticle-based sensor platform for cell tracking and status/function assessment. Sci Rep 5(1):14768. https://doi.org/10.1038/srep14768

Yousef Sheikh B (2014) Recent advances in nanotechnology: potential prospects in neuromedicine and neurosurgery. Nanosci Technol 1(3):1–13. https://doi.org/10.15226/2374-8141/1/3/00115

Zhang Y, Li M, Gao X, Chen Y, Liu T (2019) Nanotechnology in cancer diagnosis: progress, challenges and opportunities. J Hematol Oncol 12(1):137. https://doi.org/10.1186/s13045-019-0833-3

# Chapter 18
# Nanoparticles in Molecular Imaging

**Rahat Bukhari and Rahat Abdul Rehman**

**Abstract** Nanoparticles' special qualities, like their substantial surface area, adjustable size, and multifunctionality, have made them effective instruments in molecular imaging.

Because of these characteristics, nanoparticles can be used as contrast agents in a variety of imaging modalities, including computed tomography (CT), magnetic resonance imaging (MRI), and fluorescence imaging.

The sensitivity and specificity of imaging are improved by their capacity to be functionalized with targeting ligands, which enables precise delivery to certain cellular or molecular targets.

The applications of nanoparticles have been further broadened by recent developments in their production and design, which hold promise for major gains in molecular disease monitoring and diagnostics.

**Keywords** Nanoparticles · Molecular imaging · Contrast agents · Diagnostics & targeting ligands

## 18.1 Introduction

Metal and metal oxides, polymers, lipids, semiconductors, and carbon structures are the primary building blocks used to create nanoparticles (NPs) (Bukhari and Rehman 2023). Targeted distribution is made possible by nanotechnology, which also enhances solubility, bioavailability, and stability in various gastrointestinal tract circumstances (Jones et al. 2019). These characteristics are crucial for the active ingredients in fortified foods and pharmaceutical medications. Silver and gold nanoparticle-based nanomaterials have antimicrobial qualities that are utilized in the field food sector (for example in poultry farms or equipments used in storage

R. Bukhari (✉) · R. A. Rehman
University of Health Sciences, Lahore, Pakistan

 
V. Khanna et al. (eds.), *Nanotechnologies for Drug Delivery and Biopharmaceutical Development*, Progress in Drug Research 77,
https://doi.org/10.1007/978-3-032-00796-4_18

of food), personal care products, and cosmetics, as well as in medicine (e.g., in silver-coated patches and bandages). Although nanoparticles are still smaller than cells, they are larger than many proteins and tiny compounds (Luther et al. 2020). They are absorbed by the endoplasmic reticulum because of their comparatively large diameters. By applying polymers to the surfaces of nanoparticles, this behavior can be partially mitigated. Nanoparticles can be used to transport medicinal drugs, signaling elements, and target molecules (ligands) because of their high surface area to volume ratio (Woźniak et al. 2022). Recently, many researchers and physicians from a variety of professions have been very interested in molecular imaging, which is defined as quantitative, non-invasive, and repeatable imaging of specific biomolecules and monitoring of relevant biological processes in living patients (MacRitchie et al. 2020). It has historically been divided into two categories. With the ability to offer anatomical information, the first group comprises medical imaging modalities like computed tomography (CT), magnetic resonance imaging (MRI), and ultrasound imaging. The other group consists of optical imaging, single photon emission CT (SPECT), and positron emission tomography (PET), which can provide molecular or functional information as shown in Fig. 18.1.

Recent developments in imaging techniques have made it feasible to use MRI and CT in molecular imaging. It is possible to see biological functions and related molecular processes through molecular imaging (Kim et al. 2017).

Because of their better photo stability and high quantum yield, semiconductor nanoparticles (also known as quantum dots, or QDs) and in optical imaging,

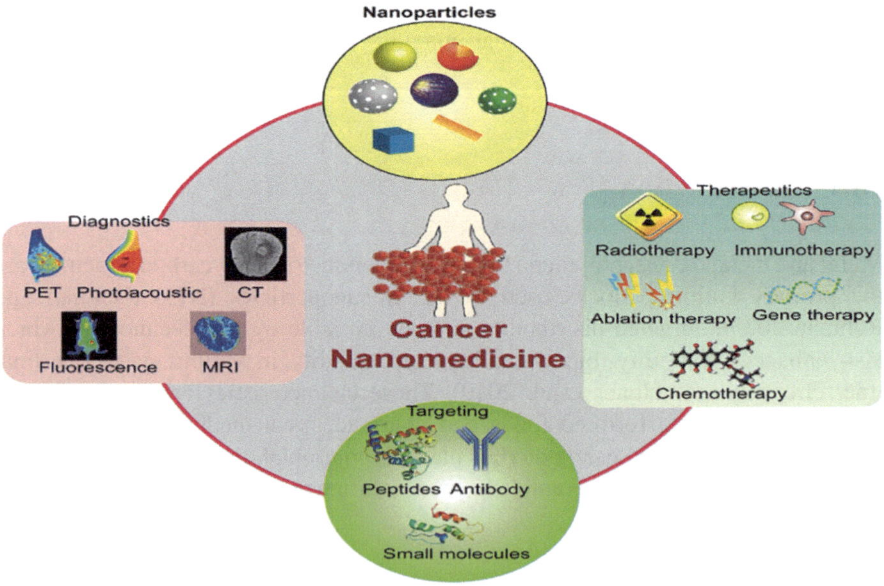


**Fig 18.1** A range of contract nano agents for in vivo stem cell tracking that are intended for use in magnetic resonance imaging, fluorescence imaging, ultrasound imaging, and photo acoustic imaging (Xu et al. 2021)

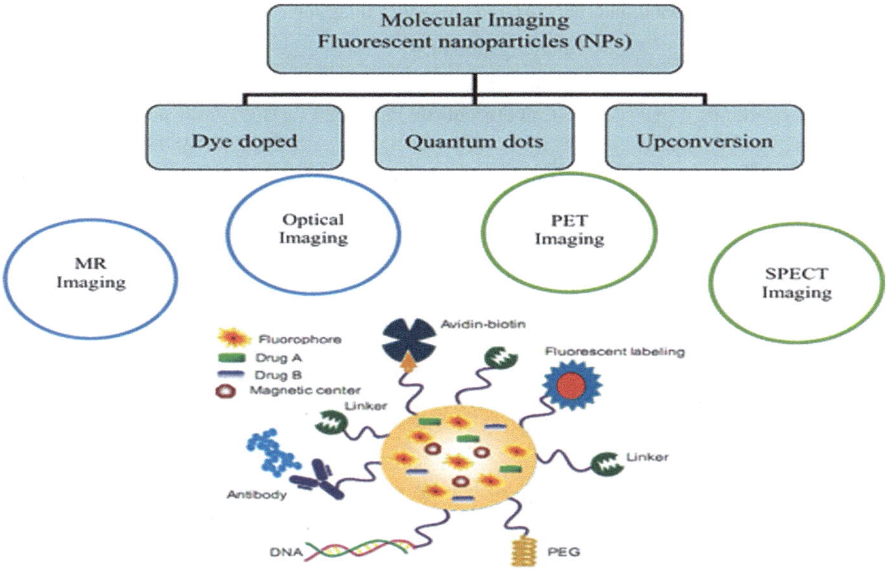


**Fig. 18.2** Fluorescence nanoparticles for molecular imaging (Jin et al. 2020)

upconversion nanoparticles (UCNPs) have been employed as fluorescent probes instead of organic dyes as shown in Fig. 18.2. Due to their higher X-ray attenuation coefficient than iodinated molecules, gold and tantalum oxide nanoparticles have been developed as CT contrast agents (Aslan et al. 2020).

## 18.2 Nanoparticle Development for Molecular Imaging

The development of a suitable, extremely sensitive contrast agent is required for molecular imaging. Nanotechnology has advanced to the point where it is now possible to accurately control the size, composition, and other properties of nanoparticles. MRI, CT, optical imaging, PET, and ultrasonic imaging are four representative imaging modalities as shown in Table 18.1.

Because of their characteristics, nanoparticles can perform exceptionally well in molecular imaging (Bouché et al. 2019) as shown in Fig. 18.3.

## 18.3 Nanoparticles for Magnetic Resonance Imaging

Since MRI is a non-invasive technique that provides anatomical, physiological, and molecular information about living people, it is the most potent imaging methods in molecular imaging (MacRitchie et al. 2020) as shown in Table 18.2. It has mostly

**Table 18.1** Types of nanomaterial's used in molecular imaging

| Nanoparticle type | Imaging modality | Advantages |
|---|---|---|
| Quantum dots (QDs) | Fluorescence imaging | Bright emission, photostability |
| Gold nanoparticles | CT, photoacoustic imaging | High X-ray absorption, biocompatibility |
| Iron oxide NPs | MRI | Superparamagnetic properties |
| Silica nanoparticles | Fluorescence, Raman | Biocompatible, tunable surface chemistry |
| Carbon nanotubes | Raman, photoacoustic | Strong optical absorption, stability |
| Upconversion NPs | NIR fluorescence | Deep tissue imaging, anti-stokes emission |

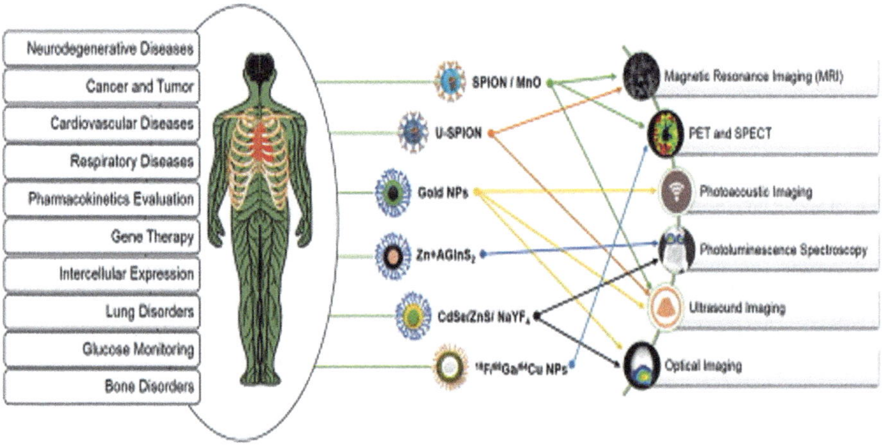


**Fig 18.3** Nanoparticles in molecular-imaging applications (Padmanabhan et al. 2016)

been modified for imaging the brain and other central nervous system organs, evaluating the performance of organs like the kidneys and heart, and identifying tumors as shown in Fig. 18.4. MRI is anticipated to become a crucial tool for molecular and cellular imaging because of its high resolution, infinite penetration depth, and superior soft tissue contrast. Even though MRI is primarily used to gather soft tissue anatomical data, molecular imaging cannot be done with it due to its low sensitivity (Bruno et al. 2019).

By altering their size and composition, nanoparticles can modify their magnetic properties, which determine their MRI contrast effect. Iron oxide nanoparticles, for instance, have a weak magnetic moment when they are less than 3 nm. With a low r2/r1 ratio, exceptionally tiny iron oxide nanoparticles (ESIONs) are suitable for use as T1 contrast agents because their magnetic properties are comparable to those of gadolinium complexes. The lengthy circulation time of ESIONs makes them beneficial for long-term, high-resolution MR imaging (Woźniak et al. 2022). Recent years have seen observable advancements in the creation of nanoparticle systems that

**Table 18.2** Types of nanoparticles used in MRI

| Nanoparticle type | Composition | Function in MRI | Applications |
|---|---|---|---|
| Iron oxide NPs | $Fe_3O_4$, $\gamma$-$Fe_2O_3$ | T2 contrast agent (darkens MRI signal) | Tumor imaging, cell tracking, brain imaging |
| Manganese-based NPs | MnO, $Mn_3O_4$ | T1 contrast agent (brightens MRI signal) | Neurological imaging, liver imaging |
| Gadolinium NPs | Gd-doped nanomaterials | T1 contrast agent | Vascular imaging, brain imaging |
| Silica-coated magnetic NPs | $Fe_3O_4$ core with $SiO_2$ shell | Biocompatible MRI contrast | Drug delivery, multimodal imaging |
| Hybrid NPs | $Fe_3O_4$-Gd, Au-$Fe_3O_4$ | Combined T1/T2 contrast | Multimodal MRI and CT imaging |

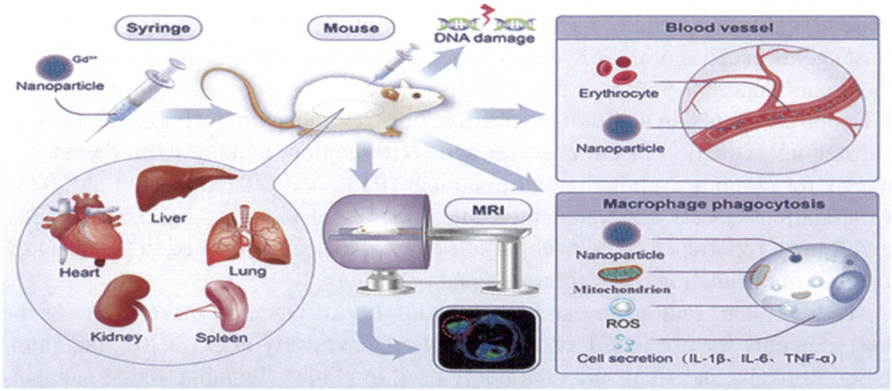


**Fig. 18.4** MRI-directed magnevist is effective in Gd-doped mesoporous carbon nanoparticle toxicity research in mice models (Zhang et al. 2023)

enable MRI imaging and diagnostics to be improved. Because nanoparticles increase contrast in MRI, it is easier to distinguish between healthy and pathologically changed tissues (Farinha et al. 2021).

Inorganic nanoparticles of iron, gold, cobalt oxide, or integrated gadolinium nanoparticles are examples of nanotechnology in this context (Woźniak et al. 2022).

## 18.4 Nanoparticles for Computed Tomography

Due to its great spatial resolution, accessibility, quick image capture, and affordability, CT is one of the most widely utilized non-invasive imaging modalities (Tanaka et al. 2019). The CT imaging principle is based on contrast chemicals or the human body's X-ray attenuation. When tissues or contrast agents absorb or scatter an irradiation X-ray beam, a detector at various angles measures the decreased X-ray intensity. It is possible to obtain three-dimensional whole-body pictures with extremely high spatial resolution by reconstructing X-ray intensity profiles (Dai et al. 2020).

Nano scale iodinated contrast agents have been developed to get beyond the drawbacks of the existing CT contrast agents. Among them are liposomal, polymeric, and micellar contrast agents. The nano-sized contrast agents have a long circulation period, which is crucial for molecular imaging, because they can avoid renal filtration (Hsu et al. 2023). Nevertheless, iodine's K-edge energy level is far lower than that of the X-ray photons utilized in CT. Therefore, heavy element-based contrast agents are needed for more sensitive CT imaging. Because of their superior biocompatibility and ease of production, gold nanomaterials in the form of spheres, rods, shells, and cages have also been researched among the high-Z elements. In general, gold nanoparticles have a greater CT contrast effect than iodine-based contrast agents because of their K-edge energy level of 80.7 keV (Oumano et al. 2021). Gold nanomaterials with a variety of shapes have been proposed as multifunctional CT contrast agents, in addition to spherical ones. Gold nanorods, for example, have substantial extinction in the near-infrared (NIR) region, making them theranostic agents for combined photothermal treatment (PTT) and CT (Nejabat et al. 2023). Conjugating several functional molecules, including targeting ligands, peptides, and Raman reporters to gold nanoparticles, has also been attempted in the development of multifunctional CT contrast agents.

The excellent radiopacity and biocompatibility of tantalum make it one of the best elements for use as CT contrast agents (Chakravarty et al. 2020). Tantalum-based implants and stents are commonly used in clinics. Tantalum oxide nanoparticles can be made by reacting tantalum ethoxide in a sol-gel process and then easily functionalising it with different silane agents. Numerous multifunctional nanomaterials can be created since the process for creating tantalum oxide nanoparticles is comparable to that of creating silica nanoparticles (Koshevaya et al. 2021). For example, tantalum oxide nanoparticles coated with PEG have an extremely extended circulation time. Apart from tantalum, nanoparticles based on bismuth or tungsten were also investigated as nanoscale CT contrast agents.

Because of their capacity to target, produce strong CT attenuation, and absorb cells, nanosized CT contrast agents have been employed in a variety of applications (Jiang et al. 2023). Red blood cell-engulfed gold nanoparticles, for instance, have been used to visualize blood flow. Tumor vasculature have been assessed using liposomal iodine, which has a lengthy circulation period and a substantial CT enhancement. To imaging prostate cancer, gold nanoparticles containing an aptamer and prostate-specific membrane antigen have been employed as shown in Fig. 18.5.

Zirconium dioxide nanoparticle accumulation has also been utilized to track drug distribution and scan tumors (Xie et al. 2019).

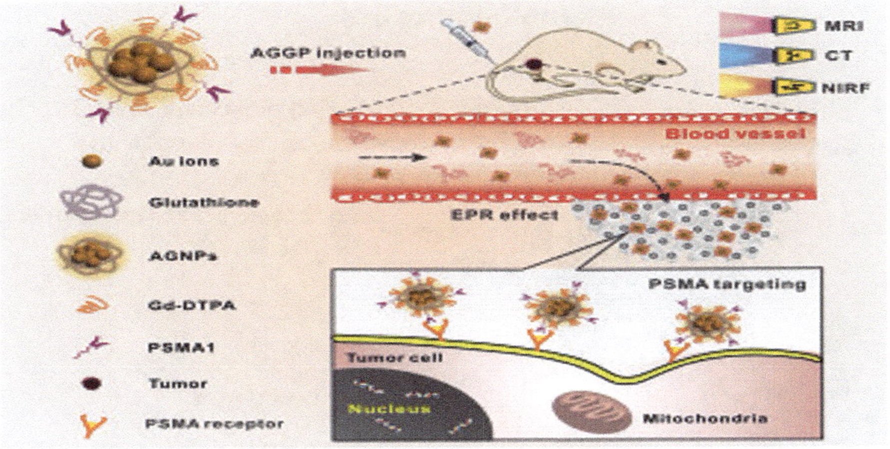


**Fig 18.5** PSMA1-mediated ultrasmall gold nanoparticles facilitate tumor targeting and MR/CT/NIRF multimodal detection of early-stage prostate cancer (Wang et al. 2023)

## 18.5 Nanoparticles for Optical Imaging

Optical imaging is quicker, less expensive, and more sensitive than MRI and CT imaging. In contrast to whole-body imaging modalities like MRI and PET, which have limited temporal and spatial resolution, optical imaging may produce high-resolution real-time images as shown in Fig. 18.6. Since optical microscopy is the only way to record subcellular occurrences, optical imaging has been extensively researched to comprehend how different nanomaterials behave in living organisms (Martynenko et al. 2017). For example, optical imaging makes it possible to see how different types of cells, including immune cells, interact with nanoparticles. Nowadays, a wide range of fluorescent dyes are widely employed as optical imaging probes. On the other hand, light scattering-induced background signals and shallow tissue penetration depth have hindered in vivo optical imaging. By using NIR light and/or anti-Stokes imaging, numerous advancements in optical imaging probes and procedures have been developed to maximise penetration depths and minimise background signals (Wu et al. 2019).

## 18.6 Nanoparticles for Positron Emission Tomography

In the last 10 years, there has been a lot of interest in the application of nanotechnology to nuclear imaging, particularly PET and SPECT. For targeted medication delivery, therapy, and patient monitoring, nanomaterial's—which are usually smaller than a few hundred nanometers—have become pioneers in nano oncology (Perez-Medina et al. 2020). In contrast to bulk material, nanoparticles have unique

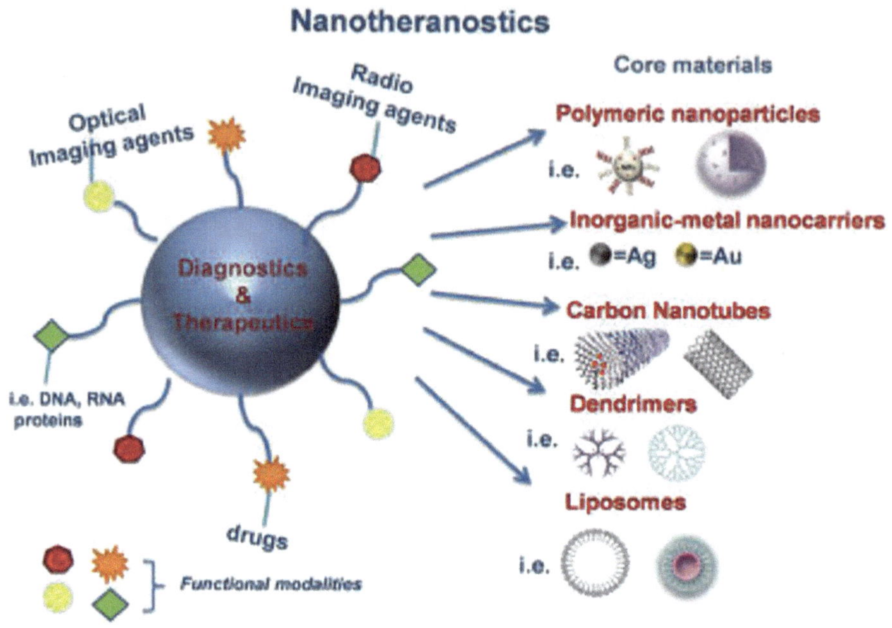


**Fig. 18.6** Using nanotheranostics and imaging modalities as medicinal and diagnostic instruments (Siafaka et al. 2021)

characteristics such as a large functional surface area, easily adjustable surface chemistry that makes it easier to attach to small molecule medicines, imaging labels, and targeted ligands like peptides, antibodies, nucleic acids, etc. Enhancement of permeability and retention (EPR) in tumor tissues and extravasation through endothelial cells are two distinctive intracellular and extracellular interactions made possible by their tiny size (about 100–10,000 times smaller than human cells). The National Institutes of Health (NIH) and National Cancer Institute (NCI) have made large investments in nanotechnology because of its enormous and distinctive potential, and some nanoparticles have even advanced to clinical trials (Henderson and Shankar 2017). In the case of head and neck cancer, for instance, the recent completion of two first-in-human phase I clinical trials (AuNP-mediated hyperthermia for therapy of refractory and/or recurrent tumors of the head and neck and AuNP-conjugated tissue necrosis factor (TNF) treatment in solid tumors) indicates that gold nanoparticles have found clinical usage as shown in Fig. 18.7.

Additionally, phase II clinical trials for lymph node mapping in patients with head and neck melanoma, breast, and cervical/uterine cancer are set to begin using fluorescent silica nanoparticles, commonly referred to as C dots (Goel et al. 2017). PET and nanotechnology work together in a symbiotic partnership that promises advantages for both parties. Unprecedented uses in molecular PET, noninvasive biological event interpretation, synergistic multimodal imaging (such as PET/CT, PET/

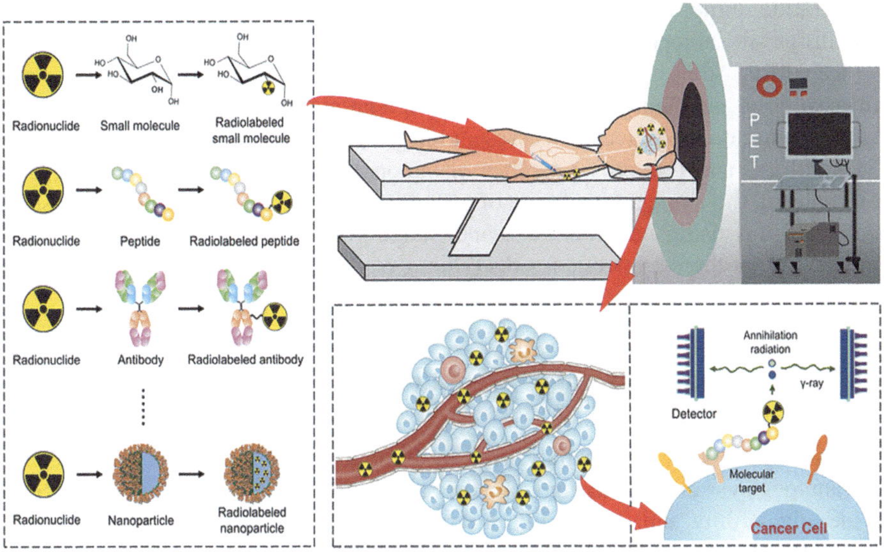


**Fig. 18.7** Principle of PET molecular imaging-based cancer phenotyping (Jin et al. 2022)

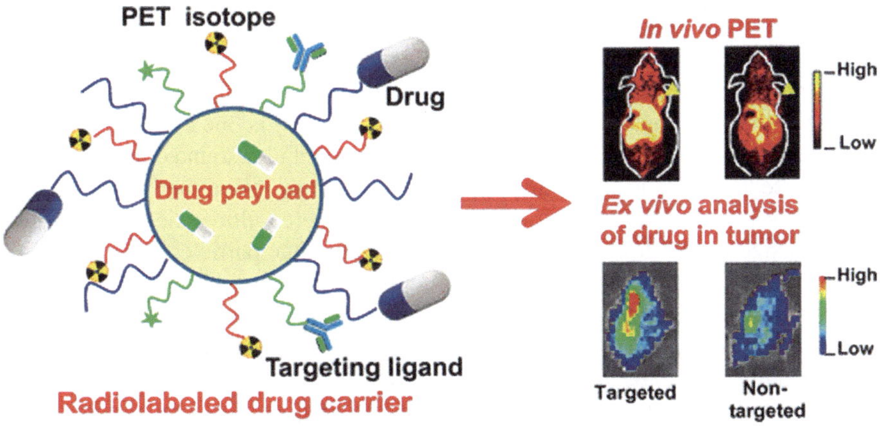


**Fig. 18.8** Positron emission tomography image-guided drug delivery (Ghosh et al. 2022)

MR, and PET/US), and theranostics are all made possible by the distinct physicochemical characteristics and multivalency of nanomaterials as shown in Fig. 18.8.

However, in the biomedical applications of nanomaterials for cancer theranostics, PET has become a powerful tool. In vivo biodistribution of nanoparticles, for instance, is known to be influenced by their physicochemical characteristics, including size, shape, content, surface charge and chemistry (PEGylation, ligand conjugation), and others (Nawaz et al. 2024). The pharmacokinetics, intratumoral penetration, tumor bioavailability, and cargo delivery mechanisms of Nano particulate drugs are subjects of investigation and optimisation.

Regarding this, PET's sensitivity, whole body imaging potential, noninvasiveness, and quantitative nature make it a great option for precisely investigating the intricate biological processes and in vivo ADME (Absorption, Distribution, Metabolism, and Excretion) profiles of nanoparticles, which have been the main obstacles in the clinical translation of nanomaterials. Second, PET has been used to deliver drugs using nanomaterials and image-guided therapy, enabling real-time monitoring of the therapeutic results (Huang et al. 2023).

## 18.7 Nanoparticles for Ultrasound

Ultrasonography (US) is widely used in clinical imaging because to its inherent advantages over other imaging techniques, including its low cost, noninvasiveness, wide adaptability and availability, nonradioactivity, and real-time monitoring (Li et al. 2020) as shown in Table 18.3. US molecular imaging can be performed using US contrast imaging based on US contrast agents. Micro bubbles are the most commonly utilized contrast agents in the United States. Their intense dispersion and great compression make them seem brilliant on US imaging. On the other hand, traditional micro bubbles are usually too big to leave the vasculature. Additionally, traditional micro bubbles are not very stable (Versluis et al. 2020).

Due to its portability, noninvasiveness, high spatial resolution, affordability, and real-time imaging capabilities, ultrasound (US) imaging is one of the most popular medical diagnostic imaging modalities. The purpose of ultrasound contrast agents is to increase the acoustic signal disparities between target lesions and normal tissues. The micro bubbles that make up the currently marketed ultrasonic contrast agents range in size from 1 to 8 μm. These comparatively large particles have a relatively limited circulation life period and low stability since they solely give blood pool contrast signals rather than cellular absorption (Tarighatnia et al. 2022). These constraints have been addressed by nanotechnology; nonetheless, US contrast nanoparticles usually need to be larger than CT or MRI contrast nanoparticles, ranging from hundreds to thousands of nanometres, to provide adequate sonic reflection.

**Table 18.3** Types of nanoparticles used in ultrasound

| Type of nanoparticle | Function | Advantages |
|---|---|---|
| Lipid-based NPs | Ultrasound contrast + drug delivery | Biocompatible, flexible design |
| Polymeric NPs | Drug delivery, targeted imaging | Long circulation, customizable surfaces |
| Gas-generating NPs | Produce microbubbles under certain triggers | Strong acoustic signals, stimuli-responsive |
| Silica or gold NPs | Improve contrast, enable photoacoustic imaging | Durable, can be used for multimodal imaging |

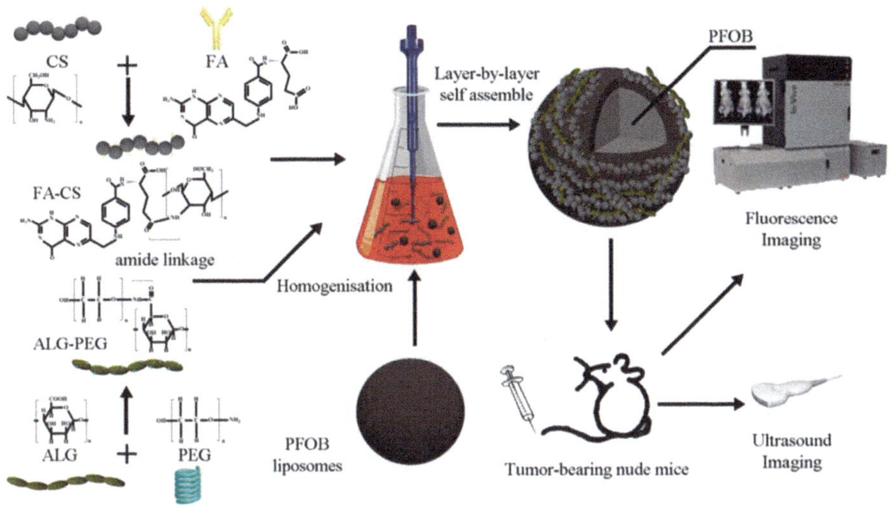


**Fig. 18.9** Folate receptor-targeted ultrasonic PFOB nanoparticles: Synthesis, characterization and application in tumor-targeted imaging (Li et al. 2017)

Ultrasound contrast agents that are nanoparticles are substantially smaller than those that are already in use (Hsu et al. 2020). Lesion targeting is made easier by the tiny size and extra surface labeling of these nanoparticles, as shown with PFOB loaded with folate-PEG-chitosan. They can produce powerful signals in local target lesions in addition to serving as blood pool contrast agents, particularly for tumor detection as shown in Fig. 18.9 (Luo et al. 2021).

## 18.8 Conclusion

Molecular imaging has been applied to a wide range of clinical applications in recent years, such as both quantitative and non-invasive illness diagnosis, disease staging, and the visualization and measurement of biological processes. Since every imaging technique has advantages and disadvantages of its own, combining various imaging modalities allows for a more precise diagnosis. Numerous imaging modalities make it easy to diagnose, prognosticate, and even treat a variety of diseases in their early stages (MacRitchie et al. 2020). As a toolkit, molecular imaging has been greatly impacted by nanotechnology, which offers further features including comprehensive molecular information. Numerous types of functional nanomaterials have been studied for use in molecular imaging. High resolution and sensitivity in molecular imaging are made possible by the unique characteristics of nanoparticles (Bukhari 2024). When compared to contrast agents based on tiny molecules, nanoparticles have outstanding biodistribution, a lengthy circulation period, and other diverse properties. Additionally, multifunctional nanomaterials can be used as theranostic or multimodal imaging agents (Bukhari and Rehman 2023).

However, compared to small molecule-based materials, the clinical translation of nanomaterials has been substantially delayed For clinical translation, a number of important issues, such as toxicity, biocompatibility, targeting effectiveness, and long-term stability of nanoparticles, need be resolved. Most importantly, it would be ideal to do long-term toxicological research on nanoparticles using big animal models. Non-human primates are considered to be crucial in determining whether nanomaterials are suitable for use in biomedical applications because of their great degree of morphological, physiological, and genetic similarities to humans (Kim et al. 2017). Before being used in clinical settings, unprecedented adverse effects on non-human primates following nanomaterial administration should be studied. Additionally, selecting the best imaging modalities and nanomaterials is crucial to obtaining the information that is needed. Multimodal imaging and theranosis, which combine various types of nanomaterials, should be used if required. The next generation of illness detection and treatment is anticipated to be driven by the sustainable development of nanotechnology in molecular imaging (Rowe and Pomper 2022).

## References

Aslan N, Ceylan B, Koç MM, Findik F (2020) Metallic nanoparticles as X-Ray computed tomography (CT) contrast agents: a review. J Mol Struct 1219:128599

Bouché M, Hsu JC, Dong YC, Kim J, Taing K, Cormode DP (2019) Recent advances in molecular imaging with gold nanoparticles. Bioconjug Chem 31(2):303–314

Bruno F, Arrigoni F, Mariani S, Splendiani A, Di Cesare E, Masciocchi C, Barile A (2019) Advanced magnetic resonance imaging (MRI) of soft tissue tumors: techniques and applications. Radiol Med 124:243–252

Bukhari R (2024) Introduction to hybrid nanomaterial: principle and application innovations and applications of hybrid nanomaterials. IGI Global, pp 183–194

Bukhari R, Rehman RA (2023) Comparative study of conventional techniques and functional nanomaterials for PMI modeling and simulation of functional nanomaterials for forensic investigation. IGI Global, pp 131–141

Chakravarty S, Hix JM, Wiewiora KA, Volk MC, Kenyon E, Shuboni-Mulligan DD et al (2020) Tantalum oxide nanoparticles as versatile contrast agents for X-ray computed tomography. Nanoscale 12(14):7720–7734

Dai X, Cheng K, Zhao W, Xing L (2020) High-speed X-ray-induced luminescence computed tomography. J Biophotonics 13(9):e202000066

Farinha P, Coelho JM, Reis CP, Gaspar MM (2021) A comprehensive updated review on magnetic nanoparticles in diagnostics. Nanomaterials 11(12):3432

Ghosh KK, Padmanabhan P, Yang C-T, Ng DCE, Palanivel M, Mishra S et al (2022) Positron emission tomographic imaging in drug discovery. Drug Discov Today 27(1):280–291

Goel S, England CG, Chen F, Cai W (2017) Positron emission tomography and nanotechnology: A dynamic duo for cancer theranostics. Adv Drug Deliv Rev 113:157–176

Henderson LA, Shankar LK (2017) Clinical translation of the national institutes of health's investments in nanodrug products and devices. AAPS J 19:343–359

Hsu JC, Nieves LM, Betzer O, Sadan T, Noël PB, Popovtzer R, Cormode DP (2020) Nanoparticle contrast agents for X-ray imaging applications. Wiley Interdiscip Rev Nanomed Nanobiotechnol 12(6):e1642

Hsu JC, Tang Z, Eremina OE, Sofias AM, Lammers T, Lovell JF et al (2023) Nanomaterial-based contrast agents. Nat Rev Methods Primers 3(1):30

Huang H, Ali A, Liu Y, Xie H, Ullah S, Roy S et al (2023) Advances in image-guided drug delivery for antibacterial therapy. Adv Drug Deliv Rev 192:114634

Jiang Z, Zhang M, Li P, Wang Y, Fu Q (2023) Nanomaterial-based CT contrast agents and their applications in image-guided therapy. Theranostics 13(2):483

Jin K-T, Yao J-Y, Ying X-J, Lin Y, Chen Y-F (2020) Nanomedicine and early cancer diagnosis: molecular imaging using fluorescence nanoparticles. Curr Top Med Chem 20(30):2737–2761

Jin C, Luo X, Li X, Zhou R, Zhong Y, Xu Z et al (2022) Positron emission tomography molecular imaging-based cancer phenotyping. Cancer 128(14):2704–2716

Jones D, Caballero S, Davidov-Pardo G (2019) Bioavailability of nanotechnology-based bioactives and nutraceuticals. In: Advances in food and nutrition research, vol 88. Elsevier, pp 235–273

Kim J, Lee N, Hyeon T (2017) Recent development of nanoparticles for molecular imaging. Philos Trans R Soc A Math Phys Eng Sci 375(2107):20170022

Koshevaya E, Krivoshapkina E, Krivoshapkin P (2021) Tantalum oxide nanoparticles as an advanced platform for cancer diagnostics: a review and perspective. J Mater Chem B 9(25):5008–5024

Li K, Liu Y, Zhang S, Xu Y, Jiang J, Yin F et al (2017) Folate receptor-targeted ultrasonic PFOB nanoparticles: synthesis, characterization and application in tumor-targeted imaging. Int J Mol Med 39(6):1505–1515

Li L, Guan Y, Xiong H, Deng T, Ji Q, Xu Z et al (2020) Fundamentals and applications of nanoparticles for ultrasound-based imaging and therapy. Nano Select 1(3):263–284

Luo D, Wang X, Burda C, Basilion JP (2021) Recent development of gold nanoparticles as contrast agents for cancer diagnosis. Cancers 13(8):1825

Luther DC, Huang R, Jeon T, Zhang X, Lee Y-W, Nagaraj H, Rotello VM (2020) Delivery of drugs, proteins, and nucleic acids using inorganic nanoparticles. Adv Drug Deliv Rev 156:188–213

MacRitchie N, Frleta-Gilchrist M, Sugiyama A, Lawton T, McInnes IB, Maffia P (2020) Molecular imaging of inflammation-current and emerging technologies for diagnosis and treatment. Pharmacol Ther 211:107550

Martynenko I, Litvin A, Purcell-Milton F, Baranov A, Fedorov A, Gun'Ko Y (2017) Application of semiconductor quantum dots in bioimaging and biosensing. J Mater Chem B 5(33):6701–6727

Nawaz T, Gu L, Fahad S, Saud S, Bleakley B, Zhou R (2024) Exploring sustainable agriculture with nitrogen-fixing cyanobacteria and nanotechnology. Molecules 29(11):2534

Nejabat M, Samie A, Ramezani M, Alibolandi M, Abnous K, Taghdisi SM (2023) An overview on gold nanorods as versatile nanoparticles in cancer therapy. J Control Release 354:221–242

Oumano M, Russell L, Salehjahromi M, Shanshan L, Sinha N, Ngwa W, Yu H (2021) CT imaging of gold nanoparticles in a human-sized phantom. J Appl Clin Med Phys 22(1):337–342

Padmanabhan P, Kumar A, Kumar S, Chaudhary RK, Gulyás B (2016) Nanoparticles in practice for molecular-imaging applications: an overview. Acta Biomater 41:1–16

Perez-Medina C, Teunissen AJ, Kluza E, Mulder WJ, Van der Meel R (2020) Nuclear imaging approaches facilitating nanomedicine translation. Adv Drug Deliv Rev 154:123–141

Rowe SP, Pomper MG (2022) Molecular imaging in oncology: current impact and future directions. CA Cancer J Clin 72(4):333–352

Siafaka PI, Okur NÜ, Karantas ID, Okur ME, Gündoğdu EA (2021) Current update on nanoplatforms as therapeutic and diagnostic tools: a review for the materials used as nanotheranostics and imaging modalities. Asian J Pharm Sci 16(1):24–46

Tanaka R, Yoshioka K, Takagi H, Schuijf J, Arakita K (2019) Novel developments in non-invasive imaging of peripheral arterial disease with CT: experience with state-of-the-art, ultra-high-resolution CT and subtraction imaging. Clin Radiol 74(1):51–58

Tarighatnia A, Fouladi MR, Nader ND, Aghanejad A, Ghadiri H (2022) Recent trends of contrast agents in ultrasound imaging: a review of the classifications and applications. Mater Adv 3(9):3726–3741

Versluis M, Stride E, Lajoinie G, Dollet B, Segers T (2020) Ultrasound contrast agent modeling: a review. Ultrasound Med Biol 46(9):2117–2144

Wang J, Li L, Li Y, Liu L, Li J, Li X et al (2023) PSMA1-mediated ultrasmall gold nanoparticles facilitate tumor targeting and MR/CT/NIRF multimodal detection of early-stage prostate cancer. Nanomedicine 47:102617

Woźniak M, Płoska A, Siekierzycka A, Dobrucki LW, Kalinowski L, Dobrucki IT (2022) Molecular imaging and nanotechnology—emerging tools in diagnostics and therapy. Int J Mol Sci 23(5):2658

Wu Y, Ali MR, Chen K, Fang N, El-Sayed MA (2019) Gold nanoparticles in biological optical imaging. Nano Today 24:120–140

Xie Y, Qi X, Xu K, Meng X, Chen X, Wang F, Zhong H (2019) Transarterial infusion of iRGD-modified ZrO2 nanoparticles with lipiodol improves the tissue distribution of doxorubicin and its antitumor efficacy. J Vasc Interv Radiol 30(12):2026–2035. e2022

Xu C, Feng Q, Ning P, Li Z, Qin Y, Cheng Y (2021) Recent advances on nanoparticle-based imaging contrast agents for in vivo stem cell tracking. Mat Matters 16:2

Zhang C, Shi C, Chang P, Bian S, Li B, Li J, Hou P (2023) MRI directed magnevist effective to study toxicity of Gd-doped mesoporous carbon nanoparticles in mice model. Int J Nanomedicine:6119–6136

# Chapter 19
# Nanomaterials: Pioneering the Future of Drug Delivery in Cancer Therapy

**Neha Atale and Vibha Rani**

**Abstract** Nanomaterials (NMs) have emerged as potential new platforms in cancer therapy to improve drug delivery, offering significant advancements in mitigating the various hurdles related to traditional treatments. Their distinctive characteristics like high volume to surface area proportion, biocompatibility and functionalization for targeted delivery, put them as highly effective candidates for enhancing the effectiveness of various cancer therapies. Nanomaterials were engineered to precisely target tumor cells, enhanced drug loading, and facilitated controlled release, thereby curtailing side effects on non-cancer cells and improving therapeutic outcomes. Polymeric nanoparticles, liposomes, dendrimers, micelles, and quantum dots, etc. have shown promising role in treating cancer owing to narrow down the drug effects to tumor sites and their multifunctionality. By targeting key components of tumor microenvironment (TME), such as immune cells, fibroblasts, tumor vasculature and the extracellular matrix (ECM), NMs can improve drug penetration, retention, and stimulate immune responses for more effective treatment. Our chapter described the summary of various nanomaterials used in cancer therapy, their role in modulating the TME and immunotherapy, as well as the challenges and future perspectives of their clinical application. Nanomaterials therefore have great potential to improve cancer treatment and influence the direction of cancer therapy in the future.

**Keywords** Cancer therapy · Dendrimers · Drug delivery · Immunotherapy · Immune checkpoint inhibitors · Liposomes · Micelles · Nanomaterials · Tumor targeting · Quantum dots

---

N. Atale (✉)
Department of Pathology, School of Medicine, University of Pittsburgh, Pittsburgh, PA, USA

V. Rani
Transcriptome laboratory, Centre of Emerging disease, Department of Biotechnology, Jaypee Institute of Information Technology Sector 62, Noida, Uttar Pradesh, India

 
V. Khanna et al. (eds.), *Nanotechnologies for Drug Delivery and Biopharmaceutical Development*, Progress in Drug Research 77,
https://doi.org/10.1007/978-3-032-00796-4_19

## 19.1 Introduction

Cancer remains the leading cause of death around the world, presenting a significant global health challenge. According to GLOBOCAN, there are currently around 20 million reported cases of cancer in the world and already nearly ten million deaths, with possibility of rising to 35 million by 2050 (Bray et al. 2024). Cancer develops as a form of primary tumor and metastasizes to other parts of the body, which makes it fatal. Traditional cancer treatments, such as chemotherapy have various side effects including toxicity, poor bioavailability and inappropriate drug delivery (Gavas et al. 2021; Liu et al. 2024). Nanomaterials (NMs) offer promising approaches in this regard by providing precisely targeted and effective strategies as well as providing new paradigms for tumor detection and therapy.

During the decades, NMs have found to be effective against tumors by two ways: active and passive targeting. These nanomaterials, such as nanoparticles and nanocarriers, possess unique physicochemical qualities due to their small size, controllable shape and surface augmentation and provide significant advantages in targeted drug delivery (Jin et al. 2020; Patra et al. 2018). NMs play a crucial role in the advancement of cancer therapies, particularly by encapsulating therapeutic agents to enhance their delivery and reduce side effects. For instance, doxorubicin encapsulated in PEGylated liposomes has demonstrated significantly reduced cardiotoxicity compared to the use of the drugs alone (O'Brien et al. 2004). Similarly, an albumin-bound paclitaxel nano particle led to fewer adverse effects and enabled better dose tolerance compared to conventionally used solvent-based taxanes (Cortés and Saura 2010).

Nanomaterials can be synthesized and applied in many different forms as part of different therapeutic agents for treating cancer, including polymeric nanoparticles, monoclonal antibody-conjugated nanoparticles, Lipid polymer hybrid nanoparticles, Liposomes, dendrimers, and carbon nanomaterials etc (Sun et al. 2023; Zhao et al. 2018). They contribute to drug retention and availability on target sites due to their inimitable size, surface properties and extended half-life (How et al. 2013). Beyond, several studies have investigated the use of nanomaterial bounded drugs to implement ablation treatments or immunotherapy in cancer (Yoon et al. 2018), it is believed that NMs can significantly enhance immunotherapy's effectiveness and the drug delivery process. They also help reverse the immunosuppressive microenvironment of tumors, thereby improving overall treatment outcomes (Zang et al. 2017). Polyphenols, widely found in various plants, possess antioxidant properties, and polyphenol-loaded nanomaterials have also shown promising potential in cancer therapy. To mitigate the inherent challenges linked with pure polyphenolic compounds like low stability, poor targeting ability, and limited solubility and bioavailability— an increasingly common and successful tactic is the use of different nano-particles such as micelles, liposomes and other nanoparticles. These nanomaterials are intended to greatly boost polyphenols' therapeutic potential by enhancing their bioavailability, stability, and delivery (Atale and Rani 2024; Atale et al. 2017; Ejiohuo et al. 2024; Kashyap et al. 2021; Rani and Yadav 2018).

Cancer nanotherapy offers significant advantages by enhancing sensitivity, specificity, and capacity for multiplexed measurements, making it a powerful tool for targeted treatment (Tiwari et al. 2020; Woodman et al. 2021). These advancements enable the development of more convenient drug formulations, lower the required concentrations of active substances, and reduce toxic effects, ultimately extending the drug's effectiveness and lifespan. Numerous NM-based drug formulations have already been developed and are undergoing clinical trials. However, each disease and therapeutic agent demands tailored delivery systems with specific modifications to optimize targeting of cancer cells (Chehelgerdi et al. 2023; He et al. 2018). As a result, the field of nanomedicine continues to improve and thrive, and a commitment to improving treatment outcomes through more precise and effective therapies.

Significant efforts have been dedicated to creating systems with precisely controlled functions and material properties. This chapter provides detailed summary of the application of different nanomaterials in immunotherapy, tumor microenvironment, and cancer treatment and their corresponding properties that help in their usage. This will also explore combined nanomaterial-assisted cancer therapies for improved therapeutic outcomes. As this field continues to progress and new innovations come forth, nanomaterials are poised to pioneer the future of cancer treatment, offering to reduce toxic side-effects, provide better survivability for cancer patients and elevate their quality of life.

## 19.2 Advancement in Nanomaterials for Cancer Therapy

NMs have taken up a renewed role in cancer therapy, offering novel strategies for targeted treatment, drug delivery, and diagnostic applications. Below are several important nanomaterials commonly used in cancer treatments. The different types of nanomaterials utilized for anti-cancer therapy are shown in Fig. 19.1. A summary of the drugs these nano materials are used to delivery and the different target cancer types are provided in Table 19.1.

### *19.2.1 Polymeric Nanoparticles (PNPs)*

The most common nanoparticles, polymeric nanoparticles (PNPs), represent a distinct class of drug delivery systems, designed with specific structural arrangements of monomers (Amreddy et al. 2018). They are colloidal macromolecules with sizes ranging from 10 nm to 1 μm. PNPs can work as efficient drug delivery systems as they can encapsulate or conjugate with drugs, allowing for their controlled release specifically at cancer sites. Polylactic-co-glycolic acid (PLGA), synthesized by copolymerization of lactic and glycolic acid, have superior biocompatibility, biodegradability, and the higher permeability into the tumor vasculature, making it suitable material for drug delivery (Acharya and Sahoo 2011; Saneja et al. 2019). In

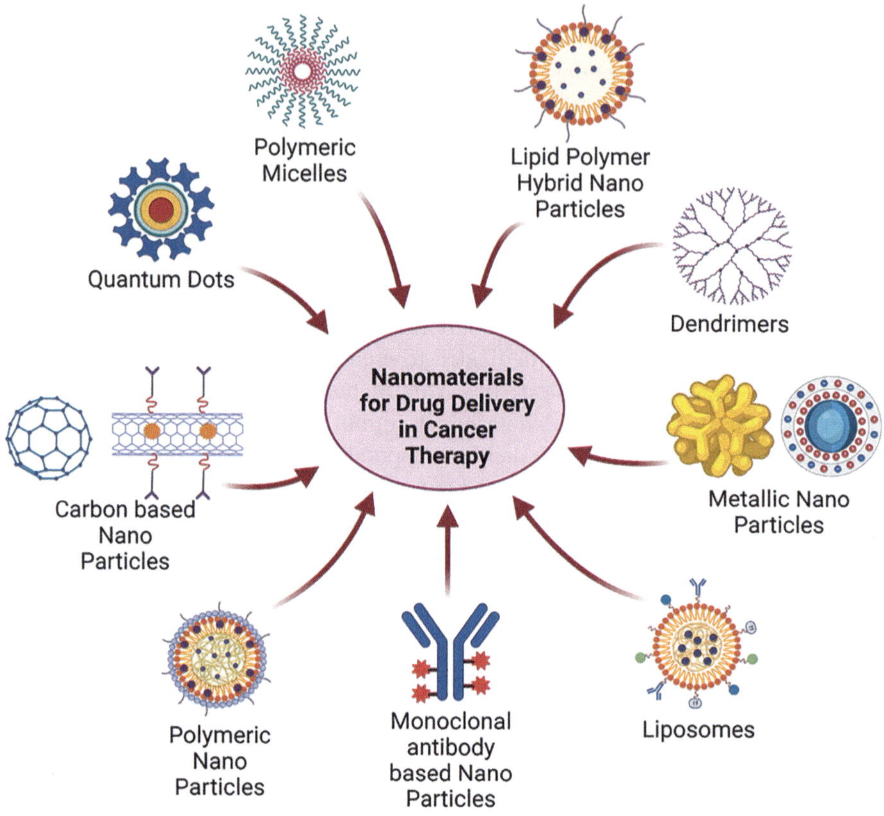


**Fig. 19.1** Different types of nanomaterials used for drug delivery in cancer therapy. Various nanomaterials have been developed over the years that exhibit different chemical, structural, electrical and magnetic properties. The commonly used NPs are Polymeric nanoparticles, Lipid polymer hybrid nanoparticles, dendrimers, liposomes, metallic NPs, carbon-based NPs such as Carbon nanotubes and graphenes, and Quantum dots etc. Owing to their varied properties, each of these different types are used in specialized applications in cancer research such as imaging, diagnosis or drug delivery

early iterations, they were formulated using non-biodegradable polymers like polymethyl methacrylate (PMMA), polystyrene and polyacrylamide (PAA) leading to concerns about their toxicity. These have been replaced by natural polymers like chitosan and biodegradable polymers including polylactic acid (PLA), poly(amino acids), poly(ε-caprolactone) (PCL) (Gagliardi et al. 2021) offering better biocompatibility and reduced toxicity, making them more suitable for medical applications (Cheng et al. 2021).

The polymeric nanoparticles have higher stability and can provide greater drug retention capacity as compared to free drugs. Enhanced efficacy of cyclodextrin polymer (CDP)-based nanoparticles was observed in drug delivery (Subhan and

**Table 19.1** Different types of nanomaterials in drug delivery in cancer therapy

| Nanomaterials | Target drug/ Bioactive compound | Treatment | Properties | References |
|---|---|---|---|---|
| Polymeric Nanoparticles (PLGA) | Fluorouracil and perfluorocarbon | Colon cancer | Biodegradable polymer co-encapsulation for enhanced anti-tumor efficacy, Anti-inflammatory activity<br>Diagnostics, Bioimaging | Wu et al. (2020a) |
| Polymeric Nanoparticles (PLGA) | Paclitaxel | HeLa cells | | Danhier et al. (2009) |
| Chitosan coated Polymeric Nanoparticles (PLGA) | Silibinin | Lung cancer, Solid tumor | Targeted drug delivery, controlled drug release, long-term stability, and non-toxicity | Raval et al. (2021) |
| Monoclonal Antibody-based Nanoparticles | Trastuzumab | HER2-positive breast cancer | High specificity and affinity, bivalent binding | Nieto et al. (2020) |
| Lipid-Polymer Hybrid NPs | Camptothecin | Breast cancer | Higher encapsulation efficiency, Superior cytotoxicity<br>Sustained release of drug, tumor targeting | Yang et al. (2013) |
| Lipid-Polymer Hybrid NPs | Docetaxel | Breast cancer | | Jadon and Sharma (2019) |
| Liposome | Paclitaxel | Breast cancer cells | Enhancing the bioavailability and stability of drug molecules, delivering them directly to the targeted site<br>Biocompatible, offering protection against drug degradation | Han et al. (2018) |
| Doxil® Liposome | Doxorubicin | Breast cancer cells | | Perez et al. (2002) |
| Myocet® Liposome | Doxorubicin | Solid tumor | | Bernardi et al. (2010) |
| Liposome MC-T-DOX | Doxorubicin | Pancreatic cancer | | Wei et al. (2020) |
| Liposome | Curcumin | Prostate cancer, Lung cancer | | Feng et al. (2017) |
| Liposome | Betulinic acid | Lung cancer, Colon cancer | | Mullauer et al. (2011) |
| Dendrimer | Trastuzumab | Breast cancer cells | Exhibiting a highest drug loading efficiency, highly branched<br>Easily modifiable surfaces | Kulhari et al. (2016) |
| PEGylated PAMAM Dendrimer | Camptothecin | Colon cancer | | Alibolandi et al. (2017) |
| PAMAM G4.5 dendrimers | Ferulic acid + Paclitaxel | Oral squamous cell carcinoma | | Anbazhagan et al. (2021) |

(continued)

**Table 19.1** (continued)

| Nanomaterials | Target drug/ Bioactive compound | Treatment | Properties | References |
|---|---|---|---|---|
| Polymeric Micelle | Paclitaxel | Solid tumor | Easily modifiable, drug loading, and release capabilities biodegradable Highly effective physiochemical characteristics, biocompatibility | Su et al. (2020) |
| Polymeric Micelle | Fisetin | Breast cancer | | Pawar et al. (2018) |
| Polymeric Micelle | Ferulic acid | Colorectal cancer | | Sweed et al. (2024) |
| Gold NPs | Doxorubicin | Breast cancer cells | Enhanced cytotoxicity to cancer cell, p-type semiconductors Photo-catalytic Photochemical Biological synthesis | You et al. (2012) |
| PEG-Gold NPs | Trastuzumab | Breast cancer | | Chattopadhyay et al. (2010) |
| Silver NPs | Camptothecin | Breast cancer cells | Optical, thermal, electrical conductivities, biological properties, anticancer potential | Elbaz et al. (2016) |
| Silver NPs | Plumbagin | Cervical cancer cells | | Appadurai and Rathinasamy (2015) |
| Iron oxide np | Paclitaxel | CT26 tumor cells | Electrical resistivity, Optical characteristics, Dynamic capabilities High magnetization, Targeted release of drugs | Schleich et al. (2015) |
| Feraheme® Iron oxide NPs | Ferumoxytol | Prostate cancer | | Kader et al. (2022) |
| Nanoscale Graphene Oxide | Doxorubicin | Oral squamous cell carcinoma | Heat and electrical conductivity, Mechanical properties | Li et al. (2021) |
| Fullerene based nano particles (DAF NPs) | – | Hela cells | | Shi et al. (2019a) |
| Single-walled carbon nanotubes | Doxorubicin | Breast cancer | Higher cytotoxicity, low ROS generation, | Unlu et al. (2018) |
| Double-walled carbon nanotubes | Survivin siRNA | Prostate cancer | Cylindrical shape offers improved stability and controlled release properties, used for gene delivery | Neves et al. (2012) |
| Graphene quantum dots | Folic acid | Breast cancer cells | Semiconductors with high thermal stability, excellent brightness, and a broad absorption spectrum, along with resistance to chemical degradation, and superior photostability | Feng et al. (2020) |
| Quantum dots | Doxorubicin | Prostate cancer | | Bagalkot et al. (2007) |
| PEG-Cdte Quantum Dots | Doxorubicin | Myeloma | | Tatar et al. (2016) |

Torchilin 2019). Additionally, transferrin-modified adamantane-Polyethylene glycol (AD-PEG) and adamantane-PEG-transferrin (AD-PEG-Tf) are effective for *in vivo* nucleic acid delivery (Bartlett and Davis 2006). Wu et al. team synthesized nanoparticles loaded with Fluorouracil (5-FU), as well as Perfluorocarbon (PFC), and functionalized with epidermal growth factor (EGF), developing in the formulation EGF-PLGA@5FU/PFC NPs for targeted cancer therapy (Wu et al. 2020a). Another study demonstrated that PEGylated PLGA nanoparticles (NPs) improved the efficacy of paclitaxel (PTX), showing increased cytotoxicity against HeLa cells. Due to enhanced permeability and retention (EPR) effect, PTX-loaded NPs reduced tumor growth and enhanced survival in mice (Danhier et al. 2009). Silibinin, encapsulated with chitosan-coated PLGA nanoparticle exhibited higher anticancer activity against lung cancer than that of silibinin alone (Raval et al. 2021). These formulations make NPs an efficient carrier in targeted drug delivery in cancer therapy.

### *19.2.2 Monoclonal Antibody Based Nanoparticles (mAb-NPs)*

Monoclonal antibody-based nanoparticles (mAb-NPs) are specialized nanoparticles which are developed by the combination of drugs and antibodies (Tsao et al. 2021; Wathoni et al. 2022). Targeting the unique antigens found on cancer cells, contrary to normal cells, they enable drug complexes to selectively target tumors, reducing toxicity. Trastuzumab (Herceptin), a monoclonal antibody, has shown improved therapeutic efficacy when incorporated into antibody-drug conjugates (ADCs) in treatment of HER2-positive breast cancer demonstrating enhanced outcomes compared to Trastuzumab alone (Nieto et al. 2020). Abedin et al. developed an ADC nanoparticle with a paclitaxel-loaded core and a trastuzumab-modified surface. Testing on HER2-positive and HER2-negative cell lines, the nanoparticle complex exhibited superior anti-tumor efficacy compared to PTX or trastuzumab individually, while also showing reduced cytotoxicity in non-cancerous human breast epithelial cells (Abedin et al. 2021). This approach highlights the possibilities of nano-particles in improving the effects of Trastuzumab in cancer therapy. Wang et al. developed microneedles (MNs) containing hyaluronic acid and dextran nanoparticles, encapsulating αPD1 and glucose oxidase enzyme. This enzyme reduced the pH, promoting nanoparticle degradation and sustained antibody release, resulted in enhanced immune responses compared to single dose αPD1 injections (Wang et al. 2016). These advancements reflect the extending potential of nanomaterials and antibody conjugates in optimizing cancer treatments through more effective and targeted therapies.

### 19.2.3 Lipid Polymer Hybrid Nanoparticles (LPHNPs)

Lipid Polymer Hybrid Nanoparticles (LPHNPs), featuring a polymeric core encased in a lipid shell, have demonstrated significant potential as drug delivery vehicles, especially for treating breast and pancreatic cancers (Gajbhiye et al. 2023; Mukherjee et al. 2019). The polymeric core provides structural stability, while the lipid shell enhances biocompatibility. Additionally, these hybrid nanoparticles were efficiently absorbed by cancer cells which enables precise medication administration to the tumor location (Sivadasan et al. 2021). Campothecin is known as a potent chemotherapy agent as it inhibits DNA topoisomerase I and can induce apoptosis in cancer cells. Yang et al. used these LPHNPs and showed promising results towards cancer therapy (Yang et al. 2013). They also demonstrated that Docetaxel (DTX)-loaded LPHNPs exhibited superior antitumor activity compared to DTX by itself. DTX-loaded LPHNPs in breast cancer also showed improved therapeutic results (Jadon and Sharma 2019). These hybridized nanoparticles were designed to optimize cancer cell targeting, enhance drug stability, and provide controlled release of the therapeutic agent.

### 19.2.4 Liposomes

Liposomes, primarily described in 1965, were self-assembling colloidal structures with a spherical shape, consisting of lipid bilayers that enclose a central aqueous compartment (Gao et al. 2019; Nsairat et al. 2022). The outer lipid bilayer covers the aqueous space, forming a platform that can be used for loading drugs and various other applications. Paclitaxel has been successfully and extensively used with liposomes in various cancer therapy (Yari et al. 2019). The D-alpha Tocopheryl Polyethylene Glycol 1000 Succinate (TPGS)-coated PTX-liposome was able to induce much higher cytotoxicity compared to the PTX alone in MCF-7/ADR breast cancer cell lines that are known to be highly resistant to many of the anti-cancer drugs (Han et al. 2018). Liposomes conjugated with anthracycline drugs, such as doxorubicin (Doxil®, Myocet®) and daunorubicin (DaunoXome®), have already been approved for clinical use. Doxil was used in metastatic breast cancer, while Myocet® was utilized for various solid tumors (Bernardi et al. 2010; Perez et al. 2002) whereas DaunoXome® loaded liposome was employed in Kaposi's sarcoma (Mukwaya et al. 1998). Onivyde™, a liposomal formulation of irinotecan (IRI), was used as a supplementary treatment method for pancreatic ductal adenocarcinoma (PDAC) and was found to increase the EPR effect leading to higher accumulation of irinotecan (IRI) within tumor sites (Bulbake et al. 2017).

A novel approach involved combining MMP14 activated Cilengitide (MC) with thermosensitive liposomes combined with doxorubicin (DOX) to form the nanovesicle MC-T-DOX (Wei et al. 2020). Additionally, Shen et al. used PEGylated liposome conjugated with mannose and levamisole hydrochloride to reduce tumor development (Shen et al. 2022). These formulations offered significant advantages, including reduced cardiotoxicity and enhanced penetration into tumor tissues. Feng et al. revealed liposome encapsulation significantly enhanced curcumin's ability to inhibit cancer cell growth (Feng et al. 2017). Another study showed betulinic acid, on delivering in liposomes, effectively reduced the growth of lung and colon cancers (Mullauer et al. 2011). With ongoing clinical evaluations, liposomal therapies hold promise for revolutionizing cancer treatment and improving patient outcomes.

### *19.2.5 Dendrimers*

Dendrimers, as their name suggests, are nano scaled molecules shaped like a tree with 3-dimensional branching structure emerging from a core (Sarode and Mahajan 2024). They are synthesized in step-by-step controlled manner, but at the same time their structure is highly repetitive like polymers and can either be divergent or convergent depending on how they are synthesized. Divergent approach starts from a core (generally ammonia) with branches synthesized from it. However, the convergent approach involves creation of branches first and then coupling them with a multivalent core (Abbasi et al. 2014). The most widely synthesized divergent dendrimers include Poly(amidoamine) (PAMAM), Polyethylene glycol (PEG) etc (Abedi-Gaballu et al. 2018). Various peripheral properties of the dendrimers can allow to conjugate with either molecules such as peptides or mAbs or different drugs. Dendrimers with hydroxylamine groups on their periphery can enhance the hydrophilic properties of anticancer drugs (Choudhary et al. 2017).

Cline et al. showed that conjugating anti-cancer drug PAX with PAMAM increases the uptake of the drug compared to other polymeric carriers (Cline et al. 2013). Kulhari et al. conjugated docetaxel, a chemotherapy drug with PAMAM dendrimers and the monoclonal antibody Trastuzumab. This was tested in targeting HER2-positive breast cancer cells (Kulhari et al. 2016). In another study, PEGylated PAMAM dendrimers were utilized for camptothecin (CPT) delivery in colon cancer treatment. AS1411-modified PAMAM dendrimers significantly boosted CPT internalization in colon cancer cells, and higher encapsulation efficiency, leading to the suppression of tumor progression (Alibolandi et al. 2017). Ferulic acid and Paclitaxel co-encapsulated within RGD-functionalized PAMAM G4.5 dendrimers represent a promising therapeutic approach for promoting apoptosis in multidrug-resistant cells that overexpress P-glycoprotein (Anbazhagan et al. 2021). Additionally, PAMAM dendrimers modified with graphene oxide represent an innovative approach to drug delivery, particularly in the context of hydrophobic anticancer agents like quercetin, which is known for its potent anticancer properties but suffers from poor solubility and bioavailability. These nanomaterials have been designed to deliver hydrophobic drugs to the targeted tumor environment as part of anticancer therapeutics (Matiyani et al. 2023).

### 19.2.6 Polymeric Micelles

Polymeric micelles are spherical structures formed by self-attracting amphiphilic molecules, typically ranging from 10 to 100 nm in diameter (Perumal et al. 2022). These are formed of hydrophilic and hydrophobic components such that when placed in an aqueous environment, the hydrophobic parts squeeze into the center to form the core whereas the hydrophilic components move to the periphery to contact the medium. Micelles can be created using both synthetic polymers and natural compounds, with properties such as critical micelle concentration (CMC), chemical groups, and the temperature below which micelles may destabilize and collapse called as Krafft temperature (Wu et al. 2020b; Zhang et al. 2014).

Block polymers, composed of a shell-forming and a core-forming component, provide micelles distinct properties depending on the materials. A range of materials can form the micellar shell, such as polyethylene glycol, polyvinyl alcohol, poly(N-vinyl-2-pyrrolidone), and polysaccharides. The core can be made from polyester, or poly(amino acid)s, each with unique characteristics (Hibino et al. 2022). Su et al. synthesized a dual-responsive PEGylated micelle that selectively delivered anti-PD-1 antibody and paclitaxel, demonstrating targeted and on-demand drug release (Su et al. 2020). The targeted formulation facilitated enhanced incorporation of the micelles by Ferulic acid (FA) receptor-positive cancer cells (Pawar et al. 2018). TPGS mixed micelles are amphiphilic structures that can solubilize hydrophobic drugs like, enabling them to be effectively delivered to target tissues. Micelles carrying FA-loaded TPGS were found to enhance the anticancer effects of FA in colorectal cancer (Sweed et al. 2024). Furthermore, micelles can also be concocted to enhance targeting by attaching ligands on their periphery, improving penetrability and retention in specific tissues.

### 19.2.7 Metallic Nanomaterials

Metallic nanomaterials have proven excellent carriers for drugs in chemotherapy owing to their volume to high surface area ratio, additional properties of metallic ions, and the ability to interact with biological systems in specific ways (Izhar et al. 2022; Xu et al. 2022). These nanomaterials, including gold, silver, and iron oxide etc. are being explored for various therapeutic strategies in cancer therapy. Gold nanoparticles (AuNPs) incorporating gold atoms on them allow for the encapsulation or attachment of anticancer drugs and controlled release. In this context, You et al. created hollow gold nanoshells (HAuNSs) with doxorubicin inside them to overcome trastuzumab resistance (You et al. 2012). Using high radiation induced HER-2 overexpression in SK-BR-3 breast cancer cells, trastuzumab-PEG-AuNPs conjugates induced more DNA double-strand breakage compared to PEG-AuNPs alone (Chattopadhyay et al. 2010). The epigallocatechin gallate (EGCG) gold nanoparticles (E-GNP) nano-formulation facilitates a sustained release of

EGCG. Additionally, further studies showed that E-GNPs resulted in a higher rate of cancer cell deaths compared to both EGCG alone or in conjunction with citrate-gold nanoparticles (C-GNPs) (Chavva et al. 2019). In addition to gold NPs, silver nanoparticles (AgNPs) have also been used as therapeutic agents and as vehicles for drugs. They can induce cytotoxicity by creating reactive oxygen species (ROS) with free radicals, disruption of cellular membranes, and interaction with DNA, leading to cell death. Additionally, when loaded with chemotherapeutic drugs, AgNPs provide controlled and targeted drug release, improving treatment while minimizing side effects. Elbaz et al. reported enhanced cytotoxicity and apoptosis in HeLa cells using a combination of AgNPs and camptothecin, with the drug attached to AgNPs coated in PEG and other polymers. Cell viability assay showed high cytotoxicity against MCF7 breast cancer cells at low concentrations due to AgNP-DOX combinations (Elbaz et al. 2016). AgNPs were utilized to encapsulate Plumbagin, a bioactive compound, with the objective of enhancing its therapeutic effectiveness, selectivity, and sensitivity in targeting cancer cells (Appadurai and Rathinasamy 2015). Iron oxide nanoparticles (IONPs) are also used for their unique magnetic properties, biocompatibility, and ability to be conjugated for targeted therapy. Scheliech et al. used Paclitaxel/superparamagnetic IONP-loaded PLGA nanoparticle formulation to demonstrate promising results in cancer therapy, exhibiting significant cytotoxicity against CT26 colon cancer cell line and delayed relapse *in vivo*. The paclitaxel-loaded nanoparticles demonstrated the ability to efficiently deliver the drug to the tumor site, where the superparamagnetic IONPs facilitated the magnetic targeting, thereby improving the therapeutic outcome and prolonging the suppression of tumor growth (Schleich et al. 2015). Feraheme® in conjugation with IONPs was investigated for its potential in prostate cancer (Kader et al. 2022). IONPs have therapeutic as well as diagnostic features, which make them highly versatile and valuable in personalized cancer treatment strategies.

### *19.2.8 Carbon Nanomaterials*

Carbon nanomaterials are nano scale material synthesized with carbon as the base element, and have specific optical, thermal and mechanical properties (Gao et al. 2024). Major carbon-based nano particles such as graphenes, carbon nanotubes etc. can be categorized based on the structure i.e. single-layer, multi-layer (Anzar et al. 2020; Singh et al. 2022). Graphene, however, has many drawbacks that lead to higher toxicity and make conjugation difficult. To mitigate these limitations of the chemically modified graphene, graphene oxide (GO) was synthesized, which consists of carbonyl and epoxy groups as the base and carboxyl and hydroxyl group on the edge. Nanoparticle for the targeted delivery of DOX to oral squamous cell carcinoma (OSCC) was developed in a novel approach by utilizing HN-1 peptide (TSPLNIHNGQKL), a targeting ligand that facilitated enhanced interaction between the nanoparticle and the OSCC cells. The encapsulation of DOX drug within DOX@NGO-PEG-HN-1 exhibited substantially increased uptake in cancer cells (Li et al. 2021). The hydrophilic groups also allow GO to attach with various types of materials, broadening their efficacy in diagnosis and targeted treatment of cancer (Sattari et al. 2021).

Fullerenes are nano-carbon structures that are formed into hollow spherical or tubular shapes and can be made up of 60–90 carbon atoms. Metals are incorporated into these structures to form metallofullerenes, which can be utilized as magnetic resonance imaging (MRI) materials, as the electrons in the encapsulated metal move to the outer cage, generating magnetic poles (Chang et al. 2024). Fullerene based nano particles (DAF NPs) with near-infrared response were used to reduce tumor growth by photothermal (PTT) and photodynamic (PDT) therapies. Compared with standard fullerenes or D-A antenna nano particles, the DAF NPs revealed higher efficiency in thermal regulation and tumor inhibition (Shi et al. 2019a).

Carbon nanotubes (CNTs) are fullerenes that are shaped as hollow tubular structures. Nano tubes can be formed by one layer or multiple concentric layers of graphene. Carbon nanotubes have optical absorption properties as they have high absorption at near infrared frequencies. In fact, owing to their excellent optical and thermodynamic properties they have become a potential avenue for all facets like imaging, diagnosis and therapy for cancer. Multi layered carbon nano particles can also cross cell membranes. Unlu et al. developed drug delivery systems using single-walled carbon nanotubes (SWNTs) to deliver the doxorubicin for NIR photothermal therapy and chemotherapy. The SWNT-DOX system demonstrated increased apoptosis in breast cancer cell models in comparison to control conditions such as DOX or photothermal therapy by themselves (Unlu et al. 2018). In another study, Neves et al. utilized peptide-coated double-walled nanotubes (DWNTs) carrying survivin siRNA to target prostate cancer. The needle-like structure of DWNTs enabled efficient siRNA delivery into cells, inhibiting survivin synthesis and inducing apoptosis. These results suggest that carbon nanotube-based delivery methods hold great potential for effective chemotherapy and gene therapy (Neves et al. 2012).

### *19.2.9 Quantum Dots (QDs)*

Quantum dots (QDs) are nano-scale structures that confine electrons in a small 3D space to keep them at discrete energy levels. They are potential biological imaging agents owing to their narrow emission bands, high photostability and broad spectrum of absorption (Gil et al. 2021). QDs offer superior photostability and intense fluorescence, enabling them to surpass traditional organic fluorescent dyes and proteins. They have improved signal transduction and cell localization and significantly aided in the early detection of cancer through their use as fluorescent probes and nanocarriers. They can be categorized as Graphene Quantum Dots (GQD) and Carbon Quantum Dots (CQD). GQDs have smaller size but have high surface area, useful electrical properties and ability to change brightness, are suitable for imaging and drug delivery (Xie et al. 2019; Zheng and Wu 2017). Folic acid-modified nitrogen-doped graphene quantum dots (FA-N-GQDs) have demonstrated cytotoxic effects against a variety of cancer cells and have been utilized for diagnostic applications (Feng et al. 2020). Magnetic quantum dots (MQDs) have been used to detect cancer, imaging and drug delivery (Karami et al. 2024). Conjugates of quantum dots and aptamer-doxorubicin (QD-Apt(Dox)) have been used to target prostate cancer (Bagalkot et al. 2007). A recent study highlighted the incorporation of DOX into cadmium-based QDs modified with PEG (PEG-CdTe QDs), enabled pH-responsive release of DOX into target cells, which significantly improved therapeutic efficacy and induced apoptosis in myeloma cells (Tatar et al. 2016).

## 19.3 Nanomaterials Targeting Cancer Cells

Drug delivery based on nanomaterials that effectively target cancer cells is essential for achieving therapeutic results, reducing dose-dependent toxicity, and protecting healthy cells from adverse effects. Anti-cancer drugs were effectively delivered to the tumor site using nanomaterials, allowing them to pass the biological barriers like phagocytic cells that usually lessen the drug's effectiveness. The interaction between nanomaterials and tumor cells is a key area of active research, focusing on advancement of NM-targeted drug delivery. NMs can play a significant role in overcoming drug resistance in cancer therapy through various mechanisms. Drug resistance, particularly in cancer treatment, occurs when cancer cells develop the ability to resist the effects of drugs that were previously effective. NMs, offer innovative strategies to address this challenge by enhancing drug delivery, overcoming drug efflux pumps, gene silencing, RNA interference and improving therapeutic outcomes (Mir et al. 2022).

Drug delivery procedures were broadly classified into two categories: passive targeting, driven by non-ligand mediated nanomaterials and active targeting, which involved ligand-conjugated nanomaterials for more precise delivery (Chen et al. 2021). Figure 19.2 demonstrates the mechanism of passive and active targeting.

### *19.3.1 Passive Targeting*

Passive targeting employs the unique biological features of tumor tissues to improve the delivery of drugs directly to cancer sites. Tumor tissues exhibit distinct characteristics compared to normal tissues, such as rapid cell proliferation, which leads to the formation of new, irregular blood vessels through a process called neovascularization. This phenomenon creates larger openings in the blood vessel walls, leading to reduced selective permeability in the tumor vasculature, which contrasts with the tighter and more controlled permeability of normal vessels (Atale and Rani 2020, 2022; Folkman 2007). These openings can allow macromolecules and nanoparticles to move from bloodstream into the tumor. Tumor environments are also found to have poor lymphatic drainage which allows the nano particles to remain in the

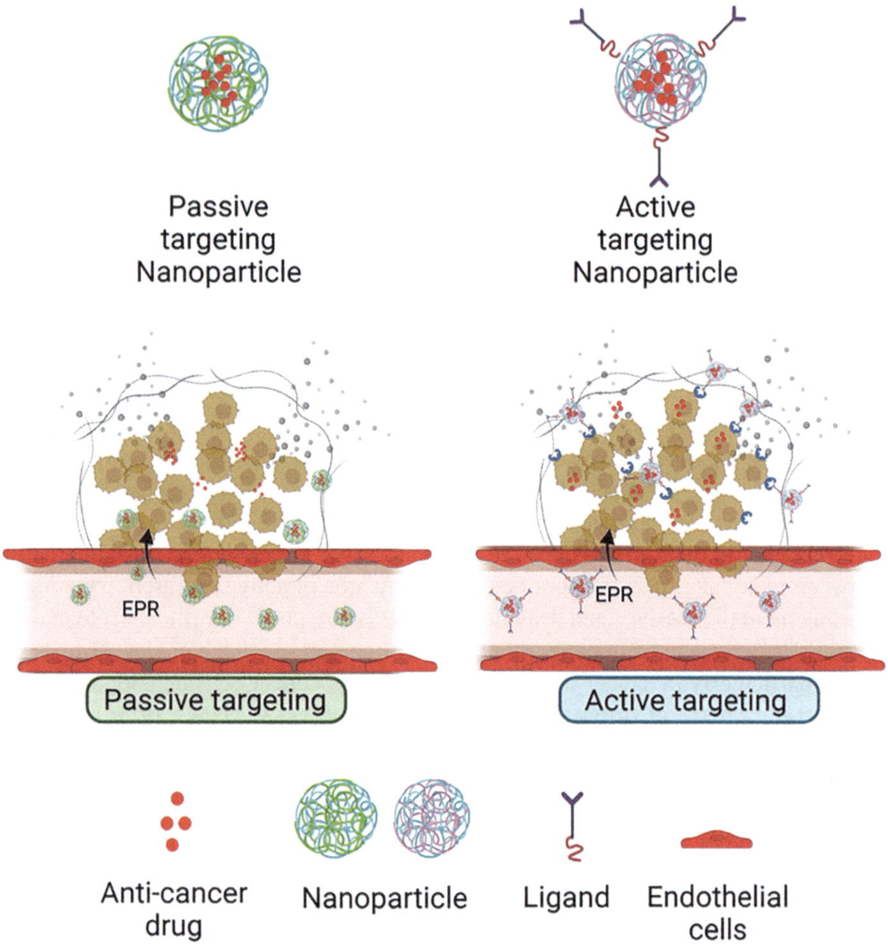


**Fig. 19.2** Passive and active targeting with nano particles. Passive nanoparticles encapsulating chemical drugs to be delivered to tumor site. Once injected into the blood stream they get absorbed into the tumor site through a process called EPR which is caused due to the gaps in endothelial layer of blood vessels. Active nanoparticles on the other hand include ligands on their periphery and drug in the core. The ligands can bind to cancer cells and therefore improve drug delivery

tumor. This improves the efficacy of the drugs and target cancer cells better. This phenomenon, referred to as EPR (Enhanced Permeability and Retention) is a key principle behind passive targeting in cancer therapy (He et al. 2020). The effects in EPR depend on the size of openings in the vasculature around the tumor and the size of nanoparticles used. Generally, smaller nano particles have better potential to penetrate tumors (Carita et al. 2018). Some drugs can also be used to first target the vasculature around the tumors and increase their permeability. For example, FENDRR, a long non-coding RNA (LncRNA), targets cell signaling to trigger growth of blood vessels around the tumor. This was shown to improve growth and angiogenesis of human retinal endothelial cells (Shi et al. 2019b).

### *19.3.2 Active Targeting*

In contrast with passive targeting, active targeting uses active compounds attached to the nano particles that target receptors specific to the cancer cells. Many active targeting nano particles have ligands that interact with receptors on cancer cells. The ligands are chosen on two criteria—they can be easily attached to the nanoparticles—and—they bind to receptors that are mainly found on cancer cells and not native cells in the tissue. This also allows the nanoparticles to enter the cancer cells easily compared to passive nano particles and therefore provide a controlled release of the drugs into the targeted cells (Attia et al. 2019). Since macromolecular drugs, like proteins and small interfering RNAs (siRNAs), need more precise targeting mechanisms to maximize their effectiveness and reduce off-target effects, active targeting is especially advantageous for their delivery. The transferrin receptor, glycoproteins, and epidermal growth factor receptor are among the receptors that are frequently overexpressed on cancer cells and can be contacted by common targeting ligands (Danhier et al. 2010). In recent studies, peptides have emerged as promising ligands for active targeting due to their small size, low immunogenicity, high stability, and strong binding affinity to drug-loaded nanoparticles (Guo et al. 2024). Human foreskin fibroblast (Hs68) cells and B16F10-Nex2 malignant melanoma cells were dual-targeted using gold nanoparticles conjugated with a combination of C7H2 and HuaL peptides. These peptide-conjugated nanoparticles demonstrate the ability to selectively target tumor cells, offering enhanced specificity and stronger binding affinities, which may ultimately lead to more effective cancer treatments (Formaggio et al. 2022). The integration of peptides with nanomaterials represents a promising avenue for improving the delivery and therapeutic outcomes of cancer therapies, as they allow for both enhanced targeting and minimal toxicity to healthy tissues.

## 19.4 Nanomaterials Targeting Tumor Microenvironment (TME)

The tumor microenvironment (TME) plays a significant role in the development of tumors. TME is a very complex amalgamation of various heterogenous components which exhibit varying properties with respect to immune response, tumor growth, and metastasis (Khalaf et al. 2021; Wang et al. 2023). In addition to tumor cells, the TME encompasses various other cell types, including T cells, the extracellular matrix, macrophages, fibroblasts etc. Tumor cells can significantly influence the immune system around them and create an immunosuppressive microenvironment which then allows the tumors to grow and circumvent or suppress all immune responses against them. These processes include an increase in tumor-associated macrophages (TAMs) and regulatory T cells ($T_{Reg}$), a reduction in lymphocyte counts, and the apoptosis of cytotoxic T cells (de Visser and Joyce 2023). Cancer

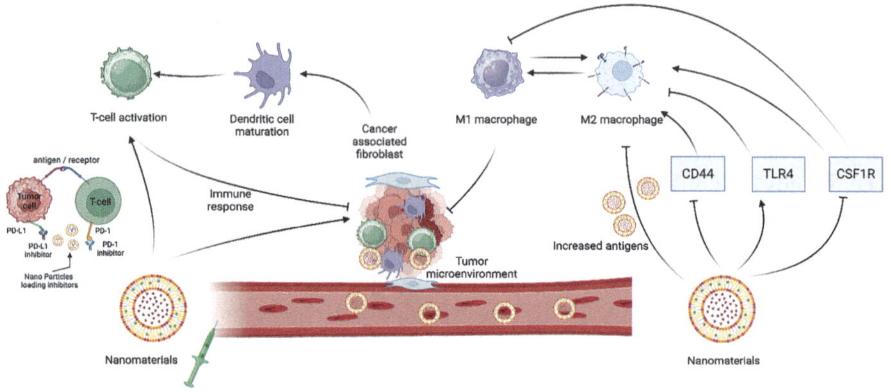


**Fig. 19.3** Nanomaterials targeting tumor microenvironment. Nano materials can be used for immune therapy by either aiding the T-cell expansion and activation or through macrophage activation. During T-cell targeting, the NPs can be used to deliver PD-1, PD-L1 inhibitors that help in T-cell activation. NPs can also directly or indirectly affect the pathways that promote M1 macrophage polarization that target the cancer cells

immunotherapy aims to overcome immune checkpoints and counteract immunosuppressive TME. As a result, targeting the TME has become a promising strategy to re-sensitize resistant tumors to current therapies. Figure 19.3 demonstrates the mechanism of nanomaterials in the tumor microenvironment.

### *19.4.1 Nanomaterials Targeting T Cells*

Nanomaterials have been used to provide significant therapeutic improvements in cancer immunotherapy by modulating T cells (Patel et al. 2022; Xiong et al. 2021). Tumor cells create immunosuppressive environment around them and improving the penetration ability of CD8+ T cells into the tumor cells environment, which could enhance the effectiveness of bodies innate immune response to counteract the tumor progression. In this regard, manganese zinc sulfide (MnZnS) nanostructures promote the Fenton reaction leading to release of reactive oxygen species (ROS) that damage cancer cells as well as induce apoptosis. Furthermore, MnZnS nanoparticles activate the cyclic GMP-AMP synthase/stimulator of interferon genes (cGAS/STING) pathway boosting the immune response (Li et al. 2022).

Glycosylated PLGA nanocarriers have been developed to deliver ovalbumin antigen and CpG adjuvants for cancer therapy. These nanocarriers are crucial for cancer vaccine development as they enhance CD8+ T cell infiltration in cancer immunotherapy increasing CD4+ T cell levels and simultaneously promoting antigen uptake and dendritic cell maturation (Chou et al. 2022). In a related development, Fengyun Shen and colleagues created a novel liposomal formulation, aNLG/Oxa(IV)-Lip, through the self-assembly of an oxaliplatin prodrug (Oxa(IV)). This formulation conjugates Oxa(IV) and phospholipids and modifies it with NLG919 (aNLG) alongside other available lipids, which significantly enhances CD8+ T cells into the tumor microenvironment, stimulates cytokine release, and amplifies the anti-tumor immune response (Shen et al. 2020).

Yin et al. developed an innovative nanoparticle, PD-1-MM@PLGA/RAPA, by combining rapamycin (RAPA)-loaded PLGA with a macrophage membrane that overexpresses PD-1. Upon administration, PD-1-MM@PLGA/RAPA effectively accumulates at tumor sites, particularly those with elevated PD-L1 expression, which is commonly seen in various cancers (Yin et al. 2022). The nanoparticle stimulates the release of TNF-α, further enhancing anti-tumor immune responses and leading to a significant reduction in tumor growth. Similarly, Mei Hu and colleagues developed the DOX@LINV system, an innovative liposomal drug delivery platform that incorporates tumor-derived nanovesicles with artificial liposomes. These tumor-derived nanovesicles, isolated from tumor tissues, are fused with liposomes to form a unique structure that efficiently delivers the doxorubicin directly to the tumor site. Additionally, the platform promotes the maturation of dendritic cells (DCs), which are critical for activating antigen-specific T cell responses. This approach makes DOX@LINV a potent tool in cancer immunotherapy overcoming the limitations of traditional chemotherapy treatments (Hu et al. 2021). These engineered nanomaterials offer a promising strategy for targeted cancer therapy, combining immune modulation with efficient drug delivery.

### *19.4.2 Nanomaterials Targeting Cancer Associated Fibroblasts (CAFs)*

Cancer-associated fibroblasts (CAFs) have long been associated with facilitation of tumor progression, metastasis, and immunity against therapy by secreting growth factors, extracellular matrix components, and cytokines that contribute to tumor cell survival, angiogenesis, and immune suppression (Wu et al. 2021). Nanoparticles have been studied recently in their role as anti-cancer agents in relation to CAFs. PLGA nanoparticles have been shown to enhance the interaction between cancer cells and CAFs, while also promoting antigen uptake by immune cells. These nanoparticles not only boost immune responses but also promote anti-CAF immunity, offering a dual strategy that targets cancer cells and modifies the TME. Similarly, peptide nanoparticles (PNPs) were used to deliver CXCL12 siRNA to CAFs using polyarginine-based cell-penetrating peptides. This nanoparticle called PNP/siCXCL12/mAb, was found to significantly inactivate CAFs, reduce metastasis and suppress prostate cancer by targeting and neutralizing CXCL12. Fibroblast activation protein (FAP), a marker commonly upregulated on CAF surfaces, serves as a key target in cancer immunotherapy (Huang et al. 2024). Nano particles targeting FAP were used by adding anti-FAP monoclonal antibodies on to the nanoparticles and improve their targeting of CAFs. These findings demonstrate the potential of targeted nanotherapy to reverse the pro-metastatic TME and inhibit CAF-mediated tumor progression (Lang et al. 2019).

FAP has been target of anti-cancer therapy in many ways. Lipid nanoparticles (LNPs) containing CpG adjuvants can be conjugated with antigenic peptide epitomes from FAP to create a nano vaccine called FAPPEP-SLNPs. This nano-vaccine effectively targets FAP-positive CAFs and activate T-cells immunity to interrupt the tumor-supportive stroma (Shin et al. 2023). Dipeptide Z-glycine proline was used to target FAP, at the same time targeting ECM components like tenascin C which are expressed by CAFs (Guo et al. 2023). CAFs can also be interrupted in other ways using nanoparticles. Previously mentioned mPEG-PLGA nano particles carrying baicalein were able to disrupt the TGF- β signaling and inhibiting CAF activation (Zheng et al. 2023). Mardhian et al. used superparamagnetic iron oxide nanoparticles (SPIONs) to deliver fibroblast growth factor 2 (FGF-2) to inhibit hPSCs (human pancreatic stellate cells). This helped to stop TGF-β1-induced CAF activation and show inhibition of tumor like spheroid growth in 3D models (Mardhian et al. 2020). These studies show the successful use of nanoparticles to target CAFs with huge potential for anti-cancer therapy.

### 19.4.3 Nanomaterials Targeting Tumor Associated Macrophages (TAMs)

Nanomaterials are being utilized to modify macrophage polarization, enhance immune responses and potentially overcoming the limitations of traditional therapies. Mesoporous magnetic nanoparticles carrying sulfasalazine and conjugated with platelets, known as Fe3O4-SAS@PLT nanostructures, were used to target the glutamate-cystine antiporter system Xc-pathway, aiming to trigger ferroptosis—a form of regulated apoptosis characterized by iron-dependent lipid peroxidation (Jiang et al. 2020). These nanomaterials promote M1 polarization, offering a promising approach to boost immunotherapy. Magnetic nanoparticles, with the ability to modify cell membranes, also influenced multiple signaling pathways, guiding macrophage polarization and inhibiting tumorigenesis (Rao et al. 2020). Plant based natural nanostructures derived from Ginseng-structures, which mimic properties of extracellular vesicle were able to trigger the TLR4/MyD88 signaling pathway, further enhanced M1 polarization of macrophages (Cao et al. 2019).

However, a significant challenge in immunotherapy is reprogramming of the macrophages into immunosuppressive M2 phenotype. Tumor-secreted MCSF (Macrophage Colony-Stimulating Factor) is a key factor in this polarization, as it increases colony stimulating factor 1 receptor (CSF1-R) expression. Additionally, elevated signal regulatory protein $\alpha$ (SIRPα) levels on myeloid cell surfaces activate SHP-1 and SHP-2, impairing phagocytosis and reducing immunotherapy efficacy. To address this, new strategies are focusing on enhancing macrophage activity, particularly phagocytosis. Nanoparticles loaded with CSF1R and SHP2 inhibitors have shown potential in boosting phagocytosis and promoting M1 macrophage polarization, thus preventing tumor growth (Ramesh et al. 2019). Feito et al. used graphene oxide nanosheets (FITC-PEG-graphene oxide) to examine macrophage polarization (Feito et al. 2019). Their findings revealed that these nanosheets increased the M2 macrophage population, which is involved in tissue repair. Hence, it can be stated that nanomaterial-based strategies modulate macrophage polarization and improve immunotherapy outcomes.

### *19.4.4 Nanomaterials Targeting Extracellular Matrix (ECM)*

The extracellular matrix (ECM) is a mechanical scaffolding providing structural integrity around the cancer cells supporting their proliferation, migration, invasion, and angiogenesis, playing a crucial role in the TME (Prakash and Shaked 2024). Key ECM components, such as lysyl oxidase, fibronectin, collagenase, hyaluronic acid (HA), and various enzymes, are essential to these processes, contributing to cancer aggressiveness and the resistance of solid tumors to therapeutic treatments. Combining traditional chemotherapies with PEGylated recombinant human hyaluronidase (PEGPH20), which specifically targets hyaluronic acid, has demonstrated therapeutic benefits for metastatic pancreatic cancer patients (Seki et al. 2022). This has made ECM target an increasingly important strategy in the design of nanocarriers for cancer therapy, with the potential to overcome drug delivery barriers and enhance treatment efficacy.

In many cancers such as breast and neck cancers, the expression of lysyl oxidase (LOX), major component of ECM, is often found to be elevated. LOX mainly provides rigidity to ECM by cross-linking collagen and elastin (Xiao and Ge 2012). The rigidity of ECM creates a protective barrier that can stop the drugs from entering the tumor cells suggesting that targeting LOX might help improve the efficacy of cancer drugs. Polymer nanoparticles can be used to deliver antibodies that inhibit LOX as shown by Kanapathipillai et al. In their study, they showed *in vitro* and xenograft models that the nanoparticles with these antibodies could effectively bind to tumor ECM and improve therapeutic responses in solid tumors (Kanapathipillai et al. 2012).

Fibronectin, another key constituent of the ECM, is also elevated in TME. It has been implicated in supporting tumor cell adhesion and migration (Yu et al. 2018). PLA nanoparticles were used to deliver paclitaxel into the TME by conjugating with peptide CLT1 (Zhang et al. 2014). Xu et al. used a nanocarrier to deliver cisplatin adjugated with collagenase. The nanocarrier incorporated a hydrophilic periphery made of poly β-amino-ester (PBAE) that disseminated the collagenase into the TME. This enhances the disintegration of ECM and improve the penetration through the tumor to deliver cisplatin (Xu et al. 2020). These different methods demonstrate the ability of nanoparticles to better target ECM to improve penetration of TME and improve drug delivery.

## 19.5 Nanomaterials and Cancer Immunotherapy

Several immunotherapeutic strategies are developed to harness the innate potential of the immune system to fight against cancer. These include immune system modulators, chimeric antigen receptor (CAR)-T cell therapy, checkpoint inhibitor therapy, and vaccines (Yan et al. 2020). These therapies utilize natural or synthetic molecules to restore immune system function or enhance its capabilities, thus exerting anti-cancer effects. A key focus in immunotherapy involves directing drugs at specific immune related proteins such as Programmed Cell Death Protein 1 (PD-1) and its ligand, Programmed Cell Death Ligand 1 (PD-L1), which are crucial in regulating immune responses. Research has demonstrated that immune checkpoint inhibitors (ICIs) can be effectively delivered using nanocarriers to target the PD-1/PD-L1 axis, improving cancer treatment outcomes (Liu and Sun 2021). A study by Bu and colleagues emphasized that uncontrolled PD-1 expression allows cancer cells to evade the immune response, and traditional PD-1/PD-L1 ICIs have shown inconsistent results. To enhance the binding between PD-L1 and ICIs, Bu et al. employed multivalent PAMAM dendrimers, which improved PD-L1 inhibition and facilitated increased drug deposition at tumor sites (Atale and Rani 2023; Bu et al. 2020; Waldman et al. 2020). These approaches involve the use of antibodies, small molecules, and proteins to modulate immune activity.

Nanoparticles targeting T-cells can promote T-cell expansion *in vivo* by delivering genes, cytokines, antibodies, and small molecules directly to tumor sites. Smith et al. demonstrated an innovative approach to enhancing T-cell-based immunotherapies by synthesizing a polymeric nanomaterial that delivers a 194-1BBz chimeric antigen receptor (CAR)-encoding plasmid and piggyBac transposase directly to T-cells. This nanomaterial was designed to promote T-cell expansion and CAR expression, essential for activating the immune system to target cancer cells. The nanomaterial was engineered with anti-CD3e F(ab′)2 fragments on its surface, ensuring that the nanomaterial would specifically target and interact with CD3+ T-cells (Smith et al. 2017) (Fig. 19.4a). To further optimize the process, sequences promoting nuclear localization and microtubule association were incorporated, which have critical role for enhancing gene transfer efficiency and ensuring successful CAR expression in T-cells.

Similarly, Alhallak et al. employed a different but equally innovative strategy by using NBiTEs (MN-based bispecific T-cell engagers) in their immunotherapy research. These nanoparticles, conjugated with antibodies, serve to form a "bridge" between T-cells and cancer cells, enabling more effective targeting of cancer cells by T-cells. NBiTEs were designed to simultaneously bind to T-cell receptors and tumor antigens, thus enhancing the T-cell's ability to localize and attack cancer cells with greater precision (Alhallak et al. 2021; Gong et al. 2021) (Fig. 19.4b). These studies highlight the promise of nanomaterials to significantly improve the effectiveness of T-cell-based immunotherapies.

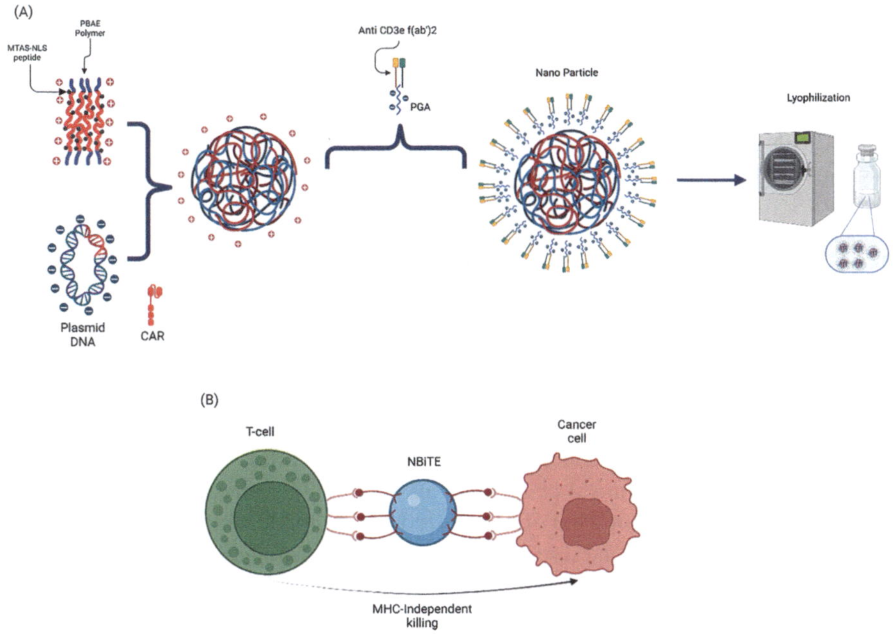


**Fig. 19.4** Synthesis of specialized nano particles for cancer immunotherapy. (**a**) Example of synthesis process of a polymeric NM to deliver a 194-1BBz chimeric antigen receptor (CAR)- encoding plasmid and a piggyBac transposase directly to T cells, effectively stimulating T-cell expansion and CAR expression (Smith et al. 2017). This nanomaterial was engineered to specifically express in CD3+ cells using anti-CD3e F(ab′)2 fragments on its surface. (**b**) Use of NBiTEs (MN-based bispecific T-cell engagers) to direct CTL functions for application in T-cell immunotherapies. NBiTEs nanoparticles conjugated with antibodies create a bridge between T- cells and cancer cells and allow for T-cells to target the cancer cells effectively (Alhallak et al. 2021; Gong et al. 2021)

## 19.6 Current Challenges in Drug Delivery in Cancer

Nanomaterials offer suitable platforms for cancer therapy, particularly due to their ability to be engineered for targeted drug delivery. However, several challenges remain that hinder their widespread clinical application in cancer treatment. The major obstacles include biodistribution, toxicity, scalability, and regulatory approval processes. To ensure the safety and efficacy of nanomaterials, careful synthesis and characterization are crucial to minimizing potential toxicity to healthy cells. During the clinical translation of nanomedicines, significant challenges arise in developing accurate pharmacokinetic models, formulating nanomedicines for practical use, and evaluating their biological properties in humans. Rational design of nanomedicines is essential for overcoming immune evasion, improving tumor targeting, and ensuring controlled drug release. However, the systematic evaluation of nanoparticle properties is complex, given the difficulty in synthesizing homogeneous nanoparticle cohorts with specific characteristics. Biocompatibility remains a critical barrier to clinical translation, as many nanomaterial-based drug delivery systems show promising results in preclinical models but encounter challenges in human clinical trials. While some nanomedicines, such as ICIs targeting CTLA-4 and PD-1/PD-L1, have been approved by the FDA for immunotherapy (e.g., atezolizumab and pembrolizumab), treatment outcomes still present significant obstacles (Finn et al. 2020; Robert et al. 2015). Moreover, responses to these therapies vary between patients, and their efficacy is often limited to specific tumor types, with only a subset of cancers responding to CTLA-4 and PD-1/PD-L1-targeting approaches (Reck et al. 2022).

ICIs require highly specific and effective targets, and the TME—characterized by hypoxia, high pressure, and acidity—further complicates the effectiveness of immunotherapeutic drugs (Yenyuwadee et al. 2022). Despite advancements in NM-based drug delivery systems for tumor immunotherapy, differences can be attributed between mouse and human tumors, which often lead to discrepancies between promising preclinical results and less favorable clinical trial outcomes. To bridge this gap, the development of humanized animal models incorporating human-derived immune systems and tumors could improve the clinical translation of nanotechnology research. Additionally, variability in the EPR effect across different patients complicates the uniform delivery of nanomaterials to tumors, further hindering therapeutic success.

Another significant challenge in immunotherapy is the failure of most drugs to effectively reactivate and proliferate T cells. To address this, promising approaches should focus on inducing the generation of tumor-specific T cells while preventing T cell dysfunction caused by excessive stimulation. Combination therapies using nanomaterials to deliver ICIs or immune-stimulating agonists hold great potential for overcoming this limitation. For example, iron-based nanomaterials could be employed to deliver anti-CD137 and anti-PD-L1 antibodies to enhance T-cell proliferation and prevent exhaustion caused by PD-L1/PD-1 interaction (Kosmides et al. 2017). Furthermore, lipid nanoparticles were being explored for delivering mRNA-based vaccines that prime T cells and trigger an antigen-specific anti-tumor immune response. *In vivo* studies using lipid nanomaterials to deliver mRNA encoding claudin 6 (CARVac) have shown promising results by enhancing the endurance of CAR T cells (Reinhard et al. 2020). Given these broad developments, we believe that combination therapies and the use of innovative nanomaterials for co-delivery of immunotherapy agents could significantly advance the clinical application of tumor immunotherapy. By addressing current limitations and enhancing the specificity and efficacy of treatments, nanomaterials have the potential to revolutionize cancer therapy.

## 19.7 Conclusion and Future Perspectives

The integration of nanotechnology into cancer treatment marks a revolutionary advancement by developing novel drug delivery systems. These nanomaterial-based systems can improve drug pharmacokinetics, and retention of therapeutic agents. Additionally, they offer improved biocompatibility, reducing the adverse reactions in healthy tissues, tumor-specific targeting capability, and minimizing systemic toxicity. Furthermore, nanomaterials can also be used to control the delivery of the drugs and improve their efficacy.

Nanomaterials are an exciting approach to enhance the effectiveness of drugs. In addition to this they can also be used to affect immune responses by targeting cancer cell receptors creating new avenues for immuno-oncology. Tumor cells are well known to modify their microenvironment and create conditions that lead to immunosuppression by forcing their surroundings to secrete immune checkpoint molecules or recruit Tregs. Nanoparticles with attached ligands can overcome these defenses and allow natural immune responses to attack cancer cells. Nanoparticles can also improve effectiveness of chemotherapies. They can be used to disintegration the ECM formed around the tumor cells, increase blood flow improving the penetration of drugs into tumor environment in both passive and active way. Modified nanoparticles that are enhanced with magnetically or electrically active components also allow improved performance of radio therapies.

While these developments with nanoparticles are promising, there are still challenges that need to be addressed. Immune therapy enhanced using nanoparticles may provide benefits but could also lead to unwanted immune response and inflammation. Moreover, the translation of nanoparticle-based therapies from *in vitro* or mice models are not always successful in humans. Thus, prolonged and rigorous clinical trials need to be performed to establish the efficacy and safety of NP based therapies. As research in this area continues to evolve, these platforms must be optimized to minimize risks and maximize therapeutic benefits.

The field of nanoparticles is still in its infancy and new methods and ideas to use nanoparticles for anti-cancer therapy in all areas—immunotherapy, chemotherapy and radiotherapy—continue to be developed. As the design of nanoparticles get more sophisticated and discoveries in materials allow to provide unique properties to the nanoparticles, we would anticipate better outcomes of cancer treatments. There might also be possibility of creating patient specific designs based on genetic profile, blood and lymphatic compositions and specific tumor characteristics of the individual. Ongoing research and development in nanotechnology hold the promise of transforming cancer treatment, potentially leading to higher survival rates, reduced side effects, and an enhanced quality of life for cancer patients around the world.

**Conflict of Interest** Authors declare no conflict of interest

## References


Abbasi E, Aval SF, Akbarzadeh A, Milani M, Nasrabadi HT, Joo SW, Hanifehpour Y, Nejati-Koshki K, Pashaei-Asl R (2014) Dendrimers: synthesis, applications, and properties. Nanoscale Res Lett 9(1):247. https://doi.org/10.1186/1556-276X-9-247

Abedi-Gaballu F, Dehghan G, Ghaffari M, Yekta R, Abbaspour-Ravasjani S, Baradaran B, Dolatabadi JEN, Hamblin MR (2018) PAMAM dendrimers as efficient drug and gene delivery nanosystems for cancer therapy. Appl Mater Today 12:177–190. https://doi.org/10.1016/j.apmt.2018.05.002

Abedin MR, Powers K, Aiardo R, Barua D, Barua S (2021) Antibody-drug nanoparticle induces synergistic treatment efficacies in HER2 positive breast cancer cells. Sci Rep 11(1):7347. https://doi.org/10.1038/s41598-021-86762-6

Acharya S, Sahoo SK (2011) PLGA nanoparticles containing various anticancer agents and tumour delivery by EPR effect. Adv Drug Deliv Rev 63(3):170–183. https://doi.org/10.1016/j.addr.2010.10.008

Alhallak K, Sun J, Wasden K, Guenthner N, O'Neal J, Muz B, King J, Kohnen D, Vij R, Achilefu S, DiPersio JF, Azab AK (2021) Nanoparticle T-cell engagers as a modular platform for cancer immunotherapy. Leukemia 35(8):2346–2357. https://doi.org/10.1038/s41375-021-01127-2

Alibolandi M, Taghdisi SM, Ramezani P, Hosseini Shamili F, Farzad SA, Abnous K, Ramezani M (2017) Smart AS1411-aptamer conjugated pegylated PAMAM dendrimer for the superior delivery of camptothecin to colon adenocarcinoma in vitro and in vivo. Int J Pharm 519(1–2):352–364. https://doi.org/10.1016/j.ijpharm.2017.01.044

Amreddy N, Babu A, Muralidharan R, Panneerselvam J, Srivastava A, Ahmed R, Mehta M, Munshi A, Ramesh R (2018) Recent advances in nanoparticle-based cancer drug and gene delivery. Adv Cancer Res 137:115–170. https://doi.org/10.1016/bs.acr.2017.11.003

Anbazhagan R, Muthusamy G, Krishnamoorthi R, Kumaresan S, Rajendra Prasad N, Lai JY, Yang JM, Tsai HC (2021) PAMAM G4.5 dendrimers for targeted delivery of ferulic acid and paclitaxel to overcome P-glycoprotein-mediated multidrug resistance. Biotechnol Bioeng 118(3):1213–1223. https://doi.org/10.1002/bit.27645

Anzar N, Hasan R, Tyagi M, Yadav N, Narang J (2020) Carbon nanotube – a review on synthesis, properties and plethora of applications in the field of biomedical science. Sens Int 1:100003. https://doi.org/10.1016/j.sintl.2020.100003

Appadurai P, Rathinasamy K (2015) Plumbagin-silver nanoparticle formulations enhance the cellular uptake of plumbagin and its antiproliferative activities. IET Nanobiotechnol 9(5):264–272. https://doi.org/10.1049/iet-nbt.2015.0008

Atale N, Rani V (2020) Angiogenesis: a therapeutic target for cancer. In: Tuli HS (ed) Drug targets in cellular processes of cancer: from nonclinical to preclinical models. Springer, Singapore, pp 165–183. https://doi.org/10.1007/978-981-15-7586-0_9

Atale N, Rani V (2022) Role of nanotherapeutics in inhibiting cancer angiogenesis: a novel perspective. In: Tuli HS (ed) Nanotherapeutics in cancer materials, diagnostics, and clinical applications. Jenny Stanford Publishing, Singapore

Atale N, Rani V (2023) Biomarkers in precision medicine with special reference to oncology. In: Tuli HS, Yerer Aycan MB (eds) Oncology: genomics, precision medicine and therapeutic targets. Springer Nature, Singapore, pp 173–187. https://doi.org/10.1007/978-981-99-1529-3_7

Atale N, Rani V (2024) In: Tuli HS (ed) Anticancer spices dietary input to health. Jenny Stanford Publishing, Singapore

Atale N, Saxena S, Nirmala JG, Narendhirakannan RT, Mohanty S, Rani V (2017) Synthesis and characterization of Sygyzium cumini nanoparticles for its protective potential in high glucose-induced cardiac stress: a green approach. Appl Biochem Biotechnol 181(3):1140–1154. https://doi.org/10.1007/s12010-016-2274-6

Attia MF, Anton N, Wallyn J, Omran Z, Vandamme TF (2019) An overview of active and passive targeting strategies to improve the nanocarriers efficiency to tumour sites. J Pharm Pharmacol 71(8):1185–1198. https://doi.org/10.1111/jphp.13098

Bagalkot V, Zhang L, Levy-Nissenbaum E, Jon S, Kantoff PW, Langer R, Farokhzad OC (2007) Quantum dot-aptamer conjugates for synchronous cancer imaging, therapy, and sensing of drug delivery based on bi-fluorescence resonance energy transfer. Nano Lett 7(10):3065–3070. https://doi.org/10.1021/nl071546n

Bartlett DW, Davis ME (2006) Insights into the kinetics of siRNA-mediated gene silencing from live-cell and live-animal bioluminescent imaging. Nucleic Acids Res 34(1):322–333. https://doi.org/10.1093/nar/gkj439

Bernardi D, Errante D, Stefani M, Salvagno L (2010) Non-pegylated liposomal doxorubicin in metastatic breast cancer patients: a valuable therapeutic option requiring caution. Breast 19(6):549–550. https://doi.org/10.1016/j.breast.2010.05.008

Bray F, Laversanne M, Sung H, Ferlay J, Siegel RL, Soerjomataram I, Jemal A (2024) Global cancer statistics 2022: GLOBOCAN estimates of incidence and mortality worldwide for 36 cancers in 185 countries. CA Cancer J Clin 74(3):229–263. https://doi.org/10.3322/caac.21834

Bu J, Nair A, Iida M, Jeong WJ, Poellmann MJ, Mudd K, Kubiatowicz LJ, Liu EW, Wheeler DL, Hong S (2020) An avidity-based PD-L1 antagonist using nanoparticle-antibody conjugates for enhanced immunotherapy. Nano Lett 20(7):4901–4909. https://doi.org/10.1021/acs.nanolett.0c00953

Bulbake U, Doppalapudi S, Kommineni N, Khan W (2017) Liposomal formulations in clinical use: an updated review. Pharmaceutics 9(2). https://doi.org/10.3390/pharmaceutics9020012

Cao M, Yan H, Han X, Weng L, Wei Q, Sun X, Lu W, Wei Q, Ye J, Cai X, Hu C, Yin X, Cao P (2019) Ginseng-derived nanoparticles alter macrophage polarization to inhibit melanoma growth. J Immunother Cancer 7(1):326. https://doi.org/10.1186/s40425-019-0817-4

Carita AC, Eloy JO, Chorilli M, Lee RJ, Leonardi GR (2018) Recent advances and perspectives in liposomes for cutaneous drug delivery. Curr Med Chem 25(5):606–635. https://doi.org/10.2174/0929867324666171009120154

Chang X, Chen M, Wei T (2024) Precision chemistry of metallofullerenes and graphene: recent advances. ChemPlusChem 89(11):e202400105. https://doi.org/10.1002/cplu.202400105

Chattopadhyay N, Cai Z, Pignol JP, Keller B, Lechtman E, Bendayan R, Reilly RM (2010) Design and characterization of HER-2-targeted gold nanoparticles for enhanced X-radiation treatment of locally advanced breast cancer. Mol Pharm 7(6):2194–2206. https://doi.org/10.1021/mp100207t

Chavva SR, Deshmukh SK, Kanchanapally R, Tyagi N, Coym JW, Singh AP, Singh S (2019) Epigallocatechin gallate-gold nanoparticles exhibit superior antitumor activity compared to conventional gold nanoparticles: potential synergistic interactions. Nanomaterials (Basel) 9(3). https://doi.org/10.3390/nano9030396

Chehelgerdi M, Chehelgerdi M, Allela OQB, Pecho RDC, Jayasankar N, Rao DP, Thamaraikani T, Vasanthan M, Viktor P, Lakshmaiya N, Saadh MJ, Amajd A, Abo-Zaid MA, Castillo-Acobo RY, Ismail AH, Amin AH, Akhavan-Sigari R (2023) Progressing nanotechnology to improve targeted cancer treatment: overcoming hurdles in its clinical implementation. Mol Cancer 22(1):169. https://doi.org/10.1186/s12943-023-01865-0

Chen L, Hong W, Ren W, Xu T, Qian Z, He Z (2021) Recent progress in targeted delivery vectors based on biomimetic nanoparticles. Signal Transduct Target Ther 6(1):225. https://doi.org/10.1038/s41392-021-00631-2

Cheng Z, Li M, Dey R, Chen Y (2021) Nanomaterials for cancer therapy: current progress and perspectives. J Hematol Oncol 14(1):85. https://doi.org/10.1186/s13045-021-01096-0

Chou PY, Lin SY, Wu YN, Shen CY, Sheu MT, Ho HO (2022) Glycosylation of OVA antigen-loaded PLGA nanoparticles enhances DC-targeting for cancer vaccination. J Control Release 351:970–988. https://doi.org/10.1016/j.jconrel.2022.10.002

Choudhary S, Gupta L, Rani S, Dave K, Gupta U (2017) Impact of dendrimers on solubility of hydrophobic drug molecules. Front Pharmacol 8:261. https://doi.org/10.3389/fphar.2017.00261

Cline EN, Li MH, Choi SK, Herbstman JF, Kaul N, Meyhofer E, Skiniotis G, Baker JR, Larson RG, Walter NG (2013) Paclitaxel-conjugated PAMAM dendrimers adversely affect microtubule structure through two independent modes of action. Biomacromolecules 14(3):654–664. https://doi.org/10.1021/bm301719b

Cortés J, Saura C (2010) Nanoparticle albumin-bound (nab™)-paclitaxel: improving efficacy and tolerability by targeted drug delivery in metastatic breast cancer. EJC Suppl 8:1–10

Danhier F, Lecouturier N, Vroman B, Jerome C, Marchand-Brynaert J, Feron O, Preat V (2009) Paclitaxel-loaded PEGylated PLGA-based nanoparticles: in vitro and in vivo evaluation. J Control Release 133(1):11–17. https://doi.org/10.1016/j.jconrel.2008.09.086

Danhier F, Feron O, Preat V (2010) To exploit the tumor microenvironment: passive and active tumor targeting of nanocarriers for anti-cancer drug delivery. J Control Release 148(2):135–146. https://doi.org/10.1016/j.jconrel.2010.08.027

de Visser KE, Joyce JA (2023) The evolving tumor microenvironment: from cancer initiation to metastatic outgrowth. Cancer Cell 41(3):374–403. https://doi.org/10.1016/j.ccell.2023.02.016

Ejiohuo O, Folami SO, Edi D, Isaac J (2024) Polyphenol encapsulated nanofibers in wound healing and drug delivery. Eur J Med Chem Rep 12:100184. https://doi.org/10.1016/j.ejmcr.2024.100184

Elbaz NM, Ziko L, Siam R, Mamdouh W (2016) Core-shell silver/polymeric nanoparticles-based combinatorial therapy against breast cancer in-vitro. Sci Rep 6:30729. https://doi.org/10.1038/srep30729

Feito MJ, Diez-Orejas R, Cicuendez M, Casarrubios L, Rojo JM, Portoles MT (2019) Characterization of M1 and M2 polarization phenotypes in peritoneal macrophages after treatment with graphene oxide nanosheets. Colloids Surf B Biointerfaces 176:96–105. https://doi.org/10.1016/j.colsurfb.2018.12.063

Feng T, Wei Y, Lee RJ, Zhao L (2017) Liposomal curcumin and its application in cancer. Int J Nanomedicine 12:6027–6044. https://doi.org/10.2147/IJN.S132434

Feng S, Pan J, Li C, Zheng Y (2020) Folic acid-conjugated nitrogen-doped graphene quantum dots as a fluorescent diagnostic material for MCF-7 cells. Nanotechnology 31(13):135701. https://doi.org/10.1088/1361-6528/ab5f7f

Finn RS, Qin S, Ikeda M, Galle PR, Ducreux M, Kim TY, Kudo M, Breder V, Merle P, Kaseb AO, Li D, Verret W, Xu DZ, Hernandez S, Liu J, Huang C, Mulla S, Wang Y, Lim HY et al (2020) Atezolizumab plus bevacizumab in unresectable hepatocellular carcinoma. N Engl J Med 382(20):1894–1905. https://doi.org/10.1056/NEJMoa1915745

Folkman J (2007) Angiogenesis: an organizing principle for drug discovery? Nat Rev Drug Discov 6(4):273–286. https://doi.org/10.1038/nrd2115

Formaggio DMD, Magalhaes JA, Andrade VM, Conceicao K, Anastacio JM, Santiago GS, Arruda DC, Tada DB (2022) Co-functionalization of gold nanoparticles with C7H2 and HuAL1 peptides: enhanced antimicrobial and antitumoral activities. Pharmaceutics 14(7). https://doi.org/10.3390/pharmaceutics14071324

Gagliardi A, Giuliano E, Venkateswararao E, Fresta M, Bulotta S, Awasthi V, Cosco D (2021) Biodegradable polymeric nanoparticles for drug delivery to solid tumors. Front Pharmacol 12:601626. https://doi.org/10.3389/fphar.2021.601626

Gajbhiye KR, Salve R, Narwade M, Sheikh A, Kesharwani P, Gajbhiye V (2023) Lipid polymer hybrid nanoparticles: a custom-tailored next-generation approach for cancer therapeutics. Mol Cancer 22(1):160. https://doi.org/10.1186/s12943-023-01849-0

Gao A, Hu XL, Saeed M, Chen BF, Li YP, Yu HJ (2019) Overview of recent advances in liposomal nanoparticle-based cancer immunotherapy. Acta Pharmacol Sin 40(9):1129–1137. https://doi.org/10.1038/s41401-019-0281-1

Gao S, Xu B, Sun J, Zhang Z (2024) Nanotechnological advances in cancer: therapy a comprehensive review of carbon nanotube applications. Front Bioeng Biotechnol 12:1351787. https://doi.org/10.3389/fbioe.2024.1351787

Gavas S, Quazi S, Karpinski TM (2021) Nanoparticles for cancer therapy: current progress and challenges. Nanoscale Res Lett 16(1):173. https://doi.org/10.1186/s11671-021-03628-6

Gil HM, Price TW, Chelani K, Bouillard JG, Calaminus SDJ, Stasiuk GJ (2021) NIR-quantum dots in biomedical imaging and their future. iScience 24(3):102189. https://doi.org/10.1016/j.isci.2021.102189

Gong N, Sheppard NC, Billingsley MM, June CH, Mitchell MJ (2021) Nanomaterials for T-cell cancer immunotherapy. Nat Nanotechnol 16(1):25–36. https://doi.org/10.1038/s41565-020-00822-y
Guo C, Zhang W, Zhang Q, Su Y, Hou X, Chen Q, Guo H, Kong M, Chen D (2023) Novel dual CAFs and tumour cell targeting pH and ROS dual sensitive micelles for targeting delivery of paclitaxel to liver cancer. Artif Cells Nanomed Biotechnol 51(1):170–179. https://doi.org/10.1080/21691401.2023.2193221
Guo S, Wang J, Wang Q, Wang J, Qin S, Li W (2024) Advances in peptide-based drug delivery systems. Heliyon 10(4):e26009. https://doi.org/10.1016/j.heliyon.2024.e26009
Han SM, Baek JS, Kim MS, Hwang SJ, Cho CW (2018) Surface modification of paclitaxel-loaded liposomes using d-alpha-tocopheryl polyethylene glycol 1000 succinate: enhanced cellular uptake and cytotoxicity in multidrug resistant breast cancer cells. Chem Phys Lipids 213:39–47. https://doi.org/10.1016/j.chemphyslip.2018.03.005
He X, Chen X, Liu L, Zhang Y, Lu Y, Zhang Y, Chen Q, Ruan C, Guo Q, Li C, Sun T, Jiang C (2018) Sequentially triggered nanoparticles with tumor penetration and intelligent drug release for pancreatic cancer therapy. Adv Sci (Weinh) 5(5):1701070. https://doi.org/10.1002/advs.201701070
He Z, Zhang Y, Feng N (2020) Cell membrane-coated nanosized active targeted drug delivery systems homing to tumor cells: a review. Mater Sci Eng C Mater Biol Appl 106:110298. https://doi.org/10.1016/j.msec.2019.110298
Hibino M, Tanaka K, Ouchi M, Terashima T (2022) Amphiphilic random-block copolymer micelles in water: precise and dynamic self-assembly controlled by random copolymer association. Macromolecules 55(1):178–189. https://doi.org/10.1021/acs.macromol.1c02186
How CW, Rasedee A, Manickam S, Rosli R (2013) Tamoxifen-loaded nanostructured lipid carrier as a drug delivery system: characterization, stability assessment and cytotoxicity. Colloids Surf B Biointerfaces 112:393–399. https://doi.org/10.1016/j.colsurfb.2013.08.009
Hu M, Zhang J, Kong L, Yu Y, Hu Q, Yang T, Wang Y, Tu K, Qiao Q, Qin X, Zhang Z (2021) Immunogenic hybrid nanovesicles of liposomes and tumor-derived nanovesicles for cancer immunochemotherapy. ACS Nano 15(2):3123–3138. https://doi.org/10.1021/acsnano.0c09681
Huang Q, Ge Y, He Y, Wu J, Tong Y, Shang H, Liu X, Ba X, Xia D, Peng E, Chen Z, Tang K (2024) The application of nanoparticles targeting cancer-associated fibroblasts. Int J Nanomedicine 19:3333–3365. https://doi.org/10.2147/IJN.S447350
Izhar F, Imran M, Izhar H, Latif S, Hussain N, Iqbal HMN, Bilal M (2022) Recent advances in metal-based nanoporous materials for sensing environmentally-related biomolecules. Chemosphere 307(Pt 4):135999. https://doi.org/10.1016/j.chemosphere.2022.135999
Jadon RS, Sharma M (2019) Docetaxel-loaded lipid-polymer hybrid nanoparticles for breast cancer therapeutics. J Drug Deliv Sci Technol 51:475–484. https://doi.org/10.1016/j.jddst.2019.03.039
Jiang Q, Wang K, Zhang X, Ouyang B, Liu H, Pang Z, Yang W (2020) Platelet membrane-camouflaged magnetic nanoparticles for ferroptosis-enhanced cancer immunotherapy. Small 16(22):e2001704. https://doi.org/10.1002/smll.202001704
Jin C, Wang K, Oppong-Gyebi A, Hu J (2020) Application of nanotechnology in cancer diagnosis and therapy – a mini-review. Int J Med Sci 17(18):2964–2973. https://doi.org/10.7150/ijms.49801
Kader A, Kaufmann JO, Mangarova DB, Moeckel J, Brangsch J, Adams LC, Zhao J, Reimann C, Saatz J, Traub H, Buchholz R, Karst U, Hamm B, Makowski MR (2022) Iron oxide nanoparticles for visualization of prostate cancer in MRI. Cancers (Basel) 14(12). https://doi.org/10.3390/cancers14122909
Kanapathipillai M, Mammoto A, Mammoto T, Kang JH, Jiang E, Ghosh K, Korin N, Gibbs A, Mannix R, Ingber DE (2012) Inhibition of mammary tumor growth using lysyl oxidase-targeting nanoparticles to modify extracellular matrix. Nano Lett 12(6):3213–3217. https://doi.org/10.1021/nl301206p
Karami MH, Abdouss M, Rahdar A, Pandey S (2024) Graphene quantum dots: background, synthesis methods, and applications as nanocarrier in drug delivery and cancer treatment: an updated review. Inorg Chem Commun 161:112032. https://doi.org/10.1016/j.inoche.2024.112032

Kashyap D, Tuli HS, Yerer MB, Sharma A, Sak K, Srivastava S, Pandey A, Garg VK, Sethi G, Bishayee A (2021) Natural product-based nanoformulations for cancer therapy: opportunities and challenges. Semin Cancer Biol 69:5–23. https://doi.org/10.1016/j.semcancer.2019.08.014

Khalaf K, Hana D, Chou JT, Singh C, Mackiewicz A, Kaczmarek M (2021) Aspects of the tumor microenvironment involved in immune resistance and drug resistance. Front Immunol 12:656364. https://doi.org/10.3389/fimmu.2021.656364

Kosmides AK, Sidhom JW, Fraser A, Bessell CA, Schneck JP (2017) Dual targeting nanoparticle stimulates the immune system to inhibit tumor growth. ACS Nano 11(6):5417–5429. https://doi.org/10.1021/acsnano.6b08152

Kulhari H, Pooja D, Shrivastava S, Kuncha M, Naidu VGM, Bansal V, Sistla R, Adams DJ (2016) Trastuzumab-grafted PAMAM dendrimers for the selective delivery of anticancer drugs to HER2-positive breast cancer. Sci Rep 6:23179. https://doi.org/10.1038/srep23179

Lang J, Zhao X, Qi Y, Zhang Y, Han X, Ding Y, Guan J, Ji T, Zhao Y, Nie G (2019) Reshaping prostate tumor microenvironment to suppress metastasis via cancer-associated fibroblast inactivation with peptide-assembly-based nanosystem. ACS Nano 13(11):12357–12371. https://doi.org/10.1021/acsnano.9b04857

Li R, Wang Y, Du J, Wang X, Duan A, Gao R, Liu J, Li B (2021) Graphene oxide loaded with tumor-targeted peptide and anti-cancer drugs for cancer target therapy. Sci Rep 11(1):1725. https://doi.org/10.1038/s41598-021-81218-3

Li Z, Chu Z, Yang J, Qian H, Xu J, Chen B, Tian T, Chen H, Xu Y, Wang F (2022) Immunogenic cell death augmented by manganese zinc sulfide nanoparticles for metastatic melanoma immunotherapy. ACS Nano 16(9):15471–15483. https://doi.org/10.1021/acsnano.2c08013

Liu YT, Sun ZJ (2021) Turning cold tumors into hot tumors by improving T-cell infiltration. Theranostics 11(11):5365–5386. https://doi.org/10.7150/thno.58390

Liu B, Zhou H, Tan L, Siu KTH, Guan XY (2024) Exploring treatment options in cancer: tumor treatment strategies. Signal Transduct Target Ther 9(1):175. https://doi.org/10.1038/s41392-024-01856-7

Mardhian DF, Vrynas A, Storm G, Bansal R, Prakash J (2020) FGF2 engineered SPIONs attenuate tumor stroma and potentiate the effect of chemotherapy in 3D heterospheroidal model of pancreatic tumor. Nano 4(1):26–39. https://doi.org/10.7150/ntno.38092

Matiyani M, Rana A, Pal M, Dokwal S, Sahoo NG (2023) Polyamidoamine dendrimer decorated graphene oxide as a pH-sensitive nanocarrier for the delivery of hydrophobic anticancer drug quercetin: a remedy for breast cancer. J Pharm Pharmacol 75(6):859–872. https://doi.org/10.1093/jpp/rgad036

Mir SA, Hamid L, Bader GN, Shoaib A, Rahamathulla M, Alshahrani MY, Alam P, Shakeel F (2022) Role of nanotechnology in overcoming the multidrug resistance in cancer therapy: a review. Molecules 27(19). https://doi.org/10.3390/molecules27196608

Mukherjee A, Waters AK, Kalyan P, Achrol AS, Kesari S, Yenugonda VM (2019) Lipid-polymer hybrid nanoparticles as a next-generation drug delivery platform: state of the art, emerging technologies, and perspectives. Int J Nanomedicine 14:1937–1952. https://doi.org/10.2147/IJN.S198353

Mukwaya G, Forssen EA, Schmidt P, Ross M (1998) DaunoXome® (liposomal daunorubicin) for first-line treatment of advanced, HIV-related kaposi's sarcoma. In: Woodle MC, Storm G (eds) Long circulating liposomes: old drugs, new therapeutics. Springer, Berlin Heidelberg, pp 147–163. https://doi.org/10.1007/978-3-662-22115-0_10

Mullauer FB, van Bloois L, Daalhuisen JB, Ten Brink MS, Storm G, Medema JP, Schiffelers RM, Kessler JH (2011) Betulinic acid delivered in liposomes reduces growth of human lung and colon cancers in mice without causing systemic toxicity. Anti-Cancer Drugs 22(3):223–233. https://doi.org/10.1097/CAD.0b013e3283421035

Neves V, Heister E, Costa S, Tilmaciu C, Flahaut E, Soula B, Coley HM, McFadden J, Silva SR (2012) Design of double-walled carbon nanotubes for biomedical applications. Nanotechnology 23(36):365102. https://doi.org/10.1088/0957-4484/23/36/365102

Nieto C, Vega MA, Martin Del Valle EM (2020) Trastuzumab: more than a guide in HER2-positive cancer nanomedicine. Nanomaterials (Basel) 10(9). https://doi.org/10.3390/nano10091674

Nsairat H, Khater D, Sayed U, Odeh F, Al Bawab A, Alshaer W (2022) Liposomes: structure, composition, types, and clinical applications. Heliyon 8(5):e09394. https://doi.org/10.1016/j.heliyon.2022.e09394

O'Brien ME, Wigler N, Inbar M, Rosso R, Grischke E, Santoro A, Catane R, Kieback DG, Tomczak P, Ackland SP, Orlandi F, Mellars L, Alland L, Tendler C, Group, C. B. C. S (2004) Reduced cardiotoxicity and comparable efficacy in a phase III trial of pegylated liposomal doxorubicin HCl (CAELYX/Doxil) versus conventional doxorubicin for first-line treatment of metastatic breast cancer. Ann Oncol 15(3):440–449. https://doi.org/10.1093/annonc/mdh097

Patel SK, Billingsley MM, Frazee C, Han X, Swingle KL, Qin J, Alameh MG, Wang K, Weissman D, Mitchell MJ (2022) Hydroxycholesterol substitution in ionizable lipid nanoparticles for mRNA delivery to T cells. J Control Release 347:521–532. https://doi.org/10.1016/j.jconrel.2022.05.020

Patra JK, Das G, Fraceto LF, Campos EVR, Rodriguez-Torres MDP, Acosta-Torres LS, Diaz-Torres LA, Grillo R, Swamy MK, Sharma S, Habtemariam S, Shin HS (2018) Nano based drug delivery systems: recent developments and future prospects. J Nanobiotechnology 16(1):71. https://doi.org/10.1186/s12951-018-0392-8

Pawar A, Singh S, Rajalakshmi S, Shaikh K, Bothiraja C (2018) Development of fisetin-loaded folate functionalized pluronic micelles for breast cancer targeting. Artif Cells Nanomed Biotechnol 46(sup1):347–361. https://doi.org/10.1080/21691401.2018.1423991

Perez AT, Domenech GH, Frankel C, Vogel CL (2002) Pegylated liposomal doxorubicin (Doxil) for metastatic breast cancer: the Cancer Research Network, Inc., experience. Cancer Investig 20(Suppl 2):22–29. https://doi.org/10.1081/cnv-120014883

Perumal S, Atchudan R, Lee W (2022) A review of polymeric micelles and their applications. Polymers (Basel) 14(12). https://doi.org/10.3390/polym14122510

Prakash J, Shaked Y (2024) The interplay between extracellular matrix remodeling and cancer therapeutics. Cancer Discov 14(8):1375–1388. https://doi.org/10.1158/2159-8290.CD-24-0002

Ramesh A, Kumar S, Nandi D, Kulkarni A (2019) CSF1R- and SHP2-inhibitor-loaded nanoparticles enhance cytotoxic activity and phagocytosis in tumor-associated macrophages. Adv Mater 31(51):e1904364. https://doi.org/10.1002/adma.201904364

Rani V, Yadav UCS (2018) Functional food and human health. Springer Nature, Singapore. https://books.google.ca/books?id=rn5yDwAAQBAJ

Rao L, Zhao SK, Wen C, Tian R, Lin L, Cai B, Sun Y, Kang F, Yang Z, He L, Mu J, Meng QF, Yao G, Xie N, Chen X (2020) Activating macrophage-mediated cancer immunotherapy by genetically edited nanoparticles. Adv Mater 32(47):e2004853. https://doi.org/10.1002/adma.202004853

Raval M, Patel P, Airao V, Bhatt V, Sheth N (2021) Novel silibinin loaded chitosan-coated PLGA/PCL nanoparticles based inhalation formulations with improved cytotoxicity and bioavailability for lung cancer. BioNanoScience 11(1):67–83. https://doi.org/10.1007/s12668-020-00797-z

Reck M, Remon J, Hellmann MD (2022) First-line immunotherapy for non-small-cell lung cancer. J Clin Oncol 40(6):586–597. https://doi.org/10.1200/JCO.21.01497

Reinhard K, Rengstl B, Oehm P, Michel K, Billmeier A, Hayduk N, Klein O, Kuna K, Ouchan Y, Woll S, Christ E, Weber D, Suchan M, Bukur T, Birtel M, Jahndel V, Mroz K, Hobohm K, Kranz L et al (2020) An RNA vaccine drives expansion and efficacy of claudin-CAR-T cells against solid tumors. Science 367(6476):446–453. https://doi.org/10.1126/science.aay5967

Robert C, Schachter J, Long GV, Arance A, Grob JJ, Mortier L, Daud A, Carlino MS, McNeil C, Lotem M, Larkin J, Lorigan P, Neyns B, Blank CU, Hamid O, Mateus C, Shapira-Frommer R, Kosh M, Zhou H et al (2015) Pembrolizumab versus Ipilimumab in advanced melanoma. N Engl J Med 372(26):2521–2532. https://doi.org/10.1056/NEJMoa1503093

Saneja A, Kumar R, Mintoo MJ, Dubey RD, Sangwan PL, Mondhe DM, Panda AK, Gupta PN (2019) Gemcitabine and betulinic acid co-encapsulated PLGA-PEG polymer nanoparticles for improved efficacy of cancer chemotherapy. Mater Sci Eng C Mater Biol Appl 98:764–771. https://doi.org/10.1016/j.msec.2019.01.026

Sarode RJ, Mahajan HS (2024) Dendrimers for drug delivery: an overview of its classes, synthesis, and applications. J Drug Deliv Sci Technol 98:105896. https://doi.org/10.1016/j.jddst.2024.105896

Sattari S, Adeli M, Beyranvand S, Nemati M (2021) Functionalized graphene platforms for anticancer drug delivery. Int J Nanomedicine 16:5955–5980. https://doi.org/10.2147/IJN.S249712

Schleich N, Danhier F, Preat V (2015) Iron oxide-loaded nanotheranostics: major obstacles to in vivo studies and clinical translation. J Control Release 198:35–54. https://doi.org/10.1016/j.jconrel.2014.11.024

Seki T, Saida Y, Kishimoto S, Lee J, Otowa Y, Yamamoto K, Chandramouli GV, Devasahayam N, Mitchell JB, Krishna MC, Brender JR (2022) PEGPH20, a PEGylated human hyaluronidase, induces radiosensitization by reoxygenation in pancreatic cancer xenografts. A molecular imaging study. Neoplasia 30:100793. https://doi.org/10.1016/j.neo.2022.100793

Shen F, Feng L, Zhu Y, Tao D, Xu J, Peng R, Liu Z (2020) Oxaliplatin-/NLG919 prodrugs-constructed liposomes for effective chemo-immunotherapy of colorectal cancer. Biomaterials 255:120190. https://doi.org/10.1016/j.biomaterials.2020.120190

Shen W, Liu T, Pei P, Li J, Yang S, Zhang Y, Zhou H, Hu L, Yang K (2022) Metabolic homeostasis-regulated nanoparticles for antibody-independent cancer radio-immunotherapy. Adv Mater 34(51):e2207343. https://doi.org/10.1002/adma.202207343

Shi H, Gu R, Xu W, Huang H, Xue L, Wang W, Zhang Y, Si W, Dong X (2019a) Near-infrared light-harvesting fullerene-based nanoparticles for promoted synergetic tumor phototheranostics. ACS Appl Mater Interfaces 11(48):44970–44977. https://doi.org/10.1021/acsami.9b17716

Shi Y, Chen C, Xu Y, Liu Y, Zhang H, Liu Y (2019b) LncRNA FENDRR promotes high-glucose-induced proliferation and angiogenesis of human retinal endothelial cells. Biosci Biotechnol Biochem 83(5):869–875. https://doi.org/10.1080/09168451.2019.1569499

Shin H, Kim Y, Jon S (2023) Nanovaccine displaying immunodominant T cell epitopes of fibroblast activation protein is effective against desmoplastic tumors. ACS Nano 17(11):10337–10352. https://doi.org/10.1021/acsnano.3c00764

Singh S, Hasan MR, Sharma P, Narang J (2022) Graphene nanomaterials: the wondering material from synthesis to applications. Sens Int 3:100190. https://doi.org/10.1016/j.sintl.2022.100190

Sivadasan D, Sultan MH, Madkhali O, Almoshari Y, Thangavel N (2021) Polymeric lipid hybrid nanoparticles (PLNs) as emerging drug delivery platform-a comprehensive review of their properties, preparation methods, and therapeutic applications. Pharmaceutics 13(8). https://doi.org/10.3390/pharmaceutics13081291

Smith TT, Stephan SB, Moffett HF, McKnight LE, Ji W, Reiman D, Bonagofski E, Wohlfahrt ME, Pillai SPS, Stephan MT (2017) In situ programming of leukaemia-specific T cells using synthetic DNA nanocarriers. Nat Nanotechnol 12(8):813–820. https://doi.org/10.1038/nnano.2017.57

Su Z, Xiao Z, Wang Y, Huang J, An Y, Wang X, Shuai X (2020) Codelivery of anti-PD-1 antibody and paclitaxel with matrix metalloproteinase and pH dual-sensitive micelles for enhanced tumor chemoimmunotherapy. Small 16(7):e1906832. https://doi.org/10.1002/smll.201906832

Subhan MA, Torchilin VP (2019) Efficient nanocarriers of siRNA therapeutics for cancer treatment. Transl Res 214:62–91. https://doi.org/10.1016/j.trsl.2019.07.006

Sun L, Liu H, Ye Y, Lei Y, Islam R, Tan S, Tong R, Miao YB, Cai L (2023) Smart nanoparticles for cancer therapy. Signal Transduct Target Ther 8(1):418. https://doi.org/10.1038/s41392-023-01642-x

Sweed NM, Dawoud MHS, Aborehab NM, Ezzat SM (2024) An approach for an enhanced anticancer activity of ferulic acid-loaded polymeric micelles via MicroRNA-221 mediated activation of TP53INP1 in caco-2 cell line. Sci Rep 14(1):2073. https://doi.org/10.1038/s41598-024-52143-y

Tatar AS, Nagy-Simon T, Tomuleasa C, Boca S, Astilean S (2016) Nanomedicine approaches in acute lymphoblastic leukemia. J Control Release 238:123–138. https://doi.org/10.1016/j.jconrel.2016.07.035

Tiwari A, Saraf S, Jain A, Panda PK, Verma A, Jain SK (2020) Basics to advances in nanotherapy of colorectal cancer. Drug Deliv Transl Res 10(2):319–338. https://doi.org/10.1007/s13346-019-00680-9

Tsao LC, Force J, Hartman ZC (2021) Mechanisms of therapeutic antitumor monoclonal antibodies. Cancer Res 81(18):4641–4651. https://doi.org/10.1158/0008-5472.CAN-21-1109

Unlu A, Meran M, Dinc B, Karatepe N, Bektas M, Guner FS (2018) Cytotoxicity of doxrubicin loaded single-walled carbon nanotubes. Mol Biol Rep 45(4):523–531. https://doi.org/10.1007/s11033-018-4189-5

Waldman AD, Fritz JM, Lenardo MJ (2020) A guide to cancer immunotherapy: from T cell basic science to clinical practice. Nat Rev Immunol 20(11):651–668. https://doi.org/10.1038/s41577-020-0306-5

Wang C, Ye Y, Hochu GM, Sadeghifar H, Gu Z (2016) Enhanced cancer immunotherapy by microneedle patch-assisted delivery of anti-PD1 antibody. Nano Lett 16(4):2334–2340. https://doi.org/10.1021/acs.nanolett.5b05030

Wang Q, Shao X, Zhang Y, Zhu M, Wang FXC, Mu J, Li J, Yao H, Chen K (2023) Role of tumor microenvironment in cancer progression and therapeutic strategy. Cancer Med 12(10):11149–11165. https://doi.org/10.1002/cam4.5698

Wathoni N, Puluhulawa LE, Joni IM, Muchtaridi M, Mohammed AFA, Elamin KM, Milanda T, Gozali D (2022) Monoclonal antibody as a targeting mediator for nanoparticle targeted delivery system for lung cancer. Drug Deliv 29(1):2959–2970. https://doi.org/10.1080/10717544.2022.2120566

Wei Y, Song S, Duan N, Wang F, Wang Y, Yang Y, Peng C, Li J, Nie D, Zhang X, Guo S, Zhu C, Yu M, Gan Y (2020) MT1-MMP-activated liposomes to improve tumor blood perfusion and drug delivery for enhanced pancreatic cancer therapy. Adv Sci (Weinh) 7(17):1902746. https://doi.org/10.1002/advs.201902746

Woodman C, Vundu G, George A, Wilson CM (2021) Applications and strategies in nanodiagnosis and nanotherapy in lung cancer. Semin Cancer Biol 69:349–364. https://doi.org/10.1016/j.semcancer.2020.02.009

Wu P, Zhou Q, Zhu H, Zhuang Y, Bao J (2020a) Enhanced antitumor efficacy in colon cancer using EGF functionalized PLGA nanoparticles loaded with 5-Fluorouracil and perfluorocarbon. BMC Cancer 20(1):354. https://doi.org/10.1186/s12885-020-06803-7

Wu S, Liang F, Hu D, Li H, Yang W, Zhu Q (2020b) Determining the critical micelle concentration of surfactants by a simple and fast titration method. Anal Chem 92(6):4259–4265. https://doi.org/10.1021/acs.analchem.9b04638

Wu F, Yang J, Liu J, Wang Y, Mu J, Zeng Q, Deng S, Zhou H (2021) Signaling pathways in cancer-associated fibroblasts and targeted therapy for cancer. Signal Transduct Target Ther 6(1):218. https://doi.org/10.1038/s41392-021-00641-0

Xiao Q, Ge G (2012) Lysyl oxidase, extracellular matrix remodeling and cancer metastasis. Cancer Microenviron 5(3):261–273. https://doi.org/10.1007/s12307-012-0105-z

Xie Y, Wan B, Yang Y, Cui X, Xin Y, Guo LH (2019) Cytotoxicity and autophagy induction by graphene quantum dots with different functional groups. J Environ Sci (China) 77:198–209. https://doi.org/10.1016/j.jes.2018.07.014

Xiong J, Wu M, Chen J, Liu Y, Chen Y, Fan G, Liu Y, Cheng J, Wang Z, Wang S, Liu Y, Zhang W (2021) Cancer-erythrocyte hybrid membrane-camouflaged magnetic nanoparticles with enhanced photothermal-immunotherapy for ovarian cancer. ACS Nano 15(12):19756–19770. https://doi.org/10.1021/acsnano.1c07180

Xu F, Huang X, Wang Y, Zhou S (2020) A size-changeable collagenase-modified nanoscavenger for increasing penetration and retention of nanomedicine in deep tumor tissue. Adv Mater 32(16):e1906745. https://doi.org/10.1002/adma.201906745

Xu JJ, Zhang WC, Guo YW, Chen XY, Zhang YN (2022) Metal nanoparticles as a promising technology in targeted cancer treatment. Drug Deliv 29(1):664–678. https://doi.org/10.1080/10717544.2022.2039804

Yan S, Luo Z, Li Z, Wang Y, Tao J, Gong C, Liu X (2020) Improving cancer immunotherapy outcomes using biomaterials. Angew Chem Int Ed Engl 59(40):17332–17343. https://doi.org/10.1002/anie.202002780

Yang Z, Luo X, Zhang X, Liu J, Jiang Q (2013) Targeted delivery of 10-hydroxycamptothecin to human breast cancers by cyclic RGD-modified lipid-polymer hybrid nanoparticles. Biomed Mater 8(2):025012. https://doi.org/10.1088/1748-6041/8/2/025012

Yari H, Nkepang G, Awasthi V (2019) Surface modification of liposomes by a lipopolymer targeting prostate specific membrane antigen for theranostic delivery in prostate cancer. Materials (Basel) 12(5). https://doi.org/10.3390/ma12050756

Yenyuwadee S, Aliazis K, Wang Q, Christofides A, Shah R, Patsoukis N, Boussiotis VA (2022) Immune cellular components and signaling pathways in the tumor microenvironment. Semin Cancer Biol 86(Pt 2):187–201. https://doi.org/10.1016/j.semcancer.2022.08.004

Yin T, Fan Q, Hu F, Ma X, Yin Y, Wang B, Kuang L, Hu X, Xu B, Wang Y (2022) Engineered macrophage-membrane-coated nanoparticles with enhanced PD-1 expression induce immunomodulation for a synergistic and targeted antiglioblastoma activity. Nano Lett 22(16):6606–6614. https://doi.org/10.1021/acs.nanolett.2c01863

Yoon HY, Selvan ST, Yang Y, Kim MJ, Yi DK, Kwon IC, Kim K (2018) Engineering nanoparticle strategies for effective cancer immunotherapy. Biomaterials 178:597–607. https://doi.org/10.1016/j.biomaterials.2018.03.036

You J, Zhang R, Zhang G, Zhong M, Liu Y, Van Pelt CS, Liang D, Wei W, Sood AK, Li C (2012) Photothermal-chemotherapy with doxorubicin-loaded hollow gold nanospheres: a platform for near-infrared light-trigged drug release. J Control Release 158(2):319–328. https://doi.org/10.1016/j.jconrel.2011.10.028

Yu Q, Xue Y, Liu J, Xi Z, Li Z, Liu Y (2018) Fibronectin promotes the malignancy of glioma stem-like cells via modulation of cell adhesion, differentiation, proliferation and chemoresistance. Front Mol Neurosci 11:130. https://doi.org/10.3389/fnmol.2018.00130

Zang X, Zhao X, Hu H, Qiao M, Deng Y, Chen D (2017) Nanoparticles for tumor immunotherapy. Eur J Pharm Biopharm 115:243–256. https://doi.org/10.1016/j.ejpb.2017.03.013

Zhang B, Shen S, Liao Z, Shi W, Wang Y, Zhao J, Hu Y, Yang J, Chen J, Mei H, Hu Y, Pang Z, Jiang X (2014) Targeting fibronectins of glioma extracellular matrix by CLT1 peptide-conjugated nanoparticles. Biomaterials 35(13):4088–4098. https://doi.org/10.1016/j.biomaterials.2014.01.046

Zhao CY, Cheng R, Yang Z, Tian ZM (2018) Nanotechnology for cancer therapy based on chemotherapy. Molecules 23(4). https://doi.org/10.3390/molecules23040826

Zheng P, Wu N (2017) Fluorescence and sensing applications of graphene oxide and graphene quantum dots: a review. Chem Asian J 12(18):2343–2353. https://doi.org/10.1002/asia.201700814

Zheng F, Luo Y, Liu Y, Gao Y, Chen W, Wei K (2023) Nano-baicalein facilitates chemotherapy in breast cancer by targeting tumor microenvironment. Int J Pharm 635:122778. https://doi.org/10.1016/j.ijpharm.2023.122778

# Chapter 20
# Quantum Dots and Nanoprobes for Bioimaging

**Rajesh Kumar, Devinder Kumar, Sunil Dutt, Pankaj Kalia, and Brijesh Kumar**

**Abstract** The development of bioimaging with novel visualisation techniques to track biological processes and enhance illness detection has been aided by quantum dots (QDs) and nanoprobes. The non-invasive analysis of anatomical and functional aspects made possible by advances in bioimaging technology has completely changed medical diagnosis. Before the discovery of quantum confinement increased the potential uses of QDs in bioimaging and sensing, their research was originally concentrated on semiconductor LED applications. Nanoprobes have altered and revolutionised biomedical/clinical diagnostics as well as medical diagnostics (biomedicine) by imaging not only at the molecular level but also genetically based biochemistry. Significant growth in nanotechnology is crucial, especially in pre-clinical imaging studies that examine the application of nanoparticles for treatment and diagnosis. The formerly impractical potential of nanomaterial applications is becoming possible attributable to these research and development advances.

**Keywords** Quantum Dots · Nanoprobes · Bioimaging · Diagnosis · Biochemistry · Nanomaterials

## 20.1 Introduction

Quantum dots (QDs) and nanoprobes have promoted the development of bioimaging with new visualization methods to monitor biological processes and improve disease diagnosis. These are small crystals made of semiconductor particles,

R. Kumar (✉) · D. Kumar · S. Dutt · P. Kalia
Department of Pharmacology, Maharaja Agrasen School of Pharmacy, Maharaja Agrasen University, Solan, Himachal Pradesh, India

B. Kumar (✉)
Department of Pharmacology, Institute of Medical Sciences, Banaras Hindu University, Varanasi, India

 
V. Khanna et al. (eds.), *Nanotechnologies for Drug Delivery and Biopharmaceutical Development*, Progress in Drug Research 77,
https://doi.org/10.1007/978-3-032-00796-4_20

typically 2–10 nm in diameter, whose optical properties differ from those of larger photodetectors and depend either on the particle size or the material (Younis et al. 2020). One of the key features of quantum dots is their size-tunable fluorescence—they emit a monochromatic light but can be precisely tuned to any wavelength by simply adjusting the dot size (very important as it allows you to design quantum dots for specific experiments) and also provides a wide range of colors when stimulated with the right lighting (Lee et al. 2018). Additionally, QDs are orders of magnitude brighter and more photostable compared to traditional organic dyes, enabling superior long-term imaging with negligible chance of rapid image-based photobleaching. Quantum dots are synthesized and there are many methods depending on the shape, size or material, such as: Colloid synthesis, chemical vapor deposition or laser ablation (Matea and Mocan 2017). Due to the simple procedures and the controlled size and surface properties of the QDs, colloid synthesis is the most common method. Functionalization—the surface modification of quantum dots—is the key to improving biological compatibility and their suitability for directed biomolecular interactions (Zrazhevskiy et al. 2010). Antibodies or peptides can be conjugated to the surface of the QDs, allowing them to bind selectively to specific cells or tissues, for example, which is crucial for many imaging and diagnostic applications (Lei and Ju 2011).

Nanoprobes, often using quantum dots as tags, are fabricated and implemented to visualize the biological markers with high affinity and sensitivity. By bringing activity detection back to molecular-level monitoring, these probes can be programmed based on the target biological environment or cellular changes themselves-making them good candidates for tracking cellular processes in real time (Ingelman-sundberg and Institutet 2004). These nanoprobes—based on quantum dots that emit different colors of light depending on fluctuations in pH, temperature or certain biomolecules built into the nanoprobe surface—provide information from which insights into cell functions and disease states can be derived (Xu et al. 2018). This property is particularly useful when studying dynamic processes such as cell signaling and gene expression cascades, which require real-time data to analyze complex relationships. Quantum dots and nanoprobes are primarily used for bioimaging, including fluorescence microscopy, in vivo imaging, and multiplex assays. In contrast, fluorescence microscopy can detect cellular structures and processes with impressive resolution and impressive detail (Belal et al. 2024). Because multiplexing can be used to label different targets with different colored QDs, it is possible to simultaneously track multiple biomarkers in the same sample. Such improvement is required for the study of complex biological systems, such as tumor microenvironments, in which multiple signaling pathways and crosstalks are involved in parallel (Trindade et al. 2024). In these animals, in vivo imaging using QDs offers the opportunity to monitor biological events non-invasively. Using QD-based imaging, researchers have imaged tumor growth and metastasis in animal models, as well as responses to treatment. Quantum dots have the property of penetrating deep into the tissue and enable the imaging of structures that cannot easily be visualized using conventional methods. For example, when QDs are seeded onto nanoparticles injected into a tumor, doctors can observe how the tumors adapt and evolve in response to treatments (Feng et al. 2017).

The applications of quantum dots and nanoprobes go far beyond pure cancer imaging. QDs have been used in cardiovascular research to study blood flow dynamics and interactions of platelets with the vascular endothelium. In neuroscience, QD-based imaging had made it possible to visualize neuronal activity and track neurotransmitter release even in real time (Fan et al. 2015). This kind of versatility speaks to the wide spectrum quantum dots, which are also useful in many other areas. Unfortunately, these quantum dots and nanoprobes could be harmful and toxic to humans when used. QDs are typically made of cadmium, lead and other potentially harmful heavy metals. Therefore, it is crucial for researchers to have sufficient data on biocompatibility and possible side effects after toxicity measurements. The development of less toxic QD materials (e.g. InP and ZnSe) as well as the optimization of surface coatings have emerged to mitigate some risks (Keshavarz et al. 2018).

## 20.2 Historical Development of Bioimaging

Advances in bioimaging technology have revolutionized medical diagnostics by enabling the non-invasive examination of anatomical and functional features. It was a journey that began with the discovery of X-rays by Wilhelm Conrad Rontgen in 1895, an astronomical event that enabled a new way to see the interior of structures. Interest in the clinical use of X-rays was initially limited to the search for broken bones and some bone-related pathologies (Kadian et al. 2020). As technology advanced, these images became more sophisticated, allowing imaging support to verify standard preparation. Radiology was born with X-ray imaging, which visualized the inside of the body in real time, but struggled with the inability to distinguish soft tissues and concerns about radiation exposure (Tufani et al. 2021).

With the introduction of non-ionizing imaging (ultrasound and magnetic resonance imaging [MRI]) in the mid-twentieth century, bioimaging changed radically. Developed in the 1950s, ultrasound is a technology that produces high-frequency sound waves and uses them to create images of internal organs. It enabled safe imaging without radiation (particularly important for use in obstetrics as an alternative to X-rays). On the other hand, MRI emerged as an excellent method for soft tissue imaging and was based on magnetic fields combined with radiofrequency pulses to provide detailed anatomical images without ionizing radiation (Montalti et al. 2015). This new development was groundbreaking and allowed doctors to better recognize soft tissue structures (brain muscles, organs) and translate this morphological data into diagnoses. The use of contrast media ushered in a new era in X-ray and MRI imaging, allowing researchers to see certain tissues (or even blood vessels) better than ever before. Introduction of modern imaging procedures such as computed tomography (CT) and positron emission tomography, which, in addition to functional monitoring, also enable 3D visualization (Feng et al. 2017). With these developments in imaging technology, the need for increasingly sensitive and multifunctional agents drove research into a revolutionary new field—nanotechnology (Kadian et al. 2020).

Quantum dots (QDs) and nanoprobes were presented as promising prospects in the field of bioimaging in the late twentieth and early twenty-first centuries. While this class of imaging agents offers different optical properties compared to traditional optics such as quantum dots (QDs; semiconductor nanoparticles that emit light at specific wavelengths when excited), the advantages are likely outweighed by the limitations when small molecules and cellular delivery are required (Niu et al. 2014). For example, their size-tunable fluorescence could enable multiplexing and simultaneously image multiple different targets in a single sample, thereby improving insights into the fundamentals of complex biological phenomena at the cellular or molecular level. In addition, quantum dots are impressively stable and bright—properties that help maintain cell specificity while ensuring fluorescence duration suitable for longer imaging time points have previously been unattainable (Ma and Su 2010). Imaging techniques have also continued to evolve to improve sensitivity solutions, such as the introduction of nanoprobes, which are often constructed with different targeting ligands. These antibody or peptide functionalized probes can be used for targeted molecular imaging of individual cells or tissues to provide early diagnosis and real-time therapeutic guidance. These nanomaterial's are versatile and expandable and have paved the way for bioimaging as they enable real-time study of dynamic biological processes such as protein interactions, cellular signaling, etc., or even the movement of nanoparticles in living organisms (Li et al. 2023).

## 20.3 Fundamentals of Quantum Dots (QDs)

In the future, fast FRFQDs may be developed for use in practical medical imaging and bioanalytical techniques using other imaging probes such as quantum dots (QDs), which are novel optical labels for QD's highly selective "two-color" molecular targeting capability. BSPs with attached evidence (Xu et al. 2018). For in vivo imaging, QDs emitting in the near infrared (NIR) are preferable due to their strong absorption, narrow emission spectra, high extinction coefficients, and resistance to photobleaching, which result from large Stokes shifts combined with low cytotoxicity. These special properties make them suitable for applications in molecular diagnostics and cellular imaging (Lei and Ju 2011). Furthermore, QDs have been functionalized with organic dyes for fluorescence resonance energy transfer (FRET) in studying protein–ligand interactions. New syntheses such as in-situ amino acid placement and surface functionalization (polymer, antibody, etc.) expand the application for molecular imaging with QD targeting (Yang et al. 2019a).

**QD Developments**

The development of QDs was initially focused on semiconductor LED applications until the discovery of quantum confinement expanded their potential application in bioimaging and sensing (Kuang et al. 2011). These QDs could potentially be useful for purposes such as biological labeling, and by adjusting their precursor ratios, researchers can achieve different optical signatures. This significant labeling property of QDs by various molecular targets makes them a good candidate for use in the field of bioimaging to productively target multiple biomolecules (El-Shabasy et al. 2021; Wang and Chen 2011).

**QD Biomedical Applications**
One of the key advantages that QDs can exploit in cell and tissue imaging is their high brightness combined with photostability (up to 20–40 times brighter than organic dyes). They find specific applications in long-term tracking and in vivo imaging of cells, enabling multicolor labeling within a cell (Wang and Chen 2011). Their tunable emission spectrum and easy functionalization with biological molecules ensure high-resolution visibility at the molecular level in imaging applications (Valizadeh et al. 2012; Singh et al. 2023a). Due to their quenching nature and the principles of quantum mechanics, QDs can be used in conjunction with gene silencing strategies such as siRNA to monitor the performance of target proteins at the molecular level (Pierobon and Cappello 2012; Kumar et al. 2022a).

**Perspectives**
Hybrid approaches using QDs and biomolecular recognition elements in bioimaging. QDs provide multicolor imaging with high spatial resolution in biological materials, providing better detectability and labeling specificity (Lee et al. 2018). The ability of QDs to distinguish between healthy and diseased cells in flow cytometry is crucial for diagnostics or monitoring the effectiveness of treatments. For example, QDs have been used to stain mouse cancer cells, enabling quantitative monitoring of gene expression and micro environmental variations, aimed at improving performance in biomedical research (Birma 2016).

## 20.4 Synthesis and Functionalization of Quantum Dots

One of them is quantum dot nanostructures (QD), the most promising nanoprobes for in vitro and/or imaging applications, designed to enable highly sensitive biomarker detection on a single living cell (whether human or bacterial) (Geszke-Moritz et al. 2013). QDs possess photophysical properties not available with typical organic dyes or fluorescent proteins, making them promising imaging agents; They have broad absorption and narrow symmetric fluorescence emissions due to the quantum confinement effect, high molar absorptivity for a wider range of excitation wavelength, and enable efficient multiplexing within a specific wavelength window without overlap compared to conventional dyes such as Cy-3 (0%–4%). The long fluorescence lifetime (>20 ns) allows filtering out Raman-Stokes from photons emitted from other probe-blinding channels in biological samples, as noise can thus lead to a reduction in auto fluorescence in the spectral region with longer emission bands due to low interrogation wavelengths (Karakoti et al. 2015; Singh et al. 2024a). It had photostability up to the maximum of the absorption band, resistance to chemical degradation and produced better signal quality than conventional

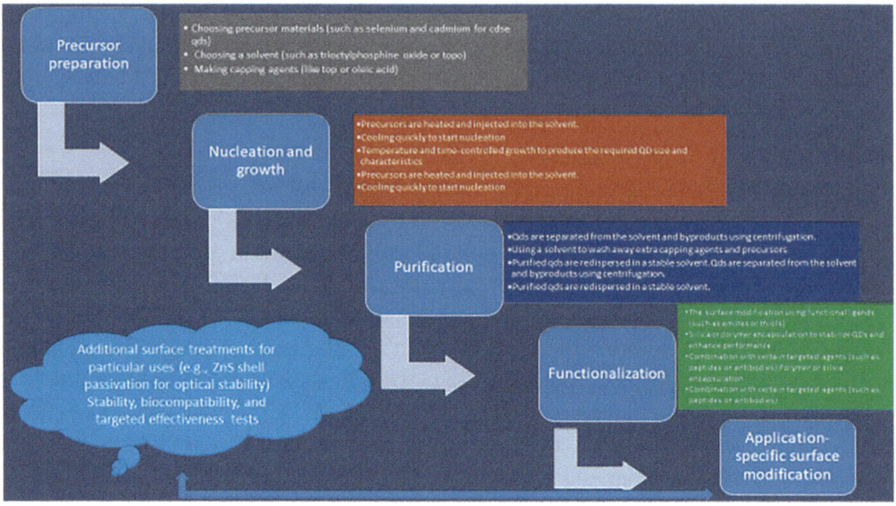


**Fig. 20.1** Synthesis and functionalization of quantum dots

chromophores with high detection limit. QDs also have adjustable maximum absorption and fluorescence emission peaks, allowing the wavelength of absorbed/emitted light to be changed as needed (Mansur et al. 2018; Dhiman et al. 2016a). This is one of the reasons why they are in great demand in biology for organic fluorescent dyes as lumiphores. This section describes the excitation principle of quantum dots and introduces its general fabrication methods for functional QDs (Fig. 20.1). This review is followed below by a brief mention of QD surface functionalization for biotic exposure and a review of the state of the art in various aspects of the production of "QD-biotic labels". Finally, improved optical properties of QD nanocomposites are addressed (Yu et al. 2018). Energy Quantification (another property of a QD) It has an important influence on the wavelength of fluorescence emission as its size increases. As a result, they are severely limited due to their small size, says Talapin: the smaller they get, the larger the band gap becomes. It is an intense and broad absorption band below the luminescence peak (higher energy light), in which this property is also present for other semiconductor nanostructures (Wang and Chen 2011; Vijaya Durga 2015).

## 20.5 Applications of Quantum Dots in Bioimaging

In recent years, new tools for imaging biological systems with particle-based probes have been developed. Metal-NP-dye conjugates can improve brightness, but these dyes also exhibit rapid fluorescence photobleaching (Martynenko et al. 2017). The 3D semiconductor nanocrystals, quantum dots (QDs), are the solution to these limitations. QDs were discovered in the 1980s and their research took off a decade later with many exciting discoveries (Fig. 20.2) (Chung et al. 2019; Kumar et al. 2021a).

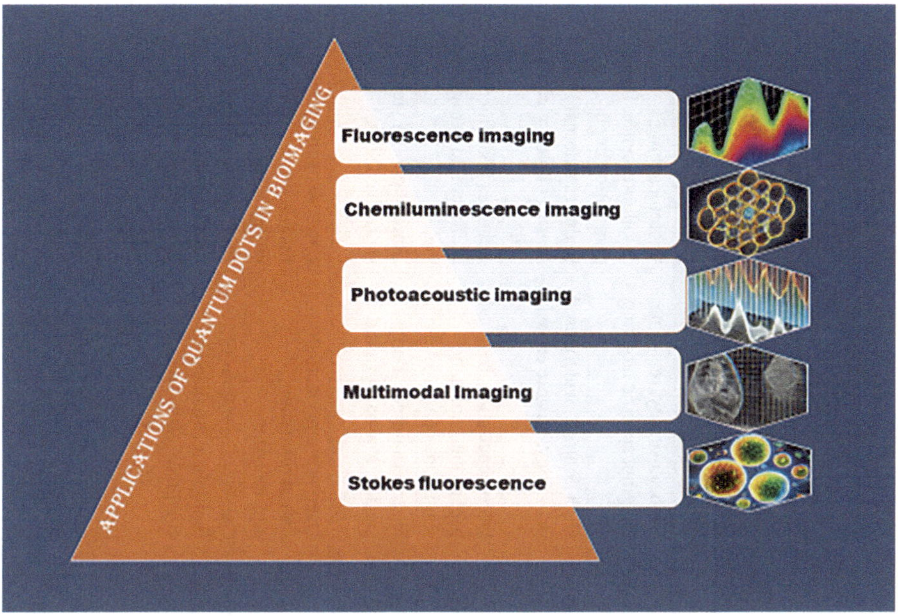


**Fig. 20.2** Applications of quantum dots in bioimaging

**Fluorescence Imaging**

Fluorescence imaging is a key technology in bioimaging. The fluorescence emission from quantum dots arises from their excitation by light of a precise wavelength (Yu et al. 2018). The emission wavelength of QDs can be changed by size and composition. Their high molar absorption coefficients ensure high fluorescence brightness with excellent stability. Quantum dots can be used to image a wide range of cellular compartments, including organelles such as mitochondria, endoplasmic reticulum, Golgi bodies, and more (Zhang et al. 2018).

Hybrid Firefly-Gold (III) quantum dot systems use different emission wavelengths to detect multiple cell components. This has the added benefit of making them sensitive to local pH changes. Scientists have developed quantum dot-based bioimaging tools for cellular and subcellular observations from single cells to entire organisms, providing important information about the expression of molecular targets in targeted regions, as well as for monitoring cell motility, protein trafficking, and drug delivery applications (Table 20.1) (Mansur et al. 2018).

**Stokes Fluorescence**

Stokes fluorescence, the ability of quantum dots to absorb light and then emit it at a longer wavelength (Birks et al. 2011). This is of great importance for bioimaging applications to avoid background signals in fluorescence microscopy. Using one wavelength for excitation and emission helps minimize interference from background photons other than the signal, enabling high quality images (Kogure et al. 2023). Such a process should be particularly useful in the context of living cells,

**Table 20.1** Biomedical applications of various techniques with advantages and challenges

| S. No. | Imaging Technique | Biomedical Applications | Examples | Advantages | Challenges |
|---|---|---|---|---|---|
| 1 | Fluorescence Imaging | Cellular and molecular imaging, cancer detection, drug delivery tracking | Fluorescence microscopy, GFP-tagged proteins in live cells | High sensitivity, real-time imaging, specific targeting | Photobleaching, autofluorescence, limited penetration depth |
| 2 | Stokes Fluorescence | Labeling biomolecules, immunofluorescence, fluorescence-guided surgery | FITC-labeled antibodies, quantum dots in diagnostics | High contrast, widely used in biology, tunable emission | Requires external light source, phototoxicity in live cells |
| 3 | Upconversion Fluorescence (Anti-Stokes) | Deep tissue imaging, biosensing, theranostics | Upconversion nanoparticles for deep tissue visualization | Deep penetration, reduced background noise, no autofluorescence | Requires high-intensity excitation, lower quantum yield |
| 4 | Chemiluminescence Imaging | Enzyme assays, oxidative stress detection, immunoassays | Luminol for reactive oxygen species detection | No external light source needed, low background noise | Limited photon yield, short-lived signal |
| 5 | Photoacoustic Imaging | Tumor imaging, blood oxygenation monitoring, brain imaging | Hemoglobin-based imaging, contrast agents for deep tissues | Deep penetration, high spatial resolution, non-ionizing | Lower sensitivity than fluorescence, requires pulsed laser |
| 6 | Multimodal Imaging | Integrating different imaging methods for enhanced diagnostics | PET/MRI, fluorescence-photoacoustic imaging | Combines strengths of multiple techniques, comprehensive data | High cost, complex instrumentation, increased processing time |

where competing signals from neighboring molecules constantly attempt to fool the receptor (Sadeghi et al. 2018; Pandey et al. 2019). The Stokes shift allows us to analyze fluorescence free of other interferences, which are completely useless for progress in fields such as medical diagnostics, drug delivery or molecular biology (Li et al. 2015; Garg et al. 2022).

**Upconversion Fluorescence (Anti-stokes)**
Anti-Stokes fluorescence or upconversion is a type of nonlinear optical transition when a material absorbs several lower energy photons and emits one higher energy photon (Dantas et al. 2002). While upconversion is less established in bioimaging, it has the potential to provide a new route to increasing the sensitivity of biological labeling (Fontenot et al. 2016). Researchers show that quantum dots doped with rare earth ions offer higher luminescence efficiency and make it possible to detect weak biological signals at lower concentrations (Avram et al. 2017; Kumar et al. 2022b).

**Chemiluminescence Imaging**
Nanochemiluminescence systems consist of luciferins, which are enzyme substrates. This is because, unlike other light production techniques, chemiluminescence is the emission resulting from an intrinsically exothermic chemical reaction that occurs without the need for an external excitation source (Koo et al. 2011). Without the need for particle conjugation, chemiluminescence imaging makes it possible to see and measure the chemical information present in biological samples (Drbohlavova et al. 2009). Chemiluminescence enables the differentiation of background signals because, unlike fluorescence-based methods, it does not depend on excitation of the sample at a broad wavelength (Niu et al. 2014; Devi et al. 2022). Compared to other tracking systems, Quantum's array for high signal-to-noise ratio and inherent photostability resulting from their longer fluorescence lifetime are enhanced by their chemiluminescent activity, which also makes them more useful (Zong et al. 2012).

**Photoacoustic Imaging**
This method, called photoacoustic (PA) imaging, combines optical and ultrasound technologies and provides new spatial information about biological systems (Shashkov et al. 2008; Kumar et al. 2022c). This advanced imaging technique involves sending short pulses of light that, when absorbed, produce sound waves to visualize tissue structures and boundaries (Guo et al. 2021; Dhiman et al. 2015a). PA imaging has shown potential for the detection and diagnosis of cancers by providing effective functional and molecular markers in many affected individuals. Consequently, PA imaging provides a new method for monitoring blood oxygen saturation and cerebral blood volume by detecting changes in tissue composition, where this can provide meaningful data for clinical diagnosis and therapy (Su et al. 2010; Binte et al. 2019).

**Multimodal Imaging**
Multimodal imaging combines various non-invasive medical imaging techniques to improve diagnostic capabilities (Koole et al. 2009). Quantum dots are used as multimodal imaging tools because they can emit different colors simultaneously (Huang

et al. 2023; Singh et al. 2024b). Multimodal techniques using fluorescence imaging (FMI) combined with photoacoustic imaging (PAI) enable a more comprehensive understanding of tumor biology by obtaining complementary molecular data (Wang et al. 2022). Integrating these capabilities into one platform enables rapid discovery and efficacy with real-time monitoring of drug delivery and therapeutic responses for cancer diagnostics, resulting in incisionless therapy (Jing et al. 2014; Dhiman et al. 2016b).

**Clinical Applications of Nanoprobes**
Imaging not only at the molecular level but also genetically based biochemistry, these nanoprobes have changed/transformed medical diagnostics (biomedicine) and biomedical/clinical diagnostics (He et al. 2022). A number of new subtypes of nanoprobes, including quantum dots, have been extensively explored, combining the best properties of high reactivity, functionality and sensitivity with the robustness inherent in all processes at a scale dictated by identity constrained to quantum mechanical laws. Therefore, they are excellent candidates for this task. Therefore, these NPs are of significant interest for a variety of preclinical and clinical procedures in imaging, cell tracking, or targeted gene/drug delivery that require biomarkers to detect early cellular molecular changes (Mé 2012). Partial imaging with deep tissue penetration and background-free images, displayed in single-cell resolution; Simultaneous multicolor emission enables tracking of multiple receptors in bioimaging applications as well as the use of multiple QDs in image-guided photothermal therapy. Quantum dots are used to treat many different diseases, such as chronic infections and cardiovascular disease; the particles are even used to treat cancer (rectal or lung cancer) (Fig. 20.3) (Wang and Chen 2011). They have also been used for in vivo and real-time molecular imaging/elevation of body temperature under room light for photothermal therapy, providing a solution for monitoring the therapeutic effect.

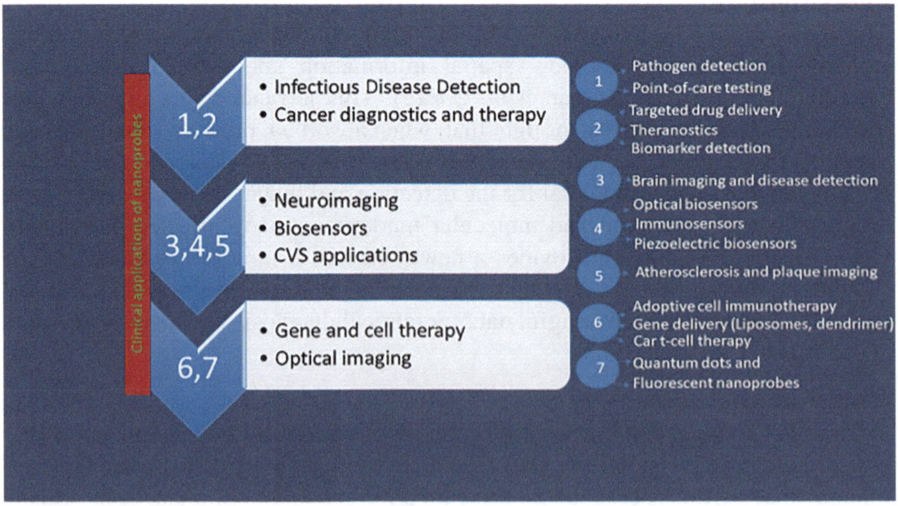


**Fig. 20.3** Clinical applications of nanoprobes

**Medicine and Disease Diagnostics**

It is important that nanotechnology continues to grow significantly, particularly with regard to preclinical imaging research into the use of nanoparticles for diagnostics and therapy. However, traditional organic fluorophores are limited by low sensitivity, poor tissue penetration, and photobleaching (Chi et al. 2012). Alternatively, quantum dots provide brighter and more photostable luminescence emission, which is of great importance for biological imaging. Highly biocompatible nanoparticles such as semiconductor quantum dots, metal nanoparticles and lanthanides are considered to have minimal risk of toxicity, while gold nanostructures improve the sensitivity of X-ray imaging, allowing lower doses to be delivered with deeper tissue penetration (Ma et al. 2021; Hamid et al. 2024).

**Molecular Imaging**

Advances in molecular imaging not only enable interactions between living cells to be observed in real time, but also make it possible to observe them dynamically (Matea and Mocan 2017). For example, traditional methods (such as PCR or immunofluorescence) only provide single snapshots, while molecular imaging with probes such as radiotracers allows visualization of biological events in REAL TIME. Imaging innovations can improve disease diagnosis, (temporal) prognosis and precision despite background activity in peripheral organs (He et al. 2022; Kumar et al. 2024a).

**Optical Imaging**

Optical endoscope imaging is typically enabled by organic dyes that are bright but not photostable. Quantum dots have emerged as a robust alternative, useful for whole-body imaging due to their ability to interact strongly with light and their resistance to photobleaching (Geszke-Moritz et al. 2013; Kumar et al. 2024b). Quantum dots have exciting potential for multiplex labeling and molecular interaction analysis, and their use in the clinic remains to be explored (Ma et al. 2021; Kumar et al. 2024c).

**Biosensors**

Its importance is demonstrated by the fact that, on the one hand, it miniaturizes crucial biosensors in order to increase their sensitivity and selectivity for real-time molecular detection (Abdellatif et al. 2022). Recent developments in probe labeling methods and the introduction of selective receptor moieties that are both highly sensitive and biocompatible have "addressed these challenges head-on." These sensor designs are still being researched to create useful diagnostic applications (Gu et al. 2018; Singh et al. 2023b).

**Therapeutics and Drug Delivery**

In drug delivery, nanoparticles enable targeted therapy by leveraging properties such as the enhanced permeability and retention (EPR) effect to reduce toxicity while increasing efficacy (He et al. 2022). Nevertheless, the clinical relevance of this so-called EPR effect in humans remains questionable, highlighting the need for systematic preclinical evaluation. They also allow imaging response to therapy, which is particularly useful in oncology, where early diagnosis and dosing accuracy can be critical (Song et al. 2010; Kumar et al. 2022d).

**Gene Therapy**

Gene therapy applications have benefited from the advent of nanoprobes that enable programmable and trackable noninvasive gene delivery (Yu et al. 2018). For imaging and therapeutic applications, superparamagnetic iron oxide nanoparticles (SPIONs) are becoming increasingly popular because they can be used as drug carriers for gene transfer in more accurate detection of sentinel lymph nodes and provide an alternative to potentially less safe methods such as viral vectors, which are currently used by used there will cause explosive cancer-invasive neurological diseases (Alaghmandfard et al. 2021; Kumar et al. 2021b).

**Photothermal and Photodynamic Therapies**

Gold nanoparticles have also been used to identify cancer cells and kill them through photothermal (PTT) or photodynamic therapies (PDT) with minimal invasion of the organism (Kurungottu et al. 2024). In particular, the use of near-infrared light to improve deep tissue delivery and treatment precision in tumor therapy has attracted considerable attention. These data will be used for further studies on biodistribution and tumor accumulation to translate these findings into clinical application (Yang et al. 2019b; Kumar et al. 2021b).

**Cancer Therapy**

In early cancer detection, nanoparticles are used to increase sensitivity at the molecular level. It turns out that they represent an alternative to standard therapy with few side effects due to their drug delivery properties (EPR effect, tumor-specific targeting) (Alaghmandfard et al. 2021). In addition to preclinical research, it is important for early detection, targeted interventions and long-term therapy monitoring (Alaghmandfard et al. 2021).

**Diagnosis and Management of Infectious Diseases**

Many infectious diseases require rapid diagnostic procedures (Chi et al. 2012). Nanoprobes enable more sensitive diagnostics faster than standard methods and allow pathogens to be identified accurately and their resistance profiles to be detected in a shorter time, which can reduce unnecessary use of antibiotics, so these tests are used to facilitate diagnosis and potentially improve treatment outcomes (Kumar et al. 2022d; Fan et al. 2024).

**Neurological Applications**

Providing functions to bypass the blood-brain barrier and monitor brain diseases, as well as potentially a substrate for cell repair, are potential applications of nanotechnology in neuroimaging (Sridhar et al. 2017). In the future, these developments could enable new non-invasive therapies for neurological diseases through combined or innovative approaches based on functionalized magnetic resonance nanoparticles (Zhang et al. 2024).

**Cardiovascular Applications**

In the area of cardiovascular health, molecular imaging could fundamentally change diagnostics and therapy. Rapid technological development, near-infrared fluorescence enables the monitoring of new generation diseases and therapeutics (Tu 2018).

## 20.6 Characterization Techniques for Quantum Dots

Quantum dots (QDs), with their unique optical and electronic properties, have attracted increasing interest due to their diverse applications in bioimaging and optoelectronics. These nanomaterials need to be characterized to better understand their behavior and improve the functionality and safety of biological systems. Various techniques are used to describe quanta, providing information about their composition, size, shape, surface properties and optical properties (Drbohlavova et al. 2009).

At the nanoscale, transmission electron microscopy (TEM) is arguably the most effective method for characterizing quantum dots. The morphology and size of QDs can then be seen in a high-quality image at atomic-scale resolutions thanks to TEM. Even the crystal structure of quantum crystals, which is essential for understanding their electrical properties, could be read with their help. Scanning electron microscopy (SEM) and TEM work together to provide a three-dimensional picture of the composition and surface morphology of quantum dots (Kadian et al. 2020). Characterizing their size distribution and aggregation is crucial, and elemental analysis confirming the composition of QDs is enabled by their ability to be combined with energy-dispersive X-ray spectroscopy (EDS). AFM (Atomic Force Microscopy) is used in bioapplications to support even higher limits of characterization.

The nearly noise-free ultrasensitivity of this technique can be exploited at the level of individual particles, and its high spatial resolution enables subnanometer imaging, enabling nm-resolution topography of QDs even in liquid environments (Bon and Cognet 2022; Kumar et al. 2017). AFM not only makes it easier to measure the mechanical properties of quantum molecules, but can also provide details about their size, shape and orientation. X-ray diffraction (XRD) is particularly important for studying the crystalline structure of quantum crystals because it helps determine the phase purity and lattice parameters required to correlate structural properties with optical behavior (Balkanloo 2023).

The optical characterization of quantum dots typically involves established methods such as ultraviolet-visible (UV-vis) spectroscopy and fluorescence spectrophotometry. When studying material with UV-Vis spectroscopy, it provides information about a material's absorption and transmission properties, while fluorescence analysis in dynamic imaging applications focuses on quantum efficiency and photostability (Drbohlavova et al. 2009; Dhiman et al. 2015b). Furthermore, the size distribution and stability of quantum dots in suspension need to be characterized using more advanced methods such as dynamic light scattering (DLS), which determines how the intensity of the scattered light varies with time upon laser irradiation of a solution of dissolved particles is important for estimating the QD interactions within biological systems (Table 20.2) (Van Veggel 2013). The surface chemistry is analyzed using Fourier transform infrared spectroscopy (FTIR), which detects characteristic vibrational transitions in molecules, thus providing information about the functional groups bound to the surface of QDs that play an important role during the bioconjugation process (Kurungottu et al. 2024).

**Table 20.2** Different techniques and their characterization

| Techniques | Description | Key Insights |
|---|---|---|
| Transmission Electron Microscopy (TEM) | High-resolution imaging; shows size and morphology of QDs | Size, shape, crystal structure |
| Scanning Electron Microscopy (SEM) | 3D images of surface morphology and composition | Size distribution, elemental composition (EDS) |
| Atomic Force Microscopy (AFM) | Topographic maps applicable in liquid environments | Size, shape, mechanical properties |
| X-ray Diffraction (XRD) | Determines crystal structure using diffraction patterns | Lattice parameters, phase purity |
| UV-Visible Spectroscopy | Analyzes absorption and transmission in the UV-Vis spectrum | Quantum yield, bandgap energy |
| Fluorescence Spectroscopy | Measures photophysical properties upon light excitation | Quantum efficiency, photostability |
| Dynamic Light Scattering (DLS) | Evaluates size distribution and stability in the suspension | Hydrodynamic radius, aggregation behavior |
| Fourier Transform Infrared Spectroscopy (FTIR) | Analyzes surface chemistry using vibrational transitions | Functional groups, surface modifications |
| Thermogravimetric Analysis (TGA) | Evaluates thermal stability based on weight changes | Thermal properties, degradation behavior |
| Mass Spectrometry (MS) | Analyzes size distribution and composition by ionization | Molecular weight, synthetic pathway insights |

Thermal stability (TS) assessment using thermogravimetric analysis (TGA) for weight changes and differential scanning calorimetry (DSC) for heat flow is widely used to assess the thermal properties of flavors and TS during storage and use (Jańczewski et al. 2011). Mass spectrometry (MS), particularly matrix-assisted laser desorption/ionization mass spectrometry, although techniques such as MALDI are used to determine the size distribution and molecular weight of quantum dots, which in turn provide an overview of how these materials were synthesized (Tang et al. 2017). The characterization of quantum dots is best applied in a broader sense, using various techniques that help create an overall picture of these unique nanomaterials. The development of new methods to characterize their quantum behavior will ensure that the applications for which they can be used in science and technology continue to expand (Chen and Liang 2014; Dhiman et al. 2015c).

## 20.7 Role of Nanoprobes in Advanced Bioimaging Techniques

Modern bioimaging techniques have transformed biology, materials science and biomedicine by providing high-resolution images that provide real-time insights into complex biological processes. Although newer techniques exist, drug discovery can greatly benefit from the use of nanoprobes such as quantum dots (QDs) (Kadian et al. 2020). Due to their unique properties, these data are still not comparable, but they provide important pharmacokinetic information for therapeutic approaches by shedding light on the way a drug interacts with its target molecules (Alaghmandfard et al. 2021).

**Synthesis of Quantum Dots**

The synthesis of quantum dots uses various techniques to tune their size, shape and composition, which can directly affect their optical properties. Synthesis techniques for spherical nanocrystals such as colloid synthesis: precursor materials are dissolved in a solvent under controlled conditions to produce uniform nanoparticles. QD growth in CVD—makes them surface functionalized (Karimian et al. 2020). Control over nanoparticle formation can also be achieved through electrochemical synthesis, in which a precursor solution is subjected to an electric current. More and more efforts to synthesize nanomaterials in an environmentally friendly manner are aimed at using non-toxic materials in low-cost, environmentally friendly large-scale production as part of sustainable practices in this field (Drbohlavova et al. 2009).

**Characterization**

The focus is on describing quantum dots and how their bioimaging works. Transmission electron microscopy (TEM) and scanning electron microscopy (SEM) are widely used to study QD size, shape, and structural morphology (Kundu 2019). By tracking the structural and spectroscopic properties of these nanoparticles, e.g. their emission wavelength or quantum yields using photoluminescence spectroscopy (PL). Additionally, methods such as dynamic light scattering (DLS) measure the stability and zeta potential of QDs in a solution and are crucial for biocompatibility (Ortiz et al. 2016). Functionalization using biomolecules (antibody conjugation to the QD surface) will expand targeted imaging capacity, enabling an almost unlimited range of potential applications (Cristian et al. 2024).

**Applications**

In the context of advanced bioimaging, particularly molecular target imaging, quantum dots play an important role in discovering cellular interactions and behaviors. QDs have been used to obtain high-contrast images of tumor microenvironments for cancer research, providing a useful tool with utility in monitoring both tumor growth and response to therapy (Arya et al. 2005). In addition, they support the tracking of transcription levels ex vivo and intervene in the in-depth analysis of cellular networks, thus revealing cellular signaling pathways, therapeutic targets, etc. Furthermore, in studying protein-protein interactions, QDs opened the possibility of also being used with Forester resonance energy transfer (FRET) by providing information about molecular mechanisms of action.

In addition, quantum dots are involved in multimodal imaging, leading to improved MRI, PET and CT for a better understanding of biological systems (Matea and Mocan 2017). By decorating these nanoparticles with specific proteins, they can be injected into the bloodstream, where their rheological properties make them preferentially adhere to vascular bifurcations rather than everywhere. With controlled surface properties, this approach could provide treatable systems that open only on selected cells and release bioactive molecules in a targeted manner when cellular signals provide appropriate permission (Jańczewski et al. 2011). Quantum dots still have a long life ahead of them in bioimaging and biotechnology as research continues to improve their properties and applications. As our nanotechnology knowledge advances, QDs are expected to become a cornerstone to advance personalized medicine and novel therapeutics at or below cellular levels/dimensions (Martynenko et al. 2017). Overall, quantum dots represent a remarkable breakthrough in bioimaging as they are synthesized and characterized on a large scale and have diverse uses, offering the opportunity to more efficiently and efficiently transform various diagnostic care modalities (Chinnathambi et al. 2013).

## 20.8 Toxicological Impact of Quantum Dots and Nanoprobes

Quantum dots (QDs) and nanoprobes have shown remarkable potential in a wide range of applications, including biomedical fields such as bioimaging, drug delivery, and therapeutic agents, due to their intriguing functional properties (Gidwani et al. 2021). Its better-known properties include resizable optical properties, high photostability, and biocompatibility, which hold great promise for real-time imaging and targeted therapeutic applications. However, in return for these advantages, the extensive use of QDs also raises specific questions regarding their potential toxicological effects on human health and the environment (Hardman 2012). Quantum dots and nanoprobes are all too often potentially toxic, so understanding this behavior is essential for human-friendly clinical/industrial use.

Quantum dots as nanoscale semiconductor particles, usually made from cadmium selenide (CdSe), cadmium telluride (CdTe) or lead sulfide (PbS). Heavy metals such as cadmium and lead are toxic and these heavy elements are found in this material (Nikazar et al. 2020). QDs are quantum objects that exhibit unique electronic and optical properties due to their nanoscale size, which is beneficial for many applications. These may be beneficial properties, but unsurprisingly they raise concerns about the toxicity of nanomaterials, particularly regarding heavy metal content and ion release (Drbohlavova et al. 2009). The toxicological effects appear to depend on their size, shape and surface chemistry as well as the chemical composition of the quantum dots. The large surface area to volume ratio of quantum dots allows for greater reactivity and, as a result, possible interactions with biological systems (Bourlinos et al. 2008).

One example is that smaller nanoparticles could more easily pass through "holes" in cell membranes directly into cells and cause unintended cellular reactions and toxicities. Some of the challenges with QDs stem from their surface functionalization. In an attempt to improve their biocompatibility and target specificity, toxic substances sometimes infiltrate the function and cause toxicity. or modify the charge by having an influence on the biological tissue interaction (Lee et al. 2018). Reactive oxygen species (ROS) are directly involved in the toxic effects of quantum dots on living systems.

After internalization into cells, QDs can generate reactive oxygen species (ROS) and cause oxidative stress, which can damage cellular components such as lipid proteins, including DNA. The increased oxidative stress caused by QDs also leads to Dysregulation of cellular homeostasis and cell cycle arrest, which ultimately leads to activation of apoptosis or programmed cell death (Zhang et al. 2015). This cytotoxic effect has been demonstrated in multiple cell types, from cancer cells to primary and stem cells. There are reports showing that exposure of cells to QDs can lead to alterations in cell viability and function, raising safety concerns regarding their use in bioapplications (Lin et al. 2023).

Another important issue related to quantum dots is their genotoxicity. It has been shown that QDs can induce DNA damage in various cell types, which can lead to a predisposition to mutations, leading to a higher risk of developing cancer. The genotoxic effects are thought to be primarily due to oxidative stress and the formation of DNA strand breaks, which prevent cells' ability to repair damaged DNA. These genotoxic effects are considered high risk, particularly when considering therapies that expose cells to QDs for prolonged periods, as long-term effects may occur that are not immediately apparent. In vivo studies have shown that QDs can be taken up by multiple organs such as liver, spleen and kidneys.

This accumulation is concerning because it is associated with chronic exposure to thallium with toxic effects and can cause organ damage over prolonged periods (Arya et al. 2005). Animal studies have shown that exposure to QDs causes histopathological changes, inflammation, and organ dysfunction. One concern is the systemic toxicity of QDs accumulating in major organs, which should be further investigated. Beyond studying the use of quantum dots and nanoprobes in human health, there is an increasingly important area, namely the study of their impact on the environment (Cristian et al. 2024). The production or mining of QDs and their release into the environment would endanger ecosystems. For example, studies have shown that QDs can be harmful to aquatic organisms such as algae, fish and invertebrates. If toxic heavy metals leak from QDs, the water source can be contaminated and harmful to aquatic life.

Cadmium-containing QDs are associated with inhibiting the growth and reproduction of fish, as well as inducing oxidative stress by disrupting lipid peroxidation processes or causing liver toxicity due to damage caused by free radicals, thereby increasing the risk of osteoporosis (Golsanamlou et al. 2021). Various behavioral changes have even been identified that can lead to ecological disruption. Another important factor that needs to be taken into account is the possible influence of quantum dots on soil organisms and plants. Studies on their potential to directly influence the food chain, particularly soil microbial communities and plant growth, are relatively new, but preliminary results already suggest that QDs may also threaten these environments. Incorporation of QDs into agricultural systems could impact their productivity in terms of crop yields and soil health in a manner that will adversely impact food security (Pacheco-liñán et al. 2020).

There are a number of solutions that can mitigate these toxicological challenges. Using artificial quantum dots can achieve this in one approach. By using less harmful materials to produce QDs such as indium phosphide (InP) or zinc selenide (ZnSe), as opposed to the heavy metal cadmium, health risks can be significantly reduced. Although these are all natural materials, the hope is that they will have similar optical properties to existing perovskites but not be as toxic (Alaghmandfard et al. 2021). Surface modification is another approach to improve the biocompatibility of QDs. This optimization of quantum dot surface chemistry can improve the biocompatibility/biofunctionality ratio, which would potentially reduce toxicity while maintaining functionality toward biological systems. It is very important that these applications of quantum dots and nanoprobes reach everyday life safely; The role of the regulatory framework is crucial. To protect human health and the environment, it is crucial to establish clear guidelines for the processing/characterization/application of nanomaterials.

Regulators need to assess the risks of nanomaterials, establish safety standards and guidelines for safe production, and deploy quantum dots in various sectors (Cristian et al. 2024). Integrated risk assessments, including conducting environmental impact studies before new processes are approved or put into safe practice. Methods are being investigated to improve the toxicological effects of quantum dots and nanoprobes, but at best they will exploit their already proven beneficial profascination properties (Zhu et al. 2020). There has been great recent progress in our understanding of the mechanisms by which nanoparticles interact with biological systems, largely due to advances in nanotechnology. This knowledge can be used to develop safer nanomaterials with better biocompatibility and reduce the health risks associated with their use (Hardman 2012).

## 20.9 Recent Progress on Clinical Application

Semiconductor nanocrystals, also known as quantum dots (QDs), are the subject of great interest in biological imaging applications due to their unique electrical and optical properties (Table 20.3). Due to their size-adjustable fluorescence, brightness and high photostability combined with broad absorption spectra, they are conceivable semiconductor nanocrystals that have achieved great success in the fields of imaging, diagnostics and targeted therapy and have made many significant advances towards clinical applications in recent years have (Abdellatif et al. 2022). Quantum dot doping is most commonly performed for bioimaging applications. Thanks to the good brightness and stability in vivo, targets can be imaged simultaneously via multiplexing. For example, cancer cell-targeting QDs modified with certain antibodies could provide high-resolution images of tumor structures and microenvironments.

This capability helps improve the accuracy of surgeries and early detection of cancer (Su et al. 2010). In addition, they increase the effectiveness of fluorescence-guided surgery (FGS) by marking tumor tissue, allowing surgeons to detect cancer

**Table 20.3** Clinical applications of quantum dots

| Application Area | Quantum Dot Type | Key Features | Clinical Outcomes/ Benefits |
|---|---|---|---|
| Bioimaging | CdSe QDs | Size-tunable fluorescence | High-resolution imaging of tumors |
| Fluorescence-Guided Surgery | QD-antibody conjugates | Tumor-targeted imaging | Improved accuracy in tumor resections |
| Targeted Drug Delivery | Polymer-coated QDs | Ligand functionalization | Enhanced drug delivery to tumor sites |
| Diagnostics | QD-based biosensors | Multiplexed detection capabilities | Early diagnosis of diseases |
| Imaging-Guided Therapy | QD-drug conjugates | Real-time therapeutic monitoring | Improved treatment efficacy and safety |

cells at their origin during a procedure and classify them as accurately as possible, leading to better patient outcomes. In addition to imaging, the QDs are also used for targeted drug delivery (Matea and Mocan 2017). Through physical encapsulation of drugs and chemical bioconjugation with targeting ligands on the surface, QDs can specifically target tumor tissue. This specific delivery method improves the effectiveness of the treatment and limits systemic side effects.

The cytotoxic effects of doxorubicin could be significantly increased with a QD-based system and there was less toxicity in normal tissue, resulting in a better quality of life for the patient (Binte et al. 2019). Another application is the use of quantum dots in various assay formats such as lateral flow assays and biosensors to detect disease biomarkers. Their high brightness and stability enable detection of low-abundance targets, which is useful in early diagnosis of diseases, including infectious diseases or cancer (Fan et al. 2024). Importantly, QD-based multiplex immunoassays in the field of personalized medicine offer the possibility of detecting multiple biomarkers from a sample simultaneously. Recent innovations include the use of quantum dots to facilitate imaging therapies, where QDs are conjugated to therapy and tracer molecules, enabling real-time tracking of therapeutic delivery and effectiveness (Cristian et al. 2024).

This feature helps tailor treatment plans to individual response profiles and toxicity, leading to better patient outcomes. However, there are challenges in the clinical implementation of quantum dots—particularly with regard to their potential toxicity and particularly those containing heavy metals. Current research results are aimed at developing biocompatible substitutes such as silicon or carbon-based QDs (Zhao et al. 2020). Another important area was the determination that QD synthesis and characterization protocols should be standardized to ensure consistent quality in clinical applications (Abdellatif et al. 2022). In addition, the clinical application of such quantum dot-based products requires clearly defined regulatory pathways so that they can be approved and used safely in clinics. This requires collaboration between researchers, regulators and industry to effectively address these issues (Yang et al. 2019a).

## 20.10 Challenges and Future Directions in Quantum Dot and Nanoprobes Bioimaging

Quantum dots (QDs) and nanoprobes have proven to be revolutionary in bioimaging, offering unique advantages such as low limit of detection, specificity, and multiplexing capacity. However, they face several challenges that need to be solved in the clinical application of these OTSs. It is important to be aware of these obstacles and explore the possible directions for further development of quantum dots and nanoprobes in biomedical imaging, diagnostics, etc (Bon and Cognet 2022). The poor biocompatibility and high toxicity of quantum dots are the biggest challenge for quantum dot bioimaging.

Traditional QDs all contain heavy metals such as cadmium and lead, both of which are highly toxic when inhaled or swallowed. The challenge now is to produce metal halide quantum dots that are not only surface coated to limit toxicity but are also biocompatible and safe for in vivo applications (Kadian et al. 2020). Efforts continue to develop alternative, safer imaging agents. Alternative options include quantum dots using non-toxic silicon or carbon-based materials that do not affect performance. In the interest of safety, future studies should investigate a detailed toxicological profile and predictive probability for these exciting materials using long-term exposure models (Korram et al. 2023).

Effectively targeting quantum dots to a specific cell or tissue is another important area where more work needs to be done. Although QDs can be decorated with targeting ligands to increase their specificity, the same biological barriers prevent them from being delivered precisely to the point of interest (sometimes due to poor access due to penetration into the ECM and cell membrane) (Kim et al. 2022). Furthermore, the biological environment can alter the behavior of nanoparticles in terms of their stability and interaction with target tissues (Li et al. 2023). To this end, future directions include the development of novel targeting strategies; For example, we can combine QDs and other therapeutic modalities in a cocktail such as antibodies or peptides to improve their target specificity. Additionally, real-time imaging should be used for dynamic biodistribution and localization. Another limitation for the clinical application of quantum dots in bioimaging is stability and photobleaching.

Although QDs are remarkably photostable compared to traditional fluorophores, they degrade with prolonged exposure to light, which in turn leads to a decrease in signal intensity and ultimately loss or corruption of images. Limitations of such systems include that this factor may limit their usefulness for certain long-term imaging studies (Fan et al. 2015). These results will motivate future studies to optimize QD formulations and surface modifications that prevent photodegradation in imaging applications. The use of protective coatings or encapsulating QDs in biocompatible compounds can improve their photobleaching resistance (Jańczewski et al. 2011).

Likewise, the complexity of biological systems leads to complications in imaging interpretation. While quantum dot-based imaging generates large data sets, interpreting the data to generate clinically relevant insights can be overwhelming. In addition, there may be idiosyncratic biological reactions leading to complications in interpreting the results. The future focus should be on merging machine learning and artificial intelligence to further improve the efficiency of image analysis (Fan et al. 2024).

These methods can help systematically reveal patterns and relationships between different variables, which will increase the diagnostic capacity of quantum dot imaging. Regulatory and ethical implications are also crucial for the further development of quantum dot nanoprobes as tools for clinical applications. Nonetheless, the emergence of nanoformulations creates a need for regulatory frameworks that ensure that these new materials are safe and always perform safely. Nanomaterials exhibit unique properties that very often become a double-edged sword and pose significant challenges to the traditional regulatory paradigms for nanomaterial synthesis and characterization as well as clinical use (Wang et al. 2022). Collaboration between researchers, regulators and industry will be paramount in overcoming these challenges in bringing together quantum dots for future clinical use. We also need standardized methods for the synthesis and characterization of quantum dots, including nanoprobes.

The differences in synthesis methods can negatively impact the size, shape, and optical properties of these spots, which are critical to their performance in bioimaging applications. Such standards for quality control of QD production and verification will improve reproducibility in research and clinical settings (Wegner and Hildebrandt 2024). It is proposed that quantum dots on the one hand and nanoprobes as a whole will significantly advance bioimaging technologies, but several challenges still need to be addressed to maximize their potential.

Important topics such as biocompatibility, target efficiency and stability of bioactivity of nanomaterials; Retrieval, data interpretation strategies used, and regulatory mechanisms require serious attention to increase research productivity in this very area. Future directions the near future of therapeutic application in liposome targeting should focus on the development of better safety materials and effective strategies for improved targeting through analytical data combined with strict regulatory guidelines (Cristian et al. 2024). This will eventually allow researchers to realize their full potential in both quantum dots and nanoprobes for diagnostic and therapeutic purposes in medicine (Bon and Cognet 2022).

## 20.11 Conclusion

This review summarizes the extraordinary capability of colloidal semiconductor nanocrystal quantum dots (QDs) for expansion and modification and thus their unique utility as emitting fluorescent probes using QD fluorescence-based NPEs. Because of their unique properties and ease of detection, these nanomaterials represent a powerful new tool to study subtle biological events in real time, such as: Selective diffusion dynamics, including ligand-receptor clustering on endocytotic vesicles or segregation of daughter DNA during cell division. To get the best results from your experiments, you need to be able to precisely control how they are carried out. By conducting a systematic examination and treating related factors, complications that may arise during research can be adequately reduced.

This allows us to unlock the true power of these cutting-edge nanoprobes. One of the critical components to improving the performance of these nanoprobes is tuning the ligand shell surrounding the QDs. Real-space imaging by moderating the composition of this protective layer allows us to overcome the low contrast limit that is a bottleneck for many other imaging techniques. Although other methods are available, none have so far shown such brilliant quality as these core-shell QD shapes doped with elements of additional types. Due to their large absorption cross section and high quantum yield during exciton generation, these novel nanoprobes show luminescence emission that is hundreds to thousands of times better than conventional QDs.

In addition, their unique stabilization properties due to the ligand shell enable them to be used even under harsh and extreme conditions. The potential of colloidal semiconductor nanocrystal quantum dots and fluorescent nanocrystal-based probes is truly amazing. These nanomaterials can study the most elusive complex processes simultaneously, bypassing all traditional limitations. With precise composition control, we are able to deliver unmatched brightness and contrast for unparalleled peak performance.

These breakthroughs in research and development are helping to realize the previously unimagined possibilities of nanomaterial applications. The future prospects lie primarily in immunoassay technology, in which the antibody specificity, in addition to the enormous sensitivity of the quantum dots, contributes to the visual detection of special proteins with the naked eye, even at very sensitive concentrations in the serum. This can be a tool for early serodiagnosis and will lead to a useful method for monitoring health conditions based on these types of innovative changes. In conclusion, colloidal semiconductor quantum dots with good photophysical properties, including ultra-broadband absorption and tunable size, have great potential for the development of bio labeled assays to elucidate the functioning of biological systems.

## References

Abdellatif AAH, Tawfeek HM, Younis MA, Alsharidah M, Al Rugaie O (2022) Biomedical applications of quantum dots: overview, challenges, and clinical potential. Int J Nanomedicine 17:1951–1970. https://doi.org/10.2147/IJN.S357980

Alaghmandfard A et al (2021) Recent advances in the modification of carbon-based quantum dots for biomedical applications. Mater Sci Eng C 120:111756. https://doi.org/10.1016/j.msec.2020.111756

Arya H, Kaul Z, Wadhwa R, Taira K, Hirano T, Kaul SC (2005) Quantum dots in bio-imaging: revolution by the small. 329:1173–1177. https://doi.org/10.1016/j.bbrc.2005.02.043

Avram D, Cojocaru B, Tiseanu I, Florea M, Tiseanu C (2017) Down- /up-conversion emission enhancement by Li addition: improved crystallization or local structure distortion? https://doi.org/10.1021/acs.jpcc.7b02897

Balkanloo PG (2023) Materials advances characterization, and application in:4272–4293. https://doi.org/10.1039/d3ma00372h
Belal A et al (2024) Nitrogen-doped carbon quantum dots as a novel treatment for black fungal bone infections (Mucormycosis): in vitro and in vivo study. Artif Cells Nanomed Biotechnol 52(1):131–144. https://doi.org/10.1080/21691401.2024.2318212
Binte A et al (2019) Photoacoustics Review article A review of clinical photoacoustic imaging: current and future trends. Photoacoustics 16:100144. https://doi.org/10.1016/j.pacs.2019.100144
Birks JB, Easterly CE, Christophorou LG, Birks JB, Easterly CE, Christophorou LG (2011) Stokes and antiStokes fluorescence of 1, 12benzoperylene in solution Stokes and anti-Stokes fluorescence of 1, 12-benzoperylene in solution. 4231(1977):10–16. https://doi.org/10.1063/1.434498
Birma I (2016) Folic acid targeted Mn: ZnS quantum dots for theranostic applications of cancer cell imaging and therapy. Int J Nanomedicine 11:413–428
Bon P, Cognet L (2022) On some current challenges in high-resolution optical bioimaging. https://doi.org/10.1021/acsphotonics.2c00606
Bourlinos AB, Stassinopoulos A, Anglos D, Zboril R, Karakassides M, Giannelis EP (2008) Carbogenic nanoparticles:455–458. https://doi.org/10.1002/smll.200700578
Chen Y, Liang H (2014) Applications of quantum dots with upconverting luminescence in bioimaging. J Photochem Photobiol B Biol. https://doi.org/10.1016/j.jphotobiol.2014.04.003
Chi X, Huang D, Zhao Z, Zhou Z, Yin Z, Gao J (2012) Biomaterials nanoprobes for in vitro diagnostics of cancer and infectious diseases. Biomaterials 33(1):189–206. https://doi.org/10.1016/j.biomaterials.2011.09.032
Chinnathambi S, Chen S, Ganesan S (2013) Silicon quantum dots for biological applications:1–20. https://doi.org/10.1002/adhm.201300157
Chung S, Revia RA, Zhang M (2019) Graphene quantum dots and their applications in bioimaging, biosensing, and therapy. 1904362:1–26. https://doi.org/10.1002/adma.201904362
Cristian RE et al (2024) In vivo assessment of hepatic and kidney toxicity induced by silicon quantum dots in mice. Nano 14(5). https://doi.org/10.3390/nano14050457
Dantas NO, Qu F, Silva RS, Ce P (2002) Anti-stokes photoluminescence in nanocrystal quantum dots. J Phys Chem B 106(30):7453–7457
Devi S, Devi S, Kumar M, Tiwari A, Tiwari V, Kaushik D (2022) Quantum dots: an emerging approach for cancer therapy. https://doi.org/10.3389/fmats.2021.798440
Dhiman R, Durga V, Rio AJ (2015a) Research Article Bioanalytical method for lornoxicam determination in human plasma by using piroxicam as internal standard by LC-MS/MS. J Chem Pharm Res 7(8):54–60
Dhiman R, Durga V, Jayasankar K, Raj S (2015b) Solid phase extraction: potential carbocisteine bioanalysis using carbocisteine 13c3 stable isotope: LC-MS/MS technique with pharmacokinetics on healthy human subjects. Der Pharm Lett 7(8):126–134
Dhiman R et al (2015c) Novel LC-MS/MS bioanalytical method validation for gabapentin in human plasma in accordance with current regulatory guidances. Int J Pharm Chem Sci 4(4):462–468, [Online]. Available: http://www.ijpcsonline.com/files/08-10-15/07-1035.pdf
Dhiman R, Kumar D, Kumar B, Pandey BL (2016a) Quantitative determination of atorvastatin, ortho-hydroxy atorvastatin, para-hydroxy atorvastatin in human plasma using rosuvastatin as internal standard by LC-MS/MS. Int J Pharm Chem Sci 4(4):487–500
Dhiman R, Kumar B, Kumar D, Kumar A (2016b) Research Article Liquid-liquid extraction method for Ziprasidone (ZRS) bioanalysis by using ZRS-D 8 (stable isotope) as internal standard. J Chem Pharm Res 8(3):722–727
Drbohlavova J, Adam V, Kizek R, Hubalek J (2009) Quantum dots — characterization, preparation and usage in biological systems:656–673. https://doi.org/10.3390/ijms10020656
El-Shabasy RM, Elsadek MF, Ahmed BM, Farahat MF, Mosleh KN, Taher MM (2021) Recent developments in carbon quantum dots: properties, fabrication techniques, and bio-applications. Processes 9(2):388
Fan J, Vankayala R, Chang C (2015) Preparation, cytotoxicity and in vivo bioimaging of highly luminescent water- soluble silicon quantum dots. Nanotechnology 26(21):1–10. https://doi.org/10.1088/0957-4484/26/21/215703

Fan Z, Liu Y, Ye Y, Liao Y (2024) Functional probes for the diagnosis and treatment of infectious diseases:1–23. https://doi.org/10.1002/agt2.620

Feng L et al (2017) A near infrared graphene quantum dots-based two-photon nanoprobe for direct bioimaging of endogenous ascorbic acid in living cells. https://doi.org/10.1021/acs.analchem.6b04943

Fontenot RS, Mathur VK, Barkyoumb JH, Mungan CE, Tran TN (2016) Measuring the anti-Stokes luminescence of CdSe/ZnS quantum dots for laser cooling applications. 9821:1–7. https://doi.org/10.1117/12.2222826

Garg R et al (2022) COVID-19: a challenge towards the sustainability of health in platform era. NeuroQuantology 20(11):2503–2519. https://doi.org/10.47750/pnr.2022.13

Geszke-Moritz M et al (2013) Thioglycerol-capped Mn-doped ZnS quantum dot bioconjugates as efficient two-photon fluorescent nano-probes for bioimaging. J Mater Chem B 1(5):698–706. https://doi.org/10.1039/c2tb00247g

Gidwani B, Sahu V, Shankar S, Pandey R, Joshi V (2021) Quantum dots: prospectives, toxicity, advances and applications. J Drug Deliv Sci Technol 61:102308. https://doi.org/10.1016/j.jddst.2020.102308

Golsanamlou Z, Soleymani J, Abbaspour S, Siahi-Shadbad M, Rahimpour E, Jouyban A (2021) Sensing and bioimaging of lead ions in intracellular cancer cells and biomedical media using amine-functionalized silicon quantum dots fluorescent probe. Spectrochim Acta – Part A Mol Biomol Spectrosc 256:119747. https://doi.org/10.1016/j.saa.2021.119747

Gu T, Zou W, Gong F, Xia J, Chen C, Chen X (2018) A specific nanoprobe for cysteine based on nitrogen-rich fluorescent quantum dots combined with $Cu^2$. Biosens Bioelectron 100:79–84. https://doi.org/10.1016/j.bios.2017.08.031

Guo T et al (2021) Boron quantum dots for photoacoustic imaging-guided photothermal therapy. https://doi.org/10.1021/acsami.0c21198

Hamid A et al (2024) Artificial intelligence (AI) based colonoscopy. https://doi.org/10.1088/978-0-7503-6134-7CH12

Hardman R (2012) Review a toxicologic review of quantum dots: toxicity depends on physicochemical and environmental factors. 114(2):165–172. https://doi.org/10.1289/ehp.8284

He H et al (2022) Molecular imaging nanoprobes for theranostic applications. Adv Drug Deliv Rev 186:114320. https://doi.org/10.1016/j.addr.2022.114320

Huang G, Luo X, Jiang G, Li Y, Dauda SA (2023) Insight into nanomaterials mediated dual-modal assay for food safety: design principles and applications. Trends Food Sci Technol 139. https://doi.org/10.1016/j.tifs.2023.104132

Ingelman-sundberg M, Institutet K (2004) Polymorphisms human drug metabolising cytochrome P450 enzymes: properties and polymorphisms:2014. https://doi.org/10.1007/s00210-003-0819-z

Jańczewski D, Tomczak N, Han MY, Vancso GJ (2011) Synthesis of functionalized amphiphilic polymers for coating quantum dots. Nat Protoc 6(10):1546–1553. https://doi.org/10.1038/nprot.2011.381

Jing L, Ding K, Kershaw SV, Kempson IM, Rogach AL (2014) Magnetically engineered semiconductor quantum dots as multimodal imaging probes:1–20. https://doi.org/10.1002/adma.201402296

Kadian S, Manik G, Das N, Roy P (2020) Targeted bioimaging and sensing of folate receptor-positive cancer cells using folic acid-conjugated sulfur-doped graphene quantum dots. Microchim Acta 187(8). https://doi.org/10.1007/s00604-020-04448-8

Karakoti AS, Shukla R, Shanker R, Singh S (2015) Surface functionalization of quantum dots for biological applications. Adv Colloid Interf Sci 215:28–45. https://doi.org/10.1016/j.cis.2014.11.004

Karimian A, Gorjizadeh N, Alemi F (2020) CRISPR/Cas9 novel therapeutic road for the treatment of neurodegenerative diseases. Life Sci:118165. https://doi.org/10.1016/j.lfs.2020.118165

Keshavarz M, Tan B, Venkatakrishnan K (2018) Multiplex photoluminescent silicon nanoprobe for diagnostic bioimaging and intracellular analysis. 1700548. https://doi.org/10.1002/advs.201700548

Kim J et al (2022) High-quantum yield alloy-typed core/shell CdSeZnS/ZnS quantum dots for bio-applications. J Nanobiotechnology 20(1):1–12. https://doi.org/10.1186/s12951-021-01227-2
Kogure T, Kawano H, Abe Y, Miyawaki A (2023) Fluorescence imaging using a fluorescent protein with a large Stokes shift dKeima. 45(2008):223–226. https://doi.org/10.1016/j.ymeth.2008.06.009
Koo H et al (2011) Nanoprobes for biomedical imaging in living systems. Nano Today 6(2):204–220. https://doi.org/10.1016/j.nantod.2011.02.007
Koole R, Mulder WJM, Van Schooneveld MM, Strijkers GJ, Meijerink A, Nicolay K (2009) Magnetic quantum dots for multimodal imaging. https://doi.org/10.1002/wnan.014
Korram J, Koyande P, Mehetre S, Sawant SN (2023) Biomass-derived carbon dots as nanoprobes for smartphone – paper-based assay of iron and bioimaging application. https://doi.org/10.1021/acsomega.3c03969
Kuang H, Zhao Y, Ma W, Xu L, Wang L, Xu C (2011) Recent developments in analytical applications of quantum dots. Trends Anal Chem 30(10):1620–1636. https://doi.org/10.1016/j.trac.2011.04.022
Kumar R, Kumar B, Kumar A (2017) Pharmacokinetic profile of antitubercular drugs and their clinical applications in therapeutic drug monitoring. Asian J Pharm 3(5):145–154, [Online]. Available: http://ajpp.in/uploaded/p71.pdf
Kumar A et al (2021a) Antiurolithiasis, antioxidant, anti-inflammatory, analgesic, and diuretic activity of ethanolic extract of seeds of Caesalpinia bonducella. Int J Pharm Investig 11(3):306–311. https://doi.org/10.5530/ijpi.2021.3.54
Kumar R et al (2021b) HPTLC and GC-MS analyses of biologically active extracts and fractions from Premna latifolia Roxb leaves reveal anti-inflammatory and analgesic effects. J Neurol Disord 12(10):811–818
Kumar A et al (2022a) Acute and subacute toxicity study of ethanolic extract of Calotropis procera (Aiton) Dryand flower in Swiss albino mice. Phytomed Plus 2(2):100224. https://doi.org/10.1016/j.phyplu.2022.100224
Kumar A et al (2022b) Toxicity (acute and subacute) assessment and in-vivo antiurolithiatic activity of ethanolic extract of Caesalpinia bonducella seed in albino Wistar rat. J Appl Pharm Sci 12(2):187–197. https://doi.org/10.7324/JAPS.2022.120220
Kumar R, Kumar B, Kumar A, Kumar A, Singh M (2022c) GC-MS analysis of phytocomponents in the methanol extract of Premna latifolia Roxb. Pharm Res 14(1):19–23. https://doi.org/10.5530/pres.14.1.4
Kumar A et al (2022d) Assessment of pain-ameliorative potential of ethanolic extract of Calotropis procera and its active components characterization by HPTLC and GC-MS analysis. 13(09):581–586. https://doi.org/10.31858/0975-8453.13.6.659-664
Kumar R, Dutt S, Goyal A, Kumar A, Kumar B (2024a) Nanotechnology in gastrointestinal endoscopy. https://doi.org/10.1088/978-0-7503-6134-7CH5
Kumar R, Singh Y, Piplani M, Bhateja P, Garg R, Kumar B (2024b) The chemistry and pharmacological study of Copaifera spp. oils. Chem Insid Spices Herbs Res Dev 4:79–116. https://doi.org/10.2174/9789815196832124040006
Kumar A, Singh P, Kumar R, Dutt S (2024c) Novel drug delivery systems for inflammatory bowel disease. https://doi.org/10.1088/978-0-7503-6134-7CH4
Kundu S (2019) Synthesis and characterization of graphene abstract: graphical abstract:1–35. https://doi.org/10.1515/psr-2019-0013
Kurungottu P et al (2024) Biodegradable nanocomposite of ZnS(Mn) quantum dots immobilized graphene oxide for bioimaging applications. Nano 8(2):150–162. https://doi.org/10.7150/ntno.87536
Lee JS, Youn YH, Kwon IK, Ko NR (2018) Recent advances in quantum dots for biomedical applications. J Pharm Investig 48(2):209–214. https://doi.org/10.1007/s40005-018-0387-3
Lei J, Ju H (2011) Fundamentals and bioanalytical applications of functional quantum dots as electrogenerated emitters of chemiluminescence. Trends Anal Chem 30(8):1351–1359. https://doi.org/10.1016/j.trac.2011.04.010

Li C et al (2015) Large stokes shift and high efficiency luminescent solar concentrator incorporated with CuInS 2/ZnS quantum dots. Nat Publ Gr:1–9. https://doi.org/10.1038/srep17777

Li S, Wei J, Yao Q, Song X, Xie J (2023) Emerging ultrasmall luminescent nanoprobes for in vivo bioimaging. Chem Soc Rev:1672–1696. https://doi.org/10.1039/d2cs00497f

Lin X, Chen T, Lin X, Chen T (2023) A review of in vivo toxicity of quantum dots in animal models a review of in vivo toxicity of quantum dots in animal models. https://doi.org/10.2147/IJN.S434842

Ma Q, Su X (2010) Near-infrared quantum dots: synthesis, functionalization and analytical applications:1867–1877. https://doi.org/10.1039/c0an00233j

Ma J, Wang X, Feng J, Huang C, Fan Z (2021) Individual plasmonic nanoprobes for biosensing and bioimaging: recent advances and perspectives. Small 17(8):1–24. https://doi.org/10.1002/smll.202004287

Mansur AAP, Mansur HS, Mansur RL, de Carvalho FG, Carvalho SM (2018) Bioengineered II–VI semiconductor quantum dot–carboxymethylcellulose nanoconjugates as multifunctional fluorescent nanoprobes for bioimaging live cells. Spectrochim Acta – Part A Mol Biomol Spectrosc 189:393–404. https://doi.org/10.1016/j.saa.2017.08.049

Martynenko IV, Litvin AP, Baranov AV, Fedorov AV, Gun YK (2017) Application of semiconductor quantum dots in bioimaging and biosensing. J Mater Chem B. https://doi.org/10.1039/c7tb01425b

Matea CT, Mocan T (2017) Quantum dots in imaging, drug delivery and sensor applications. Int J Nanomedicine 12:5421–5431

Mé J (2012) Fluorescent nanoprobes dedicated to in vivo imaging: from preclinical validations to clinical translation:5564–5591. https://doi.org/10.3390/molecules17055564

Montalti M, Cantelli A, Battistelli G (2015) Nanodiamonds and silicon quantum dots: ultrastable and biocompatible luminescent nanoprobes for long-term bioimaging. Chem Soc Rev. https://doi.org/10.1039/C4CS00486H

Nikazar S, Sivasankarapillai VS, Rahdar A (2020) Revisiting the cytotoxicity of quantum dots: an in-depth overview. Biophys Rev 12(3):703–718

Niu J, Wang X, Lv J, Li Y, Tang B (2014) Luminescent nanoprobes for in-vivo bioimaging. TrAC – Trends Anal Chem 58:112–119. https://doi.org/10.1016/j.trac.2014.02.013

Ortiz SNC, Ospino EM, Cabanzo R (2016) Spectroscopy characterization and quantum yield determination of quantum dots spectroscopy characterization determination of quantum dots and quantum yield. https://doi.org/10.1088/1742-6596/687/1/012097

Pacheco-liñán PJ et al (2020) Functionalized CdSe/ZnS quantum dots for intracellular pH measurements by fluorescence lifetime imaging microscopy functionalized CdSe/ZnS quantum dots for intracellular measurements by fluorescence lifetime imaging microscopy pH. https://doi.org/10.1021/acssensors.0c00719

Pandey H, Srivastava S, Kumar A, Kumar R, Bhusan Tripathi Y (2019) Pharmacokinetics study of Puerarin absorption in blood after consumption of Pueraria tuberosa water extract (PTWE) by rats. Explor Anim Med Res 9(2):188–196

Pierobon P, Cappello G (2012) Quantum dots to tail single bio-molecules inside living cells ☆. Adv Drug Deliv Rev 64(2):167–178. https://doi.org/10.1016/j.addr.2011.06.004

Sadeghi S et al (2018) Functional nanostructured materials (including low-D carbon) Stokes-shift-engineered indium phosphide quantum dots for efficient luminescent solar concentrators. https://doi.org/10.1021/acsami.7b19144

Shashkov EV, Everts M, Galanzha EI, Zharov VP (2008) Quantum dots as multimodal photoacoustic and photothermal contrast agents:7–9

Singh A, Prasad LB, Shiv K, Kumar R, Garai S (2023a) Synthesis, characterization, and in vitro antibacterial and cytotoxic study of Co(II), Ni(II), Cu(II), and Zn(II) complexes of N-(4-methoxybenzyl) N-(phenylethyl) dithiocarbamate ligand. J Mol Struct 1288:135835. https://doi.org/10.1016/j.molstruc.2023.135835

Singh M et al (2023b) Pharmacokinetic screening to estimate the drug likeliness characteristics of selected herbal anticancer drugs. Res J Pharm Technol 16(7):3422–3426. https://doi.org/10.52711/0974-360X.2023.00566

Singh A et al (2024a) Synthesis, crystal structure and screening for anticonvulsant and antianxiety activities of three new Ni(II), Cu(II), and Zn(II) dithiocarbamate complexes. J Mol Struct 1298(P2):137052. https://doi.org/10.1016/j.molstruc.2023.137052

Singh A et al (2024b) Dithiocarbamate-based novel anti-histaminic agents: synthesis, characterization, crystal structure and thermal study. Dalton Trans. https://doi.org/10.1039/d4dt01777c

Song S, Qin Y, He Y, Huang Q (2010) Functional nanoprobes for ultrasensitive detection of biomolecules:4234–4243. https://doi.org/10.1039/c000682n

Sridhar S, Mishra S, Gulyás M, Padmanabhan P (2017) An overview of multimodal neuroimaging using nanoprobes. https://doi.org/10.3390/ijms18020311

Su JL et al (2010) Advances in clinical and biomedical applications of photoacoustic imaging. Expert Opin Med Diagn 4(6):497–510

Tang Z, Lin Z, Li G, Hu Y (2017) Amino nitrogen quantum dots based nanoprobe for fluorescence detection and imaging for cysteine in biological samples amino nitrogen quantum dots based nanoprobe for fluorescence detection and imaging for cysteine in biological samples. https://doi.org/10.1021/acs.analchem.7b00284

Trindade FCS, De Souza IG, Pereira G, Pereira GAL, Raimundo IM, Pereira CF (2024) A surface-enhanced infrared absorption spectroscopy (SEIRA) multivariate approach for atrazine detection. Spectrochim Acta – Part A Mol Biomol Spectrosc 322:124867. https://doi.org/10.1016/j.saa.2024.124867

Tu Y (2018) Multimodality molecular imaging of cardiovascular disease based on nanoprobes. 150008:1401–1415. https://doi.org/10.1159/000492251

Tufani A, Qureshi A, Niazi JH (2021) Iron oxide nanoparticles based magnetic luminescent quantum dots (MQDs) synthesis and biomedical/biological applications: a review. Mater Sci Eng C 118:111545. https://doi.org/10.1016/j.msec.2020.111545

Valizadeh A, Mikaeili H, Samiei M, Farkhani SM, Zarghami N (2012) Quantum dots: synthesis, bioapplications, and toxicity. Nanoscale Res Lett 7(1):19–21

Van Veggel FCJM (2013) Near-infrared quantum dots, their delicate synthesis, challenging characterisation, and exciting potential applications. https://doi.org/10.1021/cm4021436

Vijaya Durga RD (2015) Determination of S-amlodipine in human plasma using tizanidine as internal standard by Lc/Ms/Ms method. J Bioanal Biomed 07(03):91–96. https://doi.org/10.4172/1948-593x.1000130

Wang Y, Chen L (2011) Quantum dots, lighting up the research and development of nanomedicine. Nanomed Nanotechnol Biol Med 7(4):385–402. https://doi.org/10.1016/j.nano.2010.12.006

Wang G et al (2022) Sensitive and specific detection of breast cancer lymph node metastasis through dual – modality magnetic particle imaging and fluorescence molecular imaging: a preclinical evaluation. Eur J Nucl Med Mol Imaging:2723–2734. https://doi.org/10.1007/s00259-022-05834-5

Wegner KD, Hildebrandt N (2024) Trends in analytical chemistry near infrared quantum dots for biosensing and bioimaging. Trends Anal Chem 180:117922. https://doi.org/10.1016/j.trac.2024.117922

Xu Q et al (2018) Two-dimensional quantum dots: fundamentals, photoluminescence mechanism and their energy and environmental applications. Mater Today Energy 10:222–240. https://doi.org/10.1016/j.mtener.2018.09.005

Yang Z et al (2019a) Recent advances in quantum dot-based light-emitting devices: challenges and possible solutions. Mater Today 24:69–93. https://doi.org/10.1016/j.mattod.2018.09.002

Yang Z et al (2019b) Advances in nanomaterials for use in photothermal and photodynamic therapeutics (Review):5–15. https://doi.org/10.3892/mmr.2019.10218

Younis MR, He G, Lin J, Huang P (2020) Recent advances on graphene quantum dots for bioimaging applications. 8:1–25. https://doi.org/10.3389/fchem.2020.00424

Yu L, Lin H, Lu X, Chen Y (2018) Multifunctional mesoporous silica nanoprobes: material chemistry–based fabrication and bio-imaging functionality. Adv Ther 1(8). https://doi.org/10.1002/adtp.201800078

Zhang D, Wei Y, Chen K, Zhang X, Xu X, Shi Q (2015) Biocompatible reactive oxygen species (ROS) – responsive nanoparticles as superior drug delivery vehicles. https://doi.org/10.1002/adhm.201400299

Zhang Q et al (2018) Cancer-targeting graphene quantum dots: fluorescence quantum yields, stability, and cell selectivity. 1805860:1–11. https://doi.org/10.1002/adfm.201805860
Zhang J, Yu L, Yu M, Yu D, Chen Y, Zhang J (2024) Engineering nanoprobes for magnetic resonance imaging of brain diseases. Chem Eng J 481:148472. https://doi.org/10.1016/j.cej.2023.148472
Zhao X, Gao W, Zhang H, Qiu X, Luo Y (2020) 20. Graphene quantum dots in biomedical applications: recent advances and future challenges. Elsevier Inc. https://doi.org/10.1016/B978-0-12-816699-4.00020-7
Zhu C et al (2020) Progress in natural science: materials international recent advances in non-toxic quantum dots and their biomedical applications. Prog Nat Sci Mater Int 29(6):628–640. https://doi.org/10.1016/j.pnsc.2019.11.007
Zong C, Wu J, Wang C, Ju H, Yan F (2012) Chemiluminescence imaging immunoassay of multiple tumor markers for cancer screening. Anal Chem 84(5):2410–2415
Zrazhevskiy P, Sena M, Gao X (2010) Designing multifunctional quantum dots for bioimaging, detection, and drug delivery:4326–4354. https://doi.org/10.1039/b915139g

# Chapter 21
# Applications of Computer-Aided Drug Design Tools in Nanotechnologies for Drug Delivery Systems; Discovery and Development

**Heyam Mohamed Ali Sid Ahmed and Mohamed Ahmed Mohamed Ali**

**Abstract** The integration of Computer-Aided Drug Design (CADD) tools with nanotechnologies has enhanced the era of drug discovery, drug delivery systems, research and development. CADD tools, including its major approaches; structure-based drug design (SBDD) and ligand-based drug design (LBDD), streamlining the process of Hit identification, lead optimization, and validation of novel therapeutic agents. These in silico methods aid to the rational drug design by predicting their molecular interactions with biological targets, thus enhancing the efficiency and accuracy of the basic steps in the drug discovery process. Nanotechnology, with its unique properties such as high surface-area-to-volume ratio and tunable surface chemistry, provides pioneering solutions for targeted drug delivery, improving the bioavailability and therapeutic index of drugs. The synergy between CADD tools; such as pharmacophore modeling and molecular dynamic simulations, with nanotechnology not only fast-tracks the drug discovery process but also boosts the accuracy and efficiency of targeted drug delivery, paving the way for innovative therapeutic solutions. This chapter explores the applications of CADD tools in the design and development of nanotechnology-based drug delivery systems, reviewing hot advancements, challenges, and future perspectives in this interdisciplinary field.

H. M. A. Sid Ahmed (✉)
Department of Pharmaceutical Chemistry, Faculty of Pharmacy, Al Neelain University, Khartoum, Sudan
e-mail: Heyam.sidahmed@neelain.edu.sd

M. A. M. Ali
Department of Computer Sciences, Faculty of Mathematical Sciences and Informatics, University of Khartoum, Khartoum, Sudan

Information Technology Department, University of Technology and Applied Sciences, Sur, Sultanate of Oman

 
V. Khanna et al. (eds.), *Nanotechnologies for Drug Delivery and Biopharmaceutical Development*, Progress in Drug Research 77,
https://doi.org/10.1007/978-3-032-00796-4_21

## Abbreviations

| | |
|---|---|
| AChE | Acetylcholinesterase |
| ADMET | Absorption, Distribution, Metabolism, Excretion, and Toxicity |
| AI | Artificial intelligence |
| AML | Myeloid leukemia |
| BBB | Blood-brain barrier |
| BChE | Butyrylcholinesterase |
| BSA | Bovine serum albumin |
| CADD | Computer-Aided Drug Design |
| CCL18 | C-C motif chemokine 18 |
| CD35 | Cluster of differentiation 35 |
| CGMD | Coarse-grained molecular dynamics simulations |
| CoMFA | Comparative Molecular Field Analysis |
| CoMSIA | Comparative Molecular Similarity Indices Analysis |
| CSCs | Cancer stem cells |
| CuO NPs | Copper oxide nanoparticles |
| cys-GnRH-I | Cysteine-tagged modification of GnRH-I |
| DL | Deep learning |
| DLS | Dynamic Light Scattering |
| DMH | Dimethyl hydantoin |
| ENMs | Engineered nanomaterials |
| EPR | Enhanced permeability and retention |
| EVs | Extracellular vesicles |
| FCC | Face-centered cubic |
| $Fe_3O_4$ NPs | Iron oxide nanoparticles |
| GnRH-I | Gonadotropin-releasing hormone I |
| hERG | Human ether-a-go-go-related gene |
| HSA | Human serum albumin |
| ICAM-1 | Intercellular Adhesion Molecule 1 |
| IL1B | Interleukin 1 beta |
| IL8 | Interleukin 8 |
| kcal/mol | The amount of energy (in kilocalories) associated with one mole of a substance. |
| LBDD | Ligand-based drug design |
| MD | Molecular Dynamics simulations |
| ML | Machine learning |
| MLR | Multiple linear regression |
| $Mn_2O_3$ NPs | Manganese oxide nanoparticles |
| MOE | Molecular operating environment |
| NF-κB | Nuclear factor kappa-light-chain-enhancer of activated B cells |
| nm | Nanometer: A unit of length equal to one billionth of a meter. |
| NMR | Nuclear Magnetic Resonance |
| NONPs | Nitric oxide-releasing nanoparticles |

| | |
|---|---|
| NPs | Nanoparticles |
| PbO | Lead oxide |
| PDB | RSCB Protein Data Bank |
| PIGF | Phosphatidylinositol-glycan biosynthesis class F protein |
| PMA-TiO2 NPs | Pyrrole and methyl anthranilate (PMA) with TiO2 nanoparticles |
| PS-80-CUR-NP | Polysorbate-80 coated, polymeric curcumin nanoparticles |
| QSAR | Quantitative structure–activity relationships |
| ReaxFF | Reactive force fields |
| RF | Random forest |
| RNAi | RNA interference |
| ROS | Reactive Oxygen Species |
| SBDD | Structure-based drug design |
| SDS | Sodium dodecyl sulfate |
| siRNA | Small interfering RNA |
| SVM | Support vector machines |
| TiO2 NPs | Titanium dioxide nanoparticles |
| ZnO NPs | Zinc oxide nanoparticles |
| °C | Degrees Celsius, a unit of temperature |
| 3D QSAR | 3D Quantitative Structure-Activity Relationship |

## 21.1 Introduction

### *21.1.1 Overview of Computer-Aided Drug Design (CADD)*

CADD has developed as a key asset in the pharmaceutical industry, renovating the processes of drug discovery and development. Since its inception in the 1970s, CADD has significantly progressive, forced by the speedy development of computer technology. Despite facing initial doubt and time of fruitless, CADD has matured into a central component of current drug discovery (Parikh et al. 2023).

Actually, the rise of CADD was driven by two pivotal advancements: the growth of structural biology field, which uncovered the crystal and three-dimensional structures of macromolecules, and the rapid growth in computational capabilities, empowering complex simulations at molecular level to be achieved more quickly. CADD is primarily divided into two main categories: structure-based drug design (SBDD) and ligand-based drug design (LBDD). SBDD utilizes the crystal three-dimensional structure of macromolecules (i.e., protein, DNA) to understand how possible drugs or small molecules can interact with it. In contrast, and in the absence of the high resolutioner 3D structure of the target, LBDD utilized known drug molecules and their pharmacological profiles as templates to design new drug candidates (Fig. 21.1) (Niazi and Mariam 2023).

In the absence of a crystal structure, homology modeling or de novo prediction models can be utilized to find small molecules with good binding affinity to the

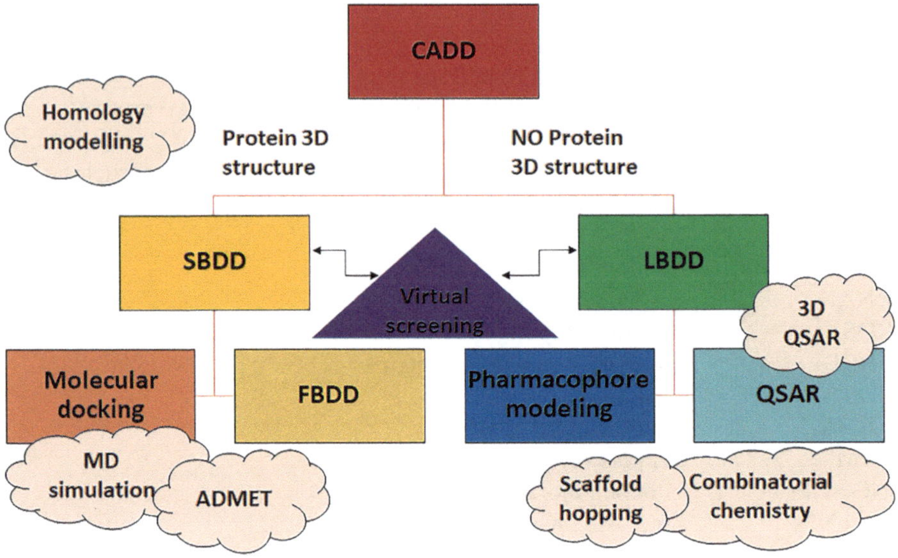


**Fig. 21.1** Schematic summary of the main computational approaches in drug discovery process

target protein, (Hughes et al. 2011). Recent years there are tremendous number of open databases accessible for drug research and development. These repositories are updated regularly with new uploaded data; some, like the ZINC and PubChem databases, include tens of millions of chemicals (Villoutreix et al. 2007). The availability of large databases fast-tracks the process of chemical identification to verify therapeutic targets, which can assist unravel numerous unmet needs in drug discovery and design. As a result, innovative software and prediction algorithms are encouraged to be developed. Moreover, quantitative structure–activity relationships (QSAR) or classification models that predict compound activities based on their structures may be found using large bioactivity data sets. These calculations can aid in drug ADMET prediction, molecular target elucidation, and possible therapeutic repurposing. (Nantasenamat and Prachayasittikul 2015). In general, it can be said that the key elements of CADD include target identification, where biological targets like proteins or enzymes involved in the progression of diseases are identified; hit identification, which focuses in identifying compounds that docked with the target through high-throughput or virtual screening; lead optimization to refine hit compounds for better efficacy, selectivity, and pharmacokinetics; SBDD using the target's three-dimensional structure to design and optimize compounds; LBDD screened chemical spaces for those similar to known ligands and their pharmacological profiles; ADMET screening to evaluate pharmacokinetics properties ensuring safety and effectiveness; computational methods like molecular docking, molecular dynamics, and QSAR modeling to predict and analyze compound- target interactions and properties; data integration, simulation, all working together to rationalize drug design (Anwar et al. 2021).

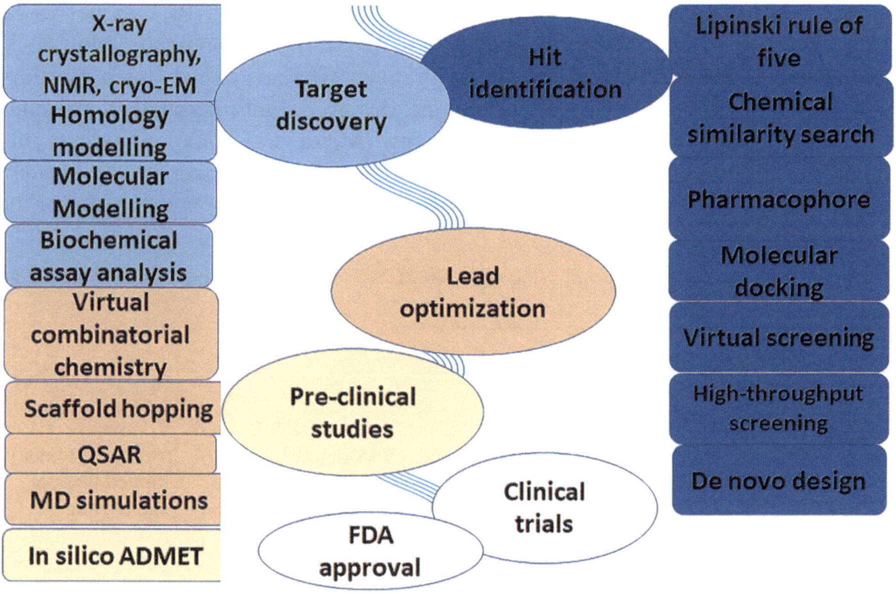


**Fig. 21.2** Conceptual map of CADD tools in drug discovery pipeline: The conceptual map provides a summary of the experimental and computational methodologies used in the drug discovery process. It highlights the integration of various techniques to streamline the efficiency of drug discovery. Target identification comprises identify disease modulating target protein and define protein 3D structure. Hit identification is the screening for ligand with pharmacological activity. Lead optimization is the stage for hit to lead conversion. The pre-clinical studies for evaluating pharmacokinetic properties. Finally, the clinical trials for evaluating safety, efficacy and adverse effects. The use of homology modeling or de novo prediction models when protein-crystal structures are unavailable

CADD is now at the cutting edge of drug discovery, becoming an indispensable tool for drug validation, optimization, and identification. The rational development of novel therapeutic agents is been achievable via the interdisciplinary research in the eras of computational chemistry, structural biology, molecular biology, and organic chemistry. (Kumari and Hasija 2024). A graphical abstract on the experimental and in silico approaches as utilized to the drug discovery process is summarized in (Fig. 21.2).

### *21.1.2 Integration of CADD with Nanotechnology Drug Delivery*

Recent advancements in nanotechnology and the application of nanomedicines for targeted drug delivery have increase more consideration to nanoparticles (NPs) for drug discovery and pharmaceutical applications. Currently, several NPs with special and extraordinary physical properties are being incorporated in pharmaceutical industry, including liposomes as well as inorganic such as gold NPs, polymer-based NPs, silica NPs, dendrimers, magnetic NPs, and carbon nanotubes (Table 21.1) (Bunker and Róg 2020). Various FDA approved NPs are already used for their unique therapeutic benefits, for instant the design of antimicrobial compounds (Fatima et al. 2021), bioimaging developments (Mirza et al. 2020), and the highly innovated formulations for anticancer therapy (Haley and Frenkel 2008). Though, one of the biggest problems in drug discovery pipeline is the non-specificity of the

**Table 21.1** Major classification of NPs, subtypes of each major class, and their highlights and challenges

| Major classification of NPs | Types of each major class | subtypes of each major class | Advantages | Disadvantages |
|---|---|---|---|---|
| Organic | Protein based | Collagen NPs, albumin NPs, gelatin NPs, casein NPs, why protein NPs, plant protein NPs | Biocompatibility and biodegradability, flexibility in design, surface modification | Complex synthesis, toxicity, limited thermal and chemical resistance |
| | Carbohydrate based | Starch NPs, dextran NPs, Cyclodextrin NPs | | |
| | Lipid based | Liposomes, solid-lipid NPs, nanostructured lipid carrier NLCs | | |
| | Polymeric | Chitosan NPs, alginate NPs, poly(lactic acid) PLA NPs and poly(lactic-co-glycolic) acid PGA NPs | | |
| Inorganic | Magnetic | Iron oxide NPs, gold NPs, silver NPs | Unique electrical, magnetic, optical properties, variability in size, structure, geometry, catalytic activity | Agglomeration and stability issues, complex synthesis and high cost, toxicity |
| | Ceramic | Silica NPs, titanium oxide NPs, alumina NPs | | |
| | Semiconductor | Zinconium dioxide NPs, zinc oxide NPs | | |
| Carbon based | Graphene | Graphene oxide NPs | Exceptional mechanical properties, thermal conductivity, chemical stability, | Toxicity, aggregation and dispersion concern, complex and costly production. |
| | Carbon quantum dots | – | | |
| | Carbon black | – | | |
| | Fullerenes | – | | |

Extracted from Joseph et al. (2023)

drugs that could bind to multiple targets leading to serious side effects and reduce their efficacy. Here is comes the rescue role of NP drug delivery system where NP can precisely reach the targeted disease cell without any effect on the healthy one (Chen and Liu 2012; Thota and Jiang 2015). Basically, the word 'nanotechnology' describes a material having unique properties of nanostructure, which is on a scale from 1 to 100 nm. Additionally, these unique properties which come from their high surface-to-volume ratios can as well render NPs highly toxic to multicellular organisms (Khorsandi et al. 2021).

The integration of CADD with nanotechnology offers a promising approach to transform drug delivery systems. Nanotechnology-based drug delivery has made significant forward steps in overcoming many important medical issues such as better improved formulation to precisely targeting the site of many disease, lessen drug toxicity, and enhanced therapeutic efficacy (Simonazzi et al. 2018). In particular, nanomedicine has been developed remarkably due to the emergence of advanced nanodrugs, which are designed to operate at the nanometer scale with unique engineered properties that offer better therapeutic effects (Mu et al. 2024). Mesoporous silica NPs are the best example for such innovation since they have various and multifunctional nano-based drug delivery system. Mesoporous silica NPs can be designed to target the cells of disease site specifically without affecting the normal and healthy cells, induce their effects as magic bullets with improved safety profile of the therapeutic agents. Likewise, a wide variety of other organic, inorganic, polymeric, and metallic nanostructures were discovered as targeted and controlled drug delivery carriers (N. Singh et al. 2023). These nanocarriers can boost the drug-likeness properties such as the solubility and bioavailability of poorly soluble drugs, rendering them to be more efficient drugs for treating various diseases.

The strategic application of CADD methods can additionally augment the capabilities of these nano-based delivery systems. Computational modeling and simulation tools can be leveraged to rationally design nanocarriers with ideal physicochemical features, such as size and complementary chemical geometry (Ahmad et al. 2023). Ultimately, the therapeutic index will be highly improved by optimizing the intended effects and at the same time reducing any unfavorable off-target or toxic effects (Sharma et al. 2024).

By integrating the powerful and advantages of the predictive capabilities of CADD with the unique properties of nanomaterials, drug developer can design targeted, controlled, and personalized drug delivery systems, and even more formulate highly engineered drug delivery system that can address many complex medical issues (Khong et al. 2020). For instant, issues like high lipophilicity or toxicity may impede the development of new drugs. Here where nanoparticles come in handy, using various types of nanoparticle delivery system to resolve such problems. The most common and significant example in this regard is patisiran, the first RNA designed to treat polyneuropathy brought on by familial transthyretin-mediated amyloidosis. Patisiran works by inhibiting the development of amyloid fibrils using lipid nanoparticles to encapsulate small interfering RNA (siRNA) (Padda et al. 2022; Suzuki et al. 2021).

## 21.2 Application of CADD Key Elements in Nanotechnology

### *21.2.1 Molecular Modeling and Structural- Based Drug Design in Nanotechnology*

Molecular modeling is the term that refer to all computational methods employed to predict the structure and behavior of the compounds. Molecular modeling is used widely in many fields including structural biology to understand and see at molecular bases different molecules in different sizes, ranging from very small molecules to macromolecules such as proteins, RNA, and DNA. To accomplish molecular modeling, the concepts of molecular mechanics are used, in particular the force fields that deal with atoms as particles with charges and potential energy. The molecules are represented in quantum mechanics by applying wave functions at the atomic and subatomic scales. This will display the simulation of a very detailed chemical structure moving within a definite period ranging from millisecond-long simulations to minutes (Sarı 2020).

The majority of drugs are hydrophobic in nature, the things that reduce their solubility and bioavailability. NPs have been developed comperhensively to resolve this issue by changing the nature of such hydrophobic drugs and making them more soluble. The compatibility and the affinity between NPs and hydrophobic drugs could be enhance and adjusted with the aid of computational molecular modelling. Such visualization can help even to assess the load process of the drug as well as the stability of the formulated NPs-drug (Huynh et al. 2012).

In terms of affordability and cost effectiveness, computer-aided drug design tools can effectually reduce the need for performing many experiments with diverse of NPs-drug formulations. One of these tools is the molecular dynamics and Monte Carlo simulations. Essentially, these simulations offer quantitative measurements that are very challenging to be obtained through conventional methods (Thota and Jiang 2015).

#### 21.2.1.1 Database Repository of NPs

Selecting the best NPs to be formulated with a drug is a challenging task. Therefore, gathering all NPs in one database will make it easier to assess and identifying computationally the suitable NP scaffold and their functional group for encapsulation releasing particular drugs. Many efforts have been made to create such database of NPs called the Collaboratory for Structural Nanobiology (CSN). This database offer access to the 3D structures, physicochemical properties of NPs, among other features (Gonzalez-Ibanez et al. 2009). Same like the Protein Data Bank (PDB), the CSN was designed to display the annotation, curation, and validation of nanoparticle structures, helping in correlation between the NPs structures with the prediction of the possible toxicity and physiochemical properties. Unfortunately, CSN is currently not exist anymore on the web. The Nanomaterial Registry is another database that gathered all available literature and data about various classes of NPs (Ostraat et al. 2013). eNanoMapper is additional NPs database that specialized in offering comprehensive data about NPs safety profile (Jeliazkova et al. 2015). Despite of many efforts to build and create repository specialized in NPs atomic 3D structure, still no such data is available yet. A resource like this would be very helpful for scientists to determine a suitable scaffold for molecular simulations (Roco et al. 2011). Certainly, defining a suitable scaffold will give rise for consequence queries; if the NPs and drug will produce a stable formulation? Whether the cargo within the target tissue will be released efficiently by the nanoparticle? How the nanoparticle will influence the loaded and the release process? All these questions could be clarified via using molecular modeling, the thing that will further clarify the nanoparticle synthesis and structure, drug-NPs interactions and release, and studying drug-NPs adsorption on membrane surfaces.

### *21.2.2 Evaluating Nanoparticle Formation and Morphology*

The shape of NPs influences more than only how they interact with drugs. The nanoparticle's geometry and structure are crucial for the cell and membrane interaction. According to Gratton's research employing NPs larger than 100 nm, Rod-shaped NPs exhibit the maximum uptake, followed by spheres, cylinders, and cubes (Gratton et al. 2008). Because excretion is influenced by nanoparticle size, the kidneys can eliminate particles with diameters less than 6 nm (Choi et al. 2007). Oversized NPs (>200 nm) accumulate within the liver and spleen, where the MPS cells break down them. It has also been shown that, in comparison to bigger NPs, smaller ones exhibit lessened ligand-to-receptor interaction. On the other hand, even bigger ones may function as clustering agents, preventing other NPs from binding. In this regard, it has been proposed that the ideal ligand density for NPs is between 30 and 50 nm (Choi et al. 2007).

To evaluate the morphology of well-researched amphiphilic block copolymers, various molecular modeling techniques have been performed. In diluted solution, those NPs may self-assemble into a variety of shapes, including micelles, bilayers, and more, based on the size and physicochemical properties of their constituent parts. By sampling across longer durations, coarse-grained techniques may aid in achieving this goal (Duarte et al. 2019). For example, the most researched method in the literature was a coarse-grained technique which was created by Marrink, and his fellows (Marrink et al. 2004); where four to one mapping was used in the MARTINI coarse grained force field, in other words, four heavy atoms are regarded as a single bead. Additionally, it accounts for various types of interaction beads, including polar, non-polar, apolar, and charged. Similar to classical force fields, interactions between beads encompass both bonded and non-bonded terms. However, MARTINI has certain limitations that need to be considered. For instance, systems tend to freeze at low temperatures, electrostatic interactions beyond 1.2 nm are not taken into account, and transitions to secondary structures in proteins cannot be addressed. On the other hand, the MARTINI model has demonstrated promising outcomes in the investigation of lipids and amphiphilic compounds, making it best suited for investigating micelles, liposomes, polymers, etc. Márquez-Miranda et al. used the MARTINI technique to study the self-assembly of amphiphilic PAMAM G2 dendrimers (Márquez-Miranda et al. 2016). This investigation showed that such dendrimers may form regular micelles and that their capacity to bind a small-interfering RNA depends on their generation and amphiphilic tail length. Other kinds of NPs can also be studied using this method.

### 21.2.3 Evaluating NP–Drug Interaction and Release

Modeling and predicting drug loading and formulation stability has mostly been handled computationally using techniques like docking, QSAR, as well as molecular dynamics simulations. In light of this, a study created by Hrkach was best example to illustrate the use of the initial dataset for QSAR research. In this work, a novel drug-loaded nanoparticle targeting solid tumors was created using chemical combinatorial libraries of 100 targeted polymeric NPs. Various formulations of the library differ in terms of particle size, targeting ligand density, surface hydrophilicity, drug loading capacity, and drug release characteristics (Hrkach et al. 2012). In 2017, Cern, Marcus, Tropsha, Barenholz, and Goldblum created a computational method to assess the quantities of active pharmacological components that might promote the development of a therapeutically effective drug-loaded nanoliposome. By taking into account the constant lipid composition and nanoparticle size, the scientists computed a score that represents the liposome's potential to load and leak a total of 13,700 medications that are accessible in the comprehensive medicinal chemistry and Drug Bank databases, according to their physicochemical properties. Leakage is recommended as a crucial characteristic that should be assessed to prevent drug-loaded liposomes from degrading their pharmacological efficacy while being stored. To date, there are 67 FDA-approved medications that satisfy the requirements to be put into these liposomes (Cern et al. 2017).

### 21.2.4 Common Docking Programs Used for NPs Modelling

Molecular docking is one of the central computational tools (Fig. 21.3). It helps in visualizing the interaction between ligand and target at molecular level. This comprehensive molecular inspection result in enhance level of virtual screening of compound databases to identify potential candidates (Pinzi and Rastelli 2019). Besides, the approach of reverse molecular docking enables the scientist to get a deep dive visual understanding aid in upgrading the ability to predict drug target and even modify the drug target interaction (Fan et al. 2019). In the field of nanotechnology, molecular docking as well helps in screening and controlling the interaction between

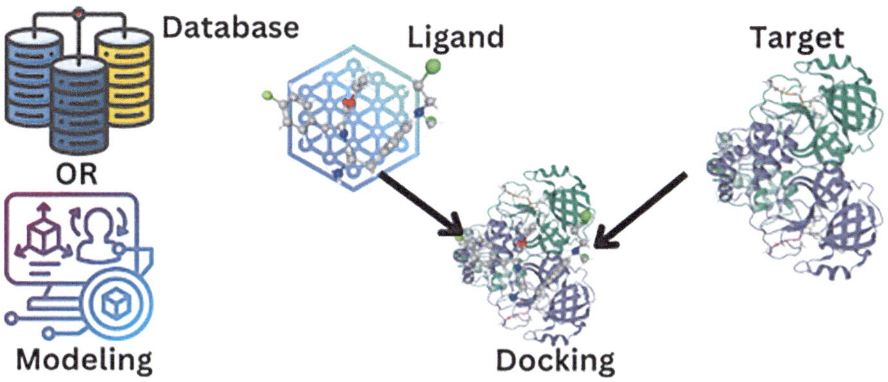


**Fig. 21.3** The molecular docking process

nanomaterials and the targeted proteins (Ali et al. 2020). Practically, docking can be used to find the best fit NPs for a given complex, minimizing the time and consumables needed for in vitro and in vivo testing. For instant, molecular docking can be performed to calculate the binding affinity and specific feature of each NPS formulations against the target, informing about the most promising design of targeted drug delivery systems (Chibber et al. 2016). The study of chitosan-curcumin NPs is the one example that highlight the utilization of molecular docking to analyze the interaction between different NPs and selected synthetic compounds to determine best combination (Yadav et al. 2018).

Molecular docking typically performed through several steps. First, the NPs need to be prepared and reproduce the 3D structure (Leifert et al. 2013). In some cases, the structure of NPs needs to be buildup from scratch using many common software such as Material Studio and Chem Draw, or it could be downloaded online database like the Cambridge Cluster Database (Arami et al. 2017). Second, the 3D structure of NPs must go for energy minimization to using several software such as Forcite, CASTEP, and Gaussian are a few examples. Gaussian is a powerful tool that can carry out molecular mechanics calculation as well as the semi empirical, DFT, and abinition calculations (Wu et al. 2017). Forcite is another potent molecular mechanic's tool that able to optimize NPs 3D structure without changing the crystal symmetry. CASTEP is consider an abinitio quantum mechanics software that is able to model many NPs with their different material states (Abdelsattar et al. 2021). Once the NPs- ligand formulation and the target are prepared, the docking experiment is performed using docking software. Till now there is no specialized docking software designed only for NPs, however the most commonly used docking software for such experiment is the AutoDock docking packages that use the Lamarckian genetic algorithm (Sarkar et al. 2024). Online server are also available for NPs docking such as Patchdock, that uses the shape complementarity-based algorithm (Futornyi et al. 2023), and HEX 6.3 that utilized the fourier transform-based algorithm(Huang 2004). Almost all docking experiments done for NPs were using a rigid docking parameter, assuming that the protein structure is fixed (Jin et al. 2024). The final step in docking is the visual inspection of the molecular interaction between NPs and the target using various software such as pymol, discovery studio (BIOVIA), and UCSF chimera (Verma et al. 2023).

It is important to mentioned that rigid docking approach is not the accurate situation, alternatively it is better to perform an induced-fit algorithm (Miller et al. 2021). Experimental data evidently display the dynamic and continuous movement of biological targets (Punjani and Fleet 2021). Therefore, recent docking studies, including NPs, used force field well-suited with flexible approach. For example, HEX software uses parameters from OPLS force fields (Kedir et al. 2024). However, the energy parameters in the force field used are not compatible with inorganic polymers and NPs, making them unsuitable to model inorganic-target interaction having different energy parameters (Hoff et al. 2022). A notable example is the INTERFACE FF, having the same energy parameters and shares common force field such as AMBER and OPLS -AA. INTERFACE FF is a suitable platform to model inorganic NPs, polymers, as well as the macromolecules (Heinz et al. 2013).

### 21.2.5 *Desolvation of the NPs Upon Protein Binding*

one of the crucial issues to consider before docking is the desolvation of NPs when react to the target. The docking searching algorithm utilize a scoring function, which accounts for the desolvation that take place before the binding with the target (Fig. 21.4). The NPs-target interaction will minimize the overall enthalpy energy. Nevertheless, the desolvation accomplished by the elimination of the bonded water molecule to the NPs, which reflect in the rise of the overall system's entropy (Ke et al. 2017).

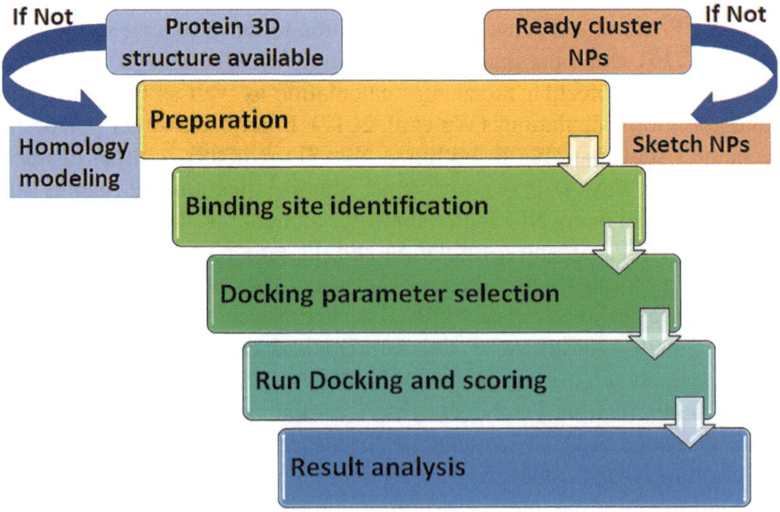


**Fig. 21.4** Standard procedures for simulating molecular docking of nanoparticles

almost all docking programs are considering the desolvation process and the resulting entropy alteration, which is integrated within their scoring function. For instant, autodock since it is the widely used docking software for NPs, it utilizes the AMBER force field. Basically, AMBER force field process a desolvation terms as well as terms accounts for the summation of the sp3 bonds to aid in calculating the unfavorable entropy of NPs-ligand interaction due to the limited rotations (Trott and Olson 2010).

## *21.2.6 Molecular Docking for Predicting Various Poses of Drugs into the Nanospheres*

Molecular docking is an ideal computational tool to determine the best binding poses of a drug inside the nanosphere. Earlier in silico study identify the highest loading of the formulated PEG and oligo nanosphere. The docking results display all the chemical bonding involved in the binding affinity calculation, importantly the hydrophobic contacts as well as the hydrogen bonds (Costache et al. 2009). Other studies performed docking experiment to assigned the binding mood and the molecular interaction between curcumin, resveratrol, and genistein against many dendrimer, which showed hydrophilic interaction mediated by hydroxyl groups (Abderrezak et al. 2012). In different studies, molecular docking was helpful in designing dendrimer using key functional groups, which boost the binding affinity. Moreover, molecular docking was handy tool to rationally place compounds in the core space within the dendrimer better than the default way, the thing that allow the compound to find its spot along the MD trajectory (Brocchini et al. 2008; Martinho et al. 2014).

## *21.2.7 Using Molecular Docking to Investigate Interactions Between Biomacromolecules and NPs*

### 21.2.7.1 Titanium Dioxide NPs

Molecular docking is used to understand the molecular interaction between TiO2 NPs, with it is tiny size of 1.09 nm, and divers of targeted proteins. TiO2 NPs revealed considerable binding with the Intercellular Adhesion Molecule 1 (ICAM-1), P-38 mitogen-activated protein kinases, Phosphatidylinositol-glycan biosynthesis class F protein (PIGF), and nuclear factor kappa-light-chain-enhancer of activated B cells (NF-κB). Docking results displayed that TiO2 NPs favorably bind to the amino acids having aliphatic groups, however they did not show any reaction with aromatic amino acids (Ranjan et al. 2017). Studying the interaction of any compounds to the human serum albumin (HAS) give a comprehensive understanding to its pharmacokinetic behavior. Molecular docking was used to calculate the TiO2

NPs half-life via calculating the binding affinity of the interaction between TiO2 NPs and the subdomain IIA of the HSA protein (Chibber et al. 2016). The results displayed a significant interaction of TiO2 NPs between subdomains IB and IIA (Wu et al. 2017). Another docking study evidenced that formulated pyrrole and methyl anthranilate (PMA) with TiO2 NPs (PMA-TiO2 NPs) unveiled boosted antibacterial and anthelmintic effect when compared to pyrrole and methyl anthranilate alone. Overall it could be said that the R-group in amino acids have a key role in the binding interaction preferences of TiO2 NPs with the targeted proteins (Abdelsattar et al. 2021).

#### 21.2.7.2 Copper Oxide NPs

copper oxide NPs (CuO NPs) have involved in many therapeutic formulations such as anti H1N1, antibacterial and antifungal effects. Therefore, studying its pharmacokinetic and toxicity profile is of high important. When compared to TiO2 that interact with HAS at subdomain IIA,. CuO NPs are found to interact to HSA at subdomain IIIA (Chibber et al. 2016).

#### 21.2.7.3 Iron Oxide and Manganese Oxide NPs

The cytotoxic effects of iron oxide NPs ($Fe_3O_4$ NPs) as well as for manganese oxide ($Mn_2O_3$) NPs are still not completely understood. Therefore, molecular docking was conducted to visualize virtually the interaction of these important NPs with vital targets such as DNA and other central proteins. Docking experiments documented that Fe NPs have a significance interactions with lysozyme (Aghili et al. 2016). In other virtual study upon bloodstream proteins, molecular docking analysis showed a binding affinity between bovine serum albumin (BSA) and MnFe2O4 nanocomplex (Gandhi and Roy 2019). Since NPs have the ability to overpass the BBB and expected to bind to multiple CNS proteins. A comprehensive study was conducted to study the molecular interaction of Mn2O3 NPs against three different fragments of tau protein. The study concluded that Mn2O3 NPs showed high binding affinities, which reform the folding behavior of tau protein into a condense structure, affecting the cell viability (Mehdizadeh et al. 2019). In the same manner, the same three different fragments of tau protein were docked against silica NPs (SiO2) showing high binding affinities (Shariati et al. 2018). Molecular docking also showed significant binding affinity of Fe3O4 NPs against Acetylcholinesterase (AChE) and butyrylcholinesterase (BChE) enzymes as key targets for Alzheimer's disease (Khalil et al. 2018). Other docking studies revealed the significant interaction between Fe NPs and human hemoglobin (Asl et al. 2018) whereas, Fe3O4/SiO2/APTS NPs were displayed bind mode within the buried active sites of the DNA duplex (Eshaghi Malekshah et al. 2020).

#### 21.2.7.4 Zinc Oxide NPs

Docking analysis documented that ZnO NPs can bind to human insulin as well as to the bovine serum albumin (BSA) at subdomain IIIA (Giannousi et al. 2020). Different docking study stated that ZnO NPs interact with IB and IIA subdomains (Ledesma et al. 2017). Compared to Fe3O4 NPs, ZnO NPs exhibit higher binding affinity towards acetylcholinesterase (AChE) and butyrylcholinesterase (BChE). However, lead oxide (PbO) NPs displayed highly interesting binding against these CNS enzymes (Khalil et al. 2018). It was found that ZnO NPs can modulate various proteins such as ICAM-1, interleukin 8, interleukin 1 beta, whereas have inhibitory effect on the a-glucosidase enzyme (Singh et al. 2018).

#### 21.2.7.5 Silver and Gold NPs

Docking studies visualized the molecular interaction between Ag NPs against three different Cytochrome P450 enzymes, showing binding affinity to CYP2C9 via Leu362, CYP2C19 via Ile362, and CYP2D6 via Val370. These interactions give insight on the metabolic-modulatory effect of Ag NPs upon the drugs substrate to these enzymes, which consequently affect their effectiveness (Wasukan et al. 2019). On the other hand, gold NPs (Au NPs) showed predicted anticancer activity via its inhibitory effect on EGFR-iRGD (Kaushik et al. 2019). Regarding Au NPs, docking studies showed that this NPs interact with ubiquitin through short-range hydrophobic interactions, forming five distinct complexes. (Brancolini et al. 2012).

## 21.3 Application of Ligand-Based Drug Design in Nanotechnology

Ligand-based drug design is a strategic approach utilized when the crystal-3D structure of the target is unavailable (Acharya et al. 2011). At the intersection of nanotechnology and pharmaceutical sciences, Ligand-based drug design entails the synthesis of molecules engineered to selectively bind to specific proteins or receptors within the body. This methodology is pivotal in the development of therapeutics that exhibit precise target interaction, thereby reducing adverse effects and enhancing therapeutic efficacy. The integration of ligand-conjugated NPs amalgamates the advantages of nanoparticle technology with ligand-based strategies, facilitating highly targeted and efficient drug delivery systems (Acharya et al. 2011). Indeed, Ligand-conjugated NPs can be formulated to target disease cells more specifically like the tumor cells with less or occasionally no effect on normal cells. Additionally, NPs can improve the solubility of the low soluble compounds, which result in improving their effectiveness. By engineering NPs for controlled release, a consistent therapeutic effect can be achieved over a specific period. Finally, targeted drug delivery using NPs can minimize the possible adverse effects that are generally arise with conventional drug delivery methods (Alkilany et al. 2019; Wang et al. 2014).

Ligand-conjugated NPs can be categorized mainly into two categories. The first and commonly one is the formulation that selectively inhibit a target that modulate the pathological pathway with no effect on the healthy tissues, the ideal example is imatinib, which formulated to suppress BCR-ABL. Imatinib target the fusion protein resulting from the translocation of the chromosome, that subsequently lead to myelogenous leukemia. This modulation is selective to cancer cells due to the presence of fusion protein in cancer cell with no existence in healthy one (Goldman and Melo 2003). The second category of ligand-conjugated NPs include an effective formulation target the desired diseased cells but are not selective modulator affecting normal cells as well. A common example for this category is vintafolide that utilized folic acid as carrier to specifically deliver complex vinca alkaloid to the folate receptor over expressed in the tumor cells of the ovary and lung tissues (Low and Kularatne 2009). Ligand-targeted therapeutics have often demonstrated greater potency and reduced toxicity compared to their non-targeted counterparts, significantly increasing consideration in their ongoing search and development.

### *21.3.1 Selecting the Most Effective Ligand for Targeted NP Delivery*

Selecting ligand to formulate with NPs is a mission required to consider many criteria. Some of these considerations are common like what applied to small molecules, whereas the rest are exclusive to NPs. The common criteria involved the need to select a target that is highly over expressed on the diseased cells, which could be modulated due to the interaction with the ligand-conjugated NPs. Following that, this target should be engulfed back to the surface of the affected cell. As a consequence, the encapsulated compound will be released out of the NPs, diffuse into the target cell membrane, ensuring the probability of the compound to modulate disease cells rather than being processed and excreted through convective flow (B. Wang et al. 2014).

Ligands those exhibited higher affinity are favored than the lower one since they allow the target to reach maximum saturation at few amounts, hence minimize the possibility of off-target toxicity. The minimum acceptable affinity for ligand-target interaction is about 10 nM. However, higher affinity, more than 10 nM, is needed if the docking experiment is conducted against one of blood vessel target where shear forces can affect the binding. Contrary, the target that located at the interstitial space, a ligand with lower affinity could be appropriate, since the surrounding circumstances is of lower shear stress (Xia and Low 2010).

Due to their significant functional and structural flexibility, NPs exhibit numerous advantageous properties not exist in small molecules. Among the most specialized functionalities of NPs is their significant capacities for loading drugs, extended time circulating in the blood stream, diverse use in therapeutic and diagnostic fields, the high ability to big molecules as well as hydrophobic one,reduced toxicity compared to some non-targeted ligands, and controlled release of therapeutic cargos (Barenholz 2012). Indeed, the integration of ligand-based drug design and nanotechnology has the ability to transform the field of therapeutic interventions, offering innovative avenues for personalized medicine and enhanced treatment outcomes.

The design of advance nanoparticle delivery system needs to comprehend the physiochemical behavior of nanoparticles, the molecular interaction between the target and the nanocomplex, and the possible physiological barrier. The intersection between the tools of LBDD and the advanced techniques in nanotechnology is notably reflect on improving drug design and therapeutic outcomes. The important methods used in the approach of LBDD are 3D QSAR and pharmacophore modeling. These virtual experiments are involved in producing predictive models that simplify the identification and optimization of lead compounds (Acharya et al. 2011).

### 21.3.2 Pharmacophore Modelling

Pharmacophores, a concept dating back over a century to Paul Ehrlich, have evolved significantly to their modern definition. Pharmacophore could be defined as a set of steric and electronic properties needed for ideal and best interactions with a defined target to stimulate or inhibit its biological response. Actually, pharmacophore is an abstract concept, not representing an actual molecule, but rather the common interaction capacities of a group of compounds towards their target. In simpler terms, it is a 3-D arrangement of molecular interactions that describes the biological activity of a ligand, typically in the context of protein-ligand binding (Fig. 21.5). Practically, Pharmacophore modeling is a powerful tool in drug discovery that identifies the essential features of a molecule necessary for its biological activity. This method is used to design and optimize new compounds that interact effectively with biological targets, such as proteins or nucleic acids. By understanding these essential features, researchers can create 3D pharmacophore models that represent the spatial arrangement required for optimal interaction with the target (Güner et al. 2014).

Pharmacophores are typically defined by features such as hydrogen bond donors and acceptors, positive and negative charges, hydrophobic regions, and aromatic rings. Additionally, excluded volumes (areas the molecule cannot occupy) and inclusion volumes (areas the molecule must occupy) provide steric constraints. Detailed interaction features, like metal interactions, can also be included. While highly specific features that directly plot each functional groups (e.g., protein side chain mimics) do not strictly fit the pharmacophore definition, they can yet been beneficial for virtual screening (Koes et al. 2012).

The spatial arrangement of features is as crucial as the features themselves in pharmacophore elucidation and matching. Molecules must be annotated with precise pharmacophore feature points. A pharmacophore model defines a series of probable geometries for these feature points, typically represented by tolerance

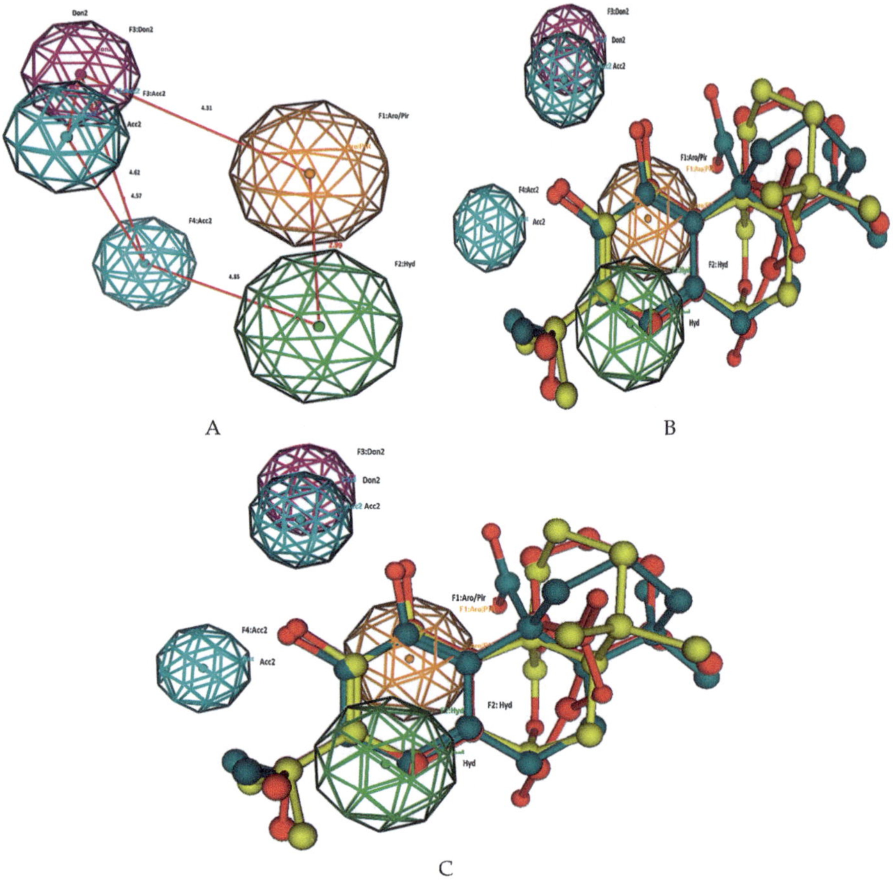


**Fig. 21.5** Pharmacophore modelrepresenting the essential chemical features of carnosol, carnosic acid and rosmanol. (**a**) Pharmacophore assembly of the compounds and distances between hot points are measured in angestrom. (**b**, **c**) superimpose of the compounds on the pharmacophore model (Etsassala et al. 2019)

spheres. A molecule matches a pharmacophore model if it can adopt a binding mode and conformation that make its crucial features positioned within these tolerance spheres, such as an aromatic ring center within an aromatic tolerance sphere (Anstead et al. 1997).

There are two strategies to build up a pharmacophore model. The first is the pharmacophore-ligand based, which allocating a library of biologically active ligands, superposing them and then determine the common features that involved in their activity. The second strategy is the pharmacophore structural based where the binding hot points between the target and the ligands are determined. Pharmacophore is a central concept used in multiple stages of drug design and development such as for virtual screening and lead optimization (Yang 2010). Interacting features can be prioritized by calculating their energetic contributions arising from their corresponding functional group. Simulation can precisely detect the positions of interacting features, which also help in considering target flexibility (Koes 2016).

One key application of pharmacophore modelling lies in the targeted delivery of drugs and other bioactive compounds using NPs, where pharmacophore modelling aids in the determination of the structural features necessary for effective binding and delivery to the target sites. The benefits of pharmacophore modelling is further demonstrated by its ability to guide the development of nanomaterials to detect and even treat complex diseases, such as the case of atherosclerosis (Nguyen et al. 2017). In this study, pharmacophore-help in engineering NPs with tailored surface properties and ligand functionalization, enabling their targeted accumulation at diseased tissue and improving their diagnostic or therapeutic efficacy. The study highlights the advantages of using pharmacophore modeling to design extracellular vesicles (EVs), which target atherosclerotic plaques. The findings determine the essential molecular features necessary for the EVs to interact effectively to the plaques. These features were drawn like a spheric dotes onto the pharmacophore model to generate a spatial arrangement to represents the best interaction with the target. Following that, the pharmacophore model was used to screen a library of bioactive ligands to select candidates that occupy the chosen features. This virtual screening approach result in narrowing the huge number of proposed molecules to find few but more promising candidates that could be incorporated into the EVs. Subsequently, the selected ligands were then incorporated into the design of the artificial EVs. This involved modifying the EV membrane to include the recognized molecular features, confirming that the EVs could effectively interact to the atherosclerotic plaques. By integrating these features into the EVs design, promising NPs has been formulated that demonstrated improved targeting and therapeutic outcomes in preclinical models. (Miao and Huang 2024).

pharmacophore modeling is used to classify the crucial structural features that ensure optimal interactions between NPs and the target, reducing the possible adverse effects and improving therapeutic outcomes. Pharmacophore model has been settled to recognize vital interaction points between the NPs and cancer cells. By utilizing these features, a generative formulation of NPs has been designed, which verified better binding to the target as well as less toxicity profiles. (Kumar et al. 2024).

Pharmacophore modeling is also applied to study the important structural features that contribute to the stabilization of NPs. Pharmacophore modeling was performed to determine the key binding points that stabilized the interaction between glycyrrhizin and a candidate NPs. The results revealed the capability of glycyrrhizin to stabilized nanosuspension better than common surfactant such as Tween 80 (Ozsoysal and Bilgili 2024). Pharmacophore model also aids in identifying the central biomolecular mechanism of action of polysorbate-80 coated, polymeric curcumin NPs (PS-80-CUR-NP) as potential antidepressant effect. The pharmacophore model detects the important steric and electronic features essential for curcumin's interaction with the specified targets, involved into its antidepressant effects. This model helps in comprehend the binding mode of how curcumin interacts with matrix metallopeptidase 12, result into its inactivation and consequent reduction in bowel inflammation (Yusuf et al. 2016).

One significant application of pharmacophore model in nanotechnology is the study of NPs toxicity. The alarming healthy risks accompanying with the increased exposure to the eco-friendly biodegradable plastics, particularly those of human origin, remain largely unexplored. A study reveals that during gastrointestinal processes, the enzymatic breakdown of polylactic acid microplastics produces nanoplastic particles by competing with triglyceride-degrading lipase. These nanoplastic oligomers form through hydrophobic self-aggregation. In a mouse model, these polylactic acid oligomers and their NPs accumulated in the vital organ such as the liver and intestine. The hydrolyzed oligomers induced intestinal damage and acute inflammation. A comprehensive pharmacophore study showed that these oligomers bind to the matrix metallopeptidase 12. The oligomers exhibited high binding affinity toward the catalytic zinc-ion finger domain, which result in inactivation of matrix metallopeptidase 12, potentially causing inflammation to the bowel following the exposure to polylactic acid oligomers. Although, the biodegradation of plastics are commonly known as a solution to environmental plastic pollution, molecular insight into their gastrointestinal fate and toxicity is crucial for assessing potential health risks (Wang et al. 2023).

In summary, using pharmacophore modeling in nanotechnology offer a promising approach to developing new and more effective NPs formulations by optimizing the interaction between drugs and their biological targets.

### 21.3.3 Quantitative Structure-Activity Relationship -QSAR

In recent years, there has been a notable rise in computational studies focusing on the activity and toxicity of NPs. Most of these studies are grounded in the fundamental chemistry principle that similar compounds exhibit similar biological properties. Among these techniques, Quantitative Structure–Activity Relationships (QSAR) modeling stands out as the most significant approach (Toropov and Toropova 2015). The classical QSAR approach, also known as Hansch Analysis (Hansch et al. 1963), is a field of molecular modeling that examines the relationships between a compound's structure and its biological activity using mathematical statistics and machine learning methods. QSAR is effectively utilized to predict the toxicity of chemical substances. A molecular descriptor is essentially a numerical value representing a specific molecular property, which can be experimentally determined (such as dipole moment), calculated (like potential energy), or derived from the chemical structure (such as the number of methyl groups). Furthermore, molecular descriptors can be mathematically derived properties, often using chemical graph theory, such as Wiener (Ivanciuc and Sciences 2000), Balaban (Thakur et al. 2004), or Randic indices. The resulting model combines set of statistically significant variables that provide deep dive insights into the interaction mechanisms under study (Puzyn et al. 2010).

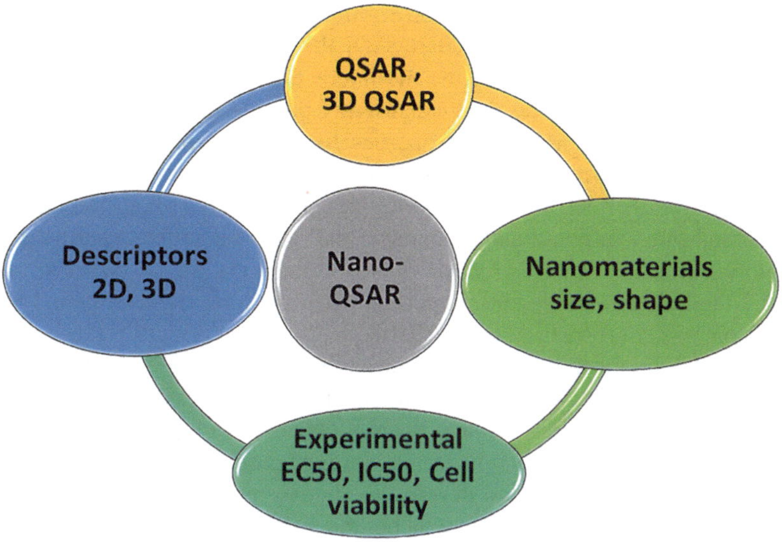


**Fig. 21.6** Main elements of QSAR in nanotechnology

QSAR models are beneficial in nanotoxicology studies due to their advantages in reducing the time and consumable needed for toxicity testing by providing reliable toxicity predictions (Fig. 21.6). QSAR models clarify the relationship between nanomaterial structures and their biological activities, which helps in formulating less toxic with high biological efficacy (Gajewicz et al. 2012). A previous study utilized QSAR models to predict the cytotoxicity of various engineered nanomaterials (ENMs) (Buglak et al. 2019). Five types of ENMs where studied those are metal oxides, metal-containing NPs, multi-walled carbon nanotubes, fullerenes, and silica NPs (Jagiello et al. 2016). The study starts by calculating molecular descriptors, which are numerical values reflecting exact molecular properties such as physicochemical properties, structural features, and quantum-chemical parameters. Following, QSAR models are build using statistical and machine learning methods, to correlate molecular descriptors with the biological activity of the nanomaterials. The models are validated using various statistical parameters, such as the determination coefficient ($R^2$) and cross-validation techniques, to make sure about their predictive accuracy and reliability. Finally, the validated models are utilized to predict the cytotoxicity of other NPs, offering detailed illustration about their potential bioactivity and guiding the formulation of safer NPs (Buglak et al. 2019). Although, QSAR models are efficient and cost-effective but they may lack the 3D spatial resolution needed for complex NPs (Sippl 2010). The evolving and the continuous development in drug discovery lead to the invention of the 3D QSAR, which was first introduce in 1988 with the Comparative Molecular Field Analysis (CoMFA) program (Cramer et al. 1988). The concept of the 3D QSAR assume that the variations in the pharmacological effects are due to alterations in the shapes and the power of the non-covalent bonding surrounding the molecules (Sippl 2010). Further methods using Comparative Molecular Similarity Indices Analysis (CoMSIA) by Klebe (Klebe et al. 1994) and the GRID/GOLPE program by Reynolds (Reynolds et al. 1989), are as well describe the 3D interactions quantitatively. CoMFA and CoMSIA specifically provide detailed insights provide detailed insights into the spatial arrangement of molecular properties like steric demand and electrostatic interactions. This information is essential for optimizing the formulation of NPs to enhance the overall effectiveness and minimize at possible the toxic effect. However, these methods are computationally intensive and require detailed experimental data (Jagiello et al. 2016). The 3D QSAR is best in studying ligand-target interaction since the model study the ligand properties when the bioactive compound is in its best fit bioactive conformation (Sippl 2010).

As the field of nanotechnology continues to flourish and the growing innovation of AI tools lead to a synergistic integration of 3D QSAR with machine learning algorithms. This integration has significantly enhanced computational methods in nanotechnology. Using highly concise and large experimental datasets, it was possible to extract key features and develop predictive models, which revolutionized NPs design and development. These technique will accelerate the risk assessment and safety evaluation of NPs, helping in unlock of high potent NPs design (Göller et al. 2022).

## 21.4 Molecular Dynamics (MD) Simulations

MD simulations have several key applications in CADD. They are crucial for studying the conformational flexibility and dynamics of drug-target complexes, helping to understand drug interactions at an atomic level, which is essential for designing effective drugs. Indeed, MD simulations is a computer simulation technique that focuses on biological molecules, such as proteins and nucleic acids, visualizing their atomic and molecular movements. MD simulations virtually mimic the structural movement of the biological macromolecules over time, providing molecular level insights into these movements, which aids in comprehend the biological functions and defining the system's dynamic evolution (Kerrigan 2013). In addition, MD simulations provide precise prediction about the thermodynamics and kinetics associated with drug-target interactions, which help in predicting binding affinity and stability of drug candidates. The development of new force-field methods has enhanced the accuracy of MD simulations, making them more reliable to observe and analyze processes difficult to study experimentally, such as drug binding and unbinding (Shukla and Tripathi 2021).

In nanotechnology, MD simulations has become instrumental in the rational design of nanoparticle-based drug delivery systems. It allows researchers to study the interactions between NPs and drug molecules, as well as their behavior within biological systems, at the nanoscale level, by offering comprehensive information about the fluctuations and flexibility of proteins and nucleic acids under study. These methods are extensively used to study the organization and dynamics of biological molecules, their complexes, and conformational changes in proteins and nucleic acids, revealing many mysteries on the femtosecond scale. Hence, provide brief understanding about the interactions between proteins and their ligands, and how these interactions influence conformational changes when mutations are introduced (Badar et al. 2022). One of the key applications of MD simulations in nanoparticle drug design is the exploration of NPs properties, such as size, shape, and surface characteristics, and their impact on drug loading, release kinetics, and targeting efficiency. By simulating the interactions between NPs and biological systems, researchers can gain valuable insights into the endocytosis, exocytosis, and clearance mechanisms of these nanomaterials, which is crucial for minimizing their potential toxicity and optimizing their therapeutic efficacy (Fig. 21.7). Moreover, molecular dynamics simulation has been instrumental in the development of novel nanomaterial-based drug delivery vehicles, such as dendrimers, micelles, and

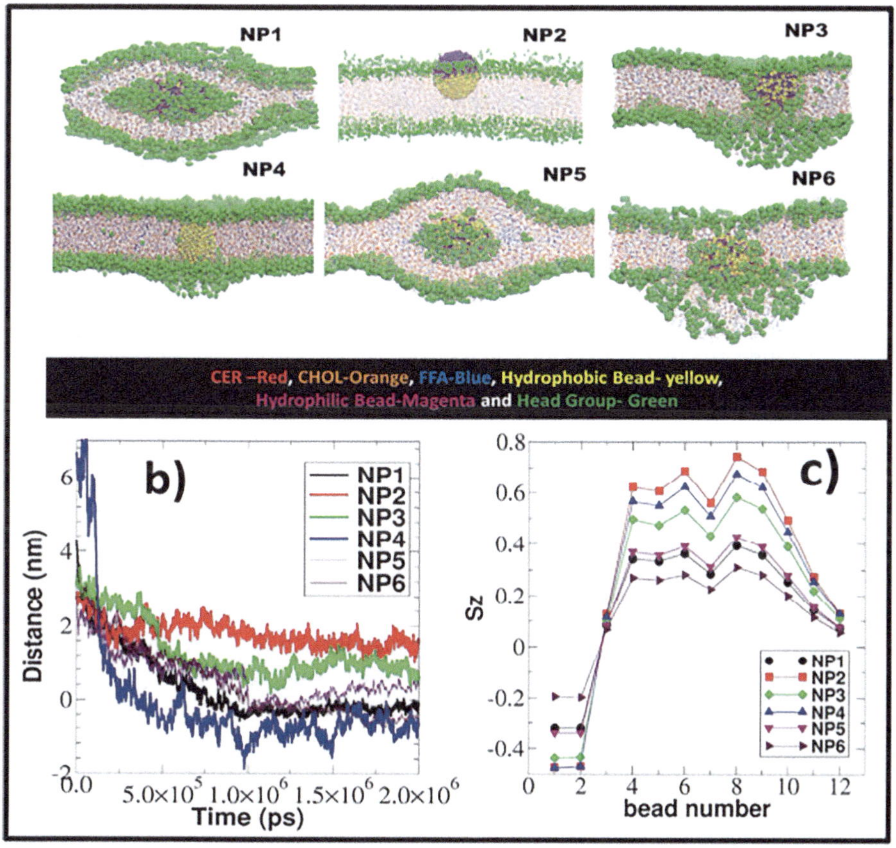


**Fig. 21.7** interaction between the lipid bilayer of the skin and nanoparticles. (**a**) The final shot of interaction between each nanoparticle and skin lipid bilayer system (**b**) fluctuation of the interaction between the lipid membrane and the center of NPs (**c**) Total order parameter of skin lipid membrane in each NP bilayer system (Gupta and Rai 2018)

liposomes. These nanostructures can be designed and engineered to achieve targeted and controlled drug release, thereby improving the bioavailability and therapeutic index of therapeutic agents (Adem et al. 2022). Many studies demonstrated that the MD simulations have been utilized to study key structural and functional properties of different NPs that are essential for their success in pharmaceutical applications. The following subsection illustrate such applications:

### 21.4.1 Structural Characteristics

In previous study, coarse-grained (CG) MD simulations were utilized to investigate the structural features of lipid-based NPs composed of stearic acid, oleic acid, and sodium dodecyl sulfate (SDS). This computational modeling verified the formation of vesicle-like structures with the water molecule encapsulated within the lipid droplet. These lipid-based NPs are exhibiting a beneficial value for delivering both hydrophilic and hydrophobic drugs (Ahmed et al. 2018).

MD simulations also aids in examining the interaction between gonadotropin-releasing hormone I and a cysteine-tagged modification of GnRH-I on simulated silica complex. The study revealed the involvement of two unique peptide-silica nanoparticle systems were caused by the variable orientation of peptides on the silica surface (Connell et al. 2018). In yet another study, MD simulations using reactive force fields (ReaxFF) was performed to inspect the constancy, clustering, and sintering of gold (Au) NPs when interacting with cysteine-based peptides (Monti et al. 2018). The findings displayed the stability of Au-NPs, due to their ability to adsorb peptide chains to their surface, which prevents particles from sintering.

### 21.4.2 Drug Loading and Release

MD simulations was used to study the loading of cyclosporin A (CyA) in lipid-based NPs made up of poly (ethylene glycol) molecules and medium-chain fatty acids at various compositions. The results find that increasing the drug loading did not induce any change observe to the NPs' nanostructure. CyA molecules often locate on the surface of the NP. Nevertheless, it was found that the CyA molecules submerged inside the NPs when the particle was overloaded with CyA. In particular, when the drug molecules loading raising from 1 to 10 (Benson and Pleiss 2014). In other study, MD simulations revealed that donepezil has a higher total drug loading capacity in chitosan NPs compared to rivastigmine when ions are present (Mousavi and Hashemianzadeh 2019).

MD simulations helps in understanding details NPs-target interactions, taking into account the NPs' diverse functional and structural characteristics, including size, shape, and surface modifications with coating polymers. For instant, the loading of hydrophobic small molecules can be enhanced by PEGylation, which entails coating NPs with polyethylene glycol (PEG) polymers (Bunker 2012). ReaxFF MD simulations was helpful to study the structure, dynamics, and drug release of gentamicin-loaded gold NPs (Au-NPs) functionalized with chitosan. The findings observed that the length of the chitosan polymer chains and the size of the carrier both affect drug loading. Besides, the release mechanism was found to be very slow, which could be explained by a delayed movement of gentamicin towards the solvent (Monti et al. 2019).

Drug release is greatly influenced by the size and internal microstructure of NPs and where it is located inside the NPs. The ligand can be placed into the NPs' core, depending on how well the ligand can attach to the NP components. The drug's release is greatly influenced by where it is located inside the NPs. The steered molecular dynamics (SMD) free energy calculations were advantages in understanding the location of a ligand inside the NPs and their release process. These approach was helpful to locate the exact position of the highly charged human vasoactive intestinal peptide and the hydrophobic bexarotene within a PEGylated nanoparticle (Vukovic et al. 2013). These free energy calculations were helpful in understanding how a nanoparticle complexed with molecules from intestinal fluid and loaded with the hydrophobic danazol were consumes a reduced amount of energy to release the danazol molecules from the surface of the nanoparticle to the membrane center (Kabedev et al. 2021).

MD simulations was also employed to test the effect of drug concentration and system pH on achieving ideal drug release and loading. Cytarabine was tested at pH values of 6.4 and 7.4 using chitosan NPs as carriers for controlled release. The results showed that cytarabine was released at a higher pH of 6.4 than at a pH of 7.4, which is consistent with the experimental data (Deepa et al. 2018).

All-atom molecular dynamics (AAMD) and coarse-grained molecular dynamics (CGMD) simulations to explore the self-assembly behavior of dual-responsive block copolymers and the capacity of the resultant micelles to load doxorubicin (DOX). The effective loading of DOX into the polymeric micelles was verified by the aid of CGMD simulations. However, it was found that when the drug concentration increased, the loading efficiency decreased, mostly probably due to the smaller particle size of DOX (Koochaki et al. 2020).

### *21.4.3 Diffusion Through Cell Protecting Layers*

MD simulations to revealed that the lipid-based NPs with intermediate stiffness had increased diffusivity through the mucus layer, in spite of their softer and harder counterparts. These stiff NPs have the capability to rotate, take on an ellipsoidal shape, and diffuse into the mucous layer. Additionally, MD simulations discovered a temperature-dependent diffusion mechanism for lipid-based NPs. Particularly, the lipid-based NPs had enhanced diffusion rate close to the phase transition temperature, which is the point at which the liposome membrane transform from a gel-like solid to a liquid crystalline phase (Yu et al. 2018). Again, MD simulations verify the diffusion of NPs over the endothelium glycocalyx. The model showed that the force was vital to indent the glycocalyx surface rises sharply as the NPs get larger. Besides, the results showed that NPs with positive charge have the ability to improve the diffusion across the glycocalyx layer (Kabedev and Lobaskin 2018).

### *21.4.4 Direct Membrane Penetration*

One of the interesting properties of NPs, particularly cationic NPs, is the direct penetration through the membrane without producing significant toxicity at lower doses. Knowing the molecular penetration process will aids in designing NPs for better cellular absorption, however the mechanisms behind this smooth penetration were undiscovered. Therefore, many studies using MD simulations to unlock this mechanisms (Nangia and Sureshkumar 2012). For instant, steered molecular dynamics (SMD) have used to examined the impact of particle size and shape, surface charge, and the characteristics of ligands employed as particle coatings on the cellular absorption of NPs (Mhashal and Roy 2016). Other work used unrestrained or regular MD simulations by calculating additional measures, like using multiple NPs, applying external electric or magnetic fields, or creating ionic concentration gradients across the cell membrane (Salo-Ahen et al. 2020). Coarse-grained molecular dynamics (CGMD) simulations were also used to investigate the interaction of many peptide-nanoparticle (NP) conjugates with the cell membrane. The model revealed that specified NPs generate transmembrane pore through which the other NPs can penetrate as well. Additionally, CGMD simulations using an external electric field were aid in studying the direct translocation of cationic gold NPs (Au-NPs). The results disclosed that NP translocation may occur in the presence of a insignificant external electric field without inducing deleterious effect on the membrane (Nakamura et al. 2019). A study utilized all-atom MD simulations to examine how a magnetic field affected the penetration of NPs across the BBB endothelial cell membrane. The gold was used as the NPs' shell whereas the iron oxide was used as the core of NPs. The results displayed that when a very mild magnetic force (<5000 pN) was applied, NPs with sizes ranging from 5 to 45 nm may pass through the cell membrane. Using CGMD simulations, Lin and Alexander-Katz disclosed that gold NPs (Au-NPs) may be induced to penetrate a cell membrane due to the effect of an ionic concentration gradient across the membrane. Interestingly, the pore was found to went for self-resealed and the ionic concentration across the membrane stabilized after the penetration process. CGMD simulations evident that NPs may directly pass through a membrane if there are gradients of several conditions, such as pH or osmotic pressure, across the membrane (Lin and Alexander-Katz 2013).

### *21.4.5 Interactions with Viruses and Biomacromolecules*

MD simulations was beneficial in understanding the interactions between NPs and viruses. All-atom MD simulations was used to investigate the potential of gold NPs (Au-NPs) coated with ligands mimicking heparan sulfate proteoglycan (HSPG) molecules that exhibit significant interaction with various virus classes and induce irreversible viral deformation. At molecular base, the simulations discovered that multivalent binding takes place between the negative sulfonate groups of the NP ligands and the positively charged L1 protein of the virus. This interaction resulted in significant forces—about −34 kcal/mol and 189 pN—sufficient to deform the L1 protein, disrupt the arrangement of L1 proteins in the virus, and eventually destabilize the virus (Cagno et al. 2018). Other study utilized MD simulations to investigate how gold NPs (Au-NPs) suppress amyloid β (Aβ) fibrillation. The model displayed that Au-NPs can influence peptide conformational changes and decrease inter-peptide interactions, which prevent Aβ peptides from forming β-sheets and slows down Aβ aggregation. Moreover, MD simulations was also used to examine the binding between Aβ40 peptides and their fibrils and NPs coated with diverse of ligands such as peptide, synthetic inhibitor, and positively and negatively charged ligands. The simulations revealed that the peptide-NPs and negatively charged NPs were interacted to the free peptides and fibril tips, respectively, that may limit the formation of fibrils and peptide fibrillization. On the other hand, deformation resulted from positively charged NPs bind to the surface of the peptide and fibril (Sen et al. 2019).

## 21.5 ADME (Absorption, Distribution, Metabolism, and Excretion) Analysis

The attractiveness of NPs for drug delivery is well-established, as they can facilitate the transport and delivery of therapeutic agents to remote or secluded areas of the body. Since, that NPs exhibit a high surface-area-to-volume ratio, chemical and geometric tunability, and the ability to interact with biomolecules, all of which can be leveraged to enhance drug targeting and controlled release (Wilczewska et al. 2012). However, the toxicity of nanoscale materials is a concern that must be addressed to ensure the safe and effective use of these drug delivery systems.

Understanding how NPs are absorbed and cleared by cells is essential to lowering the possible risks connected with them. The dispersion, biodistribution, and possible toxicity of NPs are ultimately determined by the fundamental mechanisms of endocytosis and exocytosis, which control the interface between the particles and mammalian cells. For NPs to effectively navigate the intricate biological environment and provide the intended therapeutic impact, it is essential to comprehend these cellular processes. Certainly, in order to reduce the toxicity of nanoscale materials in the body and ensure the safety of drug delivery nano-systems, it is necessary to investigate the elimination processes for NPs produced from nanoparticle-drug conjugates (Oh and Park 2014). One of the essential approaches is to assess ADMET properties, which plays an important role in the formulation of the nanoparticle-based drug design. ADME analysis assess the drug-likeness, which is crucial in drug discovery to determine if the ligands can be administered to a biological system (Umar et al. 2020). Majority of candidate drugs that failed to pass the clinical trials are documented to have either poor ADME profile or high toxicity.

Pfizer's Rule of Five, also known as Lipinski's Rule of Five (Ro5), firstly mentioned by Christopher A. Lipinski in 1997, is generally become as a threshold for evaluating drug-likeness. It aids to determine if a drug candidate with certain pharmacological activities can be taken orally (Attique et al. 2019). The rule of five is generally stated that a molecule can be absorbed orally if it does not violate two or more of these thresholds: molecular weight (Mw) < 500, octanol/water partition coefficient (iLOGP) $\leq 5$, number of hydrogen bond acceptors (nHBA) $\leq 10$, number of hydrogen bond donors (nHBD) $\leq 5$, and topological polar surface area (TPSA) < 40 Å$^2$ (Lipinski 2000). These drug-likeness parameters are reflecting the oral bioavailability of a drug candidate by calculation aqueous solubility and intestinal permeability. In-silico ADME and drug-likeness predictions have established their ability to offer vital information about the solubility, oral bioavailability, and toxicity of NPs, which effectively, guiding the design and synthesis of more potent NPs-drug candidates (Abdullahi and Adeniji 2020).

Virtually, ADME was utilized to predict the pharmacokinetic profile of the biogenic zinc oxide NPs (ZnO NPs) using AdmetSAR program that predict the human intestine absorption, Caco-2 cell permeability, blood-brain barrier (BBB) penetration, human oral bioavailability, drug metabolism, and acute oral toxicity. Organ toxicity, metabolism, toxicity endpoints (carcinogenicity, BBB penetration, and clinical toxicity), Tox21-Nuclear receptor signaling pathways (AR and ER), and molecular starting events (THRα and THRβ) were all computed using the PRO TOX-II server. According to the ADMET research, every ZnO NPs had a greater oral bioavailability in humans, was better absorbed in the gastrointestinal system, and could cross the blood-brain barrier. Even if drugs candidates had excellent effectiveness in prior in vitro and/or in vivo experiments, poor bioavailability may cause them to fail in clinical trials. Target compound excretion characteristics were anticipated by glycoprotein inhibition evaluations, which revealed that all ZnO NPs had negative P-glycoprotein inhibition values, signifying no disruption of drug absorption, permeability, or retention. Several cytochrome P450 enzymes (CYPs), including CYP3A4, CYP2C19, CYP1A2, CYP2D6, and CYP2E1, influence drug metabolism, more especially the biotransformation of drug compounds. All compounds, with the exception of chloramphenicol (CYP2C19I), did not inhibit these enzymes, according to the findings, indicating that there was no obstruction to the metabolism of medicinal substances. With the exception of chloramphenicol, which solely responded to the androgen receptor, none of the substances demonstrated any reaction to either the estrogen or androgen receptor binding. This suggests that they are not actively binding to these receptors, which prevents the endocrine system from being changed. Furthermore, the human ether-a-go-go-related gene (hERG), which is essential for the repolarization of the cardiac action potential, was strongly inhibited by all derivatives. The fact that this inhibition did not result in sudden mortality or long QT syndrome suggests that these substances may be safe. The results evident that ZnO NPs display favorable gastrointestinal absorption, oral bioavailability, and low toxicity, making them applicable for therapeutic use.

## 21.6 Successful Stories of Applying CADD Tools in Nanotechnology

Gold NPs (AuNPs') important significance in cancer treatment has been well-documented. The distinctive physicochemical characteristics of AuNPs—such as their high surface-area-to-volume ratio, chemical and geometric tunability, and potent optical absorption in the near-infrared spectrum—are emphasized. Because of these features, AuNPs are ideal in delivering therapeutic ligands and targeting cancer cells, which reduces systemic toxicity and off-target consequences while increasing therapeutic efficacy. Additionally, AuNPs' photothermal properties enable them to absorb light and transform it into heat, which efficiently kills cancer cells by producing localized hyperthermia. By optimizing these NPs size, shape, and surface chemistry through CADD design, their targeting effectiveness and therapeutic results are further enhanced (Ejiofor et al. 2023).

Many studies illustrate that NPs can be designed using CADD to enhance immune responses, deliver pharmaceuticals to target sites, and provide rapid and accurate diagnostics which highlight the key use of Nanotechnology in Infectious Diseases. Such study indicate that the therapeutic medications can be made from NPs with unique immunotherapeutic qualities. Metallic NPs and their oxides produce reactive oxygen species that have antibacterial properties when exposed to ultraviolet light. Significant antibacterial, antifungal, and antiviral properties are displayed by metallic NPs that include silver, gold, copper, titanium, magnesium, zinc, iron, or metal oxides. Numerous bacterial species, such as E. coli, S. aureus, B. subtilis, and S. typhi, are efficiently killed by silver NPs. Significant toxicity is also displayed by copper NPs; in one investigation, when S. cerevisiae was cultured on a polymer thin film laden with copper NPs, no colonies developed. Nano antibiotics are nanomaterials that have built-in antibacterial properties (Allaker et al. 2008). Nitric oxide-releasing NPs (NONPs) have many antibacterial mechanisms that work simultaneously. After interacting with superoxide, nitric oxide mostly produces reactive nitrogen oxide intermediates (RNOS), which have antibacterial properties. RNOS cause bacterial cell death by interacting with the amino acid residues of bacterial proteins and plasma membrane proteins. Through nucleotide deamination, basic site creation, and strand breakage, RNOS also directly harm bacterial DNA. Antibiotic-resistant strains of E. coli, K. pneumoniae, E. faecalis, and P. aeruginosa have been demonstrated to be inhibited in their growth by NONPs. At 1.25–5 mM, NONPs effectively eliminated E. coli, MRSA, and E. faecalis.

Compared to traditional antibiotics, nano antibiotics have the advantage of being stable for long periods of time and interacting with several bacterial biological pathways(Brar et al. 2023; Cabral et al. 2021). Furthermore, the solubility and delivery efficiency of conventional antimicrobial agents are improved by antibiotic nanoparticle polymers (Qasim et al. 2014). Strong antibacterial action was demonstrated by nylon-6 nanofiber containing 5,5-dimethyl hydantoin (DMH) as opposed to DMH alone.

## 21.7 Conclusion

The integration of CADD tools with nanotechnology has revolutionized the field of drug delivery systems, significantly enhancing the discovery and development of new therapeutic agents. CADD tools, including molecular docking, molecular dynamics simulations, and quantitative structure-activity relationship (QSAR) modeling, have provided invaluable insights into the interactions between drugs and their biological targets. These tools have empowered the rational design of NPs with enhanced physicochemical properties, improving drug solubility, bioavailability, and targeting effectiveness. Nanotechnology-based drug delivery systems, for example liposomes, dendrimers, and polymeric NPs, especially with their unique properties such as high surface-area-to-volume ratio and ability to cross biological barriers, have revealed great potential with enhanced therapeutic efficacy and minimized side effects. The successful application of CADD tools in nanotechnology has aid in the development of advanced drug delivery systems that address many serious medical challenges such as cancer, infectious diseases, and neurodegenerative disorders. These advancements have paved the way for personalized medicine, where treatments can be tailored to individual patients based on their specific genetic and molecular profiles. Molecular docking has long been a valuable tool for studying the interaction of biologically active molecules with their targets, while other techniques like MD simulation and free energy perturbation provide insights into binding modes and mechanisms of therapeutic agents. Recently, significant progress has been made in designing NPs and predicting their nanotoxicity. However, there are still shortcomings and potential for improvement in studying NP cytotoxicity. Future studies should address the solubility and redox potential of NPs, reproduce the exact shape and dimensions of NPs before molecular docking investigations, consider the effect of the protein corona, use specialized energy expressions for studying protein-metal interfaces, conduct more MD simulation studies of protein-NP complexes, and combine structure-based and ligand-based techniques, such as QSAR modeling, for studying the interaction of NPs with abundant proteins that have large and/or flexible binding sites. Although nanoparticle-based drug delivery systems have great potential, there are a number of obstacles to overcome. These include making sure the systems are biocompatible to prevent negative bodily reactions, developing scalable and economical manufacturing techniques, and negotiating the complicated regulatory environment to get these treatments approved for therapeutic use. Continued research and collaboration between computational scientists, chemists, biologists, and clinicians are essential to overcome these challenges and fully realize the potential of CADD tools in nanotechnology for drug delivery systems.

In conclusion, the synergy between CADD tools and nanotechnology has opened new avenues for the discovery and development of advanced drug delivery systems. This interdisciplinary approach holds great promise for improving patient outcomes and advancing the field of drug discovery.

## References


Abdelsattar AS, Dawoud A, Helal MAJN (2021) Interaction of nanoparticles with biological macromolecules: a review of molecular docking studies. Nanotoxicology 15(1):66–95

Abderrezak A, Bourassa P, Mandeville J-S, Sedaghat-Herati R, Tajmir-Riahi H-AJP (2012) Dendrimers bind antioxidant polyphenols and cisplatin drug. PLoS One 7(3):e33102

Abdullahi M, Adeniji SEJCA (2020) In-silico molecular docking and ADME/pharmacokinetic prediction studies of some novel carboxamide derivatives as anti-tubercular agents. Chem Africa 3(4):989–1000

Acharya C, Coop A, Polli EJ, MacKerell AD (2011) Recent advances in ligand-based drug design: relevance and utility of the conformationally sampled pharmacophore approach. Curr Comput Aided Drug Des 7(1):10–22

Adem Ş, Eyupoglu V, Ibrahim IM, Sarfraz I, Rasul A, Ali M et al (2022) Multidimensional in silico strategy for identification of natural polyphenols-based SARS-CoV-2 main protease (Mpro) inhibitors to unveil a hope against COVID-19. Comput Biol Med 145:105452

Aghili Z, Taheri S, Zeinabad HA, Pishkar L, Saboury AA, Rahimi A, Falahati MJPO (2016) Investigating the interaction of Fe nanoparticles with lysozyme by biophysical and molecular docking studies. PLoS One 11(10):e0164878

Ahmad S, Khan FN, Ramlal A, Begum S, Qazi S, Raza K (2023) Nanoinformatics and nanomodeling: recent developments in computational nanodrug design and delivery systems. In: Emerging nanotechnologies for medical applications. Elsevier, pp 297–332

Ahmed S, Govender T, Khan I, Rehman NU, Ali W, Shah SMH et al (2018) Experimental and molecular modeling approach to optimize suitable polymers for fabrication of stable fluticasone nanoparticles with enhanced dissolution and antimicrobial activity. Drug Des Devel Ther 12:255–269

Ali I, Mukhtar SD, Ali HS, Scotti MT, Scotti LJCPD (2020) Advances in nanoparticles as anticancer drug delivery vector: need of this century. *Current Pharmaceutical Design, 26*(15):1637–1649

Alkilany AM, Zhu L, Weller H, Mews A, Parak WJ, Barz M, Feliu NJADDR (2019) Ligand density on nanoparticles: A parameter with critical impact on nanomedicine. *Advanced Drug Delivery Reviews,143*, 22–36

Allaker RP, Ren GJT, Hygiene (2008) Potential impact of nanotechnology on the control of infectious diseases. *Transactions of the Royal Society of Tropical Medicines and Hygiene, 102*(1):1–2

Anstead GM, Carlson KE, Katzenellenbogen JAJS (1997) The estradiol pharmacophore: ligand structure-estrogen receptor binding affinity relationships and a model for the receptor binding site. *Steroids 62*(3):268–303

Anwar T, Kumar P, Khan AU (2021) Modern tools and techniques in computer-aided drug design. In: Molecular docking for computer-aided drug design. Elsevier, pp 1–30

Arami S, Mahdavi M, Rashidi M-R, Yekta R, Rahnamay M, Molavi L et al (2017) Apoptosis induction activity and molecular docking studies of survivin siRNA carried by

Fe3O4-PEG-LAC-chitosan-PEI nanoparticles in MCF-7 human breast cancer cells. *Journal of pharmaceutical and biomedical analysis,142,*145–154

Asl BA, Mogharizadeh L, Khomjani N, Rasti B, Pishva SP, Akhtari K et al (2018) Probing the interaction of zero valent iron nanoparticles with blood system by biophysical, docking, cellular, and molecular studies. *International journal of biological macromolecules,109*, 639–650

Attique SA, Hassan M, Usman M, Atif RM, Mahboob S, Al-Ghanim KA et al (2019) A molecular docking approach to evaluate the pharmacological properties of natural and synthetic treatment candidates for use against hypertension. *International journal of environmental research and public health,16*(6):923

Badar MS, Shamsi S, Ahmed J, Alam MA (2022) Molecular dynamics simulations: concept, methods, and applications. In: Transdisciplinarity. Springer, pp 131–151

Barenholz YCJJ (2012) Doxil®—the first FDA-approved nano-drug: lessons learned. *Journal of controlled released, 160*(2):117–134

Benson SP, Pleiss J (2014) Molecular dynamics simulations of self-emulsifying drug-delivery systems (SEDDS): influence of excipients on droplet nanostructure and drug localization. *Langmuir, 30*(28):8471–8480

Brancolini G, Kokh D, Calzolai L, Wade R, Corni S (2012) Docking of ubiquitin to gold nanoparticles. ACS Nano. 6:9863–9878

Brar B, Marwaha S, Poonia AK, Koul B, Kajla S, Rajput VDJA (2023) Nanotechnology: a contemporary therapeutic approach in combating infections from multidrug-resistant bacteria. *Archives of Microbiology, 205*(2):62

Brocchini S, Godwin A, Balan S, Choi J-W, Zloh M, Shaunak SJA (2008) Disulfide bridge based PEGylation of proteins. *Advanced drug delivery reviews, 60*(1):3–12

Buglak AA, Zherdev AV, Dzantiev BBJM (2019) Nano-(Q) SAR for cytotoxicity prediction of engineered nanomaterials. *Molecules, 24*(24):4537

Bunker AJP (2012) Poly (ethylene glycol) in drug delivery, why does it work, and can we do better? All atom molecular dynamics simulation provides some answers. *Physics procedia, 34,* 24–33

Bunker A, Róg TJF (2020) Mechanistic understanding from molecular dynamics simulation in pharmaceutical research 1: drug delivery. *Frontiers in Molecular Biosciences, 7*:604770

Cabral FV, Pelegrino MT, Seabra AB, Ribeiro MSJNO (2021) Nitric-oxide releasing chitosan nanoparticles towards effective treatment of cutaneous leishmaniasis. *Nitric Oxide, 113*, 31–38

Cagno V, Andreozzi P, D'Alicarnasso M, Jacob Silva P, Mueller M, Galloux M et al (2018) Broad-spectrum non-toxic antiviral nanoparticles with a virucidal inhibition mechanism. *Nature materials,17*(2):195–203

Cern A, Marcus D, Tropsha A, Barenholz Y, Goldblum AJJ (2017) New drug candidates for liposomal delivery identified by computer modeling of liposomes' remote loading and leakage. *Journal of Controlled Released, 252*,18–27

Chen Y, Liu LJA (2012) Modern methods for delivery of drugs across the blood–brain barrier. *Advanced drug delivery reviews, 64*(7):640–665

Chibber S, Ahmad IJB, Reports B (2016) Molecular docking, a tool to determine interaction of CuO and TiO2 nanoparticles with human serum albumin. *Biochemistry and Biophysics reports, 6*,63–67

Choi B-H, Lee H-H, Jin S, Chun S, Kim S-HJN (2007). Characterization of the optical properties of silver nanoparticle films. *Nanotechnology, 18*(7):075706

Connell DJ, Gebril A, Khan MA, Patwardhan SV, Kubiak-Ossowska K, Ferro VA, Mulheran PAJSR (2018). Rationalising drug delivery using nanoparticles: a combined simulation and immunology study of GnRH adsorbed to silica nanoparticles. *Scientific Reports, 8*(1):17115

Costache AD, Sheihet L, Zaveri K, Knight DD, Kohn J (2009) Modeling polymer-drug interactions in biodegradable Tyrosine-derived nanospheres using Molecular dynamics simulations and docking studies. Paper presented at the Nanotechnology 2009: Life sciences, medicine, diagnostics, bio materials and Composites-2009 NSTI nanotechnology conference and expo, NSTI-Nanotech 2009

Cramer RD, Patterson DE, Bunce JDJJ (1988) Comparative molecular field analysis (CoMFA). 1. Effect of shape on binding of steroids to carrier proteins. *Journal of the American Chemical Scociety,110*(18):5959–5967

Deepa G, Sivakumar K, Sajeevan TJB (2018) Molecular simulation and in vitro evaluation of chitosan nanoparticles as drug delivery systems for the controlled release of anticancer drug cytarabine against solid tumours. *3 Biotech, 8*(12):493

Duarte Y, Márquez-Miranda V, Miossec MJ, González-Nilo FJWIRN, Nanobiotechnology (2019). Integration of target discovery, drug discovery and drug delivery: a review on computational strategies. *Wiley Interdisciplinary Reviews: Nanomedicine and Nanotechnology, 11*(4):e1554

Ejiofor II, Ekeomodi CC, IlecChukwu AU, Ochiamu MC (2023) CADD for cancer therapy: current and future perspective. In: Targeted cancer therapy in biomedical engineering. Springer, pp 325–363

Eshaghi Malekshah R, Fahimirad B, Khaleghian AJI (2020). Synthesis, characterization, biomedical application, molecular dynamic simulation and molecular docking of Schiff base complex of Cu (II) supported on Fe3O4/SiO2/APTS. *International Journal of Nanomedicine, 15*:2583–2603

Etsassala NG, Adeloye AO, El-Halawany A, Hussein AA, Iwuoha EIJA (2019) Investigation of in-vitro antioxidant and electrochemical activities of isolated compounds from Salvia chamelaeagnea PJ Bergius Extract. *Antioxidants, 8*(4):98

Fan J, Fu A, Zhang LJQB (2019) Progress in molecular docking. *Quantitative Biology, 7*:83–89

Fatima F, Siddiqui S, Khan WAJBTER (2021) Nanoparticles as novel emerging therapeutic antibacterial agents in the antibiotics resistant era. *Biological Trace Element Research 199*(7):2552–2564

Futornyi D, Lozhko D, Kornelyuk OJB, Cell (2023). Computational modeling of the nanocomposite complex of EMAP II cytokine with TiO 2 nanoparticles. *Biopolymers and Cell, 39*(1):24–32

Gajewicz A, Rasulev B, Dinadayalane TC, Urbaszek P, Puzyn T, Leszczynska D, Leszczynski JJA (2012). Advancing risk assessment of engineered nanomaterials: application of computational approaches. *Advanced drug delivery reviews, 64*(15):1663–1693

Gandhi S, Roy IJJ (2019). Synthesis and characterization of manganese ferrite nanoparticles, and its interaction with bovine serum albumin: a spectroscopic and molecular docking approach. *Journal of molecular liquids, 296*:111871

Giannousi K, Geromichalos G, Kakolyri D, Mourdikoudis S, Dendrinou-Samara CJA (2020) Interaction of ZnO nanostructures with proteins: In vitro fibrillation/antifibrillation studies and in silico molecular docking simulations. *ACS Chemical Neuroscience, 11*(3):436–444

Goldman JM, Melo JVJNEJ (2003). Chronic myeloid leukemia—advances in biology and new approaches to treatment. *New England Journal of Medicine, 349*(15):1451–1464

Göller AH, Kuhnke L, Ter Laak A, Meier K, Hillisch AJA (2022). Machine learning applied to the modeling of pharmacological and ADMET endpoints. *Artificial intelligence in drug design, 2390*:61–101

Gonzalez-Ibanez AM, Gonzalez-Nilo F, Cachau RJBJ (2009). The collaboratory for structural nanobiology. *Biophysical Journal, 96*(3):49a

Gratton SE, Ropp PA, Pohlhaus PD, Luft JC, Madden VJ, Napier ME, DeSimone JMJP (2008). The effect of particle design on cellular internalization pathways. *Proceedings of the national academy of sciences, 105*(33):11613–11618

Güner OF, Bowen JPJJ, Modeling (2014). Setting the record straight: the origin of the pharmacophore concept. *Journal of chemical information and modeling, 54*(5):1269–1283

Gupta R, Rai BJN (2018) In-silico design of nanoparticles for transdermal drug delivery application. *Nanoscale, 10*(10):4940–4951

Haley B, Frenkel E (2008) Nanoparticles for drug delivery in cancer treatment. *Paper presented at the Urologic Oncology: Seminars and original investigations,* Vol .26, No 1, pp. 57-64. Elsevier.

Hansch C, Muir RM, Fujita T, Maloney PP, Geiger F, Streich MJJ (1963) The correlation of biological activity of plant growth regulators and chloromycetin derivatives with Hammett constants and partition coefficients. *Journal of the American Chemical Society, 85*(18):2817–2824

Heinz H, Lin T-J, Kishore Mishra R, Emami FSJL (2013). Thermodynamically consistent force fields for the assembly of inorganic, organic, and biological nanostructures: the INTERFACE force field. *Langmuir, 29*(6):1754–1765

Hoff SE, Liu J, Heinz HJJ, Science I (2022) Binding mechanism and binding free energy of amino acids and citrate to hydroxyapatite surfaces as a function of crystallographic facet, pH, and electrolytes. *Journal of Colloid and Interface Science, 605*:685–700

Hrkach J, Von Hoff D, Ali MM, Andrianova E, Auer J, Campbell T et al (2012). Preclinical development and clinical translation of a PSMA-targeted docetaxel nanoparticle with a differentiated pharmacological profile. *Science translational medicine, 4*(128):128ra139

Huang P-S (2004) Computational design and experimental characterization of protein oligomers. California Institute of Technology

Hughes JP, Rees S, Kalindjian SB, Philpott KLJB (2011). Principles of early drug discovery. *British journal of pharmacology, 162*(6):1239–1249

Huynh L, Neale C, Pomès R, Allen CJN, Biology, Medicine (2012). Computational approaches to the rational design of nanoemulsions, polymeric micelles, and dendrimers for drug delivery. *Nanomedicine, Nanotechnology, Biology, and Medicine, 8*(1):20–36

Ivanciuc OJJ, Sciences C (2000). QSAR comparative study of Wiener descriptors for weighted molecular graphs. *Journal of chemical information and computer sciences, 40*(6):1412–1422

Jagiello K, Grzonkowska M, Swirog M, Ahmed L, Rasulev B, Avramopoulos A et al (2016). Advantages and limitations of classic and 3D QSAR approaches in nano-QSAR studies based on biological activity of fullerene derivatives. *Journal of Nanoparticle Research, 18*:1–16

Jeliazkova N, Chomenidis C, Doganis P, Fadeel B, Grafström R, Hardy B et al (2015). The eNanoMapper database for nanomaterial safety information. Beilstein journal of nanotechnology, 6(1):1609–1634

Jin Z, Wei ZJCR, Safety F (2024). Molecular simulation for food protein–ligand interactions: a comprehensive review on principles, current applications, and emerging trends. *Comprehensive Reviews in Food Sciences and Food Safety, 23*(1):e13280

Joseph TM, Kar Mahapatra D, Esmaeili A, Piszczyk Ł, Hasanin MS, Kattali M et al (2023). Nanoparticles: taking a unique position in medicine. *Nanomaterials, 13*(3):574

Kabedev A, Lobaskin VJSR (2018). Structure and elasticity of bush and brush-like models of the endothelial glycocalyx. *Scientific reports, 8*(1):240

Kabedev A, Hossain S, Hubert M, Larsson P, Bergström CAJJ (2021). Molecular dynamics simulations reveal membrane interactions for poorly water-soluble drugs: impact of bile solubilization and drug aggregation. *Journal of Pharmaceutical Sciences, 110*(1):176–185

Kaushik AC, Wang Y-J, Wang X, Kumar A, Singh SP, Pan C-T et al (2019). Evaluation of anti-EGFR-iRGD recombinant protein with GOLD nanoparticles: synergistic effect on antitumor efficiency using optimized deep neural networks. *RSC advances, 9*(34):19261–19270

Ke PC, Lin S, Parak WJ, Davis TP, Caruso FJA (2017). A decade of the protein corona. *ACS nano,11*(12):11773–11776

Kedir WM, Li L, Tan YS, Bajalovic N, Loke DKJJ (2024). Nanomaterials and methods for cancer therapy: 2D materials, biomolecules, and molecular dynamics simulations. Journal of Materials Chemistry B, 12(47), 12141-12173.

Kerrigan JEJI (2013). Molecular dynamics simulations in drug design. *In silico models for drug discovery, 993*:95–113

Khalil AT, Ayaz M, Ovais M, Wadood A, Ali M, Shinwari ZK et al (2018). In vitro cholinesterase enzymes inhibitory potential and in silico molecular docking studies of biogenic metal oxides nanoparticles. *Inorganic and Nano-Metal Chemistry,* 48(9):441–448

Khong J, Wang P, Gan TR, Ng J, Anh TTL, Blasiak A et al (2020) The role of artificial intelligence in scaling nanomedicine toward broad clinical impact. In: Nanoparticles for biomedical applications. Elsevier, pp 385–407

Khorsandi K, Keyvani-Ghamsari S, Khatibi Shahidi F, Hosseinzadeh R, Kanwal SJJ (2021). A mechanistic perspective on targeting bacterial drug resistance with nanoparticles. *Journal of Drug Targeting, 29*(9):941–959

Klebe G, Abraham U, Mietzner TJJ (1994). Molecular similarity indices in a comparative analysis (CoMSIA) of drug molecules to correlate and predict their biological activity. *Journal of medicinal chemistry, 37*(24):4130–4146

Koes DRJC-ADD (2016). Pharmacophore modeling: methods and applications, Computer-Aided Drug Discovery, pp 167–188

Koes D, Khoury K, Huang Y, Wang W, Bista M, Popowicz GM et al (2012). Enabling large-scale design, synthesis and validation of small molecule protein-protein antagonists. *PloS one, 7*(3):e32839

Koochaki A, Moghbeli MR, Nikkhah SJ, Ianiro A, Tuinier RJR (2020). Dual responsive PMEEECL–PAE block copolymers: a computational self-assembly and doxorubicin uptake study. *RSC advances,10*(6):3233–3245

Kumar R, Dutt S, Tripathi AD, Singh AK, Chaturvedi VK, Singh SK (2024) Navigating Safety and toxicity challenges in Nanomedicine: strategies, assessment, and mitigation. In: Gautam V, Kumar R, Das Manandhar K, Kamble SC (eds) Nanomedicine: innovations, applications, and breakthroughs in the quest for health and medicine's future. Springer Nature Switzerland, Cham, pp 15–37

Kumari N, Hasija Y (2024). CADD:-exploring the digital frontier in drug designing. Paper presented at the 2024 3rd international conference on computational modelling, simulation and optimization (ICCMSO) (pp. 272-277). IEEE.

Ledesma AE, Chemes DM, de los Angeles Frías M, Torres MdPGJASS (2017). Spectroscopic characterization and docking studies of ZnO nanoparticle modified with BSA. *Applied Surface Science, 412*:177–188

Leifert A, Pan Y, Kinkeldey A, Schiefer F, Setzler J, Scheel O et al (2013). Differential hERG ion channel activity of ultrasmall gold nanoparticles. *Proceedings of the National Academy of Sciences, 110*(20):8004–8009

Lin J, Alexander-Katz AJA (2013). Cell membranes open "doors" for cationic nanoparticles/biomolecules: insights into uptake kinetics. *ACS nano, 7*(12):10799–10808

Lipinski CAJJ (2000). Drug-like properties and the causes of poor solubility and poor permeability. Journal of pharmacological and toxicological methods, 44(1):235–249

Low PS, Kularatne SAJC (2009). Folate-targeted therapeutic and imaging agents for cancer. *Current opinion in chemical biology,13*(3):256–262

Márquez-Miranda V, Araya-Durán I, Camarada MB, Comer J, Valencia-Gallegos JA, González-Nilo FDJS (2016). Self-assembly of amphiphilic dendrimers: The role of generation and alkyl chain length in siRNA interaction. Scientific reports, 6(1):29436

Marrink SJ, De Vries AH, Mark AEJTJ (2004). Coarse grained model for semiquantitative lipid simulations. *The Journal of Physical Chemistry B,108*(2):750–760

Martinho N, Florindo H, Silva L, Brocchini S, Zloh M, Barata TJM (2014). Molecular modeling to study dendrimers for biomedical applications. *Molecules, 19*(12):20424–20467

Mehdizadeh P, Fesharaki SSH, Nouri M, Ale-Ebrahim M, Akhtari K, Shahpasand K et al (2019). Tau folding and cytotoxicity of neuroblastoma cells in the presence of manganese oxide nanoparticles: biophysical, molecular dynamics, cellular, and molecular studies. *International journal of biological macromolecules,125*:674–682

Mhashal AR, Roy SJC (2016). Free energy of bare and capped gold nanoparticles permeating through a lipid bilayer. *ChemPhysChem,17*(21):3504–3514

Miao Y, Huang X (2024) Design of Artificial Extracellular Vesicles for treatments of cardiovascular diseases. In: Li Z (ed) Extracellular vesicle: biology and translational application. Springer Nature Singapore, Singapore, pp 91–115

Miller EB, Murphy RB, Sindhikara D, Borrelli KW, Grisewood MJ, Ranalli F et al (2021). Reliable and accurate solution to the induced fit docking problem for protein–ligand binding. *Journal of Chemical Theory and Computation,17*(4):2630–2639

Mirza S, Ahmad MS, Shah MIA, Ateeq M (2020) Magnetic nanoparticles: drug delivery and bio-imaging applications. In: Metal nanoparticles for drug delivery and diagnostic applications. Elsevier, pp 189–213

Monti S, Barcaro G, Sementa L, Carravetta V, Ågren HJNR (2018). Dynamics and self-assembly of bio-functionalized gold nanoparticles in solution: reactive molecular dynamics simulations. *Nano Research, 11*:1757–1767

Monti S, Jose J, Sahajan A, Kalarikkal N, Thomas SJPCCP (2019). Structure and dynamics of gold nanoparticles decorated with chitosan–gentamicin conjugates: ReaxFF molecular dynamics simulations to disclose drug delivery. *Physical Chemistry Chemical Physics, 21*(24):13099–13108

Mousavi SV, Hashemianzadeh SMJMRE (2019). Molecular dynamics approach for behavior assessment of chitosan nanoparticles in carrying of donepezil and rivastigmine drug molecules. Materials Research Express, 6(4):045069

Mu J, Luo D, Li W, Ding YJBT (2024). Multiscale polymeric fibers for drug delivery and tissue engineering. *Biomedical Technology, 5*:60–72

Nakamura H, Sezawa K, Hata M, Ohsaki S, Watano SJPCCP (2019). Direct translocation of nanoparticles across a model cell membrane by nanoparticle-induced local enhancement of membrane potential. *Physical Chemistry Chemical Physics, 21*(35):18830–18838

Nangia S, Sureshkumar RJL (2012). Effects of nanoparticle charge and shape anisotropy on translocation through cell membranes. *Langmuir, 28*(51):17666–17671

Nantasenamat C, Prachayasittikul VJEO (2015) Maximizing computational tools for successful drug discovery, vol 10. Taylor & Francis, pp 321–329

Nguyen LT, Muktabar A, Tang J, Dravid VP, Thaxton CS, Venkatraman S, Ng KWJD (2017). Engineered nanoparticles for the detection, treatment and prevention of atherosclerosis: how close are we?. *Drug discovery today, 22*(9):1438–1446

Niazi SK, Mariam ZJP (2023). Computer-aided drug design and drug discovery: a prospective analysis. *Pharmaceuticals, 17*(1):22

Oh N, Park J-HJI (2014). Endocytosis and exocytosis of nanoparticles in mammalian cells. *International journal of nanomedicine, 9*(sup1):51–63

Ostraat ML, Mills KC, Guzan KA, Murry DJI (2013). The Nanomaterial Registry: facilitating the sharing and analysis of data in the diverse nanomaterial community. *International journal of nanomedicine, 8*(sup1):7–13

Ozsoysal S, Bilgili E (2024). Non-traditional natural stabilizers in drug nanosuspensions. *Journal of Pharmaceutical and BioTech Industry, 1*(1):38–71

Padda IS, Mahtani AU, Parmar M (2022) Small interfering RNA (siRNA) based therapy. In: StatPearls Publishing, Treasure Island (FL), 20 Mar 2024. PMID: 35593797

Parikh PK, Savjani JK, Gajjar AK, Chhabria MTJB (2023). Bioinformatics and cheminformatics tools in early drug discovery. *Bioinformatics tools for pharmaceutical drug product development,* pp 147–181

Pinzi L, Rastelli GJI (2019). Molecular docking: shifting paradigms in drug discovery. *International journal of molecular sciences, 20*(18):4331

Punjani A, Fleet DJJB (2021). 3D flexible refinement: structure and motion of flexible proteins from Cryo-EM. *BioRxiv*, 2021.2004. 2022.440893

Puzyn T, Leszczynski J, Cronin MT (2010) Recent advances in QSAR studies: methods and applications. Springer Netherlands

Qasim M, Lim D-J, Park H, Na DJJ, Nanotechnology (2014). Nanotechnology for diagnosis and treatment of infectious diseases. *Journal of nanoscience and nanotechnology, 14*(10):7374–7387

Ranjan S, Dasgupta N, Chinnappan S, Ramalingam C, Kumar AJP, India Section B: Biological Sciences (2017) A novel approach to evaluate titanium dioxide nanoparticle–protein interaction through docking: an insight into mechanism of action. *Proceedings of the National Academy of Sciences, India Section B: Biological Sciences*, 87:937–943

Reynolds CA, Wade RC, Goodford PJJJ (1989). Identifying targets for bioreductive agents: using GRID to predict selective binding regions of proteins. *Journal of Molecular Graphics, 7*(2):103–108

Roco MC, Mirkin CA, Hersam MCJJ (2011). Nanotechnology research directions for societal needs in 2020: summary of international study. *Journal of nanoparticle research, 13*:897–919

Salo-Ahen OM, Alanko I, Bhadane R, Bonvin AM, Honorato RV, Hossain S et al (2020). Molecular dynamics simulations in drug discovery and pharmaceutical development. Processes, 9(1):71

Sarı SJHUJ (2020). Molecular modelling and computer aided drug design: the skill set every scientist in drug research needs and can easily get. *Hacettepe University Journal of the Faculty of Pharmacy, 40*(1):34–47

Sen S, Vuković L, Král PJS (2019). Computational screening of nanoparticles coupling to Aβ 40 peptides and fibrils. *Scientific reports, 9*(1):17804

Shariati B, Yektadoost E, Behzadi E, Azmoodeh E, Attar F, Sari S et al (2018). Interaction of silica nanoparticles with tau proteins and PC12 cells: Colloidal stability, thermodynamic, docking, and cellular studies. Scientific reports, 9 118:1963–1973

Sharma U, Khambete H, Singh N, Jain S, Jain N, & Dua KJDDSUQC(2024). Computational approaches for drug delivery of nanoparticles, *Drug Delivery Systems Using Quantum Computing*, pp 247–282

Shukla R, Tripathi TJI (2021). Molecular dynamics simulation in drug discovery: opportunities and challenges, Innovations and implementations of computer aided drug discovery strategies in rational drug design, pp 295–316

Simonazzi A, Cid AG, Villegas M, Romero AI, Palma SD, Bermúdez JM (2018) Nanotechnology applications in drug controlled release. In: Drug targeting and stimuli sensitive drug delivery systems. Elsevier, pp 81–116

Singh KP, Dhasmana A, Rahman Q (2018). Elucidation the toxicity mechanism of zinc oxide nanoparticle using molecular docking approach with proteins. Asian journal of pharmaceutical and clinical research,11(3)

Singh N, Shi S, Goel SJADDR (2023). Ultrasmall silica nanoparticles in translational biomedical research: overview and outlook. *Advanced Drug Delivery Reviews, 192*:114638

Sippl W (2010) 3D-QSAR—applications, recent advances, and limitations. In: Puzyn T, Leszczynski J, Cronin MT (eds) Recent advances in QSAR studies: methods and applications. Springer Netherlands, Dordrecht, pp 103–125

Suzuki Y, Ishihara HJDM, Pharmacokinetics. (2021). Difference in the lipid nanoparticle technology employed in three approved siRNA (Patisiran) and mRNA (COVID-19 vaccine) drugs. *Drug Metabolism and Pharmacokinetics, 41*:100424

Thakur A, Thakur M, Khadikar PV, Supuran CT, Sudele PJB, Chemistry M (2004). QSAR study on benzenesulphonamide carbonic anhydrase inhibitors: topological approach using Balaban index. *Bioorganic & medicinal chemistry,12*(4):789–793

Thota N, Jiang JJF (2015). Computational amphiphilic materials for drug delivery. Frontiers in Materials, 2:64

Toropov AA, Toropova APJC (2015). Quasi-QSAR for mutagenic potential of multi-walled carbon-nanotubes. *Chemosphere*, 124:40–46

Trott O, Olson AJJJ (2010). AutoDock Vina: improving the speed and accuracy of docking with a new scoring function, efficient optimization, and multithreading. *Journal of computational chemistry, 31*(2):455–461

Umar AB, Uzairu A, Shallangwa GA, Uba SJSAS (2020). Design of potential anti-melanoma agents against SK-MEL-5 cell line using QSAR modeling and molecular docking methods. *SN Applied Sciences, 2*(5):815

Verma SK, Nandi A, Simnani FZ, Singh D, Sinha A, Naser SS et al (2023). In silico nanotoxicology: The computational biology state of art for nanomaterial safety assessments. *Materials & Design, 235*:112452

Villoutreix BO, Renault N, Lagorce D, Sperandio O, Montes M, Miteva MAJCP, Science P (2007). Free resources to assist structure-based virtual ligand screening experiments. *Current Protein and Peptide Science, 8*(4):381–411

Vukovic L, Madriaga A, Kuzmis A, Banerjee A, Tang A, Tao K et al (2013). Solubilization of therapeutic agents in micellar nanomedicines. *Langmuir, 29*(51):15747–15754

Wang B, Galliford CV, Low PSJN (2014). Guiding principles in the design of ligand-targeted nanomedicines. *Nanomedicine, 9*(2):313–330

Wang M, Li Q, Shi C, Lv J, Xu Y, Yang J et al (2023). Oligomer nanoparticle release from polylactic acid plastics catalysed by gut enzymes triggers acute inflammation. *Nature Nanotechnology, 18*(4):403–411

Wasukan N, Kuno M, Maniratanachote RJJ (2019). Molecular docking as a promising predictive model for silver nanoparticle-mediated inhibition of cytochrome P450 enzymes. Journal of Chemical Information and Modeling, 59(12):5126–5134

Wilczewska AZ, Niemirowicz K, Markiewicz KH, Car HJP (2012). Nanoparticles as drug delivery systems. *Pharmacological reports, 64*(5):1020–1037

Wu Y, Zhang H, Wang YJJ (2017). Conformational and functional changes of bovine serum albumin induced by TiO2 nanoparticles binding. *Journal of Molecular Liquids, 243*:358–368

Xia W, Low PSJJ (2010). Folate-targeted therapies for cancer. *Journal of medicinal chemistry, 53*(19):6811–6824

Yadav P, Bandyopadhyay A, Chakraborty A, Sarkar KJC (2018). Enhancement of anticancer activity and drug delivery of chitosan-curcumin nanoparticle via molecular docking and simulation analysis. *Carbohydrate polymers, 182*:188–198

Yang S-YJD (2010). Pharmacophore modeling and applications in drug discovery: challenges and recent advances. *Drug discovery today,15*(11–12):444–450

Yu M, Xu L, Tian F, Su Q, Zheng N, Yang Y et al (2018). Rapid transport of deformation-tuned nanoparticles across biological hydrogels and cellular barriers. *Nature communications, 9*(1):2607

Yusuf M, Khan M, Khan RA, Maghrabi IA, Ahmed BJJ (2016). Polysorbate-80-coated, polymeric curcumin nanoparticles for in vivo anti-depressant activity across BBB and envisaged biomolecular mechanism of action through a proposed pharmacophore model. *Journal of microencapsulation,* 33(7):646–655

# Chapter 22
# Future Perspectives and Challenges in Nano-Biopharmaceuticals: Advancing a New Era in Medicine

**Sumanta Bhattacharya**

**Abstract** Through nano-biopharmaceuticals we are witnessing new solutions being brought in to the field of medicine which was for years suffering from problems such as, drug delivery, precise targeting and effectiveness of treatment. These therapies utilize nanoscale materials like liposomes, nanoparticles and dendrimers which allows depot formulation, drug targeting and controlled drug release apart from improving the stability greatly hence increasing positive outcomes for hard diseases such as cancer, cardiovascular diseases, neurological diseases and last but not the least infectious diseases. Thus, new opportunities are accompanied by numerous threats: legislation and toxicity, manufacturing difficulties, and costs factors that may lead to restricted availability. Technological advancement is becoming more intense with the introduction of new techniques like CRISPR gene editing, immuno-nanomedicine and AI powered DOE opening door for personalized efficient treatments. More specifically, the foremost priorities include the therapeutic's ethical implications, especially in the sphere of patient safety, and the therapeutic's societal implications concerning fairness of access. Meeting these challenges will of course necessitate interdisciplinary efforts within pharmaceutical, technological, and regulatory forms, all within the context of an innovative and transparent process.

**Keywords** Nano-biopharmaceuticals · Medica nosology's · Pharmaceutical · CRISPR gene editing · Immuno-nanomedicine

## 22.1 Introduction

Nano biopharmaceuticals is a revolutionary field in the medical science which is based on the synthesis of pharmacology and nanotechnology to develop molecular medicines where there is a proper locate targeting of largely illnesses. These

S. Bhattacharya (✉)
Asian International University, Imphal, Ghari, Manipur, India

 
V. Khanna et al. (eds.), *Nanotechnologies for Drug Delivery and Biopharmaceutical Development*, Progress in Drug Research 77,
https://doi.org/10.1007/978-3-032-00796-4_22

pharmaceuticals utilize nanoscale materials, generally in the size between 1 to 100 nanometers, which exhibit different properties from convention drugs because of the size, surface area, and level of bioactivity and reactivity. Here nanoscale biopharmaceuticals can cross cell membranes, reach areas that were not available earlier, and transport active moieties into target cells with enhanced efficiency and minimal toxicity. The field covers a broad area of applications of drug delivery systems, diagnostics, and therapeutics.This is probably one of the most exciting issues regarding nano biopharmaceuticals; enhancing the delivery and effectiveness of treatments for sophistcated diseases where other forms of therapies fail due to lack of targeting ability, including cancer, nervous disorders and reoccurring infections. These systems can be engineered to deliver drugs in response to change in pH, temperature, or the presence of certain enzymes; thus imparting a controlled and prolonged release profile that does not cause high systemic concentrations of the drug, but also maximises the therapeutically active concentration. This capability is extremely useful when dealing with chronic diseases and conditions that progress over time, and, as a result, drug released at the site of the disease is crucial. Nano biopharmaceuticals are a broad and comprehensive field as it encompasses links between medicine, biotechnology, engineering, materials science to build new therapies that address certain unmet clinical requirements. With the development of the science it is predicted that it will cover even more tailored treatments including gene therapies and immunotherapies where molecular targeting can significantly determine the effects of the treatment. Due to its rapidly evolving nature, nano biopharmaceuticals have the promising capabilities of becoming the foundation upon which future medical treatments will depend as it will make treatments accurate, efficient, and as unique as each patient (Joseph et al. 2023).

Applied in pharmacology, nano biopharmaceuticals have already started making remarkable performances in drug delivery and diseases therapies that offer new approaches to counter the classical shortcomings in the area. Among these developments, one of the most important ones is probably the increased drug bioavailability, for the nano-sized particles are taken up by cells in greater numbers and penetrate biological barriers more easily than the non-nano-sized drugs. Salient in this vein is the ability to hit target such as cancerous cells where nanoparticles can be used to transport chemotherapeutic drugs thereby minimizing harm on healthy cells and tissues which are otherwise associated with these treatments. Also, central to the development of nano biopharmaceuticals is the capacity to deliver drugs across what has been a challenging biological frontier—the blood-brain barrier, which has inhibited efforts to find effective treatments for diseases like Alzheimer's and Parkinson's (Patra et al. 2018).

Nano formulated bio-pharmaceuticals are being developed in treatment of infections by delivering the anti-infective agents more selectively thereby reducing the chances of antibiotic resistance, a growing concern in global health care. These nanoscale pharmaceuticals are aimed directly at pathogens and localized delivery of drugs enhances the localized drug concentrations and minimizes side effects in patients leading to better patient compliance and prognosis. Furthermore, the nano biopharmaceuticals also support the use of personalized medicine strategy because the products themselves can be designed to act in response to biomarker or existent mutation in a patient. This approach is radically changing standard treatments for chronic diseases, making it possible to have medicines that will perform their intended function in the body of a subject, something that was previously an extremely low possibility, and enhancing on the overall performance of drugs. The potential of NBP is not only in treatment of disease but also, detection of diseases wherein nanotechnology implies that detection instruments recognize molecular changes at stages which are often pre-symptomatic. The idea is that, early proactive approach allow improving prognosis and helps the patient to cope with the disease better. With advancing nanotechnology in biopharmaceutical industries, regenerative medicine, gene and immunotherapy, disease treatment and care becomes more personalize, specific and effective. These observations of the current effects of nano biopharmaceuticals therefore bear testimony to the reality that they can become a practical instrument in the twenty-first century medical science to transform the methods of disease treatment and management (Hetta et al. 2023).

The future of nano biopharmaceuticals is very much aligned to these two; this is because nanobiopharmaceuticals need cross disciplinary studies in order to be fully effective. Nanotechnology in its essence may be defined as working at a scale where the facility of physics, chemistry and materials science becomes implausible, where we are manipulating atoms and molecules. This produces therapeutics that are safe, efficacious, and targeted—all of which have significant implications for biology, pharmacokinetics and clinical medicine. By integrating these fields, the researchers can develop utility that can apply properties of the nanoscale materials inherently to solve a biological predicament that cannot be solved using single-discipline technique. However, one of the fields where the convergence of research between different disciplines turns out to be highly helpful is the field of targeted drug delivery. Designing nanoparticles that penetrate cancer cells, on the other hand, require material scientist to prepare the architectures of nanoparticles, chemists to surface active sites on the nanoparticles, and biologists to ascertain that these particles interact with cell in a safe fashion and deliver the drug as desired. It also makes it possible to produce strong yet selective drugs with less undesirable effects, and better therapeutic effects. Furthermore, engineering and biological simulations enhance nano biopharmaceuticals whereby scientists are able to predict the behavior of new drugs in the body before they are tested on patients.Apart from inventiveness, interdisciplinary work fosters flexibility, a valuable quality in view of the emergent area of nano biopharmaceuticals. There are always new standards, new rules, new ethical questions and new requirements set by public health—teams from different disciplines are better suited to handle such challenges. There is an incredible diverse group of people involved and all of them can predict possible obstacles and modify actions for safety, optimal effectiveness and inclusiveness. Due to advanced diseases and emerging, constant need of individualized medicine, team-based research in the fields of nanotechnology and pharmaceuticals will be crucial for the future advancements and for the realization of the potential of nano biopharmaceuticals in the further evolution of modern medicine (Farjadian et al. 2018).

## 22.2 Advances in Nano Biopharmaceutical Technology

The nanocarriers are one of the fundamental components of further development of the nanomedicine aimed at the improvement of the drug delivery and treatment of diseases. Liposomes, nanoparticles and dendrimers are commonly used nanocarriers as each of them present their benefits for delivering and releasing the drugs at the targeted site with high controlled ratio. Liposomes are spherical structures made from phospholipids thereby giving them a biomembrane like structure where they can encapsulate both amphitically and hydrophobically active drugs. These non-harmful, natural materials make liposomes suitable for multiple uses including to enhance the effectiveness of chemotherapy, where liposomes can encapsulate anti-cancer drugs and deliver them directly to the tumour thus have fewer negative impact. More recent accomplishments have made PEGylated liposomes that reduce the immunity system from recognizing the container and hence the molecules prolong the circulation time. Polymer, metal or ceramic based nanoparticles are very versatile and can be designed to release drug in response to numerical stimuli like pH and temperature. For instance, the gold nanoparticles may be employed in photo thermal therapy where they can be dates to trap light and convert the trapped light energy to heat in order to cause destruction of cancer cells. The polymer based nanoparticles can effectively encapsulate and deliver any class of drugs and provide long term delivery, which can reduce the rate of administration. Dendrimers, the third generation of nanocarriers, are highly branched, tree-like, polymeric structures possessing many functional groups in their periphery for drug encapsulation and delivery to the target tissues. Their structure allows multiple drug, imaging agent or targeting ligand to be simultaneously attached to them which makes them useful for multipart therapies. However, the best supported application of dendrimers is in gene therapy whereby dendrimers containing DNA or RNA fragments can be taken directly to the target cell. Altogether, these features of nanocarriers enhance the possibilities of controlled and refined drug delivery, thus opening the way to the application of those therapies which have not been attainable using conventional drug delivery systems (ud Din et al. 2017). Figure 22.1 depicts the impacts of advancement in nano biopharmaceutical technologies.

Pharmacology advanced and targeted drug delivery mechanisms and controlled release system are some of the recent emerging concepts with a view to enhance drug efficacy and reduce side effects. Such systems are centered at ensuring that drugs are provided in the target area while at the same time providing them in controlled release manners to treat diseases that are multifaceted. Targeted drug delivery involves use of nanocarriers that are able to afford molecular targeting, for instance via antigens or receptors expressed on cancer cells. It enables drugs to be targeted only on the affected tissues and spare healthy tissues as well as raise the local concentration of the agent of treatment, which is very useful in cases of cancer as well as in infectious diseases (Ezeako et al. 2023).

Sustained release mechanisms aim at dispensing a certain quantity of a drug with high thermaberapeutic activity over a given interval of time. An advantage of this

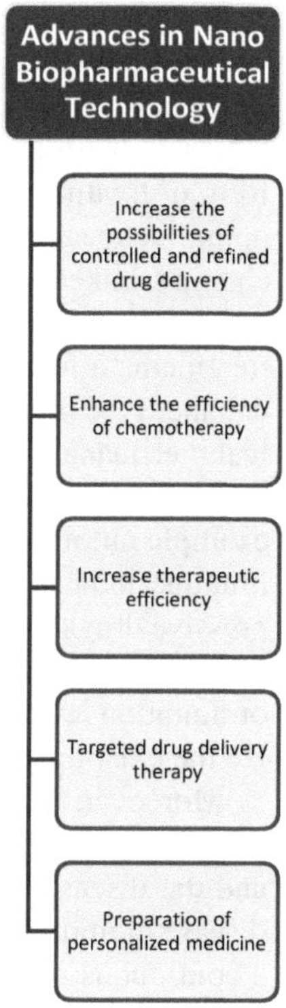


**Fig. 22.1** Impacts of advancement in Nano Biopharmaceutical Technology

process is the increase in patients' concordance since they take fewer doses in a day, which is important in chronic diseases that require medicines for a long time. That's why the use of polymer-based nanoparticles as nanocarriers, which are capable of trapping a required payload and release it in response to certain stimuli like pH level, temperature or enzymes. For example, in tumour environment, which is frequently acidic, pH-sensitive nanocarriers will deliver the drugs selectively within the tumour area thereby enhancing the therapeutic impact in the tumour tissue while having minimal effects on the surrounding normal tissue. Likewise, stimuli-responsive nanoparticles may release their cargo by temperature change of the environment and externally applied heating for hyperthermic cancer therapies (Bates 2002).

Moreover, other approaches that employed nanoparticles with response to magnetic or light stimuli are being investigated to gain better control over the release of drugs. Such improvements are highly beneficial for the diseases with complex progression dynamics and the time and concentration of the active compound release. In general, targeted drug delivery and controlled release formation emphasize the trend in medicine for targeted, molecular therapy, which is much more effective, but at the same time reflects a completely new image of medicine as a combination of breakthrough technologies targeting molecular processes in a living organism (Cheng et al. 2023).

Nanotechnology is enhancing the future of personal medicine by making it possible to target diseases depending on a patient's profile. Personalized medicine aims to identify and categorize every patient depending on the genetic makeup, the patient's environment, and patient's lifestyle in order to develop the most effective form of treatment for every patient. Nanotechnology makes this possible by developing nanocarriers that can communicate with particular biomarkers or gene changes linked to a disease. For instance, the nanoparticles can be programed to attack mutated genes like HER2 in breast cancer, or EGFR in lung cancer, whereby treatments affect the cancerous cells only with little effects on the normal cells. It enhances the treatment's success and minimizes unfavorable side effects since medical marijuana is less generic (Alghamdi et al. 2023).

Besides cancer, nanotechnology assists in the treatment of other ailments, for example autoimmune diseases, in which treatment can be formulated according to immune indices. Specific drug delivery systems designed to liberate immunosuppressive drugs to tissues can reduce side effects while improving patient experience. In the case of infections, it perfects the process of pathogen diagnosis to the strain or mutation level so that one can treat the patient with a remedy that would neutralize the existing strain (Prosperi et al. 2017).

Moreover, diagnostics based on nanotechnology possess the capability of supporting the doctrine of personalized medicine because of the early disease diagnosis and the disease status assessment. Nano-sensors and nano-biosensors can identify disease biomolecules within body fluids at very low concentrations, and this enables a continuous monitoring of the status of a patient. This ability is particularly relevant for those conditions which often present themselves in latent stages, for instance, some forms of cancer or metabolic diseases, which can then be treated before the disease has had the chance to worsen considerably. Nanotechnology thus offers the healthcare industry a new concept of the personalised therapeutic medicine to dispel with the changes in the approach to the molecular level instead of the mass approach. Over time, the role of nanotechnology in enhancing the accuracy of treatment regimens rises, thereby promoting better, efficient, and safer treatment to any patient (Ramesh et al. 2023).

## 22.3 Applications in Disease Treatment

Nanotechnology is enhancing the future of personal medicine by making it possible to target diseases depending on a patient's profile. The therapeutic effects of different nanomaterials have been depicted in Table 22.1. Personalized medicine aims to identify and categorize every patient depending on the genetic makeup, the patient's environment, and patient's lifestyle in order to develop the most effective form of treatment for every patient. Nanotechnology makes this possible by developing nanocarriers that can communicate with particular biomarkers or gene changes linked to a disease. For instance, the nanoparticles can be programmed to attack mutated genes like HER2 in breast cancer, or EGFR in lung cancer, whereby treatments affect the cancerous cells only with little effects on the normal cells. It enhances the treatment's success and minimizes unfavourable side effects since medical marijuana is less generic (Alghamdi et al. 2023).

Besides cancer, nanotechnology assists in the treatment of other ailments, for example autoimmune diseases, in which treatment can be formulated according to immune indices. Specific drug delivery systems designed to liberate immunosuppressive drugs to tissues can reduce side effects while improving patient experience. In the case of infections, it perfects the process of pathogen diagnosis to the strain or mutation level so that one can treat the patient with a remedy that would neutralize the existing strain. Moreover, diagnostics based on nanotechnology possess the capability of supporting the doctrine of personalized medicine because of the early disease diagnosis and the disease status assessment. Nano-sensors and nanobiosensors can identify disease biomolecules within body fluids at very low concentrations, and this enables a continuous monitoring of the status of a patient. This ability is particularly relevant for those conditions which often present themselves in latent stages, for instance, some forms of cancer or metabolic diseases, which can then be treated before the disease has had the chance to worsen considerably (Prosperi et al. 2017).

Nanotechnology thus offers the healthcare industry a new concept of the personalised therapeutic medicine to dispel with the changes in the approach to the molecular level instead of the mass approach. Over time, the role of nanotechnology in enhancing the accuracy of treatment regimens rises, thereby promoting better, efficient, and safer treatment to any patient (Alghamdi et al. 2023).

CVDs are among the main causes of mortality, and many of them are made worse by the coagulation of the blood and the formation of the blood clots that hinder normal blood circulation resulting in cases such as heart attacks, stroke and pulmonary embolism. Nanotechnology seems to provide the solutions for these problems, especially through the utilization of nanocarriers for selective dissolution

**Table 22.1** Therapeutic effects of different types of nanomaterials

| Different types of nanomaterials | Therapeutic functions |
|---|---|
| Nano-sensors and nanobiosensors | Monitor the activities of disease biomolecules |
| Lipid-based Nanocarriers | Enhance the rate of curability of cardiovascular diseases by selective dissolution of blood clots |
| Magnetic nanoparticles | Target based delivery of drug molecules |
| Polymeric nanomaterials | Help to cross blood brain barrier and |
| Nanocarriers | Treatment of Alzheimer's and Parkinson's diseases and other central nervous system issues |
| Cohesive nanosystems | Prevent infections by entrapping antibiotics and antiviral agents and enhance the anti-infectious activities by increasing bio-availability |
| Nano-based antiviral agents | Reduce viral infection and prevent chronic diseases like AIDS, hepatitis etc. |

of blood clots while avoiding side impacts. Standard anticoagulant therapies have been coloring ineffectively in prevention from clot formation but expose the patient to bleeding complications due to general inhibition of coagulation throughout the vascular compartment. But in the nanotechnology-based approaches, the specific clot dissolving agents are delivered where clots are forming, hence minimizing chances of formation of other blood clots (Roth et al. 2020).

Lipid based nanocarrier system for treating cardiovascular disease involves the use of nanocarrier that encapsulates thrombolytic agents like tissue plasminogen activator tPA. They can be designed as carriers with stimuli-responsive properties, and their contents can be delivered based on activated platelets or fibrin. This targeted enable the thrombolytic agents to released specifically at the clot site in order to clear it effectively without causing side effect to the other parts of the body. Another advance is the use of magnetic nanoparticles that are capable to be guided to the clot site by the magnetic field, thus providing dense concentrations of the drug with minimal affect on adjacent tissue. Other nanocarriers system developed in the recent past include those that release anticoagulants to high-risk sites like areas close to ruptured atheromatous plaques. These nanoparticles can dispense anticoagulant or anti-inflammatory agents that prevent clot formation within those areas, something that can be used to prevent major cardiovascular events. Further, nanotechnology has made it possible to deliver new drugs that address molecular factors in clotting seen to be critical in the development of cardiovascular diseases including inflammation and oxidative stress. Nanotechnology in cardiovascular disease management not only envisages exact prevention of Exenatide formation at the requisite site but also includes the management of clots formed, which may decrease acute events such as heart attacks and strokes. Thus, utilizing the advantage of having more precise control the release and targeting properties of nanocarriers for the treatment and prevention of clot-related complication in cardiovascular disease represents a safer and more effective alternative for potentially millions of patients who are at risk (Hassanpour et al. 2020).

The management of neurological conditions poses difficulties, mainly as a result of the presence of the blood brain barrier, a membrane which essentially maintains the integrity of the brain while at the same time denying free access to most drugs & macromolecules. Until now, this barrier prevents nearly all drugs from gaining access to the brain, limiting the therapy of many neurological afflictions including Alzheimer's, Parkinson's and brain tumors. We have seen that nanotechnology introduces new opportunities in dealing with such a barrier and delivering the drugs to the CNS more efficiently (Upadhyay 2014).

Polymeric nanoparticles like liposomes, polymeric nanoparticles and dendrimers can be designed to transport across the BBB via several routs. One of them is able to functionalize the surface of nanocarriers with ligands or peptides that would mimic the ability of such molecules as transferrin or insulin to penetrate the BBB. Basically, drugs can be delivered more effectively to the brain by linking these molecules to the surface of nanoparticles. Passive diffusion has also been applied with ultrasmall nanoparticles opening with a diameter of less than 10 nm that easily penetrates the BBB. These nanoparticles can accommodate drugs and deliver them incrementally over time, giving protracted treatment to the brain. Also, stimuli-responsive nanoparticles have been created to deliver drugs only if they encounter appropriate conditions inside the brain. For instance some nanoparticles were designed to release anti-inflammatory drugs when exposed to the oxidative conditions which characterise neuroinflammatory diseases. This specificity reduces the effects on other areas thereby eliminates side effects. Another potential technique relates to magnetic nanoparticles which can be guided across the BBB with the help of an imposing exterior magnetic field, which is particularly effective in localized therapy of brain tumors (Pinheiro et al. 2021).

Nanocarriers used for Alzheimer's and Parkinson's patients who require long term and constant treatment can reduce the amount of time and expense of extending treatment all over the brain since they deliver the drug in the needed concentration to the specific areas affected. It is useful not only from the point of view of improving the effectiveness of therapy but also in the context of increasing patient compliance due to a lower frequency of taking medicines. In addition to drug delivery some of the applications of nanotechnology in the brain includes imaging and diagnosing diseases, which assists the clinician to decipher the disease process and decide on a reformative strategy. The delivery of nanocarriers across the BBB is a giant leap towards effective management of neurological disorders, given that conditions of the CNS that hitherto were deemed untreatable are now in prospect, thanks to the application of nano-technology in medication delivery. As the advancement goes on, these technologies bear the promise of fundamentally changing the way a range of neurological diseases are treated, giving new chances to those patients who have narrow choices within the existing therapeutic approaches (Abbas 2021).

Infections are still a major international problem to mankind's health and are worsened by the enlightenment of antibiotic and antiviral resistance. Nanotechnology opens up opportunity for the improvement of the usage of antibiotics and antivirals for instance dealing with difficulties such as drug adaption, poor solubility and the capability for targeted delivery of drugs. Cohesive nanosystems like polymeric nanoparticles and liposomes, or metallic nanoparticles, can simultaneously entrap antibiotics or antiviral agents, enhance their stability and bioavailability and region targeted release. Repeat administrations of targeted drug delivery lower the needed quantity and boost the effectiveness of the therapy, especially in severe destroy cases (Hetta et al. 2023).

One strategy is to use nanoparticles that would have tenable release rate so that the drug concentration at the site of infection could be sustainably sustained at a therapeutically effective level. For instance, using nanoparticles developed from polymers allows antibiotics to be easily and steadily delivered and may prevent the occurrence of antibiotic resistance from the suboptimal concentration. Moreover, metal nanoparticles like NPs of silver or gold have inherent antimicrobial activity and can prevent bacterial cell membranes' cohesion on the nanoparticle's surface or inhibit viral replication. These contracted metal nanoparticles give improved action profiles when co-prescribed with conventional antibiotics or antivirals through multiple complimentary mechanisms against pathogens, self resisting strains in particular. In diseases, for example AIDS or hepatitis nanotechnology is used to introduce antiviral agents directly into infected cells to reduce the impact of the viruses while having little or no effect on other cells. It is also possible to make nanoparticles receptive to certain enzymes or condition prevailing around infected tissues so as to deliver the drug at the right time and place. This approach is particularly helpful in managing chronic infections where use of drugs results in toxicity and poor patients' compliance. Apart from performing a therapeutic function, nanotechnology also strengthens diagnostics by enabling the identification of microbial presence at minimal concentrations using nano-sensors. It helps prevent the endemics from spreading out as all is done in a day when the ailments are detected at an early stage. As nanotechnology advances in the realm of infectious diseases' treatment there are potential signs for better, more targeted, and efficient therapy against the steadily increasing problem of antimicrobial resistance and subsequently the enhanced quality of patients' quality of life all around the world. Different applications of nanotechnology in disease treatment have been depicted in Fig. 22.2 (Tang et al. 2022).

## 22.4 Challenges in Nano Biopharmaceutical Development

One of the greatest difficulties that nano biopharmaceuticals manufacturers face is the issue concerning the regulation of development and the standardization that comes along with it. Nanotechnology carries one or more properties that can significantly alter the pharmacokinetic characteristics of a drug. The challenges in Nano Biopharmaceutical Development have been shown in Fig. 22.3. Some regulatory bodies like the FDA and EMA already have guidelines that govern normal drugs but since nano-scale drug pose new sets of issues, there is need to come up with new regulations regarding nano scale drugs. They observed that the physicochemical properties of drug-loaded nanocarriers such as size, surface properties, and chemical nature can impact drug activity and toxicity and that, for each newly developed nanomedicine, these factors should be optimized.Classification of nano-drugs is one of the emerging regulatory concerns. Most nanomedicines are multi-component nano formulations: combining the therapeutic agent with the delivery system, which makes a distinction between a drug, a biologic and a medical device appear rather artificial. It influences the regulatory process because each category has their own rules when it comes to testing, approval, and subsequent monitoring. Also, the characterization of nano-drugs brings in extra challenges with regard to their size as well as the surface area to volume ratio that a Nano particle carries which renders it behave differently in living systems. There is currently no accurate means by which regulators can measure with a high degree of precision these characteristics of degree of operating market competition, switching costs, product differentiation, and buyer and seller information, using new analytical tools and methodologies are needed but are not yet fully developed (Allan et al. 2021).

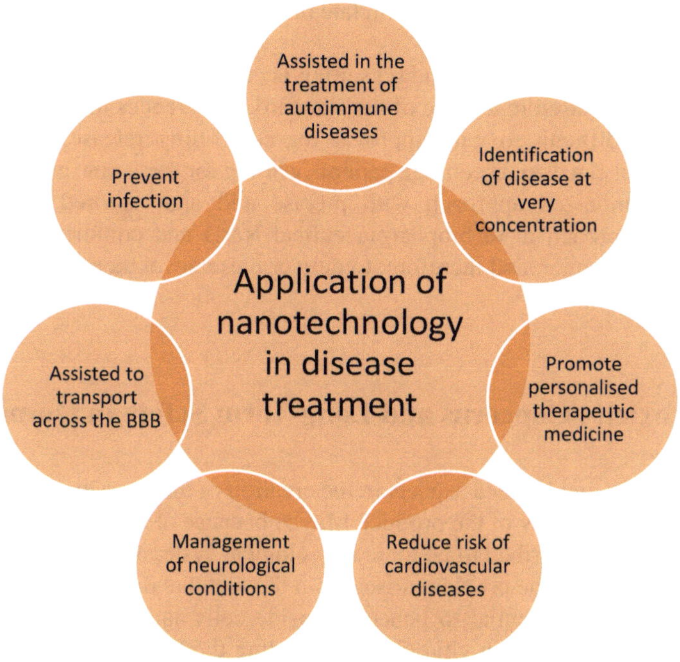


**Fig. 22.2** Application of nanotechnology in disease treatment

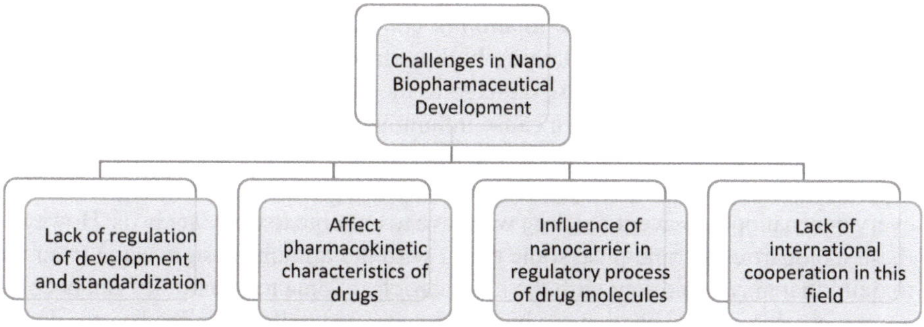


**Fig. 22.3** Challenges in Nano biopharmaceutical development

The other area where regulatory hurdles are marked is safety and efficacy testing for nano-drugs. The current traditional models of the preclinical analysis do not depict the relations of the nano-scale particles with biological systems well enough and we desperately need better models that would allow for accurate prediction of potential dangers. Prolonged safety data is highly relevant as concrete impacts of nanoparticles on human health in the long run are unknown. Authorities are gradually asking for clearance trials to be conducted for longer periods to determine whether the nano-drugs tend to accumulate in organs/tissues thus leading to toxicity effects (Đorđević et al. 2021).

Finally, harmonization of standard setting is crucial since international cooperation is largely required in anyone of the standard. Differences in rules across countries make it difficult to gain approval for controlling release mechanisms of nano-drugs and can also discourage patients waiting for these new innovative drugs. A single harmonized approach with precise and nano-tailored instructions is required to approach nano biopharmaceutical R&D and commercialization in a more efficient manner, and therefore, to deliver these novel therapies to patients all around the globe.

## 22.5 Toxicity Concerns and Long-Term Safety in Humans

One of the most emphasized aspects in the creation of the nano biopharmaceuticals is toxicity and longevity of the product. Maybe because of their size and properties, nanoparticles are described to be toxic to biological systems in ways that pure drug may not be toxic to. The problem arises from the fact that due to their size nanoparticles have a great potential to penetrate inside cells and tissues thus having the potential to be deposited in vital organs including the liver, spleen and kidneys. It means the toxin accumulates in the body that over time, may cause chronic toxicity or immune reactions which are not noticeable promptly, hence raising concerns about the safety of nano-based drugs. Among them, the possibility that nanoparticles cause oxidative stress, inflammation or cell damage may pose the most threat. Metal based nanoparticles containing high surface reactivity has the potential for producing ROS, which can cause damage to the cellular structures and thus oxidative stress. This stress can in turn cause inflammations and if prolonged will lead to more chronic health complications. The reactive toxicities of nanoparticles change with size, shape, surface charges and the materials used in the coatings and thus every new nanopharmaceutical drug will have to undergo toxicity analysis. However, these nanoparticles could pose some threat with the immune system aside from the toxicity that may be present with the nanoparticles. Some nanoparticles can become an object of immune reactions, which can cause side effects or lead to the lower effectiveness of a drug. The latter remain in circulation and can actually be missed by the immune system of the body since they are not destroyed and removed by the immune system as occur with small immune recognizable molecules but compound within the body and this can act a s a bioaccumulation that is toxic to the required level after some time. It also raises questions about scope for accidental interaction of nanoparticles with the blood brain barrier and subsequent effects on the central nervous system resulting in neurotoxicity. Such effects may not cumulatively manifest from short-term exposures and thus long-term experiments are very helpful in the research of the impacts of such substances (Zielińska et al. 2020).

Another issue which further complicates attempts to evaluate toxicity of nanoparticles is the absence of the standard protocols for testing. Conventional techniques do not adequately describe how nanoparticles engage cells and tissues as well as other complex behaviours. Hence, there is a need to develop superior in vitro and in vivo models to judge the chronic impacts of nanomedicines commensurately. Furthermore, reporting of the result is a crucial aspect as the absence of long-term toxicity information may lead to reduced acceptance of nano biopharmaceuticals. Mitigating these toxicity aspects will be equally important to develop safe nano-based therapies that can offer sustainable therapeutic impact in medicine in the advancing applications (Xuan et al. 2023).

## 22.6 Manufacturing and Scalability Issues

Challenges of manufacturing and scaling up nano biopharmaceuticals for commercialization are compelling because the nano drugs have very complicated structures and high quality standards that make production on large scale almost impossible as depicted in Fig. 22.4. Nano drugs on the other hand are much difficult to scale up from a small quantity of production because they have to be designed using atomic or molecular engineering unlike normal drugs which are just tiny chemical compounds. For example, engineering size, shape, and surface properties of nanoparticles well-maintained for proper drug behaviour and response; nevertheless, tiny differences in these characteristics are achievable during production leading to variability in the safety and efficacy of the final product (Paliwal et al. 2014).

One of the important challenges of scaling up the production of nanoparticles is the ability to replicate their synthesis. Two, techniques employed in test tubes and vials can be very different from those suitable for large scale production. Techniques such as high-pressure homogenization, microfluidization and sonication are employed especially when dealing with small quantity of nanoparticles but are often difficult to scale up while maintaining homogeneity. Third, enforcing clean room and preventing contamination in manufacturing is slightly easier said than done for

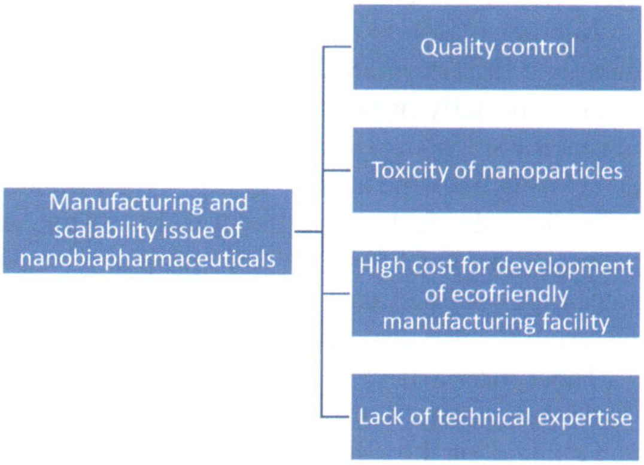


**Fig. 22.4** Manufacturing and scalability issue of nanobiapharmaceuticals

nano scale products since any unwanted particulate in the product changes the efficacy of the drug (Khairnar et al. 2022).

Quality control in nano biopharmaceuticals is another major issue in the same respect. Standard quality control approaches are sometimes ineffective for nano-drugs, and to address this issue, new methods of analysis were created. Dynamic light scattering, electron microscopy, and zeta potential analysis are applied in nanotechnology for characterization of nanoparticles; however, they are too slow to apply for massive and real-time measurement. Sustaining such quality and scaling it up to industrial production requires certain stages of automation of such processes and incorporation of the methods into production lines (Desai 2012).

Manufacturing of nano-drugs narrates some more rules and regulation that has to be followed. Given that many regulatory organizations are insisting on enhanced uniformity and safety in nanocarriers and the drugs that they delivery, the cost of enhancing the manufacturing plants to meet these requirements can be prohibitive and may involve expensive equipment. The expenses and skills necessary to construct proper plants for manufacturing nano-drugs on a big scale are usually high so that only a few companies can enter the sector. Solving these manufacturing and scalability problems is pertinent to bringing nano biopharmaceuticals into reasonable clinical practice, so patients can rely on them and receive these novel therapies with confidence (Kumari et al. 2023).

## 22.7 Cost and Accessibility of Advanced Nano-Drugs

Pursuit of nano biopharmaceuticals is capital-intensive, and these next-generation drugs are therefore expensive to produce, reducing the chances of achieving equitable accessibility. Nano-drugs call for unique investigation, designing, and production strategies that are significantly more intricate than those of standard drugs. Starting from the designing and synthesizing of nanoparticles to characterization and quality control, each step requires advanced technique and a good amount of capital. Therefore, the ultimate cost of nano-drugs can be excessively high, discouraging patient use and an economic burden to health systems (Patra et al. 2018).

Another reason is related to the introduction of the required fabrication equipment and infrastructure for manufacturing nano-drugs. Nanotechnology manufacturing especially in products manufacturing needs very careful measures, and this makes it compulsory to have controlled environmental conditions. To manufacture and closely track the nano-drugs the existing and new infrastructures must be outfitted with sophisticated equipment and software. Also, the skills needed to run such technologies and to adhere to high quality requirements raises operational costs since highly skilled staff are essential to manage the intricacies of nano-drug manufacturing (Charitidis et al. 2014).

These three aspects hitch the cost of nano-drugs: The draconian regulatory standards that they must meet. Governmental bodies insist on data, including preclinical and clinical, on the safety and effectiveness of nano biopharmaceuticals. These tests are especially expensive for nano drugs as the models required for testing and the need for research over fairly long periods for unique risks of nanoparticles are expensive. Moreover, regulatory approval for nano-drugs takes time in processing and might incur extra expenses in terms of time for development approval and compliance to the regulatory requirements (Đorđević et al. 2021).

Another factor is the high costs as they reduce the affordability aspect and will either remain being charged to the patients. Access to nano-drugs can also be problematic, especially since insurance usually provides only a limited cover for health given the high unit costs in developed countries or patient affordability in low income or underfunded health insurance in developing nations. This is why attempts to negotiate for cost subsides or Come search for better economic ways of producing the commodity are probably the best ways to overcome the problem. A few researchers are pursuing options for reducing costs of production by either employing other materials or optimizing techniques; however, these strategies are not well developed yet. It will be effective to summarize that nano-drugs' affordability and availability are slated to be more enhanced through correct apportionment of technological ingenuity, right reforms, suitable legislations and health care system support. Getting rid of such cost-restrains the field of nano biopharmaceuticals can come to millions of patients to succeed while keeping the needs of healthcare systems in mind in coping with the integration of the nano-biopharmaceuticals in the general practice (Milewska et al. 2021).

## 22.8 Ethical and Societal Considerations

The concept of patient safety is both a critical factor in the creation and evaluation of nano biopharmaceuticals and poses new ethical questions. Nanomedicine brings new materials and interactions at molecular level that affect the human body in different ways which are not well understood in the long run as depicted in Fig. 22.5. This novelty leads to difficulties in risk evaluation in the precisely manner because nanoparticles can act in an unpredictable way within the organism. Indeed, ethical

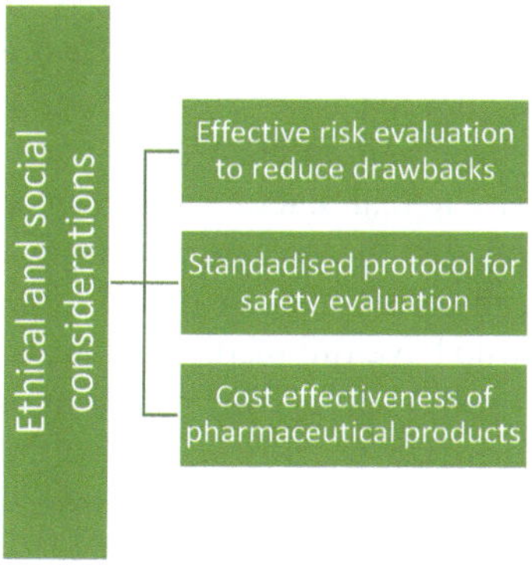


**Fig. 22.5** Societal impact of accessibility to advanced therapies

issues are most critical in phase one trials whereby the possible therapeutic value of the treatment has to be weighed against risks that are still unknown. Since nanotechnology involves highly risky drugs, it is necessary also to get true and valid consent from participants in nano-drug trials and they should be warned that what they undergo is experimentation. The major ethical consideration often observed in nano-drug trials include the question of standardized protocol for safety evaluation. The way in which nano-drugs affect cells and tissues can be substantially different from that of conventional drugs, and this fact indicates that the effectiveness of conventional models for drug tests may be limited in testing nano-drugs. This uncertainty calls for the creation of innovative approaches to preclinical/clinical assessment that would determine the behavior of nanoparticles in biological environment. Scholars need to affirm that trials failed sufficiently safe guarding measures hardships oppose any unfavorable interaction when investing in trials for the welfare of the patients. Since nano-drugs are potentially high risk, there is need to have independent monitors and an enhanced scrutiny of the regulations need to be provided at different trial phases (Wolfram et al. 2016).

Another ethical issue is concerned with reporting and such issues normally come about when there is existence of one or several of the other issues. It is the duty of the researchers to present all data from the trials, outcomes whether negative or adverse to the interest of the users of such products. Ethical concerns also apply to post-trial procedures since nano-drugs can cause side effects that are not easily noticeable immediately after the trial. Safety outcomes from trials only reflect the frequency of adverse events during the trial period; therefore, follow-up of participants in trials is important to detect delayed adverse effects and ethical guidance for the conduct of trials must incorporate provisions for extended safety monitoring. There are some ethical issues related to nanotechnology drugs' availability for trials. With nano-drug trials bringing expensive and complex procedures, it may be impractical to enrol lower-income or rural patients. There is no doubt that trial populations to be enrolled in nanoparticle conjugate development should reflect diversity since nano-drugs need to be safe for all people. If such ethical concerns are addressed, nano-drug trials may progress in a responsible and moral way to ensure that patients are protected, and their rights as participants in the trial are respected while at the same time the novel field of nanomedicine is fully developed.

An important question is related to the availability of new and effective nano-biopharmaceuticals for the treatment of diseases, which can produce better outcomes in case of application, yet which can be used by a limited number of patients since their application can be expensive and rare. Nano drugs are normally expensive to manufacture and develop because they have complex structures as well as the regulatory and manufacturing procedures that accompany them. As such, only a limited number of patients, who are in majority residing in developed nations or who have full medical insurance at their disposal, can take advantage of such innovations while many people are still dying or living with the untreated diseases. Such distinctions stir questions about fair distribution of health care services as well as overriding principles in allowing a fraction of the population to access trendy treatments (Alghamdi et al. 2023).

A potential consequence of this accessibility issue is that health care equality will be threatened. Cancer treatments, neurological disorder treatments and treatments for infectious diseases are most required in additional amount in low income earning areas, yet such regions may not be in a financial position to afford the high cost of these treatments. This situation makes vulnerable groups in health care even more disadvantaged because they are denied their rights of accessing basic health care. Unless concerted efforts are made to increase Nano-drugs affordability and availability, society is likely to build a bi-level health care delivery system where only those who can afford it or those who are insured get the best treatments while others are left with the second best or pre-Nano drugs.

Hospitals have, however, been experiencing problems in adopting expensive nano-drugs since the national or hospital health care budgets may not accommodate these innovative treatments for all the patients. Therapies are frequently reliant on insurance, but nano-drugs may be limited to insurance, owing to their high cost and the absence of information on the safety of long-term utilization. This limited access can put a lot of pressure on the care givers, especially in the health sector, as often they have to make decisions on who to treat next hence at times make some rather hard decisions that are ethically. Approaches to such accessibility issues are; calling for policies encouraging for equitable access to nano-biopharmaceuticals that may include subsidies, tiered pricing or government or philanthropist provided funding to support the treatment costs for the less fortunate. Further, developing linkages between the private and public domains may lead to lower costs associated with development and production of the treatments. Making further improvements of the availability of the nano-therapies is not only an ethical but a social demand as well since access to superior health contributes to healthier and more stable societies. This way, society can focus on achieving availability as a goal of both the development and distribution of nano-biopharmaceuticals with the view of making this resultant scientific innovation available to the entire population (Anjum et al. 2021).

## 22.9 Effective Risk Evaluation to Reduce Drawbacks

### *22.9.1 Addressing Public Perception and Awareness of Nanomedicine*

Thereby, the adoption of nano-biopharmaceuticals in health care, social acceptance and awareness about nanomedicine play a central role as rejection by the public could be a major stumbling block to the incorporation of these products into the society. Nanotechnology may appear vague or even scary to the broader population, primarily associated with the risks it poses, potential misuse, and currently underestimated consequences. The public's distrust could be attributed to an ignorance of how nano-drugs function or major scandals in other areas of biotechnology, for instance, genetically modified organisms (GMOs) or even stem cell research. In order to increase people's confidence, there is a strong need to engage in process of public enlightens and explain what nanomedicine is, what the positive consequences of its usage are and which precautions need to be taken.The public needs to be informed mainly on the rational use of nano-drugs as well as specific precaution measures taken in the development processes. This education can take the form of the collaborative development of informational material by scientific societies and educational establishments in cooperation with healthcare institutions for the purposes of effective lodging of nanomedicine concepts from different angles. The proper information that was made available to the public regarding how these nano-drugs are produced, tested, and monitored, can more or less remove the perception that they are dangerous. Also, when addressing potential risks and conducting constant research for the effects in a long term do promote trust since people appreciate safety and openness (Farjadian et al. 2018).

Social presentation through media has an imperative role in developing impression of the society. Informing journalists and media about nanomedicine can significantly limit the amount of misinformation through exposure to accurate information and of various opinions of other experts. Misinformation in addition is furthered by sensationalism in the media which might lead to fear or result in generating unrealistic expectations. With such topics, it should be critical for journalists to in possession of accurate and balanced information from science and required reference materials that would ensure appropriate coverage of nanomedicine.Also, more people discussing nanomedicine in public forums can help people share these concerns and ask questions and then the scientists and policy makers can address those concerns. To this end, engagement can be conducted offline via public forums, or informational campaigns, and online through Engaging Men for Gender Equality 27 activeness platforms. Another factor that can enhance acceptance of nanomedicine is the integration of people's input regarding policy choice about nanomedicine into lawmakers' decision making process, thus would endorse innovation as a social obligation instead of only a state's priority on possibly dangerous sciences.It is important not only for the public to understand the advantages of nano-biopharmaceuticals but also the availability of funds for research, policies endorsing the usage of the nanomedicine, and societal acceptability of the nano-biopharmaceutical products. Due to the focus on openness and engagement with the public, the field of nanomedicine has all the potential for patient, healthcare providers, and the general public to embrace nano medicine as the future of medicine (Lewandowsky et al. 2012).

## 22.10 Future Perspective

Scientifically originated as nano-biopharmaceuticals, emerging technologies specifically with CRISPR and immuno-nanomedicine are paving way for precise, personalized and effective treatments. There is a new therapeutic gene editing technique called CRISPR, which gives great potential in handling genes directly responsible for diseases. The incorporation of nanocarrier systems helps in the safe delivery of CRISPR systems into target cells, which enhance gene editing, especially with minimal off target effects. For instance, to deliver CRISPR components, nanoparticles are employed to protect them as they travel in the body to reach the target, say cancerous or genetic tissue cells. This targeted delivery makes CRISPR applications safer and more effective for treating genetic diseases and cancer and other complex diseases.Another of the promising fields is immuno-nanomedicine—utilization of nanotechnology in order to stimulate immune reactions vs diseases. Nanoparticles may be programmed to transport immunostimulants or antigens to selected immune cells resulting in better immunization against diseases, cancers, and possibly some autoimmune disorders. Through active control, systemic effects can be achieved, which is a contrary to traditional immunotherapies wherein; systemic adverse effects prevail. For example, medical scientists are working on the preparation of nanoparticles as carriers of antigens in order to target dendritic cells that are involved in initiating immune responses. It also improves the effectiveness of the drug and matters of immune response acceptable for the specific patient, as the concept of personalized medicine suggests. Nano-biopharmaceuticals can best be seen as a means of rewiring healthcare, as a way of unlocking the potential for more targeted, more inclusive and more personalized medica nosology's for an impressive range medical conditions.

In combination, CRISPR and immuno-nanomedicine show how nascent technologies are gradually improving nano-biopharmaceuticals to achieve new solutions to undruggable diseases. As these technologies develop, one can safely assume that they will change the current concept of modern medicinal practice that targets a broad spectrum of illnesses and disease with promises of more efficacious, safer and individualised therapeutic solutions.

AI is gradually being made part of nanomedicine to bring about increased accuracy, speed and full customization of the nano-biopharmaceutical. Many AI algorithms are efficient in high-throughput data analysis to predict structures of novel nano particulate forms, their interactions with biological systems, and to design exclusive treatment regimens for each patient. For example, AI models can make nanomedicine treatment decisions based on patient's genetic and disease markers today with least side effects. This integration helps the overall goals of precision medicine in that all treatments need to be individualized based on the biology of the patient. The fine example in using AI in nanomedicine is in the design and improvement of nanoparticles. Therefore the realization of efficient nanocarriers has various parameters that need to be optimized including size, shape, surface charge and release profile. AI can help here because it can model thousands of different nanoparticles involving all sorts of permutations, and quickly come up with potential new candidates for controlled drug delivery. Machine learning algorithms can also model how such nanoparticles will behave inside the human body, such as the distribution pattern, the nanoparticles' interaction with the cells, and their ageing patterns. This capability of prediction enables the researchers to achieve more enhanced formulation with nano drugs far more faster and less time consuming than just trial and error approach. Moreover, in regard to clinical application of nano-drugs, AI contributes to constant supervision that enables accurate administration changes. For instance, AI algorithms can track the nature of patient response to certain treatments so that it can recognize of patterns that indicate positive effects of the treatment or indications of negative side effects. This feedback circle makes it possible to always assess and get a close to the ideal treatment protocols for the patients to be administered with. AI tools also play another role in promoting the safety of drugs by pointing likely risks in the drugs before it goes out to be clinically used. With the development of AI technology into the nanomedicine world, there are good prospects for the further advancement into more of a tailored precise and responsive approach to treating patients that would lead to better outcomes and advancements in the nano-biopharmaceutical research and development process.

Stability and formulation of drug molecules is very critical in the use of nano-biopharmaceuticals since the properties of nanocarriers may sometimes be disadvantageous to the stability and efficacy of the entrapped drugs. Nano-drugs therefore possess a challenge because they are sensitive to temperature changes, humidity, and exposure to light when stored in vials and transported to serve their intended purpose. In response, researchers have proposed novel approaches to formulation processing of nano-drugs and stabilizing methodologies that shield nano-drugs from degradation modes and failure. For instance, the incorporation of biodegradable polymers and protective coatings on nanoparticles can greatly improve the stability of the carried drugs so that they maintain their potency for remarkably long periods of time. Nanotechnology is also coming up with other techniques in formulation to enhance bioavailability and other characteristics of drug delivery systems. Systemically administered traditional drugs may be metabolized quickly in the body or never achieve therapeutic levels at the site of disease. Nanocarriers redeem these loses by protecting the drug from metabolism allegedly immediately after the administration and targeting these specific cell or tissue. Recent products are multi-layered nanoparticles and liposomes that enable a combination of a burst released followed by another sustained release. This strategy improves the effectiveness of the treatment process, at the same time avoiding frequent doses that may be useful in chronic diseases requiring lifelong care. Further, enhancement in drug delivery system have paved way for including several pharmacological substances in a single nanocarrier package. These combination therapies may be synergistic and can affect several pathway at once as indicated by cancer therapies that involve both chemotherapy and immunomodulation. Through co-delivery, the nanocarriers also increase the chances of the multi-drug therapy on synergism, and less on antagonism. Optimizing these two factors therefore remains key to actualizing the benefits of nanomedicine nanodrugs so that patients get both better treatment response and easy to administer medicines and thus progress in treatment becomes widely attainable.

Ventures between the pharma firms and technology industries have emerged as the key forces behind the progress in nano-biopharmaceutical R&D. Nanomedicine integrates knowledge from various fields such as material science, bio technology and data analysis. Through such collaborations, the pharmaceutical firms receive access to innovative technology and knowledge that will facilitate faster development of innovative nano-drugs, while on the other side, tech firms get an entrée into the profitable health sector. The collaboration of both sectors creates a combined greater platform that amplifies each sector's resource and capability to address multifaceted issues revolving around drug delivery, formulation, and precision medicine.

The collaborations have been most effective in one particular fields which is artificial intelligence and machine learning in nanomedicine. End-user drug industries willing to embrace nanomedicine gain access to the latest technology that key-tech industries specializing in data analytic offer in terms of drug designing and high-throughput computational screening for identifying efficient nanocarriers for drug delivery or improving drug formulations. Such collaborations enable better and faster development of nano-drugs with fewer costs required to conduct research on the new drugs. Further, IT industries also provide knowledge about the sensor technology and real time monitoring systems for capturing the reaction of a patient towards the nanoparticle drugs and even for modifying the treatment procedures accordingly.

Manufacturing and regulatory assistance is also a common management area where tech companies provide sophisticated automation and quality control systems that enable companies to increase the production of nano-drugs, at the same time meeting all the required standards. For example, advancements in robotics and process automation make it possible to achieve economies of scale to produce nano-carriers thus mass production of nano–drugs. Furthermore, collaborations with technology firms enable the use of advanced qualitative assurance methods, which include monitoring techniques and machine learning, anomaly detection in the process; critical for keeping up high levels of quality in nano-pharmaceuticals. In addition to the technological factors, focus has to be placed on cost and access associated with these collaborations. This way pharmaceutical and tech companies can donate on the common pot and when the Codex is complete there is a better chance that nano-drug development cost will lower and therefore the potential of more advanced treatments per the price will increase. Cooperative R& D in nano-biopharmaceuticals can deliver a great health care, and can extend the availability of better and effective medical treatment to the large number of patients suffering from different diseases. With the development of these collaborations, these will become crucial in achieving progress in the field of nanomedicine and deliver the maximum possible benefit within numerous therapeutic areas.

## 22.11 Conclusion

nano-biopharmaceuticals are in fact considered a new frontier of medicine as a whole, as they can drastically change treatment strategies regarding many diseases: including many malignant diseases such as cancer and cardiovascular disorders, neurological disorders, and infections. Efficacies of targeted drug delivery, drug stability and precision mechanism make the nanotechnology basis of therapies safer as well as more effective and capable of being administered according to the need of the patient. Nonetheless, the existing science of external applications is not without its problems such as legal issues, toxicity, technical difficulties in production, and issues of availability and accessibility. Mitigating these challenges will thus entail multi sector stakeholders' efforts from the pharmaceutical industry players, technology industries, regulators and healthcare industries.Innovative enterprises like CRISPR, AI for Optimization, and immuno-nanomedical are expected to redefine nano-biopharmaceuticals for SMART and PERSONALIZED medicines. These developments underline the concept of multi-disciplinary investigation, and the requirement to adhere to high standards of moral responsibility, and openness in order to regain the confidence of society. This paper envisions the nano-biopharmaceutical industry to be the foundation of modern health care delivery as research goes on and collaboration deepens, which will in the process, address the emerging research gaps for improved health care delivery in developing nations. He also talks about the current challenges that nanomedicine faces and looks at the future as full bright opportunity that is capable to provide patients all over the world with effective and affordable treatments.

## References


Abbas M (2021) Potential role of nanoparticles in treating the accumulation of amyloid-Beta peptide in Alzheimer's patients; MDPI. Polymers 13(7):1051. https://doi.org/10.3390/polym13071051

Alghamdi MA, Fallica AN, Virzi N, Kesharwani P, Pittala V, Greish K (2023) The promise of nanotechnology in personalized medicine; MDPI. J Pers Med 12(5):673. https://doi.org/10.3390/jpm12050673

Allan J, Belz S, Hoeveler A, Hugas M, Okuda H, Patri A, Rauscher H, Silva P, Slikker W, Sokull-Kluettgen B, Tong W, Anklam E (2021) Regulatory landscape of nanotechnology and nanoplastics from a global perspective. Regulat Toxicol Pharmacol 122:104885. https://doi.org/10.1016/j.yrtph.2021.104885

Anjum S, Ishaque S, Fatima H, Farooq W, Hano C, Abbasi BH, Anjum I (2021) Emerging applications of nanotechnology in healthcare systems: grand challenges and perspectives. Pharmaceuticals (Basel, Switzerland) 14(8):707. https://doi.org/10.3390/ph14080707

Bates DW (2002) Sustained-release preparations and medication errors. J Gen Intern Med 17:657–658. https://doi.org/10.1046/j.1525-1497.2002.20628.x

Charitidis CA, Georgiou P, Koklioti MA, Trompeta AF, Markakis V (2014) Manufacturing nanomaterials: from research to industry. Manufact Rev 1. https://doi.org/10.1051/mfreview/2014009

Cheng X, Xie Q, Sun Y (2023) Advances in nanomaterial-based targeted drug delivery systems. Front Bioeng Biotechnol 11. https://doi.org/10.3389/fbioe.2023.1177151

Desai N (2012) Challenges in development of nanoparticle-based therapeutics; Springer. AAPS J 14:282–195. https://doi.org/10.1208/s12248-012-9339-4

Đorđević S, Gonzalez MM, Conejos-Sánchez I, Carreira B, Pozzi S, Acúrcio RC, Satchi-Fainaro R, Florindo HF, Vicent MJ (2021) Current hurdles to the translation of nanomedicines from bench to the clinic; Springer. Drug Deliv Transl Res 12:500–525. https://doi.org/10.1007/s13346-021-01024-2

Ezeako EC, Okpara OJ, Okoroafor CC, Eze SC, Kalu OL, Odoh EC, Nwadike UG, Ogbodo JO, Umeh BU, Ossai EC, Nwanguma BC (2023) Advances in drug delivery systems, challenges and future directions. Heliyon 9(6):e17488. https://doi.org/10.1016/j.heliyon.2023.e17488

Farjadian F, Ghasemi A, Gohari O, Roointan A, Karimi M, Hamblin MR (2018) Nanopharmaceuticals and nanomedicines currently on the market: challenges and opportunities. Taylor & Francis 14(1):93–126. https://doi.org/10.2217/nnm-2018-0120

Hassanpour S, Kim HJ, Saadati A, Tebon P, Xue C, van den Dolder FW, Thakor J, Baradaran B, Mosafer J, Baghbanzadeh A, de Barros NR, Hashemzaei M, Lee KJ, Lee J, Zhang S, Sun W, Cho HJ, Ahadian S, Ashammakhi N, Dokmeci MR, Mokhtarzadeh A, Khademhosseini A (2020) Thrombolytic agents: nanocarriers in controlled release. Small (Weinheim an der Bergstrasse, Germany) 16(40):e2001647. https://doi.org/10.1002/smll.202001647

Hetta HF, Ramadan YN, Al-Harbi AI, Ahmed EA, Battah B, Abd Ellah NH, Zanetti S, Donadu MG (2023) Nanotechnology as a promising approach to combat multidrug resistant bacteria: a comprehensive review and future perspectives; MDPI. Biomedicines 11(2):413. https://doi.org/10.3390/biomedicines11020413

Joseph TM, Mahapatra DK, Esmaeili A, Piszczyk L, Hasanin MS, Kattali M, Haponiuk J, Thomas S (2023) Nanoparticles: taking a unique position in medicine; MDPI. Nano 13(3):574. https://doi.org/10.3390/nano13030574

Khairnar SV, Pagare P, Thakre A, Nambiar AR, Junnuthula V, Abraham MC, Kolimi P, Nyavanandi D, Dyawanapelly S (2022) Review on the scale-up methods for the preparation of solid lipid nanoparticles. Pharmaceutics 14(9):1886. https://doi.org/10.3390/pharmaceutics14091886

Kumari R, Suman K, Karmakar S, Mishra V, Lakra SG, Saurav GK, Mahto BK (2023) Regulation and safety measures for nanotechnology-based agri-products. Front Genome Editing 5. https://doi.org/10.3389/fgeed.2023.1200987

Lewandowsky S, Ecker UKH, Seifert CM, Schwarz N, Cook J (2012) Misinformation and its correction: continued influence and successful Debiasing. Psychol Sci Public Interest 13(3):106–131. https://doi.org/10.1177/1529100612451018

Milewska S, Niemirowicz-Laskowska K, Siemiaszko G, Nowicki P, Wilczewska AZ, Car H (2021) Current trends and challenges in Pharmacoeconomic aspects of nanocarriers as drug delivery systems for cancer treatment; Dovepress. Int J Nanomedicine 16:6593–6644. https://doi.org/10.2147/IJN.S323831

Paliwal R, Babu J, Palakurthi S (2014) Nanomedicine scale-up technologies: feasibilities and challenges. AAPS PharmSciTech 15(6):1527–1534. https://doi.org/10.1208/s12249-014-0177-9

Patra JK, Das G, Fraceto LF, Campos EVR, Rodriguez-Torres MDP, Acosta-Torres LS, Diaz-Torres LA, Grillo R, Swamy MK, Sharma S, Habtemariam S, Shin HS (2018) Nano based drug delivery systems: recent developments and future prospects. J Nanobiotechnol 16(1):71. https://doi.org/10.1186/s12951-018-0392-8

Pinheiro RGR, Coutinho AJ, Pinheiro M, Neves AR (2021) Nanoparticles for targeted brain drug delivery: what do we know?; MDPI. Int J Mol Sci 22(21):11654. https://doi.org/10.3390/ijms222111654

Prosperi D, Colombo M, Zanoni I, Granucci F (2017) Drug nanocarriers to treat autoimmunity and chronic inflammatory diseases. Semin Immunol 34:61–67. https://doi.org/10.1016/j.smim.2017.08.010

Ramesh M, Janani R, Deepa C, Rajeshkumar L (2023) Nanotechnology-enabled biosensors: a review of fundamentals, design principles, materials, and applications; MDPI. Biosensors 13(1):40. https://doi.org/10.3390/bios13010040

Roth, G. A., Mensah, G. A., Johnson, C. O., Addolorato, G., Ammirati, E., Baddour, L. M., Barengo, N. C., Beaton, A. Z., Benjamin, E. J., Benziger, C. P., Bonny, A., Brauer, M., Brodmann, M., Cahill, T. J., Carapetis, J., Catapano, A. L., Chugh, S. S., Cooper, L. T., Coresh, J., Criqui, M. et. al. (2020). Global burden of cardiovascular diseases and risk factors, 1990-2019: update from the GBD 2019 study. J Am Coll Cardiol, 76(25), 2982–3021. https://doi.org/10.1016/j.jacc.2020.11.010

Tang J, Ouyang Q, Li Y, Zhang P, Jin W, Qu S, Yang F, He Z, Qin M (2022) Nanomaterials for delivering antibiotics in the therapy of pneumonia; MDPI. Int J Mol Sci 23(24):15738. https://doi.org/10.3390/ijms232415738

ud Din F, Aman W, Ullah I, Qureshi OS, Mustapha O, Shafique S, Zeb A (2017) Effective use of nanocarriers as drug delivery systems for the treatment of selected tumors. Int J Nanomedicine 12:7291–7309. https://doi.org/10.2147/IJN.S146315

Upadhyay RK (2014) Drug delivery systems, CNS protection, and the blood brain barrier. Biomed Res Int 2014:1–37. https://doi.org/10.1155/2014/869269

Wolfram J, Zhu M, Yang Y, Shen J, Gentile E, Paolino D, Fresta M, Nie G, Chen C, Shen H, Ferrari M, Zhao Y (2016) Safety of nanoparticles in medicine; Bentham Science. Curr Drug Targets 16(14):1671–1681. https://doi.org/10.2174/1389450115666140804124808

Xuan L, Ju Z, Skonieczna M, Zhou PK, Huang R (2023) Nanoparticles-induced potential toxicity on human health: applications, toxicity mechanisms, and evaluation models. MedComm 4:e327. https://doi.org/10.1002/mco2.327

Zielińska A, Carreiró F, Oliveira AM, Neves A, Pires B, Venkatesh DN, Durazzo A, Lucarini M, Eder P, Silva AM, Santini A, Souto EB (2020) Polymeric nanoparticles: production, characterization, toxicology and ecotoxicology. Molecules 25(16):3731. https://doi.org/10.3390/molecules25163731

GPSR Compliance

*The European Union's (EU) General Product Safety Regulation (GPSR) is a set of rules that requires consumer products to be safe and our obligations to ensure this.*

*If you have any concerns about our products, you can contact us on ProductSafety@springernature.com*

In case Publisher is established outside the EU, the EU authorized representative is:

Springer Nature Customer Service Center GmbH
Europaplatz 3
69115 Heidelberg, Germany

**Batch number: 09137409**

Printed by Printforce, the Netherlands